Ecological Forest Management Handbook

The second edition of *Ecological Forest Management Handbook* continues to provide forestry professionals and students with basic principles of ecological forest management and their applications at regional and site-specific levels. Thoroughly updated and revised, the handbook addresses numerous topics and explains that ecological forest management is a complex process that requires broad ecological knowledge. It discusses how to develop adaptive management scenarios to harvest resources in a sustainable way and provide ecosystem services and social functions. It includes new studies on ecological indicators, the carbon cycle, and ecosystem simulation models for various forest types: boreal, temperate, and tropical forests.

NEW IN THE SECOND EDITION

- Provides a comprehensive collection of sustainable forest management principles and their applications
- Covers new ecological indicators that can be applied to address forest environmental issues
- Includes all types of models: empirical, gap, and process-based models
- Explains several basic ecological and management concepts in a clear, easy-to-understand manner

This handbook is intended for researchers, academics, professionals, and undergraduate and graduate students studying and/or involved in the management of forest ecosystems.

Applied Ecology and Environmental Management

Series Editors: Steven M. Bartell, University of Tennessee and Sven E. Jorgensen, Copenhagen University, Denmark

The ecology is closely tied to environmental science and management. As a result, ecology has developed from a more descriptive to a more quantitative science, to be applicable as support for environmental management decisions. Several ecological sub-disciplines with emphasis on the application aspects have emerged as a result of this development:

Ecological Modeling
Ecological Indicators
Ecotechnology
Ecological Engineering
Ecohydrology
Ecological Economics
Ecological Informatics

This book series presents the state-of-the-art and an overview of these ecological sub disciplines with emphasis on the application in environmental management. It summarizes all relevant and basic knowledge needed to understand and utilize in environmental science, environmental management, applied ecology and pollution abatement.

Sustainable Development Indicators: An Exergy-Based Approach
Søren Nors Nielsen

Environmental Management Handbook, Second Edition—Six Volume Set
Edited by Brian D. Fath and Sven Erik Jorgensen

Managing Environmental Data: Principles, Techniques, and Best Practices
Gerald A. Burnette

Global Blue Economy: Analysis, Developments, and Challenges
Md. Nazrul Islam and Steven M. Bartell

Memoirs of an Environmental Science Professor
William Mitsch

Ecological Forest Management Handbook, Second Edition
Edited by Guy R. Larocque

For more information on this series, please visit: www.routledge.com/Applied-Ecology-and-Environmental-Management/book-series/CRCAPPECOENV?pd=published,forthcoming&pg=2&pp=12&so=pub&view=list

Ecological Forest Management Handbook

Second Edition

Edited by Guy R. Larocque

CRC Press
Taylor & Francis Group
Boca Raton London New York

CRC Press is an imprint of the
Taylor & Francis Group, an **informa** business

Designed cover image: © Guy R. Larocque and Jennifer A. Holm

Second edition published 2025
by CRC Press
2385 NW Executive Center Drive, Suite 320, Boca Raton FL 33431

and by CRC Press
4 Park Square, Milton Park, Abingdon, Oxon, OX14 4RN

CRC Press is an imprint of Taylor & Francis Group, LLC

First edition published by CRC Press 2016

Library of Congress Cataloging-in-Publication Data
Names: Larocque, Guy R., 1955– editor.
Title: Ecological forest management handbook / edited by Guy R. Larocque.
Description: Second edition. | Boca Raton, FL : CRC Press, [2025] | Series: Applied ecology and environmental management | Includes bibliographical references and index.
Identifiers: LCCN 2024011049 (print) | LCCN 2024011050 (ebook) | ISBN 9781032555171 (hbk) | ISBN 9781032555188 (pbk) | ISBN 9781003431084 (ebk)
Subjects: LCSH: Forest management—Handbooks, manuals, etc. | Ecosystem management.
Classification: LCC SD411 .E26 2024 (print) | LCC SD411 (ebook) | DDC 333.75/16—dc23/eng/20240617
LC record available at https://lccn.loc.gov/2024011049
LC ebook record available at https://lccn.loc.gov/2024011050

ISBN: 978-1-032-55517-1 (hbk)
ISBN: 978-1-032-55518-8 (pbk)
ISBN: 978-1-003-43108-4 (ebk)

DOI: 10.1201/9781003431084

Contents

SECTION IV Addressing Ecological and Socioeconomic Issues

Preface to the First Edition

Ecological forest management can be defined as the use of forest-specific knowledge and understanding of ecological processes to guide the development of adaptive management scenarios to harvest resources in a sustainable way and provide ecosystem services and social functions and benefits (Hummel and O'Hara 2008). The term "forest ecosystem management" is commonly used and appears synonymous to "ecological forest management" with respect to the importance of understanding ecological processes and applying sustainable development principles (see Kimmins 2003; Iftekhar 2005; Patry et al. 2013). However, there is a difference between both terms. In Chapter 1, the concept of ecosystem management is discussed in the context of the application of the knowledge of natural disturbances to ensure the maintenance of biodiversity in different management regimes. Ecological forest management is a more general term that is associated with the concept of sustainable development. Advancements in sustainable development for the management of natural resources have occurred over the last few decades during formal conferences on the establishment of global environmental policies (Holvoet and Muys 2004). Independent of the terminology used, it should be mentioned that the application of the basic principles of ecological forest management has increased in importance during the last few decades because of specific requirements originating from public pressure, which demands that policymakers or forest managers ensure the sustainability of forest management practices; consider the needs and concerns of end users; preserve the diversity of ecological structures, functions, and species richness; and support solid economic activities (Côté et al. 2010).

The application of the basic principles of ecological forest management is a complex process that requires broad ecological knowledge and the use of quantitative and qualitative metrics, such as indicators, and aid-decision tools to predict the development of forest ecosystems. For these reasons, this book covers a wide range of topics, including management concepts, forest models, and ecological indicators. The fact that this book covers a series of interrelated topics offers a significant advantage. Several books have been published on forest management, modeling, and ecological indicators, but the topics covered in each one are within their area of specialty. For instance, forest management books focus on how to manage forests to ensure long-term industrial wood supplies; forest modeling books generally focus on a specific type of forest model, either empirical growth and yield models or process-based models. This book provides readers with a solid introduction to the essential concepts that are at the foundation of ecological forest management practices, which will allow them to develop a broad perspective on the application of its basic principles. As ecological forest management is a vast subject, this book does not include all the detailed knowledge and concepts on forest management concepts, models, or ecological indicators. However, each chapter does include a comprehensive list of references that will allow readers to learn more about the different topics presented in the book.

Section I covers forest management concepts, beyond the traditional notions of sustained yield or wood supply analysis, and reviews basic concepts on management systems, wildlife management, political and social issues, and ecosystem services. Chapter 1 describes how the application of knowledge on natural disturbances contributes to the development of an ecosystem management approach. A review of ecological principles for application in boreal, temperate, and tropical forest types is presented. Triad forest management is introduced in Chapter 2. It consists of dividing a forest area into three functional zones: conservation, extensive management, and intensive wood production. The challenges and difficulties that may occur in real-world applications are reviewed for three North American forest regions. The potential benefits for worldwide application are also discussed. Chapter 3 reviews the basic principles of forest wildlife ecology and management. The impacts of disturbances and management regimes are described. As mentioned, political and social issues and ecosystem services are important components in the decision-making process for the application of the basic principles of ecological forest management. In Chapter 4, three processes that have influenced the social and political environment of forest management practices are presented: urbanization, tertiarization, and globalization. Chapter 5, on ecosystem services, reviews the different types of services that forest ecosystems provide, which are classified into provisioning, regulating, cultural, and supporting services. Linkages with ecological indicators are discussed. Two case studies are presented for Canadian ecosystems and mangrove forests.

Section II focuses on models that support ecological forest management. Chapters 6 through 8 review the basic concepts of the three categories of forest models associated with the prediction of productivity or dynamic processes: growth and yield, succession, and process-based models. The fact that there are three chapters on productivity models may appear unusual. However, the three types of models differ in characteristics, usefulness, strengths, and weaknesses (Larocque 2008). Growth and yield models (Chapter 6) require input data that can be easily obtained from inventory surveys conducted by forest agencies, including government forestry departments or forest companies. These models are very useful for performing annual allowable cut computations and simulating silvicultural treatments over short time periods (e.g., 20–50 years). Succession models (Chapter 7) are suitable for the examination of succession issues over both short- and long-time periods (e.g., 150–200 years), as they simulate the successive occurrences of tree growth and mortality and regeneration. They require, as inputs, relatively simple data that can be obtained from inventory surveys and more complex data, such as climatic data. Compared to growth and yield models, succession models have more flexibility to simulate the dynamics of uneven-aged mixed forests. Process-based models (Chapter 8) have a long-term projection time frame and are most suitable for evaluating the impacts of disturbances, such as climate change. However, they require complex input data, such as climatic data or species-specific ecophysiological parameters related to photosynthesis, respiration, transpiration, or carbon or nutrient contents. Chapter 9 introduces basic concepts on carbon budget models, which can be used to estimate greenhouse gas exchanges in forests and the forestry sector at regional or national levels. Three recognized models are described: CBM-CFS, FORCARB, and CO2FIX. Ecosystem models are also used to predict the impact of pest damage (Chapter 10). The effect of pests on the stand structure and basic concepts on pest epidemiology are reviewed. An overview of pest models is presented, and case studies for British Columbia, Canada, based on the use of model components within the TASS/TIPSY growth and yield model are described.

In the scientific literature, the word "model" is used to represent either a single mathematical equation or a system of interrelated equations integrated in applications to perform simulations. This situation creates confusion, as it is not always evident to what the word "model" refers to, particularly in titles. For consistency with the literature, the distinction between both uses of the word "model" is not made in this book; however, the confusion is noted in Chapter 6. As many systems of equations have been developed over the last few decades to perform simulations, it is appropriate to suggest highlighting more formally the difference in the use of the word "model." For systems of equations used in simulations, it would be logical to use the word "simulator." This word has been used occasionally in the literature on forest models, for example, by Hilt (1985), Salminen et al. (2005), Schultz and Matney (2006), and in the name of the official growth and yield model of the U.S. Department of Agriculture (USDA) Forest Service, FVS, which stands for Forest Vegetation Simulator.[1]

In Section III, several ecological indicators for forest ecosystems are discussed. In a paper on the development and use of ecological indicators, Dale and Beyeler (2001) highlighted the need for methods to address complex issues associated with the optimal management of natural resources and the rapid changes in environmental conditions and social values. Ecological indicators can fulfill these goals by evaluating ecosystem states, assessing their patterns of change over time, or identifying the sources of environmental problems. In theory, the combined use of groups of ecological indicators is the optimal approach and should ideally aim at providing essential knowledge about the composition, structure, and function of ecosystems (Dale and Beyeler 2001). Altogether, the ecological indicators reviewed in this section belong to one of these three categories.

The different chapters do not merely consist of lists of indicators, but, more importantly, they include discussions on how they may contribute to drawing inferences on ecosystem states, how they can be measured, and how they can be used with other tools, such as models. Chapter 11 discusses the importance of forest biomass determination and reviews some of the most widely used methodologies. Examples of applications for tropical, temperate, and boreal forests are provided. Biodiversity indicators are covered in Chapter 12. The broader concepts are explained in the context of the benefits of biodiversity and its importance for sustainable forest management. Also, environmental and anthropogenic factors that may affect biodiversity are discussed. Chapter 13 covers forest integrity. Basic concepts of stability and resilience and metrics to characterize disturbance regimes are reviewed. Examples for some Canadian forest types are discussed. Criteria and indicators that can be used to support sustainable forest management are presented in Chapter 14. The historical background that led to the use of criteria and indicators and the different processes that have been developed are reviewed. Several methodological approaches that can be used for the development of criteria and indicators are explained. In Chapter 15, different indicators of ecosystem and stand productivity are reviewed and discussed in the context of forest functions and services. Chapter 16 discusses turnover times for wood, litter, and soil organic matter, all of which are important indicators for the carbon cycle of forest ecosystems. This chapter nicely complements the review of productivity indicators in Chapter 15.

Forest ecosystem health is a broad topic with many issues. Sustainable land use also includes the maintenance of ecosystem health, which is a very broad topic in itself. For this book, the discussion is restricted to native and exotic forest insects and pathogens (Chapter 17). Extensive lists of examples of insects and diseases that affect the health of boreal, temperate, and tropical forest types are included. These lists provide examples for a discussion on how the interpretation of indicators may be used to evaluate the extent to which diseases or insect infestations may affect ecosystem function or contribute to the development of forest management strategies to decrease stand vulnerability.

Section IV contains three case study chapters. In Chapter 18, the application of surface modeling methods to estimate forest carbon stocks in China is presented. First, the basic principles of surface modeling are introduced. Then, the different steps in the methodology used to quantify carbon stocks at the continental scale are described. Chapter 19 examines forest tenure issues in relation to socioeconomic factors. The Bas-Saint-Laurent region in Québec, Canada, provided a unique opportunity for this type of study, as the spatial distribution of private and public tenures in this region is highly heterogeneous. Also, the forest sector is a very important driver of the economy in this region. Finally, Chapter 20 discusses management strategies that can be applied as adaptive measures to climate change in the temperate–boreal interface in eastern North America. Potential impacts of climate change on forest dynamics are presented. For both even- and uneven-aged forests, silvicultural adaption treatments at various evolution stages of the stands are discussed.

It is my hope that this handbook will open the possibility for readers to learn about the topics and issues in the application of the basic principles of ecological forest management and find many sources of inspiration for their own work or research.

Guy R. Larocque
Québec City, Québec, Canada

NOTE

1 https://www.fs.usda.gov/fvs/

REFERENCES

Côté, P., R. Tittler, C. Messier, D.D. Kneeshaw, A. Fall, and M.-J. Fortin. 2010. Comparing different forest zoning options for landscape-scale management of the boreal forest: Possible benefits for the TRIAD. *For. Ecol. Manage.* 259:418–427.

Dale, V.H. and S.C. Beyeler. 2001. Challenges in the development and use of ecological indicators. *Ecol. Ind.* 1:3–10.

Hilt, D.E. 1985. *OAKSIM: An individual-tree growth and yield simulator for managed, even-aged, upland oak stands*. Research Paper NE-562. Radnor, PA: USDA Forest Service.

Holvoet, B. and B. Muys. 2004. Sustainable forest management worldwide: A comparative assessment of standards. *Int. For. Rev.* 6:99–121.

Hummel, S. and K.L. O'Hara. 2008. Forest management. In *Encyclopedia of ecology*, eds. S.E. Jorgensen and B.D. Fath, pp. 1653–1662. Oxford, UK: Elsevier.

Iftekhar, M.S. 2005. From timber management to forest management: An initial discussion on forest management evolution. *J. For. Sci.* 51:412–429.

Kimmins, J.P. 2003. *Forest ecosystem management: An environmental necessity, but is it a practical reality or simply an ecotopian ideal?* Paper presented at the XII World Forestry Congress, Québec City, Québec, Canada.

Larocque, G.R. 2008. Ecological models: Models for the simulation of forest ecosystem dynamics. In *Encyclopedia of ecology*, 5 vols., eds. S.E. Jorgensen and B.D. Fath, pp. 1663–1773. Oxford, UK: Elsevier.

Patry, C., D. Kneeshaw, S. Wyatt, F. Grenon, and C. Messier. 2013. Forest ecosystem management in North America: From theory to practice. *For. Chron.* 89:525–537.

Salminen, H., M. Lehtonen, and J. Hynynen. 2005. Reusing legacy FORTRAN in the MOTTI growth and yield simulator. *Comp. Electron. Agric.* 49:103–113.

Schultz, E.B. and T.G. Matney. 2006. An integrated growth and yield simulator for predicting lob lolly pine dry weight pulp yields. *Wood Fiber Sci.* 38:672–681.

Preface to the Second Edition

I have been extremely grateful for the positive response that led to the production of the second version of *Ecological Forest Management Handbook*. Twelve chapters have been modified by introducing new concepts, upgrading knowledge on different topics, updating the literature, or adding new figures or tables. In Chapter 1, which discusses forest management in relation to natural disturbances, the introduction has been revamped. Some of the terms are better defined or clarified, and some key concepts are introduced earlier in the text. The characteristics of a suite of natural disturbances are described in more depth. In particular, the importance of improving knowledge of historical disturbances is better highlighted. Some of the sections have been shortened. Strategies to alleviate differences between natural and human disturbances have been updated. Citations have also been updated. Chapter 2 focuses on triad forest management. Additional details on the definition and application of the triad forest management and the context of the discipline of sustainability science are provided. The maintenance of multiple uses of the forests is discussed by highlighting in more depth the importance of ecological silviculture, emulation of natural disturbances, and ecosystem services. In the updated chapter, the authors have provided additional examples of the application of modeling studies to evaluate the applicability of triad management over large areas and the impacts on other forest ecosystem resources, including habitat types for bird species and potential of carbon storage. Additional real-world applications are included and discussed—one in Nova Scotia, Canada, and the other one in Oregon, USA. For both regions, the linkage with government legislation on forest management policy and the implementation steps of triad management are emphasized. The challenges associated with the application of triad management have been updated. In particular, the importance of traditional uses by Indigenous peoples is highlighted. Chapter 3 on forest wildlife management has been modified substantially. Forest succession is redefined to better reflect the importance of composition to transitions among several stages and to capture the role that management plays in driving successional change and manipulating vegetation structure within successional stages. The subsection on edges and fragmentation has been updated. In particular, distinctions between high- and low-contrast edges are better highlighted. New peer-reviewed publications on biomass case studies have been added. Additional figures have been added, and some of them in the first edition replaced. Citations of scientific papers have been updated.

The content of Chapter 5 on ecosystem services has been considerably modified. While the chapter in the first edition included a detailed review of basic concepts, the revised version focuses on understanding ecosystem services, contribution of nature to human well-being, and assessment of services provided by ecosystems. The 2023 wildfire season in Canada is used as a case study to illustrate the extent to which different ecosystem services provided by forest ecosystems can be negatively affected by this type of disturbance. The impacts on people are addressed. The historical roots that led to the development of ideas of ecosystem services are discussed. The role of the Intergovernmental Science-Policy Platform on Biodiversity and Ecosystem Services (IPBES) to develop linkage between science and policy development of ecosystem services is highlighted.

Chapter 6 focuses on empirical growth and yield models and contains one new section: process-based models. The impact of global change on forests and how to include them in growth and yield models are now explicitly detailed. Methods to model the effects of climate change in empirical growth and yield models are discussed and compared with process-based models. In particular, the interactions between climate and tree growth, tree recruitment, and tree mortality have been expanded. The contribution of remote sensing technology to improve the definition and quantification of forest structure and model development has been updated. In particular, new types of variables that could be developed using remote sensing technology are proposed. The use of artificial intelligence algorithms is also discussed. Derivation of indices to quantify the intensity of competition is provided in the section on individual-tree-level competition.

Chapter 7, which discusses forest succession models, have several new elements. The introduction to gap models now includes a description of the first individual-tree models that were developed in the early 1960s, which were either distance-dependent or distance-independent. This introduction is then followed by reasons that supported the development of ecologically based models. Regarding gap models, more details on how they represent biological processes in forest ecosystems are included. There is a new section on hybrid forest succession models that have been developed in the last two decades to assist in forest management planning. Since the first edition, several citations that have focused on evaluating the performance of gap models to predict forest growth and the effects of climate change are cited. The section on forest landscape models has also been updated considerably by explaining in more depth the concepts and adding citations on their use. In particular, several examples on the use of LANDIS are discussed.

In Chapter 9 on carbon budget models, the introduction has been modified to improve the definitions of basic terminology. Other frameworks of carbon accounting models that have been introduced since the first edition and additional study cases are included. A new subsection consisting of combinations of different model types has been added. For instance, discussions on combined forest succession and soil carbon models and carbon budget and forest harvesting models are included. The subsection bioenergy has been further developed by linking carbon extraction from harvested wood products and carbon accounting. The importance of uncertainty issues was further addressed by describing the basic approaches in more depth. The effect of climate change using carbon budget models has been updated with recent literature.

The chapter on biodiversity (Chapter 12) now includes the description of new concepts on biodiversity and covers agreements that were signed on biodiversity-related conventions and programs initiated to monitor biodiversity. Among the new concepts that are introduced, it is important to mention the state response and planetary-based frameworks, which were developed by governments or environmental organizations. The goals of the Kunming Montreal Global Biodiversity Framework, adopted during the fifteenth meeting of the Conference of the Parties (COP 15), are reviewed in depth. Several new concepts highlighting the importance of biodiversity indicators are described. The potential effects of climate change are updated. There are new sections on adaptive management approach to face the challenges of meeting biodiversity targets in the forthcoming decades, and methods to visualize changes in indicators over time are described.

The chapter on stand productivity (Chapter 15) has additional elements introduced about the effects of forest management on forest ecosystems and stand productivity. Global terrestrial primary production values have been updated and further discussed. Three new sections and four figures have been added to the chapter. A new section on competition-based mortality has been added to highlight the results of their study that consisted in analyzing 476 monospecific sample plot data throughout Europe. Several species were included in their analysis to relate mortality of individual trees to competition, age, and site conditions. A second section outlining the concepts on continuous versus cyclic stand productivity has been added. The third new section describes the effects of extreme events. In particular, the impacts of droughts that might occur under climate change are discussed. Also, additional elements on growth trends for pure and mixed stands are included.

Chapter 16 on turnover times of forest floor biomass has been considerably modified. The title has been changed by including the term modeling. This modification reflects more accurately the mathematical process involved in the development of integrated equations to represent the complexity of the relationships of the factors involved in turnover models. The literature on organic matter decomposition experiments to quantify turnover rates has been updated and reviewed in more detail. Additional modeling results are provided and discussed, with additional focus on nitrogen accumulation and variation in temperature and precipitation.

The contents of the chapter on forest ecosystem health and biotic disturbances (Chapter 17) have been notably enhanced for greater clarity. The introduction has been modified to more clearly elucidate the concept of ecosystem health, highlight the importance of considering Indigenous perspectives regarding forest health, and update relevant terminology and literature on non-native species.

Case studies of important forest insects and diseases have been updated with the most recent literature, including, but not limited to, climate change influences, range expansions, economic/ecological impacts, and surveillance and management options. Some tables and figures have been revised and improved by incorporating latest information. Scientific nomenclature has been updated, where necessary, to ensure accuracy.

The chapter on surface modeling (Chapter 18) focuses on methods of its application to estimate forest carbon stocks in China. The inventory dataset used has been updated to include inventory data from 2009 to 2013 and from 2014 to 2018. Regression analyses on the patterns of increase in annual mean carbon stock (AMCS), annual mean carbon density (AMCD), and forest area (FA) predicted by the high accuracy surface modeling (HASM) that estimate carbon stocks have been updated. The implications of the new regression analyses in AMCS and AMCD on carbon stock estimates are discussed.

Guy R. Larocque
Québec City, Québec, Canada

Editor

Guy R. Larocque, PhD, is a research scientist for the Canadian Forest Service, a sector of Natural Resources Canada, at the Laurentian Forestry Centre, Quebec City, Quebec, Canada. He is also an adjunct professor in the Faculty of Natural Resources Management at Lakehead University (Thunder Bay, Ontario, Canada). Dr. Larocque is Associate Editor of *Ecoscience*. Until recently, he served on the editorial board of *Ecological Modelling*. He has published papers on productivity and succession, carbon cycles, uncertainty analysis, and the development of empirical, succession (gap), and process-based models for forest ecosystems and was Guest Editor of special issues in *Ecological Modelling, Écoscience and Forests*.

Contributors

René I. Alfaro
Natural Resources Canada
Canadian Forest Service
Pacific Forestry Centre
Victoria, British Columbia, Canada

Madhur Anand
Global Ecological Change (GEC) and Sustainability Laboratory
School of Environmental Sciences
University of Guelph
Guelph, Ontario, Canada

Benjamin Andrieux
Institut de recherche sur les forêts
Université du Québec en Abitibi-Témiscamingue
Rouyn-Noranda, Québec, Canada

Paul A. Arp
Faculty of Forestry and Environmental Management
University of New Brunswick
Fredericton, New Brunswick, Canada

André Arsenault
Natural Resources Canada
Canadian Forest Service
Atlantic Forestry Centre
Corner Brook, Newfoundland and Labrador, Canada

Raphaël Aussenac
Institut de recherche sur les forêts
Université du Québec en Abitibi-Témiscamingue
Rouyn-Noranda, Québec, Canada

Sue Baker
School of Biological Sciences
University of Tasmania, Sandy Bay Campus
Tasmania, Australia

Jürgen Bauhus
University of Freiburg
Freiburg im Breisgau, Germany

Tanya Baycheva
Forest and Environmental Policy Group
University of Freiburg
Freiburg im Breisgau, Germany

F. Wayne Bell
Ontario Forest Research Institute
Ontario Ministry of Natural Resources and Forestry
Sault Ste. Marie, Ontario, Canada

Yves Bergeron
Institut de recherche sur les forêts
Université du Québec en Abitibi-Témiscamingue
Rouyn-Noranda, Québec, Canada

Jagtar S. Bhatti
Natural Resources Canada
Canadian Forest Service
Northern Forestry Centre
Edmonton, Alberta, Canada

Michal Bosela
National Forest Centre
Forest Research Institute Zvolen
Zvolen, Slovakia

Yan Boulanger
Natural Resources Canada
Canadian Forest Service
Laurentian Forestry Service
Québec, Canada

Luc Bouthillier
Faculté de foresterie, de géographie et de géomatique
Université Laval
Québec, Canada

Remington Bracher
Environment and Climate Change Canada
Canadian Wildlife Service
Gatineau, Québec, Canada

Ryan Burke
Environment and Climate Change Canada
Canadian Wildlife Service
Gatineau, Québec, Canada

Philip J. Burton
Ecosystem Science and Management
University of Northern British Columbia
Prince George, British Columbia, Canada

Virginia Capmourteres
Global Ecological Change (GEC)
and Sustainability Laboratory
School of Environmental Sciences
University of Guelph
Guelph, Ontario, Canada

Oleg Chertov
Institute of Physicochemical
and Biological Problems
of Soil Science
Russian Academy of Sciences
Moscow, Russia

Mohan Chinnarasu
Tropical Forest Research Institute
Jabalpur, Madhya Pradesh, India

Michael G. Cruickshank
Natural Resources Canada
Canadian Forest Service
Pacific Forestry Centre
Victoria, British Columbia, Canada

Michel Cusson
Natural Resources Canada
Canadian Forest Service
Laurentian Forestry Service
Québec, Canada

Jennifer Dacosta
Ontario Forest Research Institute
Ontario Ministry of Natural Resources
and Forestry
Sault Ste. Marie, Ontario, Canada

Victor Danneyrolles
Institut de recherche sur les forêts
Université du Québec en
Abitibi-Témiscamingue
Rouyn-Noranda, Québec, Canada

Ronnie de Camino
Centro Agronómico Tropical
de Investigación y Enseñanza (CATIE)
Turrialba, Costa Rica

Johanne Delisle
Natural Resources Canada
Canadian Forest Service
Laurentian Forestry Service
Québec, Canada

Pierre DesRochers
Natural Resources Canada
Canadian Forest Service
Laurentian Forestry Service
Québec, Canada

Mario Di Lucca
British Columbia Ministry of Forests, Land,
and Natural Resource Operations
Victoria, British Columbia, Canada

Christine Farcy
Université de Louvain (UCL)
Earth and Life Institute
Environmental Sciences
Louvain-la-Neuve, Belgium

Nicole Fenton
Forest Research Institute
Université du Québec en
Abitibi-Témiscamingue
Rouyn-Noranda, Québec, Canada

Mathieu Fortin
Natural Resources Canada
Canadian Wood Fibre Centre
Ottawa, Ontario, Canada

Tony Franceschini
Département de biologie,
chimie et géographie
Université du Québec à Rimouski
Rimouski, Québec, Canada

Laurent Gagné
Département de biologie, chimie et géographie
Université du Québec à Rimouski
Rimouski, Québec, Canada

Vincent Gauthray-Guyénet
Département de biologie, chimie et géographie
Université du Québec à Rimouski
Rimouski, Québec, Canada

Rosa C. Goodman
Yale School of Forestry and Environmental Studies
Yale University
New Haven, Connecticut, United States

Steven M. Grodsky
U.S. Geological Survey
New York Cooperative Fish and Wildlife Research Unit
Cornell University
Department of Natural Resources and the Environment
Ithaca, New York, United States

Christian Hébert
Natural Resources Canada
Canadian Forest Service
Laurentian Forestry Service
Québec, Canada

Austin Himes
Department of Forestry and Forest and Wildlife Research Center
Mississippi State University
Starkville, Mississippi, United States

Jennifer A. Holm
Climate Sciences Department
Earth Sciences Division
Lawrence Berkeley National Laboratory
Earth Sciences Division
Berkeley, California, United States

Sarah J. King
Environment and Climate Change Canada
Canadian Wildlife Service
Gatineau, Québec, Canada

Daniel Kneeshaw
Centre d'étude de la forêt/Centre for Forest Research
Département des sciences biologiques
Université du Québec à Montréal
Montréal, Québec, Canada

Alexander Komarov
Institute of Physicochemical and Biological Problems of Soil Science
Russian Academy of Sciences
Moscow, Russia

Nitin Kulkarni
Tropical Forest Research Institute
Jabalpur, Madhya Pradesh, India

Darshan Kumara
Tropical Forest Research Institute
Jabalpur, Madhya Pradesh, India

Timo Kuuluvainen
Department of Forest Sciences
University of Helsinki
Helsinki, Finland

Guy R. Larocque
Natural Resources Canada
Canadian Forest Service
Laurentian Forestry Service
Québec, Canada

Luc Lavoie
Conférence régionale des élues du Bas-Saint-Laurent
Rimouski, Québec, Canada

Markus Lier
Finnish Forest Research Institute
Joensuu, Finland

Jinxun Liu
San José State University Research Foundation
San José, California, United States

Christian Malouin
Environment and Climate Change Canada
Canadian Wildlife Service
Gatineau, Québec, Canada

Véronique Martel
Natural Resources Canada
Canadian Forest Service
Laurentian Forestry Service
Québec, Canada

Olivier Martin-Ducup
Département de biologie, chimie et géographie
Université du Québec à Rimouski
Rimouski, Québec, Canada

Inazio Martinez de Arano
European Forest Institute—Mediterranean Regional Office (EFIMED)
Sant Pau Art Nouveau Site
Barcelona, Spain

Modeste Meliho
Chaire de recherche sur la forêt habitée
Département de Biologie
Université du Québec à Rimouski
Rimouski, Québec, Canada

Christian Messier
Centre for Forest Research
Département des sciences biologiques
Université du Québec à Montréal
Montreal, Québec, Canada

Christopher E. Moorman
North Carolina State University
Department of Forestry and Environmental Resources
Fisheries, Wildlife, and Conservation Biology Program
Raleigh, North Carolina, United States

Patrick Morin
Département de biologie, chimie et géographie
Université du Québec à Rimouski
Rimouski, Québec, Canada

Niya Nijat
Environment and Climate Change Canada
Canadian Wildlife Service
Gatineau, Québec, Canada

Marco Pautasso
Animal and Plant Health Unit
European Food Safety Authority
Parma, Italy

Changhui Peng
Centre for Forest Research
Département des sciences biologiques
Université du Québec à Montréal
Montreal, Québec, Canada

Susan Preston
Environment and Climate Change Canada
Canadian Wildlife Service
Gatineau, Québec, Canada

Hans Pretzsch
School of Life Sciences
Technical University of Munich
Freising, Germany

Deepa Pureswaran
Natural Resources Canada
Canadian Forest Service
Atlantic Forestry Centre
Fredericton, New Brunswick, Canada

Danny Rioux
Natural Resources Canada
Canadian Forest Service
Laurentian Forestry Service
Québec, Canada

Eduardo Rojas Briales
Food and Agriculture Organization of the United Nations (FAO)
Viale delle Terme de Caracalla
Rome, Italy

Thomas Rötzer
School of Life Sciences
Technical University of Munich
Freising, Germany

Kevin R. Russell
College of Natural Resources
University of Wisconsin—Stevens Point
Stevens Point, Wisconsin, United States

Kishan Sambaraju
Natural Resources Canada
Canadian Forest Service
Laurentian Forestry Service
Québec, Canada

Robert Schneider
Département de biologie, chimie et géographie
Université du Québec à Rimouski
Rimouski, Québec, Canada

Gianluca Segalina
Département de biologie, chimie et géographie
Université du Québec à Rimouski
Rimouski, Québec, Canada

Simon Shamoun
Natural Resources Canada
Canadian Forest Service
Pacific Forestry Service
Victoria, British Columbia, Canada

Vladimir Shanin
Bingen University of Applied Sciences
Bingen am Rhein, Germany

Dipesh Kumar Sharma
Département de biologie, chimie et géographie
Université du Québec à Rimouski
Rimouski, Québec, Canada

Herman H. Shugart
Department of Environmental Sciences
University of Virginia
Charlottesville, Virginia, United States

Tommy Simard
Département de biologie, chimie et géographie
Université du Québec à Rimouski
Rimouski, Québec, Canada

Luc Sirois
Département de biologie, chimie et géographie
Université du Québec à Rimouski
Rimouski, Québec, Canada

Rona N. Sturrock
Natural Resources Canada
Canadian Forest Service
Pacific Forestry Centre
Victoria, British Columbia, Canada

Rebecca Tittler
Centre for Forest Research
Département des sciences biologiques
Université du Québec à Montréal
Montreal, Québec, Canada

Ruben Valbuena
European Forest Institute HQ
Yliopistokatu, Finland

Ram Keerti Verma
Tropical Forest Research Institute
Jabalpur, Madhya Pradesh, India

Yi-Fu Wang
College of Forestry
Beijing Forestry University
Beijing, PR China

Weifeng Wang
College of Ecology and Environment
Nanjing Forestry University
Nanjing, Jiangsu, China

Weimin Xi
Department of Biological and Health Sciences
Texas A&M University—Kingsville
Kingsville, Texas, United States

Tian-Xiang Yue
State Key Laboratory of Resources and Environment Information System
Institute of Geographical Sciences and Natural Resources Research
Chinese Academy of Sciences
Beijing, PR China

Section I

Forest Management Concepts

1 Ecological Forestry Derived from Knowledge on Natural Disturbances

Daniel Kneeshaw and Yves Bergeron

1.1 INTRODUCTION

In forest ecosystems, disturbances causing tree mortality, such as forest fires, insect outbreaks and windthrow, create structural variability and tree species diversity (Bergeron et al. 1999; Frelich 2002; Kneeshaw et al. 2011; Kuuluvainen et al. 2021). This structural–compositional variation is critical for the maintenance of biodiversity. Since species can be assumed to be adapted to conditions created by the dominant natural disturbances, researchers have proposed that ecological forest management should aim to maintain or recreate the conditions found following these natural disturbances. This principle is the foundation for forest ecosystem management based on emulating natural disturbances (Gauthier et al. 2009).

To evaluate whether management follows the principle of emulating natural disturbance, the following questions can be asked (Haeussler and Kneeshaw 2003): do the forest management techniques recreate the conditions necessary to maintain biodiversity and ecological processes? Do forestry practices create conditions beyond the range of natural variability found in non-managed forests? Do forestry practices recreate the complete range of forest structures and conditions observed naturally? How can forestry practices be modified to minimize the differences between natural and managed forest? (Gauthier et al. 2009; Kuuluvainen et al. 2021).

This chapter proposes an approach to answer these questions by identifying key processes and their historical bounds and then proposing practices to reduce the differences between natural and managed forests. As they structure forests at both stand and landscape scales, forest fires and insect outbreaks and windthrow are the main natural disturbances considered while designing forest management systems based on the emulation of natural disturbances (Hunter 1993; Bergeron et al. 1998, 2006). Succession after crown fires initially produce large areas of relatively homogeneous forest conditions in terms of composition and age-class structure; however, when large fires are infrequent, late-successional forest composition and structure is controlled by single tree or group tree mortality (Kneeshaw 2001; Berglund and Kuuluvainen 2021). Heterogeneous multilayered canopy conditions and a mosaic of small forest units characterize stands in such forests in different stages of development (Watt 1947; Pickett and White 1985).

Conceptually, a model for ecosystem management proposed dividing the forest into three structural types, the proportions of which could be determined based on the frequency of large-scale disturbance: (1) young, even-aged stands; (2) mature or overmature stands; and (3) old-growth stands (Bergeron et al. 1999). Each of these structural types was associated with a particular silvicultural treatment (e.g., Bergeron et al. 1999 and Figure 1.1) – clearcutting, or partial cutting, or selection cutting – to emulate forest structure linked to aging. This model has been instrumental in forcing managers to reflect on the overuse of even-aged management strategies in boreal forests. It does so by proposing a rigorous ecologically defendable method to attain a balance between even-aged and uneven-aged management techniques.

DOI: 10.1201/9781003431084-2

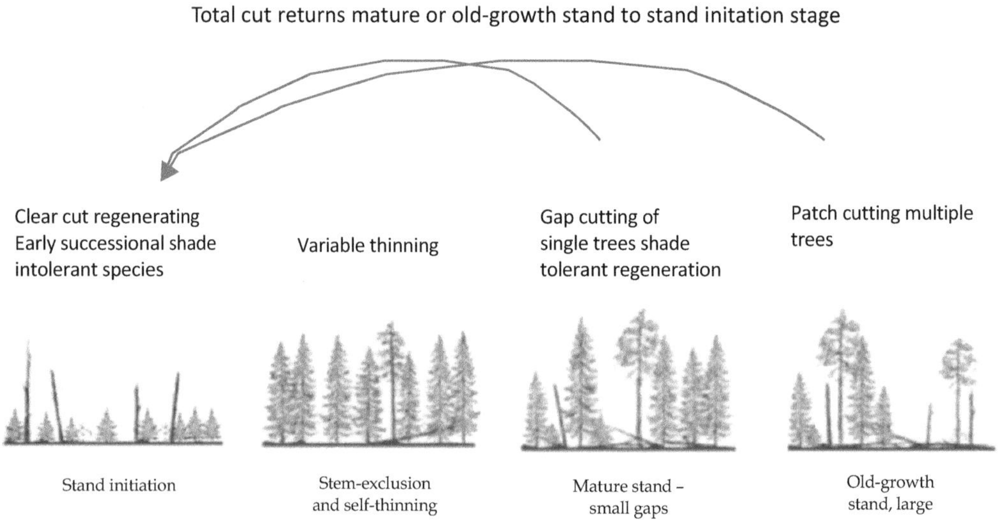

FIGURE 1.1 Conceptual diagram illustrating how natural forest development can be used to develop silvicultural approaches. (Inspired from Kuuluvainen, T., *Ambio*, 38, 309, 2009; Kneeshaw, D.D. et al., Appliquer des connaissances sur les régimes de perturbations naturelles pour développer une foresterie qui s'inspire de la nature dans le sud de la péninsule Gaspésienne, in Gauthier, S. et al. (eds.), *Aménagement forestier écosystémique en forêt*, U Québec Press, Montréal, Québec, Canada, 2008; Oliver, C.D. and Larson, B.C., *Forest Stand Dynamics*, Update Edition, John Wiley & Sons, New York, 1996, 521pp.)

Despite the fact that numerous authors have reflected on how to emulate natural disturbances (Hunter 1993; Kuuluvainen 2002; Bergeron et al. 2002; Kneeshaw et al. 2008), there is still a lack of information on the relative natural occurrence and importance of disturbance types across many forest regions. Such knowledge is critical if we are to evaluate the feasibility of using natural disturbance-based approaches to manage temperate and boreal forests. Examples from around the globe will provide an understanding of the variability in natural dynamics and their potential use in forest ecosystem management, particularly with respect to forest attributes that are profoundly affected by forest harvesting (e.g., stand age and composition, biological legacies, spatial extent of disturbances etc.).

There is urgency to acquire knowledge on the natural range of variability, as industrial forestry practices have led to a reduction in the abundance of species that were either commercially overexploited, such as white pine (*Pinus strobus* L.) and spruces (*Picea* spp.), or those that were viewed as weed species (e.g., aspen (*Populus tremul*) in Fennoscandia) and the uniformization of forest age-class structure, including decline in the area of old-growth forest and large increase in the proportion of younger stands.

In this chapter, we will use as an example, knowledge on the natural disturbance regimes in the mixedwood Gaspé region in northeast Canada. This region was chosen due to its mixed historical disturbance regime: fires, insect outbreaks, and senescence, leading to canopy gaps. Conditions following these natural disturbances will be contrasted with conditions found following the dominant forestry practices. From this exercise, guidelines will be proposed to modify forestry practices to be ecological. For each natural disturbance type, we will examine different characteristics such as the proportion of the landscape affected, disturbance frequency, composition, and structure of post-disturbance stands, as well as the different anthropological and biophysical factors affecting the disturbance regime. This will help us determine the processes driving natural variation of forest ecosystems, in both space and time. Further, examples from selected forests from around the globe (Figure 1.2) will help to describe the global variability in natural forest ecosystem dynamics as shaped by natural disturbances, identify differences between natural systems and managed forests, and subsequently, let us propose guidelines to develop forestry practices that are closer to nature.

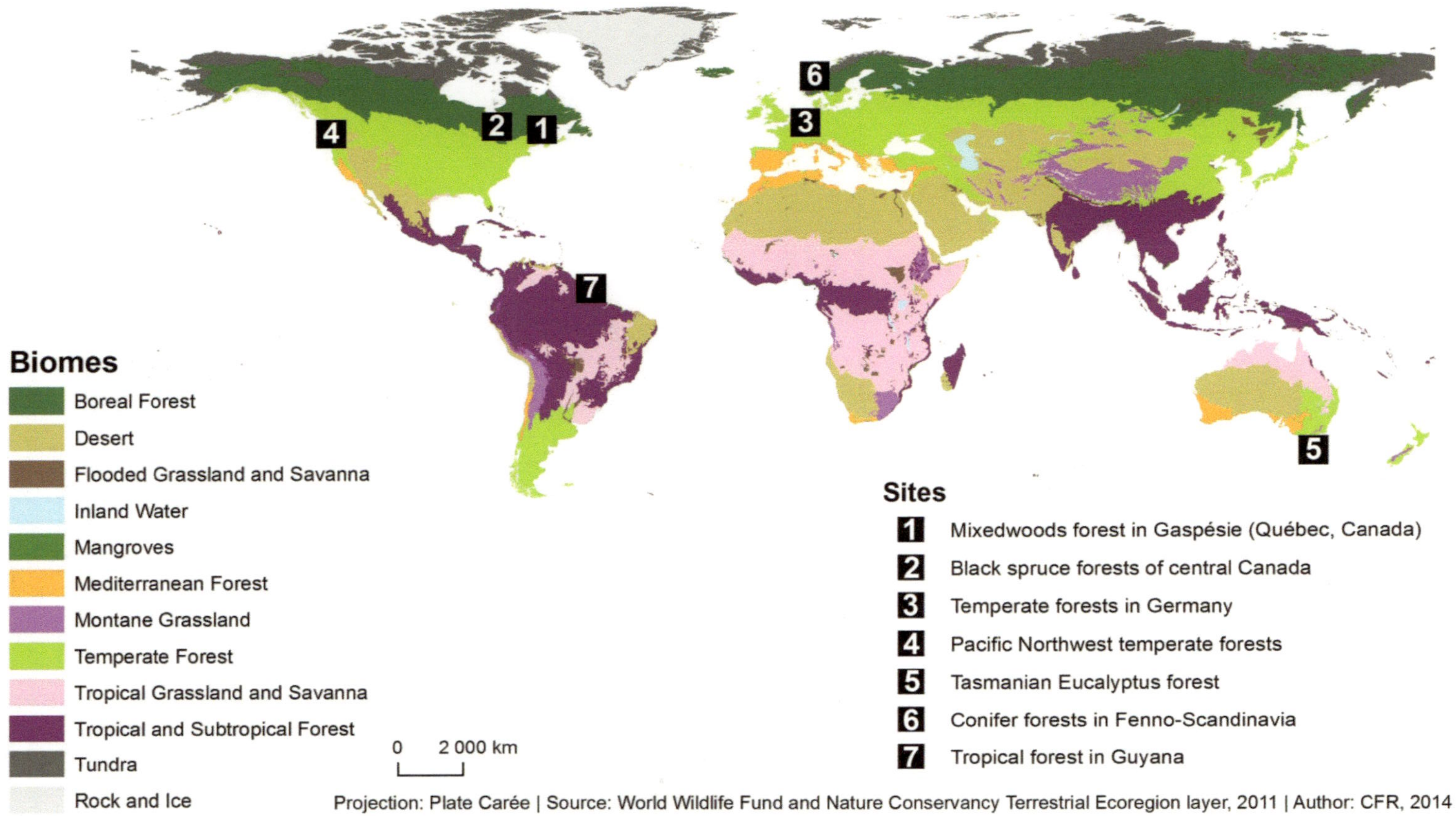

FIGURE 1.2 Location of the study areas presented in the boxes.

1.2 CONTEXT OF FOREST MANAGEMENT: CHARACTERIZING A SUITE OF NATURAL DISTURBANCES

The application of an ecosystem management approach in forestry is based on identifying the range of natural variability in conditions created naturally, which can be done by assessing different characteristics of natural disturbances (Table 1.1). We thus evaluate natural disturbance severity, frequency, specificity, and associated structural legacies (Gauthier et al. 2009; Girona et al. 2023). As with Bergeron et al. (1999), we assume that intense fire returns the forest to even-aged early successional conditions, with regeneration dominated by shade-intolerant, ruderal species. At this stage, small-scale, partial disturbances have little influence on forest development. For ensuing decades, stands are productive and quickly sequester carbon (Anderson-Teixeira et al. 2021; Thuille and Schulze 2006). As stands age, other disturbances have greater effects on forest dynamics, and in most cases, some canopy cover is maintained. Insect outbreaks, for example, even when severe rarely kill all trees, younger trees are much less vulnerable than older ones, and non-host species are not damaged (Kneeshaw et al. 2015, 2021). It is thus in the mature and old growth stages that forests are generally more influenced by insect outbreaks, windthrow, and senescence (Shorohova et al. 2012, 2023).

Four steps can guide the proportion and appropriate type of silvicultural treatments (Smith et al. 1997): (1) calculate the frequency of disturbances that lead to either even-aged or uneven-aged forests, (2) determine the spatial extent and severity of affected areas for the different disturbances,

TABLE 1.1
Determining Silviculture and Forest Management from Natural Disturbances

Application to Forest Management/Silviculture	Natural Disturbance Characteristics
Forest Management Planning	
The frequency of catastrophic disturbance can determine the relative proportions of even- and uneven-aged forest. The frequency of smaller-scale disturbances can be used to determine intervals between interventions.	Frequency (return intervals)
Environmental and habitat conditions in even-aged/total harvest management differ greatly from those in uneven-aged/continuous cover management. The proportion (and its range of variability) of forest affected by catastrophic disturbance will serve as a guideline to determine the proportion of a forest management area to be managed under different silvicultural systems. For example, the proportion of the region recently affected by fire will determine the proportion of the region to be managed by even-aged practices.	Proportion and range of variability of forested landscapes affected by different types of disturbance
Silviculture	
Severity at the canopy level (determined by the proportion of trees killed within the disturbed area) can be used to determine the silvicultural technique to be used at the stand scale. For smaller-scale disturbances, severity is often measured by the size of openings in the forest stand.	Severity (tree mortality) and conversely residual live trees or other biological legacies (e.g., snags, coarse woody debris, etc.)
For catastrophic disturbances, the size, number, and configuration of live tree residual groups can be important to ecosystem functioning. The size of the catastrophic disturbance may be important in its long-term effect on large contiguous blocks of mature and old-growth forest.	
Coupled with disturbance severity, this criterion helps to determine the silvicultural technique to employ (i.e., regeneration of shade-intolerant, pyrogenous species such as jack pine, could be used to justify clear-cutting, soil scarification, and planting (or seeding))	Regeneration composition after disturbance
Shade-tolerant species could be maintained in smaller openings.	

i.e., spatial extent to determine how much area to disturb/harvest and severity to establish the different structures associated with, e.g., uneven-aged forests, (3) determine the composition and density of regeneration after disturbance to evaluate whether forest interventions are necessary to reestablish natural stand conditions, and (4) legacies—in terms of standing and downed woody debris, germination substrates, and at larger scales connectivity. All can be easily manipulated by forestry practices.

Managers should evaluate the conditions created by natural disturbances that are important for forest regeneration, composition, or structure and ensure that these processes are emulated. The first step of our approach is, to determine the proportion of forest affected by large, catastrophic disturbances (i.e., fires). This is used to determine the proportion of the forest that could be expected to be naturally dominated by even-aged forests (sensu Bergeron et al. 1998). The remainder of the forest is older and composed of mixed-aged forests affected by disturbances of varying severity and size. Forest development in the older stages is not linear but depends on the occurrence of intermediate disturbances punctuating mortality due to natural stand senescence.

In even-aged areas of the forest, total cuts could recreate even-aged conditions. In this case, clearcutting followed by scarification and planting or seeding could be viable alternatives to establish pyrogenous pioneer species. Techniques that favor the protection of shade-tolerant advance regeneration will, in many cases, not return the forest to a composition typical of that found in burns but could be appropriate in large windthrows where advance regeneration is not destroyed. Biological legacies such as live- and dead-leave trees could also be maintained in part through different forms of variable retention. Seed tree systems may also be used as an even-aged management tool to emulate natural legacies for structure and the recruitment of the subsequent cohort.

As senescence begins, gaps are small but become larger in the old-growth stages. This understanding of structure thus suggests a reversal of Bergeron et al's. (1999) model in which their second cohort would be better emulated with small single-tree harvesting and patch cuts with larger openings being used to create structures associated with old-growth stages. Emulating insect disturbances could focus on the most devastating insects such as the spruce budworm, which attacks primarily late successional conifers (balsam fir, and white, red, and black spruce), and thus the largest openings could be used to emulate a multi-cohort structure.

Silviculturally, this multi-cohort structure could be created by patch cuts or harvesting with the protection of small merchantable stems and advance regeneration. In other words, techniques that protect shade-tolerant advance regeneration are more appropriate to emulate disturbances that primarily affect the overstory. Irregular shelterwood, group harvesting, or selection harvesting could be used to emulate smaller disturbances. In these latter cases, transitions to shade-tolerant species (i.e., directional succession regeneration) would often occur naturally and advance regeneration should be favored. Windthrow would favor advance regeneration but would also expose some mineral seedbeds that would maintain earlier successional species.

Beyond attempting to maintain and emulate forest composition and structures within historical bounds, there is also a need to consider future conditions. There is an argument that climate change will move forests outside of historical limits and that the past should not be used to inform decisions (Messier et al. 2019). However, this critique is mainly linked to the emulation of patterns rather than processes. We argue that, by understanding the processes that lead to different structures and compositions, ecosystem management through the emulation of natural disturbances is a powerful tool to ensure biodiversity is maintained in the face of global change. If the goal is to maintain biodiversity, then it should be remembered that structural attributes are important for many processes and wildlife. Large dead trees are for example important for cavities and cavity-using organisms (Cadieux et al. 2023). Finding a balance between even-aged and uneven-aged forests will be associated with structures but also different forest composition (Nolet et al. 2018). Disturbances that cause soil scarification versus those that maintain the understory should be used to ensure a balance between creating conditions for early successional species and techniques that protect advance regeneration.

Knowledge of historical disturbance regimes will be critical to highlight when cumulative disturbances, i.e., those that are natural and harvesting, will move ecosystems into a new state-space (Dakos and Kefi 2022). If forest resilience is a goal, then the identification of tipping points or limits, where forested ecosystems may be approaching tipping points, needs to be considered. However, knowledge on tipping points may be in the realm of uncertainty. In such a case, homogenization of the forest, often used to increase operational productivity, may be counterproductive to long-term resilience, and maintaining the full suite of natural variability would have a greater success in ensuring long-term resilience.

1.3 EXAMPLE: MIXEDWOODS OF THE GASPÉ PENINSULA IN NORTHEASTERN CANADA

The study area is located in the Chaleur Bay region of the Gaspé Peninsula, in southeastern Québec, between 64° 22′ and 67° 42′ W and 47° 49′ to 49° 15′ N (Site 1, Figure 1.2). The region is within the northern limit of the temperate forest zone and southern limit of the boreal forest zone. The study area is mainly located in a boreal mixedwood forest dominated by balsam fir (*Abies balsamea* L.), white birch (*Betula papyrifera* Marsh), yellow birch (*Betula alleghaniensis* Britt.), white spruce (*Picea glauca* (Moench) Voss), black spruce (*Picea mariana* (Mill.) B.S.P.), and trembling aspen (*Populus tremuloides* Michx.) (Saucier et al. 1998). The land is mostly composed of mesic sites (95%) while water represents approximately 1% of the surface area. The mean annual temperature is 2.5°C and the mean annual precipitation varies between 900 mm and 1,200 mm, with 35% occurring as snowfall.

1.3.1 Management Strategies and Proposed Silvicultural Practices

Based on the prescriptions we set out, the first step was to calculate the fire return interval that was defined as the mean time since fire in the forest. The mean time since fire of 161 years or a frequency of 0.62% (Lauzon et al. 2007) leads to almost 40% of the area being even aged (Bergeron et al. 1999). The average fire size is over 7,000 ha, although there is considerable variation from 27 ha to over 150,000 ha. Given that the largest fires also contain residual forest blocks (e.g., Bergeron et al. 2002; Dragotescu and Kneeshaw 2012), these sizes could be used to guide sizes of agglomerated harvest blocks rather than be used to determine a single cut-block size. The other important element to note is that the majority of the post-fire regeneration in these forests is composed of pyrogenous species, i.e., black spruce and paper birch. Since balsam fir is omnipresent in stands surrounding the burns as well as a dominant species in residual blocks, it seeds into the edges of burned areas although it is a minor overall component in most burns. Harvesting with advance regeneration techniques, which tend to promote greater proportions of balsam fir in the post-harvest stands (Kneeshaw et al. 2022), should NOT be used. Instead, black spruce and paper birch should be planted or favored through scarification and removal of the advance regeneration layer.

Natural fires burn small to large areas but also leave large distances between burns. Forest managers should thus ensure that even-aged stands are not distributed evenly across the landscape but should be in aggregated blocks separated by large distances. The intervening forests should be dominated by mixed-age and uneven-aged stands (Belleau et al. 2007). In other words, foresters should consider the matrix in between and surrounding even-aged cuts. From a temporal perspective, large burns eventually led to large contiguous areas of old forests. These large blocks of old growth are one of the main challenges for foresters to maintain or recreate (Kneeshaw et al. 2018).

Disturbances in these forests are windthrow and spruce budworm (SBW) outbreaks. These disturbances cover a gradient of small openings to patches as large as 100 ha for windthrows and 45 ha following SBW outbreaks (Kneeshaw et al. 2008). These disturbances could thus be emulated by small-to-large patch cuts covering about one-fifth of the total land area (Table 1.2) being created at an annual rate of 0.5% for the larger patches and 1.1% for the smaller gap cuts. Most of the regeneration

TABLE 1.2
Principal Natural Disturbances and Their Influence on Forests in the Gaspé Peninsula

Category of Opening Size	Frequency	Proportion of Region Affected	Area Disturbed/ Severity/Legacies	Post-Disturbance Regeneration
Large, catastrophic disturbances	Mean time since fire: 161 years Range: 89–250 years	Mean: 39%	Mean: 7,330 ha	Black spruce and white birch dominate
Severe effects to soils and the canopy (e.g.: fire, catastrophic windthrow)	Fire: Frequency: 0.62%/yr Range: 1.12%–0.4%/yr Severe windthrow: Range: 0.12%/yr Range: 0.16%–0.005%	Range: 27%–59%	Range: 27–158,459 ha High (but variable) severity 70%–95% overstory mortality Irregular shapes— with 5%–30% retention of single to multiple hectare islands	Balsam fir always present at low densities
Moderate-severity disturbances (e.g.: windthrow, insect outbreaks, etc.)	Partial windthrow: 1.29%/year Range: 2%–0.22%/yr SBW affects fir SBW cycle Insect outbreaks: Range: 33 years at regional scale: 42% area in fir and mortality rates from 14% to 41% Average Frequency: 0.38% Range: 0.18–0.52%	For both windthrow and insect outbreaks combined: Mean: 22% Range: 11%–29% Partial (50%–75%) mortality Mean: 10% of territory Range: 6%–14%	For both windthrow and insect outbreaks combined 0.5–100 ha Following insect outbreaks most non-host and smaller vigorous hosts survive	After windthrow, composition is variable, depends on pre-disturbance stand composition 20%–65% balsam fir After spruce budworm outbreaks, balsam fir dominates 71%–89%
Small disturbance (e.g. canopy gaps due to insects, partial windthrow, and senescence)	Mean: 1.1% year–1 Range: 0.7%–2.7% year	Mean: 39% Range: 12%–62%	75% of canopy gaps <20 m² 25% of canopy gaps >225 m² % canopy openness in forest stands: Mean: 42% Range: 18%–64%	Shade-tolerant balsam fir dominated (60%–95%)

Note: Both mean values and ranges are presented to ensure that managers are provided with knowledge on the natural variability to create prescriptions that vary across a landscape (frequency for fires is presented as both a frequency and mean time since fire for stands in the landscape which has been suggested to be a more robust value than fire cycle: frequency for insect outbreaks refers to the return interval of outbreaks since outbreaks only affect host species we consider the proportion of the trees that are vulnerable and their mortality rate; proportion of region for fires refers to first cohort of even-aged stands; disturbed areas for gaps have two components: the individual gap opening size and the proportion of the stand in canopy gaps).

in these small openings is from advance regeneration of the shade-tolerant balsam fir. In fact, approximately 80% of the regenerating stems in natural openings are balsam fir, and thus harvesting with protection of advance regeneration makes sense in these smaller openings. It is primarily in the windthrows (and not the insect disturbed stands) where there is some soil disturbance that allows other species to recruit (Reyes and Kneeshaw 2008). This observation is thus a key element for ecosystem management, where the process of soil scarification is more important than windthrow return intervals (DeGrandpré et al. 2018). Spatially, insect outbreaks are concentrated in older balsam-fir dominated forests. In other words, patch cutting of balsam fir that then regenerates back to fir would emulate natural processes (Kneeshaw et al. 2022). Windthrow, on the other hand, affects all forests and is spread out across the landscape. Windthrow is one of the disturbances that reset the clock and bring back some shade-intolerant or less tolerant species (like paper birch or white spruce).

Finally, almost as much land area (39%) is covered by small gaps as by even-aged stands (Table 1.2). Most of the openings are single- or multi-tree gaps suggesting that some form of selection or group selection or irregular shelterwood system may be appropriate. The natural turnover rates of up to 2.7 year^{-1} (average 1.1% year^{-1}) in these forests due to gap dynamics are quite high and could be an argument for frequent entries that maintain a high proportion of forest cover. These small openings greatly favor the recruitment of shade-tolerant advance regeneration, which in this region is primarily balsam fir. As an alternative to frequent harvests of small groups of trees, it is also reasonable to prescribe a systematic removal of trees in a reduced number of passes, given that at a stand scale 42% of the forest is open (58% in forest cover). For example, Beaudet et al. (2011) and Ligot et al. (2014) showed that by harvesting trees in aggregated patches or using systematic removal (e.g., of every third or fourth tree), foresters can control the proportion of canopy opening and understory light levels (to favor different regenerating species) while harvesting the same quantity of wood. Using a uniform harvest of trees, it was possible to remove 60% of the stand basal area without increasing understory light above 50% (Beaudet et al. 2011). Silvicultural practices inspired from continuous canopy forestry models in Europe could thus be applied to harvest timber fiber with minimal impact on overstory cover (Nolet et al. 2018).

All natural disturbances leave a large legacy of standing and downed wood. As the goal of forest management is the harvest of timber fiber, the maintenance of such structural legacies is in direct conflict with the goals of the forest industry. Understanding the importance of different biological legacies may help develop treatments that are compatible with both biodiversity concerns and industrial needs. For example, in conifer-dominated forests, hardwood snags may be limiting for biodiversity (Esseen et al. 1997; Brawn et al. 2001), but they may not be of economic interest or operationally viable to harvest them as the resource is not sufficient for the development of mills. Retention could thus be structured around such trees with the goal of leaving large standing hardwoods for wildlife. In other areas, mills are not able to take large-diameter trees as saws have a restricted range to improve efficiency. In such areas, some large trees could be left for biodiversity, as wildlife disproportionately uses large trees (Ibara et al. 2020). Furthermore, trees that are moribund or already dead should be left standing and become pivot points for islands of retention.

Although the forests of this region are species poor in terms of tree composition, not all forests are monocultures. Balsam fir, although naturally abundant, is resilient but not resistant to SBW disturbances where black spruce is resistant to SBW outbreak but not as resilient once a tipping point is reached (Sanchez-Pinillos et al. 2019).

1.3.2 Comparison of Harvesting and Natural Disturbances

Currently, forest management in this region is based primarily on large-scale cuts that protect advance regeneration. At the landscape level, the resulting forest structure consists of large agglomerations of mostly even-aged forests dominated by balsam fir (Sainte-Marie et al. 2015). This differs in two striking ways from natural forests, the first is that in the natural system less than 40% of the forest was even aged, whereas most managed forests are even aged. If we assume that even- and uneven-aged forests have different structures and that this is important for biodiversity and ecosystem functions, then the goal should be to ensure that human interventions do not unbalance the system.

The second is that in natural post-fire stands, the dominant regeneration was black spruce and paper birch and not balsam fir. Harvesting with the protection of advance regeneration could have led to similar composition as following moderate natural disturbances; however, Belle-Isle and Kneeshaw (2007) noted that following harvesting, there was a much greater proportion of noncommercial hardwoods (i.e., shrubs) than occurred in stands that naturally recruit following moderate disturbances. In European forests, close-to-nature forestry may in fact be far from nature if continuous-cover forestry eliminates the larger openings necessary for the recruitment of shade-intolerant species (Ligot et al. 2014). Current harvesting techniques are thus modifying forest composition at the landscape scale. In fact, it is not just composition that is being altered but the structure of the forest, as large blocks of spruce forest are broken up into smaller blocks. Small blocks of fir forest are also coalescing into larger blocks of fir. These changes radically alter forest vulnerability to future SBW outbreaks (Kneeshaw et al. 2022).

Although a few windstorms doubled (approx. 100 ha) the size of the largest SBW opening, these represent the tail of the distribution, and the same general conclusion can be drawn that harvesting creates a more homogeneous pattern of large openings across the landscape. Openings due to harvesting also differ from natural openings in that they contain far fewer elements of retention (Patry et al. 2013). Dead snags and patches of live trees are far less abundant in cut blocks than in naturally disturbed stands.

1.4 STRATEGIES TO DECREASE DIFFERENCES BETWEEN NATURAL AND HUMAN DISTURBANCES

Emulating the natural proportions of large openings due to stand-initiating disturbances and to disturbances that only partially remove the canopy is a first step in planning forest management to ensure the suite of natural conditions are maintained across the landscape. For example, group selection and continuous canopy forestry are techniques effectively absent from the forest management strategies employed in many commercial boreal forest operations (see Boxes 1.1 through 1.6) despite the fact that a large proportion of the natural forest in areas throughout the world is dominated by small openings (Box 1.5). Although these are sometimes viewed as not economically viable, our work demonstrates that much of the landscape is open and that turnover rates are high, suggesting that biomass removal could be substantial as long as forest cover is maintained.

Variable retention forestry is an alternative that permits foresters to create a variety of conditions found across most landscapes and at multiple temporal and spatial scales (Box 1.4). It should also be encouraged, as it would force foresters to reflect upon what is left behind as well as what is removed (Box 1.3). Besides maintaining a balance between partial and almost total canopy removal, greater understanding of the role of disturbances in determining stand composition is needed. Rhetorical suggestions that advance regeneration by being natural should lead to natural forest conditions should not be accepted if the dominant disturbance removes the advance regeneration layer and creates conditions for heliophilic pioneer species. Forest managers should thus use a palate of silvicultural techniques, including scarification, to create the appropriate conditions for the recruitment of desired species.

Ecosystem management based on the emulation of natural disturbances has not been accepted by all. It has been criticized for being based on the past and thus not a useful tool to prepare forests for unknown future conditions due to global change. Such an argument is based on the misconception that the primary goal is to replicate arbitrary past patterns. Instead, we remind the reader that the goal of natural disturbance-based ecosystem management is to use an understanding of natural disturbances to first determine how humans are changing forests and subsequently how to reduce this impact. In other words, it is to offset some human-induced global changes, those due to forest landscape alterations. In a best-case scenario, we will correctly identify the most important parameters to emulate and reduce negative human effects, and in a worst-case scenario, we will reduce one of the stresses that ecosystems are facing and permit species to more successfully transition through other changes such as warming temperature. In a changing world, reducing stresses caused by human-induced forest alteration is arguably a key to the long-term viability of many species.

BOX 1.1 BLACK SPRUCE (*PICEA MARIANA*)–FEATHERMOSS-DOMINATED BOREAL FOREST

An interesting area for a case study of the potential of ecosystem-based forest management is the black spruce–feathermoss-dominated boreal forest of northwestern Québec and northeastern Ontario, where a variety of natural disturbances create a mosaic of stands with both complex and simple structures (Site 1, Figure 1.2). Old-growth stands of this region have multi-storied, open structures and have a low productivity (Figure B1.1), while young post-fire stands, established following severe fires, show dense and even-sized/even-aged structures and are more productive. In contrast, current forest management emulates the effects of low severity fires, which could lead to unproductive even-aged stands. The natural disturbance and forest management regimes differ in such a way that both young productive and old-growth forests could ultimately be underrepresented on the landscape under a fully regulated forest management. An ecosystem-based management regime would be based on both of these dynamics by proposing two groups of strategies to be applied across the landscape, proportionally to the prevalence of these stand types in natural landscapes.

Despite the relatively long fire cycle in this region, large fires followed by successful regeneration of the black spruce, jack pine, and aspen strongly influenced the natural pattern, with approximately 45%–55% of the landscape in even-aged stands (Figure B1.2). Consequently, 45%–55% of the landscape could be managed with even-age techniques, such as clearcuts with appropriate biological legacy retention (as fires do not burn all trees within their perimeters; Madoui et al. 2010). Fires burn all forest types across the landscape. Clearcuts should

FIGURE B1.1 Conceptual diagram illustrating how natural forest development can be used to develop silvicultural approaches. (Inspired from Kuuluvainen, T., Ambio, 38, 309, 2009; Kneeshaw, D.D. et al., Appliquer des connaissances sur les régimes de perturbations naturelles pour développer une foresterie qui s'inspire de la nature dans le sud de la péninsule Gaspésienne, in Gauthier, S. et al. (eds.), Aménagement forestier écosystémique en forêt, U Québec Press, Montréal, Québec, Canada, 2008; Oliver, C.D. and Larson, B.C., Forest Stand Dynamics, Update Edition, John Wiley & Sons, New York, 1996, 521pp.)

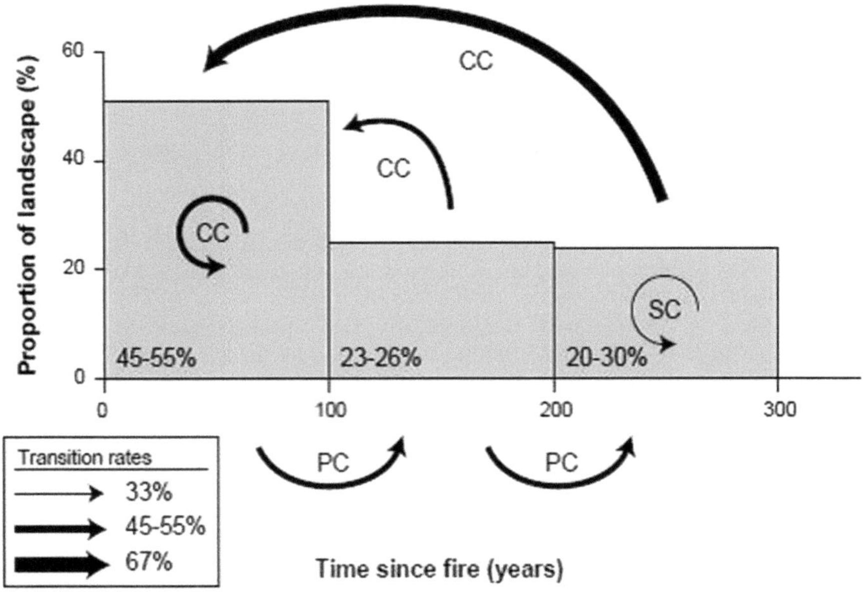

FIGURE B1.2 Locations of the study areas presented in the boxes.

be applied to all forest types present on the landscape, including stands that are currently less productive due to paludification (accumulation of organic material over mineral soil; Simard et al. 2007). However, harvesting these paludified stands without disturbing the organic layer results in less dense stands with poorer growth. Consequently, harvest should be followed by mechanical site preparation and plantation to ensure that the soil layers are disturbed as in a fire and that sufficient high-quality tree microsites are generated for the regeneration of a dense and productive forest.

The balance of this landscape is dominated by forests with complex structures and long post-fire histories. Uneven-aged management via partial harvests could be used to ensure that forests with complex structures remain outside of protected areas. Results from a series of operational tests of partial harvest in black spruce forests have indicated that most old-growth-associated species are maintained when 40–60% of preharvest basal area is removed (Fenton et al. 2013). While more trials need to be examined, younger stands seem to be a good target for partial harvest, as the total basal area is higher, leaving a denser stand postharvest. Similarly, our results indicate that partial harvests successfully maintained deadwood levels and deadwood specialist species.

We recognize that by suggesting the use of partial and selection cuts with little soil disturbance, we are encouraging paludification on the one hand, while on the other hand, suggesting the application of aggressive site preparation treatments in cutover areas to counter paludification. While these two strategies appear to be contradictory, they are in reality the logical outcome of a management strategy based on natural disturbances. Both severe stand-replacing disturbances (i.e., fire) and less severe disturbances (tree mortality) are important on this landscape, creating a complex mosaic of both even- and uneven-aged stands. As a result, any management strategy that aims at maintaining biodiversity and ecosystem function by emulating natural disturbances needs to take both of these types of disturbances into account.

To know more: Fenton, N.J., M. Simard, and Y. Bergeron. 2009. Emulating natural disturbances: The role of silviculture in creating even-aged and complex structures in the black spruce boreal forest of eastern North America. *J. For. Res.* 14:258–267.

Nicole Fenton

REFERENCES

Fenton, N.J., L. Imbeau, T. Work, J. Jacobs, H. Bescond, P. Drapeau, and Y. Bergeron. 2013. Lessons learned from 12 years of ecological research on partial cuts in black spruce forests of northwestern Québec. *For. Chron.* 89:350–359.

Madoui, A., A. Leduc, S. Gauthier, and Y. Bergeron. 2010. Spatial pattern analyses of post-fire residual stands in the black spruce boreal forest of western Québec. *Int. J. Wildl. Fire* 19:1110–1126.

Simard, M., N. Lecomte, Y. Bergeron, P.Y. Bernier, and D. Paré. 2007. Forest productivity decline caused by successional paludification of boreal soils. *Ecol. Appl.* 17:1619–1637.

BOX 1.2 CLOSE-TO-NATURE FOREST MANAGEMENT OF EUROPEAN BEECH (*FAGUS SYLVATICA*)-DOMINATED FORESTS IN CENTRAL EUROPE

European beech would naturally, without anthropogenic disturbances and modifications, dominate vast tracts of forestland in central Europe (Peters 1997). Only where site conditions become extreme (e.g., too dry or too wet) would other species gain dominance (Ellenberg and Leuschner 2010). Owing to the long period of intensive human-induced changes to these forests, very little is known about the disturbance dynamics of native beech forests in western and central Europe, despite their widespread natural occurrence. Our knowledge in this regard relies on studies of primeval beech forests in eastern Europe (e.g., Hobi et al. 2014) (Site 3, Figure 1.2). These studies show that these forests are shaped by small gap disturbances, which provide little opportunity for other, more light-demanding species to establish and grow. For example, less than 3% of the basal area is occupied by other species in the Uholka reserve in the Ukrainian Carpathians. In addition, in these unmanaged forests, trees can grow to old ages (the mean age of the dominant tree layer is estimated at 350 years), and large quantities of deadwood (ca. 160 m^3 ha^{-1}) accumulate.

The high ecological stability and small-scale disturbance patterns in these forests can be approximated by close-to-nature forest management, which is the prevailing paradigm of forest management in central Europe (Bauhus et al. 2013). This type of forest management avoids clearfelling and focuses on the continuity of ecosystem processes throughout forest development. Clearfelling is not only avoided but also prohibited by forest legislation in many jurisdictions. This reduces the impact of harvesting on nutrient pools and fluxes (e.g., Bauhus and Bartsch 1995) and optimizes the economic return from harvesting trees individually or in groups when their development in economic value culminates. In addition, the avoidance of clearfelling is associated with a reliance on natural regeneration and associated savings in establishment costs. The retained overstory following harvesting is presumed to have a positive influence not only on survival but also on the development of quality of advance regeneration (stem form and branch size). While this is generally observed for conifer advance growth, it is questionable whether such canopy shading has a similarly beneficial influence on quality development in hardwood regeneration (e.g., Leonhardt and Wagner 2006).

More recently, it has been recognized that it is not just important to maintain a canopy but that it is also important to retain structural elements typical of old forest development stages, which are relatively rare. Economically, mature beech trees are harvested typically between 100 and 160 years of age, depending on the previous thinning regime. Yet a large proportion of threatened species in European forests require deadwood as habitat (Lassauce et al. 2011). Thus, in most public forests, strategies are being implemented to increase the number of habitat trees and deadwood beyond economic production cycles. The retention of forest structures during forest harvesting has evolved as a central concept of ecologically sustainable forest management (e.g., Gustafsson et al. 2012). The nationwide forest inventory in Germany shows that these approaches are being successfully implemented. The proportion of old forests (>100 years) has increased from 2002 to 2012 by 393,000 ha, which is 3.6% of the total forest area. The average volume of deadwood increased in this period by 18% (BMEL 2014).

However, the focus on small-scale disturbances, long regeneration periods, and the continuous maintenance of mid- to late-successional conditions provide for limited regeneration options and favors a selective set of mostly shade-tolerant species (e.g., Wohlgemuth et al. 2002). Concerns have also been voiced that the proportion of light-demanding tree species (e.g., oaks) might decrease. Long-term studies in Swiss forest reserves have shown that the richness of tree species declines following cessation of management (Heiri et al. 2009). Many of these light-demanding species are of special interest, as they are also viewed as having a greater potential to adapt to future climatic conditions (e.g., Lüpke 2004).

Thus, there is an obvious tension between managing the forests in accordance with natural disturbance dynamics on the one hand, which leads to a dominance of European beech and other shade-tolerant species, and on the other hand, managing forests for greater tree species diversity and adaptability in the face of climate change. The latter would require a greater variety of management-related disturbances. Whereas larger disturbances in native hardwood forests are a rare occurrence (e.g., Pontailler et al. 1997), natural disturbances are a common phenomenon in large areas of European secondary conifer forests (Schelhaas et al. 2003), which are more susceptible to storms and bark beetle attack. Owing to the increasing average growing stock and age of Europe's forests, these disturbances are likely to increase.

FIGURE B1.3 A typical submontane European beech (Fagus sylvatica) forest on sandstone managed with target diameter harvesting with long regeneration periods of three to five decades. Advance regeneration of shade-tolerant beech is accumulating beneath the sheltering canopy of mature trees. Individual, economically mature trees with diameters of ca. 60 cm DBH are harvested ca. every 5–7 years. This process opens new and expands existing gaps. Owing to the very gradual opening of the stand, few other shade-tolerant species will regenerate. In most public forests, at least five habitat trees per ha are retained in the process of regenerating such stands. (Photo by Jürgen Bauhus.)

These natural disturbances will create ample opportunities to regenerate forests with more light-demanding and possibly more drought-adapted species. Forest management in these secondary forests could thus include a greater proportion of harvests as large as permitted by legislation. At the stand level, the close-to-nature forest management of European-dominated forests, which maintains these in a mid-to late successional state and thus close to historical reference conditions, may be most suited to protect native biodiversity under the current disturbance regime. However, largely monospecific beech forests may be more vulnerable to new disturbances (Figure B1.3).

To know more: Bauhus, J., K.J. Puettmann, and C. Kühne. 2013. Close-to-nature forest management in Europe: Does it support complexity and adaptability of forest ecosystems? In *Managing Forests as Complex Adaptive Systems: Building Resilience to the Challenge of Global Change*, eds. C. Messier, K.J. Puettmann, and K.D. Coates, pp. 187–213. New York: Routledge/The Earthscan Forest Library.

Jürgen Bauhus

REFERENCES

Bauhus, J. and N. Bartsch. 1995. Mechanisms of carbon and nutrient release and retention within beech forest gaps. I. Microclimate, water balance and seepage water chemistry. *Plant Soil* 168–169:579–584.

Bauhus, J., K.J. Puettmann, and C. Kühne. 2013. Close-to-nature forest management in Europe: Does it support complexity and adaptability of forest ecosystems? In *Managing Forests as Complex Adaptive Systems: Building Resilience to the Challenge of Global Change*, eds. C. Messier, K.J. Puettmann, and K.D. Coates, pp. 187–213. New York: Routledge/The Earthscan Forest Library.

BMEL (Bundesminister für Ernährung und Landwirtschaft). 2014. *Der Wald in Deutschland: Ausgewählte Ergebnisse der dritten Bundeswaldinventur*, p. 52. Bonn, Germany: Bundesministerium für Ernährung und Landwirtschaft.

Ellenberg, H. and C. Leuschner. 2010. *Vegetation Mitteleuropas mit den Alpen in ökologischer, dynamischer und historischer Sicht*, p. 1357. Stuttgart, Germany: Ulmer Verlag.

Gustafsson, L., S.C. Baker, J. Bauhus, W.J. Beese, A. Brodie, J. Kouki, D.B. Lindenmayer et al. 2012. Retention forestry to maintain multifunctional forests: A world perspective. *Biosci.* 62(7):633–645.

Heiri, C., A. Wolf, L. Rohrer, and H.H. Bugmann. 2009. Forty years of natural dynamics in Swiss beech forests: Structure, composition, and the influence of former management. *Ecol. Appl.* 19:1920–1934.

Hobi, M.L., B. Commarmot, and H. Bugmann. 2014. Pattern and process in the largest prime- Val beech forest of Europe (Ukrainian Carpathians). *J. Veg. Sci.* 26:323–336.

Lassauce, A., Y. Paillet, H. Jactel, and C. Bouget. 2011. Deadwood as a surrogate for forest bio-diversity: Meta-analysis of correlations between deadwood volume and species richness of saproxylic organisms. *Ecol. Indic.* 11:1027–1039.

Leonhardt, B. and S. Wagner. 2006. Qualitative Entwicklung von Buchen-Voranbauten unter Fichtenschirm. *Forst und Holz.* 61:454–457.

Lüpke, B.V. 2004. Mixed species forests as a means for risk reduction and close to nature silvi-culture: An area of conflict (in German). *Forstarchiv.* 57:43–50.

Peters, R. 1997. *Beech Forests*: *Geobotany*, vol. 24, p. 169. Dordrecht, the Netherlands: Kluwer Academic Publishers.

Pontailler, J.Y., A. Faille, and G. Lemée. 1997. Storms drive successional dynamics in natural forests: A case study in Fontainebleau forest (France). *For Ecol. Manage.* 98:1–15.

Schelhaas, M.J., G.J. Nabuurs, and A. Schuck. 2003. Natural disturbances in the European forests in the 19th and 20th centuries. *Glob. Change Biol.* 9:1620–1633.

Wohlgemuth, T., M. Bürgi, C. Scheidegger, and M. Schütz. 2002. Dominance reduction of species through disturbance—a proposed management principle for central European forests. *For. Ecol. Manage.* 166:1–15.

BOX 1.3 USING NATURAL DISTURBANCE REGIMES TO GUIDE THE MANAGEMENT OF NORTHWESTERN COASTAL FORESTS

The Pacific Coast of western North America (Site 4 in Figure 1.2) is characterized by a maritime climate and mountainous topography, which in turn influences the distribution, composition, and disturbance regimes governing some of the world's most productive forest ecosystems. Stretching from Oregon to Alaska and from sea level to alpine tundra (approximately 2,000 m elevation in the south, 700 m in the north), the region supports distinctly different forest types at low elevation and high (subalpine) elevations, and on the windward (wet) and leeward (drier) sides of mountain ranges. Trees can achieve great sizes— >90 m in height for the Douglas fir (*Pseudotsuga menziesii* var. *menziesii*) and Sitka Spruce (*Picea sitchensis*)—and great age— >1,000 years for the western redcedar (*Thuja plicata*) and Alaska yellow-cedar (*Callitropsis nootkatensis*). With many coastal species achieving great height and being quite shade tolerant, forests are also able to achieve very high standing volumes (e.g., >1,500 m^3/ha^{-1} for 100–125-year-old stands dominated by the Pacific silver fir, *Abies amabilis*; Crawford and Oliver 1990). Mild winters and plentiful moisture can support mean annual increments exceeding 8–18 m^3 ha^{-1} $year^{-1}$ (Douglas fir, 50-year rotation; Acker et al. 1998; Humphreys et al. 2003), making areas with gentle terrain highly desirable for commercial forestry. With awe-inspiring trees, undeveloped watersheds, unusual fauna, and largely intact ecological processes, this region has also become a priority for protection by biologists and environmental activists.

The disturbance regimes of coastal forest types vary according to climate, with fire prevalent in the drier interior lowlands while storm damage and gap dynamics prevail on the outer coast and at higher elevations. An overview of dominant disturbance agents, return intervals, and opening sizes is provided in Table B1.1.

Garry oak (*Quercus garryana*) savannas and meadows (now mostly urbanized or under cultivation) were historically maintained by regular prescribed burning and root harvesting by Indigenous people. In conjunction with efforts to protect remnant stands and their high biodiversity, management today primarily consists of reintroducing fire (cool surface burns) and the removal of exotic grasses and shrubs. Douglas fir forests in the lee of coastal mountains, especially over wide expanses of interior Oregon and Washington, are typically of fire origin but can have a complex structure due to thick-barked Douglas fir veterans that can survive multiple fires and can achieve magnificent size as the stands transition into old-growth status with gaps and irregular canopy structure (Franklin et al. 1981). Nonetheless, these forests are typically managed on an even-aged basis, though increasingly with the retention of biological legacies in the form of scattered live and dead canopy trees, fallen logs, and clusters of natural regeneration (Hansen et al. 1991) (Figure B1.4). Shelterwood management with adequate seedbed preparation (e.g., from machine traffic, light mechanical scarification, or surface fire) is the logical extension of the natural disturbance regime in this zone. Also, it has been demonstrated that these forests can be managed for continuous forest cover by single-tree selection (L'Anson et al. 1996; Acquah et al. 2023). To maintain forest health, the emulation of natural disturbance patterns and legacies cannot be limited to the harvesting stage but should include a program of variable-density thinning as well, leaving a within-stand patchwork of tree clusters and gaps within regenerating stands (Brodie and Harrington 2020).

Where rainfall is higher, the relatively rare forest fires tend to be severe and cover large areas, resulting in forest stands that start out as even aged. Douglas fir is intimately mixed with western hemlock and western redcedar, which can all establish after wildfire or any disturbance

TABLE B1.1
Natural Disturbance Attributes of Northwest Coast Temperate Forests and Their Incorporation in Ecologically Appropriate Forest Management

Coastal Forest Type	Natural Disturbance Regime	Estimated Disturbance Return Interval[a]	Estimated Range in Disturbance Event Size[a]	Corresponding Silvicultural Regime
Garry oak	Indigenous burning and cultivation	1–20 years	(Not known)	Threatened ecosystem, restoration treatments only: prescribed fire, removal of exotics
Douglas fir	Mixed fire regime	200 years	(Not known)	Selection or shelterwood management with natural regeneration; underburning site preparation possible; even-aged plantation forestry with structural retention
Western hemlock/ Douglas fir	Occasional fire, windstorms, landslides	200–450 years	5–118 ha	Even-aged plantation forestry with structural retention; mechanical site preparation; mixed rotation management possible
Hypermaritime western hemlock	Windstorms, landslides	345–3,000 years	<0.1–8 ha	Single-tree selection for harvesting and regeneration of high-value species; small patch cuts or group shelterwoods with mounding
Mountain hemlock/ Pacific silver fir	Single-tree mortality, snow avalanches	556–1,111 years	<0.2 ha	Single-tree selection for harvesting and regeneration of high-value species; small patch cuts or group shelterwoods on long rotations

[a] Compiled from Everett et al. (2000), Wong et al. (2003), Parminter (2012), Alaback et al. (2013).

FIGURE B1.4 Natural regeneration of Douglas fir and ponderosa pine (Pinus ponderosa) in a stand of commercially thinned Douglas fir that was subsequently treated with prescribed fire. (Photo by Rod Negrave.)

(such as landslides or windthrow) that exposes mineral soil, though hemlock and redcedar can regenerate in shade as well. Over a successional period of several centuries, these more shade-tolerant species tend to replace the fast-growing Douglas fir. Even-aged management prevails, with variable retention (Beese et al. 2019) being the means to most practically conserve old-growth attributes and respect the disturbance ecology heritage of these forests. Interesting challenges and opportunities are associated with fertile floodplains and riparian zones in this region. For example, fast-growing red alder (*Alnus rubra*), paper birch (*Betula papyrifera*), and black cottonwood (*Populus balsamifera* ssp. *trichocarpa*) rapidly establish on old logging roads, skid trails, landslides, and silt deposits, with western redcedar and western hemlock seedlings often establishing in their shade. While not widely practiced, this system lends itself to a mixedwood or mixed rotation management system, harvesting the short-lived broadleaf colonizers approximately 40–50 years after establishment, protecting the understory of conifers for released growth and harvest in another 50–60 years. On fertile floodplains, elevated substrates such as mounds or large conifer logs, under some level of shade, may be needed to successfully regenerate Sitka spruce (*Picea sitchensis*), where the colonizing hardwoods provide protection from the white pine terminal weevil (*Pissodes strobi*).

On the extremely wet outer coasts of Vancouver Island, the British Columbia mainland, Haida Gwaii, and southeast Alaska, where fires are extremely rare (Daniels and Gray 2006), Douglas fir is absent, and the disturbance regime is characterized by landslides, storm damage, and gap dynamics. Landslides and windstorms clearly initiate stand-level forest regeneration, frequently with much exposed mineral soil but typically at restricted topographic locations. The even-aged stands that develop on such sites (which may be dominated by western hemlock or Sitka spruce) are often very productive and are attractive targets for forest harvesting. Where such disturbances are absent, forest floor organic matter accumulates, paludification occurs even on some slopes, and forest stands are maintained only through single-tree replacement (Banner et al. 2005; Hoffman et al. 2021). Under such conditions, stocking and productivity deteriorate, with regeneration favoring species such as western redcedar that can reproduce by vegetative layering. Little commercial forestry occurs in those forests, with selective harvesting restricted to filling local needs or occasionally using helicopter yarding to extract individual trees or small groups of high-value species such as Alaska yellow-cedar. In most cases, the release of advance regeneration (of those same species or by Pacific silver fir) is counted on to regenerate the small gaps that have been harvested.

The prevalence of old-growth forests in Pacific coastal forests indicates that their ecology reflects a relatively stable environment and long intervals between natural disturbance events. Consequently, forest protection in the form of parks and reserves or old-growth management areas (long-term deferrals from timber harvesting) must be an integral part of forest management that sustains the biology and ecological integrity of coastal forests. This principle is reflected in ongoing public protests to stop the logging of old-growth forest. As a result of such pressure and the growing sovereignty of Indigenous peoples, some progressive land use plans have been implemented. For example, one-third of British Columbia's Great Bear Rain Forest is designated for permanent protection, and ecosystem-based management approaches must be practiced to conserve old-growth attributes in a total of 70% of this 6.4 Mha landscape while ensuring functional connectivity of old forest between upland and riparian habitats (Price et al. 2009). Variable retention harvesting and silviculture (Beese et al. 2019; Gustafsson et al. 2012) provide a flexible template for disturbance-based forest management throughout the region. Across a gradient from low to high precipitation, the emulation of natural disturbance grades from small openings with primarily understory disturbance, through large openings with green-tree patches, to progressively smaller openings at high elevations and more coastal locations. Habitat connectivity over space and time is the key to maintaining ecological integrity while still extracting resources from these very productive forest landscapes.

Philip J. Burton

REFERENCES

Acker, S.A., T.E. Sabin, L.M. Ganio, and W.A. McKee. 1998. Development of old-growth structure and timber volume growth trends in maturing Douglas-fir stands. *For. Ecol. Manage.* 104(1–3): 265–280.

Acquah, S.B., P.L. Marshall, B.N. Eskelson, and I. Barbeito. 2023. Temporal changes in tree spatial patterns in uneven-aged interior Douglas fir dominated stands managed under different thinning treatments. *For. Ecol. Manage.* 528, 120640.

Alaback, P., G. Nowacki, and S. Saunders. 2013. Natural disturbance patterns in the temper-ate rain forests of southeast Alaska and adjacent British Columbia. In *North Pacific temperate rain forests: Ecology & conservation*, eds. G.H. Orians and J.W. Schoen, pp. 73–88. Seattle, WA: University of Washington Press.

Banner, A., P. LePage, J.R. Moran, and A. de Groot. 2005. *The HyP3 project: Pattern, process and productivity in hypermaritime forests of coastal British Columbia: A synthesis of 7-year results.* Special Report Series 10, p. 142. Victoria, British Columbia, Canada: Research Branch, B.C. Ministry of Forests. www.for.gov.bc.ca/hfd/pubs/Docs/Srs/Srs10.pdf. Accessed 5 August 2023.

Beese, W.J., J. Deal, B.G. Dunsworth, S.J. Mitchell, and T.J. Philpott. 2019. Two decades of variable retention in British Columbia: A review of its implementation and effectiveness for biodiversity conservation. *Ecol. Process.* 8(1):1–22.

Brodie, L.C. and C.A. Harrington. 2020. *Guide to variable-density thinning using skips and gaps*. Gen. Tech. Rep. PNW-GTR-989, p. 37. Portland, OR: U.S. Department of Agriculture, Forest Service, Pacific Northwest Research Station. www.fs.usda.gov/pnw/pubs/pnw_gtr989.pdf. Accessed 5 August 2023.

Crawford, P.D. and C.D. Oliver. 1990. *Abies amabilis* Dougl. ex Forbes, Pacific silver fir. In *Technical coordinators: Silvics of North America*, vol. 1, eds. R.M. Burns and B.H. Honkala. Conifers. Agriculture Handbook 654, pp. 17–25. Washington, DC: USDA Forest Service.

Daniels, L.D. and R.W. Gray. 2006. Disturbance regimes in coastal British Columbia. *J. Environ. Manage.* 7:44–56. https://jem-online.org/index.php/jem/article/view/542. Accessed 5 August 2023.

Everett, R.L., R. Schellhaas, D. Keenum, D. Spurbeck, D., and P. Ohlson. 2000. Fire history in the ponderosa pine/Douglas-fir forests on the east slope of the Washington Cascades. *For. Ecol. Manage.* 129(1–3):207–225.

Franklin, J.F., K. Cromack, W. Denison, A. McKee, C. Maser, J. Sedell, F. Swanson, and G. Juday. 1981. *Ecological characteristics of old-growth Douglas-fir forests.* Gen. Tech. Rep. PNW-GTR-118, p. 48. Portland, OR: USDA Forest Service. www.fs.usda.gov/research/treesearch/5546. Accessed 5 August 2023.

Gustafsson, L., S.C. Baker, J. Bauhus, W.J. Beese, A. Brodie, J. Kouki, D.B. Lindenmayer et al. 2012. Retention forestry to maintain multifunctional forests: A world perspective. *Biosci.* 62:633–645.

Hansen, A.J., T.A. Spies, F.J. Swanson, and J.L. Ohmann. 1991. Conserving biodiversity in managed forests. *Biosci.* 41:382–392.

Hoffman, K.M., B.M. Starzomski, K.P. Lertzman, I.J. Giesbrecht, and A.J. Trant. 2021. Old-growth forest structure in a low-productivity hypermaritime rainforest in coastal British Columbia, Canada. *Ecosphere* 12(5):e03513.

Humphreys, E.R., T.A. Black, G.J. Ethier, G.B. Drewitt, D.L. Spittlehouse, E-M. Jork, Z. Nesic, and N.J. Livingston. 2003. Annual and seasonal variability of sensible and latent heat fluxes above a coastal Douglas fir forest, British Columbia, Canada. *Agric. For. Meteor.* 115:109–125.

L'Anson, B., B. Snowden, and P. Bradford. 1996. *Wildwood tree farm: 50 years of selection forestry*. Notes to the Field No. 3, p. 6. Silvicultural Systems Program. Victoria, British Columbia, Canada: B.C. Ministry of Forests. www.for.gov.bc.ca/hfp/publications/00092/note_03.pdf. Accessed 5 August 2023.

Parminter, J. 2012. *Natural Disturbance Bibliography for British Columbia*. Unpublished file Report, p. 132. Victoria, British Columbia, Canada: B.C. Ministry of Forests. www.for.gov.bc.ca/hfd/pubs/docs/Misc/Misc099_BCDisturbanceHistoryBibliography.pdf. Accessed 5 August 2023.

Price, K.A., A. Roburn, and A. MacKinnon. 2009. Ecosystem-based management in the Great Bear Rain forest. *For. Ecol. Manage.* 258:495–503.

Wong, C., B. Dorner, and H. Sandmann. 2003. *Estimating historical variability of natural disturbances in British Columbia*. Land Management Handbook 53, p. 140. Victoria, British Columbia, Canada: B.C. Ministry of Forests and B.C. Ministry of Sustainable Resource Management. www.for.gov.bc.ca/hfd/pubs/docs/lmh/lmh53.htm. Accessed 5 August 2023.

BOX 1.4 TALL, WET EUCALYPT FORESTS IN TASMANIA, AUSTRALIA

Efforts to guide management by natural disturbance ecology for wet eucalypt forests in Tasmania have been ongoing over several decades. Large-scale occasional wild-fires are the dominant disturbance regime. In the 1960s, early researchers recognized that poor regeneration outcomes followed clearcutting, and this understanding led to the development of the "clearfall, burn and sow" silvicultural system that is still frequently used today. In many regions globally, societal dislike for the smoke associated with planned burning means that fire is not used for site preparation, in spite of its apparent relevance where fire is the dominant natural disturbance regime. Widespread community opposition to both clearfelling and burning is a particular challenge in Tasmania, because the ecology of eucalypt regeneration means there is not an adequate alternative to burning for large-scale regeneration. Burning logging debris results in a seedbed with readily available nutrients, reduces initial competition for light and water from other plants, kills seed-harvesting ants, and stimulates the germination of soil-stored seeds of some understory species. Eucalypt regeneration and early growth are far greater on burnt than mechanically disturbed seedbeds, as is species diversity of native understory plants. Furthermore, previous attempts at planting the common native wet forest eucalypt species were unsuccessful and ultimately resulted in the plantations being bulldozed because they were so unproductive. Combined with burning, sowing of locally collected eucalypt seed of the same species mix that occurred in the stand preharvest results in a locally adapted "unimproved" new cohort of trees, with enough seeds sown to allow for high genetic diversity and largely natural processes of self-thinning and competition. Thus, burning and sowing result in a much more "natural" regrowth forest than would be the case if other practices were used (e.g., mechanical disturbance, herbicides, planting of genetically improved material).

Studies comparing clearcutting regeneration to sites burned by wildfires have highlighted some differences, even though, overall, species composition was mostly similar (or equivalent in the case of ground-active beetles) (Site 5, Figure 1.2). In wildfire regeneration, there was better establishment of epiphytic and moisture-sensitive plants along with better recovery of plants by coppice regeneration in the absence of harvest machinery. When comparing wildfire to clearfelling, there were differences in relation to all aspects of the disturbance regime. The area, distribution, frequency, predictability, rotation period, intensity, and severity of the two disturbance types were all found to differ, with one of the major differences being much greater degrees of variability in these metrics for wildfire than for logging regeneration.

FIGURE B1.5 Seedbed after the regeneration burn in an aggregated retention site. This is at the Warra Silvicultural Systems Trial with two clearfells in the background. (Photo by Mick Brown.)

A lot of effort has gone in to developing the "aggregated retention" variable retention approach for wet forests in Tasmania, in recognition of those aspects where clearfelling is not equivalent to wildfire. In particular, variable retention retains structural and biological legacies within sites, facilitates connectivity, and encourages variability in implementation (Figure B1.5). Forest managers were initially reluctant to adopt aggregated retention, and some said it would never work, but it has now successfully been operationalized. Elements critical to this success included (1) funding, dedicated staff, silvicultural and ecological research as an initial basis, and over time, quantitative demonstration that the system was meeting its ecological and silvicultural objectives; (3) twice yearly "Variable Retention Implementation Group" meetings that included specialists in silviculture, biodiversity, fire management, and, most importantly, forest planners and other operational staff. These meetings allowed the group to brainstorm challenges that operational and planning foresters were encountering and to share their local successes and failures with staff from around the state. Developing a new silvicultural system based only on the science and theory of natural disturbance would have failed without close ongoing liaison over several years with the foresters directly involved with its implementation.

A major recent research effort has related to understanding the concept of "forest influence." One of the defining features of variable retention implementation in Tasmania is a forest influence target, which requires that the majority (>50%) of the harvested area should be within one co-dominant tree height of long-term retention. Until our recent research, the concept of forest influence was largely theoretical; that is, it was assumed that harvested areas near to mature forest will have buffered microclimatic conditions and be closer to dispersal sources, thereby facilitating the reestablishment of species associated with older forest conditions. A large collaborative study conducted in three age classes of clearcuts immediately adjacent to unlogged old-growth forest surveyed microclimate, vascular plants, bryophytes, ground-active beetles, and birds. These studies found that there are, in practice, forest influence gradients on biodiversity but that both the depth and magnitude of forest influence change with succession from young (~7-year-old) through mid (~27-year-old) to older (~45-year old) silvicultural regeneration. The impacts of forest influence varied substantially and support from upper management to try and make it work; (2) among the taxonomic groups surveyed; however, the conclusion was that 1-tree-height is a reasonable rule of thumb for management purposes. The variation among taxa also highlighted the importance of surveying taxa with contrasting life history attributes, dispersal capabilities, and habitat requirements.

Recent efforts in Tasmania are directed at shifting focus from the site scale to consideration of the broader landscape. Eucalypts are poorly suited to aging via tree ring analysis, so knowledge of previous fire histories is much less detailed than for some other regions of the world. However, old-growth eucalypt forest and rainforest are readily distinguished on aerial photos. Detailed aerial photo assessment of the entire managed forest landscape, increasingly combined with information from LiDAR, means that good information is available about the location and amount of mature compared to regrowth forest. Mapping of these data showed that some areas of the landscape have high proportions of mature forest remaining while in other areas this can be very low, for example, as a result of previous logging, wildfires, and land clearing. This information is being used to inform a new planning tool, the "Landscape Context Planning System." This system aims to ensure that sufficient amounts of mature forests are retained and that these are well distributed across the forest landscape. The premise for this system is research showing that both the distance to nearest mature forest and the amount of mature forest in a 1-km-radius landscape impact biodiversity recovery after logging. Likewise, there is recognition that the large areas of mature forest protected in the permanent reserve network need to be supplemented with mature forest retention in production landscapes (i.e., "land sharing" rather than just "land sparing"). Under the Landscape Context Planning System, a target has been established so that at least 20% of forest should be retained, for the long term, in a 1-km-radius landscape around any proposed harvest unit. This retention should focus on mature forest where available but can also include regrowth areas that, with time, will become future mature forest.

To know more: Baker, S. 2013. Management of Tasmanian eucalypt forests as complex adaptive systems. In *Managing forests as complex adaptive systems: Building resilience to the challenge of global change*, eds. C. Messier, C. Puettmann, and K.D. Coates, pp. 269–298. New York: Routledge.

Sue Baker

BOX 1.5 BOREAL CONIFEROUS FORESTS IN FINLAND

Managed forests of boreal Fennoscandia present an interesting case for ecosystem management because of their long history of intensive utilization and modification from their natural structure and species composition (Kuuluvainen and Siitonen 2013; Kuuluvainen et al. 2019). Forest utilization has taken varied forms over time. However, currently, forests are mostly shaped by intensive forestry practiced on most of the land area in Fennoscandia. The prevailing management system is compartment-wise low-retention even-aged management based on clearcut timber harvesting and forest regeneration by tree planting or using seed trees. This plantation-like management regime manages stand density to optimize timber yield or economic profit. This management regime is efficient from a timber yield point of view, but it has resulted in major changes in natural forest ecosystem characteristics. These include reductions in the amount of natural early and late successional forests with high structural variability and abundant deadwood and related species diversity.

FIGURE B1.6 Diversifying silvicultural harvesting practices is a key to ecosystem management in Finland. Examples from forests in southern Finland where cuttings inspired by small or partial disturbances have been applied. (a) A spruce forest after single tree and group selection cutting. (b) Abundant regeneration in an uneven-aged forest treated with selection cuttings. (c) Abundant regeneration including deciduous trees in a gap in a pure spruce forest.

One definition of ecosystem management is that it should ensure the maintenance of socially important services and values. Defining what these values and their relationships are not an easy task. However, there is unequivocal evidence of the difficulties in combining current timber production-oriented management with new requirements to maintain a variety of economic, ecological, and sociocultural values of the forest. The difficulty to reconcile forestry and ecosystem management stems from the fact that in natural conditions boreal forest landscapes in Finland would be characterized by small-scale and partial disturbance dynamics that maintain a dominance of late-successional forest characteristics (Kuuluvainen 2009; Kuuluvainen and Aakala 2011; Berglund and Kuuluvainen 2021). Maintaining the key ecological features of such forested landscapes, including their biodiversity, is difficult with the current management system, which aims at homogenizing the structure of young stands and reducing or eliminating the proportion of late-successional forests that are considered to have lower timber productivity. In ecological terms, the current forest management system implements a type of disturbance regime that the ecosystem has never undergone and to which not all elements and processes adjust. With an increasingly utilized and simplified forest, we may risk the ability of the forest to deliver valuable ecosystem services and values in the future.

Applying ecosystem management in the present situation requires diversifying the way the forest is managed (Figure B1.6a through c, Site 6, Figure 1.2). Because of Finland's long history of forest use and the resulting modified and simplified forests, the implementation of ecosystem management requires patient restoration of more natural forest structures (Kuuluvainen 2009). To facilitate this, management approaches (variable density thinning in young stands, partial harvesting and variable retention harvesting in mature stands, and single tree and group selection in old-growth forests) inspired by natural forest dynamics have been suggested as a promising avenue for forest ecosystem management (Figure B1.6). In this approach, management is designed to take into account and to emulate the dynamics of natural disturbances and their outcomes on forest ecosystems. Our aim should thus be at restoring important natural ecosystem properties and ecosystem values at various scales while at the same time providing a continuous flow of timber and other ecosystem services (Koivula et al. 2014; Kuuluvainen et al. 2021).

Timo Kuuluvainen

REFERENCES

Berglund, H. and Kuuluvainen, T. 2021. Representative boreal forest habitats in northern Europe, and a revised model for ecosystem management and biodiversity conservation. *Ambio* 5:1003–1017. https://doi.org/10.1007/s13280-020-01444-3

Koivula, M., Kuuluvainen, T., Hallman, E., Kouki, J., Siitonen, J., and S. Valkonen. 2014. Forest management inspired by natural disturbance dynamics (DISTDYN)—along-term research and development project in Finland. *Scand. J. For. Res.* 29:579–592.

Kuuluvainen, T. 2009. Forest management and biodiversity conservation based on natural ecosystem dynamics in northern Europe: The complexity challenge. *Ambio* 38:309–315.

Kuuluvainen, T. and T. Aakala. 2011. Natural forest dynamics in boreal Fennoscandia: A review and classification. *Silva Fennica* 45:823–841.

Kuuluvainen, T., H. Lindberg, I. Vanha-Majamaa, P. Keto-Tokoi, and P. Punttila. 2019. Low-level retention forestry, certification, and biodiversity: Case Finland. *Ecol. Process.* 8:47.

Kuuluvainen, T. and J. Siitonen. 2013. Fennoscandian boreal forests as complex adaptive systems: Properties, management challenges and opportunities. In *Managing forests as complex adaptive systems*, eds. C. Messier, K.J. Puettmann, and K.D. Coates, pp. 244–268. New York: Routledge.

BOX 1.6 TROPICAL RAINFOREST OF GUIANA

French Guiana is a unique region from the perspective of its tropical rainforest management. Forest management in this region is based on long-term studies of forest dynamics and allows the total recovery of stand aboveground biomass between harvests. French Guiana is an overseas department of France located on the northeastern Atlantic coast of South America. The Guianese climate is humid equatorial with a mean annual temperature of 26°C and annual precipitation of up to 3,000 mm. With 84,000 km^2 and 250,000 inhabitants, French Guiana is a sparsely populated region. Furthermore, 90% of the population is concentrated in the coastal strip; thus, 95% of the territory is covered by a nearly uninhabited tropical rainforest. In Guiana, the National Forest Office (Office National des Forêts, ONF) is mandated to conserve and manage 5.3 million ha of forest, among which 2.4 million ha are managed for timber production in a 70 km strip following the coastline called the "permanent forest domain." The remaining two million ha of forest are protected by national park status.

The tropical rainforest is one of the world's most complex forest ecosystems. The aboveground biomass can reach more than 400 tons · ha^{-1}, with a complex multilayer forest structure. Alpha tree diversity ranges from 200 to 300 species · ha^{-1} along with high beta tree diversity. Forest dynamics are mostly controlled by small (300 m^2 mean area) treefall gaps, and approximately 1% of the forest area is affected by treefall each year (Riera et al. 1990).

Forest management is performed in several steps. First, based on topography and watercourse maps, the ONF defines forest units and subunits in which forest inventories are conducted. Thereafter, a network of primary services is set up. Inventories consist of locating

FIGURE B1.7 Skid trail in a harvested site. By operating only on this trail, forestry machines have a limited impact on forest soils and on the residual stand. (Photo credit by Anthony Percevaux.)

critical habitats and selecting trees to be conserved or harvested. Finally, logging operations take place following the Reduced Impact Logging (RIL) technique in which operations are conducted so as to allow timber biomass to regrow between harvests and to minimize the impact on the environment. Logging intensity thus does not exceed 5 trees · ha–1 or 25 m^3 · ha–1 with a cutting cycle of 65 years. Moreover, a minimum cutting diameter of 55 cm at breast height has been defined for most species. Although tree diversity is high, the logging operations are targeted on a small number of commercial species. Indeed, 75% of the harvested volume comes from four species.

The RIL also requires controlling the falling of trees so as to protect the remaining stand and reduce the traffic of forestry machines to protect soils (Figure B1.7), (Site 7, Figure 1.2). Indeed, the ONF and the CIRAD (Center for International Cooperation in Agronomic Research for Development) have developed the "piste" software to optimize machinery traffic.

After logging, significant changes in stand demographic parameters (mortality, growth, and recruitment) are observed. First of all, mortality increases significantly (Sist et al. 2012). Indeed, mortality due to injuries inflicted on trees during logging operations is added to natural mortality. As for residual stand growth, it is stimulated by canopy opening. It is projected that recruitment and growth will make up for biomass loss due to mortality. About 8–15 years after logging, a recruitment peak is observed, also contributing to biomass recovery (Sist et al. 2012).

Depending on the logging method, stand type, and for a logging intensity of 4–5 trees · ha–1, the total biomass recovery would take about half a century (Sist et al. 2012). However, this biomass recovery does not mean that stands will return to their initial structure and composition. Indeed, over the long term, a significant difference in structure and composition between managed forests and natural forests is expected. In particular, emerging tree species and commercial tree species are expected to be less abundant. One of the major challenges for forest managers will be to define acceptable thresholds of change for forest structure and composition. These thresholds should be defined through a holistic approach integrating local environmental and socioeconomic criteria.

Raphaël Aussenac, Benjamin Andrieux, and Victor Danneyrolles

REFERENCES

Riera, B., H. Puig, and J.-P. Lescure. 1990. La dynamique de la forêt naturelle. *Bois et forêts des tropiques* 219:69–78.

Sist, P., L. Blanc, L. Mazzei, C. Baraloto, and R. Aussenac. 2012. Current knowledge on overall post-logging biomass dynamics in Northern Amazonian forests. *Bois et forêts des tropiques* 314(4):41–49.

1.5 CONCLUSION

The research presented here (including the boxes) demonstrates that greater reflection on desired conditions is needed to orient forest management decisions on the types and proportions of silvicultural techniques to be employed. Advice that could potentially be perceived as conflicting (i.e., to conduct more partial cuts to maintain cover while implementing more soil disturbance to create appropriate seedbeds and maintain productivity (e.g., Box 1.1)) can be ecologically justified as the most justifiable ecosystem-based approach but will require managers to carefully reflect on future desired forest conditions. The "one-size fits all" mentality that is pervasive throughout resource management needs to be reevaluated (Holling and Meffe 1996), and greater attention needs to be given to a heterogeneous suite of practices to maintain diversity and ecosystem processes in managed landscapes (Gauthier et al. 2009). Thought needs also be given to creating forests that may be somewhat different than natural forests but which will be more resilient to possible future changes.

REFERENCES

Anderson-Teixeira, K.J., V. Herrmann, R.B. Morgan, B. Bond-Lamberty, S.C. Cook-Patton, A.E. Ferson, H.C. Muller-Landau, and M.M. Wang. 2021. Carbon cycling in mature and regrowth forests globally. *Environ. Res. Lett.* 16(5):053009.

Beaudet, M., B.D. Harvey, C. Messier, K.D. Coates, J. Poulin, D.D. Kneeshaw, and Y. Bergeron. 2011. Managing understory light conditions in boreal mixedwoods through variation in the intensity and spatial pattern of harvest: A modelling approach. *For. Ecol. Manage.* 26:84–94.

Belleau, A., Y. Bergeron, A. Leduc, S. Gauthier, and A. Fall. 2007. Using spatially explicit simulations to explore size distribution and spacing of regenerating areas produced by wildfires: Recommendations for designing harvest agglomerations for the Canadian boreal forest. *For. Chron.* 83:72–83.

Belle-Isle, J. and D.D. Kneeshaw. 2007. Comparison of the effects of a spruce budworm [*Choristoneura fumiferana* (Clem.)] outbreak to the combined effects of harvesting and thinning at stand and landscape scales. *For. Ecol. Manage.* 246:163–174.

Bergeron, A.L., B. Harvey, and S Gauthier. 2002. Natural fire regime: A guide for sustainable forest management in the Canadian boreal forest. *Silva Fennica* 36:81–95.

Bergeron, Y., D. Cyr, C.R. Drever, M. Flannigan, S. Gauthier, D.D. Kneeshaw, E. Lauzon, and K. Logan. 2006. Past, current, and future fire frequencies in Québec's commercial forests: Implications for the cumulative effects of harvesting and fire on age-class structure and natural disturbance-based management. *Can. J. For. Res.* 36:2737–2744.

Bergeron, Y., B. Harvey, A. Leduc, and S. Gauthier. 1998. Basing forest management on natural disturbance: Stand- and landscape-level considerations. *For. Chron.* 75(1):49–54.

Bergeron, Y., B. Harvey, A. Leduc, and S. Gauthier. 1999. Stratégies d'aménagement forestier qui s'inspirent de la dynamique des perturbations naturelles: Considérations à l'échelle du peuplement et de la forêt. *For. Chron.* 75:55–61.

Brawn, J.D., S.K. Robinson, and F.R. Thompson III. 2001. The role of disturbance in the ecology and conservation of birds. *Ann. Rev. Ecol. Syst.* 32:251–276.

Cadieux, P., P. Drapeau, U. Ouellet-Lapointe, A. Leduc, L. Imbeau, R. Deschênes, and A. Nappi. 2023. Old forest structural development drives complexity of nest webs in a naturally disturbed boreal mixedwood forest landscape. *Front. Glob. Chang.* 6:9.

Dakos, V. and S. Kéfi. 2022. Ecological resilience: What to measure and how. *Environ. Res. Lett.* 17(4):043003.

DeGrandpré, L.D., K. Waldron, M. Bouchard, S. Gauthier, M. Beaudet, J.C. Ruel, C. Hébert, and D.D. Kneeshaw. 2018. Incorporating insect and wind disturbances in a natural disturbance-based management framework for the boreal forest. *Forests* 9(8):471.

Dragotescu, I. and D.D. Kneeshaw. 2012. A comparison of residual forest following fires and harvesting in boreal forests in Québec. *Silva Fennica* 46:365–376.

Esseen, P.A., B. Ehnström, L. Ericson, and K. Sjöberg. 1997. Boreal forests. *Ecol. Bull.* 46:16–47.

Frelich, L.E. 2002. *Forest dynamics and disturbance regimes: Studies from temperate evergreen-deciduous forests*. Cambridge: Cambridge University Press.

Gauthier, S., M.-A. Vaillancourt, A. Leduc, L. DeGrandpré, D.D. Kneeshaw, H. Morin, P. Drapeau, and Y. Bergeron, eds. 2009. *Ecosystem management in the boreal forest*. Montréal, Québec, Canada: Presses Université du Québec.

Girona, M.M., H. Morin, S. Gauthier, and Y. Bergeron. 2023. *Boreal forests in the face of climate change. Sustainable management*. Cham: Springer.

Haeussler, S. and D.D. Kneeshaw. 2003. Comparing forest management to natural processes. In *Towards sustainable management of the boreal forest*, ed. P.J. Burton, pp. 307–368. Ottawa: NRC Research Press.

Hunter, M.L. 1993. Natural fire regimes as spatial models for managing boreal forests. *Biol. Cons.* 65:115–120.

Holling, C.S. and G.K. Meffe. 1996. Command and control and the pathology of natural resource management. *Conserv. Biol.* 10:328–337.

Ibarra, J.T., F.J. Novoa, H. Jaillard, and T.A. Altamirano. 2020. Large trees and decay: Suppliers of a keystone resource for cavity-using wildlife in old-growth and secondary Andean temperate forests. *Austral Ecol.* 45(8):1135–1144.

Kneeshaw, D.D. 2001. Are non-fire gap dynamics important to boreal forest dynamics? *Recent Res. Develop. Ecol.* 1:43–58.

Kneeshaw, D.D., Y. Bergeron, and T. Kuuluvainen. 2011. Forest ecosystem dynamics across the circumboreal forest. In *Handbook of biogeography*, eds. A.C. Millington, M.A. Blumler, G. MacDonald, and U. Shickhoff, Chapter 14, pp. 261–278. Washington: Sage.

Kneeshaw, D., P.J. Burton, L. De Grandpré, S. Gauthier, and Y. Boulanger. 2018. Is management or conservation of old growth possible in North American boreal forests? In *Ecology and recovery of eastern old-growth forests*, pp. 139–157. Washington, DC: Island Press. https://doi.org/10.5822/978-1-61091-891-6_8

Kneeshaw, D.D., L. De Grandpré, L. D'Orangeville, M. Marchand, J. Moisan-Perrier, L.E. Robert, and M. Bouchard. 2022. Forest structure and composition diverge following harvesting compared to a spruce budworm Choristoneura fumiferana (Clem.) outbreak. *Front. For. Glob. Change*, 5:48.

Kneeshaw, D.D., E. Lauzon, A. de Römer, G. Reyes, J. Belle-Ile, J. Messier, and S. Gauthier. 2008. Appliquer des connaissances sur les régimes de perturbations naturelles pour développer une foresterie qui s'inspire de la nature dans le sud de la péninsule Gaspésienne. In *Aménagement forestier écosystémique en forêt*, eds. S. Gauthier et al. Montréal, Québec, Canada: University of Québec Press.

Kneeshaw, D., B.R. Sturtevant, B. Cooke, T. Work, D. Pureswaran, L. DeGrandpre, and D.A. MacLean. 2015. Insect disturbances in forest ecosystems. In *Routledge handbook of forest ecology*, pp. 109–129. London: Routledge.

Kneeshaw, D.D., B.R. Sturtevant, L. DeGrandpé, E. Doblas-Miranda, P.M. James, D. Tardif, and P.J. Burton. 2021. The vision of managing for pest-resistant landscapes: Realistic or utopic? *Curr. For. Rep.* 7:97–113.

Kuuluvainen, T. 2002. Natural variability of forests as a reference for restoring and managing biological diversity in boreal Fennoscandia. *Silva Fenn.* 36:97–125. https://doi.org/10.14214/sf.552

Kuuluvainen, T., P. Angelstam, L. Frelich, K. Jõgiste, M. Koivula, Y. Kubota, B. Lafleur, and E. MacDonald. 2021. Natural disturbance-based forest management: Moving beyond retention and continuous-cover forestry. *Front. For. Glob. Chang.* 4:24.

Lauzon, È., D.D. Kneeshaw, and Y. Bergeron. 2007. Forest fire history reconstruction (1680–2003) in the Gaspesie region of eastern Canada. *For. Ecol. Manage.* 244:41–49.

Ligot, G., P. Balandier, B. Courbaud, M. Jonard, D. Kneeshaw, and H. Claessens. 2014. Managing understory light to maintain a mixture of species with different shade tolerance. *For. Ecol. Manage.* 327:189–200.

Messier, C., J. Bauhus, F. Doyon, F. Maure, R. Sousa-Silva, P. Nolet, M. Mina, N. Aquilué, M.J. Fortin, and K. Puettmann. 2019. The functional complex network approach to foster forest resilience to global changes. *For. Ecosyst.* 6(1):1–16.

Nolet, P., D.D. Kneeshaw, C. Messier, and M. Beland. 2018. Comparing the effects of even- and uneven-aged silviculture on ecological diversity and processes: A review. *Ecol. Evol.* 8(2):1217–1226.

Oliver, C.D. and B.C. Larson. 1996. *Forest stand dynamics*, p. 521, update edition. New York: John Wiley & Sons.

Patry, C., D. Kneeshaw, S. Wyatt, F. Grenon, and C Messier. 2013. Forest ecosystem management in North America: From theory to practice. *Forest. Chron.* 89:525–537.

Pickett, S.T.A. and P.S. White. 1985. *The ecology of natural disturbance and patch dynamics*. San Diego: Academic Press.

Reyes, G.P. and D.D. Kneeshaw. 2008. Moderate-severity disturbance dynamics in *Abies balsamea- Betula spp.* forests: The relative importance of disturbance type and local stand and site characteristics on woody vegetation response. *Ecosci.* 15:241–249.

Sainte-Marie, G., D.D. Kneeshaw, D.A. MacLean, and C. Hennigar. 2015. Estimating forest vulnerability to the next spruce budworm outbreak: Will past silvicultural efforts pay dividends? *Can. J. For. Res.* 45:314–324.

Sánchez-Pinillos, M., A. Leduc, A. Ameztegui, D. Kneeshaw, F. Lloret, and L. Coll. 2019. Resistance, resilience or change: Post-disturbance dynamics of boreal forests after insect outbreaks. *Ecosys.* 22:1886–1901.

Saucier, J.-P., J.-F. Bergeron, P. Grondin, and A. Robitaille. 1998. Les régions écologiques du Québec méridional (3rd edition): Un des éléments du système hiérarchique de classification écologique du territoire mis au point par le Ministère des Ressources naturelles du Québec. Supplement of *L'Aubelle*, February–March.

Shorohova, E., T. Aakala, S. Gauthier, D. Kneeshaw, M. Koivula, J.-C. Ruel, and N. Ulanova. 2023. Natural disturbances from the perspective of forest ecosystem-based management. In *Boreal Forests in the Face of Climate Change*, ed. M. Montori. pp. 89–122. Cham: Springer.

Shorohova, E., D.D. Kneeshaw, T. Kuuluvainen, and S. Gauthier. 2012. A comparison of old-growth forest dynamics across the circum-boreal forest zone. *Silva Fennica* 45:785–806.

Smith, D.M., B.C. Larson, M.J. Kelty, and P.M.S. Ashton. 1997. *The practice of silviculture: Applied forest ecology.* New York: John Wiley & Sons Inc.

Thuille, A. and E.D. Schulze. 2006. Carbon dynamics in successional and afforested spruce stands in Thuringia and the Alps. *Glob. Change Biol.* 12(2):325–342.

Watt, A.S. 1947. Pattern and process in the plant community. *J. Ecol.* 35:1–22.

2 Triad Forest Management
Local Fix or Global Solution?

Austin Himes, Rebecca Tittler, Christian Messier, and Rosa C. Goodman

2.1 INTRODUCTION: WHAT IS TRIAD FOREST MANAGEMENT?

With the 1987 Brundtland Commission report (WCED 1987) came global acceptance of the idea of sustainable development. Sustainable development involves planning for the future and developing within certain limits such that future generations will have the same access to resources as we do today, be these resources economic, social, or ecological (WCED 1987). Sustainable development has gained international prominence over the decades, culminating in 17 sustainable development goals (SDGs) outlined in the United Nations 2030 Agenda for Sustainable Development (United Nations 2015). The important role of forestry for sustainable development is captured in SDG 15, "protect, restore and promote sustainable use of terrestrial ecosystems, sustainably manage forests, combat desertification, and halt and reverse land degradation and halt biodiversity loss." Sustainably managing forests means managing the forest in a way that will allow future generations to benefit from the forest as a source of social values, ecological resources, and economic well-being (Paavlainen 1994).

The discipline of sustainability science has developed around the goal of sustainable development and aims to understanding human interactions with natural systems, such as our management of forest landscapes, through the lens of complexity (Kates et al. 2001), i.e., by considering forests complex adaptive systems which are self-organizing and characterized by emergent properties, cross-scale interactions, nonlinear threshold behavior, and feedbacks with ecological and social components that should be considered holistically as aspects of a single system (Fischer 2018; Puettmann et al. 2009). Approaching forests and their management as part of complex adaptive social-ecological systems requires careful consideration of social, ecological, and economic outcomes, touching on all three pillars of sustainability. It also emphasizes building the adaptive capacity and resilience of the system, helping to ensure forest systems will persist and continue to function and provide benefits for future generations (Messier et al. 2019). This approach has the potential to enable forestry practices to be more innovative, inclusive of the diversity of interactions between people and forests, and lead to more equitable outcomes (West et al. 2020; Himes and Dues, in press).

At a large spatial scale, different strategies for meeting the goals of sustainable forest management have been proposed. The first includes various concepts from literature and policy including "land sharing" and "integrated resource management" where we can seek to meet all goals on all land at the same time. The second is to divide the forest into multiple zones, each managed to optimize a single goal or output—called land sparing, functional zoning, or (when there are three functional zones) triad. In agriculture systems, empirical evidence suggests a general trend that producing food on the smallest area possible while sparing larger areas for native vegetation is a better strategy for protecting biodiversity and wildlife populations than land-sharing approaches (Phalan 2018; Phalan et al. 2011) and that strategies including a third zone with less intensive agricultural production may be even better (Finch et al. 2019). In forestry, functional zoning has been promoted as a theoretically sound approach (Binkley 1997; Fox 2000; Vincent and Binkley 1993), and there have been a limited number of simulation studies suggesting functional zoning can maintain levels

DOI: 10.1201/9781003431084-3

of timber harvest while supporting more protected areas and old growth (Côté et al. 2010; Montigny and MacLean 2011; Tittler et al. 2012) which reduces trade-offs between ecosystem services by promoting forest landscape multifunctionality (Blattert et al. 2023).

Although different zoning options are possible, in forestry, triad is a very promising approach to (i) accommodate the three pillars of sustainability and (ii) consider all the different services provided by the forest (see Chapter 5) and uses for forest resources (Himes et al. 2022). It was first proposed by Seymour and Hunter (1992) to balance the various goals of ecosystem management in the state of Maine and was soon after suggested by Thompson and Welsh (1993) to incorporate an ecosystem approach into the management of Canada's productive forest. Triad management involves the division of the forest into three functional zones: a conservation zone, a multi-use or extensive management zone, and an intensive wood production management zone.

In the conservation zone, the focus is on the conservation of forests of "unique ecological value" and providing benchmarks against which impacts of management can be assessed (Seymour and Hunter 1992, 1999). These concepts were later developed to include a representative sample of native ecosystems (Kcrmar et al. 2003), i.e., native species, ecosystems, and evolutionary and ecological functions, as per Grumbine (1994). Biodiversity conservation may be a main goal, and social and economic values may also be accommodated, but only insofar as they are consistent with the overarching goal of ecological sustainability. Goals of social sustainability may also be met: forests set aside for conservation may still provide recreational, spiritual, and cultural services. In this zone, goals of economic sustainability may also be addressed in the conservation of native genetic resources that may be available at a later date to bolster productivity and increase resilience in managed forests. However, the rapid pace of global change and increasing risks of catastrophic forest losses due to pests, disease, and disturbance may justify, in some circumstances, management interventions in conservation areas to protect ecological values (Adams et al. 2019; Rannow et al. 2014). Also, there are many instances where setting aside reserve areas has resulted in restricting Indigenous peoples' access to their traditional lands, which must be avoided to prevent further colonial injustice (Fletcher et al. 2021).

In the intensive wood production management zone, the focus is on timber production (Seymour and Hunter 1992) and thus mainly on the economic and social sustainability associated with a thriving economic sector. To best fulfill the production goal without negatively impacting the conservation goal, this zone should be located on productive, accessible lands with relatively little conservation value (Seymour and Hunter 1992; Himes et al. 2022). Although ecological issues such as biodiversity may be considered, they are mainly considered in this zone insofar as they do not affect timber production. For example, although there may be intensively managed plantations in this zone, these may be mixed-species plantations rather than monocultures (Paquette and Messier 2009). Mixed-species plantations may be home to more native species and be more resistant and resilient to disturbance and support additional ecosystem services compared to traditional monocultures (Yachi and Loreau 1999; Hartley 2002; Spiecker 2003; Lindenmayer and Hobbs 2004; Erskine et al. 2006; Brockerhoff et al. 2008; Felton et al. 2010; Himes et al. 2020; Messier et al. 2021) and in some cases be more productive and generate greater economic returns (e.g., Piotto 2008; Pinnschmidt et al. 2023). Plantations also provide ecosystem services, such as flood control and carbon sequestration, thus further contributing to ecological and social goals (Pawson et al. 2013), although the main goal is still timber production.

In the multiuse or extensive management zone, the focus is on balancing timber production and extraction with ecological and social goals (Seymour and Hunter 1992). This maintains multiple uses of the forest (recreation, hunting, trapping, extraction of non-timber forest resources, etc.), as well as native species, ecosystems, and evolutionary and ecological processes. Thus, forest management activities in this zone must tread lightly, focusing less on timber production alone and more on maintaining the self-organizing and adaptive capacity of the forest. Although stands may be logged and replanted, these activities should be carried out with the idea of maintaining a diverse forest structure and composition rather than simplifying it to optimize timber production. In this zone, ecological silviculture provides a sound framework for practice; it aims to sustain ecosystem functions and processes, seek complexity and diversity, support forest continuity, maintain options for the future, consider landscape context, and considers place-specific conditions and the rich array of forest ecosystem components when developing objectives and actions (Palik et al. 2020). Partial cuts and variable retention logging and emulating natural disturbances are hallmarks of ecological silviculture. Similar approaches, like natural disturbance-based management and close-to-nature forestry are also excellent paradigms for management in this zone. These practices are all attempts to limit the negative effects of sustainable timber production on the natural functioning of these forests. Thus, this zone will contribute to economic goals, but the focus is on balancing them with social and ecological goals.

The proportion of the forest designated to each zone may vary from case to case, depending on the relative value placed on conservation, timber production, and non-timber forest uses. However, even in a system focused largely on profit from timber production, the proportion of land designated as conservation or extensive management may increase with increases in production in the intensive management zone. In other words, the faster timber can be produced in the intensive management zone, the larger the proportion of the forest that may be set aside for the other two zones (Paquette and Messier 2009; Pawson et al. 2013). The arrangement of the zones within the landscape is also important (Himes et al. 2022). If possible, reserves should be larger than the largest expected disturbance in an area and placed to ensure habitat requirements of sensitive species. Connectivity between areas is also desirable, to facilitate gene flow and species migration. Ideally, the extensive management zone will provide a buffer between intensively managed forests and reserves and for many species may also provide adequate corridors for connectivity. The functional complex network approach to forest management may be useful for guiding the strategic arrangement of the zones within a given landscape (Messier et al. 2020, 2019). Similarly, ideal placement of production zones will minimize transportation distances and the extent of road networks to increase economic efficiency and minimize habitat fragmentation caused by roads.

2.2 APPLICATIONS OF THE TRIAD APPROACH

2.2.1 Modeling

Several modeling studies have pointed to triad zoning as preferable to land sharing or even to a dyad model with extensive management and conservation but with no intensive management zone. In a modeling study of a 13,000-ha forest estate in the Douglas fir–Western hemlock forests of coastal British Columbia, Krcmar et al. (2003) concluded that triad management would lead to higher harvest volumes and more unlogged forest than a dyad model without intensive management. In a nonspatial simulation study on a 190,000-ha parcel of privately owned land in the Acadian temperate forest of New Brunswick, Montigny and MacLean (2006) found triad management to be preferable to a dyad system with intensive and extensive management but no conservation in terms of the amount of old-growth forest. Furthermore, increases in intensive wood production management resulted in increased harvest volume. In spatially explicit simulations on a 390,000-ha forest management unit in the Mauricie, Quebec, Côté et al. (2010) found that triad scenarios with 12–20% conservation and 10–36% intensive management resulted in more old-growth, less fragmentation, and greater harvest volume over the long term than both the status quo (2% conservation and 5% intensive) and the government proposed (8% conservation and 7% intensive) scenarios. Further simulations in the area clearly indicated that, without sacrificing harvest volume, fragmentation was reduced, old growth forest maximized, and road construction minimized under triad management (20% intensive and 20% conservation), as opposed to a system with no zoning (100% extensive, land sharing) or a dyad system without no intensive management zone (Tittler et al. 2012). Harris and Betts (2023) conducted a spatially explicit modeling study of zoning and land-sharing scenarios for a 34,000-ha forest landscape in the coastal mountains of Oregon and considered the impacts on birds species associated with different habitat types, carbon storage, and timber production.

Results of their study were complicated and suggested that species dependent on both old growth and early seral forest stages benefited most from a land-sharing strategy, but triad-type zoning was best for mid-seral associated species. After considering 43 different scenarios, some with different levels of timber production, the authors concluded that substantial trade-offs occur as a function of sharing, sparing, and triad strategies between different species and ecosystem services. Their study emphasizes the importance of management flexibility and demonstrated that multiple strategies and allocations levels to different zones can produce efficient outcomes for a specified level of wood production representing a wide range of trade-offs between other desired outcomes.

Research on triad in the context of tropical forests has been more limited but has yielded similar results. For example, researchers with the NGO-based Science for Nature and People Partnership (SNAPP) program developed a project to use a wide array of data to model outcomes in specific research areas where the conflict between wood production and conservation (climate) goals is particularly salient (https://snappartnership.net/teams/forest-sharing-or-sparing/). A case study from the program in Indonesian Borneo found that optimal zoning for timber production and protection of habitat for key species most resembled a triad approach with a mix of intensive, extensive, and reserve zones across the landscape (Runting et al. 2019). However, a more generalized modeling exercise for tropical forests conducted by the group was less conclusive and highlights how land tenure and logging techniques may affect the impact of zoning and land-sharing approaches on biodiversity and carbon flux (Griscom et al. 2018).

The most efficient forest use (zone) in a given location will depend on the biophysical constraints, conservation value, and economic and social drivers. Optimal solutions will likely depend upon the specific landscape context, the scope of the analysis (commodity type, economic metrics, ecosystem service), the potential to improve each system, the spatial arrangement of different zones, and the long-term sustainability of a given land use regime. Defining optimal outcomes will further depend on specific targets for competing benefits like timber, carbon, and biodiversity (Harris and Betts 2023), and it may be necessary to navigate differences in stakeholders' preferences and consider the importance of local knowledge through deliberative or other means to support social sustainability (Hallberg-Sramek et al. 2023).

2.2.2 Real-World Applications

Despite decades of interest in triad forest management and multiple modeling studies, there have been very few real-world applications of the approach, and most of those are very recent or lacked adequate monitoring and data collection to support rigorous assessment. Despite the relative dearth of empirical data, triad has been adopted as the guiding framework for managing crownlands in Quebec (Tittler et al. 2016) and Nova Scotia (Lahey 2018), and the 34,000-ha Elliott State Forest in Oregon (Tollefson 2021). These examples and the few other past attempts help illustrate how such a system may work in practice, as well as point out certain strengths and potential weaknesses of the approach.

2.3 EARLY ATTEMPTS

In the 1990s, Champion International set up a triad management system in a large forest management area in the temperate broadleaf forest of Maine. However, the project was short-lived as the company changed priorities and was later acquired by another company, leading to the abandonment of the triad project, but a legacy of large protected areas is still in place (Redelsheimer 1996). Almost a decade later, Riverside Forest Products funded research to set up triad zoning on a 145,000-ha tree farm in the spruce-fir montane forest of south-central British Columbia, Canada. After a few iterations, the company landed on a plan allocating 17% of the forest land to conservation, 49% extensive management, and 34% intensive management with special consideration to have diverse ecosystems and habitat for sensitive species allocated to conservation reserves and highly accessible areas with low levels of biodiversity and higher fire risk the focus of intensive production zones (D'Eon et al. 2004; Nitschke and Innes 2008). However, the project fell through after the 2004 sale of Riverside Forest Products to Tolko Industries. While both of these early examples with private companies failed to fully materialize, future projects can learn from the process that went into planning triad management in this area. These examples also highlight the challenge of implementing triad on large privately managed forest landscapes that may change ownership on relatively short intervals. The success of future efforts to implement triad in large private management may require binding agreements tied to the land that will persist even if landownership or timber rights change hands, for instance through conservation easements. However, such agreements may also lack the flexibility necessary to adapt zoning allocations to unexpected future conditions.

2.4 THE MAURICIE TRIAD PROJECT, QUEBEC

The Mauricie Triad Project was set up through a partnership between AbitibiBowater Inc. (now called Resolute Forest Products) and academic researchers in the boreal mixedwood forest of the Haute Mauricie in central Quebec. It was spearheaded in 2003 and touted as a pilot project by a provincial government seeking to innovate its forest management policies. Although Quebec's forestland is publicly owned, at that time, it was largely managed by large industrial operators who leased rights to exploit the forest for timber. However, other stakeholders also had use of the forest. In the 890,000-ha forest management unit of this project, the area is divided into six usage zones that support 14 fishing/hunting outfitters, four sugarshack operations, two native communities (the Atikamekw of Wemontaci and Montagnais of Lac St-Jean), and two regional county municipalities (Domaine-du-Roy and the City of La Tuque). Many individual trappers, hunters, fishers, and recreationists also use the forest. Thus, logging must be done in a way that accommodates these varied users and uses.

As part of this project, almost a fifth of the forest management unit was designated for intensive wood production, over 10% as conservation, and the remaining forest as extensive management (Figure 2.1). Since this was a pilot project, extensive research was done in all three zones to maximize timber production in the intensive management zone while minimizing the negative effects of logging in the intensive management zone. In the extensive management zone, an adaptive-management-inspired process is used to set goals and minimize the differences in forest composition and structure between the pre-industrial or unmanaged and the managed forest (Figure 2.2). In the intensive management zone, the focus is on managing the forest as a complex adaptive system (CAS), incorporating mixed-species plantations, for example (Figure 2.1). In the conservation zone, the focus is on large protected areas that include representative samples of all the ecosystems of the area. After 20 years, the Mauricie project is regarded as a success based on all three pillars of sustainability—economic, social, and environmental (Messier et al. 2009).

Now in its second phase, that of implementation rather than experimentation, the success of the Mauricie Triad Project has served as the inspiration for some of the new forest management legislation that came into effect in Quebec in 2014. This legislation now calls for intensive management to be carried out in specifically designated areas of the forest (*aires d'intensification de la production ligneuse* or intensive timber production areas), for natural-disturbance-based management to be carried out in the rest of the forest and for all this to be counterbalanced by conservation in protected areas (Gouvernment du Québec 2014). The overall goal for conservation zoning has been set at 12% across the province of Quebec (MDDEP 2011). At present, decisions as to how much of the forest will be designated for intensive management and how this will be implemented have yet to be made due to the extra cost that such an implementation will require and the lack of public support for intensive forestry. New opportunities are emerging, however, to use intensive forestry to help make the forest more resilient and adapt to future environmental conditions (Messier et al. 2020). By establishing new tree species or provenances better adapted to the future environmental conditions and/or tree species with specific functional traits in well-situated locations across

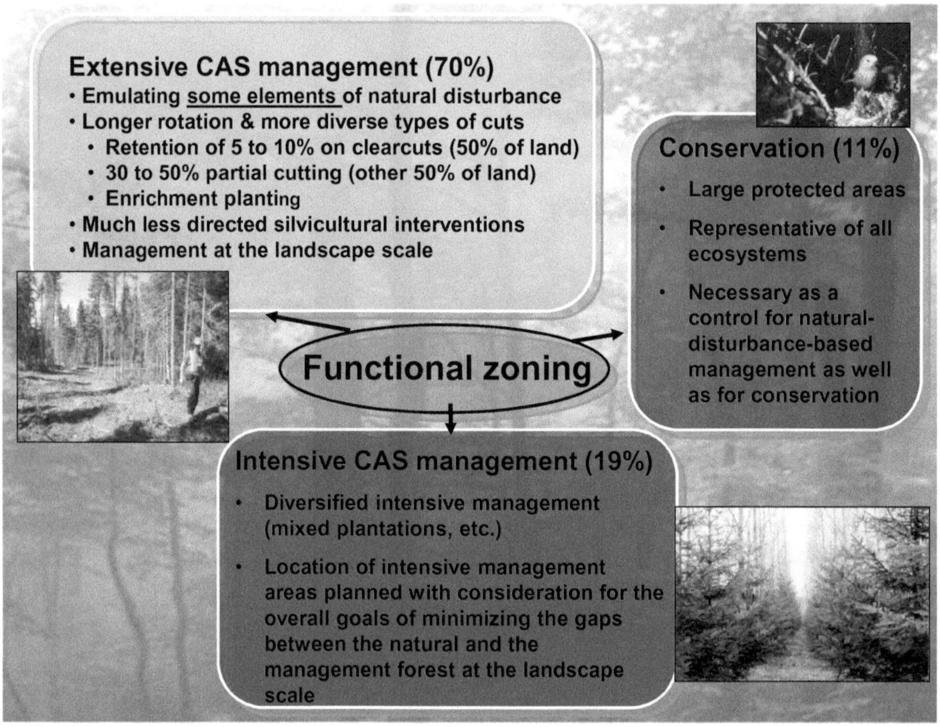

FIGURE 2.1 The three zones of the Mauricie triad project.

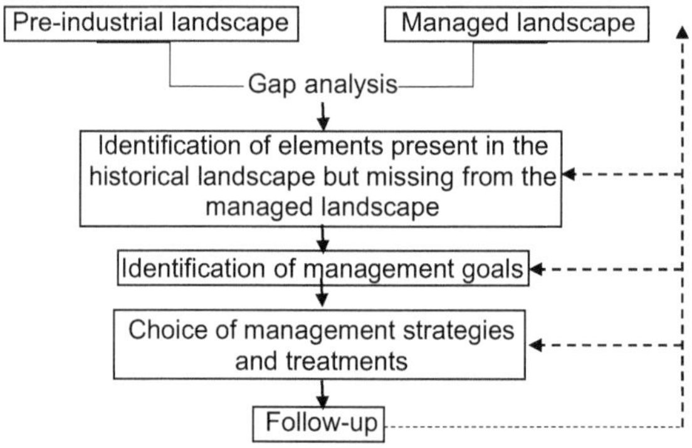

FIGURE 2.2 Schematic representation of the process used to set goals and select management strategies and treatments in the extensive management zone. (From Messier et al. 2009.)

the landscape (in centrally located nodes), following the functional complex network approach of Messier et al. (2019), intensive forestry could play a new fundamental role into the sustainable development of the forest.

In this particular case, where natural-disturbance-based management is to be practiced in the extensive management zone, as initially suggested by Seymour and Hunter (1992), the conservation zone may serve yet another purpose. It may be used as a control or reference for natural-disturbance-based management in the extensive management zone. Natural-disturbance-based management is based on the assumption that native forests are adapted to natural disturbance regimes. Therefore, the negative effects of logging can be minimized if forestry practices are designed to mimic these natural disturbances (Hunter 1990, 1993; Attiwill 1994). On a landscape scale, the goal is to maintain the managed forest within the range of natural variability in terms of structure and composition (Attiwill 1994; Hunter 1999; Landres et al. 1999). Thus, knowledge of the natural disturbance regime and of the range of natural variability is crucial to this type of management. This knowledge can be acquired through analysis of historical records (e.g., Cyr et al. 2009) sometimes combined with modeling (e.g., Tittler et al. 2012), but these records are limited both spatially and temporally. Furthermore, historical pattern may not be indicative of future patterns under climate change scenarios. It may therefore be useful to have unmanaged controls, such as protected forestlands, to inform forest management decision on nearby managed lands.

2.5 NOVA SCOTIA CROWN FOREST LAND

The Nova Scotia government passed the Environmental Goals and Climate Change Reduction Act on November 5, 2021, which included a commitment to implement an ecological forestry approach for Crown lands in the province consistent with recommendations from an independent review of forestry practices, commonly referred to as the "Lahey Report" (https://nslegislature.ca/legc/bills/64th_1st/3rd_read/b057.htm). The central recommendation of the Lahey report was to implement a triad model of forest management (Lahey 2018). As a result the triad model has been officially adopted by the Government of Nova Scotia for management of Crown forest. Since the official adoption, the government has published a report identifying 246,000 ha, ~16% of forested land, that is suitable for intensive, high-production forestry. Three primary criteria were used for identifying areas suitable for intensive management: 1) excludes areas where non-timber values take primacy, e.g., protected areas, sensitive habitats, wildlife special management zones, and remnant old growth, 2) exclusion of ecosites capable of supporting tolerant hardwood forests as intensive management would require ecologically inappropriate conversion to softwood plantations, and 3) includes land capable of supporting fast softwood tree growth (McIntyre et al. 2021). Specific sites of intensive management are in the process of being identified at the smaller scale of ecodistricts previously mapped across the province based on features such as climate, elevation, topography, bedrock formation, and vegetation. Areas already in planted forests or abandoned agricultural fields will be prioritized for intensive production, but sites within 100 meters of conservation zone boundaries, known locations for at-risk species, critical habitat areas, areas exposed to extreme winds, parcels too small or isolate to be operationally viable, sites with rare or sensitive ecosystems, and areas with poor soil fertility or drainage will be excluded (High Production Forestry in Nova Scotia Phase 2 Guidance for Implementation 2023). The reserve zone of the triad was designated as existing parks, nature reserves, wilderness areas, and officially designated old growth forests, which make up an estimated 514,000 ha or about 33% for the forested crown land. The remaining area (783,000 ha ~51%) is designated as the extensively managed zoned, referred to as the "ecological matrix," and the government has published silvicultural guidelines for that zone, which government foresters are now being trained on (McIntyre et al. 2021; McGrath et al. 2021).

Nova Scotia's implementation of the triad on Crown lands is the largest application of triad to date with around 1.5 million ha of forestland affected. The process of initiating triad at this scale has taken many years and will take many more before it is fully implemented. Challenges to triad at this scale in Nova Scotia are that ~70% of the forest land in the province is privately owned, and while the Lahey report made specific recommendations for incorporating private forest land into the triad scheme, the government has not yet made serious efforts to do so (Lahey 2021). While the adoption of triad forest management by the Novia Scotia government is a monumental benchmark, very little actual change in forest practices has occurred in the few subsequent years. Barriers to more rapid adoption have been a lack of clear articulation of the overall implementation strategy, recalcitrance within relevant government departments to adjusting mindsets and culture, and resistance to triad management from within and outside of the government (Lahey 2021). These types of challenges are likely to be present in most large jurisdictions implementing triad forestry. Forest industry interests and many trained foresters are likely to resist substantive shifts from historical forest management approaches, and environmental interests are likely to challenge intensive management practices on any portion of public forest land.

2.6 ELLIOTT STATE RESEARCH FOREST, OREGON

In 2018, the State Land Board of Oregon requested that Oregon State University and the Oregon Department of State Lands explore the potential of dedicating the Elliott State Forest, approximately 33,000 ha of forestland in the Coast Range of Oregon, to long-term research and management by the university. The proposal developed by Oregon State University and an advisory committee was to create a long-term forest study to explore the efficacy of triad forest management and land sharing/sparing alternatives. In 2022, the Oregon Legislature passed a bill creating a new public agency to oversee the Elliott State Forest, which was renamed the Elliott State Research Forest (ESRF). The current draft management plan, developed around the original proposal, is designed to provide empirical data about the multidimensional benefits and trade-offs associated with triad forest management and alternatives while meeting the tripartite goals for ecological, economic, and social sustainability (Elliott State Research Forest Forest Management Plan, October 2023 Draft). A revised management plan is expected in 2024.

The ESRF management plan begins by clearly stating a mission, vision, and guiding principles and recognizing and committing to working with Tribal Nations (the Coos, Lower Umpqua, and Siuslaw Indians) to honor their Sovereignty Rights, support the cultural values of the forest, and develop co-stewardship plans. These steps are essential to ensuring landscape-scale forest management approaches like triad, particularly on public land, are appropriately scoped to meet sustainability goals, and produce desirable outcomes that consider the diverse values provided by forests. Regulatory constraints, such as the endangered species act, and policy mandates are also outlined and are important considerations for how different activities and management options may be limited, which can help with determining the placement of zones. Both the social and ecological dimensions of the forest landscape are explored in detail prior to allocating zones. Since the ESRF is intended to be a research forest, different zoning treatments are intended to be replicated at the sub-watershed basin scale so that benefits and trade-offs of different zoning approaches can be assessed statistically, which differs from purely operational applications of triad. Within the ESRF, a large contiguous area of critical habitat and extensive older forests has been identified for conservation research watershed, which will have hands-off management and not be part of the triad experiment. The remaining area is more than 17,000 ha, divided into 40 sub-watersheds, each assigned one of four treatments: 1) *extensive,* consisting of 100% extensive management (i.e., land sharing); 2) *triad-E,* consisting of 60% extensive management, 20% intensive management, and 20% reserves; 3) *triad-I,* consisting of 20% extensive, 40% intensive, and 40% reserves; and 4) *reserves with intensive,* consisting of 50% reserves and 50% intensive management (i.e., land sparing). Each treatment zoning allocation will be replicated in ten sub-watersheds ranging in size from 160 ha to 800 ha and assigned based on stand age, protection of sensitive topographic features, configurations to reduce fragmentation, ensure protection of endangered species habitat and stakeholder priorities, and with consideration for silvicultural and operational suitability.

Basing the major experimental design for the ESRF on triad forest management has been controversial (Tollefson 2021), and the process has taken years and required public feedback, compromises, and expanded outreach and communication to affected communities, especially the Indigenous peoples who have been stewards of the forest for millennia. These challenges and subsequent compromises led to Confederated Tribes of the Coos, Lower Umpqua, and Siuslaw Indians to publicly oppose the ESRF management plan over concerns that the amount of area proposed for reserves was too large, compromising economic viability and severely limiting the ability of the tribes to actively steward the forests. Shortly following public opposition by the tribes, Oregon State

University withdrew from the project. At the time of writing, the State Land Board of Oregon intends to move forward with the ESRF. While the specific circumstances of the ESRF have made it more complicated than other attempts of implementing the triad, it is a good example of the amount of work, consideration, and public involvement that may be necessary to effectively manage publicly owned forest landscapes under triad zoning. The research focus of the ESRF is also unique but will provide a first of its kind statistically sound empirical experiment to test the efficacy of the triad to meet sustainability goals while minimizing trade-offs between timber production and other benefits provided by forests.

2.7 CHALLENGES AND OPPORTUNITIES FOR TRIAD MANAGEMENT GLOBALLY

One thing common to all these examples of triad forest management is the ownership structure of the forest. In all cases, triad is applied to large forest landscapes under single ownership by the government or large industrial interests. Triad forest management may be more challenging to apply when forest land ownership is more fragmented (i.e., when the forest is split among many small landowners). In terms of land ownerships, it is relatively easy for a large landowner to split the forest into three zones, although particularly in the case of public, multiple stakeholder concerns must also be considered so that the multiple uses of the forest may be accounted for in the zoning process. For example, people who live next to the forest may not want the intensive management zone to be in their backyard. Similarly, it may be problematic to put a very strict conservation zone in an area where non-timber use of the forest is common (e.g., hunting, trapping etc.). Nevertheless, any landowner, be it the government or industry, that has tenure over a large forest area can designate specific parts of their forest for different purposes.

On the other hand, in areas where land ownership is more fragmented, such as in Sweden or parts of the United States, the production of a large-scale zoning scheme may be more difficult. Such a process would involve small landowners cooperating, some agreeing to set their land aside for conservation, others agreeing to manage intensively for timber production, and still others managing less intensively. There would likely need to be financial mechanisms to incentivize landowners to cooperate, such as profit sharing, payments of non-timber ecosystem services, and/or easements for set aside land. Local forest owners' associations such as those in place in Sweden, or administrators of group sustainable forest management certificates, like International Paper in the U.S. Southeast might be able to organize such a system. Despite these challenges, the triad model has been suggested as an appropriate mechanism to conserve old-growth forest and biodiversity landscape contexts with fractured small private landownership, e.g., in Sweden (Ranius and Roberge 2011; Strengbom et al. 2011).

Alternatively, under these fragmented systems of forest ownership, individual landowners could seek to accommodate all three zones at the smaller scales of their properties. Although this might not result in the same savings in terms of transportation costs nor in the same benefits derived from large, protected areas (i.e., habitat for organisms such as large mammals that may need large areas of relatively untouched forest), it may nonetheless be beneficial at a smaller spatial scale, allowing for a similar balance between the various goals of sustainable resource management. For instance, a case study based in eastern Texas, United States, found splitting properties as small as 80 ha into different zones for timber production and habitat based on site quality could improve financial returns (Resch et al. 2023).

However, fragmented ownership is not an issue in much of the world since most forests are public. More than 80% of the forestlands of South America, Europe, Asia, and Africa are government owned and managed, with percentages exceeding 60% in the rest of the world (FAO 2010). Most of these public forests are also managed by governments, except for the forests of Oceania, where more than 50% of the public forests are managed by private corporations and institutions (FAO 2010). The trend of community-based management of public forest is increasing worldwide, accounting for more than 20% of the public forests of South America (FAO 2010), for example, but this type of ownership and management structure may facilitate the multi-stakeholder processes necessary for effective triad zoning on public lands. However, tenure rights and traditional uses, particularly of Indigenous peoples and local communities, should be considered when determining the types of activities and locations of different zones.

Another commonality of the existing examples of triad forest management is the relative homogeneity of the forest. In boreal or temperate systems, there are relatively few tree species and relatively little biodiversity in general compared to tropical systems. It is therefore more straightforward to set aside large, protected areas as part of a conservation zone that is representative of the forest in general. In more diverse systems, such as tropical forests, the protected areas that make up the conservation zone might have to be smaller and more dispersed across the landscape to effectively protect this diversity. Thus, the triad model might not be as beneficial in terms of reducing road construction and transportation costs in more species-rich tropical systems as it is in relatively species-poor boreal and temperate systems.

Nonetheless, triad management is still likely to be a solution to balance conservation and production values in the tropics (Putz and Romero 2014). Thus far, work in the tropical forests of Borneo indicates that, in terms of biodiversity, the total conservation value across the landscape is likely to be greater in forested landscapes split into protected areas and intensive management (Edwards et al. 2014), even if conservation values are greatly reduced or nearly lost in the intensive management zone (Edwards et al. 2010). However, this type of management, without the extensive management zone, is problematic in that it can exclude the non-timber users of the forest, benefiting only industrial forestry operators who can afford to plant and tend plantations and those who are concerned with biodiversity and carbon (i.e., often those from the global North rather than the local people). This system therefore fails to fulfill the goals of social sustainability. In addition, since local people have historically been officially excluded from forest preserves, and since governance is often weak, the protected areas that make up the conservation zone may be prone to illegal logging and other uses that may erode the conservation value of this supposedly protected forest (Nelson and Chomitz 2011). These types of exclusionary practices also lead to social justice issues, particularly for Indigenous peoples whose access to sacred sites and ability to conduct ceremonies has often been inhibited through the designation of protected areas, often leading to ecological degradation because these areas had historically been under the care of Indigenous practices for thousands of years (Fletcher et al. 2021). To maintain true ecological reserve zones, the management system must also provide space for local people and for non-timber users of the forest and may also have to consider some degree of active management to protect ecological values in the face of global change (Himes et al. 2022). Extensive management zone may also provide crucial areas where non-timber values can be provisioned.

For the triad system to remain functional, particular care must be taken to maintain this crucial extensive management zone. In some cases, in the tropics, allowing logging leads to the forest being classified as degraded and, therefore, susceptible to conversion to agriculture (Edwards et al. 2012). In more extreme cases, logging is intentionally the first step toward deforestation (Asner et al. 2006). Involvement of multiuse stakeholders in management planning, particularly in this zone, is crucial in that it provides incentives to maintain forest cover (Nelson and Chomitz 2011; Porter-Bolland et al. 2012; Griscom and Cortez 2013) and therefore helps ensure the success of the management system in meeting environmental, social, and economic goals.

2.8 TRIAD MANAGEMENT WORLDWIDE

Despite the paucity of literature on actual applications of triad management, there are indications that such zoning may be taking place worldwide in a less formally defined way. With concern about biodiversity losses and climate change, international agreements, such as the UN Convention on Biological Diversity (United Nations 1992) and the UN World Heritage Convention (United Nations

1972), have long been calling for increases in protected areas. The Convention on Biological Diversity is now calling for 30% protection by 2030. Worldwide, 18% of the world's forests are protected (FAO 2020), but there is debate as to how well these protected areas are conserving biodiversity (Butchart et al. 2012). Tropical forests are also specifically targeted by climate change mitigation projects such as REDD+ as their value as carbon stores and sinks becomes increasingly important. In addition, intensive forest management also seems to be increasing worldwide. Plantation forestry has been increasing steadily at the particularly fast rate of about four million hectares per year since 1990, most of which are planted with exotic species that are favored for their high production value (FAO 2020).

Despite these overall global trends, triad forest management is more likely to be successful if the three zones are specifically delimited in a well-thought-out, multi-stakeholder process. First, protected areas would be more effective if they were designated considering biodiversity and ecosystem services as a goal, as in the British Columbia example described earlier (D'Eon et al. 2004; Nitschke and Innes 2008). Furthermore, these areas may be more likely to be maintained and allowed to perform their conservation function if they are seen as part of a larger system that specifically allows for other uses of the forest as well. Similarly, those with conservation concerns may also be more willing to accept intensive forest management if they know that it is counterbalanced by conservation. In a focal-group study done as part of the Mauricie Triad Project in central Quebec, Berninger (2007) found a triad approach with 10–25% conservation and 13–20% intensive management (including 3% fast-growing plantations) to be more acceptable to self-identified environmentalists, foresters, and other users of the forest than scenarios with both less protected areas and no fast-growing plantations. Had the plan been presented at a smaller scale, including, for example, just the intensive management zone with fast-growing plantations, environmentalists would likely not have been as accepting. Likewise, had just the conservation zone been presented, foresters would certainly have been upset at being excluded from the forest. In the Amazon forests of Peru, Miranda et al. (2016) found that protected areas that allowed for some extraction resulted in reduced forest degradation, as illegal logging was reduced. If this extraction was instead designated to a specific location, as in a zoning scenario, forest degradation might have been reduced even further. Research is currently underway in the Mauricie Triad Project to determine what role intensive management zones could play in bringing new species better adapted to the future climate in the area. The need for careful planning, zoning, and stakeholder involvement may be particularly important in the tropics, where an intensive management zone may have perverse outcomes rather than meeting timber production goals and allowing for the protection of conservation zones. Plantation forestry, for example, has become a major driver of deforestation in the subtropics (Hansen et al. 2013) and is the primary driver of deforestation in Indonesia (Abood et al. 2014).

The process of delimiting zones for triad must also vary from one situation to another depending on the ecological, social, and economic context. The percentage of land designated for each zone will likewise vary as a result of differences in management and forest use priorities. The work of D'Eon et al. (2004) and Nitschke and Innes (2008) in central British Columbia demonstrates how the specific ecological characteristics of the forest in question may be used in deciding how much land to set aside for conservation. D'Eon et al. (2004) recommended 10–20% of the forest be set aside for conservation based on decisions about how much of each type of forest should be conserved, but more aggressive targets have been trending recently such as the 30-by-30 initiative that calls for 30% of all terrestrial and aquatic ecosystems to be protected by 2030 and even more aggressive calls to protect 50% of the terrestrial biosphere (Dinerstein et al. 2017). Decisions about the percentage of the forest to be delimited as intensive management, and about the intensity of practices to be carried out there and in the extensive management zone, may depend on market conditions. Similarly, decisions about how much of the forest to designate as extensive management, as well as the practices to be carried out in this zone, will depend on the balance between the demand for wood and non-timber uses of the forest. If retention management is to be carried out in the extensive management zone, as suggested by Gustafsson et al. (2012), the amount to be retained may vary depending on the percentage of the forest to be set aside for the other two zones. Bos (1993) and Boyland et al. (2004) present a simulated annealing method to optimize zoning, but the equation may become overly complex when climate change is taken into consideration, thus favoring an approach more like that applied by Nitschke and Innes (2008) and modeled by Krcmar et al. (2003).

2.9 CONCLUSION: TRIAD MANAGEMENT, GLOBAL SOLUTION

Much research and some real-world applications thus far indicate that the triad forest management system is a promising approach to meet the goals of sustainability and ecosystem management in different regions of the world. Such a system may allow us to meet the social, environmental, and economic goals of sustainable forest management and ecosystem management by increasing timber production and limiting transportation costs while simultaneously increasing conservation and reducing conflict between logging companies and other forest users. Although the triad system may be best suited to relatively species-poor forests under public or large industrial ownership with high governance, it is also likely to be successful in more species-rich forests and in forests under more fragmented ownership. In countries with weak governance, such as many tropical countries, the risks and opportunities of each zoning option should be especially carefully considered. As for agriculture (Phalan et al. 2011), urban development (Sushinsky et al. 2013), and road building (Jaeger et al. 2006), concentrating anthropogenic disturbance in specific areas of the forest may allow for healthier forest ecosystems (Tittler et al. 2012), as well as more sustainable forestry-based and non-forestry-based economies and societies worldwide. The addition of the extensive management zone to the dichotomy of conservation and intensive management favored particularly in the agricultural literature is appropriate in the forest because of the multiple uses to which the forest is put and the multiple services it provides.

REFERENCES

Abood, S.A., J.S.H. Lee, Z. Burivalova, J. Garcia-Ulloa, and L. Koh. 2014. Relative contributions of the logging, fiber, oil palm, and mining industries to forest loss in Indonesia. *Conserv. Lett.* 8:58–67.

Adams, V.M., G.D. Iacona, and H.P. Possingham. 2019. Weighing the benefits of expanding protected areas versus managing existing ones. *Nat. Sustain.* 2(5):404–411.

Asner, G.P., E.N. Broadbent, P.J.C. Oliveira, M. Keller, D.E. Knapp, J.N.M. Silva. 2006. Condition and fate of logged forests in the Brazilian Amazon. *P. Natl. Acad. Sci. USA* 103:12947–12950.

Attiwill, P.M. 1994. The disturbance of forest ecosystems. *For. Ecol. Manage.* 63:247–300.

Berninger, K. 2007. *Attitudes de trois groupes d'intérêt sur les forêts et la foresterie en Mauricie*. Rapport pour projet TRIADE [Online]. http://www.projettriade.ca/recherche_doc.php

Binkley, C.S. 1997. Preserving nature through intensive plantation forestry: The case for forestland allocation with illustrations from British Columbia. *For. Chron.* 73(5):553–559.

Blattert, C., M. Mönkkönen, D. Burgas, F. Di Fulvio, A. Toraño Caicoya, M. Vergarechea, J. Klein, M. Hartikainen, C. Antón-Fernández, R. Astrup, M. Emmerich, N. Forsell, J. Lukkarinen, J. Lundström, S. Pitzén, W. Poschenrieder, E. Primmer, T. Snäll, and K. Eyvindson. 2023. Climate targets in European timber-producing countries conflict with goals on forest ecosystem services and biodiversity. *Commun. Earth Environ.* 4(1):Article 1. https://doi.org/10.1038/s43247-023-00771-z

Bos, J. 1993. Zoning in forest management: A quadratic assignment problem solved by simulated annealing. *J. Environ. Manage.* 37:127–145.

Boyland, M., J. Nelson, and F.L. Bunnell. 2004. Creating land allocation zones for forest management: A simulated annealing approach. *Can. J. For. Res.* 34:1669–1682.

Brockerhoff, E.G., H. Jactel, J.A. Parrotta, C.P. Quine, and J. Sayer. 2008. Plantation forests and biodiversity: Oxymoron or opportunity? *Biodiv. Conserv.* 17:925–951.

Butchart, S.H.M., J.P.W. Scharlemann, M.I. Evans et al. 2012. Protecting important sites for biodiversity contributes to meeting global conservation targets. *PLoS One* 7.

Côté, P., R. Tittler, C. Messier, D.D. Kneeshaw, A. Fall, and M.-J. Fortin. 2010. Comparing different forest zoning options for landscape-scale management of the boreal forest: Possible benefits of the TRIAD. *For. Ecol. Manage.* 259(3):418–427. https://doi.org/10.1016/j.foreco.2009.10.038

Cyr, D., S. Gauthier, Y. Bergeron, and C. Carcaillet. 2009. Forest management is driving the eastern North American boreal forest outside its natural range of variability. *Frontiers Ecol. Environ.* 7:519–524.

D'Eon, R.G., D. Hebert, and S.L. Viszlai. 2004. An ecological rationale for sustainable forest management concepts at Riverside Forest Products, southcentral British Columbia. *For. Chron.* 80:341–348.

Dinerstein, E., D. Olson, A. Joshi, C. Vynne, N.D. Burgess, E. Wikramanayake, N. Hahn, S. Palminteri, P. Hedao, R. Noss, M. Hansen, H. Locke, E.C. Ellis, B. Jones, C.V. Barber, R. Hayes, C. Kormos, V. Martin, E. Crist, M. Saleem et al. 2017. An ecoregion-based approach to protecting half the terrestrial realm. *BioSci.* 67(6):534–545. https://doi.org/10.1093/biosci/bix014

Edwards, D.P., B. Fisher, and D. Wilcove. 2012. High conservation value or high confusion value? Sustainable agriculture and biodiversity conservation in the tropics. *Conserv. Lett.* 5:20–27.

Edwards, D.P., J.J. Gilroy, P. Woodcock et al. 2014. Land-sharing versus land-sparing logging: Reconciling timber extraction with biodiversity conservation. *Glob. Chang. Biol.* 20:183–191.

Edwards, D.P., J.A. Hodgson, K.C. Hamer et al. 2010. Wildlife-friendly oil palm plantations fail to protect biodiversity effectively. *Conserv. Lett.* 3:236–242.

Elliott State Research Forest Management Plan. 2023, October. *Draft.* www.forestry.oregonstate.edu/sites/default/files/ESRF_FMP_Public%20Comment%20Draft_10.9.23_v2.pdf

Erskine, P.D., D. Lamb, and M. Bristow. 2006. Tree species diversity and ecosystem function: Can tropical multispecies plantations generate greater productivity? *For. Ecol. Manage.* 233:205–210.

FAO (Food and Agriculture Organization of the United Nations). 2010. *Global forest resources assessment 2010.* FAO Forestry Paper 163. www.fao.org/docrep/013/i1757e/i1757e.pdf

FAO (Food and Agriculture Organization of the United Nations). 2020. *Global forest resources assessment 2020: Main report.* Rome: FAO. https://doi.org/10.4060/ca9825en

Felton, A., M. Lindbladh, J. Brunet, and O. Fritz. 2010. Replacing coniferous monocultures with mixed-species production stands: An assessment of the potential benefits for forest biodiversity in Northern Europe. *For. Ecol. Manage.* 260:939–947.

Finch, T., S. Gillings, R.E. Green, D. Massimino, W.J. Peach, and A. Balmford. 2019. Bird conservation and the land sharing-sparing continuum in farmland-dominated landscapes of lowland England. *Conserv. Biol.* 33(5):1045–1055. https://doi.org/10.1111/cobi.13316

Fischer, A.P. 2018. Forest landscapes as social-ecological systems and implications for management. *Landsc. Urban Plan.* 177:138–147. https://doi.org/10.1016/j.landurbplan.2018.05.001

Fletcher, M.-S., R. Hamilton, W. Dressler, and L. Palmer. 2021. Indigenous knowledge and the shackles of wilderness. *Proc. Natl. Acad. Sci.* 118(40). https://doi.org/10.1073/pnas.2022218118

Fox, T.R. 2000. Sustained productivity in intensively managed forest plantations. *For. Ecol. Manag.* 138(1–3):187–202.

Gouvernment du Québec. 2014. *Sustainable forest development act.* http://www2.publicationsduquebec.gouv.qc.ca

Griscom, B.W. and R. Cortez. 2013. The case for improved forest management (IFM) as a priority REDD plus strategy in the tropics. *Trop. Conserv. Sci.* 6:409–425.

Griscom, B.W., R.C. Goodman, Z. Burivalova, and F.E. Putz. 2018. Carbon and biodiversity impacts of intensive versus extensive tropical forestry. *Conserv. Lett.* 11(1):e12362. https://doi.org/10.1111/conl.12362

Grumbine, R.E. 1994. What is ecosystem management? *Conserv. Biol.* 8:27–38.

Gustafsson, L., S.C. Baker, J. Bauhus et al. 2012. Retention forestry to maintain multifunctional forests: A world perspective. *BioSci.* 62:633–645.

Hallberg-Sramek, I., E.-M. Nordström, J. Priebe, E. Reimerson, E. Mårald, and A. Nordin. 2023. Combining scientific and local knowledge improves evaluating future scenarios of forest ecosystem services. *Ecosyst. Serv.* 60:101512. https://doi.org/10.1016/j.ecoser.2023.101512

Hansen, M.C., P.V. Potapov, R. Moore et al. 2013. High-resolution global maps of 21st-century forest cover change. *Science* 342:850–853.

Harris, S.H. and M.G. Betts. 2023. Selecting among land sparing, sharing, and Triad in a temperate rainforest depends on biodiversity and timber production targets. *J. Appl. Ecol.* 60:737–750. https://doi.org/10.1111/1365-2664.14385

Hartley, M.J. 2002. Rationale and methods for conserving biodiversity in plantation forests. *For. Ecol. Manage.* 155:81–95.

High Production Forestry in Nova Scotia Phase 2 Guidance for Implementation. 2023. *Department of natural resources and renewables.* Province of Nova Scotia. ISBN:978-1-77448-432-6. https://novascotia.ca/ecological-forestry/docs/HPF-phase2-guidance-for-implementation.pdf

Himes, A., M. Betts, C. Messier, and R. Seymour. 2022. Perspectives: Thirty years of triad forestry, a critical clarification of theory and recommendations for implementation and testing. *For. Ecol. Manag.* 510:120103. https://doi.org/10.1016/j.foreco.2022.120103

Himes, A. and K. Dues. In press. Relational forestry: Transforming forestry's core principles. *Ecosystems and People*. https://doi.org/10.1080/26395916.2024.2365236

Himes, A., K. Puettmann, and B. Muraca. 2020. Trade-offs between ecosystem services along gradients of tree species diversity and values. *Ecosyst. Serv.* 44. Scopus. https://doi.org/10.1016/j.ecoser.2020.101133

Hunter, M.L., Jr. 1990. *Wildlife, forests, and forestry*. Englewood Cliffs, NJ: Prentice-Hall.

Hunter, M.L., Jr. 1993. Natural fire regimes as spatial models for managing boreal forests. *Biol. Conserv.* 65:115–120.

Hunter, M.L. 1999. *Maintaining biodiversity in forest ecosystems*. Cambridge: Cambridge University Press.

Jaeger, J.A.G., L. Fahrig, and K.C. Ewald. 2006. Does the configuration of road networks influence the degree to which roads affect wildlife populations? In *Proceedings of the 2005 international conference on ecology and transportation*, eds. C.L. Irwin, P. Garrett, and K.P. McDermott. Raleigh, NC: Center for Transportation and the Environment, North Carolina State University.

Kates, R.W., W.C. Clark, R. Corell et al. 2001. Sustainability Science. *Science* 292(5517):641–642. https://doi.org/10.1126/science.1059386

Krcmar, E., I. Vertinsky, and G.C. van Kooten. 2003. Modelling alternative zoning strategies in forest management. *Int. Trans. Op. Res.* 10:483–498.

Lahey, W. 2018. *An independent review of forest practices in Nova Scotia*. Halifax, NS: Dalhousie University. https://novascotia.ca/natr/forestry/forest_review/Lahey_FP_Review_Report_ExecSummary.pdf

Lahey, W. 2021. *Independent evaluation of implementation of the forest practices report for Nova Scotia (2018)*. https://novascotia.ca/natr/forestry/Forest_Review/Report_FPR_Evaluation_2021.pdf

Landres, P.B., P. Morgan, and F.J. Swanson. 1999. Overview of the use of natural variability concepts in managing ecological systems. *Ecol. Appl.* 9:1179–1188.

Lindenmayer, D.B. and R.J. Hobbs. 2004. Fauna conservation in Australian plantation forests: A review. *Biol. Conserv.* 119:151–168.

McGrath, T., M. Pulsifer, R. Seymour, L. Doucette, G. Forbes, R. McIntyre, R. Milton, L. Cogan, M. Retallack, and T. Crewe. 2021. *Nova Scotia silvicultural guide for the ecological matrix*. Province of Nova Scotia: Department of Lands and Forestry. ISBDN 978-1.

McIntyre, R., R. O'Keefe, K. Keys, K. George, J. Ring, J. Steenberg, M. Retallack, T. Erdle, and G. Forbes. 2021. *High production forestry in Nova Scotia: Phase 1*. Province of Nova Scotia: Department of Lands and Forestry. ISBN: 978-1-77448-216-2

MDDEP (Ministère du Développement durable, de l'Environnement et des Parcs). 2011. *Strategic guidelines for Quebec protected areas: We take growth seriously*. Québec, QC: Gouvernement du Québec.

Messier, C., J. Bauhus, F. Doyon, F. Maure, R. Sousa-Silva, P. Nolet, M. Mina, N. Aquilué, M.-J. Fortin, and K. Puettmann. 2019. The functional complex network approach to foster forest resilience to global changes. *For. Ecosyst.* 6(1):21. https://doi.org/10.1186/s40663-019-0166-2

Messier, C., J. Bauhus, R. Sousa-Silva, H. Auge, L. Baeten, N. Barsoum, H. Bruelheide, B. Caldwell, J. Cavender-Bares, and E. Dhiedt. 2021. For the sake of resilience and multifunctionality, let's diversify planted forests! *Conserv. Lett.* e12829.

Messier, C., M.G. Betts, R. Tittler, and A. Paquette. 2020. A novel TRIAD approach to increase resilience of the forest landscape to global change: Or how to make a better omelette without cracking too many eggs. In *How to balance forestry and biodiversity conservation—a view across Europe*, eds. F. Krumm, A. Schuck, and A. Rigling, p. 640. Birmensdorf: European Forest Institute (EFI); Swiss Federal Institute for Forest, Snow and Landscape Research (WSL).

Messier, C., R. Tittler, D.D. Kneeshaw, N. Gélinas, A. Paquette, K. Berninger, H. Rhealt, P. Meek, and N. Beaulieu. 2009. TRIAD zoning in Quebec: Experiences and results after 5 years. *For. Chron.* 85: 885–896.

Miranda, J.J., L. Corral, A. Blackman, G. Asner, and E. Lima. 2016. *Effects of protected areas on forest cover change and local communities: Evidence from the Peruvian Amazon. World Dev.* 78:288–307.

Montigny, M.K. and D.A. MacLean. 2006. Triad forest management: Scenario analysis of forest zoning effects on timber and nontimber values in New-Brunswick, Canada. *For. Chron.* 82:496–511.

Montigny, M.K. and D.A. MacLean. 2011. Triad forest management: Scenario analysis of forest zoning effects on timber and non-timber values in New Brunswick, Canada. *The Forestry Chronicle* 82. https://doi.org/10.5558/tfc82496-4

Nelson, A. and K.M. Chomitz. 2011. Effectiveness of strict vs. multiple use protected areas in reducing forest fires: A global analysis using matching methods. *PLoS One* 6.

Nitschke, C.R. and J.L. Innes. 2008. Integrating climate change into forest management in South-Central British Columbia: An assessment of landscape vulnerability and development of a climate-smart framework. *For. Ecol. Manage.* 256:313–327.

Paavilainen, E. 1994. The concept of sustainable management in boreal and temperate forests. *IUFRO News* 23(3):8–9.

Paquette, A. and C. Messier. 2009. The role of plantations in managing the world's forests in the Anthropocene. *Front. Ecol. Environ.* 8:27–34.

Palik, B.J., A.W. D'Amato, J.F. Franklin, and K.N. Johnson. 2020. *Ecological silviculture: Foundations and applications*. Long Grove, IL: Waveland Press.

Pawson, S.M., A. Brin, E.G. Brockerhoff, D. Lamb, T.W. Payn, A. Paquette, and J.A. Parrotta. 2013. Plantation forests, climate change and biodiversity. *Biodiv. Conserv.* 22:1203–1227.

Phalan, B.T. 2018. What have we learned from the land sparing-sharing model? *Sustain.* 10(6):1760.

Phalan, B.T., M. Onial, A. Balmford, and R.E. Green. 2011. Reconciling food production and biodiversity conservation: Land sharing and land sparing compared. *Science* 333:1289–1291.

Pinnschmidt, A., R. Yousefpour, A. Nölte, and M. Hanewinkel. 2023. Tropical mixed-species plantations can outperform monocultures in terms of carbon sequestration and economic return. *Ecol. Econ.* 211:107885. https://doi.org/10.1016/j.ecolecon.2023.107885

Piotto, D. 2008. A meta-analysis comparing tree growth in monocultures and mixed plantations. *For. Ecol. Manage.* 255:781–786.

Porter-Bolland, L., E.A. Ellis, M.R. Guariguata, I. Ruiz-Mallen, S. Negrete-Yankelevich, and V. Reyes-Garcia. 2012. Community managed forests and forest protected areas: An assessment of their conservation effectiveness across the tropics. *For. Ecol. Manage.* 268:6–17.

Puettmann, K.J., K.D. Coates, and C.C. Messier. 2009. *A critique of silviculture: Managing for complexity*. Washington, DC: Island Press.

Putz, F.E., and C. Romero. 2014. Futures of tropical forests *(sensu lato)*. *Biotropica* 46:495–505 2014.

Ranius, T., and J.-M. Roberge. 2011. Effects of intensified forestry on the landscape-scale extinction risk of dead wood dependent species. *Biodiv. Conserv.* 20:2867–2882.

Rannow, S., N.A. Macgregor, J. Albrecht, H.Q. Crick, M. Förster, S. Heiland, G. Janauer, M.D. Morecroft, M. Neubert, and A. Sarbu. 2014. Managing protected areas under climate change: Challenges and priorities. *Environ. Manag.* 54(4):732–743.

Redelsheimer, C.L. 1996. Enhancing forest management through public involvement: An industrial landowner's experience. *J. For.* 94:24–27.

Resch, B.S., A. Himes, S.C. Grado, R.K. Grala, and B. Kanieski da Silva. 2023. Functional zoning by site index to balance multiple objectives for Pine in Southern US family forests. *J. For.* 121(4):307–318. https://doi.org/10.1093/jofore/fvad015

Runting, R.K., Ruslandi, B.W. Griscom et al. 2019. Larger gains from improved management over sparing-sharing for tropical forests. *Nat. Sustain.* 2(1):Article 1. https://doi.org/10.1038/s41893-018-0203-0

Seymour, R.S., and M.L. Hunter, Jr. 1992. New forestry in eastern spruce–fir forests: Principles and applications to Maine. *Maine Ag. Exp. Stn. Misc. Pub.* 716.

Seymour, R.S., and M.L. Hunter Jr. 1999. Principles of ecological forestry. *Maintaining biodiversity in forest ecosystems*, pp. 22–64. Cambridge, UK: Cambridge University Press.

Spiecker, H. 2003. Silvicultural management in maintaining biodiversity and resistance of forests in Europetemperate zone. *J. Environ. Manage.* 67:55–65.

Strengbom, J., A. Dahlberg, A. Larsson, Å. Lindelöw, J. Sandström, O. Widenfalk, and L Gustafsson. 2011. Introducing intensively managed spruce plantations in Swedish forest landscapes will impair biodiversity decline. *Forests* 2:610–630.

Sushinsky, J.R., J.R. Rhodes, H.P. Possingham, T.K. Gill, and R.A. Fuller. 2013. How should we grow cities to minimize their biodiversity impacts? *Glob. Chang. Biol.* 19:401–410.

Thompson, I.D. and D.A. Welsh. 1993. Integrated resource management in boreal forest ecosystems—impediments and solutions. *For. Chron.* 69:32–39.

Tittler, R., C. Messier, and A. Fall. 2012. Concentrating anthropogenic disturbance to balance ecological and economic values: Applications to forest management. *Ecol. Appl.* 22:1268–1277.

Tittler, R., C. Messier, and R.C. Goodman. 2016. Triad forest management: Local fix or global solution. In *Ecological forest management handbook*, vol. 33, pp. 33–45. Boca Raton, FL: CRC.

Tollefson, J. 2021. Controversial forestry experiment will be largest-ever in United States. *Nature* 594:20–21.

United Nations. 1972. *Convention concerning the protection of the world cultural and natural heritage*. http://whc.unesco.org/en/conventiontext/

United Nations. 1992. *Convention on biological diversity*. www.cbd.int/doc/legal/cbd-en.pdf

United Nations. 2015. *Transforming our world: The 2030 agenda for sustainable development*. New York, NY: United Nations. www.cma-lifelonglearning.org/lll/wp-content/uploads/2015/001%20Transforming%20our%20world%20Agenda%20Sustainable%20Development%202030%20EngFreSpa%20AEL%20151004.doc

Vincent, J.R. and C.S. Binkley. 1993. Efficient multiple-use forestry may require land-use specialization. *Land Econ.* 69(4):370. https://doi.org/10.2307/3146454

WCED (World Commission on Environment and Development). 1987. *Our common future*. Oxford: Oxford University Press.

West, S., L.J. Haider, S. Stålhammar, and S. Woroniecki. 2020. A relational turn for sustainability science? Relational thinking, leverage points and transformations. *Ecosyst. People* 16(1):304–325.

Yachi, S. and M. Loreau. 1999. Biodiversity and ecosystem productivity in a fluctuating environment: The insurance hypothesis. *P. Natl. Acad. Sci. USA* 96:1463–1468.

3 Forest Wildlife Management

Steven M. Grodsky, Christopher E. Moorman, and Kevin R. Russell

3.1 INTRODUCTION

Just like a collection of buildings devoid of human inhabitants does not make a city, a collection of trees alone does not make a forest ecosystem. Whereas the plant world is sedentary, silent, and full of utilitarian hues of green for photosynthesis, wildlife enliven forests with movement, sound, and color. Wildlife heavily influence forests via behaviors, including pollination, seed dispersal, and herbivory. Importantly, effective management of forest wildlife is contingent on a sound understanding of the habitat requirements of focal species and how forest management actions affect habitat availability or quality for those species. In this chapter, we focus on the underlying relationships between plants and animals in forested systems and describe forest management approaches that conserve forest-dwelling wildlife.

3.2 A PRIMER IN FOREST WILDLIFE-HABITAT RELATIONSHIPS

Forests provide habitat for many wildlife species and consequently are biologically diverse terrestrial ecosystems. Habitat can be defined as the set of resources—food, cover, and water—necessary to support a population of a specific wildlife species through space and time (Hall et al. 1997). By this definition, habitat is species-specific and must meet the needs of not only individuals but entire populations of the species. In forests, vegetation is the basis of the food and cover that is habitat, so understanding the response of vegetation to disturbance is critical for predicting how management actions affect focal wildlife species.

An animal's fitness—the ability to survive and reproduce—is influenced by its ability to exploit habitat components to gain and maintain energy. *Food*—Energy flow through a forest begins with solar energy from the sun, which is photosynthesized by plants to form a major component of the bottom of the forest food pyramid. Some plant species and parts are more digestible, and thus higher quality food for herbivores, than others. In fact, forests contain lignin and cellulose, both of which are indigestible to many wildlife species. After herbivores consume digestible plant material or predators consume herbivores, some energy is lost through metabolism and in their urine and feces; the remainder is considered net energy. *Cover*—Once wildlife gain net energy from food, they may use cover to conserve it or to reduce risk of predation. Specifically, net energy from food is lost as heat, and the remainder is considered available energy—the true currency of physiological condition. Endotherms (e.g., birds and mammals) generate their own body heat and require relatively constant body temperatures for favorable metabolic function; individual survival and successful reproduction are predominantly driven by energy gains from food and exploitation of cover for energy conservation (i.e., reduction of heat loss), but access to water and nutrients also plays a role (McComb 2008). Ectotherms (e.g., reptiles, amphibians, fish) absorb their body heat from the surrounding environment. Ectotherms' fitness is mainly driven by energy gains from food, exploitation of cover to adjust ambient temperature and exposure, and in some cases, availability of water for thermoregulation and reproduction (Forsman 2000). Apart from energy conservation, wildlife also use cover for rearing young and evading predators.

The spectrum of environmental conditions in which a species occurs depends, in part, on the range of food and cover resources it can exploit. Habitat generalists can exploit a wide range of

DOI: 10.1201/9781003431084-4

environmental conditions and use a broad suite of food and cover resources. These species typically occur over large geographic ranges and inhabit a variety of forest types and ages. For example, the white-tailed deer (*Odocoileus virginianus*) is a generalist species that occurs in boreal, temperate, and tropical forests throughout the western hemisphere (Branan et al. 1985; Russell et al. 2001). On the other hand, habitat specialists are adapted to a narrow range of food and cover resources, selecting unique conditions and specific vegetation types or structural conditions. As such, these species may have a small geographic range and may be restricted to a single community type, often within a narrow range of vegetation structural conditions. For example, red-cockaded woodpecker (*Picoides borealis*) is a habitat specialist that requires live conifer trees and reduced hardwood under- and mid-stories for foraging and nesting (Walters 1991), conditions historically characteristic of fire-mediated pine (*Pinus* spp.) ecosystems of the southeastern United States (Figure 3.1).

Wildlife select habitat hierarchically based on its apparent ability to meet their food and cover requirements (Johnson 1980). First-order selection is conceptualized on a regional scale and is represented by the overall distribution of a species. For example, wildlife species adapted to tropical forests may not occur simultaneously in boreal forests. However, first-order selection can vary seasonally, as is the case for neotropical migrant birds that occur in temperate or boreal forests during the breeding season and tropical forests during the non-breeding season (Rappole 1995). Second-order selection is represented by home range establishment—the area an animal typically uses throughout the year. Larger species have larger energy demands and thus require larger home ranges for food gathering relative to smaller species (McNab 1963). Additionally, home range size may vary with habitat quality (e.g., an abundance of food resources may equate to a smaller home range). Third-order selection refers to the core-use area within an animal's home range, which may be centered on a nesting or denning site or a concentration of food or cover resources. Lastly, fourth-order selection includes specific habitat elements within an animal's core-use area (e.g., shrub for a bird's nest, a particular downed log harboring a salamander).

FIGURE 3.1 Red-cockaded woodpecker (*Leuconotopicus borealis*) is a habitat specialist that requires open, older-aged pine woodlands of the southeastern United States. (Photo by James Garabedian, used with permission.)

3.3 BASIC PRINCIPLES OF FOREST WILDLIFE ECOLOGY

Forest wildlife management involves consideration of five, interrelated, basic principles: (1) disturbance; (2) succession; (3) structure (vegetation and deadwood); (4) composition; and (5) landscape ecology. In the following section, we describe how each of these ecological principles shapes forest wildlife communities.

3.3.1 Disturbance

A forest free of severe, canopy-replacing disturbance reaches a climax state, and the trees comprising it eventually senesce and die naturally. However, forests historically were shaped by natural disturbances of different severity and periodicity (e.g., windthrow, disease, insect outbreaks, hurricane, flood, fire, landslides, volcanic eruptions; White 1979; Pickett et al. 1989). Indeed, natural disturbance is the primary driver of the development of forest structure and function (Attiwill 1994). Importantly, disturbance triggers the process of forest succession. Succession influences plant community composition and structure and thereby has profound effects on wildlife habitat (see *Succession*—3.3.2). Plants and wildlife have evolved with disturbance regimes endemic to the forest ecosystems they inhabit, and they are adapted to the size and shape, severity, and frequency at which disturbances occur.

Size and shape—The size of a disturbance can influence which wildlife species remain or recolonize after it occurs (Rosenberg and Raphael 1986). Large-scale, severe disturbances create large swaths of shrubland rich in available food and cover resources for some wildlife. For example, following large-scale, severe forest fires in the western United States, woodpeckers and ungulates can be attracted by the influx of snags for nesting and increase in herbaceous vegetation and woody sprouts as forage, respectively (Pearson et al. 1995). In most forest landscapes, there are more frequent, small natural disturbances (e.g., resulting from wind, fire, tree-fall gaps) than large ones. As such, a matrix of these small openings created through natural disturbance can be present in forest ecosystems.

Wildlife respond to variations in the size of openings and resultant vegetation conditions created by disturbances. For instance, white-footed mouse (*Peromyscus leucopus*) abundance may increase in small openings, whereas deer mice (*Peromyscus maniculatus*) may occur only in larger openings (Buckner and Shure 1985). Similarly, breeding bird habitat use in bottomland forests has been shown to increase with increasing size of forest openings for some forest bird species (Moorman and Guynn 2001). The shape of an area created by disturbance also may have implications on wildlife use of openings, particularly in reference to edge effects (see *Landscape Considerations*—3.3.6).

Severity—The severity of a disturbance directly impacts the amount of organic material (alive or dead) remaining following a disturbance. Severe disturbances such as crown fires can completely consume the forest canopy, shifting succession to early stages dominated by herbaceous plants (e.g., Turner et al. 1997). Conversely, removal of only a few trees from the canopy results in minor and ephemeral changes in vegetation structure and a small influx of legacy material (e.g., downed wood in a tree-fall gap) (Runkle 1981). Legacy material such as deadwood provides important food and cover resources for some wildlife species immediately following a disturbance (see *Structure—Deadwood*—3.3.4).

Frequency—The frequency of disturbance—or return interval—influences the floral and faunal composition of forests and the amount of organic material present over time. The frequency of disturbance varies among forest types and by type of disturbance. For example, fire return intervals may range from 2–5 years in some woodland or savannah systems (Harrell et al. 2001) to 300–400 years in northern coniferous forests (Wimberly et al. 2000). More frequent, severe disturbances may delay or preclude forest development, whereas less frequent and severe disturbances may facilitate establishment of more complex vertical vegetation structure in a stand (Franklin et al. 2002). In turn, the vegetative structure created by short and long return intervals, respectively, impacts the types of wildlife that remain or recolonize a stand following disturbance (see *Structure—Vegetation*—3.3.3).

3.3.2 Succession

Ecological succession describes the gradual replacement of plant communities over time in a specific location. The somewhat predictable changes in vegetation following a disturbance form the basis for the concept of ecological succession (Palmer et al. 1997). Importantly, each successional stage is defined by the plant species composition and vegetation structure and density. Early succession is dominated by annual and perennial grasses and forbs (i.e., herbaceous species). Shade-intolerant tree species establish and form the canopy in mid-succession, and shade-tolerant tree species eventually replace shade-intolerant species in late successional forest communities (i.e., climax forest; Bazzaz 1975). Herbaceous species may persist in the understory of mid- and late succession, especially where the canopy is open and sunlight is able to penetrate to the ground. Another way of conceptualizing the middle and later stages of succession, those dominated by trees, is to break the process down into four general physiognomic stages: (1) stand initiation; (2) stem exclusion; (3) understory reinitiation; and (4) old growth (Oliver 1980).

Wildlife diversity and composition differ among successional stages. Traditionally, wildlife diversity was thought to increase through succession and reach its peak in late succession forests because this terminal successional stage provides the greatest accumulation of biomass and thus prevalence of niches (Odum 1969). However, the richness of the wildlife community may be more a result of the structural conditions of the vegetation and legacy material within a successional stage than the stage itself (also see *Structure—Vegetation and Structure—Deadwood—3.3.3; 3.3.4*). Moreover, early successional stages can support high species richness comprising opportunists and grassland and shrubland obligates, which often use abundant organic legacy material and post-disturbance plant communities (Verschuyl et al. 2008; Swanson et al. 2010). Wildlife responses to succession may be thought of as coinciding with the cycle in which a series of plants and associated structural conditions, and thus food and cover resources, appear and disappear in between disturbances (Hunter 1990). Early successional stages typically are characterized by bare ground, abundant downed wood, and lack of canopy cover immediately following disturbance, followed by the onset of lush herbaceous cover. At this early stage, wildlife species dependent on bare ground for feeding on seeds (e.g., mourning dove (*Zenaida macroura*) (Grodsky et al. 2016a) or herbaceous vegetation (e.g., small mammals for cover; Fritts et al. 2017) are abundant. Conversely, canopy-dwelling species are absent because the trees upon which they rely have yet to grow. Over time, as trees establish and begin to shade out shade-intolerant herbaceous species during the stem exclusion stage of forest development, deadwood (i.e., snags, downed wood) becomes increasingly scarce (see U-shaped chronosequence in *Structure—Deadwood—3.3.4*). As such, cavity nesting wildlife (e.g., woodpeckers) are rare, until the forest stand ages and older trees senesce and become snags and downed deadwood. Adequate resources may be available to some wildlife species, particularly habitat generalists, through multiple successional stages or across a range of forest stand ages. On the other hand, habitat specialists tend to be linked to specific successional stages (e.g., grassland birds) or structural conditions within mid- and late succussion forests (e.g., woodland birds that require open canopy and sparse midstory).

3.3.3 Structure—Vegetation

Vegetation structure in a forest can be characterized on both the vertical and horizontal planes and is tightly linked to wildlife habitat use. Vertical structure is a measure of the stratification of forest layers starting with ground cover, spanning through the shrub and midstory layers, up to the canopy layer. The understory, including the ground and shrub layers, is especially important to wildlife that cannot fly or climb because this is the layer in forests that is accessible to them. Wildlife that can fly (e.g., birds, insects) or climb (e.g., primates, some herpetofauna) can access all forest layers and demonstrate differentiation of niches among these layers. Niche differentiation can be achieved by vertical segregation of species among forest layers, which reduces spatially explicit competition for resources. For instance, birds are well known to segregate vertically in forests and on trees while feeding to reduce competition for food (Cody 1974). In general, a more complex vertical structure (i.e., more layers) in forests leads to more niches available to wildlife (MacArthur and MacArthur 1961; Hunter 1990). However, some wildlife species thrive in forests with little vertical stratification and many fire-regulated forest ecosystems maintain relatively high species diversity despite the reduction or absence of midstory layers (Means 2006).

Horizontal patchiness refers to the distribution of vegetation along the horizontal plane of a forest; a forest stand with complex vertical structure may contain every forest layer along the vertical plane, whereas within-stand horizontal patchiness pertains to forest layers independently existing along the horizontal plane (e.g., ground cover without canopy cover) (Noss 1999). Within-stand patchiness naturally occurs after canopy closure when forest openings are created (e.g., tree-fall gaps) and can be created or enhanced through forest management practices such as thinning and group selection (see *Thinning—3.4.5* and *Uneven-Aged Systems and Wildlife—3.4.3*, respectively). At the landscape-level, horizontal patchiness can be achieved by interspersing even-aged stands of various ages and structural conditions with more lightly managed stands and riparian corridors (see *Landscape Considerations—3.3.6*).

3.3.4 Structure—Deadwood

When trees die due to competition, disease, insect outbreaks, lighting strikes, or senescence, among other factors, they become snags (i.e., standing deadwood) and ultimately fall to the ground to become downed wood (Figure 3.2). Snags and downed wood provide valuable food and cover for a plethora of wildlife (Harmon et al. 1986). Wildlife, including primary cavity users (wildlife capable of excavating cavities in trees) like woodpeckers and secondary cavity users (wildlife reliant on primary cavity users for cavity production) like flying squirrels, use snags for nesting, roosting, and feeding. Downed wood is predominantly used by ground-dwelling wildlife for cover, escape from predators, travel corridors, and in the case of amphibians, its presumably favorable microclimate (Lanham and Guynn 1996; Loeb 1996; Whiles and Grubaugh 1996; Fritts et al. 2015b, 2016, 2017). Additionally, predators of all sorts feed on abundant arthropods harbored by downed wood (Castro and Wise 2010; Lohr et al. 2002; Ulyshen and Hanula 2009; Grodsky et al. 2019). A vast array of invertebrates associate with downed wood, spanning downed wood-obligate (saproxylic) insects (Grove 2002) to habitat generalists (Hagan and Grove 1999).

FIGURE 3.2 Downed wood is an important habitat component for many ground-dwelling, forest wildlife species. (Photo by Christopher E. Moorman, used with permission.)

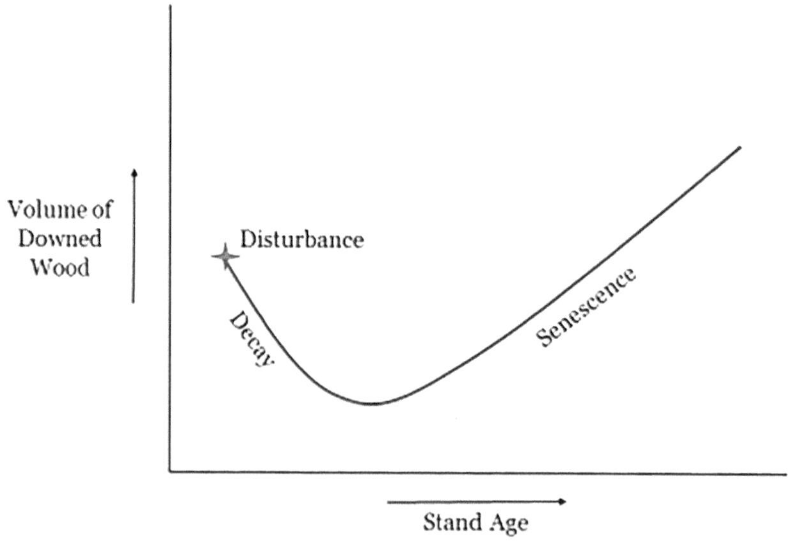

FIGURE 3.3 An interpretation of the U-shaped chronosequence (Harmon et al. 1986) illustrating downed wood dynamics in forests. Volume of downed wood (represented by the "U-shaped" line) is high immediately following a disturbance, decreases with stand age as it decays, and slowly increases as mature trees senesce and die. (Schematic by Steven M. Grodsky.)

In general, volume of downed wood on the forest floor of a stand follows a U-shaped chronosequence (Harmon et al. 1986; Figure 3.3). Immediately following a disturbance, the volume of downed wood is great. Over time, the downed wood decays, leading to a decrease in its volume. A slight but steady increase in downed wood volume then follows as mature trees senesce and die. Both the role and dynamics of downed wood vary regionally. For example, downed wood in the southeastern United States historically was consumed by frequent fires and otherwise decays rapidly due to damp, warm climatic conditions (Moorman et al. 1999). As such, many wildlife species in the southeastern United States have adapted to forest floors with little available downed wood. On the other hand, wildlife in cooler, less disturbed areas of the United States like the Pacific Northwest are adapted to historically high volumes of downed wood on the forest floor (Bunnell et al. 1999; Butts and McComb 2000).

3.3.5 Composition

Plant species composition influences availability of resources to wildlife. Wildlife diversity generally increases with increasing plant diversity within a stand (also see *Landscape Considerations—3.3.6*). A mixture of plant species leads to more complex vertical structure, which, in turn, has positive implications for niche partitioning among wildlife in many forests (MacArthur and MacArthur 1961). High plant diversity specifically increases the availability of food and cover. For example, in temperate forests, a mixture of coniferous (e.g., *Pinus* spp.) and deciduous (e.g., *Quercus* spp.) trees may help balance winter cover and food availability to wildlife throughout the year. Conifers are particularly important as winter cover for ungulates (Moen 1973) and as winter roosting sites for eastern wild turkey (*Meleagris gallopavo*; Kilpatrick et al. 1988) in colder climates. Additionally, high plant diversity helps to ensure food resources are available to wildlife year-round, especially in the absence of primary food sources.

Coniferous and deciduous forests differ in the wildlife species they support. The foliage of coniferous trees is less palatable than that of deciduous trees because it is rich in distasteful compounds like terpenes (Longhurst et al. 1968). The decreased palatability of coniferous foliage has ecological implications beginning at the bottom of the food web, with fewer invertebrates and herbivores eating conifer litter and foliage, respectively, and consequentially fewer predators that eat herbivores (Kennedy and Southwood 1984). As such, deciduous forests often are more biologically diverse than coniferous forests. However, many coniferous forests support unique communities (Wiens 1975) and contribute to wildlife diversity at larger spatial scales.

3.3.6 Landscape Considerations

Spatial Heterogeneity—Wildlife diversity at the landscape level is affected by the arrangement (i.e., interspersion and juxtaposition) of vegetation types and stand ages on the landscape, in addition to unique landscape features (e.g., bluffs, prairie fragments, caves). In general, dispersal of many forest types, stand ages, and stand sizes across the landscape helps to maximize landscape-level wildlife diversity. Many wildlife species require multiple forest types or stand ages because of the varying vegetation composition and structure necessitated by their life histories (Hunter 1990). Although more mobile species (e.g., birds) can easily access different forest types or stand ages, less mobile species may require some level of connectivity to exploit varied habitat conditions across the landscape (Todd et al. 2009) (see *Connectivity* discussed later).

Edges and Fragmentation—Edge is the location where two landcover types (e.g., forest ages, land uses) meet. Aldo Leopold (1933) described the "edge effect" after noticing wildlife, particularly game animals, increased in diversity and abundance near edges. At an edge, animals have simultaneous access to two or more vegetation types, along with the associated unique resources in each type. Additionally, vertical vegetation structure at high contrast edges (i.e., hard edges) may be more complex than in the adjacent interior of forest patches because the absence of canopy cover on one side of the edge (e.g., young forest, cropland) allows sunlight to penetrate into the forest on the other side. Many forest wildlife species have been labeled edge specialists, but these associations are because the closed canopy and sparse understory in forests do not provide appropriate food and cover for wildlife; instead, individuals are relegated to forest edges where greater sunlight penetration creates more abundant food and cover resources. Additionally, small and irregularly shaped stands have more edge than larger and more uniformly shaped stands. Importantly, hard edges between forest and non-forested land uses (e.g., urban, agriculture) differ from the edges (i.e., soft edges) between different forest types or ages. With the former, true forest fragmentation occurs, which more often leads to negative edge effects (Wade et al. 2003). For instance, edges between forests and non-forest land uses may attract avian nest predators and brood parasites (e.g., brown-headed cowbirds (*Molothrus ater*)), thereby lowering nest success of songbirds in forest patches (Patton 1994).

Connectivity—Animals may need to move among habitat patches in search of food and cover, to access seasonal life-history requirements, such as breeding or overwintering sites, to locate available resources, to disperse and establish territories, and to maintain genetic connections with adjacent populations. In landscapes where habitat patches are fragmented, connectivity that facilitates movement of less mobile wildlife from one patch to others may be important (Haddad et al. 2003). For example, linear rainforest remnants have been shown to be particularly important for providing habitat connectivity for arboreal mammals in tropical forests (Laurence and Laurence 1999). However, corridors designed to increase connectivity also have been shown not to facilitate movement between habitat patches for some wildlife (e.g., salamanders—Rosenberg et al. 1998). Instead, less mobile species may be conserved by increasing landscape permeability, which is a measure of how easily an animal can move through a landscape. Permeability varies with a species' dispersal ability, the degree of habitat specialization the species exhibits, and the arrangement of available habitat across the landscape. At the landscape scale, permeability can be increased simply by increasing the percent cover of habitat. Understanding the ecology of target species for which connectivity is a concern is important when considering corridor design (e.g., width, length, connectivity—Lindenmayer and Nix 1993) and permeability (i.e., what is habitat for the target species, what vegetation characteristics facilitate movement or reduce predation risk).

3.3.7 Aquatic Habitats

Wetlands typically are characterized by hydrology, occurrence of hydric soils, and presence of plant species adapted to hydric soils. Wetlands can have high plant diversity, complex vegetative structure, and a diversity of wildlife foods, including arthropods, fruits, and nuts. Examples of wetlands in forested environments include river bottoms, stream corridors, and depressional ponds. Wetlands are especially important for amphibians, insects, and reptiles that rely on water for one or more parts of their life cycles (Batzer and Wissinger 1996; Semlitsch and Bodie 2003). Specifically, many forest amphibians rely on ephemeral pools to breed because they hold water only temporarily and thus do not harbor fish predators that typically consume their eggs (DiMauro and Hunter 2002). Waterfowl and wading birds also rely on wetlands for food and cover, often year-round (Isola et al. 2000). In drier regions, wetlands may provide important sources of drinking water for wildlife.

Given their high value to wildlife and water quality, wetland management is an important consideration during forest harvest activities. Buffers parallel to rivers and streams or surrounding small, isolated wetlands established prior to timber harvest typically are effective means by which to reduce potentially negative impacts on wetlands during timber harvest (Semlitsch and Bodie 2003). These forested buffers provide shade to maintain water temperatures and help prevent sediment deposition into wetlands during timber harvest. Buffer specifications vary based on target wildlife species, stream width, harvest intensity, slope, forest ecosystem, and the type of natural and anthropogenic disturbances (Russell et al. 2002, 2005). In the United States, each state has developed Best Management Practices (BMPs) for forest management that include protection of wetlands and water quality during forest harvest activities; BMPs are tailored for each state's local conditions and, when implemented properly, are effective means by which to manage and protect forest wetlands (Cristan et al. 2016).

3.4 FOREST MANAGEMENT FOR WILDLIFE

The central thesis of game management is this: game can be restored by the creative use of the same tools which have heretofore destroyed it—axe, plow, cow, fire, and gun.

Aldo Leopold, Game Management (1933)

Aldo Leopold, a forester and wildlife conservationist, postulated that people have a right to use and manage forests wisely *and* the responsibility of doing so in a way that recognizes the intrinsic value people share with other species and the ecosystems that contain them (Leopold 1933). By this definition, Leopold captured the true essence of forest wildlife management; he identified that conservation of forest products and wildlife could be simultaneously achieved. Indeed, forest management, when appropriately implemented, may provide practical means by which to manage forest wildlife habitat.

Forest management, especially timber harvest, changes plant structure and composition, which, in turn, affects wildlife habitat (Hunter 1990). The ecological changes that accompany forest management in some part simulate the effects of natural disturbance on forest ecosystems, and therefore may similarly affect wildlife (Long 2009). Importantly, long-standing natural disturbance regimes have been altered by humans through means such as wildfire suppression, introduction of invasive and non-native species, and urbanization; as such, the natural disturbance regimes to which many wildlife species adapted over millennia are far less common in today's modern world (Johnstone et al. 2016).

As described earlier, different wildlife species associate with different forest stand ages or structural conditions. Hence, changes brought upon by timber harvest create or enhance habitat for some species while simultaneously diminishing habitat for others. In other words, there are always "winners" and "losers." For example, large-scale removal of trees in a mature stand via clearcutting shifts the vegetation biomass from the overstory to the understory, in turn benefiting wildlife adapted to the conditions therein. Concurrently, mature forest associates, especially canopy dwellers, may be eliminated in the stand until the canopy redevelops. Because habitat requirements vary among wildlife species, providing habitat for all wildlife species at small spatial and temporal scales is implausible. Therefore, forest wildlife managers can identify focal wildlife species or communities before initiating forest management practices and account for local variations in wildlife-habitat relationships, or habitat planning for wildlife can be conducted at large spatial scales (i.e., thousands of hectares). Table 3.1 includes responses of selected wildlife to changes in vegetation structure that may occur following forest management practices described in the following sections.

3.4.1 Final Harvest

Forest harvest and regeneration are achieved using one of two silvicultural systems—*even-aged* or *uneven-aged* (summarized in McComb 2008). Even-aged systems regenerate an entire stand at once and contain trees of either one or two age classes. Single-aged stands are created by clearcutting, whereas multi-aged stands are created when some overstory trees are retained during harvest (i.e., reserve shelterwood or seed tree regeneration harvests). Uneven-aged systems create stands with three or more age classes of trees by employing frequent, repeated harvests of scattered individual trees (single-tree selection harvests) or small patches of trees (group-selection harvests). Even-aged forest stands usually are dominated by shade-intolerant plant species. Meanwhile, uneven-aged stands commonly promote regeneration of shade-tolerant and intermediate shade-tolerant plants.

Environmental conditions created by even- and uneven-aged systems vary both spatially and temporally (summarized in McComb 2008). Even-aged harvests may yield large areas of early successional vegetation that persist for several years. However, stump sprouts from many tree species dominate following clearcut harvests, thereby maintaining or rapidly shifting the vegetation to mid- or late succession post-harvest. Additionally, artificial regeneration following clearcutting accelerates succession by favoring the planted shade-intolerant crop tree species over herbaceous species that are quickly shaded out by the canopy of the planted trees (Figure 3.4). Once even-aged stands of timber mature, they offer large areas of vegetation with relatively consistent within-stand vegetative structure. Dispersed, even-aged harvests may lead to increased vegetation diversity at the landscape scale because a variety of stand ages may be present among stands. Meanwhile, uneven-aged harvests often yield heterogeneous within-stand environmental conditions, particularly by promoting complex vertical vegetation structure within and adjacent to harvest openings, mimicking small-scale natural canopy disturbance created by treefall.

3.4.2 Even-Aged Systems and Wildlife

Wildlife response to implementation of even-aged systems largely is influenced by reduction of forest canopy cover and release of understory vegetation. Following clearcutting, overstory trees are typically removed, and canopy-dwelling wildlife are reduced or displaced. For example, clearcutting may result in nearly complete turnover from mature forest bird species to shrubland bird species (Conner and Adkisson 1977; King and DeGraaf 2000). However, individual canopy trees, clumps of trees, and buffers on streams and waterbodies are often retained during clearcut harvests (i.e., clearcut with reserves). Reductions in canopy-dwelling bird species can be short-lived, especially where forest regeneration is rapid following harvest. Mature forest birds, including some canopy dwellers, use young forest (i.e., 5–10 years following even-aged harvests) during the post-fledging period (Champlin et al. 2009; Chandler et al. 2012; Porneluzi et al. 2014). Whereas clearcutting removes most of the canopy, seed tree and shelterwood regeneration harvests temporarily maintain some live overstory, thereby retaining denser canopy structure and, in some cases, continued overstory mast production (e.g., acorns). Seed tree and shelterwood cuts provide vegetation structure in both the understory and canopy until overstory trees are harvested. Hence, these regeneration methods provide food and cover for diverse wildlife that use multiple forest layers.

Even-aged timber harvests may maintain lush herbaceous layers for several years post-harvest, providing habitat for grassland or shrubland wildlife species. Herbaceous vegetation provides abundant forage and seeds; indeed, many herbivorous and granivorous wildlife are associated with herbaceous vegetation, whereas insectivores are associated with older forest with midstory and canopy

TABLE 3.1
Responses of Selected Wildlife "Winners" and "Losers"[1] to Changes in Habitat Structure Resulting from Select Forest Management Practices

Management Practice	Post-Management Vegetation Structure[2]	Availability of Deadwood	Percent of Stand Influenced	"Winners"	"Losers"	Citations[3]
Even-aged harvest (clearcut)[4]	Dense herbaceous and shrub/scrub cover; canopy cover absent	High (downed wood); low (snags)	High	Early successional and shrubland species requiring large disturbances; downed-wood associated species; granivores; herbivores	Canopy-dwelling species; species requiring a cool, moist micro-climate; leaf-litter obligates	Annand and Thompson 1997; deMaynadier and Hunter Jr. 1995; Kirkland 1990; Moorman et al. 2011; Niemelä 1997
Even-aged harvest (shelterwood/ seed tree)[4]	Dense herbaceous cover; moderately dense shrub/scrub cover; low canopy cover	High	High	Similar to clearcut, although retained canopy benefits overstory species and hard-mast consumers	Similar to clearcut, although retained canopy lessens effect on canopy-dwelling species and leaf-litter obligates	Rosenvald and Lõhmus 2008
Uneven-aged harvest	Patchy herbaceous and shrub/scrub cover; high canopy cover	Moderate	Low–moderate	Gap-specialists; hard-mast consumers; species using both mature and young forest	Early successional species requiring large disturbances	Bouget and Duelli 2004; Lempckert 1999; Moorman and Guynn 2001
Herbicide site prep (after clearcut)[5]	Low to moderate herbaceous cover; low shrub/scrub cover; canopy cover absent	High	High	Downed-wood associated species; early successional species requiring large disturbances	Herbivores and granivores if reduced herbaceous cover; leaf litter obligates; shrub-nesting birds if reduced shrub cover; species requiring a cool, moist micro-climate	Guynn Jr. et al. 2004; Miller and Miller 2004; Miller and Wigley Jr. 2004
Mechanical site prep (after clearcut)[5]	Moderate herbaceous cover; low shrub/scrub cover; canopy cover absent	Low (especially if burned)	High	Early successional species requiring large disturbances; granivores and herbivores (relative to herbicide site prep)	Downed-wood associated species; fossorial species (e.g., invertebrates, snakes); leaf litter obligates; species requiring a cool, moist micro-climate; shrub-nesting birds	O'Connell and Miller 1994

Thinning	Moderately dense herbaceous and shrub/scrub cover; moderate canopy cover	Moderate	High	Species benefiting from increased vertical structure (e.g., songbirds) and understory cover (e.g., shrub-nesting birds, some small mammals)	Species requiring a cool, moist micro-climate	Verschuyl et al. 2011
Prescribed burning (high intensity)	Dense herbaceous cover; moderately dense shrub/scrub cover; moderate canopy cover due to some overstory mortality	Moderate	Moderate	Cavity nesters (if overstory mortality occurs); grazing herbivores; species associated with herbaceous cover; species requiring hot, dry microclimate (e.g., lizards); understory species	Species requiring a cool, moist micro-climate; leaf-litter obligates	deMaynadier and Hunter Jr. 1995; Moorman et al. 2011; Russell et al. 1999; Schowalter 2012
Prescribed burning (low intensity)	Moderately dense herbaceous, shrub/scrub, and canopy cover	Moderate	Moderate	Browsing herbivores; shrub-nesting birds	Leaf-litter obligates but effects are short-term	

Note:

1 *There are always exceptions to the "winners" and "losers" wildlife groups following management. Furthermore, generalist species may be present following management practices, whereas specialist species may show more unique responses to management practices or variations within a management practice.*

2 *Refers to vegetation conditions within the first five years after management practice is implemented. Vegetation structure changes over time following management; for example, see Succession (section 3.3.2).*

3 *This list of citations is not exhaustive.*

4 *Effects on vegetation and deadwood are for scenarios with no site preparation. For effects of site preparation, see Herbicide site preparation and Mechanical site preparation (section 3.4.7).*

5 *Effects of site preparation, especially with herbicides, on wildlife are relatively short-lived (e.g., < 3 years post treatment; Guynn et al. 2004). Mechanical and herbicide site preparation may be concurrently implemented. "Winners" and "losers" include clearcut effects.*

FIGURE 3.4 Lead author and average-sized human among artificially regenerated loblolly pine (*Pinus taeda*) in an intensively managed forest plantation, North Carolina, USA. (Photo by Chris Moorman, used with permission.)

layers (Constantine et al. 2004). Dense understory in young, even-aged stands provides nesting and brooding cover for gallinaceous birds, such as wild turkey (Dickson 2001) and northern bobwhite (*Colinus virginianus*; Burger 2001). Herbaceous groundcover also provides important escape cover for small animals, including sparrows (Lima 1993) and small mammals, among other wildlife species. As woody shrubs and sprouting hardwoods develop, other wildlife species colonize recently harvested even-aged stands. For instance, shrub-nesting birds are especially adapted to nesting in harvested timber stands several years after harvest (Brawn et al. 2001). Even-aged stands also are characterized by a hotter, drier environment on the forest floor because of the greatly reduced canopy cover, which may be unfavorable for some amphibians prone to desiccation, such as salamanders (Figure 3.5), but favorable for ectothermic wildlife (e.g., reptiles) because hotter conditions facilitate their thermoregulation (Moorman et al. 2011).

FIGURE 3.5 Northern red salamander (*Pseudotriton ruber*) requires moist conditions to persist in forests, which may be promoted by some canopy cover and availability of downed wood. (Photo by Chris Moorman, used with permission.)

3.4.3 Uneven-Aged Systems and Wildlife

Uneven-aged regeneration harvests promote a mosaic of small openings within a matrix of older trees. Given the generally small size of openings, wildlife species adapted to large-scale disturbances typically are excluded in uneven-aged systems. For example, some shrubland bird species (e.g., prairie warbler (*Dendroica discolor*)) did not colonize small, experimental gaps created via group and single-tree selection (Robinson and Robinson 1999; Moorman and Guynn 2001). However, creating larger group-selection openings (> 0.5 ha) may increase use by shrubland wildlife species, as has been seen with birds in the southeastern United States (Moorman and Guynn 2001). Lemckert (1999) found that selective logging had little impact on forest-dwelling frogs, which may be related to the relatively low impact of small disturbances brought upon by uneven-aged management. Within-stand forage (e.g., hard mast) and cover (e.g., early successional herbaceous layer) are interspersed in uneven-aged stands. For example, white-tailed deer may bed and forage in openings and eat acorns in mature forest adjacent to openings, all within the same stand (Murray et al. 2016).

3.4.4 Regeneration and Site Preparation

Stand regeneration following forest harvest can be accomplished either through planting or naturally via stump sprouts, advanced regeneration, or seed from residual trees or the seed bank. Shelterwood and seed tree regeneration harvests in even-aged systems and single-tree selection and group-selection regeneration harvests in uneven-aged systems usually depend on natural regeneration. Planting typically follows clearcutting, though some tree species such as aspen (*Populus* spp.) or oaks (*Quercus* spp.) regenerate naturally from stump sprouts following clearcut harvests (Atwood et al. 2009). Prior to planting, sites may be prepared with heavy machinery, prescribed burning, herbicides, or a combination of these practices (the latter two are covered in their respective sections; *Prescribed Burning 3.4.6* and *Herbicide Application—3.4.7*). Specifically, site preparation is implemented to help control vegetation that may suppress planted crop tree seedlings and is categorized as either mechanical (e.g., shearing, burning, disking, bedding) or chemical (i.e., herbicides). O'Connell and Miller (1994) reported that small mammal capture rates were greater in mechanically prepared sites than on those sites that were chemically prepared, indicating that mechanical site preparation alone may be less detrimental to small mammal populations in some cases. However, soils may be disturbed during mechanical site preparation, which may, in turn, impact fossorial (underground-dwelling) wildlife species (Moorman et al. 2011). The intensity of site preparation (e.g., chemical only, mechanical only, chemical and mechanical) may influence wildlife response. However, Lane et al. (2011a) showed that the addition of chemical site preparation had little effect on birds beyond planted pine spacing, indicating that avian abundance was not proportional to site preparation intensity. Winter bird abundance has been shown to decrease with increased mechanical site preparation intensity (Hanberry et al. 2013a), but site preparation intensity had no effect on small mammal abundance (Hanberry et al. 2013b). Planting seedlings at high stocking rates with little spacing may facilitate rapid canopy closure, reducing the time early successional vegetation is available to wildlife. However, wider spacing between planted trees may allow sunlight to reach the forest floor and extend early successional stages, thus increasing herbaceous vegetation available as wildlife forage and cover.

3.4.5 Thinning

As canopy closure occurs during stem exclusion, understory vegetation is shaded out and vertical vegetation structure declines (Figure 3.6). Thinning reduces the number of overstory trees in a stand, which can simultaneously improve growing conditions for residual crop trees through decreased

(a) (b)

FIGURE 3.6 (a) An unthinned forest stand with a closed canopy and shaded understory that lacks vertical vegetation structure; and (b) a thinned stand with more diverse vertical vegetative structure due to increased light penetration to the forest floor. (Photos by Chris Moorman, used with permission.)

competition (Barbour et al. 1997) and increase sunlight penetration to the forest floor, promoting understory development (Bailey and Tappeiner 1998). Thinning during early stand development (i.e., pre-commercial thinning) helps maintain forest herbaceous layers and extends the time until crown closure. Intensified thinning typically results in increased understory vegetation cover as light reaching the forest floor increases with greater reductions in canopy cover (Thomas et al. 1999; Figure 3.6).

Thinning increases structural complexity in forests and may improve habitat for many wildlife species (Hayes et al. 1997). Indeed, a meta-analysis of the effects of forest thinning reported that forest thinning treatments had generally positive or neutral effects on forest biodiversity (Verschuyl et al. 2011). As described earlier, wildlife diversity tends to increase with vertical vegetation structure because the presence of multiple layers in the vertical plane of forests provides a high number of niches. Indeed, short-term increases in species richness and densities of songbirds have been recorded following experimental thinning (DeGraaf et al. 1991; Hagar et al. 2004). Similarly, some small mammal populations positively respond to forest thinning (Converse et al. 2006a, 2006b). Effects of thinning may become immeasurable as time passes since the initial thinning, as has been seen with bird communities in temperate rainforests 20 years after thinning (Matsuoka et al. 2012) and insect communities in plantations three years after thinning (Taki et al. 2010). In fire-mediated forests, thinning may be an effective forest wildlife management tool when used in conjunction with prescribed burning (Lashley et al. 2011; also see *Prescribed Burning—3.4.6*).

3.4.6 Prescribed Burning

Forest communities around the globe have been shaped by fire as a natural or anthropogenic disturbance, making prescribed burning a valuable forest wildlife management tool in historically fire-mediated systems. Historical fire frequencies have helped form plant communities in traditionally fire-mediated systems present today (Frost 1998). In turn, many plant species within these communities are dependent on fire for reproduction. For example, several pine species exhibit serotiny—the ability to retain seeds in the plant canopy (Lamont et al. 1991)—and their cones remain closed until exposed to high temperatures such as those resulting from fire (Gauthier et al. 1996). Fire also promotes understory plant communities by removing leaf litter, exposing mineral soil, and releasing nutrients for seedbed preparation (Russell et al. 1999).

As a management tool, prescribed burning typically reduces forest fuels and diminishes the potential for uncharacteristic wildfire (Agee and Skinner 2005), suppresses competing vegetation, encourages regeneration of certain plant species on the forest floor, reverts successional growth, and, in many cases, improves wildlife habitat (Pyne et al. 1996; Figure 3.7). Effects of prescribed burning on plant communities vary with frequency, intensity, and seasonality (i.e., growing-season versus dormant-season burns). For example, intense, hot fires reduce leaf litter and kill overstory trees, whereas lower intensity fires have minimal effects on overstory trees but may reduce leaf litter and shrub and midstory vegetation in the short term (Greenberg et al. 2023). Wildlife species shift concomitantly with vegetative response to fire seasonality, frequency, and intensity (Harper et al. 2016; Arthur et al. 2021). For example, reptile abundance can increase following severe fires that result in canopy mortality and more open, hot conditions on the forest floor (Matthews et al. 2010), and shrubland birds similarly may increase because of the pulse of woody understory plant species (Greenberg et al. 2023). Conversely, low-intensity burns, especially those under a closed canopy, typically do not result in a change in the wildlife community (Greenberg et al. 2018a, 2018b). Moreover, fire may consume snags and downed wood, thereby reducing these habitat elements for some wildlife; however, new snags resulting from fire reduce net snag loss and are in turn used by cavity nesting birds (Bagne et al. 2008). Similarly, Hanula et al. (2012) determined that frequent fires had little effect on the availability of downed wood and consequently abundance of termites (*Reticulitermes* spp.). However, insect response to prescribed fire may vary over longer time periods (e.g., 15 years), as previously documented for beetles (Campbell et al. 2018) and ants (Campbell et al. 2019).

Prescribed burning can result in direct mortality of wildlife, especially when fires are intense. Yet, given the coevolution of wildlife species and fire, wildlife can be well adapted to its occurrence on the landscape (Harper et al. 2016). Although some wildlife may perish during fires, some species exhibit behavioral adaptations that allow them to survive. Larger animals and birds may escape fire by running and flying away, respectively, whereas amphibians, reptiles, and small mammals often remain underground or seek cover under rocks or downed wood (Russell et al. 1999).

Following prescribed burning, wildlife forage may be more prolific, nutritious, and palatable (Harris and Covington 1983; Masters et al. 1993). For example, post-fire nutrition in the diets of ungulates has been shown to be elevated for bighorn sheep (*Ovis canadensis*) and mule deer (*Odocoileus hemionus*) (Hobbs and Spowart 1984) and white-tailed deer (Lashley et al. 2011). Similarly, prescribed burning may promote fruit production (Penman and Penman 2010) and encourage the growth of seed-producing herbaceous plants, which, in turn, may increase wildlife forage availability. However, effects on fruit and seed production vary depending on the season and

FIGURE 3.7 Prescribed fire is an important management tool to maintain plant diversity and understory structure for wildlife. (Photo by M. Colter Chitwood, used with permission.)

frequency of prescribed fire. For example, fire return intervals less than three years may greatly limit production of understory soft mast (Lashley et al. 2017). Many insects that provide important forage for vertebrate and invertebrate wildlife alike are favored by the flush of understory plant growth following fire (Campbell et al. 2022), especially in open forest (Schowalter 2012).

Twentieth-century fire suppression practices have significantly modified historic fire regimes throughout the world (Ryan et al. 2013). Excluding fire from historically fire-mediated systems changes forest structure, composition, and integrity across multiple scales (Varner et al. 2005). The fate of prescribed fire as a management tool may rely on increasing public awareness of its positive benefits through education and public involvement (Ryan et al. 2013).

3.4.7 Herbicide Application

Technological advancements in the chemical industry have led to widespread use of herbicides in forest management, especially in plantations. Often, herbicides are used to release crop trees from competing vegetation in young plantations but also may be used later in the rotation to suppress understory or midstory competitors. Forest herbicides typically target biochemical pathways unique to plants and thus have few acute toxic effects on wildlife (Tatum 2004). Rather, herbicides indirectly affect wildlife by changing vegetation structure and composition (Miller and Wigley 2004; Kroll et al. 2020). Research studies indicate that the impact of herbicides on plant communities may be relatively short-lived (e.g., < three years; Miller and Miller 2004). Similarly, negative effects of herbicide application on plants and thereby wildlife may be temporary (Guynn et al. 2004). Herbicide application to recent clearcuts has been shown to reduce vegetation complexity, resulting in the temporary decline of some species of songbirds, for instance (Santillo et al. 1989), and, to some degree, avian species richness (Stokely et al. 2022). However, some studies have shown herbicide use has no appreciable effect on wildlife species or communities or, in some cases, positive effects via shifted successional trajectories and plant communities (Guynn et al. 2004). For example, an "Intensive Forest Management" study in the Pacific Northwest (United States) tested experimental gradients of herbicide applications in industrial forests at stand scales; they determined that herbicide applications at current industry standards can increase wood production without major effects on avian diversity (Kroll et al. 2020) and moths (Root et al. 2017), for example. Indeed, many forest herbicides are selective and when applied can favor plants beneficial to wildlife (e.g., fruit producers) while eliminating undesirable plant species (e.g., non-native, invasive plants; Miller 2003). In forests where snags and downed wood are lacking, herbicides can be injected into live trees to artificially create snags (McComb and Rumsey 1983), which, in turn, increases structural complexity and habitat suitability for cavity-associated wildlife species.

Herbicides can be useful for restoring potentially desirable open forest conditions (i.e., woodland), especially when used in conjunction with prescribed fire and thinning, and for eradicating invasive plant species. For example, herbicides were used to remove the hardwood midstory in pine woodlands to enhance red-cockaded woodpecker habitat (Conner 1989). Maintenance of early successional vegetation through herbicide applications targeting removal of woody stems also may improve forage availability for herbivores like white-tailed deer (Edwards et al. 2003) and vegetative structure for northern bobwhite (Jones and Chamberlain 2004).

Until recently, most studies pertained to the effects of single herbicides applied one time and focused on a relatively small window of time (Guynn et al. 2004; Iglay et al. 2010). However, Botten et al. (2021) determined that sublethal doses of glyphosate remain in forest plant tissues 12 years after application. Forest management may also entail multiple applications of herbicides and simultaneous use of several herbicides (i.e., tank mixes—Miller and Miller 2004). Studies involving combinations of herbicides applied to regenerating clearcuts showed that these tank mixes had little effect on small mammal abundance (Hanberry et al. 2013b) or breeding birds (Lane et al. 2011a). In a concurrent study, Lane (2011b) also found tank mixes had short-lived (i.e., < four year) effects on herbaceous vegetation but had long-term effects on woody vegetation. Additionally, Tatum et al.

(2012) documented no synergistic toxicity to aquatic forest wildlife from commonly used forestry herbicide tank mixes.

3.5 CASE STUDIES

The following section consists of case studies that reinforce and apply principles of ecological forest management covered in this chapter. These case studies also illustrate the pivotal role wildlife play as ecological indicators to inform adaptive forest-wildlife management.

3.5.1 Boreal Forests

3.5.1.1 Palm Warblers as Indicators of Peatland Composition and Structure within Boreal Forests

Palm warblers (*Dendroica palmarum*) are short-distance migratory birds that breed primarily in boreal forests of North America and winter in the southeastern United States and Caribbean region (Wilson 1996). They breed in Canada's boreal peatlands (Calmé and Desrochers 1999), but the southern limit of the species' breeding range includes scattered populations in the northern Great Lakes and New England regions of the United States (Wilson 1996; Cutright et al. 2006). Palm warblers are among the bird species most strongly and perhaps exclusively associated with boreal peatlands during the breeding season (Wilson 1996; Calmé et al. 2002; Lachance et al. 2005; Figure 3.8). Within these boreal forest landscapes, the peatlands used by palm warblers are often distributed as "islands" that are clearly distinguished from the surrounding upland forests in which they are embedded (Calmé and Desrochers 2000).

Specifically, breeding sites of this species are associated with muskegs or open bogs containing deep beds of sphagnum moss (*Sphagnum* spp.) and small copses of black spruce (*Picea mariana*) surrounded by open areas, a situation intermediate between open and closed canopy (Lachance et al. 2005; Cutright et al. 2006). Palm warblers typically nest on the ground but use trees within the copse for singing and surveillance (Welsh 1971). Zolkowski (2008) showed that palm warblers were more frequently encountered and more abundant in peatlands containing primarily conifer trees but with low amounts of overall overhead canopy, consistent with dense but scattered copses of conifers within otherwise open sites (Calmé and Desrochers 2000). Palm warblers were associated with peatlands containing abundant moss cover and larger numbers of snags (Zolkowski 2008). Snags may provide additional singing perches but may also reflect the mature, open canopy bogs used by palm warblers. Overall, palm warblers appear particularly sensitive to changes in the vegetative

FIGURE 3.8 Peatlands embedded within boreal forests provide important breeding habitat for palm warblers (*Setophaga palmarum*). (Photo by Stephanie Mueller, used with permission.)

structure of peatlands that increase forest cover at the expense of open patches (Lachance et al. 2005). In addition, palm warbler presence appears to be strongly influenced by peatland area, and individuals may actively avoid small, isolated or fragmented peatlands (Zolkowski 2008).

Accordingly, palm warblers may be an important indicator of avian habitat conditions within boreal forest peatlands (Calmé and Desrochers 2000; Lachance et al. 2005; Cutright et al. 2006). Afforestation in response to wetland drainage may be among the primary vegetative changes recently observed in North American peatlands (Lachance et al. 2005). Because saturated conditions inhibit the production of commercially important tree species, large areas of peatlands within boreal forests have been drained to support commercial forestry operations (Turetsky and St. Louis 2006). In addition to direct alteration for commercial forestry, peatland ecosystems may be at risk of changes or loss from climate change (Weltzin et al. 2003). The actual effects of climate change on peatlands are unknown but potentially include increased densities of shrub and tree cover and decreased amounts of graminoid species (Weltzin et al. 2003). The frequency of large-scale wildfires has increased across the boreal forests of North America during the last 50 years, and fires are occurring later in the growing season (Kasishcke and Tureksky 2006; Flannigan et al. 2009). These changes in fire frequency appear to be climate influenced (Flannigan et al. 2009) and are expected to significantly alter vegetation dynamics within boreal peatlands.

Changes in the vegetative composition and structure of peatlands certainly could be expected to significantly alter the distributions of peatland-associated birds within boreal forests (Lachance et al. 2005). Because of their unique composition and structure, peatlands at lower latitudes form "refugia" of boreal forest surrounded by more temperate communities, thus supporting distinct avifaunas (Calmé et al. 2002). This refugia dynamic is particularly true for peatlands at the southern extent of their distribution, where they support plant and animal species common at more northern latitudes (Lachance et al. 2005). Populations of species that are currently at the low-latitude edge of their distributional ranges (e.g., Great Lakes and New England populations of palm warblers) may be among the first species to exhibit changes as a result of anthropomorphic alternations to boreal forest peatlands (Hampe and Petit 2005).

3.5.1.2 Invertebrate Biodiversity within Fennoscandian Boreal Forests

Boreal coniferous forest is the dominant biome in Finland and the Scandinavian Peninsula (Esseen et al. 1997). Fennoscandian forests are among the most intensively managed forests in the world, with less than 5% virgin forest remaining (Linnell et al. 2000). Most of these forests are clearcut on a 70–80-year rotation using mechanized harvesting and dense road networks (Linnell et al. 2000). Increasingly, post-harvest woody debris and stumps are removed and used for bioenergy production (Siira-Pietikäinen and Haimi 2009). Regeneration is achieved either through artificial planting or through natural regeneration. According to Esseen et al. (1997), the most important ecosystem change in Fennoscandian boreal forests has been the replacement of natural fires with timber harvesting as the predominant disturbance regime. As a result of more than 100 years of fire suppression, early post-fire successional stands have become extremely rare (Esseen et al. 1997; Rolstad et al. 2017). In addition, fire suppression is contributing to the increasing dominance of spruce in previously mixed stands of pine, birch, and aspen (Esseen et al. 1997).

The combined effects of intensive timber harvest, removal of post-harvest residue, and fire suppression have had substantial effects on the biodiversity of Fennoscandian boreal forests, including forest invertebrates. Modern forest management practices are considered a major force influencing the landscape-level composition of boreal forest invertebrate communities (Väisänen and Heliövaara 1994). Accordingly, numerous studies have evaluated the impacts of forest management practices on Fennoscandian boreal invertebrates (Esseen et al. 1997; Niemelä 1997; Niemelä et al. 2007; Paillet et al. 2010). For example, most studies of European forest-dwelling *Coleoptera* have been conducted in boreal forests (Paillet et al. 2010). Traditional methods of clearcutting that do not incorporate structural retention (i.e., on-site retention of live and dead vegetation after harvest) have influenced the distribution and abundance of a large proportion of red-listed Fennoscandian forest invertebrates (Esseen et al. 1997; Niemelä 1997). Changes in forest microclimates after clearcutting can cause significant shifts in boreal invertebrate richness and abundance. The drier, warmer conditions on clearcut sites have been linked to the declines of several specialist species associated with spruce-dominated mires or swamp forests in late-successional, unmanaged forests, including carabid beetles, mollusks, and noctuid moths (Esseen et al. 1997; Niemelä 1997; Niemelä et al. 2007; Hylander 2011). Wood-inhabiting species associated with old-growth boreal forests in particular experienced marked declines (Heliövaara and Väisänen 1984; Esseen et al. 1997). In contrast, several studies have documented greater richness and abundance of generalist, open-canopy, and disturbance-tolerant species of carabid beetles, spiders, ants, and butterflies in logged stands when compared to unlogged forests (Niemelä 1997; Niemelä et al. 2007). Although invertebrate species diversity may increase after logging, these communities typically are different than those present in old growth Fennoscandian forests (Niemelä 1997).

Invertebrate species dependent on snags and downed woody debris also have experienced significant declines in Fennoscandian boreal forests. For example, several studies reported declines of saproxylic beetles in managed boreal forests when compared to unmanaged forests (Esseen et al. 1997; Niemelä et al. 2007; Paillet et al. 2010). Large logs and snags in different stages of decay are key habitat elements for these taxa (Paillet et al. 2010). However, these structures are almost completely lacking in Fennoscandian managed forests as a result of successive rotations of intensive fiber removal and fire suppression (Hyvärinen et al. 2009). Additionally, some tree- and snag-living species of beetles require open and dry conditions in early successional stages (Muona and Rutanen 1994). Historically, these early successional conditions were created by wildfires. Although clearcutting approximates some early successional conditions created by natural burning, residual live and dead trees traditionally are removed during harvesting. Additionally, fire suppression has resulted in the declines of many Fennoscandian invertebrates that depend on burned substrates (Muona and Rutanen 1994; Niemelä et al. 1997; Hyvärinen et al. 2009).

Because of the significant reduction of mature or old growth forests on the one hand and the virtual disappearance of young forests regenerated by fire on the other, many specialist Fennoscandian invertebrates have experienced significant population declines and, in some cases, local extinctions (Esseen et al. 1997; Niemelä 1997; Niemelä et al. 2007). To mitigate this situation, forest reserves were established across Fennoscandia with the purpose of maintaining invertebrates and other taxa dependent on old growth forests (Esseen et al. 1997; Niemelä 1997; Lundström et al. 2011). Some saproxylic beetles require snags decayed by specific polypore fungi that occur only in older forests (Kaila et al. 1994; Niemelä 1997). However, forest reserves in Fennoscandia are considered too small and unrepresentative of the range of forest conditions to adequately maintain the biodiversity of boreal forests (Esseen et al. 1997; Niemelä 1997; Lundström et al. 2011). Forest reserves constitute only about 1% of the productive forests of Norway and Sweden and approximately 2% of Finnish forests (Esseen et al. 1997; Lundström et al. 2011). Accordingly, the incorporation of silvicultural techniques that more closely mimic natural disturbances into production forest management is considered critical for the conservation of Fennoscandian invertebrate diversity (Niemelä 1997; Hyvärinen et al. 2009; Siira-Pietikainen and Haimi 2009; Lundström et al. 2011). Specifically, the use of green-tree retention and prescribed burning in managed forests have been suggested as means of more closely approximating structural conditions associated with natural disturbances and providing habitat for specialist invertebrate taxa in boreal forests (Hyvarinen et al. 2006, 2009; Siira-Pietikäinen and Haimi 2009; Lundström et al. 2011). Although the results of some studies are equivocal (e.g., Siira-Pietikäinen and Haimi 2009), other studies indicate that retention of green trees within clearcuts, pre- and post-harvest prescribed burning, and combinations thereof can provide habitat for declining and specialist boreal invertebrates (Hyvärinen et al. 2006, 2009). Important structural features that are almost completely lacking in the managed boreal forests in Fennoscandia, such as deadwood or live overstory trees within young forests, are produced by green tree retention and the burning of harvested sites with retention trees (Hyvärinen et al. 2009). In the short term, prescribed burning may temporarily decrease suitability for herbivore invertebrates and those associated with leaf litter (Hyvärinen et al. 2009). However, saproxylic and pyrophilous species that have declined from successive short rotations in Fennoscandian forests appear to significantly benefit from these measures (Niemelä 1997; Hyvärinen et al. 2006, 2009). This case study highlights the deep intersection of invertebrate conservation and forest management in intensively managed forests.

3.5.2 Temperate Forests

3.5.2.1 Multi-Scale Conservation Strategies and Bachman's Sparrow

The flora and fauna associated with longleaf pine (*Pinus palustris*) forests comprise one of the most diverse communities in North America (Van Lear et al. 2005). As of 2006, this ecosystem occupies a mere 3–5% of its historical range (Landers et al. 1995; Frost 2006), although it has increased in extent due to proactive conservation efforts. Historical longleaf pine forest experienced frequent low-intensity fire and was characterized by widely spaced overstory longleaf pine trees, scattered hardwood patches, and diverse understory vegetation (Frost 2006). Longleaf pine forests once dominated the Atlantic and Gulf Coastal Plains in the southeastern United States, interrupted only by rivers and scattered swamps, whereas today these forests persist as isolated patches within a matrix of pasture, row crop agriculture, fire-suppressed pine plantations, and urban development. This fundamental shift in landscape context has important implications for the many plant and animal species that evolved in a landscape comprising vast, contiguous tracts of open-canopy longleaf pine forest, including the Bachman's sparrow (*Peucaea aestivalis*; Figure 3.9).

Bachman's sparrow selects specific vegetation conditions that reflect fire-maintained forest stands, particularly the presence of a diverse herbaceous groundcover that results from an open overstory and frequent low-intensity fire (Taillie et al. 2015; Winiarski et al. 2017a; Fish et al. 2018). In fact, Bachman's Sparrow is one of six surrogate species used by the United States Fish and Wildlife Service for strategic habitat conservation of longleaf pine communities at the landscape scale (Shelton 2014). The range of preferred conditions of Bachman's Sparrows is so narrow that they abandon habitat patches after as little as three years without fire when grasses begin to form a dense thatch that can restrict movement along the ground (Brooks and Stouffer 2010; Jones

FIGURE 3.9 Bachman's sparrow (*Peucaea aestivalis*) is a specialist bird species that benefits from multi-scale conservation efforts that encourage large and connected patches of fire-maintained longleaf pine (*Pinus palustris*) woodlands with lush groundcover of native grasses and forbs. (Photo by Alexander Fish, used with permission.)

et al. 2013), and less fire-tolerant woody species begin to replace existing herbaceous vegetation (Engstrom et al. 1984; Dunning and Watts 1990; Rutledge and Conner 2002; Tucker et al. 2004).

Although local vegetation conditions, particularly the herbaceous groundcover, are emphasized as the most critical aspects of Bachman's sparrow habitat (Dunning and Watts 1990; Haggerty 1998), sparrow occurrence also is strongly influenced by the percent of sparrow habitat in the surrounding landscape (Taillie et al. 2015; Winiarski et al. 2017b). Patch isolation affects colonization of habitat patches by Bachman's Sparrow, possibly due to poor dispersal through the matrix of non-habitat (Dunning et al. 1995; Dunning and Kilgo 2000; Winiarski et al. 2017b). Individual sparrows will abandon a patch either because it burned recently (Seaman and Krementz 2001) or because the lack of fire permitted unsuitable vegetation conditions to develop (Engstrom et al. 1984; Tucker et al. 2004), requiring potentially multiple dispersal events in a lifetime (Cox and Jones 2010). Isolation of a habitat patch may amplify the challenges associated with each of these dispersal events (Lampila et al. 2005). Hence, increased landscape connectivity can facilitate successful dispersal to previously unoccupied habitat (Dunning et al. 1995).

Restoration of the longleaf pine ecosystem may involve both efforts to establish and maintain vegetation conditions that once characterized fire-maintained upland woodland and larger-scale coordinated efforts focused on landscapes comprising at least 20% habitat (i.e., open canopy pine woodland, including longleaf pine and other pine species, with a diverse herbaceous plant layer); this is particularly important as restoration efforts continue to shift to private lands that typically are more fragmented than public lands in the eastern United States (Riitters et al. 2012). Specifically, private lands adjacent to large patches of forest managed for longleaf pine woodland, such as those on many public lands, may serve to more efficiently conserve some species as such properties could still be functionally connected from the prospective of dispersing wildlife. Additionally, frequent burning can maintain a diverse groundcover layer and prevent the encroachment of woody vegetation (Glitzenstein and Streng 2003; Heuberger and Putz 2003). The logistical and financial resources associated with prescribed fire currently restrict the widespread and frequent use of prescribed fire by landowners across the historical range of the longleaf pine ecosystem (Moorman et al. 2002; Kupfer et al. 2022). Overcoming these logistical hurdles may be essential for conserving biodiversity depend on functional longleaf pine ecosystems.

3.5.3 Tropical Forests

3.5.3.1 Neotropical Migrant Birds and Coffee Agroecosystems

In tropical forests around the globe, agricultural development is in direct conflict with the conservation of forests and the wildlife therein (Gliessman 1992). Agroecosystems define agricultural systems as ecosystems, enabling a framework for examination of the complex interactions between overlapping ecological, environmental, and cultural components of tropical agriculture (Conway 1985; Francis 1986). Traditional agroecosystems in the tropics focused on long-term sustainability of the system rather than maximizing profits and yield (Gliessman 1992). Yet increased economic and political pressures for the export of cash crops have led many tropical agroecosystems to become production driven and thereby potentially less sustainable (Conway 1985).

Coffee (*Coffea* spp.) has been the dominant agroecosystem in the tropics of Latin America, second only to oil in value in many parts of the region (Rice and Ward 1997). Coffee production is particularly prevalent in some of the biodiversity hotspots of Latin America (e.g., Brazil, Columbia, Mexico) and occurs in areas with few remaining forests (Mittermeier et al. 1998; Hardner and Rice 2002; Philpott et al. 2007). In its native form, coffee is a shade-tolerant understory tree. As such, traditional (i.e., primitive) coffee plantations were established and maintained by placing young coffee plants under canopy provided by one or more native tree species ("shade coffee"; Rappole et al. 2003; Figure 3.10). Efforts in the Latin American tropics to increase coffee production and combat devastating coffee diseases (e.g., coffee leaf rust (*Hemileia vastatrix*)) led to the development of

FIGURE 3.10 Shade-grown coffee plantations are characterized by more complex vegetation structure than sun-grown coffee, and as such, typically maintain higher levels of biodiversity. (Photo by Chris Moorman, used with permission.)

coffee varieties tolerant to direct sunlight ("sun coffee"; Rappole et al. 2003). Compared to shade coffee, sun-grown coffee produces greater yields, is amenable to mechanized agricultural practices, and exhibits greater resistance to disease, making it an attractive agroindustry throughout much of modern-day Latin America (Rice and Ward 1997). Despite the cultural and economic benefits of sun coffee, the system is typically far less biologically diverse than shade coffee (Rappole et al. 2003). Indeed, the ecological and economic trade-offs of these competing agroecosystems defines the true challenge of implementing conservation-minded coffee production throughout Latin America.

Neotropical migrant birds (herein birds) have become ecological indicators informing the sustainability of coffee production in the Latin American tropics. Birds are especially useful for assessing the conservation value of shade versus sun-grown coffee production because they serve many ecological functions (e.g., seed dispersal, secondary consumers), span across both temperate and tropical forests of the western hemisphere, and respond to changes in vegetation structure (MacArthur and MacArthur 1961; Hernandez-Aquilera et al. 2019). Researchers have noted a high diversity and abundance of bird species in both primitive and modern (i.e., emulating primitive production) shade coffee plantations (Griscom 1932; Greenberg et al. 1997; Williams et al. 2021). Shade coffee inherently maintains more structurally complex vegetation in comparison to sun coffee, where the coffee plants themselves or coffee and a single tree species provide vertical structure (Perfecto et al. 1996). The vertical complexity of shade coffee supports more diverse fauna as a whole (MacArthur and MacArthur 1961; Williams et al. 2021), and increased prevalence of tall, taxonomically and structurally diverse canopies are especially important for supporting bird diversity (Greenberg et al. 1997). The importance of shade coffee as habitat and/or corridors for wildlife also has been demonstrated for ants (Perfecto and Snelling 1995), butterflies (Mas and Dietsch 2004), and monkeys (Williams-Guillén et al. 2006), among other wildlife groups. In general, shade coffee provides important refugia for biodiversity, especially birds, in the increasingly deforested landscape throughout much of Latin America (Mas and Dietsch 2004; Fritz et al. 2022).

Coffee certification programs in Latin America exemplify the complex series of trade-offs between economic and ecological forest management inherent to forest certification programs in boreal, temperate, and tropical forests alike. Coffee certification programs operate on the basic premise of branding ecologically beneficial coffee production practices as sustainable (e.g., "green consumerism"—Albertin and Nair 2004), passing some of the economic burden of such production onto the consumer and, in turn, providing economic incentives to coffee growers (Rainforest Alliance 2000). Coffee certification programs may offer a way in which to protect biodiversity while simultaneously maintaining the economic livelihood of coffee farmers (Philpott et al. 2007). However, there is significant contention among ecologists and economists regarding the ability of coffee certification programs to achieve these goals, and variation among the viability of specific certification programs (e.g., organic, fair trade, shade grown) is thought to be high (Philpott et al. 2007). For example, the premium offered to shade coffee growers may not always compensate for the low yields of the most shaded (i.e., most biologically diverse) plantations (Perfecto et al. 2005). Importantly, maintaining the economic viability of shade coffee is paramount to keeping coffee farmers interested in sustainable coffee production.

Cultural and societal inputs in the framework of coffee certification programs illustrate the value of land manager involvement (e.g., coffee farmers). In the early 2000s, coffee prices were comparatively low (Food and Agricultural Organization 2002), in part due to global economic policies and coffee overproduction, putting extreme economic pressure on predominantly, small, rural coffee farmers willing to participate in coffee certification programs. During this time, a philosophical disconnect occurred between conservation coffee certification programs and economic stability of rural farmers. In response to this situation, Philpott and Dietsch (1993) suggested the unification of historically independent coffee certification programs linked to price premiums for coffee producers could provide an all-inclusive, long-term solution to meet conservation and economic goals. Importantly, coffee farmers' perception of the role of shade trees in coffee plantations overlaps with the conservation objectives of coffee certification programs. For example, farmers in Costa Rica considered tree height an important characteristic of their plantations for coffee production (Albertin and Nair 2004). Meanwhile, Greenberg et al. (1997) documented tree height as a significant contributing factor to increased biodiversity in coffee plantations. Most of those same Costa Rican coffee farmers (88%) also expressed interested in adding more trees to their plantations, especially via intercropping fruit trees (Albertin and Nair 2004). In turn, intercropping could increase vegetation structure and potentially improve wildlife habitat. Given the connection between coffee farmers and the agroecosystems they manage in tropical Latin America, a sound understanding of their perception of conservation goals in addition to their financial needs could contribute to conserving coffee agroecosystems into the future. This message can be extended to a suite of agroforest systems and the land managers therein (e.g., forests for timber production).

Coffee is just one of a number of cash crops in Latin America. Yet many of the same land use, forest management, and conservation concerns and strategies (e.g., certification programs) apply to Latin American cash crops. In the case of agroforests, conservation issues surrounding cacao (Schroth and Harvey 2007) and banana/plantain (Harvey et al. 2006) production are especially similar to that of coffee. Oil palm (*Elaeis guineensis*; see *Oil Palm Plantations in Tropical Forests—3.6.2*) production also invokes similar global tropical conservation (Fitzherbert et al. 2008) and societal (Rist et al. 2010) concerns as monocultures of sun coffee, although oil palm plantations may have even more severe impacts on land cover.

3.6 CONTEMPORARY ISSUES IN FOREST WILDLIFE MANAGEMENT—FOREST BIOENERGY

The interface between forest-based renewable energy development and wildlife conservation is an increasingly relevant research area as the energy transition from fossil fuels to renewables reshapes conservation and energy targets (Moorman et al. 2019; Grodsky 2021, 2023). Forests are capable of producing vast amounts of woody biomass (i.e., dead and live trees), which, in turn, can be used to produce traditional wood products and generate forest bioenergy in the form of heat, electricity, and liquid fuels (Parikka 2004). While high-value wood products (e.g., timber) from forests remain the primary driver of decisions regarding forest management, by-products of timber production such as woody biomass are an important feedstock for forest bioenergy markets. Further, bioenergy production in facilities with carbon capture and storage may address emissions concerns generated from burning wood for energy, with potential traction for maintaining ecosystems services (Donnison et al. 2020). European Union renewable energy mandates still heavily rely on imported wood pellets as feedstock (Goh et al. 2013; Sgarbossa et al. 2020)

3.6.1 Woody Biomass Harvest in Boreal and Temperate Forests

3.6.1.1 Harvest Residues

Forest harvest operations (e.g., clearcutting for timber harvest) typically yield large amounts of harvest residue collectively referred to as fine (foliage, small limbs and trees) and coarse (downed logs) woody material (Riffell et al. 2011a). Harvest residues, particularly coarse woody material, may be gleaned following harvest (i.e., woody biomass harvest; Figure 3.11) and used as a feedstock for the following forest-bioenergy technologies: (1) wood pellets (Sgarbossa et al. 2020); (2) woody biomass and co-fired coal power production (Picciano et al. 2022); and (3) liquid biofuels (Nigam and Singh 2011). While coarse woody material may be a prospective feedstock for forest-bioenergy, it also provides important food and cover resources to many wildlife species (Harmon et al. 1986; Janowiak et al. 2010; see *Structure—Downed Wood*—3.3.4).

Wildlife ecologists have made steady strides to elucidate effects of harvest residue removal for forest bioenergy on wildlife. Riffell et al. (2011a) forecasted potential, negative effects of intensified harvest residue removal on abundance but not diversity of forest wildlife based on available literature at the time. Recent studies with experimental manipulations of harvest residues based on previously established Biomass Harvesting Guidelines (Perschel et al. 2012; Fritts et al. 2014) conducted in clearcuts following harvest residue removal for forest bioenergy among plantations of the southeastern United States yielded novel insights on forest-bioenergy-wildlife interactions with empirical underpinnings. Through this body of work, researchers determined that forest wildlife largely responded to changes in vegetation structure and cover following clearcut harvests more so than downed wood retention in clearcuts (e.g., shrews—Fritts et al. 2015a, herpetofauna—Fritts et al. 2017; breeding shrubland birds, Grodsky et al. 2016a), with the exception of invertebrates (Grodsky et al. 2018a). However, studies revealed that vertebrate wildlife do in fact use harvest residues for food and cover, including southern toads (*Anaxyrus terrestris*) (Fritts et al. 2015b) and wintering birds (Grodsky et al. 2016b). Removal of harvest residues for forest bioenergy harvests may negatively affect forest invertebrates in clearcuts, including wood roaches (Ectobiidae), fungus gnats (Mycetophilidae and Sciaridae), cave crickets (Rhaphidophoridae), and millipedes (Diplopoda) (Grodsky et al. 2018a), and impact interactions among invasive ants in regenerating stands (Grodsky et al. 2018b). Retention of at least 15% of harvest residues in a stand can minimize negative effects on forest invertebrate communities (Grodsky et al. 2018a) and ground beetles specifically (Grodsky et al. 2019). Loy et al. (2020) also found that harvest residue removal in clearcuts reduced bee diversity.

FIGURE 3.11 Following clearcutting, harvest residues may be chipped on site and transported to forest-bioenergy facilities to be converted to renewable energy. (Photo by North Carolina State University Forestry Extension, used with permission.)

Recent modeling suggests that some forecasted bioenergy scenarios may spatially overlap with areas of high biodiversity value in the state of North Carolina in the United States (Costanza et al. 2016). Studies explicitly linking harvest residue retention and its effects on wildlife to forest bioenergy remain relatively rare. Woody biomass harvesting may be a more prevalent issue in Europe, for example, because even heavily forested states in the United States are trending toward solar energy development (SEIA 2023). Homyack and Verschuyl (2019) provide a summary of published research on effects of harvesting forest-based biomass on terrestrial wildlife across the globe.

3.6.1.2 Thinning

Forest thinning is a well-established silvicultural practice that generates downed wood, which, in turn, can be used in all of the same forest bioenergy technologies that we outlined for harvest residues earlier. Thinning increases structural complexity, understory vegetation, and consequently wildlife habitat for a variety of wildlife species (see *Thinning—3.4.5*). A review by Verschuyl et al. (2011) reported that forest thinning treatments (pre-commercial, commercial, and fuels reduction) in North America generally had positive or neutral effects on wildlife, although the magnitude of wildlife response varied with thinning intensity and type and the review did not include any thinning treatments explicitly meant for woody biomass harvest. Homyack and Verschuyl (2019) conducted an updated meta-analysis of forest thinning in North America; they documented no forest thinning operations specifically designed for bioenergy production within the published literature. In general, the effects of thinning for forest bioenergy on wildlife, if they occur independently or in conjunction with other forest management objectives, generally increase diversity and abundance of a variety of taxa across forest types and thinning intensities (Homyack and Verschuyl 2019). Further, Gottlieb et al. (2017) determined that thinning had positive effects on avian occupancy, whereas harvesting pine residues and short-rotation woody crops had relatively negative effects in the southeastern United States; meanwhile, all pine biomass extractions were better for birds relative to corn production.

3.6.1.3 Short-rotation Woody Cropping

While forest bioenergy from harvest residues and thinning generally originates as a by-product of traditional forest management, short-rotation woody cropping (SRWC) involves specific cultivation of high-yielding, fast-growing woody crops (e.g., cottonwood and poplar (*Populus* spp.) and *Eucalyptus*, using intensive silvicultural techniques to produce woody biomass across short harvest cycles (e.g., 1–15 years; Dickman 2006). Short-rotation woody crops typically are burned as feedstock for electricity production (Parikka 2004). Riffell et al. (2011b) reviewed the effects of cottonwood SRWC on birds and mammals relative to reference woodlands in North America. Their results indicated that overall bird and mammal abundance was lower in SRWC plantations relative to reference woodlands, but shrub-associated birds were more abundant. Another meta-analysis based on available literature from studies in the southeastern United States, which still largely lacks forest-bioenergy specific studies, corroborated the negative relationship between bird and mammal diversity and SRWC relative to unmanaged forests (Griffiths et al. 2018). Potentially significant land use conversion from agriculture to SRWC may lead to landscape-level effects on some wildlife populations, such as birds common in agricultural landscapes (Dhont et al. 2004). Greene et al. (2019) features an updated review of short-rotation woody crops and wildlife conservation.

3.7 CONCLUSION

Land use change (rather than specific forest management actions) may be the greatest single near-term threat to the combined conservation of forest products and forest dependent wildlife species—in both temperate and tropical systems. However, where forests are conserved, proper management is contingent on a sound understanding of forest wildlife ecology. Forests provide habitat—food, cover, and water—for many wildlife species and consequentially are among the most biologically diverse terrestrial ecosystems. Understanding wildlife habitat requirements is key to predicting forest wildlife response to natural or anthropogenic disturbances, including forest management activities. Specifically, forest wildlife management can be related to five, interrelated ecological principles: (1) disturbance; (2) succession; (3) structure (vegetation and deadwood); (4) plant composition; and (5) landscape ecology. Following Aldo Leopold (1933), the true essence of forest wildlife management is captured by the idea that conservation of forest products for human consumption and forest wildlife could be simultaneously achieved. Forest management practices, including final harvest, regeneration and site preparation, thinning, prescribed burning, and herbicide application, have direct effects on the five ecological principles outlined earlier and consequentially influence forest wildlife habitat. Successful forest wildlife management is achieved by implementing forest management practices that facilitate habitat conditions for target wildlife species. The fundamental concepts presented in this chapter generally apply to all forests. However, we presented case studies illustrating the application of forest wildlife management in boreal, temperate, and tropical forests and the use of forest wildlife as ecological indicators.

3.8 ACKNOWLEDGMENTS

Drs. Jessica Homyack and Jake Verschuyl reviewed an earlier version of this chapter. Any use of trade, firm, or product names is for descriptive purposes only and does not imply endorsement by the U.S. government.

REFERENCES

Agee, J.K. and C.N. Skinner. 2005. Basic principles of forest fuel reduction treatments. *For. Ecol. Manage.* 211:83–96.

Albertin, A. and P.K.R. Nair. 2004. Farmers' perspectives on the role of shade trees in coffee production systems: An assessment of the Nicoya Peninsula, Costa Rica. *Hum. Ecol.* 32:443–462.

Arthur, M.A., J.M. Varner, C.W. Lafon, H.D. Alexander, D.C. Dey, C.A. Harper, S.P. Horn, T.F. Hutchinson, T.L. Keyser, M.A. Lashley, C.E. Moorman, and C.J. Schweitzer. 2021. Fire ecology and management in Eastern Broadleaf and Appalachian forests. In *Fire ecology and management: Past, present, and future of US forested ecosystems*, eds. C.H. Greenberg and B. Collins, pp. 105–148. New York, NY: Springer.

Attiwill, P.M. 1994. The disturbance of forest ecosystems: The ecological basis for conservative management. *For. Ecol. Manage.* 63:247–300.

Atwood, C.J., T.R. Fox, and D.L. Loftis 2009. Effects of alternative silviculture on stump sprouting in the southern Appalachians. *For Ecol. Manage.* 257:1305–1313.

Bagne, K.E., K.L. Purcell, and J.T. Rotenberry. 2008. Prescribed fire, snag population dynamics, and avian nest site selection. *For. Ecol. Manage.* 25:99–105.

Bailey, J.D. and J.C. Tappeneir. 1998. Effects of thinning on structural development in 40 to 100-year old Douglas-fir stands in western Oregon. *For. Ecol. Manage.* 108:99–113.

Barbour, R.J., S. Johnston, J.P. Hayes, and G.F. Tucker. 1997. Simulated stand characteristics and wood product yields from Douglas-fir plantations managed for ecosystem objectives. *For. Ecol. Manage.* 91:205–219.

Batzer, D.P. and S.A. Wissinger. 1996. Ecology of insect communities in nontidal wetlands. *Annu. Rev. Entomol.* 41:75–100.

Bazzaz, F.A. 1975. Plant species diversity in old-field successional ecosystems in Southern Illinois. *Ecology* 56:485–488.

Botten, N., L.J. Wood, and J.R. Werner. 2021. Glyphosate remains in forest plant tissues for a decade or more. *For. Ecol. Manage.* 493:119259.

Branan, W.V., M.C.M. Werkhoven, and R.L. Marchinton. 1985. Food habits of brocket and white-tailed deer in Suriname. *J. Wildl. Manage.* 49:972–976.

Brawn, J.D., S.K. Robinson, and F.R. Thompson III. 2001. The role of disturbance in the ecology and conservation of birds. *Ann. Rev. Ecol. Syst.* 32:251–276.

Brooks, M.E. and P.C. Stouffer. 2010. Effects of Hurricane Katrina and salvage logging on Bachman's Sparrow. *Condor Ornith. Apps.* 112:744–753.

Buckner, C.A. and D.J. Shure. 1985. The response of Peromyscus to forest opening size in the Southern Appalachian Mountains. *J. Mammal.* 66:299–307.

Bunnell, F.L., L.L. Kremsater, and E. Wind. 1999. Managing to sustain vertebrate richness in forests of the Pacific Northwest: Relationships with stands. *Environ. Rev.* 7:97–146.

Burger, L.W., Jr. 2001. *Northern Bobwhite in wildlife of Southern forests: Habitat and management*, ed. J.G. Dickson, p. 133. Blaine, Washington, DC: Hancock House Publishers, Ltd.

Butts, S.R. and W.C. McComb. 2000. Associations of forest-floor vertebrates with coarse woody debris in managed forests of Western Oregon. *J. Wildl. Manage.* 64:95–104.

Calmé, S. and A. Desrochers. 1999. Nested bird and micro-habitat assemblages in a peatland archipelago. *Oecologia* 118:361–370.

Calme, S. and A. Desrochers. 2000. Biogeographic aspects of the distribution of bird species breeding in Quebec's peatlands. *J. Biogeogr.* 27:725–732.

Calme, S., A. Desrochers, and J-P.L. Savard. 2002. Regional significance of peatlands for avifaunal diversity in Southern Quebec. *Biol. Conserv.* 107:273–281.

Campbell, J.W., S.M. Grodsky, O. Keller, C. Vigueira, E. Waite, P. Vigueira, and C. Greenberg. 2018. Response of beetles (Coleoptera) to repeated applications of prescribed fire and other fuel reduction techniques in the southern Appalachian mountains. *For. Ecol. Manage.* 429:294–299.

Campbell, J.W., S.M. Grodsky, D. Halbritter, P. Vigueira, C. Vigueira, O. Keller, and C.H. Greenberg. 2019. Asian needle ant (*Brachyponera chinensis*) and woodland ant responses to repeated applications of fuel reduction methods. *Ecosphere* 10:1–12. https://doi.org/10.1002/ecs2.2547

Campbell, J.W., S.M. Grodsky, M. Milne, P. Viguiera, C.C. Viguiera, E. Stern, and C.H. Greenberg. 2022. Prescribed fire and other fuel-reduction treatments increase spider abundance in a Southern Appalachian hardwood forest. *For. Ecol. Manage.* 510:120127.

Castro, A. and D.H. Wise. 2010. Influence of fallen coarse woody debris on the diversity and community structure of forest-floor spiders (Arachnida:Araneae). *For. Ecol. Manage.* 260:2088–2101.

Champlin, T.B., J.C. Kilgo, and C.E. Moorman. 2009. Food abundance does not determine bird use of early-successional habitat. *Ecology* 90:1586–1594.

Chandler, C.C., D.I. King, and R.B. Chandler. 2012. Do mature forest birds prefer early-successional habitat during the post-fledging period? *For. Ecol. Manage.* 264:1–9.

Cody, M.L. 1974. *Competition and the structure of bird communities*. Princeton, NJ: Princeton University Press.

Conner, R.N. 1989. *Injection of 2,3-D to remove hardwood midstory within red-cockaded woodpecker colony areas*. Nacogdoches, TX: United Sates Department of Agriculture Forest Service, Southern Forest Experiment Station, Research Paper SO-251.

Conner, R.N. and C.S. Adkisson. 1977. Effects of clearcutting on the diversity of breeding birds. *J. For.* 73:781–785.

Constantine, N.L., T.A. Campbell, W.B. Baughman, T.B. Harrington, B.R. Chapman, and K.V. Miller. 2004. Effects of clearcutting with corridor retention on abundance, richness, and diversity of small mammals in the Coastal Plain of South Carolina, USA. *For. Ecol. Manage.* 202:293–300.

Converse, S.J., W.M. Block, and G.C. White. 2006a. Small mammal population and habitat responses to forest thinning and prescribed fire. *For. Ecol. Manage.* 228:263–273.

Converse, S.J., G.C. White, and W.M. Block. 2006b. Small mammal response to thinning and wildfire in Ponderosa Pine-dominated forests of the southwestern United States. *J. Wildl. Manage.* 70:1711–1722.

Conway, G. 1985. Agroecosystem analysis. *Agric. Adm.* 20:31–55.

Costanza, J.K., R.C. Abt, A.J. McKerrow, and J.A. Collazo. 2016. Bioenergy production and forest landscape change in the southeastern United States. *GCB Bioener.* 9:924–939.

Cox, J.A. and C.D. Jones. 2010. Estimating probabilities of survival of male Bachman's Sparrows from plot-based, mark-resighting, off-plot surveys and multi-strata models. *Condor Ornith. Apps.* 112:663–669.

Cristan, R., W.M. Aust, M.C. Bolding, S.M. Barrett, J.F. Munsell, and E. Schilling. 2016. Effectiveness of forestry best management practices in the United States: Literature review. *For. Ecol. Manage.* 360:133–151.

Cutright, N.J., B.R. Harriman, and R.W. Howe, eds. 2006. *Atlas of the breeding birds of Wisconsin*. Waukesha, WI: Wisconsin Society for Ornithology, Inc.

DeGraaf, R.M., W.M. Healy, and R.T. Brooks. 1991. Effects of thinning and deer browsing on breeding birds in New England oak woodlands. *For. Ecol. Manage.* 41:179–191.

Dhont, A.A., P.H. Wrege, K.V. Sydenstricker, and J. Cerretani. 2004. Clone preference by nesting birds in short-rotation coppice plantations in Central and Western New York. *Biomass Bioener.* 27:429–435.

Dickson, J.G. 2001. *Wild Turkey in wildlife of Southern forests: Habitat and management*, ed. J.G. Dickson, p. 114. Blaine, Washington, DC: Hancock House Publishers, Ltd.

Dickman, D.L. 2006. Silviculture and biology of short-rotation woody crops in temperate regions: Then and now. *Biomass Bioener.* 30:696–705.

DiMauro, D. and M.L. Hunter, Jr. 2002. Reproduction of amphibians in natural and anthropogenic temporary pools in managed forests. *For. Sci.* 48:397–406.

Donnison, C., R.A. Holland, A. Hastings, L.M. Armstrong, F. Eigenbrod, and G. Taylor. 2020. Bioenergy with carbon capture and storage: Finding the win-wins for energy, negative emissions, and ecosystem services—size matters. *GCB Bioener.* 12:586–604.

Dunning, J.B. Jr., R. Borgella Jr., K. Clements, and G.K. Meffe. 1995. Patch isolation, corridor effects, and colonization by a resident sparrow in a managed pine woodland. *Cons. Biol.* 9:542–550.

Dunning, J.B. and J.C. Kilgo. 2000. Avian research at the Savannah River Site: A model for integrating basic research and long-term management. *Stud. Avian Biol.* 21:1–170.

Dunning, J.B. Jr., and B.D. Watts. 1990. Regional differences in habitat occupancy by Bachman's Sparrow. *Auk.* 107:463–472.

Edwards, S.L., S. Demarais, B. Watkins, and B.K. Strickland. 2003. White-tailed deer forage production in managed pine stands and summer food plots. *Southeast Deer Study Group Meeting* 26:22–23.

Engstrom, R.T., R.L. Crawford, and W.W. Baker. 1984. Breeding bird populations in relation to changing forest structure following fire exclusion: A 15-year study. *Wilson Bull.* 96:437–450.

Esseen, P.A., B. Ehnström, L. Ericson, and K. Sjöberg. 1997. Boreal forests. *Ecol. Bull.* 46:16–47.

Fish, A.C., C.E. Moorman, C.S. DePerno, J.M. Schillaci, and G.R. Hess. 2018. Predictors of Bachman's sparrow occupancy at its northern range limit. *Southeast. Natur.* 17:104–116.

Fitzherbert, E.B., M.J. Struebig, A. Morel, F. Danielsen, C.A. Brühl, P.F. Donald and B. Phalan. 2008. How will oil palm expansion affect biodiversity? *Trends Ecol. Evol.* 23:538-545. 10.1016/j.tree.2008.06.012

Flannigan, M., B. Stocks, M. Turetsky, and M. Wotton. 2009. Impacts of climate change on fire activity and fire management in the circumboreal forest. *Glob. Change Biol.* 15:549–560.

Food and Agricultural Organization (FAO). 2002. *Agricultural production statistics.* Rome, Italy: FAO. http://apps.fao.org/. Accessed December 2014.

Forsman, A. 2000. Some like it hot: Intrapopulation variation in behavioral thermoregulation. *Evol. Ecol.* 14:25–38.

Francis, C.A. 1986. *Multiple cropping systems.* New York: Macmillan.

Franklin, J.F., T.A. Spies, R. Van Pelt, A.B. Carey, D.A. Thornburgh, D.R. Berg et al. 2002. Disturbances and structural development of natural forest ecosystems with silvicultural implications, using Douglas-fir as an example. *For. Ecol. Manage.* 155:339–423.

Fritts, S.R., S.M. Grodsky, D.W. Hazel, J.A. Homyack, S.B. Castleberry, and C.E. Moorman. 2015b. Quantifying multi-scale habitat use of woody biomass by southern toads. *For. Ecol. Manage.* 346:81–88.

Fritts, S.R., C.E. Moorman, S.M. Grodsky, D.W. Hazel, J.A. Homyack, C.B. Farrell, and S.B. Castleberry. 2015a. Shrew response to variable woody debris retention: Implications for sustainable forest bioenergy. *For. Ecol. Manage.* 336:35–43.

Fritts, S.R., C.E. Moorman, S.M. Grodsky, D.W. Hazel, J.A. Homyack, S.B. Castleberry, K.H. Pollock, and C.B. Farrell. 2016. Can biomass harvesting guidelines sustain herpetofauna following harvest of logging residues for renewable energy? *Ecol. Appl.* 26:926–939.

Fritts, S.R., C.E. Moorman, S.M. Grodsky, D.W. Hazel, J.A. Homyack, C.B. Farrell, and S.B. Castleberry. 2017. Rodent response to harvesting woody biomass for bioenergy production in the Southeastern United States. *J. Wildl. Manage.* 81:1170–1178.

Fritts, S.R., C.E. Moorman, D.W. Hazel, and B.D. Jackson. 2014. Biomass Harvesting Guidelines affect downed woody debris retention. *Biomass Bioenerg.* 70:382–391.

Fritz, S. et al. 2022. A continental assessment of the drivers of tropical deforestation with a focus on protected areas. *Front. Conserv. Sci.* 3:830248.

Frost, C.C. 1998. *Presettlement fire frequency regimes of the United States: A first approximation.* Tall Timbers Fire Ecology Conference Proceedings, vol. 22, pp. 207–210. Tallahassee, FL: Tall Timbers Research Station.

Frost, C.C. 2006. History and future of the longleaf pine ecosystem. In *The longleaf pine ecosystem—ecology, silviculture, and restoration*, eds. S. Jose, E.J. Jokela, and D.L. Miller, pp. 9–48. New York: Springer.

Gauthier, S., Y. Bergeron, and J.P. Simon. 1996. Effects of fire regime on the serotiny level of jack pine. *J. Ecol.* 84:539–548.

Gliessman, S.R. 1992. Agroecology in the tropics: Achieving a balance between land use and preservation. *Environ. Manag.* 16:681–689.

Glitzenstein, J.S., and D.R. Streng. 2003. Fire frequency effects on longleaf pine vegetation in South Carolina and Northeast Florida, USA. *Nat. Areas J.* 23:22.

Goh, C.S. et al. 2013. Wood pellet market and trade: A global perspective. *Biofuel Bioprod Biorefinery.* 7:24–42.

Gottlieb, I.G., R.J. Fletcher Jr., M.M. Nuñez-Regueiro, H. Ober, L. Smith, and B.J. Brosi. 2017. Alternative biomass strategies for bioenergy: Implications for bird communities across the Southeastern United States. *GCB Bioener.* 9:1606–1617.

Greenberg, C.H., C.E. Moorman, K.J. Elliott, K. Martin, M. Hopey, and P.V. Caldwell. 2023. Breeding bird abundance and diversity greatest in high-severity wildfire patches in eastern hardwood forests. *For. Ecol. Manage.* 529. https://doi.org/10.1016/j.foreco.2022.120715

Greenberg, C.H., C.E. Moorman, C.E. Matthews-Snoberger, T.A. Waldrop, D. Simon, A. Heh, and D. Hagan. 2018a. Long-term herpetofaunal response to repeated fuel reduction treatments. *J. Wild. Manage.* 82:553–565.

Greenberg, C.H., J. Tomcho, A. Livings-Tomcho, J.D. Lanham, T.A. Waldrop, D. Simon, and D. Hagan. 2018b. Long-term avian response to fire severity, repeated burning, and mechanical fuel reduction in upland hardwood forest. *For. Ecol. Manage.* 424:367–377.

Greenberg, R., R. Bichier, A.C. Angon, and R. Reitsma. 1997. Bird populations in shade and sun coffee plantations in central Guatemala. *Cons. Biol.* 11:448–459.

Greene, R., J.A. Martin, and T.B. Wigley. 2019. Short-rotation woody crops and wildlife conservation. In *Renewable energy and wildlife conservation*, eds. C.E. Moorman, S.M. Grodsky, and S.P. Rupp, pp. 13–22. Baltimore: Johns Hopkins University Press.

Griffiths, N.A., B.M. Rau, K.B. Vaché, G. Starr, M.M. Bitew, D.P. Aubrey, J.A. Martin, E. Benton, and C.R. Jackson. 2018. Environmental effects of short-rotation woody crops for bioenergy: What is and isn't known. *GCB Bioener.* 11:554–572.

Griscom, L. 1932. The distribution of bird-life in Guatemala. *Bull. Am. Mus. Nat. Hist.* 64.

Grodsky, S.M. 2021. Matching renewable energy and conservation targets for a sustainable future. *One Earth.* 4:924–926.

Grodsky, S.M. 2023. Join the revolution: For wildlifers, renewable energy offers unprecedented challenges—and opportunities. *The Wildlife Professional* 26–27, May–June Issue. Bethesda.

Grodsky, S.M., J.W. Campbell, S.R. Fritts, T.B. Wigley, and C.E. Moorman. 2018b. Variable responses of non-native and native ants to coarse woody debris removal following forest bioenergy harvests. *For. Ecol. Manage.* 427:414–422.

Grodsky, S.M., R.R. Hernandez, J.W. Campbell, K.R. Hinson, O. Keller, S.R. Fritts, J.A. Homyack, and C.E. Moorman. 2019. Ground beetle (Coleoptera: Carabidae) response to harvest residue retention: Implications for sustainable forest bioenergy production. *Forests* 11:48.

Grodsky, S.M., C.E. Moorman, S.R. Fritts, J.W. Campbell, M.A. Bertone, C.E. Sorenson, S.B. Castleberry, and T.B. Wigley. 2018a. Invertebrate community response to coarse woody debris removal for bioenergy production from intensively managed forests. *Ecol. Appl.* 28:135–148.

Grodsky, S.M., C.E. Moorman, S.R. Fritts, S.B. Castleberry, and T.B. Wigley. 2016b. Breeding, early-successional bird response to forest harvests for bioenergy. *PLoS One* 11(10):e0165070.

Grodsky, S.M., C.E. Moorman, S.R. Fritts, D.W. Hazel, J.A. Homyack, S.B. Castleberry, and T.B. Wigley. 2016a. Winter bird use of harvest residues in clearcuts and the implications of forest bioenergy harvest in the southeastern United States. *For. Ecol. Manage.* 379:91–101.

Grove, S.J. 2002. Saproxylic insect ecology and the sustainable management of forests. *Annu. Rev. Ecol. Evol.* Syst. 33:1–23.

Guynn Jr., D.C., S.T. Guynn, B.T. Wigley, and D.A. Miller. 2004. Herbicides and forest biodiversity—what we do know and where do we go from here? *Wildl. Soc. Bull.* 32:1085–1092.

Haddad, N.M., D.R. Browne, A. Cunningham, B.J. Danielson, D.J. Levey, S. Sargent, and T. Spira. 2003. Corridor use by diverse taxa. *Ecol.* 84:906–615.

Hagan, J.M. and S.L. Grove. 1999. Coarse woody debris: Humans and nature competing for trees. *J. For.* 97:6–11.

Hagar, J., S. Howlin, and L. Ganio. 2004. Short-term response of songbirds to experimental thinning of young Douglas-fir forests in the Oregon Cascades. *For. Ecol. Manage.* 199:333–347.

Haggerty, T.M. 1998. Vegetation structure of Bachman's Sparrow breeding habitat and its relationship to home range. *J. Field Ornithol.* 69:45–50.

Hall, L.S., P.R. Krausman, and M.L. Morrison. 1997. The habitat concept and a plea for standard terminology. *Wildl. Soc. Bull.* 25:173–182.

Hampe, A. and R.J. Petit: 2005. Conserving biodiversity under climate change: The rear edge matters. *Ecol. Lett.* 8:461–467.

Hanberry, B.B., S.L. Edwards, S. Demarais, and J. Jones. 2013b. Small mammal responses to intensively established pine plantations in Coastal Plain Mississippi. *South. J. Appl. For.* 37:53–58.

Hanberry, B.B., P. Hanberry, S.K. Riffell, S. Demarais, and J.C. Jones. 2013a. Wintering birds in intensively established pine plantations of Coastal Plain Mississippi. *South. J. Appl. For.* 37:91–96.

Hanula, J.L., M.D. Ulyshen, and D.D. Wade. 2012. Impacts of prescribed fire frequency on coarse woody debris volume, decomposition and termite activity in the Longleaf Pine Flatwoods of Florida. *Forests* 3:317–331.

Hardner, J. and R. Rice. 2002. Rethinking green consumerism. *Sci. Am.* 286:88–95.

Harmon, M.E. et al. 1986. Ecology of coarse woody debris in temperate ecosystems. *Adv. Ecol. Res.* 15:133–302.

Harper, C.A., W.M. Ford, M.A. Lashley, C.E. Moorman, and M. Stambaugh. 2016. Fire effects on wildlife in the Central Hardwoods and Appalachian regions. *Fire Ecol.* 12:127–159.

Harrell, W.C., S.D. Fuhlendorf, and T.G. Bidwell. 2001. Effects of prescribed fire on sand shinnery oak communities. *J. Range Manag.* 54:685–690.

Harris, G.R. and W.W. Covington. 1983. The effect of a prescribed fire on nutrient concentration and standing crop of understory vegetation in ponderosa pine. *Can. J. For. Res.* 13:501–507.

Harvey, C.A., J. Gonzalez, and E. Somarriba. 2006. Dung beetle and terrestrial mammal diversity in forests, indigenous agroforestry systems and plantain monocultures in Talamanaca, Costa Rica. *Biodivers. Conserv.* 15:555–585.

Hayes, J.P., S. Chan, W.H. Emmingham, J.C. Tappeiner, L.D. Kellogg, and J.D. Bailey. 1997. Wildlife responses to thinning in young forests of the Pacific Northwest. *J. For.* 95:28–33.

Heliövaara, K. and R. Väisänen. 1984. Effects of modern forestry on northwestern European forest invertebrates: A synthesis. *Acta Forestalia Fennica* 189:1–32.

Heuberger, K.A. and F.E. Putz. 2003. Fire in the suburbs: Ecological impacts of prescribed fire in small remnants of longleaf pine (*Pinus palustris*) sandhills. *Restor. Ecol.* 11:72–81.

Hernandez-Aquilera, J.N., J.M. Conrad, N.I. Gómez, and A.D. Rodewald. 2019. The economics and ecology of shade-grown coffee: A model to incentivize shade and bird conservation. *Ecol. Econ.* 159:110–121.

Hobbs, N.T. and R.A. Spowart. 1984. Effects of prescribed fire on nutrition of mountain sheep and mule deer during winter and spring. *J. Wildl. Manage.* 48:551–560.

Homyack, J. and J. Verschuyl. 2019. Effects of harvesting forest-based biomass on terrestrial wildlife. In *Renewable energy and wildlife conservation*, eds. C.E. Moorman, S.M. Grodsky, and S.P. Rupp, pp. 22–41. Baltimore: Johns Hopkins University Press.

Hunter, M.L., Jr. 1990. *Wildlife, forests, and forestry: Principles of managing forests for biological diversity.* Englewood Cliffs, NJ: Prentice Hall, Inc.

Hylander, K. 2011. The response of land snail assemblages below aspens to forest fire and clear-cutting in Fennoscandian boreal forests. *For. Ecol. Manage.* 261:1811–1819.

Hyvärinen, J. Kouki, and P. Martikainen. 2006. Fire and green-tree retention in conservation of red-listed and rare deadwood-dependent beetles in Finish boreal forests. *Cons. Biol.* 20:1711–1719.

Hyvärinen, J. Kouki, and P. Martikainen. 2009. Prescribed fires and retention trees help to conserve beetle diversity in managed boreal forests despite their transient negative effects on some beetle groups. *Insect Conserv. Divers.* 2:93–105.

Iglay, R., B. Leopold, D.A. Miller, and L.W. Burger. 2010. Effect of plant community composition on plant response to fire and herbicide treatments. *For. Ecol. Manage.* 260:543–548.

Isola, C.R., M.A. Colwell, O.W. Taft, and R.J. Safran. 2000. Interspecific differences in habitat use of shorebirds and waterfowl foraging in managed wetlands of California's San Joaquin Valley. *Waterbirds* 23:196–203.

Janowiak, M.K. and C.R. Webster. 2010. Promoting ecological sustainability in woody biomass harvesting. *J. For.* 108:16–23.

Johnson, D.H. 1980. The comparison of usage and availability measurements for evaluating resource preference. *Ecology* 6:65–71.

Johnstone, J.F. et al. 2016. Changing disturbance regimes, ecological memory, and forest resilience. *Front. Ecol. Environ.* 14:369–378.

Jones, C.D., J.A. Cox, E. Toriani-Moura, and R.J. Cooper. 2013. Nest-site characteristics of Bachman's Sparrows and their relationship to plant succession following prescribed burns. *Wilson. J. Ornithol.* 125:293–300.

Jones, J.D.J. and M.J. Chamberlain. 2004. Efficacy of herbicides and fire to improve vegetative conditions for Northern bobwhites in mature pine forests. *Wildl. Soc. Bull.* 32:1077–1084.

Kaila, L., P. Martikainen, P. Punttila, and E. Yakolev. 1994. Saproxylic beetles (Coleoptera) on dead birch trunks decayed by different polypore species. *Ann. Zool. Fenn.* 31:97–107.

Kasischke, E.S. and M.R. Turetsky. 2006. Recent changes in the fire regime across the North American boreal region—spatial and temporal patterns of burning across Canada and Alaska. *Geophys. Res. Lett.* 33:L09703. https://doi.org/10.1029/2006GL025677

Kennedy, C.E.J. and T.R.E. Southwood. 1984. The number of species of insects associated with British trees: A re-analysis. *J. Anim. Ecol.* 53:445–478.

Kilpatrick, H.J., T.P. Husband, and C.A. Pringle. 1988. Winter roost site characteristics of eastern wild turkeys. *J. Wildl. Manage.* 52:461–463.

King, D.I. and R.M. DeGraaf. 2000. Bird species diversity and nesting success in mature, clearcut and shelterwood forest in northern New Hampshire, USA. *For. Ecol. Manage.* 129:227–235.

Kroll, A.J., A. Springford, and J. Verschuyl. 2020. Conservation and production responses vary by disturbance intensity in a long-term forest management experiment. *Ecol. Appl.* 30:e02148.

Kupfer, J.A., K. Lackstrom, J.M. Grego, K. Dow, A.J. Terando, and J.K. Hiers. 2022. Prescribed fire in longleaf pine ecosystems: Fire managers' perspective on priorities, constraints, and future prospects. *Fire Ecol.* 18:27.

Lachance, D., C. Lavoie, and A. Desrochers. 2005. The impact of peatland afforestation on plant and bird diversity in Southeastern Quebec. *Ecosci.* 12:161–171.

Lamont, B.B., D.C. Le Maitre, R.M. Cowling, and N.J. Enright. 1991. Canopy seed storage in woody plants. *Bot. Rev.* 57:277–317.

Lampila, P., M. Mönkkönen, and A. Desrochers. 2005. Demographic responses by birds to forest fragmentation. *Cons. Biol.* 19:1537–1546.

Landers, J.L., D.H. Van Lear, and W.D. Boyer. 1995. The longleaf pine forests of the Southeast: Requiem or renaissance? *J. For.* 93:38–44.

Lane, V.R., K.V. Miller, S.B. Castleberry, R.J. Cooper, D.A. Miller, T.B. Wigley, G.M. Marsh, and R.L. Mihalco. 2011a. Bird community responses to a gradient of site preparation intensities in pine plantations in the Coastal Plain of North Carolina. *For. Ecol. Manage.* 262:1668–1678.

Lane, V.R., K.V. Miller, S.B. Castleberry, D.A. Miller, T.B. Wigley, G.M. Marsh, and R.L. Mihalco. 2011b. Plant community responses to a gradient of site preparation intensities in pine plantations in the Coastal Plain of North Carolina. *For. Ecol. Manage.* 262:370–378.

Lanham, J.D. and D.C. Guynn, Jr. 1996. Influences of coarse woody debris on birds in southern forests. In *Biodiversity and coarse woody debris in southern forests*, ed. J.W. McMinn and D.A. Crossley, Jr. U.S. Forest Service Technical Report SE-94, pp. 101–107. Washington, DC: U.S. Forest Service.

Lashley, M.A., M.C. Chitwood, C.S. DePerno, and C.E. Moorman. 2017. Frequent fires eliminate understory fruit production. *For. Ecol. Manage.* 405:9–12.

Lashley, M.A., C.A. Harper, G.E. Bates, and P.D. Keyser. 2011. Forage availability for white-tailed deer following silvicultural treatments in hardwood forests. *J. Wildl. Manage.* 75:1467–1476.

Laurence, S.G. and W.F. Laurence. 1999. Tropical wildlife corridors: Use of linear rainforest remnants by arboreal mammals. *Biol. Conserv.* 91:231–239.

Lemckert, F. 1999. Impacts of selective logging on frogs in a forested area of Northern New South Wales. *Biol. Conserv.* 89:321–328.

Leopold, A. 1933. *Game management.* New York: Scribner.

Lima, S.L. 1993. Ecological and evolutionary perspectives on escape from predatory attack: A survey of North American birds. *Wilson Bull.* 105:1–47.

Lindenmayer, D.B. and H.A. Nix. 1993. Ecological principles for the design of wildlife corridors. *Cons. Biol.* 7:627–630.

Linnell, J.D.C., J.E. Swenson, and R. Andersen. 2000. Conservation of biodiversity in Scandinavian boreal forests: Large carnivores as flagships, umbrellas, indicators, or keystones? *Biodivers. Conserv.* 9:857–868.

Loeb, S.C. 1996. The role of coarse woody debris in the ecology of southeastern mammals. In *Biodiversity and coarse woody debris in southern forests*, eds. J.W. McMinn and D.A. Crossley, Jr. U.S. Forest Service Technical Report SE-94, pp. 108–118.Washington, DC: U.S. Forest Service.

Lohr, S.M., S.A. Gauthreaux, and J.C. Kilgo. 2002. Importance of course woody debris to avian communities in loblolly pine forests. *Cons. Biol.* 16:767–777.

Long, J.N. 2009. Emulating natural disturbance regimes as a basis for forest management: A North American view. *For. Ecol. Mange.* 257:1868–1873.

Longhurst, W.M., K.K. Oh, M.B. Jones, and R.P. Kepner. 1968. A basis for the palatability of deer forage plants. *Trans. N. Am. Wildlife Confer.* 33:181–192.

Loy, X. et al. 2020. The impacts of bioenergy pine plantation management practices on bee communities. *J. Appl. Ecol.* 57:952–962.

Lundström, J., K. Öhman, K. Perhans, M. Rönnqvist, and L. Gustafsson. 2011. Cost-effective age structure nad geographical distribution of boreal forest reserves. *J. Appl. Ecol.* 48:133–142.

MacArthur, R.H. and J.W. MacArthur. 1961. On bird species diversity. *Ecology* 42:594–598.
Mas, A.H. and T.V. Dietsch. 2004. Linking shade coffee certification to biodiversity conservation: Butterflies and birds in Chiapas, Mexico. *Ecol. Appl.* 14:642–654.
Masters, R.E., R.L. Lochmiller, and D.M. Engle. 1993. Effects of timber harvest and prescribed fire on white-tailed deer forage production. *Wildl. Soc. Bull.* 21:401–411.
Matsuoka, S.M., J.A. Johnson, and D.A. Dellasala. 2012. Succession of bird communities in young temperate rainforests following thinning. *J. Wildl. Manage.* 76:919–931.
Matthews, C.E., C.E. Moorman, C.H. Greenberg, and T.A. Waldrop. 2010. Response of reptiles and amphibians to repeated fuel reduction treatments. *J. Wildl. Manage.* 74:1301–1310.
McComb, B.C. 2008. *Wildlife habitat management: Concepts and applications in forestry*. Boca Raton, FL: CRC Press.
McComb, W.C. and R.L. Rumsey. 1983. Bird density and habitat use in forest openings created by herbicides and clearcutting in the central Appalachians. *Brimleyena* 9:83–95.
McNab, B.K. 1963. Bioenergetics and the determination of home range size. *Am. Nat.* 97:133–140.
Means, D.B. 2006. Vertebrate faunal diversity of longleaf pine ecosystems. In *The longleaf pine ecosystem: Ecology, silviculture, and restoration*, eds. S. Jose, E.J. Jokela, and D.L. Miller. New York, NY: Springer.
Miller, D.A. and T.B. Wigley. 2004. Introduction: Herbicides and forest biodiversity. *Wildl. Soc. Bull.* 32:1016–1019.
Miller, J.H. 2003. Nonnative invasive plants of southern forests: A field guide for identification and control. Revised. United States Department of Agriculture Forest Service, Southern Research Station, General Technical Report SRS-62, Washington, DC: USA.
Miller, K.V., and J.H. Miller. 2004. Forestry herbicide influences on biodiversity and habitat in southern forests. *Wildl. Soc. Bull.* 32:1049–1060.
Mittermeier, R., N. Mayers, J.B. Thomsen, and G.A.B. Olivieri. 1998. Biodiversity hot spots and major tropical wilderness areas: Approaches to setting conservation priorities. *Cons. Biol.* 12:516–520.
Moen, A.N. 1973. *Wildlife ecology*. San Francisco, CA: W.H. Freeman.
Moorman, C.E., S.M. Grodsky, and S.P. Rupp, eds. 2019. *Renewable energy and wildlife conservation*. Baltimore, MD: Johns Hopkins University Press.
Moorman, C.E., and D.C. Guynn, Jr. 2001. Effect of group-selection opening size on breeding bird habitat use in a bottomland forest. *Ecol. Appl.* 11:1680–1691.
Moorman, C.E., K.R. Russell, and C.H. Greenberg. 2011. Reptile and amphibian response to hardwood forest management and early-successional habitats. In *Sustaining Young Forest Communities: Ecology and Management of Early Successional Habitats in the Central Hardwood Region, USA*, ed CH Greenberg, BS Collins, and FR Thompson III, 191–208. New York: Springer.
Moorman, C.E., K.R. Russell, M.A. Megalos, and D. Drake. 2002. The role of non-industrial private forest lands in the conservation of southern fire-dependent wildlife. In *The role of fire for nongame wildlife management and community restoration: Traditional uses and new directions from eastern and western perspectives*, ed W.M. Ford, K.R. Russell, and C.E. Moorman, 116–123. USDA Forest Service GTR-NE-288.
Moorman, C.E., K.R. Russell, G.R. Sabin, and D.C. Guynn Jr. 1999. Snag dynamics and cavity occurrence in the South Carolina Piedmont. *For. Ecol. Manage.* 118:37–48.
Muona, J. and I. Rutanen. 1994. The short-term impact of fire on the beetle fauna in boreal coniferous forests. *Ann. Zool. Fenn.* 31:109–122.
Murray, B.D., C.R. Webster, M.A. Jenkins, M.R. Saunders, and G.S. Haulton. 2016. Ungulate impacts on herbaceous-layer plant communities in even-aged and uneven-aged managed forests. *Ecosphere*. 7:e01378.
Niemelä, J. Invertebrates and boreal forest management. 1997. *Cons. Biol.* 11:601–610.
Niemelä, J., M. Koivula, D.J. Kotze. 2007. The effects of forestry on carabid beetles (Coleoptera: Carabidae) in boreal forests. *J. Insect Conserv.* 11:5–18.
Nigam, P.S., and A. Singh. 2011. Production of biofuels from renewable resources. *Prog. Energy Combust. Sci.* 37:52–68.
Noss, R.F. 1999. Assessing and monitoring forest biodiversity: A suggested framework and indicators. *For. Ecol. Manage.* 115:135–146.
O'Connell, W.E., and K.V. Miller. 1994. Site preparation influences on vegetative composition and avian and small mammal communities in the South Carolina Upper Coastal Plain. *Proceedings of the Annual Conference of Southeast Fish and Wildlife Agencies* 48:321–330.
Odum, E.P. 1969. The strategy of ecosystem development. *Science* 164:262–270.
Oliver, C.D. 1980. Forest development in North America following major disturbances. *For. Ecol. Manage.* 3:153–168.

Paillet, Y., L. Bergès, J. Hjältén, P. Ódor, C. Avon, M. Bernhardt-Römermann, R.-J. Bijsma, L. De Bruyn, M. Fuhr, U. Grandin, R. Kanka, L. Lundin, S. Luque, T. Magura, S. Matesanz, I. Mészáros, M.-Teresa Sebastià, W. Schmidt, T. Standovár, B. Tóthmérész, A. Uotila, F. Valladares, K. Vellak, and R. Virtanen. 2010. Biodiversity differences between managed and unmanaged forests: Meta-analysis of species richness in Europe. *Cons. Biol.* 24:101–112.

Palmer, M.A., R.F. Ambrose, and N.L. Poff. 1997. Ecological theory and community restoration ecology. *Restoration Ecology* 5:291–300.

Parikka, M. 2004. Global biomass fuel resources. *Biomass Bioenergy* 27:613–620.

Patton, P.W. 1994. The effect of edge on avian nest success: How strong is the evidence? *Cons. Biol.* 8:17–26.

Pearson, S.M., M.G. Turner, L. L Wallace, and W.H. Romme. 1995. Winter habitat use by large ungulates following fire in northern Yellowstone National Park. *Ecol. Appl.* 5:744–755.

Penman, T.D., and S.H. Penman. 2010. Influence of prescribed burning on fruit production in Proteaceae. *Pacific Cons. Biol.* 16:46–53.

Perfecto, I., R.A. Rice, R. Greenberg, and M.E. Van der Voort. 1996. Shade coffee: A disappearing refuge for biodiversity. *Bioscience* 46:598–608.

Perfecto, I., and R. Snelling. 1995. Biodiversity and the transformation of a tropical agroecosystem: Ants in coffee plantations. *Ecol. Appl.* 5:1084–1097.

Perfecto, I, J. Vandermeer, A. Mas, and L.S. Pinto. 2005. Biodiversity, yield, and shade coffee certification. *Ecol. Econ.* 54:435–446.

Perschel, B., A. Evans, and M. DeBonis. 2012. *Forest biomass retention and harvesting guidelines for the Southeast Forest Guild.* Forest Guild Southeast Biomass Working Group. Forest Guild. Santa Fe, New Mexico.

Philpott, S.M., P. Bichier, R. Rice, and R. Greenberg. 2007. Filed-testing ecological and economic benefits of coffee certification programs. *Cons. Biol.* 21:975–985.

Philpott, S.M., and T. Dietsch. 1993. Coffee and conservation: A global context and the value of farmer involvement. *Cons. Biol.* 17:1844–1846.

Picciano, P., F.X. Aguilar, D. Burtraw, and A. Marzaee. 2022. Environmental and socio-economic implications of woody biomass co-firing at coal-fired power plants. *Resour. Energy Econ.* 68:101296.

Pickett, S.T.A., J. Kolasa, J.J. Armesto, and S.L. Collins. 1989. The ecological concept of disturbance and its expression at various hierarchical levels. *Oikos* 54:129–136.

Porneluzi, P.A., R. Brito-Aguilar, R.L. Clawson, and J. Faaborg. 2014. Long-term dynamics of bird use of clearcuts in post-fledging period. *Wilson J. Ornithol.* 126:623–634.

Pyne, S.J., P.L. Andrews, and R.D. Laven. 1996. *Introduction to wildland fire.* New York, NY: Wiley Press.

Rainforest Alliance. 2000. *The conservation coffee campaign organizer's kit.* New York, NY: Rainforest Alliance.

Rappole, J.H. 1995. *The ecology of migrant birds: A neotropical perspective.* Washington, DC: Smithsonian Institution Press.

Rappole, J.H., D.I. King, and J.H. Vega Rivera. 2003. Coffee and conservation. *Cons. Biol.* 17:334–336.

Rice, R.A., and J.R. Ward. 1997. *Coffee, conservation, and commerce in the Western Hemisphere.* Smithsonian Migratory Bird Center and Natural Resources Defense Council. Washington, DC.

Riffell, S., J. Verschuyl, D. Miller, and T.B. Wigley. 2011**a**. Biofuel harvests, coarse woody debris, and biodiversity—A meta-analysis. *For. Ecol. Manage.* 261:878–887.

Riffell, S., J. Verschuyl, D. Miller, and T. B Wigley. 2011**b**. A meta-analysis of bird and mammal response to short-rotation woody crops. *GCB Bioenergy* 3:313–321.

Riitters, K.H., J.W. Coulston, and J.D. Wickham. 2012. Fragmentation of forest communities in the eastern United States. *For. Ecol. Manage.* 263:85–93.

Rist, L., L. Feintrenie, and P. Levang. 2010. The livelihood impacts of oil palm: Smallholders in Indonesia. *Biodivers. Conserv* 19:1009–1024.

Robinson, W.D., and S.K. Robinson. 1999. Effects of selective logging on forest bird populations in a fragmented landscape. *Cons. Biol.* 13:58–66.

Rolstad, J., Y. Blanck, and K.O. Storaunet. 2017. Fire history in a western Fennoscandian boreal forest as a influenced by human land use and climate. *Ecol. Monogr.* 87:219–245.

Root, H T., J. Verschuyl, T. Stokely, P. Hammond, M. Scherr, and M.G. Betts. 2017. Plant diversity enhances moth diversity in an intensive forest management experiment. Ecol. Appl. 27:134–142.

Rosenberg, D.K., B.R. Noon, J.W. Megahan, and E.C. Meslow. 1998. Compensatory behavior of *Ensatina eschscholtzii* in biological corridors: A field experiment. *Can. J. Zool.* 76:117–133.

Rosenberg, K.V., and M.G. Raphael. 1986. Effects of forest fragmentation on vertebrates in Douglas-fir forests. In *Wildlife 2000: Modeling Habitat Relationships of Terrestrial Vertebrates*, ed J. Verner, M.L. Morrison and C.J. Ralph. Madison, WI: The University of Wisconsin Press.

Runkle, J.R. 1981. Gap regeneration in some old-growth forests of the eastern United States. *Ecology*. 62:1041–1051.

Russell, F.L., D.B. Zippin, and N.L. Fowler. 2001. Effects of white-tailed deer (Odocoileus virginianus) on plants, plant communities and communities: A review. *Am. Mid. Natur*. 146:1–26.

Russell, K.R., H.G. Hanlin, T.B. Wigley, and D.C. Guynn, Jr. 2002. Responses of isolated wetland herpetofauna to upland forest management. *J. Wildl. Manage*. 66:603–617.

Russell, K.R., T.J. Mabee, M.B. Cole, and M.J. Rochelle. 2005. Evaluating biotic and abiotic influences on torrent salamanders in managed forests of western Oregon. *Wildl. Soc. Bull*. 33:1413–1424.

Russell, K.R., D.H. Van Lear, and D.C. Guynn Jr. 1999. Prescribed fire effects on herpetofauna: Review and management implications. *Wildl. Soc. Bull*. 27:374–384.

Rutledge, B.T. and M.L. Conner. 2002. Potential effects of groundcover restoration on breeding bird communities in longleaf pine stands. *Wildl. Soc. Bull*. 30:354–360.

Ryan, K.C., E.E. Knapp, and J.M. Varner. 2013. Prescribed fire in North American forests and woodlands: History, current practice, and challenges. *Front. Ecol. Environ*. 11:25–34.

Santillo, D.J., P.W. Brown, and D.M. Leslie, Jr. 1989. Response of songbirds to glyphosphate-induced habitat changes in clearcuts. *J. Wildl. Manage*. 53:64–71.

Schowalter, T.D. 2012. Insect responses to major landscape-level disturbance. *Annu. Rev. Entomol*. 57:1–20.

Schroth, G., and C.A. Harvey. 2007. Biodiversity conservation in cocao production landscapes: An overview. *Biodivers. Conserv*. 16:2237–2244.

Seaman, B.D. and D.G. Krementz. 2001. Movements and survival of Bachman's Sparrows in response to prescribed summer burns in South Carolina. *Proceedings of the Annual Conference of the Southeastern Association of Fish and Wildlife Agencies* 54:227–240.

Semlitsch, R.D. and J.R. Bodie. 2003. Biological criteria for buffer zones around wetlands and riparian habitats for amphibians and reptiles. *Cons. Biol*. 17:1219–1228.

Sgarbossa, A., M. Boschiero, F. Pierobon, R. Cavalli, and M. Zanetti. 2020. Comparative life cycle assessment of bioenergy production from different wood pellet supply chains. *Forests*. 11:1127.

Shelton, S. 2014. *Birds of a feather, linked to an open pine forest*. US Fish and Wildlife Service. www.fws.gov/landscape-conservation/stories/birds-of-a-feather-linked-to-an-open-pine-forest.html. Accessed December 2014.

Siira-Pietkäinen, A., and J. Haimi. 2009. Changes in soil fauna 10 years after forest harvestings: Comparison between clear felling and green-tree retention methods. *For. Ecol. Manage*. 258:332–338.

Stokely, T.D. et al. 2022. Experimental evaluation of herbicide use on biodiversity, ecosystem services and timber production trade-offs in forest plantations. *J. Appl. Ecol*. 59:52–66.

Swanson, M.E., J.F. Franklin, R.L. Beschta, C.M. Crisafulli, D.A. DellaSalla, R.L. Hutto, D.B. Lindenmayer, and F.J. Swanson. 2010. The forgotten stage of forest succession: Early-successional ecosystems on forest sites. *Front. Ecol. Environ*. 9:117–125. https://www.seia.org/research-resources/solar-market-insight-report-2023-year-review

Taillie, P.J., C.E. Moorman, and M.N. Peterson. 2015. The relative importance of multi-scale factors on the distribution of Bachman's Sparrow and the implications for ecosystem conservation. *Condor Ornithol. Apps*. 117:137–146.

Taki, H., T. Inoue, H. Tanaka, H. Makihara, M. Sueyoshi, M. Isono, and K. Okabe. 2010. Responses of community structure, diversity, and abundance of understory plants and insect assemblages to thinning in plantations. *For. Ecol. Manage*. 259:607–613.

Tatum, V.L. 2004. Toxicity, transport, and fate of forest herbicides. *Wildl. Soc. Bull*. 32:1042–1048.

Tatum, V.L., D.L. Borton, W.R. Streblow, J. Louch, and J.P. Shepard. 2012. Acute toxicity of commonly used forestry herbicide mixtures to *Ceriodaphnia dubia* and *Promelas pimephales*. *Environmental Toxicology* 27:671–684.

Thomas, S.C., C.B. Halpern, D.A. Falk, D.A. Liguori, and K.A. Austin. 1999. Plant diversity in managed forests: Understory responses to thinning and fertilization. *Ecol. Appl*. 9:864–879.

Todd, B.D., T.M. Luhring, B.B. Rothermel, and J.W. Gibbons. 2009. Effects of forest removal on amphibian migrations: Implications for habitat and landscape connectivity. *J. Appl. Ecol*. 46:554–561.

Tucker, J.W. Jr., W.D. Robinson, and J.B. Grand. 2004. Influence of fire on Bachman's Sparrow, an endemic North American songbird. *J. Wildl. Manage*. 68:1114–1123.

Turetsky, M.R., and V. St. Louis. 2006. Disturbance in Boreal Forests. In *Ecological Studies, Volume 188. Boreal Peatland Ecosystems*, ed. R.K. Wieder and D.H. Vitt. Berlin: Springer-Verla.

Turner, M.G., W.H. Romme, R.H. Gardnerm and W.W. Hargove. 1997. Effects of fire size and pattern on early succession in Yellowstone National Park. *Ecol. Monogr*. 67:411–433.

Ulyshen, M.D., and J.L. Hanula. 2009. Responses of arthropods to large-scale manipulations of dead wood in loblolly pine stands of the southeastern United States. *Environ. Entomol.* 38:1005–1012.

Väisänen, R., and K. Heliövaara. 1994. Assessment of insect occurrence in boreal forests based on satellite imagery and field measurements. *Acta Forestalia Fennica* 243.

Van Lear, D.H., W.D. Carroll, P.R. Kapeluck, and R. Johnson. 2005. History and restoration of the longleaf pine-grassland ecosystem: Implications for species at risk. *For. Ecol. Manage.* 211:150–165.

Varner, J.M., et al. 2005. Restoring fire to long-unburned *Pinus palustris* systems: Novel fire effects and consequences for long-unburned ecosystems. *Restoration Ecol.* 13:536–544.

Verschuyl, J., A.J. Hansen, D.B. McWethy, R. Sallabanks, and R. L Hutto. Is the effect of forest structure on bird diversity modified by forest productivity? *Ecol. Appl.* 2008. 18:1550–1770.

Verschuyl, J., S. Riffell, D. Miller, and T.B. Wigley. 2011. Biodiversity response to intensive woody biomass production from forest thinning in North American forests—a meta-analysis. *For. Ecol. Manage.* 261:221–232.

Wade, T.G., K.H. Riitters, J.D. Wickham, and K.B. Jones. 2003. Distribution and causes of global forest fragmentation. *Cons. Biol.* 7:7.

Walters, J.R. 1991. Application of ecological principles to the management of endangered species: The case of the red-cockaded woodpecker. *Annu. Rev. Ecol. Syst.* 22:505–523.

Welsh, D.A. 1971. Breeding and territoriality of the Palm Warbler in a Nova Scotia bog. *Can. Field Nat.* 85:31–37.

Weltzin, J.F., S.D. Bridgham, J. Pastor, J. Chen, and C. Harth. 2003. Potential effects of warming and drying on peatland plan community composition. *Glob. Change Biol.* 9:141–151.

Whiles, M.R., and J.W. Grubaugh. 1996. Importance of coarse woody debris to southern forest herpefauna. In *Biodiversity and coarse woody debris in southern forests, proceedings of the workshop on coarse woody debris in southern forests: Effects on biodiversity*, ed J.W. McMinn and D.A. Crossley, 94–100. USDA Forest Service GT-SE-94.

White, P.S. 1979. Pattern, process, and natural disturbance in vegetation. *Bot. Rev.* 45:229–299.

Wiens, J.A. 1975. Avian communities, energetics, and functions in coniferous forest habitat. In D.R. Smith (tech. coor.), Proceedings from the symposium on management of forest and range habitats for nongame birds. USDA Forest Service. General Technical Report Wo-1. Washington, DC.

Williams, A., A.A. Dayer, N. Hernandez-Aquilera, T.B. Phillips, H. Faulkner-Grant, M.I. Gómez, and A.D. Rodewald. 2021. Tapping birdwatchers to promote bird-friendly coffee consumption and conserve birds. *Peopl. Nature.* 3:312–324.

Williams-Guillén, K., C. McCann, J.C. Martinez Sánchez, and F. Koontz. 2006. Resource availability and habitat use by mantled howling monkeys in a Nicaraguan coffee plantation: Can agroforests serve as core habitat for a forest mammal? *Anim. Conserv.* 9:331–338.

Wilson, W.H., Jr. 1996. Palm Warbler (*Dendroica palmarum*). In *The Birds of North America No. 228.* Ed. Poole, A. and F. Gill. Philadelphia: The Academy of Natural Sciences.

Wimberly, M.C., T.A. Spies, C.J. Long, and C. Whitlock. 2000. Simulating historical variability in the amount of old forests in the Oregon Coast Range. *Cons. Biol.* 14:167–118.

Winiarski, J., J. Carpenter, and C.E. Moorman. 2017a. Bachman's Sparrows at the northern periphery of their range: Home range size and microhabitat selection. *J. of Field Ornith.* 88:250–261.

Winiarski, J., C.E. Moorman, J. Carpenter, and G. Hess. 2017b. Reproductive consequences of habitat fragmentation for a declining resident bird of the longleaf pine ecosystem. *Ecosphere* 8(7):e01898.10.1002/ecs2.1898

Zolkowski, S.B. 2008. Habitat relationships of bird communities in Wisconsin peatlands. M.S. Thesis, University of Wisconsin-Stevens Point, WI.

4 External Drivers of Changes Challenging Forestry

Political and Social Issues at Stake

Christine Farcy, Ronnie de Camino, Inazio Martinez de Arano, and Eduardo Rojas Briales

4.1 INTRODUCTION

This chapter focuses on forestry that embraces the theory and practice of all that constitutes the management and use of forests. The intention is to present some of the main challenges that forestry is facing today from a social and political point of view.

We will first recall and briefly discuss the basic principles of the thesis of the "wake effect," which is a framework that has been one of the main conceptual pillars of forestry for centuries and until recently. Conceived and designed in Western countries, it has been spread beyond their borders. The intention is not to further allow a comparative panorama nor a historical overview of the other paradigms that have been guiding forestry until today (Schmithüsen 2013). This would indeed lead too far from the expected objective. The idea with the presentation of the "wake effect" is indeed rather to offer some introductive departure point and reading grid to better understand the issues at stake today that will be presented through the prism of three global processes having been identified as the main external drivers of change.

4.1.1 The "Wake Effect"

From a conceptual point of view, the management of forests has been based for a long time, especially in the Western countries, on the thesis of the "wake effect." This meant that managing forests for sustained wood production would imply the implicit achievement of other purposes, such as soil protection, water quality, or public access (Dieterich 1953; Rupf 1960). In this scheme, developed in particular by the German and French schools of forestry, the function of wood production is thus the main implicit strategic objective that is pulling the other functions and is making them possible from a technical and economical point of view.

Although numerous examples demonstrate the interest that foresters have had in other functions than wood production, forest management has indeed often been based on the theoretical framework underpinning wood production and has been built on specific underlying principles that are linked with silviculture and the framework of neoclassical economics. In this economics thinking mode, forests are considered as resources that are inputs available to humans for their economic activities (Barre 1964). This consideration is the result of an explicit choice and made according to the system of value in place as adopted by the human society concerned rather than to immanent needs (Weber et al. 1990). Resources and explicit human utilization are indeed conceptually closely linked in this scheme (Boude and Chaboud 1995). For Kant (2003), the neoclassical framework is also

> "characterized by the utility (or profit) maximizing rational (homo economicus) agent and the "invisible hand" leading to an efficient general equilibrium. In this framework, people's preferences are static,

DOI: 10.1201/9781003431084-5

> society is a mathematical aggregation of homogeneous rational agents, public inputs are through market signals, and there is no role for any institution other than the market".

Forest stands are basic units of this system, where the balance that is being sought is a sustained yield of wood, basically at ownership scale. The underlying decision-making process is quite linear and exclusive. It is based on hierarchical modalities according to the "command and control" scheme (Harrington and Morgenstern 2007) in force in forest policy, where a central role of inspection and supervision is given by forest laws to national forest services and where initiatives were taken to reduce traditional users' rights and secure supplies for strategic uses such as navy and military foundries (Farcy and Devillez 2005; Michel and Gil 2013). The main leading actors in this framework are those of the wood production sector acting in a closed system implying not only institutional and social partitioning but also spatial partitioning, forest being considered "as an island, that is, a world apart, isolated from the rest of the rural landscape" (Plaisance 1979). Certainty and long-term planning are other keywords of such an approach, which are embodied in the formal discipline of forest management planning developed since the 1850s by the German and French schools of forestry (Huffel 1926).

In Europe, this scheme has allowed the recovery of forest area, stocks, and increasing volumes of wood supply in some sustainable and affordable way. However, the expansion of the neoclassic model has seen resistance or has been slower in areas with strong community links, cultural heritage, or emotional links to the territory (i.e., the Indigenous and communities in Latin America). Also, it has been facing increased criticism in Western postmodern societies that are giving more relevance to aesthetic or welfare values.

While the validity of those principles is indeed not conceptually called into question for industrial societies, their relevance for forest-dependent people (Chao 2012) or in agrarian or postindustrial societies (Nair 2004), and as main components of a global paradigm for forestry, has clearly been challenged for several years (Ormerod 1997; Kant 2003; Kennedy and Koch 2004; Piketty 2013).

The framework also appears too limited and too restrictive for facing late and forthcoming political and social issues and for responding to current human and social demands and needs when higher priorities are given to more holistic objectives and to the full collection of material and immaterial goods and ecosystem services provided by forests, including wood and nonwood forest products.

In any case, the validity of those principles is not unique and fully viable anymore not only for forestry but for any sector related to natural resources, particularly when the growing paradigm begins to be one of the sustainable human developments.

4.1.2 External Drivers of Change

Three main current global processes have been identified as main external drivers of changes challenging forestry: urbanization, tertiarization, and globalization. The last two are recent. The first one is developing since the first step of industrialization but can be considered contemporary because of the global rate of more than 50% of the population living in cities that has been recently crossed.

Those processes, which are deeply interdependent, were considered as mainstream and as the main levers of changes in the social and political environment that forests and forestry are facing today. They are indeed significantly impacting the relationship not only between humans and forests but also between humans themselves when being related to forests.

The topic of external drivers of changes in forestry is quite large. So this chapter is not intended to be exhaustive or detailed; it will focus on the global trends and will not be considering the explanation of the spectrum of possible effects of the global processes identified, which are depending on national or local specific socioeconomical or cultural contexts and on the level of interactions between them or with other macroprocesses.

4.2 URBANIZATION

4.2.1 Findings

The world in which we live has become urban, and this constitutes one of the major changes in the contemporary era (Seto et al. 2011; Bonnet 2014). Urbanization is defined here as the increase in the proportion of population living in cities and settlements versus rural areas (United Nations 2014). The continued increase in the human population means that urban areas tend to physically grow and expand both vertically and horizontally at the expense of rural areas. According to the last World Urbanization Prospects published by the United Nations (2014), 54% of the world's population lived in urban areas in 2014. In 1950, this proportion was 30% and is expected to increase to 66% by 2050.

About 50% of urban people are living in small settlements (less than 500,000 inhabitants) while one in seven lives in a city with one million inhabitants or more (Bonnet 2014; United Nations 2014). The most urbanized regions are North America, Latin America and the Caribbean, and Europe with 82%, 80%, and 73%, respectively. Africa and Asia still remain mostly rural, with 40% and 48%, respectively, of their population living in urban areas. The urbanization trend is expected to continue in all regions but should be faster in Africa and Asia, where 56% and 64% of the population are expected to be urban by 2050 (United Nations 2014). However, the concepts of rural and urban settlements are not equivalent in the different regions or continents. The infrastructure and quality of services in a rural community in Europe or North America are by far of higher quality than the ones in Latin America or Africa.

4.2.2 Consequences

What are the consequences of such a global, continuous, and ineluctable process for forestry? Environmental consequences of urbanization, such as habitat fragmentation, isolation of fauna and flora populations, or loss of forests areas for the benefit of the development of cities and infrastructures, are well and often documented (Nowak and Walton 2005; Pauchard et al. 2006; Devictor et al. 2007; Shen et al. 2008), showing that they influence the landscape beyond peri-urban areas. This requires connecting the urban dimension of forestry well beyond the peri-urban areas. Unfortunately, studies on political and social consequences of urbanization are rare and the information is scattered.

While it is quite risky to state that ongoing urbanization leads to a global social transformation from a rural to an urban culture (Scott et al. 2007), we could nevertheless consider that having higher daily distance to access natural processes could generate an evolution in the perception that urban population could have on forests. Virtual feelings would be replacing background and own concrete experiences. In particular, this perception is more important in the new generation, as it is globally more urban than the previous one. Changes in the school systems and lifestyle have been reinforcing this trend (Pergams and Zaradic 2008; Gubbels et al. 2011). Data from the United States show indeed that the average outdoor time in 2002 was 50% of that in 1982, whereas time at school was up to 20%, and passive leisure increased to 400% (Juster et al. 2004).

Studies also suggest that forest issues are not well understood outside the small forestry community or in sparsely wood area and that there is a significant gap between public understanding and reality (Rametsteiner and Kraxner 2003; European Commission 2009). Such social perception can have high impact on forest management decisions, where urban citizens and authorities make decisions that affect forest resources and rural population, or where people from developed countries put a global agenda that affects developing countries in forestry issues. The case of the wrong social perception of deforestation, as its rate than the process itself, in developed countries, which is generating inadequate response and finally creating controversial reality, is an interesting illustrative case in this regard (Fabra-Crespo and Rojas-Briales 2015).

Whereas examples of citizens or children resistance movements against the felling of trees or against the use of "exotic species" considered harmful are becoming common and often reported in the media, some sociological studies also show that a significant amount of young urban people is having concerns with rural activities, such as hunting and tree felling, which are considered as a priority to be banned in forests (Bodson 2005). This trend constitutes a challenge for a sector that needs social legitimacy and a "license to operate" in a context where it is becoming a huge ally of humans in their intents to mitigate and adapt to the effects of expected climate change, in particular through the development of a greener economy besides many other vital environmental services (water, soil), biodiversity, and other social services (livelihood in rural areas, recreation, landscape) (IPCC 2011; European Commission 2012). This situation still remains paradoxical and requires particular attention.

Communication efforts are being implemented by the forest sector to try to curb or even reverse the process (Rametsteiner et al. 2009; European Commission 2011; IUFRO 2014; Fabra-Crespo and Rojas-Briales 2015). The launch of the Year of Forests in 2011 and of the International Day of Forests through a dedicated United Nations process has to be understood as part of those efforts. Nevertheless, they are far from being powerful enough to change the trends. While some countries, such as Switzerland, are monitoring the evolution of public perception on forests (Hunziker et al. 2013), some research in communication sciences has slowly developed since a few years to understand such gaps in social representations and to contribute to conceive more efficient communication strategies that take into account the ability of people to understand the complexity of forest issues instead of focusing mainly on the message sent that is moreover often too simplistic (European Commission 2009; Fastrez et al. 2013).

From a political point of view, with the current trend of urbanization, rural affairs would be losing ground and power to the benefit of urban concerns while forest affairs have always been weak in relation to agricultural affairs even when they were in the same ministry. In fact, since the mid-1980s, in many countries, forests have moved from agriculture to environmental ministries. Less attention would be given by the political world to traditional rural areas and institutions and as a consequence less resources in particular to the detriment of public services. This might be reasonable in developed countries but not in developing countries, where the poverty levels in rural areas are much higher.

In postindustrial societies, such as Spain or Italy, urbanization induces progressive social desertification and aging of the population in rural areas while society seems to have lost the capacity to generate value in large proportions of a landscape that is dramatically changing as extensive grazing and family subsistence farming are abandoned, except where replacement by cash crops is possible. Those changes in the land use are, in addition, inducing less interaction not only with rural populations but also with nature itself. Those evolutions are consistent with the current process of natural expansion of forests in some regions linked with dereliction of forest management (FAO 2011). Forests and shrubs are indeed often the last stage in the vegetation dynamics that impose a forest factual situation in agriculture dereliction (Álvarez-Martínez et al. 2014) and which is likely to increase the risk and intensity of forest fires in Mediterranean climates. Such a situation where abandonment of management increases the chances of catastrophic megafires raises new issues such as rights and duties linked to ownership. In contrast, in developing Mediterranean countries (i.e., southern countries) remaining rural populations are still exercising significant pressures on relatively scarce resources extracting firewood or seeking new pasturelands. In the absence of benefit sharing and perceived value for local populations that have been moved out of the forests, thanks to classical forest regulations still in place, forests are indeed jeopardized. Both quite different cases would require revisiting tenure and exploring innovative and specific governance arrangements.

At the same time, urban forestry becomes an emerging field in many situations, including developing countries, and green leisure for urban population is being raised with growing new opportunities. Urbanization is indeed inducing a growing need for green spaces in and around cities, particularly the biggest ones (Konijnendijk et al. 2005; Nail 2006; Arnberger 2012; Kloek et al. 2013). In particular, Mitchell and Popham (2008) showed that green areas in cities mitigate health inequalities among neighborhoods of different income levels.

While green conversion of towns is an old and persistent goal of urban planners (Johnston 1996; Forrest and Konijnendijk 2005), foresters have to face with ongoing growing pressures, new expectations, and more varied demands from the society related, in particular, to urban lifestyle and tertiarization (Nowak and Walton 2005). Whereas an urgent need for anticipating urbanization effects appears significant, countries or cities are developing specific urban forest strategies and new specific forest management models (Forestry Commission Scotland 2008; HCEFLCD 2010; Bobrowski and Biondi 2012; City of Melbourne 2012). However, this is not yet a mainstream, and efforts should be taken in urban planning to better take into account the urban hinterland and the environmental services provided by the landscapes where cities are inserted.

In addition to outdoor recreation and resourcing activities, more attention is also given to human health. While the use of the forest for pharmacopoeia by traditional forest societies, such as Pygmies, is well documented and being valorized (Motte-Florac 1980; Brisson 2011; Nke Ndih 2014), initiatives inducing positive effects on the physical and psychological health of stressed city dwellers are developing in countries like Japan, Korea, and Finland (Morita et al. 2007; Annerstedt et al. 2010; Karjalainen et al. 2010; Li 2010; Nilsson et al. 2011) (Box 4.1). Such actions could be recognized in the future as positively contributing to social security budgets.

4.3 TERTIARIZATION

4.3.1 Findings

For the last 30 years, there has been a substantial shift from the primary and secondary sectors to the tertiary sector, which is becoming the largest and fastest growing sector of the economy (Elfring 1989). The 2008 crisis became nevertheless nuance and revealed the higher resilience of primary and secondary sectors.

Vulgarized by Clark (1940) and Fisher (1939), the tertiary sector is characterized by producing services (i.e., immaterial goods) instead of end products. Due to its quite open and fuzzy content, this sector is also often divided into three: the tertiary sector would consider services provided to consumers and business, such as transportation, health care, or tourism. The quaternary sector would consist of intellectual activities undertaken by governments, science, universities, nonprofits, health care, culture, and the media while the quinary sector would include the senior management level of the quaternary sector (Lisch 2014).

Tertiarization is defined here as the increase in the proportion of the contribution of the tertiary sector to the world economy versus primary (i.e., agriculture) and secondary sectors (i.e., industry and mining). The concept is often considered together with theories developed by social scientists linking it with a series of historical structural transformations of human societies (Fourastié 1958; Bell 1973; Timmer and Akkus 2008). Our intention is not to go into this debate nor fuel a controversy that is beyond the scope of this chapter.

According to the World Bank (2014), in 2012, the contribution to the GDP of the added value of primary, secondary, and tertiary sectors was 3%, 27%, and 70%, respectively, at the global level, while for the same period, the employment in those sectors was 4%, 30%, and 66%, respectively. The global data are for hiding specific situations, such as the contribution of informal and subsistence activities (Lescuyer et al. 2009; FAO 2014) or the decisive role currently played by agriculture in the economy of various low-income countries. In 1995, the percentages of GDP attributable to primary, secondary, and tertiary sectors were 26%, 38%, and 26%, respectively, in low-income countries. The equivalent percentages were 12%, 35%, and 53% in middle-income countries and 2%, 32%, and 66% in high-income countries (World Bank 2014).

BOX 4.1 FOREST FOR WELL-BEING AND HEALTH OF STRESSED URBAN DWELLERS

Following Ulrich's (1984) first demonstration of a link between recovery from surgery and a window view of a natural setting, Korea and Japan are pioneers in developing the concept of therapy forests.

The Korean Forest Service (KFS) is giving a special focus to forest for people and in particular to stressed urban dwellers.

> Korean society is aging, being industrialized and modernized at a rapid pace. There is a resulting tendency that environmental diseases, chronic illness, and senile diseases are increasing. It is said that fundamental treatments for those diseases are limited, so it draws more attention to nature treatment. Forests are believed to have a variety of natural elements good for health, such as phytoncide, anions, comforting scenery, and natural soils. To promote people's health, using nature's healing power, the KFS has created therapy forests. They are now on operation equipped with convenience facilities, simple sports facilities, and so forth. There is also a visitor center, meditation spaces, rest areas, and healing forest trails.
>
> **(KFS 2015)**

Meanwhile, Japan has designed 48 official centers dedicated to forest therapy attended each year by millions of Japanese and would have 100 by 2024. Those sites are forests specially managed by the Japanese Forest Agency to offer to stressed urban populations the opportunity to practice *Shinrin-Yoku* (Park et al. 2010), which is a green "bath" in forest, that is, a walk, aiming to leave nature entering through the five senses. Companies are already offering such trips to their executives or are planning to give their workers time to practice such a traditional medicine precautionary measure. "Forest therapy" sessions are also proposed. They consist of two hours of leisurely stroll through the forest, with breathing exercises led by monitors. Before and after the trek, blood pressure and other physiological variables of the participants are measured to test the effectiveness of the treatment (Morita et al. 2007; Li 2010). Li (2010) showed that a walk through a forest or a park significantly increases the concentration of NK cells in the blood, a type of white blood cell that fights against infections and cancer. According to Li (2010), the volatile compounds emitted by trees are primarily responsible for the beneficial effect on the immune system. Many of these natural aromatics are already used in aromatherapy and holistic medicine.

Studies have also shown, thanks to advanced techniques in the field of neurobiology, that interaction with nature decreases the activity of the prefrontal cortex, the part of the brain where cognitive and executive functions, such as planning, problem-solving, and decision-making, reside. In contrast, the activity moves to other parts of the brain involved in emotion, pleasure, and empathy, which are closer to creativity than productivity.

Research also shows that living in a wooden house or studying in a wooden classroom reduces the heart rhythm (de Camino 2012); further research on the interaction, forest–wood–health–well-being, is thus strongly needed.

4.3.2 Consequences

Despite those differences, the global trend presented is nevertheless enough straightforward and significant to be considered as a lever of change and to deserve a specific analysis of possible consequences on forestry. This can be performed from three tracks: (1) from the decreasing role of the primary sector, which includes agriculture, forestry, fishing, and hunting; (2) from the development of a tertiary/postindustrial economy; and (3) from opportunities arising, thanks to the tertiarization of those primary activities through the creation, production, and exchange of services and amenities together or thanks to them.

Apart from economic impacts, which are obvious, the main consequences and challenges of the decreasing role of the primary sector, including forestry, are linked with the need for decompartmentalization, in particular, institutional one. In most countries, forest research institutions and academies, forest laws, ministries, forest services, and forest associations have been for decades conceived, structured, and organized almost entirely to protect forests and to secure sustained wood supply to meet domestic or external demand (Nair 2004) and some primitive and modest delivery of the other functions. While forest issues and concerns are opening from commodities to supply services, the challenge would be to expand and decompartmentalize institutional horizon and framework. Tertiarization is indeed not at all synonymous of slowing or even halting a material production, which is at the contrary still growing in volume but not proportionally in value. Moreover, services are often produced on the base of or thanks to tangible goods acting as a physical support for the production of services. Air travel requires planes, the internet needs computers, health care dispenses pharmaceuticals, and water production needs forests. This is the case for the forest sector that is still living and expected to grow, thanks to its supporting role in the production of new services, in particular, social and environmental ones. There is indeed a chance to link material use with other environmental co-benefits like water, soil, and biodiversity and cultural values, especially in the area of nonwood forest products. When trying to reach this goal, the adequacy of social, spatial, and institutional scales of concomitant production types remains a challenge.

Forest sector definition and scope should consequently expand through relevant integration of postindustrial forestry. Reports on the economic and social benefits of forests, such as the last State of the World's Forest published by FAO (2014), are welcome for arguing in such a way and induce required political awareness. With dematerialization of production, tertiary economy characteristics are offering a quite different environment and framework for action (Fuchs 1968). First, there is a shift from mass production to meet the needs of smaller and more specialized markets and an increased emphasis given to personalization and customization. Then, information and knowledge that are crucial inputs in production of services are becoming the critical resources conditioning costs (Nair 2004). Indeed, it is no longer the duration for producing a good which is affecting the production costs but the time for conceiving the service; the first unit produced is thus the more expensive, the second and successive one being less costly, even almost free. More generally, the information is much more expensive to develop than the physical content, which is then given to it.

Finally, producing services would require less natural capital and more human capital than producing agricultural or industrial goods. In the context of the transition toward a bio-based economy, for example, the move to high-added-value products (engineered wood products, specialty chemicals) would allow for the creation of more jobs and wealth in tertiary sectors for every unit of resource produced, thus alleviating the pressure on natural resources. This is in sharp contrast with low-added value, resource-intensive uses (i.e., co-firing biomass).

Tertiarization also concerns forestry itself and new opportunities are being created. In the context of bioeconomy, for example, new multistory wood-based buildings are creating most of the value through services (e.g., engineering, design, and research) and in related sectors (Näyhä et al. 2014). There is a vast understanding of the important role of forests in the production of environmental services, but we are still far from the moment when services, such as scenic beauty, water, biodiversity, and carbon, will be truly recognized by the markets. The environmental services of the forests are still positive externalities, subject to international conventions, but not massively recognized by the markets.

Moreover, the current legal frame is not adapted to a tertiary economy; it has indeed a strong primary rooting with civil law and some secondary elements. It is thus crucial to transform the economic frame of forest to unlock its tertiary potential.

Tourism, and in particular, green or rural tourism, where forest plays an important role, is growing all over the world. Initiatives are also taken due to a growing need for outdoor recreation and environmental facilities (Ammer and Pröbstl 1991), where niche markets are developing, particularly through forest-based tourism around towns and in the countryside (Mantau et al. 2001), which is responding to a growing need in all regions and continents (Mermet and Moquay 2002; Konijnendijk et al. 2005; Moigneu 2005; Nail 2006; Granet and Dobré 2009) (Box 4.2). In some poor regions, this implies a rapid and sometimes difficult cohabitation or hybridization between agrarian and postindustrial societies.

Tertiarization is also questioning the linearity of the production process that is often associated with material goods. In the case of forests, as in many other sectors, the primary, secondary, and tertiary sectors need to be connected as part of the same system and not as consecutive stages.

BOX 4.2 WALLONIA: STRATEGIC FRAMEWORK FOR TOURISM DEVELOPMENT (FROM BODSON ET AL. 2012)

Long sought for the production of wood and hunting, and more recently, for the protection or conservation of nature, the forest of Wallonia, a densely populated region (198 in./km^2) located in the southern and well-forested part of Belgium, is facing a rising social demand for recreation and tourism in forests. While considered away from social life for a long time, it is becoming an area associated with daily living space, just as the weekend is associated with the week: on a playful and secondary mode. Moreover, from a tourist point of view, the green field of Walloon forests is also a real asset since it is located near the major conurbations of west/center of northern Europe. Based on this finding, the region decided to try to anticipate the rising social pressure. The idea was to turn it into an opportunity for rural development while encouraging wood production and ensuring environmental functions of forests where appropriate. In this context, an interdisciplinary study was launched in 2008 and has laid out the foundations for a strategic framework for tourism development of Walloon forests.

As shown by the supply analysis, the Walloon forest, if it is widely used as a selling point on the tourist market, is in fact no more than pleasant scenery or beautiful backdrop for outdoor activities. While too often considered like a field made available, it seems to be time to move from such a *deposit logic* to a *product logic* that would be based on a structured, equipped, and marketed offer.

For contributing efficiently and sustainably to this component of rural development, the idea pursued is to design a comprehensive framework, to split the region in around ten territorial entities that are coherent from a tourist point of view, to set up common rules, and to frame common practices allowing the respect of the integrity of forest ecosystems. To achieve this purpose, each forest territory is organized into four areas distinguishing access modalities, in which the type of tourist activities is well defined. In the common frame designed, local actors—both public and private—are then calling to develop their creativity in implementing coordinated and collaborative projects and initiatives at the territorial level previously identified.

While now at the implementing stage through pilot projects, the projects have to face two main challenges. The first one concerns the organization of the Walloon administration, which is still sectorally compartmentalized, tourism and forest being in different departments. The second one is related to the "legitimate use" of the forests, which is a stumbling block for tourism development. Indeed, this legitimacy seems so far to be the preservation of traditional actors that are close to wood production and hunting and are still considered too often as symbolic owners of the forest, in spite of the access priority given by the new Forest Code adopted in 2008 to soft users, such as walkers and bikers. The inertia of those symbolic owners is contrasting with the dynamism of local actors.

A particular forest produces wood and nonwood products, supplies industries (sawmills, board plants, construction, furniture), and provides concrete services like water, pollinators, wildlife, and vulnerability reduction. A sector of activity is not replacing another, and a conscious balance of woods and services must be pursued by the management taking into account social context and concerns.

The motivation of the producers is then quite crucial, and studies are suggesting diverse schemes and drivers. We note two examples. Experiences are showing in rural areas of developing countries, such as in the Mayan Biosphere Reserve in Guatemala, that motivation for producing environmental services is higher when associated with grant of community concessions for wood production (de Camino and Breitling 2007; Villalobos et al. 2013). Moreover, studies are taking place in France to design incentives taking into account motivations of prosocial or altruistic order such as reputation or seeking for social prestige (Bénabou and Tirole 2006).

There are some key challenges like the transition phase that, for a small and weak sector, might be seen as a difficult burden to overcome. This very understandable, collective hesitation should be recognized and faced. The other challenge is the strong elasticity of forests due, on the one hand, to their geographical and temporary extension and high stocks, and, on the other hand, to the high number of stakeholders that benefit from them despite the small number that is requested or willing to return a part of the value added, the goods and services of forests provide to them.

4.4 GLOBALIZATION

4.4.1 Findings

Globalization is a controversial word having as many definitions as authors (Scholte 2000). The main definition comes from the economic sphere and refers to the global process of market integration resulting from the liberalization of trade and rising travel as well as exchange of people, goods, data, and information. Other definitions are embracing internationalization of concerns, such as the protection of the environment or are focusing on politics, culture, or laws, justifying pluralizing the word "globalization" (Rocher 2001).

For our purpose, globalization is understood through a political point of view and is considered as the extension at the global level of challenges previously limited to national or regional scale and the emergence of related governance arrangements. Those challenges are now also related to public goods and being of general interest. The Brundtland Report (1987) on Our Common Future calls for the integration of efforts toward sustainability all over the world while key focuses stated in the 1990s in the Rio Conventions, such as climate change, biological diversity, and desertification, are some good examples of forest-related concerns justifying such a global approach. More recently, a joint program established at the global level between UNESCO and CBD* intended to integrate biological and cultural diversity, both components where forests are playing a meaningful role.

Deliberately, this chapter does not deal with the impacts generated by market integration, liberalization of trade, and the growing demand for food, fiber, and feedstock that results from the combination of demographic growth and socioeconomic development. Evidently, this growing demand for material and immaterial uses of forests (and competing lands uses) will accelerate and multiply impacts and contradictions described in the following paragraphs.

This evolution toward globalization, which implies a less sharp boundary between public and private rights and responsibilities (internationally and nationally), is leading to huge changes in the decision-making process, policymaking, and policy implementation, and the word "governance" becomes a cornerstone of a new conceptual umbrella. Bas and Visseren-Hamakers (2012) refer to governing arrangement schemes for provisioning of public *goods* and solving public *bads*, distinguishing different roles for the state including its absence. According to Scholte (2000),

> "Authority has become increasingly "multilevel" or "multi-scalar" across substate (municipal and provincial) bodies and suprastate (macroregional and transworld) agencies as well as state organs. In

addition, various private sector and civil society actors have taken on regulatory roles. On this line of argument, states survive under globalization, but they are no longer the sole—and in some cases not even the principal—site of governance".

4.4.2 Consequences

A word of warning about globalization is that, in the case of forestry, the global system is practically (through the international conventions approved by UNCED in Rio in 1992) claiming rights and putting conditions to forests and the way to make forestry.[1] The risk exists in particular that globalization imposes patterns to developing countries that are postponing a forest development that is closer to their own livelihoods than to the ones of the effluent world. This is particularly the case in regions where populations are living inside the forests and/or are still depending on forests (Nke Ndih 2014). There is indeed a dilemma about whether to globalize or not to globalize as presented in the report of the Millennium Ecosystems Assessment (MEA 2005), which shows four possible sustainable development scenarios. The most promising scenario is the so-called Adapting Mosaic that corresponds to a landscape/territorial approach, thus a nonglobalized view, more community oriented.

While legal instruments and their respective control units are no longer seen as the only and most effective way to regulate human activities, such a trend implies a diversification of a process long considered linear through a progressive shift from a command-and-control regulation framework (Harrington and Morgenstern 2007) toward a broader one embracing a set of diversified decision-making actors and related policy instruments and processes.

A visible consequence is the continuous diversification of policy instruments dedicated to enhance the sustainable and multifunctional management of forests (Cubbage et al. 2007). Besides traditional regulation tools with regional scope such as the EU Timber Regulation, new types of policy instruments are emerging as, for example, instruments of intervention on private market like certification schemes and development of private/public market such as payment for environmental services (Box 4.3) or carbon offset payments. Innovative approaches like reflexive governance where policy process is considered from its social component and by the way of a shared problem construction and collective solutions are also experimented (Brousseau et al. 2012).

A direct implication can also be seen in the field of forest planning, which is an interface process between forest policy and forest management and which has been longtime formally conceived in a restrictive mode: for wood production, at ownership level, and for decades. News approaches are formally expanding its scope to other social demands, such as biodiversity, human health, or recreation. They are proposing to add a strategic regional level to try to balance different interests and give coherence between local initiatives. While the landscape approach (Sayer et al. 2013) is somehow familiar for users where agrarian societies, including the watershed management strongly linked with the Alpine and southern European areas since the beginning of professional forestry in the early 19th century, were prevailing as, for example, in the Mediterranean region, the concept is becoming widespread and even a prerequisite embedded in the expanding concept of strategic forest planning (Bettinger et al. 2009). New approaches also consider iterative and adaptive framework to correct and react in a continuous process depending upon the level of implementation and the changing context instead of waiting for decades as stated in the adopted forest plan (Lessard 1998). Stakeholders are more often directly implicated in the decision-making process and participatory approaches are becoming concepts and wording that foresters are progressively brought to tame (Buchy and Hoverman 2000). A good illustration of contradiction generated by this evolution is the military uniform paces worn by many forest service agents; heir to early days of forestry when the police mission was predominant to ensure a first sustainability, it is now less suitable for rising environmental mediation activities that focus on participation of stakeholders. Note, however, that this enlarged scope of activities is not the case everywhere, and in some regions, such as Italy or some part of Spain, there is a recent tendency to confine the forester in law enforcement activities.

BOX 4.3 COSTA RICA—PAYMENTS FOR ENVIRONMENTAL SERVICES (FROM PORRAS ET AL. 2013)

Costa Rica is a pioneer country in the implementation of Payment for Environmental Services (PES) schemes as an innovative tool included in its Forestry Law in 1996 for recovering and protecting its forests while contributing to international conservation agenda. The law also provided the institutional framework and the initial funds to start the process.

PES is a social negotiation market-based and voluntary approach based on a transaction for a well-defined ecological service where the buyer(s) pay(s) only if the provider(s) continue(s) to deliver the defined service over time (Wunder 2005). Four main services provided by ecosystems to people are recognized: carbon sequestration, protection of watersheds, biodiversity conservation, and preservation of scenic beauty. Practically, smallholders of forests are receiving fair and direct payments for undertaking management practices whose implementation is contributing to maintain environmental services that these forests provide to Costa Rican society and the world at large.

After decades of continuous deforestation with the forest cover rate falling to around 20% in the 1980s, a process of forest transition is now well consolidated. The current forest cover rate is over 50%.

> Between 1997 and 2012, Costa Rica protected more than 860,000 ha of forest, reforested 60,000 ha, and supported sustainable forest management in almost 30,000 ha. More recently, natural regeneration was promoted on almost 10,000 hectares. This totals nearly 1 million hectares under the PES scheme at one time or another as well as 4.4 million trees planted under agroforestry systems since 2003. This is a substantial achievement for a developing country of just 51,100 km^2. By 2010, roughly 52% of the territory was under some sort of forest cover: a great victory for the environmental movement, especially considering the 1983 low 21%.
>
> **(Klein et al. 2002 cited by Porras et al. 2013)**

But in parallel with PES, the law also defined the term "forest" and prohibited the change of use of areas with a natural forest cover. It was the interaction of the prohibition and the PES that allowed the restoration and conservation of the forest cover.

After 20 years, it was time to assess such a process and take lessons to analyze its impacts on people and its cost-effectiveness. Four main areas for improvement were identified: use of simple indicators for environmental impact; target of a specific social group, in particular, properties belonging to private individuals, areas of low social development index, and small properties (less than 50 ha); consideration of approaches to increase cost-effectiveness, in particular, payment levels better tuned to local costs; and definition of an impact evaluation tool for continuous monitoring.

4.5 CONCLUSION

While human societies are ever and always under changing contexts, some recent evolutions are having deep and long-lasting impacts on the relationship between humans and forests. The subject is large and complex. The chapter is not intended to be exhaustive nor thorough. By the way of a quick outline of three main global processes, that is, urbanization, tertiarization, and globalization, the intention was to show how usual boarders and traditional comfort zones of foresters are obliging those professionals to review or upgrade some basic principles. A more extended definition of globalization than the one used here that would consider it in its economic component, together with demographic and socioeconomic changes, would make even more relevant and urgent, the challenges identified in the chapter.

The chapter is an introduction to a large topic that would for sure deserve further and more in-depth development. Among the many points raised, we see useful to highlight a few that seem to be particularly crucial and that are considered as prerequisite to move forward and toward a progressive adaptation of forestry to the current and coming challenges.

An important further future step would be to undertake a semantic work of clarification and mapping of concepts to better connect forest-related disciplines and sectors between themselves and with other non-forest-related disciplines. Forests have been long time seen for its functions, afterward through the goods and services provided, and recently by the way of ecosystem or even environmental service benefiting to the society. There is often confusion that is generating misunderstanding and also difficulties to maintain coherence in the actions.

From an institutional point of view, while intersectoral links are more needed than ever to respond to the rising diversity of social needs, the forest sector has also to be reinforced due to the growing importance of its output in combating climate change and contributing to the development of a green/bioeconomy. Forests and forest-dependent people would thus benefit from a complementary binary approach combining better networking together with effective reinforcement of the sector.

As established by Nair (2004), forestry, which has been mainly monolithically wood oriented for a long time, has to consider different social contexts, such as forest-based society, rural society, industrial society, and postindustrial society including hybrid situations, where forests are playing quite different roles in the economy. Those distinct socioeconomic situations not only require specific innovative approaches and tools but also distinguish spatial and social scales and arrangements, provided that those approaches also mean that forests have economic value for improving the livelihoods of local people.

In such a context and looking to the various controversies still persisting around forest issues and expressing divergent views, it would be appropriate to try innovative decision-making processes where the ethical and cultural context would be made explicit (Huybens 2011) instead of remaining largely encrypted for the benefit of discourses based only on science and economics (Honadle 2000).

While in some contexts, foresters are still responsible for the management of the ecosystem itself, in an increasing number of others, the responsibility is moving to the management of the relationships between humans and the ecosystem (Kennedy and Koch 2004). There is indeed a shift in focus from “what to do in the forests” to a focus on “how, who” on the social and economic frame that makes management possible, and this implies participatory and bottom-up approaches. This also requires quite new skills and competences, in particular, in human and social sciences and also another positioning toward the society; while the challenge of the first one is to implement the best technical tools, the second one is to be able to solve concrete problems encountered by the society. This needs in any case a sound technical and scientific knowledge and evidence basis.

Finally, difficult communication between humans themselves on forest issues remains a stumbling block too frequent that needs to be better managed, even solved. A forest communication that takes on board the complexity of forest issues, and that relies on media skills, changes in a relevant way the address from forests to an indirect path through media and decision-makers and society as a tool to ensure favorable frame conditions for a sustainable forestry.

4.6 ACKNOWLEDGMENT

This chapter was prepared thanks to the support of the Walloon minister responsible for forest. The chapter is dedicated to our colleague and friend Franz Schmithüsen, Emeritus Professor at ETH Zurich, who passed away in Baden (Switzerland) on April 14, 2015.

NOTE

1 http://www.unesco.org/new/en/natural-sciences/special-themes/biodiversity-initiative/biodiversity-culture/unesco-cbd-joint-programme/.

REFERENCES

Álvarez-Martínez, J.M., Suárez-Seoane, S., Stoorvogel, J.J., and E. de Luis Calabuig. 2014. Influence of land use and climate on recent forest expansion: A case study in the Eurosiberian—Mediterranean limit of north-west Spain. *Journal of Ecology* 102(4):905–919.

Ammer, U. and U. Pröbstl. 1991. *Freizeit und Natur*. Hamburg, Germany: Verlag Paul Parey.

Annerstedt, M., Norman, J., Boman, M., Mattsson, L., Grahn, P., and P. Währborg. 2010. Finding stress relief in a forest. *Ecological Bulletins* 53:33–42.

Arnberger, A. 2012. Urban densification and recreational quality of public urban green spaces—A Viennese Case Study. *Sustainability* 4(4):703–720.

Barre, R. 1964. *Economie Politique*. Paris, France: PUF, Coll. Thémis.

Bas, A. and I.J. Visseren-Hamakers. 2012. Forest governance: Mainstream and criticals views. In *Moving Toward with Forest Governance*, eds. G. Broekhoven, H. Savenije, H., and S. von Scheliha, pp. 3–10. Wageningen, the Netherlands: Tropenbos International.

Bell, D. 1973. *Coming of Post-Industrial Society*. New York: Basic Books.

Bénabou, R. and J. Tirole. 2006. Incentives and prosocial behavior. *American Economic Review* 96(5):1652–1678.

Bettinger, P., Boston, K., Siry, J.P., and D.L. Grebner. 2009. *Forest Management and Planning*. Burlington, VT: Elsevier/Academic Press.

Bobrowski, R. and G. Biondi. 2012. Characterization of planting pattern adopted for urban afforestation in the city of Curitiba, Parana state, Brazil. *REVSBAU Soc. Bras. de Arborização Urbana Piracicaba* 7(3):18–27.

Bodson, D. 2005. Comprendre les perceptions, les usages et les significations de la forêt en 2005. *Forêt Wallonne* 79:19–28.

Bodson, D., Farcy, C., and A. Robert. 2012. *Strategic framework for tourist valorization of forests in Wallonia (Belgium): From conception to implementation*. Paper presented at the IUFRO Conference Forest for People, 22–24/05/2012, Alpach, Austria.

Bonnet, J. 2014. Ville—Le fait urbain dans le monde. Encyclopædia Universalis. www.universalis.fr/encyclopedie/ville-le-fait-urbain-dans-le-monde/ (accessed August 14, 2014).

Boude, J.P. and C. Chaboud. 1995. Le concept de ressource naturelle en économie. In *Actes du colloque Forum halieumétrique. Les recherches françaises en évaluation quantitative et modélisation des ressources et des systèmes halieutiques*, eds. D. Gascuel, and J.L. Durand, pp. 269–281. Rennes, France: Orstom.

Brisson, R. 2011. *Utilisation des plantes par les Pygmées Baka*. Paris, France: L'Harmattan.

Brousseau, E., Dedeurwaerdere, T., and B. Siebenhüner. 2012. *Reflexive Governance for Global Public Goods*. Cambridge, MA: MIT Press.

Bruntland, G.H. 1987. *Our Common Future*. New York: Commission mondiale sur l'environnement et le développement de l'Organisation des Nations unies.

Buchy, M. and S. Hoverman. 2000. Understanding public participation in forest planning: A Review. *Forest Policy and Economics* 1:15–25.

Chao, L. 2012. *Forests People: Numbers across the World*. Moreton-in-Marsh, UK: Forest Peoples Programme.

City of Melbourne. 2012. Urban forest strategy. Making a great city greener 2012–2032. www.melbourne.vic.gov.au/Sustainability/UrbanForest/Documents/Urban_Forest_Strategy.pdf (accessed August 12, 2014).

Clark, C. 1940. *The Conditions of Economic Progress*. London, UK: Macmillan.

Cubbage, F., Harou, P., and E. Sills. 2007. Policy instruments to enhance multi-functional forest management. *Forest Policy and Economics* 9:833–851.

de Camino, R. 2012. Es probable que el uso de madera en construcciones nos alargue la vida. <ins cite=mailto: pg3459datetime=2015-07-04T22:56>http://finnfor.catie.ac.cr/categories/categoria_ 1329336588 (accessed April 10, 2015).

de Camino, R. and J. Breitling. 2007. *El cambio es posible: 20 años de experiencias innovadores en los recursos naturales en Guatemala*. San José, CA: University for Peace, Producciones ALPI.

Devictor, V., Julliard, R., Couvet. D., Lee, A., and F. Jiguet. 2007. Functional homogenization effect of urbanization on bird communities. *Conservation Biology* 21:741–751.

Dieterich, V. 1953. *Forstwirtschaftspolitik. Eine Einführung*. Hamburg, Germany: Paul Parey.

Elfring, T. 1989. New evidence on the expansion of service employment in advanced economies. *Review of Income and Wealth* 35:409–440.

European Commission (EC). 2009. *Shaping Forest Communication in the European Union: Public Perceptions of Forests and Forestry*. Brussels, Belgium: DG Agriculture and Rural Development.

European Commission (EC). 2011. EU Forest Communication Strategy. http://ec.europa.eu/agriculture/fore/publi/communication-strategy_en.pdf (accessed August 14, 2014).

European Commission (EC). 2012. Innovating for Sustainable Growth: A Bioeconomy for Europe. http://ec.europa.eu/research/bioeconomy/pdf/official-strategy_en.pdf (accessed August 12, 2014).

Fabra-Crespo, M. and E. Rojas-Briales. 2015. Comparative analysis on the communication strategies of the forest owners' associations in Europe. *Forest Policy and Economics* 50:20–30.

Farcy, C. and F. Devillez. 2005. New orientations of forest management planning from an historical perspective of the relations between man and nature. *Forest Policy and Economics* 7:85–95.

Fastrez, P., Meurice, C., De Smedt, T., and J. Matagne. 2013. Assessing the cognitive autonomy of audiences towards environmental media messages. *Proceedings of COCE 2013: Conference on Communication and Environment*, Uppsala, Sweden: IECA.

Fisher, A.G.B. 1939. Production, primary, secondary and tertiary. *Economic Record* 15:24–38.

Food and Agriculture Organization (FAO). 2011. *State of the World's Forests 2011*. Rome, Italy: FAO.

Food and Agriculture Organization (FAO). 2014. *State of the World's Forests 2014*. Rome, Italy: FAO.

Forestry Commission Scotland. 2008. Forests for people access, recreation & tourism on the national forest estate. www.forestry.gov.uk/pdf/fcfc114.pdf/$FILE/fcfc114.pdf (accessed August 12, 2014).

Forrest, M. and C. Konijnendijk. 2005. A history of urban forests and trees in Europe. In *Urban Forests and Trees*, eds. C.C. Konijnendijk, K. Nilsson, T.B. Randrup, and J. Schipperijn, pp. 23–48. Heidelberg, Germany: Springer.

Fourastié, J. 1958. *Le grand espoir du XXème siècle. Progrès technique, progrès économique, progrès social.* Paris, France: PUF.

Fuchs, V.R. 1968. *The Service Economy*. New York: Colombia University Press.

Granet, A.M. and M. Dobré. 2009. Les citadins et la forêt en France. *Revue Forestière Française* 61: 521–534.

Gubbels, J.S., Kremers, S.P.J., van Kann, D.H.H., Stafleu, A., Candel, M.J., and P.C. Dagnelie. 2011. Interaction between physical environment, social environment, and child characteristics in determining physical activity at child care. *Health Psychology* 30(1):84–90.

Harrington, W. and R.D. Morgenstern. 2007. Economic incentives versus command and control: What's the best approach for solving environmental problems? In *Acid in the Environment: Lessons Learned and Future Prospects*, eds. G.R. Visgilio and D.M. Whitelaw, pp. 233–240. New York: Springer.

Haut Commissariat aux Eaux et forêts et à la Lutte Contre la Désertification (HCEFLCD). 2010. Guide des forêts urbaines et périurbaines. www.eauxetforets.gov.ma/admin/telechargement/fr/Guide_For%C3%AAts_Urbaines_HCEFLCD.pdf (accessed August 14, 2014).

Honadle, G. 2000. *How Context Matters: Linking Environmental Policy to People and Place*. Bloomfield, NJ: Kumarian Press.

Huffel, G. 1926. *Les Méthodes de l'aménagement forestier en France, étude historique*. Nancy, France: Berger-Levrault.

Hunziker, M., Frick, J., Bauer, N., von Lindern, E., and O. Graf. 2013. *La population suisse et sa forêt. Rapport relatif à la deuxième enquête menée dans le cadre du monitoring socioculturel des forêts (WaMos2).* Berne, Switzerland: OFEV et WSL.

Huybens, N. 2011. Comprendre les aspects éthiques et symboliques de la controverse socio- environnementale sur la forêt boréale du Québec. *Vertigo* 11(2), http://vertigo.revues. org/11119.

Intergovernmental Panel on Climate Change (IPCC). 2011. *IPCC Special Report on Renewable Energy Sources and Climate Change Mitigation. Prepared by Working Group III of the Intergovernmental Panel on Climate Change*. New York: Cambridge University Press.

International Union of Forest Research organizations (IUFRO). 2014. *Manual of Public Relations in Forest Sciences*. Vienna, Austria: IUFRO.

Johnston, M. 1996. A brief history of urban forestry in the United States. *Arboricultural Journal* 20:257–278.

Juster, F.T., Ono, H., and F.P. Stafford. 2004. Changing time of American Youth: 1981–2003. Research report, Ann Arbor, MI: Institute for Social Research/University of Michigan.

Kant, S. 2003. Extending the boundaries of forest economics. *Forest Policy and Economics* 5:39–56.

Karjalainen, E., Sarjala, T., and H. Raitio. 2010. Promoting human health through forests: Overview and major challenges. *Environmental Health Prevention Medicine* 15:1–8.

Kennedy, J.J. and N.E. Koch. 2004. Viewing and managing natural resources as human-ecosystem relationships. *Forest Policy and Economics* 6:497–504.

Kloek, M.A., Buijs, A.E., Boersema, J.J., and M.G.C. Schouten. 2013. Crossing Borders: Review of Concepts and Approaches in Research on Greenspace, Immigration and Society in Northwest European Countries. *Landscape Research* 38:117–140.

Konijnendijk, C.C., Nilsson, K., Randrup, T.B., and J. Schipperijn. 2005. *Urban Forests and Trees*. Heidelberg, Germany: Springer.

Korean Forest Service (KFS). 2015. Therapy forests. http://english.forest.go.kr/newkfsweb/html/EngHtmlPage.do?pg=/esh/policy/UI_KFS_0102_040600.html&mn=ENG_02_04_06 (accessed April 10, 2015).

Lescuyer, G., Eba'a Atyi, R., and P. Cerutti, P. October 2009. National consumption of timber in Central Africa: A major challenge for sustainable forest management. Paper presented at *the World Forest Congress*, pp. 18–23, Buenos Aires, Argentina.

Lessard, G. 1998. An adaptive approach to planning and decision-making. *Landscape and Urban Planning* 40:81–87.

Li, Q. 2010. Effect of forest bathing trips on human immune function. *Environmental Health Prevention Medicine* 15:9–17.

Lisch, R. 2014. *Measuring Service Performance: Practical Research for Better Quality*. Burlington, VT: Gower Applied Business Research.

Mantau, U., M. Merlo, W. Sekot, and B. Welcker. 2001. *Recreational and Environmental Markets for Forest Enterprises*. Wallingford, CT: CABI.

Mermet, L. and P. Moquay. 2002. *Accès du public aux espaces naturels: Outils d'analyse et méthodes de gestion*. Paris, France: Hermès Science Publications, Lavoisier.

Michel, M. and L. Gil. 2013. *La transformación histórica del paisaje forestal en la Comunidad Autónoma de Euskadi*, p. 18. Vitoria, Spain: Servicio de publicaciones del Gobierno Vasco, Colección LUR.

Millennium Ecosystem Assessment (MEA). 2005. *Ecosystems and Human Well-Being: Synthesis*. Washington, DC: Island Press.

Mitchell, R. and F. Popham. 2008. Effect of exposure to natural environment on health inequalities: An observational population study. *The Lancet* 372(9650):1655–1660.

Moigneu, T. 2005. *Gérer les forêts périurbaines*. Paris, France: Office National des Forêts.

Morita, E., Fukuda, S., Nagano, J., Hamajima, N., Yamamoto, H., Iwai, Y., Nakashima, T., Ohira, H., and T. Shirakawa. 2007. Psychological effects of forest environments on healthy adults: Shinrin-yoku (forest-air bathing, walking) as a possible method of stress reduction. *Public Health* 121:54–63.

Motte-Florac, E. 1980. *Les Plantes chez les Pygmées Aka et les Monzombo de la Lobaye (Centrafrique)*. Paris, France: Société d'Etudes Linguistiques et Anthropologiques de France.

Nail, S. 2006. *Bosques urbanos en América Latina: Usos, funciones, representaciones*. Bogota, Colombia: Universidad Externado de Colombia.

Nair, C.T.S. 2004. What does the future hold for forestry education? *Unasylva* 216(55):3–9.

Näyhä, A., Pelli, P., and L. Hetemäki. 2014. Forest-based services outlook. In *Future of the European Forest-Based Sector: Structural Changes Towards Bioeconomy*, ed. L. Haternaki, pp. 55–62. Joensuu, Finland: What Science Can Tell Us 6, European Forests Institute.

Nilsson, K., Sangster, M., Gallis, C., Hartig, T., de Vries, S., Seeland, K., and J. Schipperijn. 2011. *Forests, Trees and Human Health*. Heidelberg, Germany: Springer.

Nke Ndih, J. 2014. Gestion des écosystèmes forestiers par les Pygmées Bakola/Bagielli et voisins Bantu au Sud-Ouest du Cameroun face à l'exploitation néolibérale, Thèse de doctorat, Université catholique de Louvain (UCL), Louvain, Belgium.

Nowak, D.J. and J.T. Walton. 2005. Projected urban growth (2000–2050) and its estimated impact on the U.S. forest resource. *Journal of Forestry* 103(8):383–389.

Ormerod, P. 1997. *The Death of Economics*. Wiley, New York.

Park, B.J., Tsunetsugu, Y., Kasetani, T., Kagawa, T., and Y. Miyazaki. 2010. The physiological effects of Shinrin-yoku (taking in the forest atmosphere or forest bathing): Evidence from field experiments in 24 forests across Japan. *Environmental Health Prevention Medicine* 15(1):18–26.

Pauchard, A., Aguayo, M., Peña, E., and R. Urrutia. 2006. Multiple effects of urbanization on the biodiversity of developing countries: The case of a fast-growing metropolitan area (Concepción, Chile). *Biological Conservation* 127:272–281.

Pergams, O.R. and P.A. Zaradic. 2008. Evidence for a fundamental and pervasive shift away from nature-based recreation. *Proceedings of the National Academy of Sciences of the United States of America* 105(7):2295–2300.

Piketty, T. 2013. *Le Capital au XXIème siècle*. Paris, France: Le Seuil.

Plaisance, G. 1979. *La forêt française*. Paris, France: Denoël.

Porras, I., Barton, D.N., Miranda, M., and A. Chacón-Cascante. 2013. *Learning from 20 Years of Payments for Ecosystem Services in Costa Rica*. London, UK: International Institute for Environment and Development.

Rametsteiner, E., Eichler, L., and J. Berg. 2009. *Shaping Forest Communication in the European Union: Public Perceptions of Forests and Forestry*. Rotterdam, the Netherlands: Ecorys Research & Consulting.

Rametsteiner, E. and F. Kraxner. 2003. *Europeans and Their Forests: What Do Europeans Think About Forests and Sustainable Forest Management?* Vienna, Austria: FAO/UNECE Forest Communicators Network.

Rocher, G. 2001. La mondialisation: Un phénomène pluriel. In *Une société-monde? Les dynamiques sociales de la mondialisation*, ed. D. Mercure, pp. 17–21. Montréal, Québec, Canada: Presses de l'Université Laval et De Boeck Université.

Rupf, H. 1960. Wald und Mensch im Geschehen der Gegenwart. *Allgemeine Forstzeitschrift* 38:545–552.

Sayer, J., Sunderland, T., Ghazoul, J., Pfund, J-L., Sheil, D., Meijaard, E., Venter, M. et al. 2013. Ten principles for a landscape approach to reconciling agriculture, conservation, and other competing land uses. *Proceedings of the National Academy of Sciences of United States of America* 110(21):8349–8356.

Schmithüsen, F. 2013. Three hundred years of applied sustainability in forestry. *Unasylva* 240(64):3–11.

Scholte, J.A. 2000. *Globalization: A Critical Introduction*. New York: Palgrave Mac Millan.

Scott, A., Gilbert, A., and A. Gelan. 2007. *The Urban-Rural Divide: Myth or Reality*? Policy Brief Series, No 2. Aberdeen, UK: The Macaulay Institute.

Seto, K.C., Fragkias, M., Güneralp, B., and M.K. Reilly. 2011. A meta-analysis of global urban land expansion. *PLoS ONE* 6(8):e23777. doi:10.1371/journal.pone.0023777.

Shen, W., Wu, J., Grimm, N.B., and D. Hope Shen. 2008. Effects of urbanization-induced environmental changes on ecosystem functioning in the Phoenix metropolitan region, USA. *Ecosystems* 11:138–155.

Timmer, P. and S. Akkus. 2008. The structural transformation as a pathway out of poverty: Analytics, empirics and politics. Working Paper, Number 150, Centre for Global Development. http://www.cgdev.org/publication/structural-transformation-pathway-out-poverty-analytics-empirics-and-politics-working (accessed September 15, 2014).

Ulrich, R.S. 1984. View through a window may influence recovery from surgery. *Science* 224:420–421.

United Nations (UN). 2014. World Urbanization Prospects: The 2014 Revision, Highlights/United Nations, Department of Economic and Social Affairs, Population Division (ST/ESA/SER.A/352). http://esa.un.org/unpd/wup/ (accessed August 12, 2014).

Villalobos, R., Carrera, F., de Camino, R., Morales, J., and W. Flores. 2013. Construcción de cultura forestal para el desarrollo. In *Escalando innovaciones rurales*, eds. Á. Paz, M. Paz Montoya, and R.H. Asensio, pp. 209–233. Lima, Peru: Estudios de la Sociedad Rural 43.

Weber, J., Betsch, J.-M., and P. Cury. 1990. A l'interface Hommes-Nature: Les ressources renouvelables. Papier présenté *au Colloque Recherche et Environnement*, pp. 24–25. Strasbourg, France: CNRS Programme Environnement.

World Bank (WB). 2014. World development indicators. http://data.worldbank.org/indicator (accessed August 14, 2014).

Wunder, S. 2005. *Payments for Environmental Services: Some Nuts and Bolts*. Occasional Paper No. 42. Jarkarta, Indonesia: CIFOR.

5 Ecosystem Services in Ecological Forest Management

Sarah J. King, Susan Preston, Christian Malouin, Ryan Burke, Remington Bracher, and Niya Nijat

5.1 INTRODUCTION

Humans are embedded within nature. People need healthy natural ecosystems to survive and thrive. Often, many people, including managers and decision-makers, take nature's benefits for granted, treating the natural world as an infinite source of resources and a limitless sink of pollution. As the non-negotiable limits of ecosystems on the planet become irrefutably clear, scientists have sought ways to assess and communicate human reliance on nature and, in this context, have developed the idea of ecosystem services. Ecosystem services, or nature's benefits to humans, "support life (by providing air, water, food, raw materials, medicines), security (e.g., by mitigating extreme weather events, spread of vector-borne diseases) and quality of life (e.g., by supporting mental and physical health, cultural identity, recreation), among many other things" (Value of Nature to Canadians Study Taskforce (VNCST) 2017). Put another way, ecosystem services demonstrate that humans are utterly dependent on ecosystems for life, security, and well-being.

This chapter explores the idea of ecosystem services; outlines key conversations, tools, and approaches to assessing ecosystem services; and identifies current challenges and limitations. Section 5.2 begins with a *case study* of the 2023 wildfire season to illustrate the complexity of ecosystem services and human impacts on them, outlines how the *idea of ecosystem services* has developed, profiles the leading global initiative for scientific assessment of biodiversity and ecosystem services *IPBES*, and explores the connections between *biodiversity, ecosystem integrity, and ecosystem services*. Section 5.3 outlines the complexity of *socio-ecological systems* using the IPBES model and explores the critical relevance of *Indigenous knowledge*. Section 5.4 argues for an equitable, interdisciplinary approach to the assessment of ecosystem services, profiling the *Ecosystem Services Toolkit* and the various sub-fields of integrated assessment (*Biophysical Assessment, Social Assessment*, and *Economic Assessment*) including challenges and limitations to specific methodologies.

For foresters, conservation biologists, and all who are engaged in ecological forest management, the idea of ecosystem services provides an important tool for understanding the complex ways that humans are embedded within and dependent upon the natural systems they work to protect. Equitable and interdisciplinary approaches to the assessment of ecosystem services are critical to the development of accurate information for decision-making. These approaches are increasingly possible because of the proliferation of new research, tools, and models to support environmental professionals, many of which are discussed here.

5.2 UNDERSTANDING ECOSYSTEM SERVICES

5.2.1 Case Study: 2023 Wildfire Season

In the summer of 2023, unprecedented wildfires burned across Canada, affecting local communities and regional economies and generating unforeseen air pollution across the continent. On a national scale, 17.6 million hectares burned in 2023, a season that vastly surpassed the previous

DOI: 10.1201/9781003431084-6

worst documented wildfire season of 1995, when 7.1 million hectares burned (CIFFC 2023). The impacts of even a single wildfire can be immense; the unprecedented effects of the 2023 wildfire season illustrate the complex ways that humans rely on and benefit from forests. In 2023, the local impact of large wildfires was enormous, as people were evacuated, communities burned, businesses were destroyed, and timber was lost. In previous years, these effects seemed significant only to those who live close to a fire. The 2023 fires were on such a large scale that they had a broad impact. For example, they caused air pollution across the continent with profound effects on human health far from the fire, impacting those with respiratory vulnerabilities, straining hospitals, causing deaths, and limiting the ability of most people to participate in daily activities. Even among those who were unaffected by pollution from wildfire smoke, research shows that news of these events had significant mental health impacts (Lowe and Garfin 2023). While forest fires cause enormous local destruction, the 2023 fires reinforce that their impact on people is far greater than this, disrupting human lives in ecological, sociocultural, and economic terms. The benefits that forests provide to humans are far more important than those typically ascribed to a "healthy natural resource." Forests are essential to the survival of humans in a broad range of less tangible ways: they provide clean air, water, species habitat, and immense biodiversity; they support human physical and mental health, meaningful livelihoods, and cultural practices like hiking and hunting; they are the foundation of human economies and trade practices.

Wildfires have increased in both frequency and magnitude in several forest ecosystem types because of increasing anthropogenic pressures and human-induced environmental change over generations. In the Canadian boreal forest, high-severity fires are a natural part of the ecosystem (Hall et al. 2008; Soverel et al. 2011; Whitman et al. 2018), and resident propagule banks are usually unaffected if they have enough time to replenish between fire intervals (Boucher et al. 2020; Whitman et al. 2019). Often, healthy fire regimes were supported by Indigenous cultural burning practices, which varied widely depending on the characteristics of different regions (Christianson et al. 2022). Indigenous fire management ensured fuel loads remained low to limit subsequent fire intensity (Lake and Christianson 2019). These practices also increased the fire-resistant vegetation, increasing the biodiversity and resilience of the ecosystem to fire disturbance (Mistry et al. 2016). However, during colonization of North America, the displacement of Indigenous communities from their lands, settler fear and misunderstanding of fire, and the colonial prioritization of timber value and the logging economy led to centuries of active fire suppression that altered fire regimes and contributed to recent fire emergencies.

The impact of wildfires extends far beyond the fire season, as increasingly frequent and more powerful fires change the ecosystems upon which humans rely. The lengthening of the wildfire season and increased likelihood of extreme fire weather due to climate change, coupled with ongoing fire suppression (Gavin et al. 2007), mean that fires are not only increasing in severity and frequency but are increasingly changing the ecological composition of forest ecosystems. Traditionally, in the boreal forest of North America, black spruce has influenced forest flammability and carbon storage for thousands of years (Baltzer et al. 2021). However, climate change and increased wildfires have resulted in a shift in forest dominance to aspen and, in some cases, grassland ecosystem types, leading to potential shifts in ecosystem dynamics (Baltzer et al. 2021). Data from across North America shows that while resilience is the most common outcome, drier conditions and severe fires often lead to complete regeneration failure, with potential consequences for this crucial biome's functioning, as warming and drying persist (Baltzer et al. 2021). Even in the more sensitive Rocky Mountain ecosystems found in British Columbia and Alberta, there is an increase in fire severity and intervals comparative to historic levels (Steel et al. 2015; Abatzoglou and Williams 2016). This could result in newly emerging fire regimes that will have significant implications for ecosystems and the humans who depend on them for their ecological, sociocultural, and economic well-being (Higuera et al. 2021).

As the wildfire season of 2023 makes clear, nature's benefits to humans are broad and complex, and humans rely fundamentally on the natural world for survival. When air is polluted by the burning of the forests that normally purify it, human lives are threatened, and mental and physical health is at risk. Further, the ecological impacts of fires are compounded as fires change ecosystems, with profound potential impacts for people. Ecosystem services offer a way to articulate human dependence on nature, assess nature's benefits to people, and support people in addressing human-induced environmental change.

5.2.2 The Idea of Ecosystem Services

The recognition that nature supports human beings in many ways, including through the provision of services, is not new (Brauman et al. 2007). Leopold (1949) highlighted human reliance on natural systems and the importance of their maintenance as early as the 1940s. Moreover, Alexander von Humboldt recognized the fundamental intertwining of humanity and the natural world in the 1700s (Wulf 2016). In fact, Humboldt and Leopold were articulating for a Western audience something that Indigenous and place-based knowledge systems had recognized much earlier: that humans are an inextricable part of natural systems. The Anthropocene discussion (Steffen 2015; Ellis 2018) focuses attention on that observation: humanity and nature are deeply intertwined. Humans have been influencing the natural world at local, regional, and perhaps global scales for many millennia. The changing natural world has had reciprocal influences on human welfare and activities, though with a worrying time lag due to nature's capacity to absorb and buffer impacts until these capacities are overwhelmed.

Beginning in the 1970s (e.g., Holdren and Ehrlich 1974), the idea of Ecosystem Services gained traction as a Western framework for articulating the ways in which nature supports humans in all aspects of their lives. Over time, this idea has matured and has been applied in diverse fields of analysis, such as ecosystem accounting, trade-off/scenario assessment, forest certification (Meijaard et al. 2014), environmental impact assessment (e.g., NAS 2013), integrated ecosystem assessment, risk assessment, costs of climate change, and payment for ecosystem services (PES) policies (Daily and Matson 2008; Naeem et al. 2015). The ecosystem services cascade model, first proposed by Haines-Young and Potschin (2010) and used/updated widely since that time, demonstrates how ecosystem services arise from ecosystems and become important to human well-being. The Value of Nature to Canadians Study Task Force refined this model as a central component of their *Ecosystem Services Toolkit* (VNCST 2017) (Figure 5.1).

This model makes clear that ecosystem services are possible because of the foundational biophysical structures and processes of ecosystems (van Oudenhoven et al. 2012). These structures and

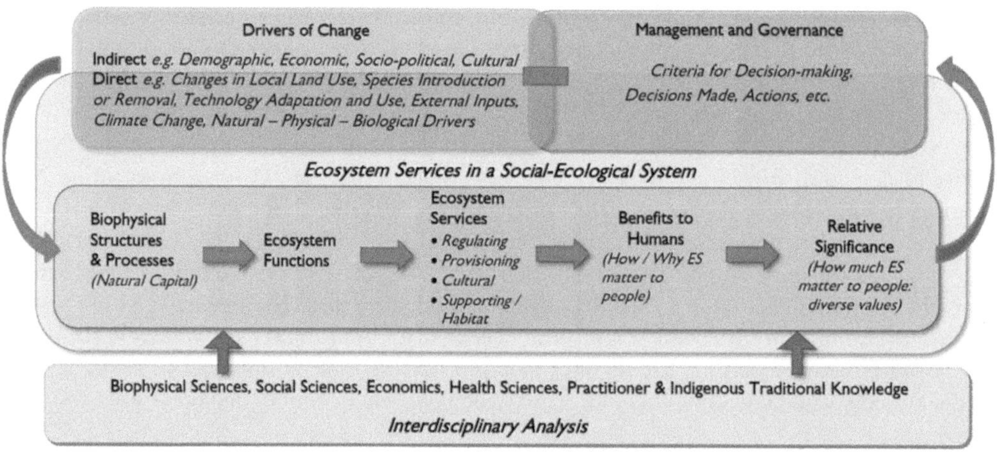

FIGURE 5.1 Ecosystem Services Toolkit Conceptual and Analytical Framework. (VNCST 2017.)

processes give rise to ecosystem functions. Ecosystem processes and functions are often mediated by human interventions that can contribute to the production or reduction of ecosystem services. These services provide benefits to humans, and the services and benefits have significance to human well-being (VNCST 2017).

Experts have articulated specific typologies for ecosystem services that make it feasible to identify, assess, and make decisions around individual services. While there are some variations in the list of service types, a typical list includes approximately 30 services grouped under four classes: provisioning (producing material goods), regulating (moderating ecosystem conditions), cultural (informing mental and physical health, identity and development), and supporting or habitat (the underlying processes necessary for all others).

Ecosystem integrity and resilience are essential to the production of ecosystem services and of sustainable forest management. Biodiversity is the cornerstone of ecosystem integrity and therefore of ecosystem services. In 2022, Parties to the Convention on Biological Diversity (CBD) adopted a new Global Biodiversity Framework to 2030. This framework builds on the Convention's previous Strategic Plans, with a vision to achieve a world living in harmony with nature by 2050. Its 23 targets explicitly address the need for change, for example, to achieve genuinely sustainable use of natural resources, elimination of subsidies harmful to biodiversity, and ensure the resilience of ecosystems and their capacities to support biodiversity and human well-being.

5.2.3 Key Initiative: IPBES

The Intergovernmental Science-Policy Platform on Biodiversity and Ecosystem Services (IPBES) was established in 2012 as an intergovernmental initiative to build knowledge and strengthen the interface between science and policy on biodiversity and ecosystem services. IPBES consists of close to 140 member States that convene in a Plenary to commit to the independent expert completion of scientific assessments. While it is not strictly a UN body, it is closely aligned with the UNEP, UNFCCC, and the Parties to the UN Convention on Biological Diversity. Member States joined in the creation of IPBES to build capacity to take meaningful action in light of the findings from the UN Millennium Ecosystem Assessment and their commitments to the conservation and sustainable use of biodiversity under the CBD. Engaging the independent expertise of scientists and knowledge holders, IPBES performs regular assessments of knowledge on biodiversity and ecosystem services on a regional or global scale that help support decision-makers in integrating science into policy (IPBES 2013, 2019, 2022; Schmeller et al. 2017). IPBES plays a crucial role in shaping global policy agreements, developing frameworks related to biodiversity and ecosystem services, and advancing understanding of the interrelationships between biodiversity, ecosystem services, and human well-being.

IPBES assessment teams are painstakingly assembled based on broad interdisciplinary representation so that they have the expertise to bridge natural and social sciences with policy, thereby advancing its core mandate to halt biodiversity loss and its attendant detrimental impacts on human societies. By publishing scientific assessments and reports, such as the *Methodological assessment regarding the diverse conceptualization of multiple values of nature and its benefits, including biodiversity and ecosystem functions and services* (also known as the *Values Assessment* 2022), IPBES provides policymakers with valuable guidance, incorporating comprehensive scientific information, analysis, and policy recommendations. The ground-breaking IPBES *Global Assessment Report on Biodiversity and Ecosystem Services* (2019)—assessing the well-being of the variety of life on Earth at all levels, from genes to ecosystems—was delivered to the public with a call for transformative change. According to that assessment, overall trends in health and functioning of species and ecosystems and the Earth's support systems are overwhelmingly negative and permeate the world everywhere. These deteriorating trends have been accelerating and intensifying since about the 1970s, at a remarkable pace and rate of change relative to the last ten million years and are projected to continue or worsen into the future under business-as-usual scenarios (Ray et al. 2021). The first key message of IPBES's *Values Assessment* outlines this clearly:

> KM1. The causes of the global biodiversity crisis and the opportunities to address them are tightly linked to the ways in which nature is valued in political and economic decisions at all levels.
>
> Unprecedented climate change and decline of biodiversity are affecting ecosystem functioning and negatively impacting people's quality of life. An important driver of the global decline of biodiversity is the unsustainable use of nature, including persistent inequalities between and within countries, emanating from predominant political and economic decisions based on a narrow set of values (e.g., prioritizing nature's values as traded in markets). Simultaneously, access to and distribution of the benefits from nature's many contributions to people are highly inequitable. Yet, a consolidated global consensus reflected by the 2030 Agenda for Sustainable Development and the 2050 Vision for Biodiversity has established a shared vision of prosperity for people and the planet. Achieving this vision depends on system-wide transformative change that incorporates the diverse values of nature and is aligned with the mutually supportive goals of justice and sustainability and their intertwined economic, social and environmental dimensions.
>
> **(IPBES 2022, p. xiv)**

Consistent with its core mandate, the expansive body of expert scientific analysis produced through its assessments has led to IPBES becoming an "agent of transformation" to support the system-wide structural changes required to achieve the 2050 Biodiversity vision. The World Wildlife Fund recently urged governments to adopt the suggestions put forth in the IPBES *Values Assessment*, noting that the *Values Assessment* and the subsequent signing of the Kunming-Montreal Global Biodiversity Framework create "a critical opportunity to ensure that nature's values are properly accounted for within political and economic decisions and can help set us on the pathway to a nature-positive world by 2030" (WWF 2022).

5.2.4 Biodiversity, Ecosystem Integrity, and Ecosystem Services

Biodiversity underpins both resistance and resilience to natural and anthropogenic disturbances and thus the overall stability of ecosystems (Holling 1973). The concept of ecosystem integrity, however, is broader in scope, as it includes the interactions between the living and nonliving components of an ecosystem. Ecosystem integrity is the notion that the maintenance of functioning ecosystems, and thus of the derived flow of ecosystem services, is dependent on sustaining and conserving all appropriate elements and occurrence of all processes at appropriate rates (Angermeier and Karr 1994; Kohm and Franklin 1997). The concept implies a state of an ecosystem that is unimpaired and representative of its natural region, including biological communities and the rate and maintenance of ecological processes within the range of natural variability (Karr 1996; Parks Canada Agency 2000). Supporting functionally unimpaired systems is critical, but challenging, since ecosystems tend to have built-in redundancies in terms of the functions of elements (including biological organisms), and scientific knowledge of the diverse functions that each organism has, and how their interactions generate system functions, is still incomplete. Redundancies are key to ensuring resilience because they create the opportunity for adaptability. Efforts to conserve biodiversity may also conserve these important redundancies.

The relationship between biodiversity and most ecosystem functions and services (including those from forests) is quite clear and has been demonstrated widely at multiple scales (IPBES 2019; See also Tilman et al. 1997; Paquette and Messier 2011; Hooper et al. 2012; Gamfeldt et al. 2013; Isbell et al. 2014; Hautier et al. 2015; Thompson et al. 2009, 2011, 2013; Hicks et al. 2014). As the planet loses biodiversity, it loses ecosystem functions, and humans suffer the degradation and even loss of ecosystem services (IPBES 2019). Moreover, erosion of biodiversity assets is affecting humanity's resilience to environmental changes now and in the future. As Dasgupta (2021) argues in his influential review *The Economics of Biodiversity*: "No amount of technological progress can make economic growth as conventionally measured an indefinite possibility. Ours is inevitably a finite economy, as is the biosphere of which we are part." Recognition of this reality has led some communities to expansive and forward-thinking acts of conservation, such as when the cities of Boston, Hartford, and New York made the decision more than a century ago to protect their water supplies by protecting forests in the watersheds that are the source of their drinking water (Barnes et al. 2009). More recently, in Canada, the Natural Assets Initiative supports municipalities in identifying, conserving, and/or restoring the natural assets that underpin all (NAI | Natural Assets Initiative—Making Nature Count (mnai.ca)). Protecting ecological systems such as aquifers, forests, streams, riparian areas, and foreshores fosters the green infrastructure that will support these communities as they navigate climate change and global environmental change over time. In essence, accounting and managing for a number of ecological functions and biodiversity underpinning ecosystem services is a central principle of all planning for resilient communities.

5.3 NATURE'S CONTRIBUTIONS TO HUMAN WELL-BEING

Nature is the essential foundation of all human success and well-being, although its contributions to human life are often taken for granted. For example, humans rely on nature for clean air, water, and food, essential elements of life. But the benefits of nature are far more vast and complex than this—the natural world is the framework within which all human economic activities exist; it is the source of most medicine and pharmaceuticals; it supports mental well-being, structures social relationships, and fosters diverse cultural practices. Ecosystem Services is one way to understand and articulate these contributions, as people make decisions about how to use nature and how to manage human impacts on it. Several classifications, or typologies, of ecosystem services have been proposed (Costanza et al. 1997; de Groot et al. 2002, 2010; MEA 2005; Wallace 2007; Fisher et al. 2009; TEEB 2010; Landers and Nahlik 2013; Haines-Young and Potschin 2010; EEA 2015), but there is no formal consensus on one in particular (Burkhard et al. 2012; Seppelt et al. 2012). It can be argued that the development of a single universally agreed upon typology is not desirable or practical (e.g., Costanza 2008; Haines-Young and Potschin 2010; Lamarque et al. 2011) because of the need for diverse types of analysis and varied end uses for ecosystem services information. For example, the need to understand ecosystem dynamics that underpin mitigation of flood or drought requires the explicit inclusion of analysis of the "regulating" category of ecosystem services, whereas the need to generate economic statistics about crops that rely on hydrological ecosystem dynamics may view regulating services as secondary inputs to other, "final" services (the crops) and are therefore assumed to be included in the value of those final services. The adoption of a global standard (such as the Common International Classification of Ecosystem Services (CICES), Haines-Young and Potschin 2013) should be informed by the users' needs as it is largely question-dependent and context-related. Noting the complex and intricate interactions between nature and human well-being, it is essential to be inclusive and embrace the variety of perspectives held by people on nature's benefits (Diaz et al. 2018; Anderson et al. 2022).

5.3.1 Socio-Ecological Systems

The Millennium Ecosystem Assessment (MEA 2005) explores *nature's benefits to people* and defines ecosystems services as benefits people obtain from ecosystems. In developing their Conceptual Framework, IPBES aimed to provide an inclusive and holistic approach—from science to other knowledge systems—to assess and understand the complex interrelationships between nature, human well-being, and decision-making processes (Diaz et al. 2015). Besides *nature's benefits to people*, the IPBES conceptual framework includes five other interlinked elements: *nature, good quality of life, anthropogenic assets, institutional and governance and other indirect drivers*, and *direct drivers*, constituting a social–ecological system that operates at various scales in time and space (IPBES 2013).

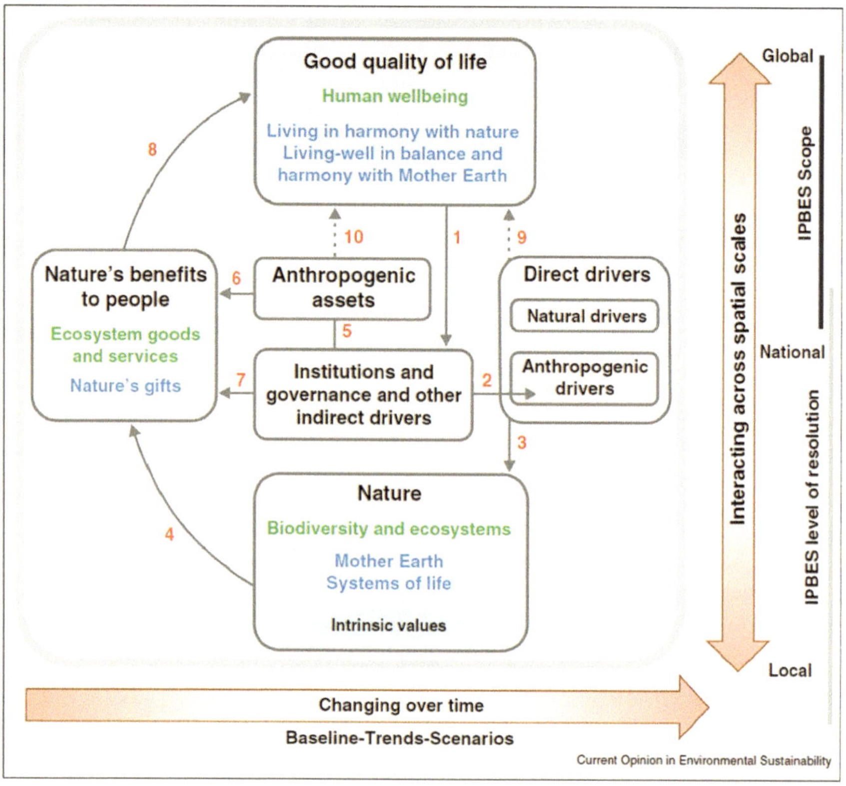

FIGURE 5.2 IPBES Conceptual Framework—connecting nature and people. (Diaz et al. 2015.)

The IPBES Conceptual Framework aims to provide a globally relevant tool for talking about the complex relationships between people and nature, one that makes space for the diverse ways that people understand and experience these relationships in cultures and ecosystems around the world (Figure 5.2). In this context, IPBES suggests a capacious framework for ecosystem services:

- *Nature* refers to the natural world with a focus on the diversity of living organisms, ecosystems, and the complex, interconnected, and dynamic interactions within and between them. *Nature* also includes categories such as evolution, biocultural diversity, and the biosphere. In other knowledge systems, it refers to Mother Earth and systems of life (IPBES 2013). This element takes into consideration nature's intrinsic values independent of human uses in addition to providing essential resources and services that sustain life.
- The definition of *good quality of life* is diverse and multidimensional (Diaz et al. 2015). However, at its core, the fundamental goal is to achieve human well-being, which involves components such as basic material for a good life, health, good social relations, security, and freedom of choice and action (MEA 2005). In non-Western knowledge systems, this is often expressed as living in harmony with nature or living well in balance and harmony with Mother Earth.
- *Nature's benefits to people* encompass a wide range of benefits humans derive from the natural world. One example of this element is the ecosystem services, including provisioning, regulating, and cultural services that people obtain from nature (MEA 2005). Within other knowledge systems, these benefits are often referred to as nature's gifts, underscoring their role in fostering a *good quality of life* (IPBES 2013). When it comes to understanding

nature's benefits to people, it is worth noting that perception of what is beneficial or undesirable can vary among individuals with different backgrounds and perspectives, and there is not a single universal way to interpret the relationship between humans and nature (Diaz et al. 2018).

People from diverse socioeconomic and cultural backgrounds have varying experiences and understandings of the role of nature (IPBES 2022). For example, nature provides many people meaningful work, whether directly such as to those who are harvesting timber or working in mines, or indirectly to those laboring outside who depend on a stable, healthy environment such as construction workers or childcare providers, or to those trading on the value of the companies built on such labor. Nature contributes to human health and well-being, as people harvest medicines, spend time outdoors to support their mental health, breathe air, drink water, and eat food. Nature supports the creation of human homes, as people harvest materials for constructing them, and when people are able to choose their preferred places to build and live. Where and how people live, work, and thrive shapes what they know and how they understand their reliance on the natural world. While some individuals may prioritize the intrinsic value of nature, others may place greater importance on instrumental values. These different perspectives are shaped by how people perceive, experience, and interact with the natural world (Anderson et al. 2022). Unfortunately, decision-makers sometimes prioritize only the "marketable" values of *nature's benefits to people*, such as those with direct economic benefits, while overlooking the "non-market" values during the decision-making processes (IPBES 2022; see section 5.4.4). When integrating nature's many diverse values into decision-making, a holistic approach to valuation helps articulate the range of nature's values surrounding the issue, including health value, social value, cultural value, and economic value. Valuation processes are implemented all around the world, in various countries, regions, Indigenous and local communities, each designed based on their distinct worldviews and ecological knowledge. Currently, many policy decisions are overly narrow, focusing primarily on short-term benefits of nature and economic or monetary gains. However, incorporating the diverse values of nature can lead to more comprehensive, sustainable, and just policy decisions that consider long-term implications (Pascual et al. 2017; IPBES 2022).

- *Institutional and governance and other indirect drivers* of change have a central role in the IPBES framework as they comprise sociopolitical, economic, technological, and cultural factors that can regulate and legislate human relationships with nature (IPBES 2013; Diaz et al. 2015). Institutional and governance policies, such as those developed through land use planning, play a significant role in environmental and landscape changes. Land use planning involves allocating the spatial and temporal distribution of activities in each environment to achieve various social, ecological, and economic objectives (CBD 2022). However, land use decisions can lead to environmental changes that result in habitat degradation or pollution. These human-induced changes, or anthropogenic drivers, are currently at the center of many discussions as they have increasingly allowed unsustainable human impacts on a global scale.
- *Direct drivers* of environmental change include the things that directly affect the natural world and affect the humans who rely upon it (IPBES 2013). This direct damage can come from natural drivers of environmental change, factors present regardless of human existence or influence, and anthropogenic drivers, such as habitat conversion, exploitation, climate change, pollution, and species introductions. In the last 75 years, anthropogenic drivers have begun to increase at an exponential rate (Steffen et al. 2015), far surpassing natural drivers in their impacts on natural systems, and sometimes triggering unexpected "natural" activities such as when global heating triggers higher numbers of wildfires, hurricanes, flooding, and droughts. These drivers can have direct impacts on *anthropogenic assets* and the *good quality of life*, for instance, an intense tornado cutting off electrical

power in built infrastructures may result in financial loss or threat to human well-being. On the other hand, sometimes resources or properties created by humans, such as built infrastructure, technological innovations, knowledge (including Indigenous knowledge and science), can increase the value of *nature's benefits to people.* IPBES refers to these as *anthropogenic assets.*

5.3.2 Case Study: Indigenous Knowledge

Indigenous perspectives and worldviews are highly place-based. Rather than understanding nature as a collection of taxa in the form of discrete species of flora, fauna, and fungi, place-based approaches recognize a web of integral and interdependent relationships bound together (Robinson et al. 2021). Local places, or ecosystems, have unique ecological relationships and distinct impacts on the sociocultural identities of the communities within them.

Indigenous nations of the Northwest Coast such as the Spokane, Coeur d'Alene, and Colville are defined in many ways by the coastal forests and waters that their communities live within. Anadromous fish such as salmon are celebrated with ceremonies and feasts during migration and spawning. The food value alone of salmon within these communities cannot be understated. Historically, Spokane tribal members were estimated to collectively consume 1.4–2.4 million pounds of salmon annually, which makes up only approximately 20% of the total Indigenous harvest within the area (US Columbia Basin Tribes and Canadian First Nations 2014). Salmon embody traditional teachings of respect and renewal that are conveyed through storytelling and cultural practices such as ensuring that all bones, fins, and meat not used from an individual are returned to the water—ensuring the fish's spirit can return home and nourishing the waters where it lived. Similarly, caribou were once so numerous they were referred to like bugs on the landscape by West Moberly Elders and represented food security when other game species were not available. The caribou of the north are intrinsically braided into the Gwich'in creation story that highlights how their lives and well-being are interconnected to one another and they together agreed that the land would sustain the caribou, which would then sustain the people who steward the land.

Culturally significant species such as salmon and caribou could be understood as cultural keystone species based on their innate importance and dependence to the persistence and resilience of those unique cultures (Garibaldi and Turner 2004). These species often become intertwined with Indigenous cultural identity as shown through the creation and continuation of art, ceremonies, traditions, and narratives that are intwined with species. Within many Indigenous communities, people self-identify by cultural affiliations with a particular species of plant or animal (Garibaldi and Turner 2004) such as "Salmon People," "Buffalo People," "Caribou People," and so on. These cultural keystone species represent vast and unique value systems that integrate ecologistic–scientific, humanistic, utilitarian, symbolic, and spiritual value (Kellert 1993) into unified systems of scientific knowledge. Western ways of thinking that categorize, quantify, contextualize, and evaluate these values often misrepresent or misunderstand these knowledge systems, distorting them to fit them within the current management toolkits, strategies, and policies. However, through the unique lens of Indigenous cultural perspectives, the ecosystem services supported by cultural keystone species are invaluable and critical for the continuation of cultural traditions that depend on those relationships to water, land, plants, fish, and all life.

When ceremonial releases of adult Chinook Salmon occurred with the Spokane and Coeur d'Alene nations, a host of revitalizing cultural ceremonies also took place, reconnecting people and communities again through this cultural keystone species (Baldwin et al. 2022). The most powerful of all the stories from these releases is the story of Nucucšnetk ("She Who Retraces Her Steps"). This adult salmon was released in 2017 and was detected in the Bonneville Dam fish ladder, which was just one stop on her journey, where she overcame the odds and passed through seven hydroelectric dams on her way home (Baldwin et al. 2022). Many of these dams did not have fish ladders. For many Indigenous peoples, this story shows hope for this declining cultural keystone species. If we can pave the way to welcome the salmon homecoming, we will then see the return of salmon culture.

Cultural valuation of ecosystem services is critical for understanding communities with holistic relationships with the natural world. Sacred sites, natural landmarks, and specific ecosystems hold significant spiritual value, serving as places of worship, ceremony, and cultural expression. These areas contribute to the sense of place and identity for Indigenous peoples. The landscapes, biodiversity, and natural resources within their traditional territories often define these communities. Preserving and restoring natural ecosystems is crucial for fostering a sense of belonging and continuity of Indigenous cultural systems. Indigenous rights and title to land, paired with knowledge of that land, make Indigenous leadership and governance critical to the stewardship of global biodiversity and ecosystem services.

—Remington Bracher

5.4 ASSESSMENT

As scientific understanding of ecosystem services becomes more and more well developed, and as frameworks for exploring nature's contributions to human well-being are defined, it becomes more possible to assess ecosystem services in particular places and cases. Assessment requires consideration of ecosystem structure and functions, how those generate the services to produce benefits, and how those benefits are distributed to society. It is therefore a broadly interdisciplinary, technical activity, requiring an interdisciplinary expert team to complete. This approach identifies the consequences of environmental change and how environmental management decisions can enhance, diminish, or maintain the flow of nature's benefits (VNCST 2017).

There are a wide variety of examples illustrating diverse approaches to study, analyze, and assess ecosystem services at various scales and in different contexts. Despite the diversity of benefits that ecosystems create for human well-being, most policymaking approaches adopt a very narrow discussion of ecosystem services and often ignore Indigenous and local communities' views (IPBES 2022, p. 4). What is learned in a valuation process depends on the process itself—what is asked, observed, and prioritized in any scientific process shapes what is learned. This emphasizes the importance of thinking carefully about the approach to assessment in each specific case—valuation processes can, with intention, be designed in a more equitable way, to take into account the various views and experiences of people. "Any strategy to achieve valuation beyond the individual needs to consider challenges of representation, inequity (e.g. within and between generations) and asymmetric power relations (e.g. predominance of certain worldviews) among the actors affected" (p. 5).

Effective assessment processes generally include a variety of types of information and are interdisciplinary in methodology. To capture the variety of ways that people value and rely upon ecosystems, ecosystem service assessments typically include biophysical, sociocultural, and economic analyses. Using a structured, planned approach that involves a wide variety of people is an important element of developing an assessment strategy that addresses equity challenges. Social scientists are often highly skilled in approaches such as participatory and deliberative methodologies. Effective assessments are also interdisciplinary. That is, they use a number of different methods (e.g., qualitative and quantitative, ecological, economic, social scientific) to gather data, carry out analysis, and develop the assessment. Interdisciplinary methods lead to more robust and verifiable analyses in which decision-makers may have confidence, especially given the limitations of some individual assessment approaches. Finally, ecosystem services assessments are challenged by multiple sources of uncertainty, such as data scarcity, ecological complexity, knowledge gaps, demand variability, social trade-offs, and normative and value-laden arguments (Jacobs et al. 2015). Assessments that consider a variety of ways people benefit from nature, and a diversity of people's experiences, using several different methodologies, are more robust and reliable in the face of such uncertainty.

5.4.1 Resource: Ecosystem Services Toolkit

Completing and Using Ecosystem Service Assessment for Decision-Making: An Interdisciplinary Toolkit for Managers and Analysts

The Ecosystem Services Toolkit is a dynamic technical guide offering practical, step-by-step guidance for understanding and using ecosystem services assessment (Figure 5.3). Developed by the Value of Nature to Canadians Study Taskforce and released in 2017, the toolkit is meant to assist in addressing the need to build capacity to reflect ecosystem services in decision-making.

The toolkit is fully interdisciplinary, clearly laying out the roles for different kinds of knowledge. Ecosystem services assessment is necessarily an interdisciplinary effort, requiring expertise from across the biophysical and social sciences, Indigenous science, and practitioner knowledge to identify the ways that ecosystem structures and processes (natural capital) generate outcomes (ecosystem services) that humans depend on every day, how they matter to society and individual well-being, and how they are affected by human activity.

The approach can be adapted as needed to each unique context. Decision-makers may want to know whether a set of ecosystem services is being managed sustainably or if any ecosystem services are close to collapse. In other cases, they may want to know which are important to local populations, in what ways they are important, and their relative significance.

The toolkit can be used by any government, community, organization, or group. It was written for Canada's federal, provincial, and territorial governments but can be used by anyone wishing to identify and analyze the ecosystem services associated with a particular place and the people who rely on those services, and the potential effects on those services from changes to the ecosystem.

Advice regarding Indigenous peoples as participants in ecosystem services assessment is provided in the chapters and tools, but most important among these is *Tool Tab 3*. This is unique as the first advice of its kind in Canada. The section was developed in collaboration with an anthropologist based on many years of work with Indigenous communities in western Canada and benefited greatly from review and suggestions provided by representatives of Indigenous organizations.

Chapter 1 sets the foundations: the types of ecosystem services and the conceptual and analytical framework used to assess them.

Chapter 2 explains the six steps of assessment, from clearly identifying the reasons and context for the work to communicating the final results. Links to key tools and resources in the *Tool Tabs* are integrated to help complete each step.

Chapter 3 offers advice on how to address ecosystem services considerations in different policy and decision contexts such as area-based planning, regulatory decision analysis, environmental damages assessment, environmental management, and conservation instruments. Canadian examples are featured.

Ten "Tool Tabs" provide resources for completing an assessment, including, among other things, descriptions for each type of ecosystem service; practical worksheets; indicators; clear advice about economic and sociocultural values and valuation; and factsheets on data sources and analysis methods. User-friendly features include an assessment progress tracker and step overviews, key messages and tip boxes, internal and external links, FAQs, a glossary, and more.

Value of Nature to Canadians Taskforce of the Federal, Provincial, and Territorial Governments of Canada, 2017.

—Susan Preston

Ready-to-use web-based packages or toolkits designed especially for ecosystem services assessment at various spatial scales have also been developed. The best known among these models include Artificial Intelligence for Ecosystem Services (ARIES) and Integrated Valuation of Ecosystem Services and Tradeoffs (InVEST). Both of these models address various analytical elements of ecosystem services

FIGURE 5.3 Cover page of the ecosystem services toolkit report developed by the Value of Nature to Canadians Study Taskforce.

(Source: VNCST 2017.)

assessment, and the latter is widely noted for providing economic values as an output. A team of experts undertook an extensive methodological assessment of ecosystem services models for IPBES (2016), although with cloud-based applications this field of analysis is rapidly evolving and the models have changed significantly since that time.

5.4.2 Biophysical Analysis

Biophysical analysis of ecosystem services focuses on ecosystem structures, processes, and resulting functions. The measurement, modeling, and monitoring of ecosystem functions are the foundation for ecosystem services assessments. This will at least involve analysis of physical characteristics of "natural capital" and the status and trends of ecosystem condition. In some cases, it may be necessary to analyze in more detail biodiversity and ecological resilience, functional dynamics related to the production and spatial flows of ecosystem services, how a change in an ecosystem will affect the natural processes and functions that generate ecosystem services and by extension how such a change would affect the flow of services, and of course the impact of cumulative effects. Indeed, those activities are necessary for making informed science-based decisions toward sustainability goals in general (Carpenter et al. 2009; de Groot et al. 2010a, 2010b; Burkhard and Maes 2017). Biophysical *values* normally refer to measures of extent, condition, integrity, and resilience.

In addition to field science, a broad range of methods and tools have been developed to assess ecological functions at various geographical scales and for different purposes. They range from the identification of indicators or proxies of key processes or structures known to have a causal relationship with a targeted function to the development of models (e.g., statistical, empirical, process-based, or mechanistic). Method selection typically considers the targeted ecosystem services, data availability, and spatial scale. It is often the case that measures of the biophysical environment will serve as an indicative proxy for the ecosystem services that it yields, because it can be more challenging to measure services directly.

To provide useful information for policy and decision making, the biophysical assessment needs to be credible, replicable, and scalable (Daily et al. 2009), as well as sensitive to changes in land management (van Oudenhoven et al. 2012). However, as the Millenium Ecosystem Assessment (MEA 2005) highlights, the establishment of ecosystem services-based sustainable management strategies should not wait for final levels of certainty and precision.

Some of the tools used for biophysical assessment of ecosystem services are profiled here.

5.4.2.1 Ecosystem Service Indicators

The use of indicators (surrogates or proxies) is valuable for the establishment of direct or indirect (qualitative or quantitative) linkages between ecosystem services and an ecosystem's structures, components, or processes (de Groot et al. 2010b, see various other chapters of this volume for more details on ecological indicators). In a review of ecosystem services indicators,[1] Egoh et al. (2012) organized their inventory according to primary and secondary indicators. Primary indicators are the proxy used to measure ecosystem services, whereas secondary indicators are the necessary information used to compose the primary indicators. For example, for the provisioning service of "raw material," Grêt-Regamey et al. (2008) used the primary indicator "timber production," with the following secondary indicators: digital elevation maps (DEM), land cover, and temperature. Overall, Egoh et al. (2012) identified 107 indicators (46 primary, 61 secondary) for regulating services, 62 (21, 41) for provisioning services, and 74 (35, 39) for cultural services, demonstrating not only the complexity of the biophysical but also the interconnections of these with other services.

To facilitate the integration of ecosystem services considerations into decision-making, de Groot et al. (2010a) argued that two types of indicators are needed: state and performance indicators. State indicators are akin to primary or secondary indicators (*sensu* Egoh et al. 2012) and aim at describing which ecosystem process or component is providing the service and associated production level. Performance indicators describe sustainable level of human use (e.g., maximum sustainable forest harvesting). Here we urge caution. The developing interdisciplinary science of planetary boundaries conclusively demonstrates that human use of resources, on a global scale, has far surpassed the sustainable level, largely because some humans are using far more than their share of planetary resources (Richardson et al. 2023; Rockstrom et al. 2023). Notions of ecological thresholds or the natural range of variability might be useful to define sustainable levels, keeping in mind the complex interplay between local and global needs and use in this age of global environmental change.

Although many ecosystem services assessments are conceptually grounded on known factors directly controlling the supply of targeted service(s), they often rely on indirect measures. These indicators are based on hypothesized, but largely untested, relationships between ecosystem services and widely available spatial data products (Andrew et al. 2014). One such product is land use and land cover (LULC) datasets that have been extensively used to measure and map ecosystem services (Costanza et al. 1997; Burkhard and Maes 2017; Burkhard et al. 2009, 2012; Petz and van Oudenhoven 2012; Thomas and Geller 2013). In addition to their relative availability and usefulness in measuring the spatial distribution and supply of ecosystem services (MEA 2005), LULC (and associated change) is recognized as a key driver of human-induced global environmental change (Rockstrom et al. 2023). Land cover information is typically extracted from land classification products (e.g., based on remote-sensing imagery) or ecosystem or watershed classifications that integrate additional environmental data (e.g., vegetation, precipitation, temperature, DEM). Such land or ecological units form the geospatial assessment base and provide the "service-providing units" (Syrbe and Walz 2012).

Linking LULC indicators or ecological units with ecosystem services (types and quantity) is usually done using opinions of experts (or "expert elicitation" *sensu* Jacobs et al. 2015), literature review, empirical research, or often in the case of cultural services, public opinion research. Lookup tables or matrices (also known as the matrix model) are often developed to weight the ecosystem potential or capacity of a land cover class to deliver a selected ecosystem service (Burkhard et al. 2009; Jacobs et al. 2015). It consists in determining the level at which land or ecosystem classes contribute to a particular landscape function (e.g., using binary or quantitative, expert opinion or empirical measures) by assigning a discrete value.

Such a technique is appealing for a number of reasons: it is technically simple, time and cost-effective, and rapidly conveys ecosystem-related data and information into maps (Jacobs et al. 2015). Furthermore, LULC information is commonly available in most areas of the world (at various spatial resolutions), and a variety of methods exist to create linkages from community-based interviews to process-based modeling. Available empirical and quantitative evidence for a given link sometimes exists, further facilitating and improving the reliability of such a technique. For example, Lenz et al. (2014) identified relationships between 13 ecological regions and wood trait signatures (e.g., wood quality). However, the approach is not without criticism, with key concerns related to restricted capacity to detect spatial variability of ecosystem services, accuracy and reliability of the classification method, lack of a mechanistic basis for each score, generalization error, the inability to integrate effects of landscape structure and configuration on service capacity, and the relative subjectivity of the expert estimates (Eigenbrod et al. 2010; Andrew et al. 2014; Jacobs et al. 2015).

Based on data availability, additional information on the local context may be used to modify the lookup score (under a matrix approach) or to develop indices. Indicators of ecosystem integrity are often selected due to the strong linkages between ecosystem integrity and ecosystem services (e.g., Stoll et al. 2015, see previous section). They include species diversity, habitat, forest structure and composition, and landscape heterogeneity and connectivity. Indicators of pressures (or driving forces) on ecosystems known to have negative impacts on their integrity can also be considered, such as drought, fragmentation, deforestation or other land conversion, natural disturbance regimes (pests, fire), soil erosion, invasive alien species, and impacts of climate change.

Another indicator-based approach consists in mapping the presence and distribution of organisms that provide ecosystem services and the area over which these operate (Kremen 2005), for example, species that perform pollination, which is a type of regulating ecosystem service. This work is underpinned by studies demonstrating the link between the presence of species that express certain traits related to performing ecosystem functions or roles and ecosystem service provision (Balvanera et al. 2006; Hooper et al. 2005, 2012, Cadotte et al. 2008; Elmqvist et al. 2010; Cardinale et al. 2011, 2012). There are, however limitations to this approach because many ecosystem services are a result of more complex ecological dynamics involving the actions of living organisms and their environments. For example, carbon sequestration and storage are regulating ecosystem services; in forests these occur as a function of dynamic relationships between the atmosphere, vegetation, soil microbiota, hydrology, and soil structure.

5.4.2.2 Modeling Tools and Techniques

In addition to elementary, indicator-based models, ecosystem services modeling can often benefit from existing and more elaborated environmental modeling efforts from various fields. In forestry, the development of modeling tools and simulation models went from the widespread use of yield tables in the early 1700s (in Europe) to increasingly complex, computer-based, process-based, and spatially explicit modeling techniques (Messier et al. 2003; Larocque 2008). With the expanding set of values that foresters and forest scientists had to manage throughout the evolution of forestry, the forest management modeling toolbox includes models on productivity, carbon, water, species (e.g., habitat and population modeling), genes (e.g., gene flow), natural disturbances (e.g., fire and pest), and climate. Clearly, the traditional forest modeling tools can support ecosystem services assessments, sometimes with some postprocessing steps required. Coupled with increasing data availability from remote sensing (e.g., LiDAR, Landsat) and forest inventories throughout the world, they offer tested approaches for biophysical analyses that can be easily repurposed to use ecosystem services as an assessment end point. Obviously, traditional models from other fields can also contribute to ecosystem services assessment (e.g., hydrology).

The production function approach (Chan et al. 2006; Naidoo et al. 2008; Willemen et al. 2008; Nelson et al. 2009; Tallis and Polasky 2009; Kareiva et al. 2011; Maes et al. 2011) consists of developing quantitative models of ecosystem services using measured ecosystem properties. The models attempt to estimate ecosystem services based on known ecological processes and environmental correlates to ecosystem services. The models can incorporate biophysical data, including climate, topography, structure, and configuration in addition to, or instead of, land cover data. This approach has been found to provide more nuanced mapping than the indicator-based approach and to better assess the multifunctionality of landscapes (Willemen et al. 2010). Disadvantages of this approach include the requirement for more detailed ecological knowledge for accurate mechanistic modeling of many ecosystem services, which may not be available.

5.4.3 Sociocultural Analysis

Sociocultural valuation of ecosystem services has gained increasing attention among foresters, conservation biologists, and ecologists relatively recently (Scholte et al. 2015; IPBES 2022). At the same time, there is a long and highly developed body of research in the social sciences (including anthropology, cultural geography, sociology, and environmental studies), which articulates and explores the various ways that humans value and rely upon the natural world. More recent initiatives such as the IPBES Methodological Assessment on Values (2022) and the Ecosystem Services Toolkit (VNCST 2017) provide models for the integration of social scientific knowledge within social and cultural assessments for Ecosystem Services Assessment. For the purpose of this chapter, the various complex ways in which the natural world creates value for humans and in which humans derive value from nature are the focus, including in the discussion of sociocultural valuation here.

All people attribute importance (values) to things on the basis of their experiences, beliefs, and understandings, which are influenced by their society and their culture. The process of attributing value occurs both within the consciousness of an individual and collectively within groups of people through shared experience. These are the subjects of sociocultural valuation approaches. Understanding both the benefits and values of ecosystem services to people involves understanding the sociocultural context of their knowledge and experience. Sociocultural valuation methods can be qualitative or quantitative; there are many methods for each. They can be designed to capture statistically representative results, and they can also focus on in-depth understanding among individuals and groups within a population.

Sociocultural values can apply to *all* types of ecosystem services and are not limited to the class of "cultural ecosystem services" (Scholte et al. 2015; VNCST 2017). Most ecosystem services are directly experienced by people in some way, for example, the service of crop production results in food that people can eat. Cultural ecosystem services are a classification that is characterized by direct personal interaction with nature and often involve an emotional, cognitive, or physical experience for the beneficiaries. This quality is sometimes confused or combined with the *values* that are attributed to them. In simplest terms, *cultural ecosystem services* are *things* (like all other ecosystem services) and *sociocultural values* are the *importance* of things.

Data on the actual use of ecosystem services can be derived from statistics, modeling, socioeconomic monitoring, or interviews (Burkhard et al. 2012). Flow models, that is, the characterization of the actual use of an ecosystem service in a given area, have been developed by Schröter et al. (2014) for moose hunting, sheep grazing, timber harvest, snow slide prevention, residential amenity, and recreational hiking. The multiple methods for completing sociocultural valuation of ecosystem services have become increasingly well known due to publications by Scholte et al. (2015) and Gómez-Baggethun et al. (2014), among others. The most comprehensive coverage of values and valuation to date has come from IPBES expert guidance (2015) and its Methodological Assessment on Values and Valuation (IPBES 2022). The Ecosystem Services Toolkit provides detailed explanations and practical advice about sociocultural analysis of values with factsheets outlining a breadth of methods to complete the analyses (VNCST 2017). Examples of methods for identifying sociocultural values include surveys, interviews, focus groups, collaborative consultations, participant observation, and action research (Bagstad et al. 2015, Cresswell and Poth 2017). A number of other methods and tools have been developed, such as the public participation geographic information system (see Bagstad et al. 2015), and Social Values for Ecosystem Services (SolVES—Sherrouse et al. 2022).

Ecosystem change that prevents access to important cultural ecosystem services can have a devastating impact on communities. For example, hunting not only provides food for people—through which it is closely related to the provisioning class of services—but for some cultures, it also maintains intergenerational relationships and knowledge systems, retaining shared identities especially in Indigenous communities. Direct experience of healthy natural environments is well-demonstrated to support psychological and physical well-being. During the summer of 2023, as wildfires raged across Canada, their impact was felt far beyond the burn zone. Locals evacuated and experienced profound disruption and uncertainty while wildfire smoke also inundated communities far from the burn zones. People across the continent were unable to leave their homes comfortably to socialize, exercise, or participate in community life, jeapordizing wellbeing, and increasing social isolation. Many experienced significant health effects from breathing the smoke-filled air, and some died (Shewaga 2023). Humans rely on healthy intact forests and the clean air they provide to thrive physically, socially, and culturally. The three are interconnected—when physical health is harmed, it also harms social and cultural lives.

The loss of natural ecosystems can have far-reaching and complex impacts on human societies and cultures. In their study exploring the impact of the slaughter of the bison on Indigenous peoples in the Great Plains, Feir et al. (2022) demonstrated that the near-extinction of bison resulted in immediate impacts which became entrenched over time including significant reduction in height; child mortality increased by 16 percentage points; employment decreased; and income per capita remained 28% lower in bison-reliant nations even into the present (2022). Sociocultural analysis of ecosystem services offers a way to begin to understand the complex ways that every aspect of human culture and society relies on intact healthy ecosystems and to try to anticipate potential consequences of disruptions to these relationships.

5.4.4 Economic Analysis

There is a large demand for economic analyses with estimates for the value of ecosystem services in monetary metrics, both from decision-makers who are trying to assess trade-offs and from environmental advocates who want to frame their concerns in ways that may convince new allies. There are several well-established methods for achieving these estimates, and each has its own associated assumptions, utilities, and limitations. While such analyses can estimate monetary values for some aspects of benefits from ecosystem services, they are unable to represent the complexity of human relationships to the natural world and the extent of our dependence upon it. As a result, it is widely recognized that monetary figures are to be understood as lower-level estimates, not comprehensive or precise measures that capture all relevant variables. Opportunities for these metrics to be misunderstood and misused pose challenges even when they are produced and communicated in the most transparent and responsible ways. Importantly, any exercise in valuation is inescapably political—all such exercises "risk to continue to overrepresent the values of those who hold power and dominate the valuation space, and to perpetuate the discrimination of the views and values of nondominant stakeholders" (Jacobs et al. 2023; IPBES 2022). Contemporary society's lack of awareness of the extent of human dependence on ecosystem services has led most people to take for granted the "free" benefits they receive every day. Remarkably, estimates of monetary value of ecosystem services, and the costs resulting from their degradation, are typically so high that they are met with disbelief. Nature is almost always undervalued, and the need to address this undervaluation is more urgent than ever (Pascual et al. 2023; Dasgupta 2021).

Finally, there are many important values provided by forests that cannot be logically represented with a monetary metric, nor replaced with money. For example, forests are essential to the moderation of global climate change as they sequester carbon, control erosion and flooding, and play a significant role in the hydrologic cycle. While the monetary costs of global climate change are immense, some impacts are complex and highly personal. The loss of a home and its contents in a flash flood can be partially expressed in dollars (real estate and land loss, vehicle loss etc.) but in addition, some of what is lost is profoundly irreplaceable to people—family photos, mementos, even the ability to ever return home (Porteous and Smith 2001). These losses cannot be fully represented in dollars, nor can any amount of money purchase an equivalent substitute.

Understanding the key approaches to economic valuation is highly useful, especially as some losses can be compensated, and avoided costs associated with natural disaster risk and impacts can help to guide environmental management decisions toward greater resilience. Given the uncertainty inherent in these evaluations, and the limits, they are most effective when they are included in interdisciplinary assessments that articulate important non-monetary values for variables, such as clean air, human health and well-being and meaningful work, and grounded in biological, sociocultural, and participatory assessments.

5.4.4.1 Monetary Valuation

Monetary valuation often has appeal for those who see currency (e.g., dollars) as a practical common denominator, particularly, as manufactured or built capital, expressed in currency, is one of the required contributors of production, and most people understand value expressed in currency (Costanza et al. 2014). This approach often attempts to incorporate the monetization of ecosystem services or natural capital into decision-making, in an effort to include the externalities associated with those decisions in the cost–benefit framework (Alexander and McDonald 2014). Ecological economics has emerged and evolved greatly as a research approach intended to address such issues (Costanza 1996) and better inform decision-making related to sustainability. At the same time, there are significant limitations to the accuracy and completeness of such approaches (see earlier), particularly, given the challenges of conveying the complexity of ecological interrelationships in simple numerical terms.

The methods relevant for estimating the value of ecosystem services generated a vast amount of debate and raised a number of technical, ethical, and philosophical issues (including those outlined earlier). A vast body of literature exists on the conceptual and methodological aspects of ecosystem services valuation, including original research and case studies (e.g., IPBES 2015, 2019, 2022; Costanza et al. 1997, 2006, 2014; Alexander and McDonald 2014) as well as reviews and guidance (e.g., TEEB 2010; Dupras et al. 2013; Cordier et al. 2014; Richardson et al. 2014; Keleman et al. 2014; Jacobs et al. 2016, 2023; Pascual et al. 2017, 2023).

A number of tools and resources have been developed to support economic valuation of ecosystem services. The Environmental Valuation Reference Inventory database, hosted by Canada, contains over 5,000 valuation study records to facilitate literature review, environmental valuation, and the application of benefit transfer method (Environment Canada 2011). Some ecosystem services modeling platforms, such as ARIES and InVEST, include economic valuation packages. Regardless of the method or tool used, however, there is typically a high level of uncertainty and complexity in such estimates, and variability in results from one method/tool, even for the same area (Bagstad et al. 2013). Costanza et al. (2014) suggest that the purpose of the analysis informs the method selection and provide guidelines based on the targeted valuation use. The concept of total economic value (TEV) is used to classify and integrate the different values that an ecosystem may provide according to the economic concepts of "use" and "non-use" values categories.

Use values represent the values expressed by economic markets. They can be divided into direct use (e.g., timber harvesting, game animals), indirect use (e.g., water filtration), and option value (i.e., future potential uses). Non-use values from ecosystems represent the value derived from nature that affects human well-being without being transacted on traditional markets (i.e., unpriced services). They include the existence value (i.e., related to the satisfaction from simply knowing that species and ecosystems continue to exist), altruist values (i.e., in relation to intragenerational equity concerns), and bequest values (when concerned with intergenerational concerns) (TEEB 2010). In the TEV framework, an economic valuation methodology is chosen that suits the type of value being measured. A brief overview of three general groups of valuation methodologies is presented in the following text.

5.4.4.2 Market-Based Valuation

The first suite of methods is relying on existing markets to identify the environmental value of interest. They include market price-based approaches (i.e., when a market already exists), cost-based approaches (e.g., the cost associated with replacing ecosystem services with technology or for ecosystem restoration), and production function approaches (i.e., the value associated with an ecosystem service's contribution to the production of a marketed good or service). An example to illustrate the production function method is a change in wetland size that leads to a change in water quality that reduces the quantity of fish caught. This lost market value can be estimated using market price information (OECD 2002).

5.4.4.3 Direct and Indirect Valuation

A second suite of methods exists for nonmarket services, which are classified into two categories: direct and indirect valuation. Contingent valuation is probably the best-known direct valuation technique. It consists of a survey-based approach to estimating the monetary equivalent of nonmarket values (Macdonald and McKenney 1996). Indirect methods, like hedonic pricing and travel cost, rely on observed behavior in related markets for valuing the ecosystems services (OECD 2002). For example, travel cost models are based on the willingness of an individual to incur costs in terms of travel and time to benefit from an ecosystem service, such as outdoor recreational activities.

5.4.4.4 Benefit Transfer

Finally, another valuation technique is benefit transfer (e.g., Plummer 2009). It involves "borrowing" an estimate of economic value, such as willingness to pay (WTP), from one site (the "study site") and applying it to another (the "policy site") (OECD 2002). The following are two main approaches to benefit transfer (Statistics Canada 2013, and see additional guidance in Richardson et al. 2014):

(1) "Unit value transfer" transfers a monetary value from a study site to a new site (e.g., \$ per hectare). This has generally been done by using estimate from a single study in the

literature that best matches the characteristics of the policy site, applying an average value from several studies, or applying administratively approved values (e.g., access to national parks within a country).

(2) "Function value transfer" consists of developing a demand or WTP function. For example, WTP is expressed as a function of the quantity or quality of the ecosystem service provided and socioeconomic characteristics of the population originally surveyed. This approach can be based on results from a single study or a number of studies. For example, in the case of meta-analysis transfer, data are consolidated from a large number of studies.

5.5 CONCLUSIONS

In this age of human-induced global environmental change, assessing human embeddedness in nature, including nature's benefits for humans, is essential to decision-making on all scales. Ecosystem services assessment can be a critical tool for managers and decision-makers when carefully executed with attention to people's social, economic, biophysical, and ecological integration with natural ecosystems. The relatively recent development of interdisciplinary values assessment tools, such as those from IPBES, and the Canadian *Ecosystem Services Toolkit* (2017), is an important contribution. Emphasizing and understanding the importance of "diverse values for sustainability" (as in IPBES, and Pascual et al. 2023) is a critical step. But as Jacobs et al. (2023) remind us, attention to the implicit politics of the valuation process is equally important. For foresters concerned with ecological forest management, this means that ecosystem services assessment requires attention to biophysical, economic and social dimensions of value, and also to who has the power: who decides the questions and criteria, whose experience is included, who is carrying out the analysis. As the best practices outlined by IPBES and the Value of Nature to Canadians Taskforce illustrate, ecosystem services assessments are most complete, accurate, and equitable when they involve interdisciplinary and multi-party approaches at all stages. These challenges are not unique to ecosystem services assessment, or ecological forest management, and they illustrate the important responsibility of all environmental professionals to use available tools with care and precision and to prioritize meaningful engagement with the public and decision-makers as an ongoing and integral part of the work.

NOTE

1 The Ecosystem Servies Toolkit (VNCST 2017) provides a listing of many possible indicators for each type of ES, for both biophysical and cultural variables, synthesized from the literature (see Tool Tab 5, pp. 127–136).

REFERENCES

Abatzoglou, J.T. and P. Williams. 2016. Impact of Anthropogenic Climate Change on Wildfire across Western US Forests. *Proc. Natl. Acad. Sci. U.S.A.* 113(42):11770–11775.

Alexander, C. and C. McDonald. 2014. *Urban forests: The value of trees in the city of Toronto*. Special Report, TD Economics. http://www.td.com/document/PDF/economics/special/UrbanForests.pdf (accessed May 6, 2015).

Anderson, C.B., Athayde, S., Raymond, C.M., Vatn, A., Arias-Arévalo, P., Gould, R.K., Kenter, J., Muraca, B., Sachdeva, S., Samakov, A., Zent, E., Lenzi, D., Murali, R., Amin, A., & Cantú-Fernández, M. 2022. Chapter 2. Conceptualizing the diverse values of nature and their contributions to people. (Version 3). Zenodo.

Andrew, M.E., M.A. Wulder, and T.A. Nelson. 2014. Potential contributions of remote sensing to ecosystem service assessments. *Prog. Phys. Geogr.: Earth Environ.* 38(3):328–353. https://doi.org/10.1177/0309133314528942

Angermeier, P.L. & J.R. Karr. 1994. Biological integrity versus biological diversity as policy directives: Protecting biotic resources. BioScience 44:690–697.

Bagstad, K.J., J.M. Reed, D.J. Semmens, B.C. Sherrouse, and A. Troy. 2015. Linking biophysical models and public preferences for ecosystem service assessments: A case study for the Southern Rocky Mountains. *Reg. Environ. Change* 16:2005–2018: DOI 10.1007/s10113-015-0756-7.

Bagstad, K.J., D.J. Semmens, S. Waage, and R. Winthrop. 2013. A comparative assessment of decision-support tools for ecosystem services quantification and valuation. *Ecosys. Serv.* 5:e27–e39. www.sciencedirect.com/science/article/abs/pii/S221204161300051X

Baldwin, C., Giorgi, C., & Biladeau, T. 2022. Cultural and educational releases of salmon in areas blocked by major hydroelectric projects on the Columbia River. Aquatic Ecosystem Health & Management 25(1):16–26.

Baltzer, J.L., Day, N.J., Walker, X.J., Greene, D., Mack, M.C., Alexander, H.D., Arseneault, D., Barnes, J., Bergeron, Y., Boucher, Y., Bourgeau-Chavez, L., Brown, C.D., Carrière, S., Howard, B.K., Gauthier, S., Parisien, M-A., Reid, K.A., Rogers, B.M., Roland, C., Sirois, L., Stehn, S., Thompson, D.K., Turetsky, M.R., Veraverbeke, S., Whitman, E., Yang, J. & Johnstone, J.F. 2021. Increasing fire and the decline of fire adapted black spruce in the boreal forest. Proceedings of the National Academy of Sciences 118(45), e2024872118.

Balvanera, P., A.B. Pfisterer, N. Buchmann, J.S. He, T. Nakashizuka, D. Raffaelli, and B. Schmid. 2006. Quantifying the evidence for biodiversity effects on ecosystem functioning and services. *Ecol. Lett.* 9:1146–1156.

Barnes, M.C., A.H. Todd, R.W. Lilja, and P.K. Barten. 2009. *Forests, water and people: Drinking water supply and forest lands in the Northeast and Midwest United States.* Newton Square, PA: United States Department of Agriculture, Forest Service, Northeastern Area State and Private Forestry. https://ice.ucdavis.edu/node/133.html

Bond, W.J., Woodward, F.I., Midgley, G.F. 2005. The global distribution of ecosystems in a world without fire. New Phytol 165(2):525–538.

Boucher, D., Gauthier, S., Thiffault, N., Marchand, W., Girardin, M., & Urli, M. 2020. How Climate Change might affect Tree Regeneration following Fire at Northern Latitudes: A Review. New For. 51:543–571.

Brauman, K.A., G.C. Daily, T. Ka'eo Duarte, and H.A. Mooney. 2007. The nature and value of ecosystem services: An overview highlighting hydrologic services. Annu. Rev. Environ. Resourc. 32:67–98.

BSR. 2013. *Private sector uptake of ecosystem services concepts and frameworks: The current state of play.* 27 p. http://www.bsr.org/reports/BSR_Private_Engagement_With_Ecosystem_Services_2014.pdf

Burkhard, B., F. Kroll, F. Müller, and W. Windhorst. 2009. Landscapes' capacities to provide ecosystem services—A concept for land-cover based assessments. *Landsc. Online* 15:1–22.

Burkhard, B., F. Kroll, S. Nedkov, and F. Müller. 2012. Mapping ecosystem service supply, demand and budgets. *Ecol. Indic.* 21:17–29.

Burkhard, B. and J. Maes (eds.) 2017. *Mapping ecosystem services.* Advanced Books. https://doi.org/10.3897/ab.e12837

Cadotte, M.W., B.J. Cardinale, and T.H. Oakley. 2008. Evolutionary history and the effect of biodiversity on plant productivity. *Proc. Natl. Acad. Sci. USA* 1055:17012–17017.

Cardinale, B.J., J.E. Duffy, A. Gonzalez et al. 2012. Biodiversity loss and its impact on humanity. *Nature* 486:59–67.

Cardinale, B.J., K.L. Matulich, D.U. Hooper, J.E. Byrnes, E. Duffy, L. Gamfeldt, P. Balvanera, M.I. O'Connor, and A. Gonzalez. 2011. The functional role of producer diversity in ecosystems. *Am. J. Bot.* 98:572–592.

Carpenter, S.R., H.A. Mooney, J. Agard et al. 2009. Science for managing ecosystem services: Beyond the millennium ecosystem assessment. *Proc. Natl. Acad. Sci. USA* 106:1305–2131.

Chan, K.M.A., M.R. Shaw, D.R. Cameron, E. Underwood, and G.C. Daily. 2006. Conservation planning for ecosystem services. *PLoS Biol.* 4:2138–2152.

Christianson, A.C., Sutherland, C.R., Moola, F., Bautista, N.G., Young, D. & MacDonald, H. 2022. Centering Indigenous Voices: The Role of Fire in the Boreal Forest of North America. Curr Forestry Rep 8:257–276. https://doi.org/10.1007/s40725-022-00168-9

CIFFC. 2023. Fire Statistics. Canadian Interagency Forest Fire Centre. https://ciffc.net/statistics

Convention on Biological Diversity (CBD) 2022. *The biodiversity plan for life on earth: Target 1.* https://www.cbd.int/gbf/targets/1

Cordier, M., J.A. Pérez Agúndez, W. Hecq, and B. Hmaide. 2014. A guiding framework for ecosystem services monetization in ecological-economic modeling. *Ecosyst. Serv.* 8:86–96.

Costanza, R. 1996. Ecological economics: Reintegrating the study of humans and nature. *Ecol. Appl.* 6:978–990.

Costanza, R. 2008. Ecosystem services: Multiple classification systems are needed. *Biol. Conserv.* 141:350–353.

Costanza, R., R. d'Arge, R. de Groot et al. 1997. The value of the world's ecosystem services and natural capital. *Nature* 387:253–260. https://doi.org/10.1038/387253a0

Costanza, R., R. De Groot, P. Sutton, S. Van der Ploeg, S.J. Anderson, I. Kubiszewski, S. Farber, and R.K. Turner. 2014. Changes in the global value of ecosystem services. *Glob. Environ. Change* 26:152–158.

Costanza, R., M. Wilson, A. Troy, A. Voinov, S. Liu, and J. D'Agostino. 2006. *The value of New Jersey's ecosystem services and natural capital.* New Jersey Department of Environmental Protection.

Cresswell, J.W. and C.N. Poth. 2017. *Qualitative inquiry and research design: Choosing among five approaches.* SAGE Publications Inc. Thousand Oaks, California, USA.

Daily, G.C. and P.A. Matson. 2008. Ecosystem services: From theory to implementation. *Proc. Natl. Acad. Sci.* 105:9455–9456.

Daily, G.C., S. Polasky, J. Goldstein, P.M. Kareiva, H.A. Mooney, L. Pejchar, T.H. Ricketts, J. Salzman, and R. Shallenberger. 2009. Ecosystem services in decision making: time to deliver. *Front. Ecol. Environ.* 7:21–28. https://doi.org/10.1890/080025

Dasgupta, P. 2021. *The economics of biodiversity: The Dasgupta review.* Abridged Version. London: HM Treasury.

de Groot, R.S., R. Alkemade, L. Braat, L. Hein, and L. Willemen. 2010a. Challenges in integrating the concept of ecosystem services and values in landscape planning, management and decision making. *Ecol. Complex.* 7(3):260–272. https://doi.org/10.1016/j.ecocom.2009.10.006

de Groot, R.S., B. Fisher, M. Christie et al. 2010b. Integrating the ecological and economic dimensions in biodiversity and ecosystem service valuation. In *The economics of ecosystems and biodiversity: Ecological and economic foundations*, ed. P. Kumar, pp. 1–40. London: Earthscan.

de Groot, R.S., M.A. Wilson, and R.M.J. Boumans. 2002. A typology for the classification, description and valuation of ecosystem functions, goods and services. *Ecol. Econ.* 41(3):393–408. https://doi.org/10.1016/S0921-8009(02)00089-7

Díaz, S., Demissew, S., Carabias, J., Joly, C., Lonsdale, M., Ash, N., Larigauderie, A., Adhikari, J.R., Arico, S., Báldi, A., Bartuska, A., Baste, I.A., Bilgin, A., Brondizio, E., Chan, K.M.A., Figueroa, V.E., Duraiappah, A., Fischer, M., Hill, R., Koetz, T., & Zlatanova, D. 2015. The IPBES Conceptual Framework—connecting nature and people. Current Opinion in Environmental Sustainability 14:1–16.

Diaz, S., Pascual, U., Stenseke, M., Martín-López, B., Watson, R., Molnár, Z., Hill, R., Chan, K., Baste, I., Brauman, K., Polasky, S., Church, A., Lonsdale, M., Larigauderie, A., Leadley, P., van Oudenhoven, A., Plaat, F., Schröter, M., Lavorel, S., & Shirayama, Y. 2018. Assessing Nature's Contributions to People. Science 359:270–272.10.1126/science.aap8826.

Dupras, J., J.-P. Revéret, and J. He. 2013. *L'évaluation économique des biens et services écosystémiques dans un contexte de changements climatiques: Un guide méthodologique pour une augmentation de la capacité à prendre des décisions d'adaptation.* Ouranos. http://www.ouranos.ca/media/publication/273_RapportReveret2013.pdf

Egoh, B., E.G. Drakou, M.B. Dunbar, J. Maes, and L. Willemen. 2012. *Indicators for mapping ecosystem services: A review.* JRC Scientific and Policy Reports. Luxembourg: Publications Office of the European Union, EUR 25456 EN.

Eigenbrod, F, P.R. Armsworth, B.J. Anderson, A. Heinemeyer, S. Gillings, D.B. Roy, C.D. Thomas, and K.J. Gaston. 2010. The impact of proxy-based methods on mapping the distribution of ecosystem services. *J. Appl. Ecol.* 47:377–385.

Ellis, Erle. 2018. Anthropocene: A Very Short Introduction. 10.1093/actrade/9780198792987.001.0001.

Elmqvist, T., E. Maltby, T. Barker, M. Mortimer, C. Perrings, J. Aronson, R. deGroot, A. Fitter, G. Mace, J. Norberg, I. Sousa Pinto, and I. Ring. 2010. Biodiversity, ecosystems and ecosystem services. Chapter 2. In *The economics of ecosystems and biodiversity: Ecological and economic foundations,* ed. P. Kumar, pp. 44–111. London: Earthscan Ltd.

Environment Canada. 2011. *Environmental Valuation Reference Inventory (EVRI).* www.evri.ca

European Environment Agency (EEA). 2015. Towards a common classification of ecosystem services: CICES 4.3 spreadsheet. http://cices.eu/

Feir, D.L., Gillezeau, R., Jones, M.E.C. 2022. The Slaughter of the Bison and Reversal of Fortunes on the Great Plains, IZA Discussion Papers, No. 15498, Institute of Labor Economics (IZA), Bonn

Fisher, B., R.K. Turner, and P. Morling. 2009. Defining and classifying ecosystem services for decision making. *Ecol. Econ.* 68(3):643–653. https://doi.org/10.1016/j.ecolecon.2008.09.014

Gamfeldt, L., T. Snälll, R. Bagchi, M. Jonsson, L. Gustafsson, P. Kjellander, M.C. Ruiz-Jaen et al. 2013. Higher levels of multiple ecosystem services are found in forests with more tree species. *Nat. Comm.* 4:1–8. https://doi.org/10.1038/ncomms2328

Garibaldi, A. & Turner, N. 2004. Cultural Keystone Species: Implications for Ecological Conservation and Restoration. Ecology & Society, 9(3).

Gavin, D.G., Hallett, D.J., Hu, F.S., Lertzman, K.P., Prichard, S.J., Brown, K.J., Lynch, J.A., Bartlein, P. & Peterson, D.L. 2007. Forest Fire and Climate Change in Western North America: Insights from

Sediment Charcoal Records. Frontiers in Ecology and the Environment 5(9):499–506. www.jstor.org/stable/20440746

Gómez-Baggethun, E., B. Martín-López, D. Barton, L. Braat, H. Saarikoski, Kelemen, M. García-Llorente, E., J. van den Bergh, P. Arias, P. Berry, L., M. Potschin, H. Keene, R. Dunford, C. Schröter-Schlaack, and P. Harrison. 2014. *State-of-the-art report on integrated valuation of ecosystem services.* European Commission FP7.

Hall, R.J., J.T. Freeburn, W.J. De Groot, J.M. Pritchard, T.J. Lynham, and R. Landry. 2008. Remote sensing of burn severity: Experience from western Canada boreal fires. Int. J. Wildland Fire, 17(4):476–489.

Haines-Young, R. & M. Potschin. 2010. The links between biodiversity, ecosystem services and human well-being. In Ecosystem Ecology: A New Synthesis, BES Ecological Reviews Series, eds. D. Raffaeilli and C.L.J. Frid, pp. 110–139. Cambridge, UK: Cambridge University Press/British Ecological Society.

Haines-Young, R. and M. Potschin. 2013. *Report to the European Environment Agency.* Common International Classification of Ecosystem Services (CICES). www.cices.eu

Hautier, Y., D. Tilman, F. Isbell, E.W. Seabloom, E.T. Borer, and P.B. Reich. 2015. Anthropogenic Environmental Changes affect Ecosystem Stability via Biodiversity. Science 348:336–340.

Hicks, C., S. Woroniecki, M. Fancourt, M., Bieri, M., Garcia Robles, H., Trumper, K., Mant, R. 2014. The Relationship between Biodiversity, Carbon Storage and the Provision of Other Ecosystem Services: Critical Review for the Forestry Component of the International Climate Fund. Cambridge, UK: United Nations Environment Programme (UNEP).

Higuera, P.E., Shuman, B.N. and Wolf, K.D. 2021. Rocky Mountain Subalpine Forests now Burning More than any Time in Recent Millennia. Proceedings of the National Academy of Sciences 118(25), e2103135118.

Holdren, J.P. & P.R. Ehrlich. 1974. Human Population and the Global Environment. Am. Sci. 62:282–292.

Holling, C.S. 1973. Resilience and Stability of Ecological Systems. Annu. Rev. Ecol. Syst. 4:1–24.

Hooper, D.U., C. Adair, B.J. Cardinale, J.E.K. Byrnes, B.A. Hungate, K.L. Matulich, A. Gonzalez, J.E. Duffy, L. Gamfeldt & M.I. O'Connor. 2012. A global synthesis reveals biodiversity loss as a major driver of ecosystem change. Nature 486:105–108.

Hooper, D.U., F.S. Chapin III, J.J. Ewel et al. 2005. Effects of biodiversity on ecosystem functioning: A consensus of current knowledge. *Ecol. Monogr.* 75:3–35.

IPBES (Intergovernmental Science Policy Platform on Biodiversity and Ecosystem Services). 2013. *Decision IPBES-2/4: Conceptual framework for the intergovernmental science-policy platform on biodiversity and ecosystem services.* Cape Town, South Africa: Multidisciplinary Expert Panel, 25–26 August.

IPBES (Intergovernmental Science Policy Platform on Biodiversity and Ecosystem Services). 2015. Preliminary guide regarding diverse conceptualization of multiple values of nature and its benefits, including biodiversity and ecosystem functions and services. Deliverable 3d to Plenary 4 (IPBES/4/INF/13). Available at www.ipbes.net/sites/default/files/downloads/IPBES-4-INF-13_EN.pdf.

IPBES. 2016. The methodological assessment report on scenarios and models of biodiversity and ecosystem services. In *Secretariat of the intergovernmental science-policy platform on biodiversity and ecosystem services*, eds. S. Ferrier, K.N. Ninan, P. Leadley, R. Alkemade, L.A. Acosta, H.R. Akçakaya, L. Brotons, W.W.L. Cheung, V. Christensen, K.A. Harhash, J. Kabubo-Mariara, C. Lundquist, M. Obersteiner, H.M. Pereira, G. Peterson, R. Pichs-Madruga, N. Ravindranath, C. Rondinini, and B.A. Wintle, p. 348. Bonn: IPBES.

IPBES (Intergovernmental Science Policy Platform on Biodiversity and Ecosystem Services). 2019. *Summary for policymakers of the global assessment report on biodiversity and ecosystem services of the intergovernmental science-policy platform on biodiversity and ecosystem services*, eds. S. Díaz, J. Settele, E.S. Brondízio, H.T. Ngo, M. Guèze, J. Agard, A. Arneth, P. Balvanera, K.A. Brauman, S.H.M. Butchart, K.M.A. Chan, L.A. Garibaldi, K. Ichii, J. Liu, S.M. Subramanian, G.F. Midgley, P. Miloslavich, Z. Molnár, D. Obura, A. Pfaff, S. Polasky, A. Purvis, J. Razzaque, B. Reyers, R. Roy Chowdhury, Y.J. Shin, I.J. Visseren-Hamakers, K.J. Willis, and C.N. Zayas. Bonn, Germany: IPBES Secretariat. https://doi.org/10.5281/zenodo.3553579

IPBES (Intergovernmental Science Policy Platform on Biodiversity and Ecosystem Services). 2022. *Methodological assessment report on the diverse values and valuation of nature of the intergovernmental science-policy platform on biodiversity and ecosystem services*, eds. P. Balvanera, U. Pascual, M. Christie, B. Baptiste, and D. González-Jiménez. Bonn, Germany: IPBES Secretariat. https://doi.org/10.5281/zenodo.6522522

Isbell, F., D. Tilman, S. Polasky, and M. Loreau. 2014. The biodiversity-dependent ecosystem service debt. *Ecol. Lett.* 18:119–134.

Jackson, S.J. 2019. Humboldt for the anthropocene. *Science* 365(6458):1074–1076. https://doi.org/10.1126/science.aax7212

Jacobs, S., B. Burkhard, T. Van Daele, J. Staes, and A. Schneiders. 2015. 'The Matrix Reloaded': A review of expert knowledge use for mapping ecosystem services. *Ecol. Modell.* 295:21–30. https://www.sciencedirect.com/science/article/abs/pii/S0304380014004141

Jacobs, S., N. Dendoncker, B. Martín-López, D.N. Barton, E. Gomez-Baggethun, F. Boeraeve, F.L. McGrath, K. Vierikko, D. Geneletti, K.J. Sevecke, N. Pipart, E. Primmer, P. Mederly, S. Schmidt, A. Aragão, H. Baral, R.H. Bark, T. Briceno, D. Brogna, P. Cabral, R. De Vreese, C. Liquete, H. Mueller, K.S.-H. Peh, A. Phelan, A.R. Rincón, S.H. Rogers, F. Turkelboom, W. Van Reeth, B.T. van Zanten, H.K. Wam, and C.-L. Washbourne. 2016. A new valuation school: Integrating diverse values of nature in resource and land use decisions. *Ecosyst. Serv.* 22(B):213–220. https://doi.org/10.1016/j.ecoser.2016.11.007.

Jacobs, S., E. Kelemen, P. O'Farrell, A. Martin, M. Schaafsma, N. Dendoncker, R. Pandit, T.H. Mwampamba, I. Palomo, A.J. Castro, M.A. Huambachano, A. Filyushkina, and H. Gunimeda. 2023. The pitfalls of plural valuation. *Curr. Opin. Env. Sust.* 64:101345. https://doi.org/10.1016/j.cosust.2023.101345

Kareiva, P., H. Tallis, T.H. Ricketts, G.C. Daily, and S. Polasky. 2011. *Natural capital: Theory and practice of mapping ecosystem services*. Oxford: Oxford University Press.

Karr, J.R. 1996. Ecological integrity and health are not the same. In *Engineering within ecological constraints*, ed. P.C. Schulze, pp. 97–109. Washington, DC: National Academy of Engineering, National Academy Press.

Kelemen, E., M. García-Llorente, G. Pataki, B. Martín-López, and E. Gómez-Baggethun, 2014. Non-monetary techniques for the valuation of ecosystem service. *OpenNESS Reference Book. EC FP7 Grant Agreement, 308428*(4).

Kellert, S.R. 1993. *The biological basis for human values of nature: The biophilia hypothesis*, pp. 42–69. Washington, DC: Island Press.

Kremen, C. 2005. Managing ecosystem services: What do we need to know about their ecology? *Ecol. Lett.* 8:468–479.

Kohm, K.A. and J.F. Franklin. 1997. *Creating a forestry for the 21st century: The science of ecosystem management*. Washington, DC: Island Press.

Lake, F.K. and Christianson, A.C. 2019. Indigenous fire stewardship. In *Encyclopedia of wildfires and wildland-urban interface (WUI) fires*, ed. S. Manzello. Cham: Springer. https://doi.org/10.1007/978-3-319-51727-8_225-1

Lamarque, P., F. Quétier, and S. Lavorel. 2011. The diversity of the ecosystem services concept and its implications for their assessment and management. *C.R. Biol.* 334:441–449.

Landers, D. and A.M. Nahlik. 2013. *Final Ecosystem Goods and Services Classification System (FEGS-CS)*. Corvallis, OR: US Environmental Protection Agency Office of Research and Development. Version 2.8a. http://gispub4.epa.gov/FEGS/.

Larocque, G.R. 2008. Ecological models: Models for the simulation of forest ecosystem dynamics. In *Encyclopedia of ecology*, 5 vols., eds. S.E. Jorgensen and B.D. Fath, pp. 1663–1773. Oxford, UK: Elsevier.

Leopold, A. 1949. *A Sand County Almanac, and sketches here and there*. New York: Oxford University Press.

Lenz, P., M. Deslauriers, and C.-H. Ung et al. 2014. What do ecological regions tell us about wood quality? A case study in Eastern Canadian white spruce. *Can. J. For. Res.* 44:1383–1393.

Lowe, S.R. and D.R. Garfin. 2023. Crisis in the air: The mental health implications of the 2023 Canadian wildfires. *Lancet Planet Health.* 7(9):e732–e733. https://doi.org/10.1016/S2542-5196(23)00188-2. PMID: 37673541.

Macdonald, H. and D. McKenney. 1996. Varying levels of information and the embedding problem in contingent valuation: The case of Canadian wilderness. *Can. J. For. Res.* 26:1295–1303.

Maes, J., M.L. Paracchini, and G. Zulian. 2011. *European assessment of the provision of ecosystem services: Towards an atlas of ecosystem services*. European Commission Joint Research Centre Institute for Environment and Sustainability EUR 24750 EN–2011.

Mamuji, A.A. and J.L. Rozdilsky. 2019. Wildfire as an increasingly common natural disaster facing Canada: Understanding the 2016 Fort McMurray wildfire. *Nat. Hazards* 98:163–180. https://doi.org/10.1007/s11069-018-3488-4

Meijaard, E., S. Wunder, M.R. Guariguata, and D. Sheil. 2014. What scope for certifying forest ecosystem services? *Ecosyst. Ser.* 7:160–166.

Messier, C., M.-J. Fortin, F. Schmiegelow et al. 2003. Modelling tools to assess the sustainability of forest management scenarios. In *Towards sustainable management of the boreal forest*, eds. P.J. Burton, C. Messier, D.W. Smith, and W.L. Adamowicz, pp. 531–580. Ottawa, ON: NRC Research Press.

Millennium Ecosystem Assessment (MEA). 2005. *Ecosystems and human well-being: Synthesis*. Washington, DC: Island Press.

Ministry of Forestry and Parks, Alberta. 2023. *Alberta wildfire Status dashboard.* ArcGIS dashboards. www.arcgis.com/apps/dashboards/3ffcc2d0ef3e4e0999b0cf8b636defa3

Mistry, J., B.A. Bilbao, and A. Berardi. 2016. Community owned solutions for fire management in tropical ecosystems: Case studies from indigenous communities of South America. *Phil. Trans. R Soc. B* 371(1696):20150174.

Naeem, S., J.C. Ingram, A. Varga, T. Agardy, P. Barten, G. Bennett, E. Bloomgarden, L.L. Bremer, P. Burkill, M. Cattau, C. Ching, M. Colby, D.C. Cook, R. Costanza, F. Declerck, C. Freund, T. Gartner, R. Goldman-Benner, J. Gunderson, D. Jarrett, A.P. Kinzig, A. Kiss, A. Koontz, P. Kumar, J.R. Lasky, M. Masozera, D. Meyers, F. Milano, L. Naughton-Treves, E. Nichols, L. Olander, P. Olmsted, E. Perge, C. Perrings, S. Polasky, J. Potent, C. Prager, F. Quetier, K. Redford, K. Saterson, G. Thoumi, M.T. Vargas, S. Vickerman, W. Weisser, D. Wilkie & S. Wunder 2015. Get the science right when paying for nature's services. *Science* 347:1206–1207.

Naidoo, R., A. Balmford, R. Costanza et al. 2008. Global mapping of ecosystem services and conservation priorities. *PNAS* 105:9495–9500.

National Academy of Sciences (NAS). 2013. An ecosystem services approach to assessing the impacts of the Deepwater Horizon oil spill in the Gulf of Mexico. Committee on the Effects of the Deepwater Horizon Mississipi Canyon-252 Oil Spill on Ecosystem Services in the Gulf of Mexico, Ocean Studies Board; Division on Earth and Life Studies. Washington DC: National Research Council. www.nap.edu/catalog.php?record_id=18387.

Nelson, E., G. Mendoza, J. Regetz et al. 2009. Modeling multiple ecosystem services, biodiversity conservation, commodity production, and tradeoffs at landscape scales. *Front. Ecol. Environ.* 7:4–11.

OECD. 2002. *Handbook of biodiversity valuation: A guide for policy makers.* Paris: OECD.

Paquette, A. and C. Messier. 2011. The effect of biodiversity on tree productivity: From temperate to boreal forests. *Glob. Ecol. Biogeogr.* 20:170–180.

Parks Canada Agency. 2000. Unimpaired for Future Generations? Protecting Ecological Integrity with Canada's National Parks. Vol. I A Call to Action. Vol. II Setting a New Direction for Canada's National Parks. Report of the Panel on the Ecological Integrity of Canada's National Parks, Ottawa, Ontario, Canada.

Pascual, U., P. Balvanera, C.B. Anderson *et al.* 2023. Diverse values of nature for sustainability. *Nature* 620: 813–823. https://doi.org/10.1038/s41586-023-06406-9

Pascual, U., P. Balvanera, S. Díaz, G. Pataki, E. Roth, M. Stenseke, R.T. Watson, E.B. Dessane, M. Islar, E. Keleman, V. Maris, M. Quaas, S.M. Subramanian, H. Wittmer, A. Adlan, S. Ahn, Y.S. Al-Hafedh, E. Amankwah, S.T. Asah, P. Berry, A. Bilgin, S.J. Breslow, C. Bullock, D. Cáceres, H. Daly-Hassen, E. Figueroa, C.D. Golden, E. Gómez-Baggethun, D. González-Jiménez, J. Houdet, H. Keune, R. Kumar, K. Ma, P.H. May, A. Mead, P. O'Farrell, R. Pandit, W. Pengue, R. Pichis-Madruga, F. Popa, S., and N. Yagi. 2017. Valuing nature's contributions to people: The IPBES approach. *Curr. Opin. Env. Sust.* 26:7–16.

Petz, K. and A.P.E. van Oudenhoven. 2012. Modelling land management effect on ecosystem functions and services: A study in the Netherlands. *Int. J. Biodivers. Sci. Ecosyst. Serv. Manage.* 8:135–155.

Plummer, M.L. 2009. Assessing benefit transfer for the valuation of ecosystem services. *Front. Ecol. Environ.* 7:38–45.

Porteous, J.D. and S.E. Smith. 2001. *Domicide: The global destruction of home.* Montreal: MCGill-Queens University Press.

Ray, J.C., J. Grimm, and A. Olive. 2021. The biodiversity crisis in Canada: Failures and challenges of federal and sub-national strategic and legal frameworks. *FACETS* 6:1044–1068. https://doi.org/10.1139/facets-2020-0075

Richardson, K. et al. 2023. Earth beyond six of nine planetary boundaries. *Sci. Adv.* 9:37. https://doi.org/10.1126/sciadv.adh2458

Richardson, L., J. Loomis, T. Kroeger, and F. Casey. 2014. The role of benefit transfer in ecosystem service valuation. *Ecol. Econ.* 115:51–58.

Robinson, J.M., Gellie, N., MacCarthy, D., Mills, J.G., O'Donnell, K. & Redvers, N. 2021. Traditional ecological knowledge in restoration ecology: A call to listen deeply, to engage with, and respect Indigenous voices. Restoration Ecology, 29(4), e13381.

Rockström, J., Gupta, J., Qin, D. *et al.* 2023. Safe and just Earth system boundaries. Nature 619:102–111. https://doi.org/10.1038/s41586-023-06083-8

Schmeller, D.S., Niemelä, J. & Bridgewater, P. 2017. The Intergovernmental Science-Policy Platform on Biodiversity and Ecosystem Services (IPBES): Getting involved. Biodiversity Conservation 26: 2271–2275. https://doi.org/10.1007/s10531-017-1361-5

Scholte, S.S.K., A.J.A. van Teeffelen, and P.H. Verburg. 2015. Integrating socio-cultural perspectives into ecosystem service valuation: A review of concepts and methods. *Ecol. Econ.* 114:67–78.

Schröter, M., D.N. Barton, R.P. Remme, and L. Hein. 2014. Accounting for capacity and flow of ecosystem services: A conceptual model and a case study for Telemark, Norway. *Ecol. Ind.* 36:539–551.

Seppelt, R., B. Fath, B. Burkhard et al. 2012. Form follows function? Proposing a blueprint for ecosystem service assessments based on reviews and case studies. *Ecol. Indic.* 21:145–154.

Sherrouse, B.C., D.J. Semmens & Z.H. Ancona. 2022. Social Values for Ecosystem Services (SolVES): Open-source spatial modeling of cultural services. Environmental Modelling & Software 148:105259. https://doi.org/10.1016/j.envsoft.2021.105259

Shewaga, J. 2023. USask experts sound warning over wildfire smoke, drought conditions. University of Saskatchewan News. USask experts sound warning over wildfire smoke, drought conditions—News—University of Saskatchewan accessed Dec. 22., 2023.

Soverel, N.O., N.C. Coops, D.D. Perrakis, L.D. Daniels, and S.E. Gergel. 2011. The transferability of a dNBR-derived model to predict burn severity across 10 wildland fires in western Canada. Int. J. Wildland Fire, 20(4):518–531.

Statistics Canada. 2013. *Measuring ecosystem goods and services in Canada.* Human Activity and the Environment: Annual Statistics 2013. Statistics Canada Catalogue no. 16–201-X. www.statcan.gc.ca/pub/16–201-x/16–201-x2013000-eng.pdf.

Steel, Z.L., H.D. Safford, and J.H. Viers. 2015. The Fire Frequency-Severity Relationship and the Legacy of Fire Suppression in California Forests. *Ecosphere* 6(1):8–23.

Steffen, W., W. Broadgate, L. Deutsch, O. Gaffney, and C. Ludwig. 2015. The trajectory of the anthropocene: The great acceleration. *Anthropocene Rev.* http://dx.doi.org/10.1177/2053019614564785.

Stoll, S., M. Frenzel, B. Burkhard et al. 2015. Assessment of ecosystem integrity and service gradients across Europe using the LTER Europe network. *Ecol. Model.* 295:75–87.

Syrbe, R.-U. and U. Walz. 2012. Spatial indicators for the assessment of ecosystem services: Providing, benefiting and connecting areas and landscape metrics. *Ecol. Ind.* 21:80–88. https://www.sciencedirect.com/science/article/abs/pii/S1470160X12000593

Tallis, H. and S. Polasky. 2009. Mapping and valuing ecosystem services as an approach for conservation and natural-resource management. *Year Ecol. Conserv. Biol.* 1162:265–283.

TEEB. 2010. *The economics of ecosystems and biodiversity ecological and economic foundations*, ed. P. Kumar. London and Washington, DC: Earthscan.

Thomas, K. and L. Geller. 2013. *Urban forestry: Toward an ecosystem services research agenda: A workshop summary.* National Academies Press, 2013. https://books.google.ca/books?hl=en&lr=&id=TgF1AgAAQBAJ&oi=fnd&pg=PA5&dq=thomas+and+geller+2013+ecosystem+services&ots=QUoaBjnBEc&sig=wuFlvERa9s47bIlUxkdOblvb6fQ#v=onepage&q=thomas%20and%20geller%202013%20ecosystem%20services&f=false

Thompson, I.D., M.R. Guariguata, K. Okabe, C. Bahamondez, R. Nasi, V. Heymell & C. Sabogal. 2013. An Operational Framework for Defining and Monitoring Forest Degradation. Ecol. Soc. 18(2):20.

Thompson, I.D., B. Mackey, S. McNulty & A. Mosseler. 2009. Forest Resilience, Biodiversity, and Climate Change. A Synthesis of the Biodiversity/Resilience/Stability Relationship in Forest Ecosystems. Secretariat of the Convention on Biological Diversity, Montréal. Technical Series no. 43.

Thompson, I.D, K. Okabe, J.M. Tylianakis, Kumar, P., Brockerhoff, E.G., Schellhorn, Parotta, J.A., Nasi, R. 2011. Forest Biodiversity and the Delivery of Ecosystem Goods and Services: Translating Science into Policy. BioScience 61:972–981. http://dx.doi.org/10.1525/bio.2011.61.12.7.

Tilman, D., J. Knops, D. Wedin, P. Reich, M. Ritchie, and E. Siemann. 1997. The influence of functional diversity and composition on ecosystem processes. *Science* 277:1300–1302.

US Columbia Basin Tribes and Canadian First Nations. 2014. *Fish passage and reintroduction into the US and Canadian upper Columbia River.* An Interim Joint Paper of the US Columbia Basin Tribes and Canadian First Nations, February.

Value of Nature to Canadians Study Taskforce (VNCST). 2017. *Completing and using ecosystem service assessment for decision-making: An interdisciplinary toolkit for managers and analysts.* Ottawa, ON: Federal, Provincial, and Territorial Governments of Canada. https://static1.squarespace.com/static/613fb778a76e244eef08775d/t/61c3973a898e7e19c3ccf9d8/1640208222620/2017_Ecosystem_Services_Toolkit.pdf

van Oudenhoven, A.P.E., K. Petz, R. Alkemade, L. Hein, and R.S. de Groot. 2012. Framework for systematic indicator selection to assess effects of land management on ecosystem services. *Ecol. Indic.* 21:110–122. https://doi.org/10.1016/j.ecolind.2012.01.012

Wallace, K.J. 2007. Classification of ecosystem services: Problems and solutions. *Biol. Conserv.* 139(3–4): 235–246. https://doi.org/10.1016/j.biocon.2007.07.015

Whitman, E., M.A. Parisien, D.K. Thompson, and M.D. Flannigan. 2019. Short-interval wildfire and drought overwhelm boreal forest resilience. Sci. Rep. 9(1):1–12.

Whitman, E., M.A. Parisien, D.K. Thompson, R.J. Hall, R.S. Skakun, and M.D. Flannigan. 2018. Variability and drivers of burn severity in the northwestern Canadian boreal forest. *Ecosphere* 9(2):e02128.

Willemen, L., L. Hein, M.E.F. van Mensvoort, and P.H. Verburg. 2010. Space for people, plants, and livestock? Quantifying interactions among multiple landscape functions in a Dutch rural region. *Ecol. Ind.* 10:62–73.

Willemen, L., P.H. Verburg, L. Hein, and M.E.F. van Mensvoort. 2008. Spatial characterization of landscape functions. *Landsc. Urban Plan.* 88:34–43.

World Wildlife Fund. 2022. WWF reaction: IPBES value of nature report. https://wwf.panda.org/wwf_news/?6023941/WWF-reaction-IPBES-Value-of-Nature-report

Wulf, A. 2016. *The invention of nature: Alexander von Humboldt's new world.* New York, NY: Vintage.

Section II

Forest Models

6 Growth and Yield Models for Predicting Tree and Stand Productivity

Robert Schneider, Modeste Meliho, Gianluca Segalina, Mathieu Fortin, Tommy Simard, Tony Franceschini, Olivier Martin-Ducup, Vincent Gauthray-Guyénet, Dipesh Kumar Sharma, and Guy R. Larocque

6.1 INTRODUCTION

Evaluating different management options for forest stands is a complex task. Forests are dynamic systems, and the rate of change is determined by many factors, including species composition, climate, site, and both natural and anthropogenic disturbances. A common approach when making decisions in a complex situation is to simplify the problem. This can be done for forest stands by using growth and yield models. Such models can be divided into (1) empirical or statistical models, (2) gap models, (3) process-based models, and (4) hybrid models.

Confusion can arise in the use of the term model. "Model" can be used to represent a mathematical equation as well as the set of interrelated equations used to predict growth and yield of a stand. Some people use the term "simulator" to represent the set of equations. However, to remain consistent with much of the growth and yield literature, we will use "model" to refer to either a given single equation or a set of equations used for simulations.

This chapter will focus on empirical growth and yield models which are based on statistical relationships between indicators of forest growth and yield and a selection of explanatory variables (e.g., tree size, competition intensity, stand density, or site variables), even though process-based models (i.e., models that are based on the modeling of physiological processes to quantify tree or stand development) have the same objectives but use more theoretical approaches. The use of empirical growth and yield models in forestry will be discussed (Section 6.2), with respect to different spatial and temporal horizons. The different components of growth and yield models will then be described by looking at how traditional models work and how new technologies can be included (Sections 6.3 and 6.4). Section 6.5 will explore how empirical models have been modified to integrate the representation of the effects of global change by looking at each component of net stand yield: net yield = tree growth + tree recruitment − mortality. As some effects are hard to capture using empirical relationships, a summary of how process-based models tackle the problem will also be presented. Section 6.6 will describe some basic approaches to quantifying uncertainty in model predictions. The final section (6.7) will conclude on perspectives on the development of growth and yield models.

6.2 MANAGEMENT AND SILVICULTURE MODELS

Foresters use growth and yield models for multiple purposes. The three traditional applications are for updating forest inventories, supporting forest management decisions, and comparing the effects of silvicultural treatments. In the last decade, models have increasingly been used to evaluate the effects of global change on forest dynamics.

DOI: 10.1201/9781003431084-8

Forest management decisions generally involve estimating current stand volume and determining allowable annual cut (AAC). The methodology generally consists of conducting some type of optimization or simulation using yield curves to project the growth of individual stands or groups of similar stands. Growth and yield models also help silviculturists to explore different scenarios and evaluate which treatments will meet the management objectives prior to implementing the treatments in the field. Moreover, with the changes in management paradigms, such as the emergence of ecosystem-based forest management (Barrette et al. 2014) or increasing resistance of forests to global change (Achim et al. 2021), novel silvicultural treatments are being proposed without any data from past field experiments. Appropriate growth and yield models can help silviculturists to evaluate the short- and long-term impacts of such treatments. Growth and yield models designed to support these four primary uses generally have different data requirements, principally spatial and temporal resolution, as well as output.

6.2.1 Growth and Yield Models for Forest Management

Forest management decisions are generally made at large spatial scales, on forest management units, landscapes, or forest estates (i.e., hundreds to millions of hectares). Forest management issues relate to the structure and dynamics of the management unit. The inputs generally include a combination of forest inventory data and stand maps. Depending on the jurisdiction, not all forest stands are sampled during a forest inventory. Generally, the forest stands are grouped into strata according to ecological and silvicultural criteria. For instance, in Quebec, Canada, forest stands are grouped into inventory strata by dominant species composition, dominant height class, age class, and several key biophysical characteristics, including surface deposit, drainage class, and ecological type (Ministère des Forêts, de la Faune et des Parcs 2014). These strata are then used as the basis for the sampling design and the desired precision per strata can be set during the planning stages. Furthermore, measurements taken during the inventory will concentrate on establishing a portrait of each stratum (e.g., volume or basal area per species). Generally, inventory strata that share the same natural dynamics are then grouped into management strata that will have the same silvicultural scenario applied with the same operational constraints (Raulier et al. 2009). These management strata are then used as the basis for AAC determinations.

Growth and yield models used for supporting forest management decisions must have the capacity to accept broad scale (i.e., forest inventory) data as input. Their spatial resolution is low, with strata average yield through time being the primary focus. Recent efforts have been made to include both sampling and modeling errors to evaluate both strata average yield and some type of confidence interval (Fortin et al. 2022a). Although yield varies with the silviculture regime through density and composition management (Assmann 1970), the objective is not to obtain specific site-level predictions with the smallest errors. Often, simulations over 150–200 years, in 5–10-year steps, are made to establish a yield curve, generally in volume (e.g., $m^3\ ha^{-1}$), for a given management stratum. If a tree-level model (see predicting growth and yield section) is used, this output is aggregated to provide stand-level information.

The AAC determinations will be used to establish a management strategy, which will outline the steps to reach the objectives defined for the management unit. The objectives will encompass the desired structure of the forest and identify the methods to achieve them.

6.2.2 Growth and Yield Models for Silviculture

Silviculturists rely on growth and yield models to make site-specific decisions. Generally, models that are designed to support silviculture decisions are used to compare the growth and yield of different treatments for a given site at a given time. Thus, the development of such models needs more intensive input data than models designed for forest management decisions (stems per ha by diameter class, crown lengths etc.). The silviculturist often requires site-specific predictions, with a need for high sensitivity to site variables such as fertility, species composition, and stand density. Generally, the simulation horizon will be shorter than for management models, where the values at the time of the next entry (e.g., 15–30-year horizon) are generally predicted in 1–5-year increment steps. These models are usually tree-level (see predicting growth and yield section).

The silviculturist can use models to compare different treatments for given stand conditions and choose one of the treatments that meet the desired objectives. The decision can be based on volume yield, or on additional criteria, such as stand composition, diameter distributions, or lumber grades.

6.3 PREDICTING GROWTH AND YIELD

Growth and yield predictions are based on one or more mathematical models (equations) that represent different aspects of stand dynamics. The parameters of each equation may be estimated using a variety of statistical tools (Robinson and Hamann 2010; Weiskittel et al. 2011). Multiple sources explain how growth and yield models may be built (e.g., Vanclay 1994; Pretzsch 2009; Weiskittel et al. 2011). This section will briefly cover the different types of growth and yield models, based on their spatial resolution: whole stand, size class, and individual trees.

6.3.1 Whole Stand

Whole stand growth models reduce the forest to a few key stand-level variables that can be divided into the state variables to be predicted and the stand characteristics influencing the temporal trajectories of the state variables. The state variables generally quantify the amount of timber in a particular forest at a given time t and often include stand volume (m^3 ha^{-1}), stand basal area (m^2 ha^{-1}), mean quadratic diameter at breast height (cm), and stand density (stems ha^{-1}). These state variables are generally predicted in 5–10-year time steps, although stand volume yield (m^3 ha^{-1}) is more often estimated at the end of the period rather than during the growth period. Due to the simplicity of this approach, stand-level models have been adapted to a wide variety of species and biomes (e.g., Bermejo et al. 2004; Anta et al. 2006; García 2011; Pothier and Auger 2011).

Yield trajectories vary with both site fertility and competition within the stand. Details on how site fertility can be quantified can be found in the section on quantifying site variations. The effect of competition on stand growth and yield is generally characterized by variables, such as stand density, basal area or relative stand density, with details found in the section on quantifying competition.

Whole-stand models rely on the principle that the dominant height of the stand is a good predictor of the standing volume (Pretzsch 2009). Consequently, the construction of such a model is achieved in two steps:

(1) Predicting the stand dominant height using tree age and site fertility; and
(2) Estimating stand yield from the dominant height and a measure of tree density to account for competition.

This approach was the basis for the first yield tables in the 1700s (cf. Pretzsch 2009), although variations in stand density were not necessarily taken into account. Over the last several decades, with increased computational power combined with complex statistical modeling methods, whole-stand models can be considerably more complex than the early yield tables.

Another approach to simulating the development of forest stands is to rely on stand density diagrams, which are related to Reineke's stand density index (Reineke 1933) and Yoda's self-thinning rule (Yoda et al. 1963). The stand density on one axis and average tree size (either average tree volume or quadratic mean dbh) on the other axis are used to place the stand in the diagram. Allometric relationships can also be used to position the stand in the diagram using other metrics, such as stand dominant height in combination with stand density. These metrics can then be extracted for any stand. North American stand density diagrams tend to place stand density on the x axis (e.g., Newton 1997; Wilson et al. 1999), whereas European ones place it on the y axis (e.g., Castedo-Dorado et al. 2009; Ningre et al. 2016).

Depending on which axis stand density is placed, stands develop vertically or horizontally in the diagram (i.e., mean tree size increases without any change in stand density) until the stand reaches the self-thinning induced mortality zone (i.e., a relative density greater than 0.65) (Figure 6.1). Within the mortality zone, the stand then follows a line parallel to the maximum density relationship. Survivorship/mortality curves calibrated empirically can also be included in the stand density management diagram (Newton 1997; Penner et al. 2006). The speed at which the stand progresses through the diagram can be established using site index curves (Jack and Long 1996; Vacchiano et al. 2008). The volume of the stand at a given age is then established by multiplying the average tree volume by stand density.

Whole-stand growth models have the advantage of rapidly providing simple stand metrics for even-aged conditions. Moreover, their use does not necessitate very detailed information, other than what is needed to establish the input variables. These types of models are ideal for supporting management decisions, where the need for detailed information is not important and when simulations

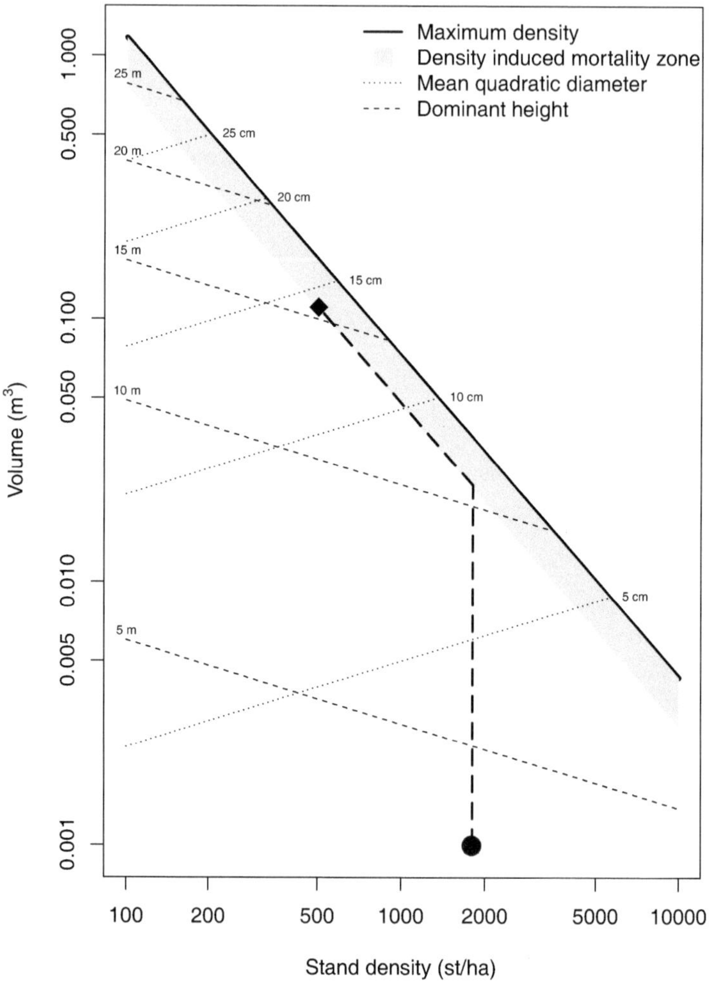

FIGURE 6.1 Stand density management diagram based on Wilson et al. (1999). The bold dashed line represents the development of a stand with a mean tree volume of 0.001 m^3 and a density of 1,800 st ha^{-1} (solid round dot) to a final density of 500 st ha^{-1} (solid diamond).

are conducted over long time periods. In addition, whole-stand models implicitly take into account growth and mortality dynamics, with recruitment assumed to be negligible in even-aged stands. However, these are difficult to use in mixed stands and cannot be used for irregular or uneven-aged stand simulations. Moreover, they may be inconsistent for use in the simulation of different silvicultural scenarios. Even stand density management diagrams, which were developed mainly for silvicultural decisions, are only valid for simulating thinning from below, as this type of thinning mimics self-thinning mortality.

6.3.2 Size Class

The size class approach groups similar-sized individuals together. The transition between each size class can be estimated by different methods (cf. Pretzsch 2009; Weiskittel et al. 2011): differential equations, matrix-based, transition probabilities, and diameter distribution methods.

In its simplest and oldest form, stand tables were used to project future stand development using one of several assumptions. An example of such an assumption is that the trees are evenly distributed in the size classes and that each tree in each class has the same growth rate on average. The number of trees that change classes in each period is thus determined by the size class interval and the growth rate. Such approaches assume that growth is strongly density independent; recruitment and mortality are not easily integrated.

The stand table approach can be modified to take into account the complexity of forest dynamics through matrix-based approaches to predict transitions between size classes. The transition functions can be time-independent (e.g., Solomon et al. 1995), linear (e.g., Liang et al. 2005), or nonlinear (e.g., Liang 2010). Matrix-based models can be applied to even- or uneven-aged, pure and mixed stands. As with the simplest form, ingrowth and mortality dynamics can be hard to incorporate, making them more adapted to silvicultural models with short temporal simulation horizons than for use to simulate stand development over long time periods such as those required in forest management.

Finally, stand dynamics can be simulated using diameter distribution models. This approach consists of fitting a probability distribution, such as the Weibull distribution, to the diameter distributions observed in permanent sample plot measurements. The parameters of the probability distribution are then related to productivity indicators and/or stand age (e.g., Hyink and Moser 1983; Návar 2014). The premise for such models is that the probability distribution parameters are sensitive to both changes in time and silvicultural inputs, which is not always the case (Schneider et al. 2013).

6.3.3 Individual Trees

The last class of growth and yield models considers individual trees. These models can work either on each tree in a sample plot or on a tree list, where the trees can be scaled to the stand level by using the plot to stand expansion factor (e.g., a tree in a 400-m^2 plot represents 25 stems ha^{-1}).

This type of model has submodels for each component of stand dynamics: tree growth, tree mortality, and ingrowth as described in the section on forest dynamics. Tree growth is generally predicted by using a height, dbh or volume increment model, where tree increment is determined by the size of the tree at time t, site factors, and competition. Quantifying competition can be done using spatial or aspatial metrics (see the section on quantifying competition). Modeling tree increment is the easiest part of developing an individual-tree growth model if good data are available.

Tree mortality is stochastic and often episodic with a low average rate (ranging between 0.5% and 4.0% across Canada, Peng et al. 2011), making its modeling difficult. Moreover, it is well known that the probability of mortality increases with decreasing growth (Wyckoff and Clark 2002) and that small and large trees have higher mortality probabilities than medium-sized trees (Platt et al. 1988). These issues are more problematic for ecosystems with a large number of species and low number of stems per species, such as tropical rainforests (Vanclay 1994). Statistical approaches to modeling mortality probabilities, such as Bayesian statistics, can help in such situations (Metcalfet al. 2009). Otherwise, logistic regressions are generally used.

Growth and yield models generally apply to stems above a certain size. Recruitment models predict the number of stems, or the probability of observing new trees, within a given plot at every time step that reach the minimum size to be considered. Ingrowth is generally predicted by the model time step, stand composition, and site variables (Weiskittel et al. 2011).

Individual-tree models have the finest spatial resolution of the different classes of models and can be used either for management (e.g., Fortin and Langevin 2010) or for silvicultural (e.g., Pretzsch et al. 2002) decisions. In addition, they can be developed for complex stands and can be used to simulate a wide variety of thinning regimes. Many individual-tree models were developed for application in numerous regions around the world.

6.3.4 Growth Model Formulation

A growth model can be either linear or nonlinear. Information on building growth models and advantages and disadvantages of linear and nonlinear approaches can be found in Pretzsch (2009) and Weiskittel et al. (2011). Here, we will briefly describe the basics of each type of model and present some of their advantages and disadvantages.

6.3.4.1 Linear Models

A linear model is defined as a linear combination of a series of parameters (Quinn and Keough 2002); hence the first derivative of the model with respect to a given parameter does not contain that parameter (Robinson et al. 2010):

$$y_i = \beta_0 + \beta_1 x_{i,1} + \beta_2 x_{i,2} + \ldots + \beta_j x_{i,j} + \varepsilon_i \tag{6.1}$$

where:

y_i: is the growth of the ith stand or tree i;
β_0 to β_j: parameters to be estimated;
$x_{i,1}$ to $x_{i,j}$: the independent variables, containing metrics of competition and site fertility and possibly other variables;
ε_i: the residual error.

Ordinary least squares regressions are commonly used to estimate the model parameters, under the assumption that each covariate is independent. The residual errors are assumed to be normally and independently distributed, with homogeneous variances. If some of these assumptions are not met, they can be relaxed using a generalized least squares estimator to account for heteroscedasticity or error correlation (Pinheiro and Bates 2000).

The advantage of linear models is that the parameters are easily estimated. However, such models do not often represent our conception of the biological processes underlying growth. Moreover, any observed linear trend between growth and the covariates might be highly dependent on the data range.

To obtain curvilinear predictions within a linear model framework, variable transformations are needed, or polynomial covariates must be used. Variable transformations, especially if the dependent variable is transformed, entails computing a correction factor to back-transform the response variable (e.g., Beauchamp and Olson 1973).

6.3.4.2 Nonlinear Models

Nonlinear models are defined as models where the parameters are nonlinearly related to the in dependent variables. The parameters can thus be powers, exponentials, and/or raised to the power of a covariate.

Growth models can take many nonlinear forms and have several relevant properties, such as asymptotes (Ratkowsky and Giles 1990). Common nonlinear function forms for growth and yield models include the power, the Chapman-Richard's (Richards 1959), generalized logistic, and the Gompertz functions.

An important nonlinear model based on the biological conception of tree growth is the potential growth reduction model (Canham et al. 2006):

$$Growth = Growth_{max} \times red_{size} \times red_{site} \times red_{competition} \tag{6.2}$$

where:

- *Growth* is the increment in individual tree size (either in diameter or basal area), and can be either in relative (increment/size) or absolute values (increment);
- $Growth_{max}$ is the maximum growth observed for the species;
- red_{size}, red_{site} and $red_{competition}$ are nonlinear reduction functions for tree size, site fertility, and competition, with values for each function ranging between 0 and 1.

In models that rely on potential diameter reduction, a problem can arise in estimating the maximum potential tree growth. Generally, the potential growth is estimated independently from the modifier function parameters (Wykoff and Monserud 1988), or else the parameter estimates may be biased (Lessard et al. 2001). Furthermore, such a model can underestimate tree growth, especially if the number of reduction functions is high (Weiskittel et al. 2011).

The parameters of nonlinear models can be estimated using nonlinear least-squares techniques or likelihood estimation. As with linear models, covariates must be independent, and residual errors must have homogeneous variances, be independent, and normally distributed. Variance modeling and error correlation can also be included when fitting the model.

Nonlinear models are more complex than linear models and can lead to some practitioners losing confidence in the model due to its complexity. Although extrapolating outside the range of data used to calibrate the model is never recommended, a biologically realistic (and thus likely nonlinear) model form helps reduce the probability and magnitude of erroneous predictions compared to models with no restriction on model form. Also, nonlinear models may have fewer parameters to estimate compared to linear models with the same shape, making them more parsimonious. The choice of the functional form is a key step in model building and must be performed meticulously.

6.3.4.3 Process-Based Models

Process-based models are based on the cause–effect mathematical representation of different ecological and physiological mechanisms that occur in forest ecosystems, including carbohydrate regulation, water and nutrient cycles, synthesis of hormones, or carbon partitioning. Most process-based models describe the effects of environmental factors that affect these processes, such as the interception of photosynthetic active radiation (PAR) at different canopy levels, nutrient and water uptake rates, and climatic conditions. Their development began in the 1980s when the need to better understand and simulate the impacts of environmental disturbances was identified. The disturbances issues considered have included the effects of acid rain, air pollution, and climate change.

In the last few decades, several process-based models differing in type and number of processes and level of details have been developed. Process-based models may consist of single models to simulate particular processes. A good example is the photosynthesis model developed by Farquhar et al. (1980) and Farquhar and von Caemmerer (1982). This model has been widely used to simulate photosynthesis (e.g., Larocque 2002b; Von Caemmerer 2013) and has been integrated in more complex models to simulate carbon cycles in forest ecosystems (e.g., Zhou et al. 2008; Yan et al. 2019). The FOREST-BGC model developed by Running et al. (1991) established a benchmark that influenced the subsequent development of complex biogeochemical models. For instance, BIOME-BGC (Running and Hunt 1993), a descendant of FOREST-BGC, included the photosynthesis model developed by Farquhar et al. (1980) and Farquhar and von Caemmerer (1982). CENTURY (Parton et al. 1988, 1993) is another process-based model that gained wide recognition since its development. It focuses on soil carbon, nitrogen, phosphorus, and sulfur cycles and includes a vegetation component.

Process-based models have been used mainly to simulate above- and below-ground carbon cycles in forest ecosystems. Thus, typical predicted variables have consisted mostly of turnover rates and pools of carbon, nutrients, water, or biomass. However, there are examples of process-based models that compute forest productivity variables, such as increase in diameter at breast height (dbh), basal area, or wood volume. This approach has been identified as essential to integrate the effects of climate change on stand productivity (Franklin et al. 2016). A good example of this approach is the work by Collalti et al. (2014), which relied on the 3D-CMCC model to predict changes in dendrometric variables, including diameter, height, crown diameter, and leaf area index. It is composed of several submodels that simulate ecological and physiological processes, such as soil water circulation, evapotranspiration, PAR interception, and carbon allocation. The 3D-CMCC model includes a detailed description of forest structure, which is used to compute new biomass produced that is allocated to stem, root, and foliage compartments. Gupta and Sharma (2019) reviewed different studies based on the 3-PG model to predict forest productivity in terms of traditional productivity indices used for forest management planning.

More details on process-based models are provided in Chapter 8.

6.3.5 Quantifying Competition

Competition affects the amount of resources that a tree can access. Together with site quality, it is one of the most important model components used to predict tree or stand growth. Competition indices can be established at the stand or individual-tree level. At the tree level, these indices can be subdivided into distant-independent and distant-dependent (Munro 1974). Moreover, with the advent of new technologies, inter-tree competition can be defined more precisely. This section will summarize the different types of competition indices and will introduce how competition can be evaluated using new technologies.

6.3.5.1 Stand-Level Competition

A competition index is designed to assess the average competitive pressure a tree receives. This can be done by simply using some measure of stand density. However, this approach has some drawbacks (Zeide 2004). A major one is that this type of competition index does not provide information on the spatial variation (e.g., clustered versus random) of the trees within the stand. Moreover, if the number of stems per ha (or basal area) is used as the indicator of stand density, it is necessarily dependent upon the sizes of the trees present. For instance, a large number of small trees may exert less competitive pressure than a considerably smaller number of larger trees.

To compensate for the effect of size, relative stand density, calculated from the self-thinning rule (Yoda et al. 1963) or Reineke's index (Reineke 1933) have also been used to quantify stand competition (Pothier and Savard 1998). This approach leads to a ratio of the observed stand density to the maximum stand density of the stand. Once again, this value expresses the competition faced by an average tree, without regard to spatial distribution of the trees within the stand.

To internalize spatial distribution, the variability of the diameter distribution has also been used in conjunction with stand density and/or basal area (Pothier and Auger 2011). Stand structure and its dbh distribution are related. Stands with a large variation in dbh are more likely to have an irregular spatial distribution, where small trees will be clumped together and more distant from the large trees (i.e., a multimodal or inverse-J distribution). Conversely, a managed stand, with regularly spaced trees, will have a dbh distribution centered around its mean (i.e., a unimodal distribution).

6.3.5.2 Individual Tree-Level Competition

Distant-independent competition indices (DICs) rely on the relative size of the tree with respect to the other trees of the stand, whereas distant-dependent competition indices (DDCs) use the spatial position of the trees to estimate competition intensity. A first approximation of the competition can be done using stand-level competition indices. However, this provides only the competition for an average tree and does not make use of more detailed information available, such as a tree list.

DICs can be divided into three different categories (Prévosto 2005), where the indices are based on (i) basal area or stand density; (ii) size ratios; and (iii) tree vigour or crown measures (Table 6.1). An example of the first category would be the basal area of the trees with a DBH greater (BAL) than the tree of interest (Wykoff et al. 1982); trees with a high BAL value will be subject to more competition. The second group of DICs attempts to evaluate the social position of the tree within the stand. For instance, the ratio of the basal area or dbh of the tree of interest to the basal area or dbh of the average tree of the stand can be used (Tome and Burkhart 1989). The third class generally relies on crown dimensions, such as surface area or volume (Prévosto 2005). Such tree metrics give an indication of past competition and can thus be used to assess the likelihood of competition in the present. DICs were at the center of the development of simulation systems composed of interrelated model components to simulate tree and stand growth and the effects of silvicultural treatments. Some of them gained strong recognition for operational use. For instance, the Forest Vegetation Simulator (FVS) has become the official growth model of the USDA forest service (Dixon 2002).

To use DDCs, more detailed information is needed, such as the location of every tree. DDC can be subdivided into three classes (Biging and Dobbertin 1992; Weiskittel et al. 2011): (i) area-based DDCs, (ii) distance-based DDCs, and (iii) light interception-based DDCs (Table 6.1). Larocque (2008) suggested a similar classification, except that the type of DDC was based on crown interference.

Area-based DDCs rely on the fact that the available growing space for a tree depends on its competitors. The first sub-category considers the available space around trees and essentially divides the stand into polygons around each tree using methods such as Voronoi tessellation (Daniels et al. 1986). The area of these polygons represents the space that is available for tree development. The second sub-category considers the area that is shared by neighboring trees and assumes that the radius of the competition zone is proportional to the maximum radius of the crown of an open-grown tree of the same size. This approach, based on the application of the crown competition factor (Krajicek 1961), defines influence zones, and the competition index corresponds to the degree of overlap between influence zones (Pretzsch 2010).

DDCs often weigh the distances between trees using tree size or ratios of distance between the trees and their size (dbh and/or total height). The assumption behind such indices is that the competition to which a tree is subjected is proportional to the competitor's size and inversely proportional to the competitor's distance. Identifying the competitors that should be included in the computation of the index is an important step. The competitors can be defined either by a fixed radius around the target tree (e.g., all trees at a distance less than r meters are considered as competitors) or by using the closest neighboring trees (e.g., trees where the crowns touch). A few indices have been derived by relating the intensity of competition to change in crown development. Hatch et al. (1975) derived a competition index based on crown interference, but the dynamic nature of crown development was not considered. Larocque (2002a) proposed a competition index based on the effect of the crown dimensions of competitors on the crown development of subject trees. The changes in crown width and length resulting from the influence of competitors were related to individual tree growth. Many such indices exist (Weiskittel et al. 2011), and some of which are listed in Table 6.1.

DDCs based on light interception can use hemispherical photographs and ray tracing algorithms to characterize canopy openness (ratio of visible sky to leaf surface as seen through the fisheye image) or the effective leaf area (Mailly et al. 2003; Groot 2004). Another choice would be to directly compute light extinction through the canopy (Duursma and Mäkelä 2007). These indices aim at quantifying the actual light available for tree growth once the neighboring shade is considered.

When using DDCs, competitors must be identified. Several methods have been proposed for this purpose. For instance, a radius proportional to tree height around the target tree (Lorimer 1983) or trees where their apex overlaps an upside down cone based either at stump or at crown base of the subject tree (Pretzsch et al. 2002). Larocque (2002a) evaluated the amount of 3D crown overlap to identify competitors, and his model differentiated between asymmetric and symmetric competition.

TABLE 6.1
List of the Major Distant-Independent and Distant-Dependant Competition Indices

	Name of Indices	Description	References
Distance-Independent Indices			
Stand level	Basal area ($m^2.ha^{-1}$)		
	Stand density ($nbst.ha^{-1}$)		
	Stand density index	$SDI = N \cdot \left(\frac{DBH}{25.4}\right)^{1.604}$	Reineke (1933)
Tree level	Basal area of trees larger than the tree	$BAL_i = \sum_{j\forall(DBH_j \geq DBH_i)} BA_j$	Wykoff *et al.* (1982)
	Basal area index for tree	$BAI_i = \frac{BA_i}{BA}$	Tome & Burkhart (1989)
	Crown dimensions	Crown length Crown surface	
Distance-Dependent Indices			
Area-based	Area potentially available	Area determined by polygons, through Voronoi tessellation or weighted by the DBH of a tree and its competitors	Moore *et al.* (1973), Daniels *et al.* (1986)
	Bella's influence zone competition index	$IC_i = \sum_{j \neq i}\left[\frac{O_{ij} \cdot DBH_j}{Z_i \cdot DBH_i}\right]$	Bella (1971)
Distance-based	Distance and size weighted ratio	$IC_i = \sum_{j \neq i}\left[\left(\frac{D_j}{D_i \cdot Dist_{ij}}\right)\right]$	Hegyi (1974)
	Negative exponential distance and size weighted ratio	$IC_i = \sum_{j \neq i}\left[\frac{D_j}{D_i} \cdot e^{\frac{16 Dist_{ij}}{D_i + D_j}}\right]$	Martin & Ek (1984)
	Distance and crown dimension weighted ratio	$IC_i = \sum_{j \neq i}\left[\left(\frac{CD_j}{CD_i \cdot Dist_{ij}}\right)\right]$	Biging & Dobbertin (1992)
	Point density index	$IC_i = \frac{75.625}{n}\sum_{j=1}^{n}\left[(j+0.5)\frac{D_j^2}{Dist_{ij}^2}\right]$	Spurr (1962)
Light interception based	Canopy openness	Percentage of open sky visible from a hemispherical image	Mailly *et al.* (2003)
	Leaf area index	Leaf area index computed from a hemispherical image	Mailly *et al.* (2003)
	Light interception models	Models of light interception taking into account the heterogeneity of the canopy	Duursma & Mäkelä (2007)

Tree maps are needed for model calibration. However, methods exist to impute the location of a tree from readily available measurements (Pretzsch 1997).

A number of authors concluded that DICs are better to predict tree growth, or at the very least, DDCs are not better enough, considering the extra cost of obtaining tree maps (e.g., Martin and Ek 1984; Biging and Dobbertin 1995). However, Pretzsch (2009) demonstrated that DDCs are similar to DICs for predicting growth in regularly distributed stands and superior for predicting growth in clustered stands. Mailly et al. (2003) showed that DDCs were better for predicting black spruce (*Picea mariana* (Mill) B.S.P.) growth. In white spruce plantations, Bérubé-Deschênes et al. (2017) found that DIC and DDC were hard to differentiate, and the choice should rest on the outputs sought by the users. With the pressure on silviculturists to use varied treatments leading to more irregular stands (Barrette et al. 2014), the use of DDCs becomes more important for simulating the growth of the residual stands (Schneider et al. 2021).

6.4 USE OF REMOTE SENSING

Enhanced forestry inventories are becoming more widely available (Lister et al. 2020) and can contribute to the development of ecological models. Forest structure can be better defined by using three-dimensional (3D) metrics (Lechner et al. 2020). For example, these methods can rely on LiDAR (Light Detection And Ranging) systems (Coops et al. 2021), photogrammetry (Goodbody et al. 2019), or a combination of sensors (White et al. 2016). LiDAR scanners utilize lasers to register coordinates of solid objects, creating 3D point cloud representations of scanned areas. These scanners can be ground-based (TLS, MLS), mounted on drones (UAVLS) or in airplanes (ALS), or deployed on satellites (DAP). Photogrammetry uses georeferenced pictures to create 3D representations of forests from an aerial perspective that are cost-effective when compared to ALS (Kangas et al. 2018; Iglhaut et al. 2019). Combining TLS or UAVLS and photogrammetry allows for improved estimates and mapping of the forest (St-Onge et al. 2008; Jayathunga et al. 2018).

The estimation of tree height, crown size, and stand structural complexity can now be performed at regional and even national scale with ALS data that are increasingly available worldwide (Marczak et al. 2020; Guo et al. 2021). At the plot scale, TLS, and now MLS, enables quick acquisition of basic dendrometric variables (dbh or height) as well as more detailed crown attributes and even tree volume to enhance forest inventories (Vandendaele et al. 2022). The estimation of vertical profile of leaf-area density (LAD) from LiDAR data through radiative transfer modeling have received a lot of attention and research effort in the last decade and can be estimated at both local and regional scales (Pimont et al. 2018; Bouvier et al. 2015). All these improvements in vegetation characterization at different scales are very promising for modeling forest growth and yield through process-based models with biophysical variables (i.e., LAD) or through empirical models using classic variables (i.e., tree height and diameter, tree location, crown size). Tompalski et al. (2016, 2018) used an area-based approach to evaluate standing volume and future growth. They first divided plots into grids and then estimated various metrics, such as basal area, timber volumes, canopy structures, dbh or tree height. These were then related to growth and yield model parameters to project growth of each forest data pixel (i.e. forel). Growth metrics can also be directly estimated using multitemporal measures (Hopkinson et al. 2008).

Remote sensing efficiently automates data acquisition and can thus lead to large datasets that can be used to train deep or machine learning algorithms. However, as deep learning is relatively new, at least in the forestry sector, work is still needed to optimize them as they are data-driven (Hamedianfar et al. 2022). Deep learning models greatly improved aboveground biomass predictions by combined LiDAR and satellite data (Zhang et al. 2019). Machine learning was also proven effective to predict growth of eucalyptus in plantations with neural networks, random forest, or vector suppression, all using ALS data (Görgens et al. 2015).

6.5 SITE VARIATIONS IN GROWTH

Differences in site characteristics associated with climatic conditions and soil properties influence stand productivity. Methods for quantifying this variability in growth and yield models have focused on using simple metrics. This section focuses on three general approaches: site index, soil characteristics, and ecological classification.

6.5.1 Site Index

De Perthuis in 1788 derived tree height-age curves that were site-dependent (Batho and García 2006). From these curves, the height at a reference age, called site index, can be determined for pure, even-aged stands (Curtis 1964; Hägglund 1981; Vanclay 1994). Site index has long been known to be intimately related to forest productivity (e.g., Eichhorn 1902). In Eastern North America, a reference age of 50 years is generally used for most species while in Europe the reference age varies among species. The major assumptions behind the use of site index are that (i) the stand is monospecific and even-aged; (ii) the height of the dominant trees in a stand is independent from stand density; and (iii) site index is time-independent (Skovsgaard and Vanclay 2008). The last assumption implies that site characteristics, such as climatic conditions or soil drainage and texture, do not change appreciably over time.

Site indices are widely used to classify and compare potential site productivity because they are easy to derive when height-age equations are available. Site index equations are available for a large number of commercial species in North America (e.g., Central US, Carmean 1972; Northwest US, Monserud 1984; Ontario, Carmean and Lenthall 1989; Montana, Milner 1992; Quebec, Pothier and Savard 1998; British Columbia, Chen, Klinka et al. 1998), in Europe (e.g., Sweden, Johansson 1999; Spain, Diéguez-Aranda et al. 2005; Denmark, Nord-Larsen 2006), in Australia (e.g., Rayner 1991; Haywood 2009), and for tropical conditions (e.g., Kenya, Schonau 1973; Ethiopia, Teshome and Petty 2000; India, Tewari and Kishan Kumar 2002; Mexico, Corral Rivas et al. 2004).

Data from temporary or permanent sample plots or stem analysis are used to calibrate tree height-age relationships. With temporary sample plots (i.e., plots without re-measurements), a statistical model is fitted to the point data to yield height-age equations (Brickell 1968). The derivation of bias-free site index equations using data from temporary sample plots requires large datasets with the range of site qualities balanced across stand ages.

Stem analysis consists of obtaining a number of disks along the main stem of a dominant tree, on which the age of each disk is determined. The height at each age can then be interpolated (Carmean 1972; Newberry 1991). A height-age equation can then be constructed from individual tree height-age data. This approach assumes that the sample trees were always dominant throughout their life.

Finally, height-age equations can be fitted using remeasured data from permanent sample plots. In these plots, the dominant height is available for each measurement year, as well as the average age of the dominant trees within the plot. The height-age equations are then fitted to these repeated measures.

It has been shown that height-age equations established using stem analysis data overestimate dominant stand height growth (Raulier et al. 2003). In other words, stand dominant height increment does not mirror the height increment of the dominant trees selected at some ages. Site indices are calculated for survey plots, and thus the height-age equations should be calibrated using similar data with which they are used. Furthermore, the time-independent assumption of site index measures has also been questioned (Charru et al. 2010) as well as its link to forest productivity (Boisvenue and Running 2006; Isbell et al. 2013; McMahon et al. 2010).

Shifts from even-aged, monospecific forest management to uneven-aged mixed-species stands are being suggested in a number of locales (Barrette et al. 2014). The use of site index in uneven-aged mixed-species stands is difficult because the basic assumptions for site indices are not met (Bontemps and Bouriaud 2014). To use site index equations in uneven-aged stands, attempts have been made to modify the equations by integrating measures of stand structure in the models (Anyomi et al. 2014; Ouzennou et al. 2008) or by explicitly integrating the effect of tree age in the site index calculation (Hamel et al. 2004). In mixed-species stands, a promising approach is to build site index conversion equations to adapt site index equations estimated for pure stands (Johansson 2006; Kahriman et al. 2013; Nigh 1995, 2002; Wang 1998). Others have also suggested using different modeling techniques to encompass species mixtures (e.g., hierarchical cross-scale modeling, Anyomi et al. (2013); explicit integration of mixture, Eriksson et al. (1997).

Finally, the climate-independent site index assumption has been shown to not hold (Pau et al. 2022; Cheng et al. 2023). This thus opens the possibility of including climate change in growth and yield models by using climate-sensitive site indices (see section 6.6.2.2).

6.5.2 Soil Properties and Tree Growth

Many studies have examined the link between forest productivity and soil properties (Weiskittel et al. 2011). Soil parameters that could influence tree productivity are numerous (Duchaufour 2001) and may consist of physical (e.g., soil depth, texture, soil moisture) or chemical (e.g., pH, C/N ratio, nutrient availability) characteristics (Schoenholtz et al. 2000). However, due to the high costs of obtaining soil data, especially chemical characteristics, relationships between tree and stand growth and such variables are seldom established (Doran et al. 1994, 1996; Zornoza et al. 2007). Using soil variables to assess stand or tree growth is a balance between the use of sufficient variables to describe the soil characteristics and the usefulness and ease of use of the models by practitioners (Schoenholtz et al. 2000). Therefore, only a limited number of soil variables is generally used to ensure the applicability of the model.

In boreal forests, nitrogen is the key element in forest conifer productivity (Lupi et al. 2013) while in tropical forests, the interaction between nitrogen and phosphorus is very important for plant growth and nutrition (Vitousek et al. 2010). However, only weak linkages have been found between site index and soil characteristics (Weiskittel et al. 2011), probably because soil variables are numerous and characterized by complex interactions (Doran et al. 1996; Schoenholtz et al. 2000). Despite the high spatial variability in soil properties, some attempts have been made to scale up the relationships between forest productivity and soil properties (Cambardella et al. 1994). Although difficult, assessing long-term ecosystem sustainability and health must also rely on soil dynamics (Schoenholtz et al. 2000).

In practice, high-resolution soil maps are needed to obtain the soil characteristics of a stand. However, such maps are often incomplete or not available (Weiskittel et al. 2011), although recently some progress has been made to improve soil mapping (Bulmer et al. 2019; Duchesne and Ouimet 2021).

6.5.3 Ecological Classification

Several ecological classifications have been proposed. The basis for such classifications is that a limited number of biophysical characteristics can be used to group stands. Ecological classification systems are hierarchical, ranging from very broad climatic or vegetation constraints to site-specific factors. For example, in Quebec, Eastern Canada, the ecological classification relies on 11 hierarchical levels (Saucier et al. 1998). The highest levels consider large-scale variations in climate and vegetation associations and are divided into vegetation zones and subzones (e.g., boreal zone and continuous boreal forest subzone), bioclimatic domains and sub domains (e.g., spruce-moss domain and eastern spruce-moss subdomain). On a finer scale, the lowest levels consider the potential vegetation (e.g., black spruce-moss potential vegetation) and ecological type (e.g., black spruce-moss on very thin soils). Similar approaches have been developed for other regions, such as the ecological classification used in Finland based on the presence of herbaceous and shrub species (Cajander 1949) and the biogeoclimatic ecosystem classification system used in British Columbia (Meidinger and Pojar 1991).

Ecological classification can be used instead of site index to account for site fertility based on the hypothesis that two different stands, with the same soil type, geomorphological deposits, and floristic communities, have the same productivity. In Quebec, information from ecological classification is used instead of site index in the Artemis-2009 and Natura-2009 growth and yield models (Fortin and Langevin 2010; Pothier and Auger 2011). Larocque et al. (1996) derived polymorphic site productivity functions that were related to different ecological types for black spruce in Northern Ontario. Also, in Ontario, Zhang et al. (2002) associated site indices with each ecotype, which were then used as input for a growth and yield model. However, work remains to precisely evaluate site index variability within a given ecological type (Laflèche et al. 2013). For example, the 95% confidence interval for site index of balsam fir for the balsam fir-yellow birch forest type on well-drained thick soils ecological type (MJ12) within the Nominingue Lake Hills ecological district varies from

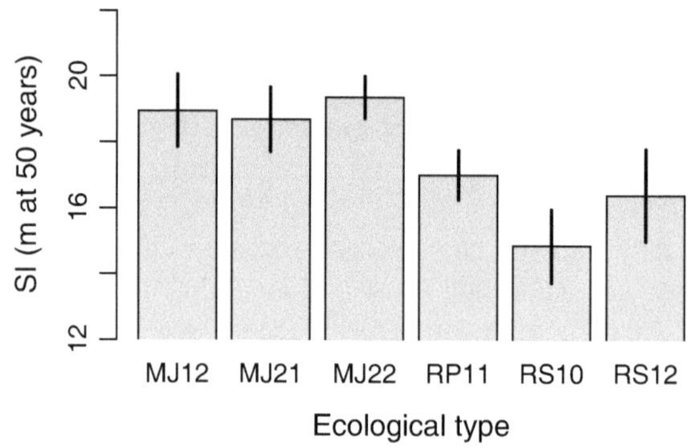

FIGURE 6.2 Balsam fir site index (dominant height at age 50) per ecological type in the Nominingue Lake Hills ecological district of Quebec (Laflèche et al. 2013). (MJ12: balsam fir-yellow birch forest type on well-drained thick soils with medium texture; MJ21: balsam fir-white spruce forest type on quickly drained thin-to-thick soils with gross texture; MJ22: balsam fir-white spruce forest type on well-drained thin-to-thick soils with coarse texture; RP11: white or red pine on quickly drained thin-to-thick soils with coarse texture; RS10: balsam fir-eastern white cedar forest type on thin, quickly drained soils; RS12: balsam fir-eastern white cedar forest type on thick, well-drained soils.)

18.95 to 20.05 m (Figure 6.2). Using the growth curves from Pothier and Savard (1998), this translates to a range in merchantable volume at age of harvesting (e.g., 45 years) between 164 and 182 m^3 ha^{-1}. Further efforts are needed to better assess the observed variations in potential stand yield within a given ecological type.

6.6 CLIMATIC CHANGE IN GROWTH AND YIELD MODELS

6.6.1 Climate Change and Forests

The Intergovernmental Panel on Climate Change (IPCC) established that climatic conditions at the earth's surface have changed and will continue to change during the 21st century (Stocker 2014). The global surface temperature has increased faster since 1970 than in any other 50-year period over at least the last 2,000 years, leading to increases of the global surface temperature of 1.09°C between 1850 and 2020, with larger increases over land (1.59°C) than over the ocean (1.01°C) (IPCC 2023). Climate change has thus become the most important factor influencing forest ecosystems, either directly by influencing temperature and precipitations (Peng et al. 2002), or indirectly by changing the occurrence of natural disturbances such as forest fires and insect outbreaks (USGCRP 2018). Direct effects of climate change (e.g., shifts in temperature and precipitation) will be explored in the following subsections.

Changes in forest fire severity, size, and return intervals with increases in temperature and droughts have the most important indirect impacts of climate change. Climate is a major factor of fire regimes through its control of fire weather, fuel availability, and flammability. The probability of ignition along with the rate of fire spread increases, as both live and dead fuels become drier and flammable (Scott 2005). In recent years, a significant increase of forest fire events has been observed across the globe (Ansmann et al. 2018; Ponomarev et al. 2016; Boer et al. 2020), with the area and severity of forest fires across Canada having surpassed historic levels in 2023. Forest fires not only interfere with timber supply and reset the regeneration process but also influence ecosystem productivity, biodiversity loss, air pollution, soil fertility, and water quality (Pérez-Cabello et al. 2012; Yin et al. 2019). However, a detailed review of the effects of fire on forest ecosystems is out of the scope of this chapter. We will thus explore how climate change influences directly tree growth, recruitment, and mortality.

6.6.2 Tree Growth

6.6.2.1 Tree Growth and Climate

Tree growth is a result of how trees allocate photosynthates to their different components and depends on environmental factors that influence the metabolic and physiological processes (Deslauriers et al. 2016). For example, species-specific ranges of temperature and amount of water and light are needed for germination after which, growth is highly dependent on climatic conditions (Lambers et al. 2008). Moreover, tree growth follows a periodicity marked by periods of activity and dormancy, depending on annual cycles of temperature, photoperiod, and water availability. Dendrochronology assumes that tree-ring growth contains climatic information for previous and current growing seasons, making radial growth a reliable indicator of historical climate variability and change.

In temperate and boreal regions, trees have an annual cycle of dormancy and activity. They enter dormancy in late autumn to cope with winter cold, with a resting phase controlled by internal factors and a quiescence phase maintained by unfavorable environmental conditions (Rossi et al. 2006). During this dormancy period, low temperatures play an essential role in regulating the physiological activities of trees. Higher temperatures, combined with a longer photoperiod, enable trees to initiate cambial cell differentiation in spring. When the temperature reaches the necessary threshold (which varies according to species and site) to trigger the onset of cambial cell division (Rossi et al. 2006), tree growth resumes. D'Orangeville et al. (2022) reported that tree growth follows a typical logistic pattern, with growth rates peaking in June and then declining until the end of the process. The same authors reported that diffuse-porous species take 16–18 days less on average than other types of woody structured species to reach 50% of their annual growth in diameter, with their maximum growth rate occurring almost a full month later than ring-porous species and conifers over an average period of 24 days. Therefore, changes in temperature impact tree growth by modulating the onset of cambium cell division. Consequently, warmer springs may result in a longer growing season for tree growth.

The xylogenesis process resulting in xylem formation is characterized by a period of slow initiation, a period of rapid growth followed by a period of slow termination (Rossi et al. 2006). Initial and final wood production depends on climate. For example, the initial wood production in balsam fir lasted 9–10 weeks in 1999, which was warmer than average, compared to 6–7 weeks during cooler years 1998 and 2000 (Deslauriers et al. 2003). Similarly, final wood formation is highly sensitive to temperature and water stress. In boreal forests, the period of final wood production generally extends from late June to mid-August, with fluctuations depending on the species (Zhai et al. 2012). In the dry Mediterranean region, the increase in temperature in August led to a reduction in photosynthesis, and consequently a negative effect on the formation of the final wood (Campelo et al. 2007). Thus, all xylogenesis processes are affected by climate during the growing season.

In the temperate and boreal forests, global warming can improve carbon sequestration due to greater radial growth (Wang et al. 2023). Furthermore, the fertilization effect of increased CO_2 in the context of climate change may enhance photosynthesis and water-use efficiency (Dusenge et al. 2019). On the other hand, increased vapour pressure deficit associated with climate warming can inhibit photosynthesis and carbon sequestration (Grossiord et al. 2020). Similarly, increased evaporation due to global warming can slow down photosynthesis and carbon sequestration (Berner et al. 2013). Projections of the effects of climate change on tree growth indicate an increase in growth at higher latitudes associated with higher rates of global warming (D'Orangeville et al. 2018). Greater increases in tree growth are projected in wetter eastern ecozones compared with drier central ecozones, suggesting that drier parts of the boreal biome will become more vulnerable to growth reductions under warming (D'Orangeville et al. 2016).

It is thus clear that climate regulates tree growth, and must be considered in growth and yield models. The effects of climate on tree growth can be considered in both empirical and process-based models.

6.6.2.2 Adjusting Empirical Models for Climate Change

In their reviews on climate-sensitive growth and yield models in Canada, Metsaranta et al. (2024) defined three possible implementations of climate sensitivity:

(1) An indirect implementation occurs when one of the model covariates is expressed as a function of some climate variables. Typically, this happens when the model relies on a climate-sensitive site index.
(2) An average climate implementation is based on average climate variables. Typically, one or many model components include 30-year normals in their equations.
(3) An annual implementation is based on annual or intra-annual climate variables.

The indirect implementation can be considered as an early approach to the integration of climate sensitivity in empirical growth and yield models. An appreciable number of climate-sensitive site indices have been published over the last decade (e.g., Nothdurft et al. 2012; Sharma et al. 2012; Antón-Fernández et al. 2016). SILVA (Pretzsch et al. 2002), the Forest Vegetation Simulator (FVS; Crookston et al. 2010), and the Open Stand Model (OSM; Hennigar et al. 2017) are examples of growth and yield models that benefited from an indirect implementation at some point in their development.

The indirect implementation has the advantage of being an easy way to account for climate effects in existing growth and yield models, especially in stand-level models as per Porté and Bartelink (2002). The indirect implementation does not require any change in the model because it is assumed that climate affects a single model covariate, which in turn affects the model behaviour. However, this assumption is subject to criticisms. It can be anticipated that climate has multiple and contrasted effects on growth. For instance, a warmer temperature could be beneficial to diameter increment but detrimental to regeneration. Should it be the case, the indirect implementation might not fully capture the climate effect on growth.

With the greater accessibility to climate data (Régnière et al. 1995; Wang et al. 2016; Fortin et al. 2022b), the average climate implementation has gained in popularity over the last decade. Many growth models now include 30-year normals in their equations and especially those related to tree growth. The Mixed wood Growth Model (MGM; Bokalo et al. 2013) applies to Canadian boreal forests, and its diameter increment predictions were shown to be dependent on the 30-year climate moisture index and frost-free period (Oboite and Comeau 2021). In France, Manso et al. (2015) found that the mean temperature of the vegetative period over the period 1958–2007 had a significant effect on diameter increments of sessile oak (*Quercus patraea* Liebl.) and European beech (*Fagus sylvatica* L.).

Although it requires more adjustment than the indirect implementation, the average climate implementation can be easily implemented, or at least tested, in any model component. A major criticism is that it neglects the year-to-year variability in climate effects (Oboite and Comeau 2021). The annual climate implementation seems preferable to the average climate implementation as it can account for the year-to-year variability. However, the annual climate implementation is often limited by the nature of the data. In permanent sample plots, for instance, tree growth is observed over intervals of several years. Unfortunately, regular statistical regression techniques do not allow for considering the annual effect of climate variables on tree growth measured over a multi-year interval.

The annual climate implementation is well suited for tree ring data. Many authors have evaluated the effect of climate variables on annual radial increments and found significant effect of monthly temperatures and precipitations of the current and previous years (e.g., Girardin et al. 2008; Giebink et al. 2022; Heilman et al. 2022). Efforts have been made to integrate these annual climate implementation of diameter or radial increment models into existing growth and yield models (e.g., Giebink et al. 2022).

Regardless of the climate sensitivity implementation, there is no consensus on the sign and magnitude of the different climate effects in the literature. The species and the location within the species distribution range can partly explain the differences. In their literature review, Marchand et al. (2018) outlined that methodological biases related to the source of data may also account for some of the discrepancies.

6.6.2.3 Incorporating Climatic Variables in Process-Based Models to Predict Tree Growth

To predict above-ground biomass accumulation at both the stand and tree levels, process-based models generally start by estimating carbon uptake through light interception and photosynthesis. Net productivity is then obtained by some type of carbon balance (Mäkelä 1997), whereby total carbon assimilation is reduced by the maintenance respiration. Finally, net productivity is allocated to the different tree compartments (i.e., stem, branches, foliage, roots, etc.). The effect of climatic control can be represented for all these steps.

Let us use the 3-PG model as a case study of how process-based models incorporate climatic variables. 3-PG is a model that is used around the world for various cover types (Landsberg et al. 2003; Gupta and Sharma 2019). In the original model, the absorbed photosynthetically active radiation is conditioned by the vapor pressure deficit, the amount of water available in the soil (current and water balance), the occurrence of frost, and the age of the forest (Landsberg and Waring 1997). The allocation to the roots is dependent on these same environmental factors, and allocation to the stem and foliage is expressed in terms of allocation to the roots. The effect of temperature has since then been modified and can include variables such as monthly mean air temperature and minimum, maximum, and optimum temperature for tree growth (Peng et al. 2002). Another example can be found in Luckai and Larocque (2002) in which two process-based models, CENTURY and FOREST-BGC, were compared to predict the effects of climate change on the carbon cycle in a black spruce ecosystem in the boreal forest of northern Ontario. More detailed information on process-based models can be found in Chapter 8.

6.6.3 Mortality

6.6.3.1 Mortality and Climate

A tree dies when its access to resources (e.g., light, water, nutrients) is limited and has used all its reserves. Although conceptually simple, understanding and modeling tree mortality remains a challenge (Das et al. 2016). The most important driver of mortality that has been widely studied is inter-tree competition. As previously stated, the self-thinning rule (Yoda et al. 1963) or Reineke's index (Reineke 1933) can be used to evaluate the number of stems a stand should have and can thus be used to evaluate the number of trees that should die. Environmental stresses caused by climate change such as high temperatures and drought can increase mortality (Anderegg et al. 2019; Hartmann et al. 2018). In fact, mortality rates could serve as a reliable indicator of global forest well-being on a global scale (McDowell et al. 2020).

Rising surface temperatures exacerbate the effects of drought on trees due to increased vapor pressure deficit, plant transpiration demand, and soil evaporative demand (Grossiord et al. 2020). Trees have thus to decide between the lesser of two evils (Sevanto et al. 2014): hydraulic failure or carbon starvation. When under hydraulic stress, trees can either close their stomata and reduce water loss or maintain their stomata open and increase the risk of embolism. There is therefore a link between a tree's water and carbon balance, as the maintenance of hydraulic function depends in part on the use of carbohydrates for the growth of conductive tissue, cellular osmoregulation, and production of defense compounds against biotic attack (McDowell 2011). The risk of increased tree mortality due to hydraulic failure and/or carbon deficiency is proportional to the increase in vapor pressure deficit. However, increasing atmospheric CO_2 concentration has been found to either mediate stomatal closure or increase the probability of tree mortality (De Kauw et al. 2021). Increased levels of CO_2 can allow plants to reduce stomatal conductance without negative feedback on photosynthetic rates. On the other hand, reduced evaporative cooling in the face of heat can lead to early leaf senescence (Warren et al. 2011). The carbon gain associated with elevated CO_2 concentration can then be allocated to root production during drought or regeneration of damaged tissues after drought (Ruehr et al. 2019), which can lead to structural overshoot (Jump et al. 2017), predisposing trees to greater mortality during subsequent drought stress (Esquivel-Muelbert et al. 2020). In conclusion, the physiological mechanisms of tree mortality and their responses to the conflicting factors of increased CO_2 and vapour pressure deficit are complex. However, the increased mortality rates observed in many parts of the world (McDowell et al. 2020), combined with theory, observations, and model predictions, can give indications of future risks of drought-related tree mortality (Liu et al. 2021, 2023).

6.6.3.2 Empirically Predicting Mortality in a Changing Climate

Evaluating mortality in growth and yield models is often performed at the tree level, with very little models evaluating it at the stand level. Generally, individual-tree mortality models include stand (i.e., some type of density measurement) and tree level (i.e., tree size and/or tree growth) covariates. Competition is thus either directly included, or indirectly if past growth is used as a predictor. This leads to only the last two possible implementations of climate variables as defined by Metsaranta et al. (2024) and were previously discussed, e.g., either use 30-year normals, or annual/intra-annual climate variables. The inclusion of climate variables was found to increase model performance for three of the 11 species studied by Hülsmann et al. (2018) in Central Europe. In the ARTÉMIS model used in the province of Québec, Canada, the fit of the logistic regression to predict tree mortality was improved in 11 out of 25 forest types, when 30-year average annual precipitation and temperature were tested (Fortin and Langevin 2010). Certain authors, nevertheless, suggest that climate change strongly influence forest dynamics and specifically tree mortality (Taccoen et al. 2019).

Empirical-based predictions will often have large variances due to the stochasticity of mortality. Slight differences in model specification can lead to very large differences in forest dynamics (Bircher et al. 2015). Several authors thus have proposed to use either process-based models or multimodel inferences to reduce the uncertainty around the predictions (e.g., Adams et al. 2013). The next subsection will explore how process-based models can be used to better understand the underpinnings of mortality.

6.6.3.3 Process-Based Models and Mortality

The simplest method to predict tree mortality is to compare the carbon uptake of a tree to the demands due to respiration (Pedersen 1998); if there is insufficient energy to sustain the tree's infrastructure, it dies. In the widely used 3-PG model, tree mortality is estimated using the self-thinning rule described in the whole stand growth model section (Figure 6.1). In TRIPLEX (Peng et al. 2002), a hybrid model based on 3-PG, TREEDYN3 (Bossel 1996) and CENTURY (Parton et al. 1988), tree mortality is expressed as a function of competition. The mortality function of TRIPLEX was then modified to evaluate the contributions of both carbon starvation and hydraulic failure (Liu et al. 2021). The latter is predicted using soil water potential and xylem conductivity to evaluate if there is hydraulic failure. The former uses a non-structural carbohydrate pool as a buffer, and when it is used up, carbon starvation mortality is forecast. The results indicated that hydraulic failure is more important than carbon starvation when drought-induced mortality is observed (Liu et al. 2021).

6.6.4 Recruitment

6.6.4.1 Recruitment and Climate

Forest succession trajectories that determine forest structure and composition depend on tree regeneration, growth, and mortality (Huber et al. 2018). Tree regeneration is driven by an ensemble of sub-processes such as seed production, seed dispersal, seed storage, post-dispersal seed predation, germination, seedling establishment, seedling and sapling growth, and recruitment of small trees above a certain height or diameter threshold. Regeneration is a complex process that depends on multiple factors (Canham and Murphy 2016). The multiplicity, multidimensionality, and interactions between the factors involved in the regeneration process make regeneration a highly stochastic process that is difficult to predict (Shoemaker et al. 2020). Recruitment refers to trees that exceed a given size threshold for the first time within a defined time interval and is therefore an indicator of regeneration success.

Temperature and water availability play an important role in all stages of development, from seed production to seedling growth and recruitment. Seed production at the northern limit of a tree's range is generally limited by low temperatures (Morin et al. 2007). Increasing temperature during the growing season leads to an increase in carbon uptake and carbon allocation to reproduction (Gallego Zamorano et al. 2018). However, temperature extremes can lead to early fruit and flower mortality and a reduction in the number of viable seeds. Temperature plays an important role in seed germination (Manso et al. 2013), which generally occurs when temperatures reach a certain threshold. Water availability also plays an important role in germination, which differs from species to species.

In cold regions, where temperature is generally the main limitation (Rossi et al. 2008), increasing temperature generally leads to increased seedling survival and growth, mainly due to direct effects on photosynthesis and hence assimilation rates (Pallardy 2010). Temperature also indirectly affects growth through nutrient and water uptake due to increased metabolism and evapotranspiration under warmer conditions (Bonan 1992). Predicted global warming may increase the length of the growing season in colder regions, which may favor seedling growth (Rossi et al. 2011). However, the beneficial effects of such climatic warming on boreal tree growth may be short-lived, given concomitant decreases in soil moisture storage and availability (D'Orangeville et al. 2018). On the one hand, high temperatures can induce summer water stress that attenuates or prevents temperature-related growth increases (Ols et al. 2016). On the other hand, an increase in winter temperatures could have different effects than summer increases (Vitasse et al. 2018). For example, at the treeline in Quebec, warmer winter temperatures and reduced duration of snow cover resulted in reduced seedling survival due to early dormancy breakup, exposing seedlings to freezing temperatures (Renard et al. 2016).

Given the limited seed dispersal, the natural migration rate is estimated at less than 1 km yr^{1} for most plant species (Corlett et al. 1998). In comparison, the climatic niches of tree species are expected to move north at a pace of 7 km yr^{1} under Canadian latitudes (McKenney et al. 2007). The difference suggests that most species will not be able to keep up with their climatic niche as temperature increases over the 21st century.

In response to this issue, some authors have proposed assisted migration as a potential climate change adaptation strategy. To be effective, assisted migration requires provenance data and modeling tools to determine where seedling should be planted (Williams and Dumroese 2013). Recent work on provenance data shows that genetic adaptation can be a limiting factor to assisted migration (Montwé et al. 2018). Effective assisted migration strategies are currently under development. The implementation of these strategies into day-to-day practice is also a challenge for decision-makers (McLachlan et al. 2007).

6.6.4.2 Empirical Approaches Including Climatic Shifts

As previously stated, recruitment is the result of the regeneration process, combining seedling survival and growth until they have a certain diameter at breast height such a merchantable dbh. As with mortality, recruitment is a highly stochastic process, making it a challenge to model. The presence of recruitment and the number of new recruits are often modeled using zero-inflated models (Fortin and DeBlois 2007). Recruitment studies are hard to compare due to the different analytical approaches used. There are however several common trends between the studies, such as the fact that recruitment is negatively related to competition, often quantified by basal area and stem density.

As with mortality, these types of models rarely have covariates that can be expressed as a function of climate variables, thus requiring the direct use of climate variables. Käber et al. (2021) found that stand structure and shade tolerance of a tree were as important or more important predictors than climate variations in European temperate forests. Similar results were reported by Zell et al. (2019) for Switzerland, where climate variables were found to be less important. In the case of recruitment in eastern United States, the presence of seedlings was found to be influenced by local climate, whereas their density is invariant to it (Canham and Murphy 2016). More recently, Fortin et al. (2023) found that growing degree-days was an important driver for recruitment.

The translation of seedling dynamics in relation to climate toward recruitment of trees of a given size (e.g., merchantable-sized trees) will have a lag of up to several decades, notwithstanding the shifts in sapling growth and survival. Even though climate change has been present for several decades, the acceleration of the changes and extreme events are only starting to be observed (IPCC 2023). This places modelers with a huge task in trying to consider these changes without any long-term data. The use of process-based models may bridge this gap.

6.6.4.3 Recruiting Trees in Process-Based Models with Climatic Effects

When trying to predict the recruitment of trees or regeneration establishment in process-based models, there is always a step that has to be empirically driven (Hartig et al. 2014). In the ZELIG-CFS model (Larocque et al. 2006), the number of seedlings reaching 1.3 m in height depends on light availability, soil moisture, fertility, and growing degree-days. In the HETEROFOR model, no climatic variables are used in the regeneration module, but seedling development is climate dependent (de Wergifosse et al. 2020; Guignabert et al. 2023).

There is however a growing interest to cover all the steps of recruitment in more detail. For example, Hanbury-Brown et al. (2022) proposed tree recruitment scheme that follows the seed to sapling process. The first step is to evaluate how the trees allocate carbon to reproduction, based on tree size and available carbon when the balance is positive. This carbon pool is then divided into seed and non-seed reproductive biomass. Seed mortality (climate-invariant), seedling emergence (climate-variant), and mortality (climate-variant) are then used with a seedling-to-sapling recruitment rate to estimate the number of saplings.

Modelers have to well be aware that the feedback loop between forest composition and regeneration and recruitment can lead to bias in model forecasts (Lischke and Löffler 2006). Furthermore, disturbances such as predation and herbivory must also be taken into account when trying to evaluate tree recruitment (Price et al. 2001; Desgagnés et al. 2022).

6.7 MODEL UNCERTAINTY

6.7.1 Definitions

Model uncertainty arises from different sources of errors. These errors have been traditionally grouped into four categories: (i) model misspecification, (ii) errors in the independent variables, including sampling and measurement errors, (iii) residual errors, and (iv) errors in the parameter estimates (Kangas 1999; McRoberts and Westfall 2014). Model misspecification may occur when the model is used in a context that differs from the one in which the model was fitted (Breidenbach et al. 2014). Errors in the independent variables frequently happen in forestry. They can be due to a measurement error. For example, both airborne LiDAR and conventional field measurement are known to provide tree height with a measurement error that varies depending on the context (Andersen et al. 2006; Sibona et al. 2016). Another source of measurement error occurs when the trees are tallied in diameter classes instead of having their diameter precisely measured. It must be stressed that these two measurement errors are fundamentally different, and they are respectively referred to as classical errors and Berkson errors (Berkson 1950). Modelers should be aware that the impacts on the statistical inferences differ depending upon the type of measurement errors (cf. Berkson 1950).

The third category of errors, residual errors, is inherent to any model. Regardless of the distributional assumptions, the residual error is defined as the difference between the observed response variable Y and its expectation conditional on the covariate X as defined by the model, i.e., $\varepsilon_{RES} = Y - E[Y \mid X]$.

Finally, errors in parameter estimates arise from the estimation process. Because model parameters are estimated from a sample of population units, estimators of the model parameters have sampling variability. In most cases, it can be assumed that errors in the parameter estimates are asymptotically normally distributed, even if the residual errors are not normally distributed. Asymptotically refers to the fact that the empirical distribution tends toward a normal distribution as the sample size increases, as stated by the central limit theorem (cf. Rosenblatt 1956).

In the next sections, two different techniques for assessing model uncertainty are presented. These two techniques allow the propagation of the errors in a single model or in a chain of models. The first one, the Taylor series, is analytical, whereas the second one, the Monte Carlo technique, is based on resampling.

6.7.1.1 Taylor Series Technique

The Taylor series technique is often used to approximate a model and obtain a more tractable form. For instance, Lindstrom and Bates's (1990) estimators for nonlinear mixed-effects models rely on a Taylor series to obtain an approximate linear mixed model. Any mathematical function $f(\cdot)$ can be approximated as a series of terms:

$$\begin{aligned} f(X+\varepsilon) &= f(X) + \frac{\partial f(X)}{1!\partial X}\varepsilon + \frac{\partial^2 f(X)}{2!\partial^2 X}\varepsilon^2 + \frac{\partial^3 f(X)}{3!\partial^3 X}\varepsilon^3 + \ldots \\ &= \sum_{n=0}^{\infty} \frac{\partial^n f(X)}{n!\partial^n X}\varepsilon^n \end{aligned} \tag{6.3}$$

Theoretically, the series is infinite unless the derivative becomes null at some point. However, in most practical applications, the series is stopped at the first or the second derivative, being called either a first- or a second-order Taylor expansion. If the subsequent derivatives are non-null, then the series is not exact, and it remains an approximation of the original function.

6.7.1.2 Propagating the Residual Error of Diameter Predictions into Basal Area Calculation

Consider the following model that predicts dbh for tree i given a vector of explanatory variables $\boldsymbol{x}_i$ and a vector of model parameters $\boldsymbol{\beta}$:

$$dbh_i = f(\boldsymbol{x}_i; \boldsymbol{\beta}) + \varepsilon_i \tag{6.4}$$

where ε_i is the residual error, and $E[\varepsilon_i \mid \boldsymbol{x}_i] = 0$.

Tree basal area (ba_i) is easily derived as: $ba_i = \pi dbh_i^2 / 4 \equiv g(dbh_i)$. If dbh is observed, then the true tree basal area is obtained through function $g(dbh_i)$. If $f(\mathbf{x}_i ; \boldsymbol{\beta}) + \varepsilon_i$ is substituted for dbh_i in $g(dbh_i)$, then a second-order Taylor series makes it possible to propagate the error ε_i in the basal area calculation:

$$\begin{aligned} ba_i &= g(dbh_i) \\ &= g(f(\mathbf{x}_i;\beta) + \varepsilon_i) \\ &= \frac{\pi(f(\mathbf{x}_i;\beta) + \varepsilon_i)^2}{4} \\ &= \frac{\pi}{4}\left((f(\mathbf{x}_i;\beta))^2 + \frac{2f(\mathbf{x}_i;\beta)}{1!}\varepsilon_i + \frac{2}{2!}\varepsilon_i^2\right) \end{aligned} \tag{6.5}$$

Note that the second-order Taylor series in Eq. 6.5 is exact since the third-order derivative is null. The expectation of the basal area is then:

$$\mathrm{E}[ba_i|\mathbf{x}_i] = \mathrm{E}\left[\frac{\pi}{4}\left((f(\mathbf{x}_i;\beta))^2 + \frac{2f(\mathbf{x}_i;\beta)}{1!}\varepsilon_i + \frac{2}{2!}\varepsilon_i^2\right)\right]$$

$$= \frac{\pi}{4}\left((f(\mathbf{x}_i;\beta))^2 + \mathrm{Var}(\varepsilon_i)\right) \tag{6.6}$$

If the basal area is calculated from the expectation of dbh without consideration of the residual error, then this quantity is underestimated, and the bias is

$$\mathrm{Bias} = E\left[g\left(f\left(x_i;\beta\right)\right) - E[ba_i \mid x_i]\right] = -\frac{\pi}{4}\mathrm{Var}\left(\varepsilon_i\right) \tag{6.7}$$

Note that no distributional assumption has been made regarding ε_i; only the variance is needed. The same developments could be used to calculate the bias related to using the medians of dbh classes instead of true values in the calculation of basal area in a common forest inventory.

6.7.1.3 Feasible Estimators

The Taylor series is based on the true parameters and variances, which are all unknown. To obtain feasible estimators, these parameters and variances are replaced by their estimates, which may result in a biased estimator. Only a few feasible estimators can be easily corrected for this bias. One of those is a feasible estimator of Eq. 6.6 if and only if the diameter model is linear, i.e., $f\left(x_i;\beta\right) = x_i\beta$. If the parameters are replaced by their estimates, this estimator is:

$$\mathrm{E}\left[\hat{\mathrm{E}}\left[ba_i|x_i\right]\right] = \frac{\pi}{4}\mathrm{E}\left[\left(x_i\hat{\beta}\right)^2 + \hat{\mathrm{V}}\mathrm{ar}(\varepsilon_i)\right] = \frac{\pi}{4}\left(x_i\beta + x_i\Omega x_i^T + \mathrm{Var}(\varepsilon_i)\right) \tag{6.8}$$

where Ω is the variance-covariance matrix of the parameter estimates. The estimator $\hat{\mathrm{E}}\left[ba_i|x_i\right]$ overestimates the true basal area by $(\pi/4)x_i\Omega x_i^T$. Subtracting the estimate of this quantity from the estimator yields an unbiased estimator:

$$\hat{E}_{\mathrm{CORR}}[ba_i \mid x_i] = \frac{\pi}{4}\left((x_i\hat{\beta})^2 + \hat{V}\mathrm{ar}\left(\varepsilon_i\right) - x_i\hat{\Omega}x_i^T\right) \tag{6.9}$$

This unbiased estimator is extensively described in Gregoire et al. (2008). An unbiased estimator for normal errors propagated into an exponential function also exists (Bradu and Mundlak 1970; Beauchamp and Olson 1973).

6.7.1.4 Monte Carlo Technique

With growth models, the Taylor series can be used to propagate the errors over a single growth step. When the predictions are reinserted in the model, the developments become tedious. Sambakhe et al. (2014) succeeded in propagating errors into a transition matrix model over many growth steps using complex algebra developments. For more complex models, such as individual-tree models, a Taylor expansion is impossible. In such contexts, the Monte Carlo technique can prove an effective alternative.

The Monte Carlo technique consists of drawing random deviates from the assumed error distributions to reproduce their variability (Vanclay 1994). This is possible only through a large number of realizations (e.g., 10 000 or more). The mean and the variance of the realizations are then used as estimators (Rubinstein and Kroese 2008, p. 100). Since the deviates are drawn at random, the estimated mean and variances are likely to vary from one Monte Carlo simulation to another. The variability tends toward 0, as the number of realizations tends to infinity. In practice, it should be expected that the mean and variance estimators show some variability even with a 50,000-realization Monte Carlo simulation. The Monte Carlo technique has been used by many authors to assess the uncertainty of growth model or national biomass and volume estimates (e.g., Mowrer and Frayer 1986; Gertner 1987; Fortin et al. 2009; Mäkinen et al. 2010; Fortin and Langevin 2012; McRoberts and Westfall 2014; Breidenbach et al. 2014).

6.7.1.4.1 Mean, Variance, and Confidence Envelopes

Distributional assumptions are essential to the Monte Carlo technique. When it comes to assessing the propagation of errors in the parameter estimates, random deviates from multivariate normal distributions are often required. If $\hat{\boldsymbol{\beta}} \sim N\left(\boldsymbol{\beta},\hat{\boldsymbol{\Omega}}\right)$, then a random vector of parameter estimates for realization b is obtained through:

$$\hat{\boldsymbol{\beta}}_b = \boldsymbol{\beta} + \boldsymbol{C}^T \boldsymbol{r}_b \tag{6.10}$$

where $\boldsymbol{C}$ is the upper triangle of the Cholesky factorization of $\boldsymbol{\Omega}$ (i.e., the matrix that satisfies $\boldsymbol{\Omega} = \boldsymbol{C}^T\boldsymbol{C}$), and $\boldsymbol{r}_b$ is a vector of independent deviates drawn from a standard normal distribution, $N(0,1)$. In practice, the true $\boldsymbol{\beta}$ and $\boldsymbol{\Omega}$ are unknown, and their estimates need to be substituted for the true value to produce the random deviates in Eq. 6.10.

Confidence limits are estimated based on the percentile method (Efron and Tibshirani 1994), which has the benefit of not requiring any distributional assumption. The 95% confidence limits are given by the 2.5th and 97.5th percentiles of the realized distribution of the random variable while the 0.5th and the 99.5th percentiles provide the limits of a 99% confidence envelope. Given that they are based on the percentiles, confidence limits are calculable even for highly skewed or irregular distributions, which is a major advantage of the Monte Carlo technique. Some improved versions of the percentile method exist (see Efron and Tibshirani 1994).

6.7.1.5 Stochastic Single Tree-Based Growth Models

Although the Monte Carlo technique requires a large amount of computing capacity, it makes it possible to propagate multiple error sources in stochastic individual-tree growth models and to produce confidence envelopes for the predictions (Fortin et al. 2009; Fortin and Langevin 2012). Basically, the input tree list is replicated B times before starting the simulation. Each copy of the list is simulated independently with different deviates to account for the errors in the parameter estimates, the residual errors, and eventually the measurement errors. Consider the simulation of a single copy as a growth realization.

Care must be taken when drawing the random deviates as they must follow the hierarchical structure of the model. Random deviates in the parameter estimates are drawn once and only once for each growth realization. If the submodels of the model were fitted using a mixed-model approach, then random deviates need to be drawn for the random effects as well. The lifetime of these deviates in the simulation depends on their hierarchical level. Once the random deviate of a plot random effect is drawn, it remains constant throughout the growth realization. Likewise, deviates for tree random effects are drawn once for each tree in a particular growth realization. On the other hand, interval random effects need new random deviates at the beginning of each growth step within a growth realization. Finally, random deviates are drawn to account for the residual errors in the submodels every time these are called during the simulation. If the residual errors follow a covariance structure, it is possible to account for the multivariate distribution through Eq. 6.10.

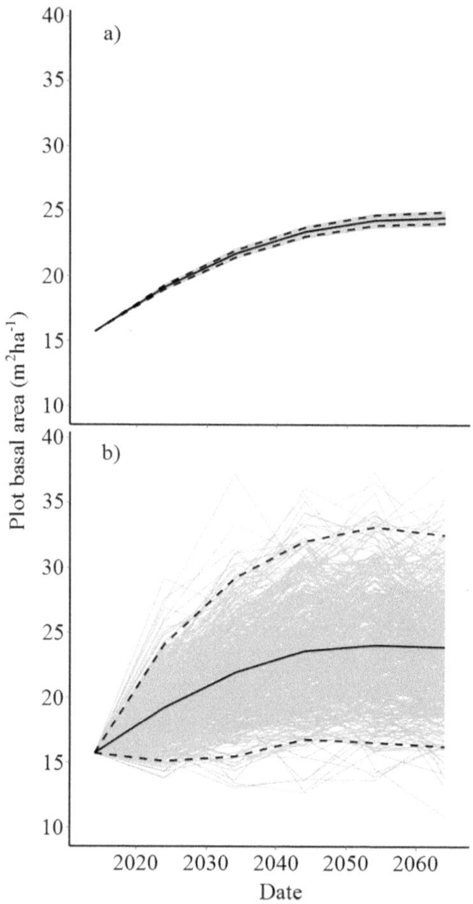

FIGURE 6.3 Growth simulation of a 400-m² plot over a 50-year period with the Artemis model (Fortin and Langevin 2012). The gray lines are the realizations, the plain black line is the estimated mean, and the dashed lines are the 0.95 confidence limits. In (a), the simulation accounts for the errors in the parameter estimates only. In (b) the simulation accounts for the errors in the parameter estimates, plot and tree random effects, and the residual errors.

Two examples of error propagation in an individual-tree model are shown in Figure 6.3. Given that the submodels were fitted on a large dataset, the variance of the parameter estimates is rather small, and the errors in these estimates result in a narrow confidence envelope for plot basal area predictions (Figure 6.3a). On the other hand, when the residual errors as well as the plot and tree random effects are propagated in addition to the errors in the parameter estimates, the predictions are more imprecise, and the confidence envelope is much wider (Figure 6.3b). The plot size induces a large component of the variability in this example since the values are reported per hectare, and the plot is only 400 m² in area. The death of a single tree causes a large drop in basal area, whereas recruitment causes a sharp increase.

6.8 PERSPECTIVES

Growth and yield models have come a long way since the first standard yield tables proposed by European foresters in the late 1700s. Model development must necessarily be accompanied by model evaluation. As pointed out by Robinson and Ek (2003), a model must not only be validated with independent data, it has to be also verified for ecological likelihood. Models have become more complex to take advantage of the better understanding of forest ecology, new statistical techniques, and the development of new inventory techniques. This has led to models that perform downscaling from the stand to the tree, use more detailed competition indices, account for prediction uncertainties, and better quantify site quality.

Recent results have shown that forest diversity can positively influence forest productivity in several biomes (Paquette and Messier 2011; Pretzsch et al. 2013). This is likely to have consequences on forest management, as it will be necessary to establish what level of mixed species to encourage and when the mixing should be favored. These processes could be better integrated into growth and yield models by considering crown allometries (Larocque 2002a; Pretzsch et al. 2012) or a more precise characterization of soil processes.

As our knowledge of the biological and ecological processes increases, growth and yield models must also continue to evolve. One means of achieving this is through Bayesian statistics, a powerful tool that is gaining popularity (McMahon et al. 2010). Bayesian statistics have many advantages: (i) they can consider prior knowledge on growth and yield processes, and (ii) they allow for fine modeling of the errors. These techniques are of particular interest for tree mortality modeling as they are able to overcome the small number of tree deaths that are observed (Metcalf et al. 2009) or enable better parameter estimation for complex model (Hartig et al. 2014).

Besides these considerations, a lot of effort has been made to deliver models on user-friendly platforms. For example, the CAPSIS platform user interface provides a generic base for growth and yield model delivery to forest practitioners (Dufour-Kowalski et al. 2012). This platform is of great interest as it presently includes models from boreal to tropical biomes and is used for forest management decisions in several countries. The development of modeling software platforms must continue to improve further model delivery in a decision-making framework and facilitate the complex analyses of simulation results, such as uncertainty analysis. For model users, undertaking complex analyses of simulation results may be viewed as a daunting task. The modeling software platform AMSIMOD (Application for the Management of SImulation MODels) was originally developed to manage the execution of a gap model and display results on graphs and Quantum GIS, but modifications were completed to facilitate the integration of different models and automate the complex analyses of simulation results, such as uncertainty or sensitivity analyses (Larocque et al. 2014). Models can also be transferred to practitioners through R packages that include the model's code (Nuske et al. 2022; Trotsiuk et al. 2020). No matter the transfer method, two gold rules of the user interface must be followed: the interface must be intuitive and explicit (i.e., the user should understand the underlying relationships of the model). Model delivery to the practitioners will also greatly evolve, as hand-held devices are becoming increasingly popular. Foresters will be able to input survey data directly and simulate different silvicultural treatments while still in the field.

Moreover, changes in the forest management paradigm (i.e., normal, even-aged forestry to ecosystem-based forestry) require foresters to integrate several objectives when making decisions, including, for example, present and future mill wood supply volume and quality, carbon sequestration within the forest, and the entire wood product chain. With efforts to optimize the whole forest value chain, better knowledge of future wood characteristics is needed. As wood properties are intimately related to tree and crown size and growth rate, many studies were conducted to predict tree form (Li and Weiskittel 2010), tree branchiness (Beaulieu et al. 2011) and knot size (Duchateau et al. 2013), wood density (Auty et al. 2014), and mechanical (Sattler et al. 2013) and anatomical properties (Franceschini et al. 2012) in relation to key variables available from growth and yield model outputs. When such models are integrated into growth and yield models, wood properties can be predicted as well as tree and/or stand development. Simultaneous predictions of growth and yield and wood properties will become increasingly important as just-in-time mill supply structures are implemented.

The complexity of the needs of the end-users must not eclipse the new forest dynamics that are being observed in the face of climate change. Although every modeler likes to think that their model is the best, one must not forget that a model is a simplification of reality. A model thus has its strengths and weaknesses and cannot be used for every purpose. Multi-model inference could be used to combine several models to yield a better picture (Adams et al. 2013; McDowell et al. 2013; Bosela et al. 2022; Boulanger et al. 2022). Even with more than two centuries of modeling, there are never-ending challenges that still need to be tackled.

REFERENCES

Achim, A., G. Moreau, N.C. Coops, J.N. Axelson, J. Barrette, S. Bédard, K.E. Byrne, J. Caspersen, A.R. Dick, L. D'Orangeville, G. Drolet, B.N.I. Eskelson, C.N. Filipescu, M. Flamand-Hubert, T.R.H. Goodbody, V.C. Griess, S.M. Hagerman, K. Keys, B. Lafleur, M.M. Girona, D.M. Morris, C.A. Nock, B.D. Pinno, P. Raymond, V. Roy, R. Schneider, M. Soucy, B. Stewart, J.-D. Sylvain, A.R. Taylor, E. Thiffault, N. Thiffault, U. Vepakomma, and J.C. White. 2021. The changing culture of silviculture. *Forestry* 95(2), 143–152. https://doi.org/10.1093/forestry/cpab047.

Adams, H., A Williams, C. Xu, S. Rauscher, X. Jiang, and N. McDowell. 2013. Empirical and process-based approaches to climate-induced forest mortality models. *Front. Plant Sci.* 4. www.frontiersin.org/articles/10.3389/fpls.2013.00438.

Anderegg, W.R.L., L.D.L. Anderegg, K.L. Kerr, and A.T. Trugman. 2019. Widespread drought-induced tree mortality at dry range edges indicates that climate stress exceeds species' compensating mechanisms. *Glob. Change Biol.* 25(11):3793–3802. https://doi.org/10.1111/gcb.14771

Andersen, H.-E., S.E. Reutebuch, and R.J. McGaughey. 2006. A rigorous assessment of tree height measurements obtained using airborne lidar and conventional field methods. *Can. J. Remote Sens.* 32(5):355–366.

Ansmann, A., H. Baars, A. Chudnovsky, I. Mattis, I. Veselovskii, M. Haarig, P. Seifert, R. Engelmann, and U. Wandinger. 2018. Extreme levels of Canadian wildfire smoke in the stratosphere over Central Europe on 21–22 August 2017. *Atmos. Chem. Phys.* 18:11831–1184.

Anta, M.B., F.C. Dorado, U. Diéguez-Aranda, J.G. Álvarez González, B.R. Parresol, and R. Rodríguez Soalleiro. 2006. Development of a basal area growth system for maritime pine in Northwestern Spain using the generalized algebraic difference approach. *Can. J. For. Res.* 36(6):1461–1474.

Antón-Fernández, C., B. Mola-Yudego, L. Dalsgaard, and R. Astrup. 2016. Climate-sensitive site index models for Norway. *Can. J. For. Res.* 46(6):794–803. https://doi.org/10.1139/cjfr-2015-0155

Anyomi, K.A., F. Raulier, Y. Bergeron, and D. Mailly. 2013. The predominance of stand composition and structure over direct climatic and site effects in explaining Aspen (Populus Tremuloides Michaux) site index within Boreal and temperate forests of Western Quebec, Canada. *For. Ecol. Manage.* 302:390–403, August. https://doi.org/10.1016/j.foreco.2013.03.035

Anyomi, K.A., F. Raulier, Y. Bergeron, D. Mailly, and M.P. Girardin. 2014. Spatial and temporal heterogeneity of forest site productivity drivers: A case study within the Eastern Boreal Forests of Canada. *Landsc. Ecol.* 29(5):905–918. https://doi.org/10.1007/s10980-014-0026-y

Assmann, E. 1970. *The Principles of Forest Yield Study: Studies in the Organic Production, Structure, Increment and Yield of Forest Stands*. Pergamon Press.

Auty, D., A. Achim, E. Macdonald, A.D. Cameron, and B.A. Gardiner. 2014. Models for predicting wood density variation in Scots Pine. *Forestry* 87(3):449–458. https://doi.org/10.1093/forestry/cpu005

Barrette, M., M. Leblanc, N. Thiffault, A. Paquette, L. Lavoie, L. Bélanger, F. Bujold, L. Côté, J. Lamoureux, R. Schneider, J.-P. Tremblay, S. Côté, Y. Boucher, and M.-È. Deshaies, 2014. Issues and solutions for intensive plantation silviculture in a context of ecosystem management. *For. Chron.* 90(6):748–762. https://doi.org/10.5558/tfc2014-147

Batho, A. and O. García. 2006. De Perthuis and the origins of site index: A historical note. *Forest Biometry, Modelling and Information Sciences* 1:1–10.

Beauchamp, J.J. and J.S. Olson. 1973. Corrections for bias in regression estimates after logarithmic transformation. *Ecology* 1403–1407.

Beaulieu, E., R. Schneider, F. Berninger, C.-H. Ung, and D.E. Swift. 2011. Modeling Jack Pine branch characteristics in Eastern Canada. *For. Ecol. Manage.* 262(9):1748–1757.

Bella, I.E. 1971. A new competition model for individual trees. *For. Sci.* 17(3):364–372.

Berkson, J. 1950. Are there two regressions? *J. Am. Stat. Assoc.* 45(250):164–180.

Bermejo, I., C. Isabel, and A. San Miguel. 2004. Growth and yield models for teak plantations in Costa Rica. *For. Ecol. Manage.* 189(1):97–110.

Berner, L.T., P.S.A. Beck, A.G. Bunn, and S.J. Goetz. 2013. Plant response to climate change along the forest-tundra ecotone in Northeastern Siberia. *Glob. Change Biol.* 19(11):3449–3462. https://doi.org/10.1111/gcb.12304

Bérubé-Deschênes, A., T. Franceschini, and R. Schneider. 2017. Quantifying competition in White Spruce (Picea Glauca) plantations. *Ann. For. Sci.* 74(2):15. https://doi.org/10.1007/s13595-017-0624-3

Biging, G.S. and M. Dobbertin. 1992. A comparison of distance-dependent competition measures for height and basal area growth of individual conifer trees. *For. Sci.* 38(3):695–720.

Biging, G.S. and M. Dobbertin. 1995. Evaluation of competition indices in individual tree growth models. *For. Sci.* 41(2):360–377.

Bircher, N., M. Cailleret, and H. Bugmann. 2015. The agony of choice: Different empirical mortality models lead to sharply different future forest dynamics. *Ecol. Appl.* 25(5):1303–1318. https://doi.org/10.1890/14-1462.1

Boer, M.M., V. Resco de Dios, and A. Bradstock. 2020. Unprecedented burn area of Australian mega forest fires. *Nat. Clim. Change* 10(3):171–172. https://doi.org/10.1038/s41558-020-0716-1

Boisvenue, C. and S.W. Running. 2006. Impacts of climate change on natural forest productivity—evidence since the middle of the 20th century. *Glob. Change Biol.* 12(5):862–882. https://doi.org/10.1111/j.1365-2486.2006.01134.x.

Bokalo, M., K.J. Stadt, P.G. Comeau, and S.J. Titus. 2013. The validation of the mixedwood growth model (MGM) for use in forest management decision making. *Forests* 4(1):1–27.

Bonan, G.B. 1992. Soil Temperature as an ecological factor in boreal forests. In *A systems analysis of the global boreal forest*, eds. G.B. Bonan, H.H. Shugart, and R. Leemans, pp. 126–143. Cambridge: Cambridge University Press. https://doi.org/10.1017/CBO9780511565489.005.

Bontemps, J.-D. and O. Bouriaud. 2014. Predictive approaches to forest site productivity: recent trends, challenges and future perspectives. *Forestry* 87(1):109–128. https://doi.org/10.1093/forestry/cpt034.

Bosela, M., K. Merganičová, C. Torresan, P. Cherubini, M. Fabrika, B. Heinze, M. Höhn, M. Kašanin-Grubin, M. Klopčič, I. Mészáros, M. Pach, K. Střelcová, C. Temperli, G. Tonon, H. Pretzsch, and R. Tognetti. 2022. Modelling future growth of mountain forests under changing environments. In *Climate-smart forestry in mountain regions*, eds. R. Tognetti, M. Smith, and P. Panzacchi, pp. 223–262. Managing Forest Ecosystems. Cham: Springer International Publishing. https://doi.org/10.1007/978-3-030-80767-2_7.

Bossel, H. 1996. TREEDYN3 forest simulation model. *Ecol. Model.* 90(3):187–227.

Boulanger, Y., J. Pascual, M. Bouchard, L. D'Orangeville, C. Périé, and M.P. Girardin. 2022. Multi-model projections of tree species performance in Quebec, Canada under future climate change. *Glob. Change Biol.* 28(5):1884–1902. https://doi.org/10.1111/gcb.16014.

Bouvier, M., S. Durrieu, R.A. Fournier, and J.P. Renaud. 2015. Generalizing predictive models of forest inventory attributes using an area-based approach with airborne LiDAR data. *Remote Sens. Environ.* 156:322–334.

Bradu, D. and Y. Mundlak. 1970. Estimation in lognormal linear models. *J. Am. Stat. Assoc.* 65(329):198–211.

Breidenbach, J., C. Antón-Fernández, H. Petersson, R.E. McRoberts, and R. Astrup. 2014. Quantifying the model-related variability of biomass stock and change estimates in the Norwegian national forest inventory. *For. Sci.* 60(1):25–33.

Brickell, J.E. 1968. *A method for constructing site index curves from measurements of tree age and height—its application to inland Douglas-Fir*. Research Paper INT-47. Ogden, UT: Intermountain Forest & Range Experiment Station, Forest Service, US Department of Agriculture.

Bulmer, C., D. Paré, and G.M. Domke. 2019. A new era of digital soil mapping across forested landscapes. In *Developments in soil science*, vol. 36, pp. 345–371. Oxford: Elsevier.

Cajander, A.K. 1949. *Forest types and their significance*. https://helda.helsinki.fi/handle/10138/17982.

Cambardella, C.A., T.B. Moorman, T.B. Parkin, D.L. Karlen, J.M. Novak, R.F. Turco, and A.E. Konopka. 1994. Field-scale variability of soil properties in Central Iowa soils. *Soil Sci. Soc. Am. J.* 58(5):1501. https://doi.org/10.2136/sssaj1994.03615995005800050033x.

Campelo, F., C. Nabais, H. Freitas, and E. Gutiérrez. 2007. Climatic significance of tree-ring width and intra-annual density fluctuations in pinus pinea from a Dry Mediterranean Area in Portugal. *Ann. For. Sci.* 64(2):229–238.

Canham, C.D. and L. Murphy. 2016. The demography of tree species response to climate: seedling recruitment and survival. *Ecosphere* 7(8):e01424. https://doi.org/10.1002/ecs2.1424.

Canham, C.D., M.J. Papaik, M. Uriarte, W.H. McWilliams, J.C. Jenkins, and M.J. Twery. 2006. Neighborhood Analyses of canopy tree competition along environmental gradients in New England forests. *Ecol. Appl.* 16(2):540–554.

Carmean, W.H. 1972. Site index curves for Upland Oaks in the Central States. *For. Sci.* 18(2):109–120.

Carmean, W.H. and D.J. Lenthall. 1989. Height-growth and site-index curves for jack pine in North Central Ontario. *Can. J. For. Res.* 19(2):215–224. https://doi.org/10.1139/x89-030.

Castedo-Dorado, F., F. Crecente-Campo, P. Álvarez-Álvarez, and M. Barrio-Anta. 2009. Development of a stand density management diagram for radiata pine stands including assessment of stand stability. *Forestry* 82(1):1–16. https://doi.org/10.1093/forestry/cpm032

Charru, M., I. Seynave, F. Morneau, and J.-D. Bontemps. 2010. Recent changes in forest productivity: An analysis of national forest inventory data for common beech (fagus sylvatica L.) in North-Eastern France. *For. Ecol. Manage.* 260(5):864–874. https://doi.org/10.1016/j.foreco.2010.06.005.

Chen, H.Y.H., K. Klinka, and R.D. Kabzems. 1998. Height growth and site index models for trembling aspen (populus tremuloides michx.) in Northern British Columbia. *For. Ecol. Manage.* 102(2–3):157–165. https://doi.org/10.1016/S0378-1127(97)00154-0.

Cheng, R., J. Zhang, X. Wang, Z. Ge, and Z. Zhang. 2023. Predicting the growth suitability of larix principis-rupprechtii mayr based on site index under different climatic scenarios. *Front. Plant Sci.* 14. www.frontiersin.org/articles/10.3389/fpls.2023.1097688.

Collalti, A., L. Perugini, M. Santini, T. Chiti, A. Nolè, G. Matteucci, and R. Valentini. 2014. A process-based model to simulate growth in forests with complex structure: Evaluation and use of 3D-CMCC forest ecosystem model in a deciduous forest in Central Italy. *Ecol. Model.* 272 (January):362–378. https://doi.org/10.1016/j.ecolmodel.2013.09.016.

Coops, N.C., P. Tompalski, T.R.H. Goodbody, M. Queinnec, J.E. Luther, D.K. Bolton, J.C. White, M.A. Wulder, O.R. van Lier, and T. Hermosilla. 2021. Modelling lidar-derived estimates of forest attributes over space and time: A review of approaches and future trends. *Remote Sens. Environ.* 260 (July):112477. https://doi.org/10.1016/j.rse.2021.112477.

Corlett, T.R. and V.J. Lafrankie. 1998. *Potential impacts of climate change on Tropical Asian forests through an influence on phenology*. Heidelberg, ALLEMAGNE: Springer.

Corral, R., J. Javier, J.G. Álvarez González, A.D. Ruíz González, and K. von Gadow. 2004. Compatible height and site index models for five pine species in El Salto, Durango (Mexico). *For. Ecol. Manage.* 201(2–3): 145–160. https://doi.org/10.1016/j.foreco.2004.05.060.

Crookston, N.L., G.E. Rehfeldt, G.E. Dixon, and A.R. Weiskittel. 2010. Addressing climate change in the forest vegetation simulator to assess impacts on landscape forest dynamics. *For. Ecol. Manage.* 260(7):1198–1211.

Curtis, R.O. 1964. A stem-analysis approach to site-index curves. *For. Sci.* 10(2):241–256.

Daniels, R.F., H.E. Burkhart, and T.R. Clason. 1986. A comparison of competition measures for predicting growth of loblolly pine trees. *Can. J. For. Res.* 16(6):1230–1237. https://doi.org/10.1139/x86-218.

Das, A.J., N.L. Stephenson, and K.P. Davis. 2016. Why do trees die? Characterizing the drivers of background tree mortality. *Ecology* 97(10):2616–2627. https://doi.org/10.1002/ecy.1497.

De Kauwe, M.G., B.E. Medlyn, and D.T. Tissue. 2021. To what extent can rising [CO_2] ameliorate plant drought stress? *New Phytol.* 231(6):2118–2124. https://doi.org/10.1111/nph.17540.

Desgagnés, J.-F., R. Schneider, and M.-H. St-Laurent. 2022. Winter browsing in absence of an apical predator: Do high moose densities compromise tree regeneration? *For. Ecol. Manage.* 520:120403.

Deslauriers, A., J.-G. Huang, L. Balducci, M. Beaulieu, and S. Rossi. 2016. The contribution of carbon and water in modulating wood formation in black spruce saplings. *Plant Physiol.* 170(4):2072–2084.

Deslauriers, A., H. Morin, C. Urbinati, and M. Carrer. 2003. Daily weather response of balsam fir (Abies Balsamea (L.) Mill.) stem radius increment from dendrometer analysis in the boreal forests of Québec (Canada). *Trees* 17(6):477–484. https://doi.org/10.1007/s00468-003-0260-4.

Diéguez-Aranda, U., H.E. Burkhart, and R. Rodríguez-Soalleiro. 2005. Modeling dominant height growth of radiata pine (pinus radiata D. don) plantations in North-Western Spain. *For. Ecol. Manage.* 215(1–3):271–284. https://doi.org/10.1016/j.foreco.2005.05.015.

Dixon, G.E. 2002. *Essential FVS: A user's guide to the Forest Vegetation Simulator*. Internal Rep. Fort Collins, CO: US Department of Agriculture, Forest Service, Forest Management Service Center. 248 p. (Revised: June 27, 2012)

Doran, J.W., D.C. Coleman, D.F. Bezdicek, B.A. Stewart, J.W. Doran, and T.B. Parkin. 1994. Defining and assessing soil quality. In *SSSA Special Publication*. Soil Science Society of America and American Society of Agronomy. https://dl.sciencesocieties.org/publications/books/abstracts/sssaspecialpubl/definingsoilqua/1.

Doran, J.W., A.J. Jones, R.F. Harris, D.L. Karlen, and D.J. Mulla. 1996. A conceptual framework for assessment and management of soil quality and health. In *SSSA Special Publication*. Soil Science Society of America. https://dl.sciencesocieties.org/publications/books/abstracts/sssaspecialpubl/methodsforasses/61.

D'Orangeville, L., L. Duchesne, D. Houle, D. Kneeshaw, B. Côté, and N. Pederson. 2016. Northeastern North America as a potential refugium for boreal forests in a warming climate. *Science* 352(6292):1452–1455. https://doi.org/10.1126/science.aaf4951.

D'Orangeville, L., D. Houle, L. Duchesne, R.P. Phillips, Y. Bergeron, and D. Kneeshaw. 2018. Beneficial effects of climate warming on boreal tree growth may be transitory. *Nat. Commun.* 9(1):1–10.

D'Orangeville, L., M. Itter, D. Kneeshaw, J.W. Munger, A.D. Richardson, J.M. Dyer, D.A. Orwig, Y. Pan, and N. Pederson. 2022. Peak radial growth of diffuse-porous species occurs during periods of lower water availability than for ring-porous and coniferous trees. *Tree Physiol.* 42(2):304–316.

Duchateau, E., F. Longuetaud, F. Mothe, C. Ung, D. Auty, and A. Achim. 2013. Modelling knot morphology as a function of external tree and branch attributes. *Can. J. For. Res.* 43(3):266–277.

Duchaufour, P. 2001. *Introduction à La Science Du Sol (6ème Éd.)*. Sciences Sup. Paris: Dunod.

Duchesne, L. and R. Ouimet. 2021. Digital mapping of soil texture in ecoforest polygons in Quebec, Canada. *PeerJ* 9:e11685, June. https://doi.org/10.7717/peerj.11685

Dufour-Kowalski, S., B. Courbaud, P. Dreyfus, C. Meredieu, and F. De Coligny. 2012. Capsis: An open software framework and community for forest growth modelling. *Ann. For. Sci.* 69(2):221–233.

Dusenge, M.E., A.G. Duarte, and D.A. Way. 2019. Plant carbon metabolism and climate change: Elevated CO_2 and temperature impacts on photosynthesis, photorespiration and respiration. *New Phytol.* 221(1):32–49. https://doi.org/10.1111/nph.15283.

Duursma, R.A. and A. Mäkelä. 2007. Summary models for light interception and light-use efficiency of non-homogeneous canopies. *Tree Physiol.* 27(6):859–870.

Efron, B. and R.J. Tibshirani. 1994. *An introduction to the bootstrap*. CRC press. http://books.google.ca/books?hl=en&lr=&id=gLlpIUxRntoC&oi=fnd&pg=PR14&dq=An+introduction+to+the+Bootstrap&ots=A8wzU5RcA-&sig=5l8doqkC0ZdphEfXNAcrS3oC2yY.

Eichhorn, F. 1902. *Ertragstafeln Für Die Weisstanne*. Berlin: Verlag Julius Springer. http://ir.nmu.org.ua/handle/123456789/21194.

Eriksson, H., U. Johansson, and A. Kiviste. 1997. A site-index model for pure and mixed stands of betula pendula and betula pubescens in Sweden. *Scand. J. For. Res.* 12(2):149–156. https://doi.org/10.1080/02827589709355396.

Esquivel-Muelbert, A., O.L. Phillips, R.J.W. Brienen, S. Fauset, M.J.P. Sullivan, T.R. Baker, K.-J. Chao, T.R. Feldpausch, E. Gloor, and N. Higuchi. 2020. Tree mode of death and mortality risk factors across amazon forests. *Nat. Commun.* 11(1):5515.

Farquhar, G.D. and S. Von Caemmerer. 1982. Modelling of photosynthetic response to environmental conditions. In *Physiological plant ecology II: Water relations and carbon assimilation*, eds. O.L. Lange, P.S. Nobel, C.B. Osmond, and H. Ziegler, pp. 549–587. Encyclopedia of Plant Physiology. Berlin, Heidelberg: Springer. https://doi.org/10.1007/978-3-642-68150-9_17.

Farquhar, G.D., S. Von Caemmerer, and J.A. Berry. 1980. A biochemical model of photosynthetic CO_2 assimilation in leaves of C_3 species. *Planta* 149(1):78–90. https://doi.org/10.1007/BF00386231.

Fortin, M., S. Bédard, J. DeBlois, and S. Meunier. 2009. Assessing and testing prediction uncertainty for single tree-based models: A case study applied to northern hardwood stands in Southern Québec, Canada. *Ecol. Model.* 220(20):2770–2781.

Fortin, M. and J. DeBlois. 2007. Modeling tree recruitment with zero-inflated models: The example of hardwood stands in Southern Québec, Canada. *For. Sci.* 53(4):529–539.

Fortin, M. and R. Langevin. 2010. *ARTEMIS-2009 : Un Modèle de Croissance Basé Sur Une Approche Par Tiges Individuelles Pour Les Forêts Du Québec. Mémoire de recherche forestière n°156*. Québec, Canada: Ministère des Ressources naturelles et de la Faune, Direction de la recherche forestière.

Fortin, M. and L. Langevin. 2012. Stochastic or deterministic single-tree models: Is there any difference in growth predictions? *Annals of For. Sci.* 69(2):271–282.

Fortin, M., D. Sattler, and R. Schneider. 2022a. An alternative simulation framework to evaluate the sustainability of annual harvest on large forest estates. *Can. J. For. Res.* 52(5), 704-715.

Fortin, M., J.-F. Lavoie, J. Régnière, and R. Saint-Amant. 2022b. A web API for weather generation and pest development simulation in North America. *Environ. Model. Softw.* 157:105476.

Fortin, M., H. Power, R. Van Couwenberghe, and B.N.I. Eskelson. 2023. The effect of climate on the occurrence and abundance of tree recruitment in the Province of Quebec, Canada. *Forestry*, cpad029.

Franceschini, T., S.-O. Lundqvist, J.-D. Bontemps, T. Grahn, L. Olsson, R. Evans, and J.-M. Leban. 2012. Empirical models for radial and tangential fibre width in tree rings of Norway Spruce in North-Western Europe. *Holzforschung* 66(2). https://doi.org/10.1515/HF.2011.150.

Franklin, J., J.M. Serra-Diaz, A.D. Syphard, and H.M. Regan. 2016. Global change and terrestrial plant community dynamics. *Proc. Natl. Acad. Sci. USA* 113(14):3725–3734. https://doi.org/10.1073/pnas.1519911113.

Gallego Zamorano, J., T. Hokkanen, and A. Lehikoinen. 2018. Climate-driven synchrony in seed production of masting deciduous and conifer tree species. *J. Plant Ecol.* 11(2):180–188.

García, O. 2011. A parsimonious dynamic stand model for interior spruce in British Columbia. *For. Sci.* 57(4):265–280.

Gertner, G. 1987. Notes: Approximating precision in simulation projections: An efficient alternative to Monte Carlo methods. *For. Sci.* 33(1):230–239.

Giebink, C.L., R.J. DeRose, M. Castle, J.D. Shaw, and M.E.K. Evans. 2022. Climatic sensitivities derived from tree rings improve predictions of the forest vegetation simulator growth and yield model. *For. Ecol. Manag.* 517:120256.

Girardin, M.P., F. Raulier, P.Y. Bernier, and J.C. Tardif. 2008. Response of tree growth to a changing climate in Boreal Central Canada: A comparison of empirical, process-based, and hybrid modelling approaches. *Ecol. Model.* 213(2):209–228. https://doi.org/10.1016/j.ecolmodel.2007.12.010.

Goodbody, T.R.H., N.C. Coops, and J.C. White. 2019. Digital aerial photogrammetry for updating area-based forest inventories: A review of opportunities, challenges, and future directions. *Curr. For. Rep.* 5(2):55–75. https://doi.org/10.1007/s40725-019-00087-2.

Görgens, E.B., P. Packalen, A.G.P. da Silva, C.A. Alvares, O.C. Campoe, J.L. Stape, and L.C.E. Rodriguez. 2015. Stand volume models based on stable metrics as from multiple ALS acquisitions in Eucalyptus plantations. *Ann. For. Sci.* 72:489–498.

Gregoire, T.G., Q.F. Lin, J. Boudreau, and R. Nelson. 2008. Regression estimation following the square-root transformation of the response. *For. Sci.* 54(6):597–606.

Groot, A. 2004. A model to estimate light interception by tree crowns, applied to black spruce. *Can. J. For. Res.* 34(4):788–799.

Grossiord, C., T.N. Buckley, L.A. Cernusak, K.A. Novick, B. Poulter, R.T.W. Siegwolf, J.S. Sperry, and N.G. McDowell. 2020. Plant responses to rising vapor pressure deficit. *New Phytologist* 226(6):1550–1566. https://doi.org/10.1111/nph.16485.

Guignabert, A., Q. Ponette, F. André, C. Messier, P. Nolet, and M. Jonard. 2023. Validation of a new spatially explicit process-based model (HETEROFOR) to simulate structurally and compositionally complex forest stands in Eastern North America. *Geosci Model Dev* 16(6):1661–1682. https://doi.org/10.5194/gmd-16-1661-2023.

Guo, Q., Y. Su, T. Hu, H. Guan, S. Jin, J. Zhang, X. Zhao, K. Xu, D. Wei, M. Kelly, and N.C. Coops. 2021. Lidar boosts 3D ecological observations and modelings: A review and perspective. *IEEE Geosci. Remote Sens. Mag.* 9(1):232–257. https://doi.org/10.1109/MGRS.2020.3032713.

Gupta, R. and L.K. Sharma. 2019. The process-based forest growth model 3-PG for use in forest management: A review. *Ecol. Model.* 397(April):55–73. https://doi.org/10.1016/j.ecolmodel.2019.01.007.

Hägglund, B. 1981. Evaluation of forest site productivity. *For. Abstr.* 42(1):515–527.

Hamedianfar, A., C. Mohamedou, A. Kangas, and J. Vauhkonen. 2022. Deep learning for forest inventory and planning: A critical review on the remote sensing approaches so far and prospects for further applications. *Forestry* 95(4):451–465. https://doi.org/10.1093/forestry/cpac002.

Hamel, B., N. Bélanger, and D. Paré. 2004. Productivity of black spruce and jack pine stands in Quebec as related to climate, site biological features and soil properties. *For. Ecol. Manag.* 191(1–3):239–251. https://doi.org/10.1016/j.foreco.2003.12.004.

Hanbury-Brown, A.R., T.L. Powell, H.C. Muller-Landau, S.J. Wright, and L.M. Kueppers. 2022. Simulating environmentally-sensitive tree recruitment in vegetation demographic models. *New Phytologist* 235(1):78–93. https://doi.org/10.1111/nph.18059.

Hartig, F., C. Dislich, T. Wiegand, and A. Huth. 2014. Technical note: Approximate Bayesian parameterization of a process-based tropical forest model. *Biogeosciences* 11(4):1261–1272. https://doi.org/10.5194/bg-11-1261-2014.

Hartmann, H., C.F. Moura, W.R.L. Anderegg, N.K. Ruehr, Y. Salmon, C.D. Allen, S.K. Arndt, D.D. Breshears, H. Davi, D. Galbraith, K.X. Ruthrof, J. Wunder, H.D. Adams, J. Bloemen, M. Cailleret, R. Cobb, A. Gessler, T.E.E. Grams, S. Jansen, M. Kautz, F. Lloret, and M. O'Brien. 2018. Research frontiers for improving our understanding of drought-induced tree and forest mortality. *New Phytol.* 218(1):15–28. https://doi.org/10.1111/nph.15048.

Hatch, C.R., D.J. Gerrard, and J.C. Tappeiner II. 1975. Exposed crown surface area: A mathematical index of individual tree growth potential. *Can. J. For. Res.* 5(2):224–228.

Haywood, A. 2009. Estimation of height growth patterns and site index curves for pinus radiata plantations in New South Wales, Australia. *South. For.* 71(1):11–19. https://doi.org/10.2989/SF.2009.71.1.2.739.

Hegyi, F. 1974. A simulation model for managing jack-pine stands. In *Growth models for tree and stand simulation*, ed. J. Fries, pp. 74–76. Stockholm: Royal College of Forestry.

Heilman, K.A., M.C. Dietze, A.A. Arizpe, J. Aragon, A. Gray, D. Shaw, A.O. Finley, S. Klesse, R.J. DeRose, and M.E.K. Evans. 2022. Ecological forecasting of tree growth: regional fusion of tree-ring and forest inventory data to quantify drivers and characterize uncertainty. *Glob. Change Biol.* 28(7):2442–2460. https://doi.org/10.1111/gcb.16038.

Hennigar, C., A. Weiskittel, H.L. Allen, and D.A. MacLean. 2017. Development and evaluation of a biomass increment based index for site productivity. *Can. J. For. Res.* 47(3):400–410. https://doi.org/10.1139/cjfr-2016-0330.

Hopkinson, C., L. Chasmer, and R.J. Hall. 2008. The uncertainty in conifer plantation growth prediction from multi-temporal lidar datasets. *Remote Sens. Environ.* 112(3):1168–1180. https://doi.org/10.1016/j.rse.2007.07.020.

Huber, N., H. Bugmann, and V. Lafond. 2018. Global sensitivity analysis of a dynamic vegetation model: Model sensitivity depends on successional time, climate and competitive interactions. *Ecol. Model.* 368:377–390.

Hülsmann, L., H. Bugmann, M. Cailleret, and P. Brang. 2018. How to kill a tree: Empirical mortality models for 18 species and their performance in a dynamic forest model. *Ecol. Appl.* 28(2):522–540. https://doi.org/10.1002/eap.1668.

Hyink, D.M. and J.W. Moser. 1983. A generalized framework for projecting forest yield and stand structure using diameter distributions. *For. Sci.* 29(1):85–95.

Iglhaut, J., C. Cabo, S. Puliti, L. Piermattei, J. O'Connor, and J. Rosette. 2019. Structure from motion photogrammetry in forestry: A review. *Curr. For. Rep.* 5(3):155–168. https://doi.org/10.1007/s40725-019-00094-3.

IPCC. 2023. Climate Change 2023: Synthesis Report. Contribution of Working Groups I, II and III to the Sixth Assessment Report of the Intergovernmental Panel on Climate Change [Core Writing Team, H. Lee and J. Romero (Eds.)]. *IPCC, Geneva, Switzerland, 184 Pp.* https://doi.org/10.59327/IPCC/AR6-9789291691647.

Isbell, F., P.B. Reich, D. Tilman, S.E. Hobbie, S. Polasky, and S. Binder. 2013. Nutrient enrichment, biodiversity loss, and consequent declines in ecosystem productivity. *Proc. Natl. Acad. Sci. USA* 110(29):11911–11916. https://doi.org/10.1073/pnas.1310880110.

Jack, S.B. and J.N. Long. 1996. Linkages between silviculture and ecology: An analysis of density management diagrams. *For. Ecol. Manag.* 86(1–3):205–220. https://doi.org/10.1016/S0378-1127(96)03770-X.

Jayathunga, S., T. Owari, and S. Tsuyuki. 2018. The use of fixed–wing UAV photogrammetry with LiDAR DTM to estimate merchantable volume and carbon stock in living biomass over a mixed conifer–broadleaf forest. *Int. J. Appl. Earth Obs. Geoinf.* 73(December):767–777. https://doi.org/10.1016/j.jag.2018.08.017.

Johansson, T. 1999. Site index curves for common alder and grey alder growing on different types of forest soil in Sweden. *Scand. J. For. Res.* 14(5):441–453. https://doi.org/10.1080/02827589950154140.

Johansson, T. 2006. Site index conversion equations for picea abies and five broadleaved species in Sweden: Alnus glutinosa, alnus incana, betula pendula, betula pubescens and populus tremula. *Scand. J. For. Res.* 21(1):14–19. https://doi.org/10.1080/02827580500526015.

Jump, A.S., P. Ruiz-Benito, S. Greenwood, C.D. Allen, T. Kitzberger, R. Fensham, J. Martínez-Vilalta, and F. Lloret. 2017. Structural overshoot of tree growth with climate variability and the global spectrum of drought-induced forest dieback. *Glob. Change Biol.* 23(9):3742–3757. https://doi.org/10.1111/gcb.13636.

Käber, Y., P. Meyer, J. Stillhard, E. De Lombaerde, J. Zell, G. Stadelmann, H. Bugmann, and C. Bigler. 2021. Tree recruitment is determined by stand structure and shade tolerance with uncertain role of climate and water relations. *Ecol. Evol.* 11(17):12182–203. https://doi.org/10.1002/ece3.7984.

Kahriman, A., H. Yavuz, and İ. Ercanli. 2013. Site index conversion equations for mixed stands of scots pine (pinus sylvestris L.) and oriental beech (fagus orientalis lipsky) in the Black Sea Region, Turkey. *Turk J Agric For* 37(4):488–494.

Kangas, A.S. 1999. Methods for assessing uncertainty of growth and yield predictions. *Can. J. For. Res.* 29(9):1357–1364.

Kangas, A.S., T. Gobakken, S. Puliti, M. Hauglin, and E. Naesset. 2018. Value of airborne laser scanning and digital aerial photogrammetry data in forest decision making. *Silva Fennica* 52(1). www.silvafennica.fi/article/9923.

Krajicek, J.E., K.A. Brinkman, and S.F. Gingrich. 1961. Crown competition—A measure of density. *For. Sci.* 7(1):35–42.

Laflèche, V., S. Bernier, J.-P. Saucier, and C. Gagné. 2013. *Indices de Qualité de Station Des Principales Essences Commerciales En Fonction Des Types Écologiques Du Québec Méridional.* Québec: Ministère des Ressources naturelles, Direction des inventaires forestiers.

Lambers, H., F.S. Chapin, and T.L. Pons. 2008. *Plant Physiological Ecology*. www.springer.com/life+sciences/ecology/book/978-0-387-78340-6.

Landsberg, J.J. and R.H. Waring. 1997. A generalised model of forest productivity using simplified concepts of radiation-use efficiency, carbon balance and partitioning. *For. Ecol. Manag.* 95(3):209–228. https://doi.org/10.1016/S0378-1127(97)00026-1.

Landsberg, J. J, R. H Waring, and N. C Coops. 2003. Performance of the forest productivity model 3-PG applied to a wide range of forest types. *For. Ecol. Manag.* 172(2–3):199–214. https://doi.org/10.1016/S0378-1127(01)00804-0.

Larocque, G.R. 2002a. Examining different concepts for the development of a distance-dependent competition model for red pine diameter growth using long-term stand data differing in initial stand density. *For. Sci.* 48(1):24–34.

Larocque, G.R. 2002b. Coupling a detailed photosynthetic model with foliage distribution and light attenuation functions to compute daily gross photosynthesis in sugar maple (acer saccharum marsh.) stands. *Ecol. Model.* 148(3):213–232. https://doi.org/10.1016/S0304-3800(01)00442-2.

Larocque, G.R. 2008. Ecological Models: Models for the Simulation of Forest Ecosystem Dynamics. In *Encyclopedia of Ecology*, Oxford: Elsevier, 1663–1773. S.V. Jorgensen and B.D. Fath.

Larocque, G.R., L. Archambault, and C. Delisle. 2006. Modelling forest succession in two Southeastern Canadian mixedwood ecosystem types using the ZELIG model. *Ecol. Model.* 199(3):350–362. https://doi.org/10.1016/j.ecolmodel.2006.05.010.

Larocque, G.R., J. Bhatti, and A. Arsenault. 2014. Integrated modelling software platform development for effective use of ecosystem models. *Ecol. Model.* www.sciencedirect.com/science/article/pii/S0304380014003809.

Larocque, G.R., W.J. Parton, and D.J. Archibald. 1996. *Polymorphic Site Productivity Functions for Black Spruce in Relation to Different Ecological Types in Northern Ontario. NEST Technical Report TR-033.* Vol. 119. www.cfs.nrcan.gc.ca/publications/?id=16720.

Lechner, A.M., G.M. Foody, and D.S. Boyd. 2020. Applications in Remote Sensing to Forest Ecology and Management. *One Earth* 2(5):405–412. https://doi.org/10.1016/j.oneear.2020.05.001.

Lessard, V.C., R.E. McRoberts, and M.R. Holdaway. 2001. Diameter growth models using minnesota forest inventory and analysis data. *For. Sci.* 47(3):301–310.

Li, R. and A. R Weiskittel. 2010. Comparison of model forms for estimating stem taper and volume in the primary conifer species of the North American Acadian Region. *Ann. For. Sci.* 67(3).

Liang, J. 2010. Dynamics and management of alaska boreal forest: An all-aged multi-species matrix growth model. *For. Ecol. Manag.* 260(4):491–501.

Liang, J., J. Buongiorno, and R.A. Monserud. 2005. Growth and yield of all-aged douglas-fir western hemlock forest stands: A matrix model with stand diversity effects. *Can. J. For. Res.* 35(10):2368–2381.

Lindstrom, M.J. and D.M. Bates. 1990. Nonlinear mixed effects models for repeated measures data. *Biometrics* 673–687.

Lischke, H. and T.J. Löffler. 2006. Intra-specific density dependence is required to maintain species diversity in spatio-temporal forest simulations with reproduction. *Ecol. Model.* 198(3):341–361. https://doi.org/10.1016/j.ecolmodel.2006.05.005.

Lister, A.J., H. Andersen, T. Frescino, D. Gatziolis, S. Healey, L.S. Heath, G.C. Liknes, R. McRoberts, G.G. Moisen, M. Nelson, R. Riemann, K. Schleeweis, T.A. Schroeder, J. Westfall, and B.T. Wilson. 2020. Use of remote sensing data to improve the efficiency of national forest inventories: A case study from the United States national forest inventory. *Forests* 11(12):1364. https://doi.org/10.3390/f11121364.

Liu, Q., C. Peng, R. Schneider, D. Cyr, Z. Liu, X. Zhou, and D. Kneeshaw. 2021. TRIPLEX-mortality model for simulating drought-induced tree mortality in boreal forests: model development and evaluation. *Ecol. Model.* 455(September):109652. https://doi.org/10.1016/j.ecolmodel.2021.109652.

Liu, Q., C. Peng, R. Schneider, D. Cyr, N.G. McDowell, and D. Kneeshaw. 2023. Drought-induced increase in tree mortality and corresponding decrease in the carbon sink capacity of Canada's boreal forests from 1970 to 2020. *Glob. Change Biol.* 29(8):2274–2285. https://doi.org/10.1111/gcb.16599.

Lorimer, C.G. 1983. Tests of age-independent competition indices for individual trees in natural hardwood stands. *For. Ecol. Manag.* 6(4):343–360.

Luckai, N. and G.R. Larocque. 2002. Challenges in the application of existing process-based models to predict the effect of climate change on C pools in forest ecosystems. *Clim. Change* 55:39–60.

Lupi, C., H. Morin, A. Deslauriers, S. Rossi, and D. Houle. 2013. Role of soil nitrogen for the conifers of the boreal forest: A critical review. *Int. J. Plant Soil Sci.* 2:155–189.

Mailly, D., S. Turbis, and D. Pothier. 2003. Predicting basal area increment in a spatially explicit, individual tree model: A test of competition measures with black spruce. *Can. J. For. Res.* 33(3):435–443.

Mäkelä, A. 1997. A carbon balance model of growth and self-pruning in trees based on structural relationships. *For. Sci.* 43(1):7–24.

Mäkinen, A., M. Holopainen, A. Kangas, and J. Rasinmäki. 2010. Propagating the errors of initial forest variables through stand-and tree-level growth simulators. *Eur. J. For. Res.* 129(5):887–897.

Manso, R., M. Fortin, R. Calama, and M. Pardos. 2013. Modelling seed germination in forest tree species through survival analysis. The pinus pinea L. case study. *For. Ecol. Manag.* 289(February):515–524. https://doi.org/10.1016/j.foreco.2012.10.028.

Manso, R., F. Morneau, F. Ningre, and M. Fortin. 2015. Effect of climate and intra- and inter-specific competition on diameter increment in beech and oak stands. *Forestry* 88(5):540–551.

Marchand, W., M. Girardin, S. Gauthier, H. Hartmann, O. Bouriaud, F. Babst, andY. Bergeron. 2018. Untangling methodological and scale considerations in growth and productivity trend estimates of Canada's forests. *Environ. Res. Lett.* https://doi.org/10.1088/1748-9326/aad82a.

Marczak, P.T., K.Y. Van Ewijk, P.M. Treitz, N.A. Scott, and D.C.E. Robinson. 2020. Predicting carbon accumulation in temperate forests of Ontario, Canada using a LiDAR-initialized growth-and-yield model. *Remote Sens.* 12(1):201. https://doi.org/10.3390/rs12010201.

Martin, G.L. and A.R. Ek. 1984. A comparison of competition measures and growth models for predicting plantation red pine diameter and height growth. *For. Sci.* 30(3):731–743.

McDowell, N.G. 2011. Mechanisms linking drought, hydraulics, carbon metabolism, and vegetation mortality. *Plant Physiol.* 155(3):1051–1059.

McDowell, N.G., C.D. Allen, K. Anderson-Teixeira, B.H. Aukema, B. Bond-Lamberty, L. Chini, J.S. Clark, M. Dietze, C. Grossiord, A. Hanbury-Brown, G.C. Hurtt, R.B. Jackson, D.J. Johnson, L. Kueppers, J.W. Lichstein, K. Ogle, B. Poulter, T.A.M. Pugh, R. Seidl, M.G. Turner, M. Uriarte, A.P. Walker, and C. Xu. 2020. Pervasive shifts in forest dynamics in a changing world. *Science* 368(6494):eaaz9463. https://doi.org/10.1126/science.aaz9463.

McDowell, N.G., R.A. Fisher, C. Xu, J.C. Domec, T. Hölttä, D.S. Mackay, J.S. Sperry, A. Boutz, L. Dickman, N. Gehres, J.M. Limousin, A. Macalady, J. Martínez-Vilalta, M. Mencuccini, J.A. Plaut, J. Ogée, R.E. Pangle, D.P. Rasse, M.G. Ryan, S. Sevanto, R.H. Waring, A.P. Williams, E.A. Yepez, and W.T. Pockman. 2013. Evaluating theories of drought-induced vegetation mortality using a multimodel–experiment framework. *New Phytol.* 200(2):304–321. https://doi.org/10.1111/nph.12465.

McKenney, D.W., J.H. Pedlar, K. Lawrence, K. Campbell, and M.F. Hutchinson. 2007. Potential impacts of climate change on the distribution of north american trees. *BioScience* 57(11):939–948.

McLachlan, J.S., J.J. Hellmann, and M.W. Schwartz. 2007. A framework for debate of assisted migration in an era of climate change. *Conserv. Biol.* 21(2):297–302.

McMahon, S.M., G.G. Parker, and D.R. Miller. 2010. Evidence for a recent increase in forest growth. *Proc. Natl. Acad. Sci. USA* 107(8):3611–3615.

McRoberts, R.E. and J.A. Westfall. 2014. Effects of uncertainty in model predictions of individual tree volume on large area volume estimates. *For. Sci.* 60(1):34–42.

Meidinger, D. and J. Pojar. 1991. Ecosystems of British Columbia. *Special Report Series-Ministry of Forests, British Columbia*, no. 6. www.cabdirect.org/abstracts/19950613205.html.

Metcalf, C.J.E., S.M. McMahon, and J.S. Clark. 2009. Overcoming data sparseness and parametric constraints in modeling of tree mortality: A new nonparametric bayesian model. *Can. J. For. Res.* 39(9):1677–1687.

Metsaranta, J., M. Fortin, J.C. White, D. Sattler, W. Kurz, J. Penner, J. Edwads, W. Hays-Byl, R. Comeau, and V. Roy. 2024. Climate sensitive growth and yield models in canadian forestry: Challenges and opportunities. *For. Chron.* 100(1): 88–106.

Milner, K.S. 1992. Site index and height growth curves for ponderosa pine, western larch, lodgepole pine, and douglas-fir in Western Montana. *West. J. Appl. For.* 7(1):9–14.

Ministère des Forêts, de la Faune et des Parcs. 2014. *Norme d'inventaire Écoforestier. Placettes-Échantillons Temporaires*. Normes techniques, Direction des inventaires forestiers, Ministère des Forêts, de la Faune et des Parcs.

Monserud, R.A. 1984. Height growth and site index curves for inland douglas-fir based on stem analysis data and forest habitat type. *For. Sci.* 30(4):943–965.

Montwé, D., M. Isaac-Renton, A. Hamann, and H. Spiecker. 2018. Cold adaptation recorded in tree rings highlights risks associated with climate change and assisted migration. *Nat. Commun.* 9(1):1574.

Moore, J.A., C.A. Budelsky, and R.C. Schlesinger. 1973. A new index representing individual tree competitive status. *Can. J. For. Res.* 3(4):495–500.

Morin, X., C. Augspurger, and I. Chuine. 2007. process-based modeling of species' distributions: What limits temperate tree species' range boundaries? *Ecology* 88(9):2280–2291. https://doi.org/10.1890/06-1591.1.

Mowrer, H.T. and W.E. Frayer. 1986. Variance propagation in growth and yield projections. *Can. J. For. Res.* 16(6):1196–1200.

Munro, D.D. 1974. Forest Growth Models—a Prognosis. In *Growth Models for Tree and Stand Simulation*, 30:7–21. Research Note 30. Department of Forest Yield Research, Royal College of Forestry, Stockholm.

Návar, J. 2014. A Stand-class growth and yield model for mexico's northern temperate, mixed and multiaged forests. *Forests* 5(12):3048–3069.

Newberry, J.D. 1991. Note: A note on Carmean's estimate of height from stem analysis data. *For. Sci.* 37(1):368–369.

Newton, P.F. 1997. Stand density management diagrams: Review of their development and utility in stand-level management planning. *For. Ecol. Manag.* 98(3):251–265. https://doi.org/10.1016/S03781127(97)00086-8.

Nigh, G.D. 1995. The geometric mean regression line: A method for developing site index conversion equations for species in mixed stands. *For. Sci.* 41(1):84–98.

Nigh, G.D. 2002. Site index conversion equations for mixed trembling aspen and white spruce stands in Northern British Columbia. *Silva Fennica* 36(4):789–797.

Ningre, F., J.M. Ottorini, and N. Le Goff. 2016. Modeling size-density trajectories for even-aged beech (Fagus silvatica L.) stands in France. *Ann. For. Sci.*, *73*(3), 765–776. https://doi.org/10.1007/s13595-016-0567-0

Nord-Larsen, T. 2006. Developing dynamic site index curves for European beech (fagus sylvatica L.) in Denmark. *For. Sci.* 52(2):173–181.

Nothdurft, A., T. Wolf, A. Ringeler, J. Böhner, and J. Saborowski. 2012. Spatio-temporal prediction of site index based on forest inventories and climate change scenarios. *For. Ecol. Manag.* 279:97–111.

Nuske, R., K. Staupendahl, and M. Albert. 2022. *et.Nwfva: Forest Yield Tables for Northwest Germany and Their Application_. R Package Version 0.1.1.* https://CRAN.R-Project.Org/Package=et.Nwfva

Oboite, F.O. and P.G. Comeau. 2021. Climate sensitive growth models for predicting diameter growth of Western Canadian boreal tree species. *Forestry* 94(3):363–373.

Ols, C., A. Hofgaard, Y. Bergeron, and I. Drobyshev. 2016. Previous growing season climate controls the occurrence of black spruce growth anomalies in boreal forests of Eastern Canada. *Can. J. For. Res.* 46(5):696–705. https://doi.org/10.1139/cjfr-2015-0404.

Ouzennou, H., D. Pothier, and F. Raulier. 2008. Adjustment of the age–height relationship for uneven-aged black spruce stands. *Can. J. For. Res.* 38(7):2003–2312. https://doi.org/10.1139/X08-044.

Pallardy, S.G. 2010. *Physiology of woody plants*. Academic Press.

Paquette, A. and C. Messier. 2011. The effect of biodiversity on tree productivity: From temperate to boreal forests. *Glob. Ecol. Biogeogr.* 20(1):170–180. https://doi.org/10.1111/j.1466-8238.2010.00592.x.

Parton, W.J., J.M.O. Scurlock, D.S. Ojima, T.G. Gilmanov, R.J. Scholes, D.S. Schimel, T. Kirchner, J.-C. Menaut, T. Seastedt, and E. Garcia Moya. 1993. Observations and modeling of biomass and soil organic matter dynamics for the grassland biome worldwide. *Glob. Biogeochem. Cycles* 7(4):785–809.

Parton, W.J., J.W.B. Stewart, and C.V. Cole. 1988. Dynamics of C, N, P and S in grassland soils: A model. *Biogeochemistry* 5(1):109–131. https://doi.org/10.1007/BF02180320.

Pau, M., S. Gauthier, R.D. Chavardès, M.P. Girardin, W. Marchand, and Y. Bergeron. 2022. Site index as a predictor of the effect of climate warming on boreal tree growth. *Glob. Change Biol.* 28(5):1903–1918. https://doi.org/10.1111/gcb.16030.

Pedersen, B.S. 1998. Modeling tree mortality in response to short-and long-term environmental stresses. *Ecol. Model.* 105(2–3):347–351.

Peng, C., J. Liu, Q. Dang, M.J. Apps, and H. Jiang. 2002. TRIPLEX: A generic hybrid model for predicting forest growth and carbon and nitrogen dynamics. *Ecol. Model.* 153(1–2):109–130. https://doi.org/10.1016/S0304-3800(01)00505-1.

Peng, C., Z. Ma, X. Lei, Q. Zhu, H. Chen, W. Wang, S. Liu, W. Li, X. Fang, and X. Zhou. 2011. A drought-induced pervasive increase in tree mortality across Canada's boreal forests. *Nat. Clim. Change* 1(9):467–471. https://doi.org/10.1038/nclimate1293.

Penner, M., D.E. Swift, R. Gagnon, and J. Brissette. 2006. A stand density management diagram for balsam fir in New Brunswick. *For. Chron.* 82(5):700–711.

Pérez-Cabello, F., A. Cerdà, J. de la Riva, M.T. Echeverría, A. García-Martín, P. Ibarra, T. Lasanta, R. Montorio, and V. Palacios. 2012. Micro-scale post-fire surface cover changes monitored using high spatial resolution photography in a semiarid environment: A useful tool in the study of post-fire soil erosion processes. *J. Arid Environ.* 76(January):88–96. https://doi.org/10.1016/j.jaridenv.2011.08.007.

Pimont, F., D. Allard, M. Soma, and J.L. Dupuy. 2018. Estimators and confidence intervals for plant area density at voxel scale with T-LiDAR. *Remote Sens. Environ., 215*, 343–370.

Pinheiro, J.C. and D.M. Bates. 2000. *Mixed-effects models in S and S-PLUS.* New-York, United-States: Springer Verlag. http://books.google.ca/books?hl=en&lr=&id=RFDe_BKxvRIC&oi=fnd&pg=PR7&dq=pinheiro+mixed&ots=mNx1ED0RI7&sig=_ZZlUcu9Jc5a10TYUPQeZw56rok.

Platt, W.J., G.W. Evans, and S.L. Rathbun. 1988. the population dynamics of a long-lived conifer (pinus palustris). *Am. Nat.* 491–525.

Ponomarev, E.I., V.I. Kharuk, and K.J. Ranson. 2016. Wildfires dynamics in siberian larch forests. *Forests* 7(6):125. https://doi.org/10.3390/f7060125.

Porté, A. and H.H. Bartelink. 2002. Modelling mixed forest growth: A review of models for forest management. *Ecol. Model.* 150(1–2):141–188. https://doi.org/10.1016/S0304-3800(01)00476-8.

Pothier, D. and I. Auger. 2011. *NATURA-2009 : Un Modèle de Prévision de La Croissance à l'échelle Du Peuplement Pour Les Forêts Du Québec.* Mémoire de recherche forestière n°163. Québec: Ministère des Ressources naturelles et de la Faune, Direction de la recherche forestière.

Pothier, D. and F. Savard. 1998. *Actualisation Des Tables de Production Pour Les Principales Espèces Forestières Du Québec.* Québec: Ministère des ressources naturelles, Forêt Québec.

Pretzsch, H. 1997. Analysis and modeling of spatial stand structures. Methodological Considerations based on mixed beech-larch stands in Lower Saxony. *For. Ecol. Manag.* 97(3):237–253.

Pretzsch, H. 2009. *Forest dynamics, growth and yield: From measurement to model.* Springer. http://books.google.ca/books?hl=fr&lr=&id=ZLNZMCSuUAQC&oi=fnd&pg=PA1&dq=forest+dynamics+growth+and+yield&ots=_I7CMadoYQ&sig=ocqhGvHWsakfz2OOUNfJZ1LcMd8.

Pretzsch, H. 2010. Re-Evaluation of Allometry: State-of-the-Art and Perspective Regarding Individuals and Stands of Woody Plants. In *Progress in Botany 71*, eds. Ulrich Lüttge, Wolfram Beyschlag, Burkhard Büdel, and Dennis Francis, pp. 339–369. Progress in Botany 71. Berlin, Heidelberg: Springer. http://link.springer.com/chapter/10.1007/978-3-642-02167-1_13.

Pretzsch, H., P. Biber, and J. Ďurský. 2002. The single tree-based stand simulator silva: construction, application and evaluation. *For. Ecol. Manag.* 162(1):3–21. https://doi.org/10.1016/S0378-1127(02)00047-6.

Pretzsch, H., K. Bielak, J. Block, A. Bruchwald, J. Dieler, H.-P. Ehrhart, U. Kohnle, J. Nagel, H. Spellmann, M. Zasada, and A. Zingg. 2013. Productivity of mixed versus pure stands of Oak (Quercus Petraea (Matt.) Liebl. and Quercus Robur L.) and European Beech (Fagus Sylvatica L.) along an ecological gradient. *Eur. J. For. Res.* 132(2):263–280. https://doi.org/10.1007/s10342-012-0673-y.

Pretzsch, H., C. Matthew, and J. Dieler. 2012. Allometry of Tree Crown Structure. Relevance for Space Occupation at the Individual Plant Level and for Self-Thinning at the Stand Level. In *Growth and Defence in Plants*, eds. Rainer Matyssek, Hans Schnyder, Wolfgang Oßwald, Dieter Ernst, Jean Charles Munch, and Hans Pretzsch, 287–310. Ecological Studies 220. Springer Berlin Heidelberg. http://link.springer.com/chapter/10.1007/978-3-642-30645-7_13.

Prévosto, B. 2005. Les Indices de Compétition En Foresterie: Exemples d'utilisation, Intérêts et Limites. *Rev. For. Fr.* 5:413–430.

Price, D.T., N.E. Zimmermann, P.J. van der Meer, M.J. Lexer, P. Leadley, I.T.M. Jorritsma, J. Schaber, D.F. Clark, P. Lasch, S. McNulty, J. Wu, and B. Smith. 2001. Regeneration in gap models: Priority issues for studying forest responses to climate change. *Clim. Change* 51(3):475–508. https://doi.org/10.1023/A:1012579107129.

Quinn, G.P. and M.J. Keough. 2002. *Experimental design and data analysis for biologists.* Cambridge University Press.

Ratkowsky, D.A. and D.E.A. Giles. 1990. *Handbook of Nonlinear Regression Models.* Marcel Dekker New York. www.sidalc.net/cgi-bin/wxis.exe/?IsisScript=BIBEST.xis&method=post&formato=2&cantidad=1&expresion=mfn=000313.

Raulier, F., M.-C. Lambert, D. Pothier, and C.-H. Ung. 2003. Impact of dominant tree dynamics on site index curves. *For. Ecol. Manag.* 184(1):65–78.

Raulier, F., A. Leduc, and M.-A. Vaillantcourt. 2009. Aménagement de La Forêt. In *Manuel de Foresterie. Nouvelle Édition.*, Éditions Multimondes, 1510. Québec, QC: Ordre des ingénieurs forestiers du Québec.

Rayner, M.E. 1991. Site index and dominant height growth curves for regrowth karri (eucalyptus diversicolor F. muell.) in South-Western Australia. *For. Ecol. Manag.* 44(2–4):261–283. https://doi.org/10.1016/0378-1127(91)90013-L.

Régnière, J., B.J. Cooke, and V. Bergeron. 1995. *BioSIM: A Computer-Based Decision Support Tool for Seasonal Planning of Pest Management Activities. User's Manual. Information Report LAU-X-116. Natural Resources Canada, Canadian Forest Service. Laurentian Forestry Centre.* https://scf.rncan.gc.ca/publications?id=16681&lang=fr_CA.

Reineke, L.H. 1933. Perfecting a stand-density index for even-aged forests. *J. Agric. Res.* 46(7):627–638.

Renard, S.M., J. Eliot, B. McIntire, and A. Fajardo. 2016. Winter conditions—not summer temperature—Influence establishment of seedlings at white spruce alpine treeline in Eastern Quebec. Edited by Kerry Woods. *J. Veg. Sci.* 27(1):29–39. https://doi.org/10.1111/jvs.12347.

Richards, F.J. 1959. A flexible growth function for empirical use. *J. Exp. Bot.* 10(29):290–300. https://doi.org/10.1093/jxb/10.2.290.

Robinson, A.P. and A.R. Ek. 2003. Description and validation of a hybrid model of forest growth and stand dynamics for the Great Lakes Region. *Ecol. Model.* 170(1):73–104.

Robinson, A.P. and J.D. Hamann. 2010. *Forest Analytics with R: An Introduction.* Springer Science & Business Media. http://books.google.ca/books?hl=en&lr=&id=z2m57dY4vgMC&oi=fnd&pg=PR8&dq=forest+analytics+with+R&ots=nZSbxXO9dN&sig=kC9Amf_zLq_6b-hyJp5HzoomKWQ.

Rosenblatt, M. 1956. A central limit theorem and a strong mixing condition. *Proc. Natl. Acad. Sci. USA* 42(1):43.

Rossi, S., A. Deslauriers, T. Anfodillo, H. Morin, A. Saracino, R. Motta, and M. Borghetti. 2006. Conifers in cold environments synchronize maximum growth rate of tree-ring formation with day length. *New Phytol.* 170(2):301–310. https://doi.org/10.1111/j.1469-8137.2006.01660.x.

Rossi, S., A. Deslauriers, J. Griçar, J.-W. Seo, C.B.K. Rathgeber, T. Anfodillo, H. Morin, T. Levanic, P. Oven, and R. Jalkanen. 2008. Critical temperatures for xylogenesis in conifers of cold climates. *Glob. Ecol. Biogeogr.* 17(6):696–707. https://doi.org/10.1111/j.1466-8238.2008.00417.x.

Rossi, S., H. Morin, A. Deslauriers, and P.-Y. Plourde. 2011. Predicting xylem phenology in black spruce under climate warming. *Glob. Change Biol.* 17(1):614–625. https://doi.org/10.1111/j.1365-2486.2010.02191.x.

Rubinstein, R.Y. and D.P. Kroese. 2008. *Simulation and the Monte Carlo method*, 2nd edn. Hoboken, NJ: John Wiley & Sons.

Ruehr, N.K., R. Grote, S. Mayr, and A. Arneth. 2019. Beyond the extreme: Recovery of carbon and water relations in woody plants following heat and drought stress. *Tree Physiol.* 39(8):1285–1299.

Running, S.W. and S.T. Gower. 1991. FOREST-BGC, A general model of forest ecosystem processes for regional applications. II. Dynamic carbon allocation and nitrogen budgets. *Tree Physiol.* 9(1–2):147–160. https://doi.org/10.1093/treephys/9.1-2.147.

Running, S.W. and E. Hunt. 1993. Generalization of a Forest Ecosystem Process Model for Other Biomes, BIOME-BCG, and an Application for Global-Scale Models. *Scaling Physiological Processes: Leaf to Globe: A Volume in Physiological Ecology*, January, 141–158. http://dx.doi.org/10.1016/B978-0-12-233440-5.50014-2.

Sambakhe, D., M. Fortin, J.-P. Renaud, C. Deleuze, P. Dreyfus, and N. Picard. 2014. Prediction bias induced by plot size in forest growth models. *For. Sci.* 60(6):1050–1059.

Sattler, D.F., P.G. Comeau, and A. Achim. 2013. Within-tree patterns of wood stiffness for white spruce (picea glauca) and trembling aspen (populus tremuloides). *Can. J. For. Res.* 44(2):162–171. https://doi.org/10.1139/cjfr-2013-0150.

Saucier, J.-P., J.-F. Bergeron, P. Grondin, and A. Robitaille. 1998. Les Régions Écologiques Du Québec Méridional (3e Version): Un Des Éléments Du Système Hiérarchique de Classification Écologique Du Territoire Mis Au Point Par Le Ministère Des Ressources Naturelles Du Québec. *Supplément de l'Aubelle*, 12.

Schneider, R., J. Bégin, A. Danet, and R. Doucet. 2013. Effect of selective precommercial thinning on balsam fir stand yield and structure. *For. Chron.* 89(6):759–768. https://doi.org/10.5558/tfc2013-138.

Schneider, R., T. Franceschini, E. Duchateau, A. Bérubé-Deschênes, L. Dupont-Leduc, S. Proudfoot, H. Power, and F. de Coligny. 2021. Influencing plantation stand structure through close-to-nature silviculture. *Eur. J. For. Res.*, January. https://doi.org/10.1007/s10342-020-01349-6.

Schoenholtz, S. H, H. Van Miegroet, and J. A Burger. 2000. A review of chemical and physical properties as indicators of forest soil quality: Challenges and opportunities. *For. Ecol. Manag.* 138(1–3):335–356. https://doi.org/10.1016/S0378-1127(00)00423-0.

Schonau, A.P.G. 1973. Height growth and site index curves for Acacia Mearnsii on the Uasin Gishu Plateau of Kenya. *The Commonwealth Forestry Review* 52(3(153)):245–253.

Scott, J.H. 2005. *Standard Fire Behavior Fuel Models: A Comprehensive Set for Use with Rothermel's Surface Fire Spread Model.* US Department of Agriculture, Forest Service, Rocky Mountain Research Station.

Sevanto, S., N.G. Mcdowell, L.T. Dickman, R. Pangle, and W.T. Pockman. 2014. How do trees die? A test of the hydraulic failure and carbon starvation hypotheses. *Plant Cell Environ.* 37(1):153–161. https://doi.org/10.1111/pce.12141.

Sharma, R.P., A. Brunner, and T. Eid. 2012. Site index prediction from site and climate variables for Norway spruce and scots pine in Norway. *Scand. J. For. Res.* 27(7):619–636. https://doi.org/10.1080/02827581.2012.685749.

Shoemaker, L.G., L.L. Sullivan, I. Donohue, J.S. Cabral, R.J. Williams, M.M. Mayfield, J.M. Chase, C. Chu, W.S. Harpole, A. Huth, J. Hille Ris Lambers, A.R.M. James, N.J.B. Kraft, F. May, R. Muthukrishnan, S. Satterlee, F. Taubert, X. Wang, T. Wiegand, Q. Yang, and K.C. Abbott. 2020. Integrating the underlying structure of stochasticity into community ecology. *Ecology* 101(2):e02922. https://doi.org/10.1002/ecy.2922

Sibona, E., A. Vitali, F. Meloni, L. Caffo, A. Dotta, E. Lingua, M. Garbarino et al. 2016. Direct measurement of tree height provides different results on the assessment of LiDAR accuracy. *Forests* 8(1), 7. https://doi.org/10.3390/f8010007

Skovsgaard, J.P. and J.K. Vanclay. 2008. Forest site productivity: A review of the evolution of dendrometric concepts for even-aged stands. *Forestry* 81(1):13–31. https://doi.org/10.1093/forestry/cpm041.

Solomon, D.S., D.A. Herman, W.B. Leak et al. 1995. *FIBER 3.0: An Ecological Growth Model for Northeastern Forest Types.* Vol. 204. USDA Forest Service, Northeastern Forest Experiment Station. www.fs.fed.us/ne/newtown_square/publications/technical_reports/pdfs/scanned/gtr204.pdf.

Spurr, S.H. 1962. A measure of point density. *For. Sci.* 8:85–96.

Stocker, T. 2014. *Climate Change 2013: The Physical Science Basis: Working Group I Contribution to the Fifth Assessment Report of the Intergovernmental Panel on Climate Change.* Cambridge University Press.

St-Onge, B., C. Vega, R.A. Fournier, and Y. Hu. 2008. Mapping canopy height using a combination of digital stereo-photogrammetry and lidar. *Int. J. Remote Sens.* 29(11):3343–3364. https://doi.org/10.1080/01431160701469040.

Taccoen, A., C. Piedallu, I. Seynave, V. Perez, A. Gégout-Petit, L.-M. Nageleisen, J.-D. Bontemps, and J.-C. Gégout. 2019. Background mortality drivers of European tree species: Climate change matters. *Proc. R. Soc. B* 286(1900):20190386. https://doi.org/10.1098/rspb.2019.0386.

Teshome, T. and J.A. Petty. 2000. Site index equation for cupressus lusitanica stands in Munessa Forest, Ethiopia. *For. Ecol. Manag.* 126(3):339–347. https://doi.org/10.1016/S0378-1127(99)00105-X.

Tewari, V. P, and V. S Kishan Kumar. 2002. Development of top height model and site index curves for Azadirachta Indica A. Juss. *For. Ecol. Manag.* 165(1–3):67–73. https://doi.org/10.1016/S0378-1127(01)00625-9.

Tome, M. and H.E. Burkhart. 1989. Distance-dependent competition measures for predicting growth of individual trees. *For. Sci.* 35(3) p. 816–831 (September). http://agris.fao.org/agris-search/search/display.do?f=1990/US/US90272.xml;US9014822.

Tompalski, P., N.C. Coops, P.L. Marshall, J.C. White, M.A. Wulder, and T. Bailey. 2018. Combining multi-date airborne laser scanning and digital aerial photogrammetric data for forest growth and yield modelling. *Remote Sens.* 10(2):347. https://doi.org/10.3390/rs10020347.

Tompalski, P., N.C. Coops, J.C. White, and M.A. Wulder. 2016. Enhancing forest growth and yield predictions with airborne laser scanning data: Increasing spatial detail and optimizing yield curve selection through template matching. *Forests* 7(11):255. https://doi.org/10.3390/f7110255.

Trotsiuk, V., F. Hartig, and D.I. Forrester. 2020. r3PG—An R package for simulating forest growth using the 3-PG process-based model. Edited by Sarah Goslee. *Meth. Ecol. Evol.* 11(11):1470–1475. https://doi.org/10.1111/2041-210X.13474.

USGCRP. 2018. *Impacts, Risks, and Adaptation in the United States: Fourth National Climate Assessment*, vol. II, eds. Reidmiller, D.R., C.W. Avery, D.R. Easterling, K.E. Kunkel, K.L.M. Lewis, T.K. Maycock, and B.C. Stewart, p. 1506. Washington, DC: U.S. Global Change Research Program. https://doi.org/10.7930/NCA4.2018

Vacchiano, G., R. Motta, J.N. Long, and J.D. Shaw. 2008. A density management diagram for scots pine (pinus sylvestris L.): A tool for assessing the forest's protective effect. *For. Ecol. Manage.* 255(7):2542–2554. https://doi.org/10.1016/j.foreco.2008.01.015.

Vanclay, J.K. 1994. *Modelling Forest Growth and Yield: Applications to Mixed Tropical Forests.* CAB International. Wallingford, Australia: CAB International.

Vandendaele, B., O. Martin-Ducup, R.A. Fournier, G. Pelletier, and P. Lejeune. 2022. Mobile laser scanning for estimating tree structural attributes in a temperate hardwood forest. *Remote Sens.* 14(18):4522.

Vitasse, Y., L. Schneider, C. Rixen, D. Christen, and M. Rebetez. 2018. Increase in the risk of exposure of forest and fruit trees to spring frosts at higher elevations in Switzerland over the last four decades. *Agric. For. Meteorol.* 248:60–69.

Vitousek, P.M., S. Porder, B.Z. Houlton, and O.A. Chadwick. 2010. Terrestrial phosphorus limitation: mechanisms, implications, and nitrogen–phosphorus interactions. *Ecol. Appl.* 20(1):5–15. https://doi.org/10.1890/08-0127.1.

Von Caemmerer, S. 2013. Steady-state models of photosynthesis. *Plant Cell Environ.* 36(9):1617–1630. https://doi.org/10.1111/pce.12098.

Wang, G.G. 1998. An ecologically based model for site index conversion among species. *Can. J. For. Res.* 28(2):234–238. https://doi.org/10.1139/x97-214.

Wang, J., A.R. Taylor, and L. D'Orangeville. 2023. Warming-induced tree growth may help offset increasing disturbance across the canadian boreal forest. *Proc. Natl. Acad. Sci. U.S.A.* 120(2):e2212780120. https://doi.org/10.1073/pnas.2212780120.

Wang, T., A. Hamann, D. Spittlehouse, and C. Carroll. 2016. Locally downscaled and spatially customizable climate data for historical and future periods for North America. *PloS One* 11(6):e0156720.

Warren, J.M., R.J. Norby, and S.D. Wullschleger. 2011. Elevated CO_2 enhances leaf senescence during extreme drought in a temperate forest. *Tree Physiol.* 31(2):117–130.

Weiskittel, A.R., D.W. Hann, J.A. Kershaw Jr, and J.K. Vanclay. 2011. *Forest Growth and yield modeling.* West Sussex: John Wiley & Sons.

Wergifosse, L. de F.A., N. Beudez, F. de Coligny, H. Goosse, F. Jonard, Q. Ponette, H. Titeux, C. Vincke, and M. Jonard. 2020. HETEROFOR 1.0: A spatially explicit model for exploring the response of structurally complex forests to uncertain future conditions—Part 2: Phenology and water cycle. *Geosci. Model Dev.* 13(3):1459–1498. https://doi.org/10.5194/gmd-13-1459-2020.

White, J.C., N.C. Coops, M.A. Wulder, M. Vastaranta, T. Hilker, and P. Tompalski. 2016. Remote sensing technologies for enhancing forest inventories: A review. *Can. J. Remote Sens.* 42(5):619–641. https://doi.org/10.1080/07038992.2016.1207484.

Williams, M.I. and R.K. Dumroese. 2013. Preparing for climate change: Forestry and assisted migration. *J. For.* 111(4):287–297.

Wilson, D.S., V. Monleon, and A Weiskittel. In press. Bayesian probabilistic platform for ecological modelling. *Ecol. Model.* In press.

Wilson, D.S., R.S. Seymour, and D.A. Maguire. 1999. Density management diagram for northeastern red spruce and balsam fir forests. *North. J. Appl. For.* 16(1):48–56. https://doi.org/10.1093/njaf/16.1.48

Wyckoff, P.H. and J.S. Clark. 2002. The relationship between growth and mortality for seven co-occurring tree species in the southern appalachian mountains. *J. Ecol.* 90(4):604–615.

Wykoff, W.R., N.L. Crookston, A.R. Stage, Intermountain Forest et al. 1982. *User's guide to the stand Prognosis model.* General Technical Report INT-133. United States: Department of Agriculture, Forest Service, Intermountain Forest and Range Experiment Station. www.fs.fed.us/rm/pubs_int/int_gtr133.pdf

Wykoff, W.R. and R.A. Monserud. 1988. *Representing site quality in increment models: A comparison of methods.* USDA Forest Service General Technical Report NC-North Central Forest Experiment Station (USA). http://agris.fao.org/agris-search/search.do?recordID=US8871951

Yan, M., Z. Li, X. Tian, L. Zhang, and Y. Zhou. 2019. Improved simulation of carbon and water fluxes by assimilating multi-layer soil temperature and moisture into process-based biogeochemical model. *For. Ecosyst.* 6(1):12. https://doi.org/10.1186/s40663-019-0171-5.

Yin, S., X. Wang, X. Zhang, M. Guo, M. Miura, and Y. Xiao. 2019. Influence of biomass burning on local air pollution in mainland southeast asia from 2001 to 2016. *Environ. Pollut.* 254(November):112949. https://doi.org/10.1016/j.envpol.2019.07.117.

Yoda, K, T Kira, H. Ogawa, and K. Hozumi. 1963. Self-thinning in overcrowded pure stands under cultivated and natural conditions. *J. Biol. Osaka City Univ.* 14:107–129.

Zeide, B. 2004. How to measure stand density. *Trees* 19(1):1–14. https://doi.org/10.1007/s00468-004-0343-x.

Zell, J., B. Rohner, E. Thürig, and G. Stadelmann. 2019. Modeling ingrowth for empirical forest prediction systems. *For. Ecol. Manage.* 433(February):771–779. https://doi.org/10.1016/j.foreco.2018.11.052.

Zhai, L., Y. Bergeron, J.-G. Huang, and F. Berninger. 2012. Variation in intra-annual wood formation, and foliage and shoot development of three major canadian boreal tree species. *Am. J. Bot.* 99(5):827–837. https://doi.org/10.3732/ajb.1100235.

Zhang, L., C. Peng, S. Huang, and X. Zhou. 2002. Development and evaluation of ecoregion-based jack pine height-diameter models for Ontario. *For. Chron.* 78(4):530–538.

Zhang, L., Z. Shao, J. Liu, and Q. Cheng. 2019. Deep learning based retrieval of forest aboveground biomass from combined LiDAR and Landsat 8 Data. *Remote Sens.* 11(12):1459. https://doi.org/10.3390/rs11121459.

Zhou, X., C. Peng, Q.-L. Dang, J. Sun, H. Wu, and D. Hua. 2008. Simulating carbon exchange in Canadian boreal forests: I. Model structure, validation, and sensitivity analysis. *Ecol. Model.* 219(3):287–299. https://doi.org/10.1016/j.ecolmodel.2008.07.011.

Zornoza, R., J. Mataix-Solera, C. Guerrero, V. Arcenegui, F. García-Orenes, J. Mataix-Beneyto, and A. Morugán. 2007. Evaluation of soil quality using multiple lineal regression based on physical, chemical and biochemical properties. *Sci. Total Environ.* 378(1–2):233–237. https://doi.org/10.1016/j.scitotenv.2007.01.052.

7 Forest Succession Models

Guy R. Larocque, Herman H. Shugart, Weimin Xi, and Jennifer A. Holm

7.1 INTRODUCTION

Forest succession models occupy a special niche among models that simulate the dynamics of forest ecosystems. While empirical models (see Chapter 6) focus on the prediction of tree and forest productivity to optimize forest management from an economic point of view, succession models predict the changes in species composition over long-term periods to examine different forest successional pathways. A fundamental difference between empirical and succession models resides in their theoretical foundation. Empirical models are generally derived from statistical relationships using data that can be obtained from forest inventories. On the other hand, succession models are based on relationships that represent the essential underlying mechanisms of seedling establishment and individual-tree growth and mortality. However, the modeling of abiotic and biotic processes is simpler than process-based models (see Chapter 8).

There are two broad categories of succession models: ecosystem- and landscape-based models (FLMs). In this chapter, the basic concepts and characteristics of both broad types of succession models are reviewed, along with some of their strengths and limitations. Their development history is discussed and the applicability of models commonly used is described and compared. Several key characteristics, including modeled processes, forest types, main drivers, spatial scaling, and contributions, are summarized. Lastly, the use of succession models in relation to ecological indicators and their application in ecological forest management, the modeling of regeneration dynamics and the application of remote sensing are presented.

7.2 TYPES AND STRUCTURE OF SUCCESSION MODELS

7.2.1 Ecosystem-Based Models

7.2.1.1 Gap Models

7.2.1.1.1 Initial Developments in Individual-Based Models

The first individual-based models (IBMs) were developed in forestry and were closely associated with prediction needs for stand growth and involved stand yield tables under novel conditions. Models were either distance-independent or distance-dependent. For distance-dependent models, the location of trees was used to compute competition indices, which altered growth of individual trees. The first model of this type was developed by Newnham (1964). These models are based on the derivation of empirical statistical relationships and are used for traditional forestry applications.

In the mid-1960s, foresters developed individual-based dynamic models of forests or IBMs (mostly for commercial forests). They did not realize at the time that their computational innovation was being paralleled in other fields, notably in astronomy, physics, and several engineering sciences (Huston et al. 1988). Despite their independent origin in multiple disciplines, IBMs are strongly ecological models. Of the thousands of such models developed across all sciences, about half of them originated in ecology (Grimm et al. 2005). In forest ecology, these models came from forestry applications as early as 1964 (reviewed in Shugart and Woodward 2011) and have been characterized as a major innovation uniting ecological theory (Huston et al. 1988). The IBMs have been nurtured by the increase in computation infrastructure over the past decades, and the issues in

 DOI: 10.1201/9781003431084-9

model development have transitioned from computational issues to the richer theoretical problem of attaining the sweet spot (the "Medawar zone") between the utility of a model to solve a problem and its degree of realism (Grimm et al. 2005). Perhaps, it is this search for the Medawar zone that has led the development of forest IBMs to cycle from high complexity in their early development to simplifications and then back to complexity (Figure 7.1).

Starting in the 1960s, remarkably complex 3D dynamic models were developed primarily by forestry researchers to investigate problems in stand spacing and timber dimensions in novel forestry practices. Several quasi-independent forestry challenges propelled these developments: forestry yield tables, essentially calibrations-based long-term remeasurement of forest stand attributes (productivity, diameter increment, biomass, average diameters of trees for mono-aged, mono-species stands, needed revision in the face of changes in forestry practice—forest fertilization, genetically improved tree cropping, introduction of exotic trees in national forestry programs, *etc.*). In the United States, the possible conversion of standard British measurement for lumber (feet, inches, board-feet) to metric standards implied recalibration of yield tables. Monterey Pine, *Pinus radiata*, which is native to the central coast of California and Guadalupe Island and Cedros island in the Mexican Pacific was planted worldwide. Endangered in its native range, *P. radiata* is the leading introduced tree in Australia, New Zealand, and Spain. Also, it is a major species for forestry plantations in Argentina, Chile, Uruguay, Kenya, and South Africa. Extensive forest models were developed to predict the yield of forest products for *P. radiata* plantations in novel environments and under the continuing selection for genetically superior plantation stock.

For forest ecologists, gap models, which were simpler models than many of the individual-based forestry models, were developed to simulate change in diameters for individual trees competing for light on a sample plot. These were used for ecological applications on the dynamics of mixed-species, mixed-aged forests. Gap models were simplified for landscape-scale simulation of forest change. With the steadily increasing computer capability, gap models became relatively more complex with a return to 3D simulations and the addition of more mechanistic plant physiology along with the biogeochemical and hydrological features and eventually cycled back to models with the complexity of the original forest IBMs. The continued increase in computational muscle has allowed continental-scale simulations with relatively complex simulators.

Forest IBMs have much to offer for assessing the potential effects of climate change over large regions, including global applications (Ma et al. 2017; Shugart et al. 2018), through their ability to

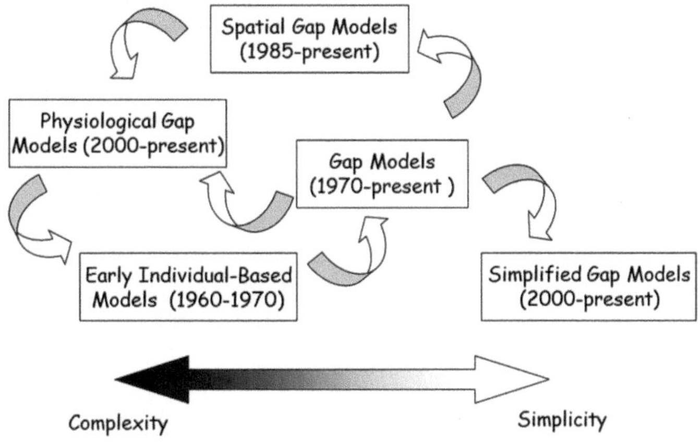

FIGURE 7.1 A cycle of development for individual-based models (IBMs). (From Shugart, H.H. and F.I. Woodward, 2011. *Global change and the terrestrial biosphere: Achievements and challenges.* Chichester: John Wiley & Sons Ltd.)

project standard state variables, such as leaf area, biomass, and productivity, and also to simulate changes in forest vertical, horizontal, and 3D structure and in species composition over time and space. One advantage of individual-based models is that two implicit assumptions associated with the more traditional state variable approach used in ecological modeling of populations are not necessary. The unique features of individuals are not assumed to be identical, nor is the population assumed to be "perfectly mixed" with no important spatial interactions. These assumptions seem particularly inappropriate for trees which are sessile and vary greatly in size over their lifespan. For this reason, tree-based forest IBMs models are among the earliest and most widely elaborated models of this genre. Botkin et al. (1972), with the JABOWA model, introduced an important simplification of early forestry work to ecologists, but the earlier forestry work remains unknown to most ecologists.

7.2.1.1.2 Concepts of Gap Models

Gap models, a term coined by Shugart and West (1980), are a category of individual-tree models that have been widely used in ecological studies. In addition to predicting individual-tree growth, they predict tree mortality and regeneration, which allows them to predict forest succession. First developed in the early 1970s and seemingly a simplification of earlier individual-based models, they have been further developed ever since. The JABOWA model (Botkin et al. 1972) was developed for forest types in New England. Since then, other models were developed with different types of modifications and ecosystem (Table 7.1), but their basic formulation, at least initially, was close to that of the JABOWA model. The models listed in Table 7.1 consist of examples of development and/or application for the three main forest types of the world: boreal, temperate, and tropical. The gap models that have been derived using JABOWA as a framework are sometimes called "JABOWA-type" models.

For many JABOWA-type models, the basic form of the annual dbh growth rate equation of individual trees is:

$$\frac{\Delta D_{real}}{\Delta t} = \frac{\Delta D_{pot}}{\Delta t} \times f(\text{Limiting site factors}) \tag{7.1}$$

The left-hand term is the realized annual dbh growth rate computed from the species-specific potential growth rate reduced by the effect of site factors, which may include the effects of temperature, intra- and inter-specific competition, precipitation, and basic soil properties. The species-specific potential dbh growth rate, which represents the maximum dbh growth rate that a species can achieve under optimal conditions within its area of distribution, is computed as:

$$\frac{\Delta D_{pot}}{\Delta t} = GD\left(1.0 - \frac{D\,H}{D_{max}\,H_{max}}\right) \times \frac{1}{f(D)} \tag{7.2}$$

where D is dbh, H is stem height, G is a growth rate parameter, D_{max} and H_{max} are the maximum dbh and height that a species can reach under optimal conditions, and f(D) is an allometric relationship between dbh and height.

The effect of limiting or controlling site factors varies among models, but their application is similar in concept to JABOWA, which consists of using derived response functions that represent how species with different ecological properties behave to site conditions (Figure 7.2). In the original version of JABOWA, the effect of light interception on growth is modeled as an annual insolation factor adjusted by a shading leaf area factor obtained from the sum of the leaf areas from all the trees within the plot by applying Beer's law. Two classes of species were recognized with respect to the response to light conditions: shade-tolerant and shade-intolerant species. ZELIG (Urban 1990), a derivative of JABOWA, computes the effect of light interception on individual-tree dbh growth

TABLE 7.1
Main Characteristics of Individual-Based Gap Models

Model Name	References	Main Processes Modeled and Equation Types	Forest Type(s)/Region(s) of the World	Main Drivers and Response Variables	Spatial Scaling	Main Contributions
JABOWA	Botkin et al. (1972)	Growth, mortality, effects of temperature and precipitation, mortality and regeneration. Semi-empirical equations	Northern hardwood forest/North America	Tree dbh[a], species-specific ecological parameters (min & max dbh, degree-days, min & max values for evapotranspiration index)	Individual-based model (IBM) with individual trees on a 0.01 ha plot updated annually using a Monte Carlo simulation of many plots	First gap model. Simulation of the succession and altitudinal transition between hardwood and conifer forests for Hubbard Brook Watershed (New Hampshire, USA)
FOREST	Ek and Monserud (1974)	Individual tree growth equations with competition driven by 3D canopy encroachment between trees. Seed rain computed for each tree based on tree height and seed aerodynamics. "Windowing" used to increase computational efficiencies. Empirical equations	Northern hardwood forests in Wisconsin/ North America	Tree coordinates, tree diameter and height, clear bole length, reproduction options	3D map of individual trees on a plot of arbitrary size	Uneclipsed in model detail, even though FOREST was an early 3D vegetation model
FORET	Shugart and West (1977)	Derived from JABOWA model. Species-specific regeneration conditions control germination. Sprouting included. Change in computations involving vertical light competition to increase computational efficiency. Semi-empirical equations	Appalachian deciduous forest/North America	Tree dbh, species-specific ecological parameters (min & max dbh, degree-days, min & max values for evapotranspiration index). Growth altered by degree-days and water availability. Shading and approach to plot biomass maximum also reduce tree growth	Individual trees on a 0.083 ha plot with Monte Carlo simulation	Simulations of elimination of American Chestnut by blight. Reconstructions of forests over the past 18,000 years. Simulation of songbird habitat dynamics. Theoretical investigations of long-term dynamics of forests. Simulations of pollution effects on forests

(Continued)

TABLE 7.1 (*Continued*)
Main Characteristics of Individual-Based Gap Models

Model Name	References	Main Processes Modeled and Equation Types	Forest Type(s)/Region(s) of the World	Main Drivers and Response Variables	Spatial Scaling	Main Contributions
KIAMBRAM	Shugart et al. (1980)	Derived from the FORET model with options to harvest trees by dbh and species. Mortality increases with felling trees included as a source of mortality. Stochastic variation in seed rain. Semi-empirical equations	Australian subtropical rainforest/Australia	Selective harvest by species and size with increase in harvest mortality from empirical experiments. Competition for light and crowding according to stand basal area.	Individual trees on a 0.05 ha plot with Monte Carlo simulation of hundreds of plots	Very diverse forest (125 species in the model). Inclusion of strangler life form. Simulation of selective harvest in rainforests. Application with new species parameters to southern African montane rainforests
FORTNITE	Aber and Melillo (1982)	Derived from JABOWA, Inclusion of nitrogen dynamics. Semi-empirical equations	Northern hardwood forest/North America	Same as JABOWA, with the addition of nitrogen cycling and nitrogen uptake. Used to evaluate the effects of different harvest rotations	Individual trees on a 0.01 ha plot with Monte Carlo simulation of multiple plots	Inclusion of detailed nitrogen cycle in individual-based models
LINKAGES	Pastor and Post (1985), Post and Pastor (1996)	Based on the development of soil processes merged with components from JABOWA & FORET. Semi-empirical equations	Northern hardwood forest/Eastern North America	Tree dbh, species-specific ecological parameters (min & max dbh, degree-days, min & max values for evapotranspiration index). Soil moisture capacity, Litter content, initial soil organic matter and nitrogen contents	Individual trees on a 0.083 ha plot with Monte Carlo simulation	Demonstration of multiple stable states in simulated northern forests mediated by high or low nitrogen availability

FORSKA	Leemans and Prentice (1987)	Derived with considerable alteration from the FORET model. Mitscherlich curve for maximum tree height and tree growth. Explicit consideration of canopy foliage distributed for each tree's crown depth. Two-year time step. Semi-empirical equations	Swedish *Pinus-Picea* forest/Europe	Similar to FORET, but the initial model was simplified and used no stochasticity in the driving variable, which provided clear results (no stochastic variation) but limits simulations to the calibration site. Later, applications varied climatic drivers	Individual trees on a 0.5 ha plot with Monte Carlo simulation	Reproduction of composition for an ancient uncut forest in Sweden. Model was adapted to different locations in Europe and elsewhere for climate-related simulations
SORTIE	Pacala et al. (1993)	Individual-based gap model with empirical parameterizations for birth, growth, dispersal, and mortality. Empirical equations	Northern Hardwood Forest/USA	Tree coordinates, more complex light extinction calculations, basic site characteristics, individual tree data	Spatially explicit in 2D (spatial locations of trees)	Demonstration of statistical/empirical methodologies for gap model parameterization
ZELIG	Urban (1990)	FORET model applied to a map of trees on a forest plot of arbitrary size using windowing. Resolution can be adjusted for greater computational speed or spatial detail. Canopy structure simulated and light computed with a ray tracing approach through a 3D canopy. Growth altered by degree-days and water availability. Shading and approach to plot biomass maximum also reduce tree growth. Semi-empirical equations	Appalachian deciduous forest and other forests, notably Douglas fir forests/North America	Tree dbh, species-specific ecological parameters (min & max dbh, degree-days, min & max values for evapotranspiration index), nutrient tolerance factor, species-specific regeneration probabilities	Can be implemented in 3D as a gap model, as a transect model, or over a multiple ha plot	Used to simulate expected spatial patterns in remote sensing for high-resolution panchromatic optical as well as radar (microwave) instruments

(Continued)

TABLE 7.1 (*Continued*)
Main Characteristics of Individual-Based Gap Models

Model Name	References	Main Processes Modeled and Equation Types	Forest Type(s)/Region(s) of the World	Main Drivers and Response Variables	Spatial Scaling	Main Contributions
ZELIG-CFS	Larocque et al. (2011), Larocque and Bell (2021), Searle et al. (2021)	Adapted from ZELIG. Modifications include new algorithms for crown interaction, mortality, and regeneration. Semi-empirical equations	South Canadian boreal mixed hardwood forest types; boreal forest types	Same as ZELIG	Implementation at the plot level	Used to test new competition algorithms at the canopy level. Integration of the Stand Visualization System and the geographic information system Quantum GIS
ZELIG-TROP	Holm et al. (2012, 2014)	Adapted from ZELIG. Modifications include a new exogenous disturbance routine, optional basal-sprouting routine, adjustment to natural mortality equation for tropical trees. Semi-empirical equations	Tropical forests; both subtropical dry and Amazonian wet rainforests/South America	Same as ZELIG.	Can be implemented in 3D as a gap model, as a transect model, or over a multiple ha plot	Used to simulate diverse and complicated tropical forests and evaluate changes due to multiple disturbance types
FORCLIM I and FORCLIM II	Bugman (1994), Fischlin et al. (1995)	Derived from the FORECE model (Kienast and Kuhn 1989), which is derived from FORET. Considerable development of functional responses to environmental variables. Semi-empirical equations	Swiss alpine forest/ Central European forests	Heat sums, minimum mean monthly temperature, carrying capacity constrained by NPP equation, nutrient constraints on growth, water balance, and dry days	Individual trees on a 0.083 ha plot with Monte Carlo simulation	Capable of simulating forests in multiple locations with different locations in Switzerland. Produces alpine patterns and regional patterns on forest change. Sensitivity analyses using equilibrium as starting point

FORMIX, FORMIX 2, FORMIX 3	Bossel and Krieger (1994)	Functional type-based mortality, seed banking, and growth equations with shading through canopy geometry (Beer's Law). Tree increment uses a photosynthesis equation decremented by tissue-specific respiration and growth respiration. Forest harvest and cutting included process-based equations	Tropical rainforest on several continents and islands/South America	Driven primarily by the light environment interacting with physiologically based tree performance equations	Spatial: a hectare or larger plot tiled with 0.04 ha gaps, details of canopy, form factors, etc. Use of a functional type approach to allow implementation in a number of rainforests	Many good applications in diverse tropical ecosystems with a number of independent model tests
FORMOSAIC	Liu and Ashton (1998)	Growth function calculated for 502 species based on tree growth rate in diameter, basal area of neighbors for inter-tree competition, slope, distance from water, and an error term. Four levels of recruitment distributed among species by "guilds." Mortality by size and longevity classes. Semi-empirical equations	Malaysian rainforest/ Southeastern Asia	Tree coordinates. Parameters were fit to half the data using an empirical approach. Environmental couplings through site conditions: (basal area, slope, etc.)	Tree locations in grid scales (0.01 ha) with internal and external seed dispersal. Scaled up to a 50 ha spatial plot	Initially implemented for the Pasoh forest of Malaysia with some landscape complexity. Parameterized on half the 50 ha plot and tested on the other half

(*Continued*)

TABLE 7.1 (*Continued*)
Main Characteristics of Individual-Based Gap Models

Model Name	References	Main Processes Modeled and Equation Types	Forest Type(s)/Region(s) of the World	Main Drivers and Response Variables	Spatial Scaling	Main Contributions
DRYADES	Mailly et al. (2000)	JABOWA/FORET growth equations, but using a Chapman-Richards Equation for height. Use of a cross- section of sapwood to determine leaf area. More detailed crown geometry. Semi-empirical equations	Conifer forests of the North American Pacific Coast/northwestern America	Probability of wildfire. Stochastic variation in seed rain. Light competition, moisture, and dry days computation. Includes harvesting	3D spatial model	Compared with measured stand data after fire and clearcut
PICUS	Lexer and Honninfer (1998)	Tree growth, reproduction and death. Semi-empirical equations	Austrian forests/Europe	Growing degree-days, winter minimum temperatures. Drought and site conditions (pH, C/N ratio of upper soil layer, soil texture)	0.01 ha plots tiled onto 1 ha grid elements	Climate change assessment for Austrian forests from 2000 to 2065. Change in forest composition, biomass and zonation with altitude
FAREAST	Xiaodong and Shugart (2005), UVAFME model (Shuman et al. 2014)	Fusion of the FORET model modified through several Chinese forest models and then merged with the FORSKA model with several modifications, notably dynamics of soil organic carbon and nitrogen cycling. Semi-empirical equations	Forest of northern China subsequently expanded to simulate forests of northern Eurasia/East Asia	Tree data. Monthly temperature and precipitation and derived heat and water balance indices. Nitrogen cycling data	0.083 ha plots. Largest simulation is 20 km on a side lattice over Russia (31,000+ grid points with 200 replications at each point)	Applications in climate change projections over continental-scale regions

TreeMig	Lischke et al. (2006)	Local tree dynamics from the FORCLIM model to produce ingrowth and outgrowth from height classes Semi-empirical equations	Central European forests/ Europe	Same environmental drivers as in FORCLIM	0.083 ha grid elements on a larger landscape, typically 1 km^2. Grids interact through seed dispersal	Local pattern formation in forests, tree species migration rates in the alpine region of Central Europe
FORSITE-BPM	Caplat and Anand (2009)	Growth uses a Gompertz function. Mortality with two components: one actuarial, one from disturbances. Seed dispersal function, also sprouting regeneration. Semi-empirical equations	Birch–Pine–Maple forest of Minnesota/North America	Tree data and habitat cells. Disturbance regime	Cell diameter of 20 m simulated over 400 cells	Investigation of the resprouting tendency of trees to alter rate and patterns of ecological succession
ForestMAS	Kolmanic et al. (2014)	Light, temperature with indicators for soil moisture, acidity, nutrients/nutrient requirements. Regeneration uses a seed rain approach. Approximation of Yoda's Law (-3/2 thinning law for mortality). Semi-empirical equations	Forests of Slovenia/ Europe	Tree data. Ray tracing for light as a function of topography. Site quality indicators: moisture, incoming solar radiation, soil pH, and nutrient availability	3D IBM[b] with topographically mapped fields of underlying driving variables	Reforestation and ecological succession in Slovenia. Use of classic Ellenberg indicator values from European phytosociological studies

[a] Diameter at breast height
[b] Individual based model

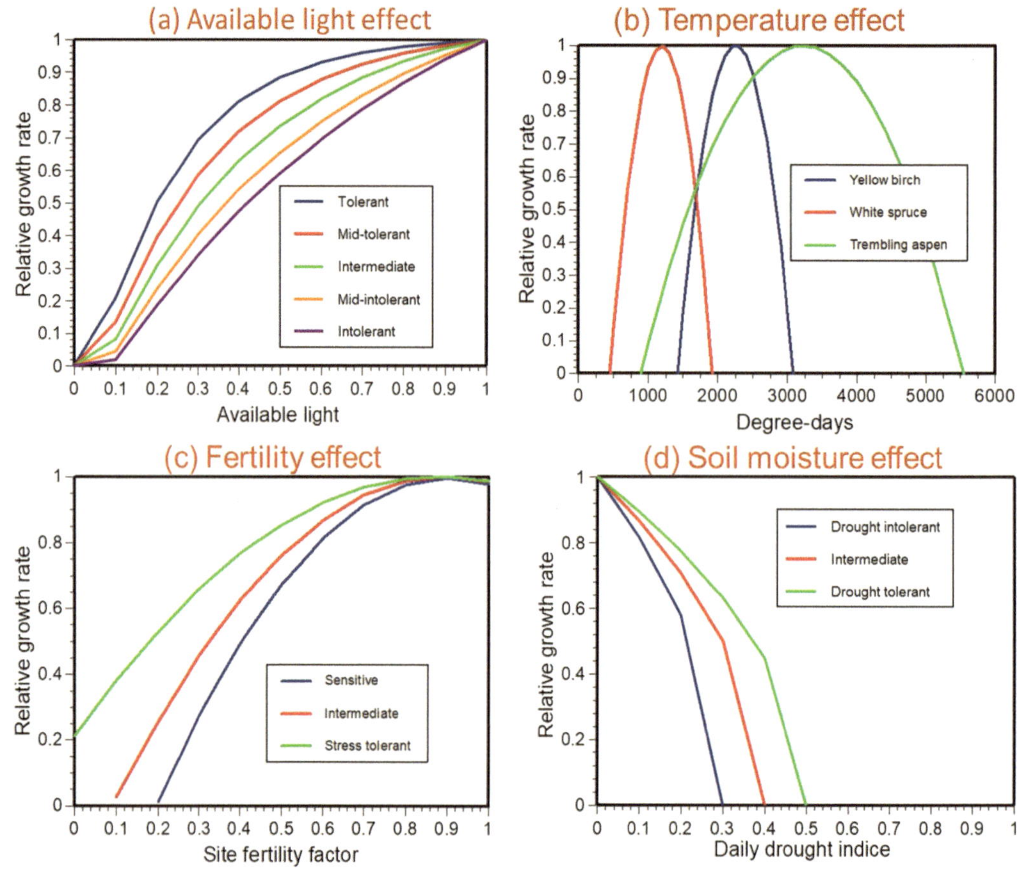

FIGURE 7.2 Response functions to site or environmental factors used in ZELIG that represent how species with different ecological properties respond to site conditions. (Adapted from Urban 1990.)

rate by calculating the available light growing factor (ALGF) as a function of the summation of the leaf areas of all trees within a plot divided by canopy length. Similarly, as in JABOWA, the Beer's law is used to compute available light within stands, but ZELIG recognizes five shade-response classes to light conditions (Figure 7.2a). The most commonly used method to take account of the effect of temperature is based on parabolic functions constrained by species-specific minimum and maximum degree-days within which they can grow (Figure 7.2b). This approach is based on the assumption that, for each species, photosynthesis occurs in physiological optimum conditions at a specific temperature, which is converted in degree-days, and then decreases above and below this optimum level. Parabolic functions for three typical North American species—yellow birch (*Betula alleghaniensis* Britton), white spruce (*Picea glauca* (Moench) Voss), and trembling aspen (*Populus tremuloides* Michx.)— are provided as examples in Figure 7.2 to illustrate the degree to which species may differ in minimum and maximum degree-days.

Different methods to represent the effect of variation in soil fertility and moisture have been derived in gap models. A detailed review is beyond the scope of this chapter, but, for illustrative purposes, the methods developed by Urban (1990) for ZELIG were among the most broadly applied. For soil fertility, site-specific maximum annual productivity in terms of above-ground woody biomass production (Mg ha^{-1} yr^{-1}) must be provided. The response of each species within a forest is then modulated by its specific response category, which may be sensitive, intermediate, or stress tolerant (Figure 7.2c). For soil moisture, a daily drought index is first computed, which is linked to the number of days with low soil water content during the growing season. Each species will have its growth rate affected depending upon whether they are drought intolerant, intermediate, or drought tolerant (Figure 7.2d).

Individual-tree mortality is usually modeled as a random process. Typical JABOWA-type gap models include model components to represent the effects of both natural mortality related to age and stress originating from site factors or suppression. For natural mortality, it is assumed that 1% of individual trees within a plot will reach the species-specific maximum age, but the death probability remains constant with age:

$$P_m = 1 - e^{\frac{-4.605}{AGEMX}} \tag{7.3}$$

where P_m is the probability of mortality, −4.605 is the natural logarithm of 0.01 (1%), and AGEMX is the maximum age that a species can live. Stress-related mortality occurs when trees fail to reach a threshold minimum dbh growth rate. The implementation of stress mortality algorithms varies among models. Some models set a specific value, such as 1 mm $year^{1}$ in dbh growth rate; other models specify a minimum value, such as a percentage of optimum growth rate for a given tree size. Trees that do not meet these conditions see their death probability increased for a given number of years.

As these equations are based on the application of basic ecological principles, they may be considered as semi-empirical because they are a compromise between empirical models and process-based models. Some of their components are based on empirical knowledge while other components have an ecological foundation. For instance, the species-specific dbh growth rate equation is similar in concept to the Gause's model for competition (Gause and Witt 1935), the classic competition model in population dynamics, as it is computed by estimating potential growth rate (absence of competition) reduced by the intensity of competitive interactions. Also, the potential growth rate component of the dbh growth model includes species-specific ecological parameters, such as the minimum and maximum growing degree-days within the area of distribution. The models that are considered as semi-empirical in Table 7.1 generally are based on the approach of computation of potential growth rate reduced by a competitive interaction factor.

The more recent gap models feature relatively simple protocols for estimating the model's parameters (Shugart 1998). For instance, Moore (1989) developed a relatively simple model formulation after noting that the JABOWA model implied that the rate of carbon loss in large trees increased proportionally more than their rate of increase in stem volume. The new diameter increment equation, also used in FORCLIM (Bugmann 1994), was formulated as:

$$\frac{\Delta D}{\Delta t} = GD\left(\frac{1-\frac{H}{H_{max}}}{274+3b_2-4b_3}\right) \tag{7.4}$$

Pacala et al. (1993) identified the need to relate the parameters of gap models directly to empirical field observations because they believed that the high degree of complexity of JABOWA-type models required information difficult to obtain from the scientific literature. Their model, SORTIE, is based on the development of statistical relationships using field data. The growth rate equation for diameter increment was developed as follows:

$$\Delta r = r\left(\frac{P_1 GLI}{\frac{P_1}{P_2}+GLI}\right)+\alpha \tag{7.5}$$

where r is the radius of a tree, Δr is the annual increment rate (ring width), GLI is a "general light index" that is computed from changing sun angles, ratios of diffuse and direct radiation, and canopy openness to represent an integrated measure of annual ltocal light availability, P_1 and P_2 are regression parameters, and α is a random variable normally distributed with a mean = 0. The variance of α is computed as C(predicted ring width)^D, where C and D are estimated parameters.

7.2.1.2 Hybrid Forest Succession Models

Hybrid forest succession models have been developed in the past 20 years to meet the need of forest management situations and problems (Monserud 2003). A hybrid model is typically a mixture of the best features of a process-based physiological model and growth and yield model at the same hierarchical level (Johnsen et al. 2001). Hybrid models hold good promise, because they are highly interbred and may provide more accurate and powerful prediction of forest dynamics and can integrate operational processes with useful products on yield for the manager (Liu and Ashton 1995; Monserud 2003).

Among them, the Forest Vegetation Simulator (FVS) is a stand-level, hybrid forest succession model used for forest management planning (Taylor et al. 2009). FVS is an individual-tree, distance-independent, growth and yield model (Dixon 2002; Larocque 2008) and was initially designed and developed for modeling multiple tree species, even- and uneven-aged forest stand development under alternative management prescriptions as Prognosis in the 1970s. As one of the earliest examples of a hybrid model, it simulates development of individual trees and the deviation-amplifying aspect of the growth process by a unique procedure for introducing the stochastic elements in a deterministic computing algorithm. Lodgepole pine, in the presence of an infestation of mountain pine beetles, was the first forest type for which this growth prognosis model was implemented (Stage 1973).

As the modern version of Prognosis, FVS represents an amalgamation of the original Prognosis model. It couples empirical modeling techniques with a process-based approach, and the forest processes that are described and represented by empirical relationships (e.g., growth) are modified and interact in a mechanistic manner. Individual trees, however, are still modeled aspatially, thus maintaining the model's distance-independent nature (Taylor et al. 2009).

The development of FVS was facilitated by the US Department of Agriculture Forest Service to provide improved quantitative understanding and prediction of stand development in pure and mixed forests under alternative management regimes through integration of known silvicultural knowledge and empirical tree growth data. It is extensively used throughout the United States to support strategic-level forest management planning and has been calibrated for 20 specific geographic areas (variants) of the United States (Crookston and Dixon 2005) and was calibrated for the boreal and Great Lakes–St. Lawrence forest regions of Ontario (FVS-Ontario) (Lacerte et al. 2004, 2006a, 2006b).

FVS has been linked to visualization systems (Stand Visualization System, SVS) and was developed in 1997 to add spatial and visual capacity to its simulation outputs (McGaughey 1997). The SVS generates graphic images depicting stand conditions represented by a list of individual stand components, e.g., trees, shrubs, and down material. The images produced by SVS, while abstract, provide a readily understood representation of stand conditions. Images produced using SVS help communicate silvicultural treatments and forest management alternatives to a variety of audiences (Roth and Finley 2007).

FVS developed over the years with various model extensions to enhance representation of stand characteristics, e.g., understory vegetation, wildlife habitat, disturbance influences and forest treatments. The Fire and Fuels Extension (FFE) to the FVS simulates to model fuel treatments, fuel dynamics, and potential fire behavior over time, in the context of stand development and management. Ager et al. (2020) developed LSim as a hybrid forest model, which integrated FVS with the FSim, a wildfire simulation model to conduct research on long-term management and wildfire feedback. LSim has the functionality to simulate spatially coordinated forest management over time under a background of large stochastic wildfires with models that have undergone decades of field application (Ager et al. 2020). The Douglas Fir Beetle Impact Model (DFBIM) simulates the epidemic impacts of Douglas fir beetle (*Dendroctonus pseudotsugae Hopk*) on its principal host, Douglas fir (*Pseudotsugae menziesii*). The Climate Extension to the Forest Vegetation Simulator (Climate-FVS) provides forest managers a tool for considering the effects of climate change on forested ecosystems.

In recent years, FVS has been linked with GIS capabilities and geospatial function. ArcFVS allows users to select spatially or by attribute the forest plots to be projected by FVS, create the FVS input files for the selected plots, and display in a geospatial environment different types of FVS output (FVS output files, tables with variables of interest, and visualization image files). Advantages of ArcFVS are enhanced format translation functions and the standardized programming environment (Chivoiu et al. 2006). ArcFuels is another example as an interface for fuel treatment planning and wildfire risk assessment (Vaillant and Ager 2014).

An additional example is the hybridization or linking of the forest yield model PTAEDA2 with the process model MAESTRO for loblolly pine. Other hybrid forest succession models include SORTIE, ZELIG, PipeQual, and Stand-BGC (Monserud 2003; Taylor et al. 2009).

7.2.1.3 Applications

For many of the more common temperate and boreal forest species, there is a considerable body of information on the performance of individual trees (growth rates, establishment requirements, and height/diameter relations) that can be used directly to estimate the parameters for such models. The models have simple rules for interactions among individuals (e.g., shading, competition for limiting resources etc.) and equally simple rules for the birth, death, and growth of individuals. The simplicity of the functional relationships in the models has positive and negative consequences. The positive aspects are largely involved in the ease of estimating model parameters for a large number of species; the negative aspects highlight a desire for more physiologically or empirically "correct" functions (Shugart 1984, 1998; Shugart and Woodward 2011). The representation of interactions among trees may also be influenced by the location factor. Similar to distance-dependent models, some gap models require tree coordinates to compute the intensity of competition, but the majority of them do not require location data (Table 7.1). Trees may be positioned in a random pattern within a delimited zone or along a transect.

The availability of extensive information in terms of tree measurement for temperate and boreal forest species theoretically provides a major advantage for model evaluation. The importance of using existing datasets to evaluate the performance of models has been stressed several times in the literature (see Bugmann et al. 2001; Didion et al. 2009; Ngugi et al. 2011; Schuman et al. 2014). In particular, evaluating model predictions against observations contributes to improving confidence in gap model predictions, which is essential to better predict the effects of management practices on long-term forest dynamics or global change. The evaluation of the performance using repeated measurements was performed for several models, such as SORTIE-ND, FAREAST, FORSKA-2V. BOREALIS, ForClim, SIBBORK, or ZELIG-CFS (Bugmann et al. 2001; Shuman et al. 2014; Stadt et al. 2007; Thorpe et al. 2010; Beaudet et al. 2011; Rasche et al. 2012; Brazhnik and Shugart 2016; Bose et al. 2015; Maleki et al. 2020; Larocque and Bell 2021; Searle et al. 2021). These studies provided invaluable results, but there is still a need to continue validation efforts. Only some of these studies were based on the use of long-term historical remeasurement datasets. In some studies, for instance, many sample plot data were used, but the predictions were compared with observations at only one point in time. For other studies, datasets consisted of historical repeated measurements, but only a few sample plots or few species were included.

In many cases, the recent gap models in many cases have functional relationships that are different from those used in earlier gap models (Table 7.1). These models differ in their inclusion of processes that may be important for the dynamics of the particular forests being simulated (e.g., hurricane disturbance, flooding, formation of permafrost etc.). However, the models share a common set of characteristics. Each individual tree is simulated as an independent entity with respect to the processes of establishment, growth, and death. This unifies individual tree-based forest models and allows the computation of species- and size-specific tree demography. Also, gap models emphasize two features that are important for a dynamic description of the vegetation pattern: the response of an individual plant to the prevailing environmental conditions and the modification of those environmental conditions by the individual tree. Gap models are hierarchical in that the higher-level patterns observed (i.e., population, community, and ecosystem) represent the integration of plant responses to the environmental constraints defined at the level of the individual. Recent increases in the capability of remote sensing instruments allow the detection of some of the structural patterns in forests generated by individual tree processes (Shugart et al. 2010). The possibility of testing and implementing individual-based models over very large areas has increased the potential to apply these biologically rich models over continental-scale areas.

Due to the remarkable drop in the cost of computation and the development of object-oriented computer languages, the computational environment of today's gap models is much more conducive to individual-based model applications than in the past. Initial individual-based forest models were extremely complex and implemented with 3D spatially explicit models. We have seen a complexity to simplicity and back to complexity cycle over the past four decades with the developers intending to improve the models at each step of the process. With a cycle from complexity-to-simplicity and back again in gap models (Figure 7.1), there has also been a significant push to expand the application of gap models to subcontinental representations of landscapes. These involve understanding global and regional carbon cycles and representing the dynamics of vegetation responding to climate change. Initial simplified versions of gap models ranged from automatic programs that converted gap model output into Markov models to computationally fast models that assessed the probability of transition of the vegetation from one type to another (Moore and Noble 1993).

Alternative approaches to developing computationally fast representations of gap models involve reducing the species in a gap model to a collective of species called functional types or roles. Both Acevedo et al. (1996) and Shao et al. (1996) treated the problem of deriving a computationally fast role-based model based on four abstract tree-species roles. Acevedo et al. (1996) developed a Markovian model directly from a gap model and then inspected the behavior of the resultant model against dynamic simulations of forest succession from a gap model. Shao et al. (1996) used a somewhat more simulation-based approach and developed regional applications for the four northern provinces of China.

Another important such application to tropical rainforests is that of Moorcroft et al. (2001), who parameterized a gap model for an Amazonian rainforest using a postulated relationship between growth rate, photosynthesis, and wood density. The resultant model was used to parameterize a statistical model of the size distribution of trees across the Amazon Basin using an approach pioneered by Kohyama (1993) to represent forest structural dynamics as a partial differential equation in time, numbers of trees, and tree size. The partial differential equation model provides canopy structure to constrain the results produced by a photosynthesis model. This model, called the Ecosystem Demography (ED) model, was used to summarize the demography of forest mosaics over large areas (Amazonia). The ED model represented Amazonian forests as a Kohyama-type partial differential equation of numbers versus tree size and used a gap model to estimate the growth (the forward vector with respect to the size axis) and mortality (the downward vector) under the constraint of cohort size-specific photosynthesis rates.

7.2.1.4 Future Developments

There are several promising directions for gap model development. These involve the following.

1. Continued development of the models for novel ecosystems and inclusion of new environmental phenomena in the models as well updating internal model functions;
2. Application of the models to a diverse array of topics, including paleo-reconstruction, climate change applications, and simulations of forest under altered management regimes;

3. Theoretical developments derived from the models, including a better understanding of landscape and regional dynamics.

In particular, the application of gap models to predict the effects of climate change on forest ecosystems deserves greater attention. It is expected that climate change will affect forest biodiversity, succession, and productivity at the ecosystem level due to local variation in climatic conditions and characteristics and water-holding capacity of soils and species-specific ecological properties and responses (Dale et al. 2010; Foster et al. 2017; Shugart et al. 2018). Significant changes in forest ecosystem development have been predicted in several modeling studies, such as LINKAGES (Dale et al. 2010), FAREAST (Huo et al. 2010), UVAFME (Foster et al. 2017, 2018), or ForCEEPS (Morin et al. 2021). However, gap models that simulate the effects of global change have been tested and used mostly for forest types in the northern hemisphere, including temperate and boreal forests. The use of gap models to simulate the effects of global change for forest types in the southern hemisphere needs to be conducted.

In Figure 7.1, these are illustrated as branches into topics, such as ecological theory and spatial ecology. There are also several branches of application and development of the models as simulators for non-forest ecosystems. The hope is for a synthesis that seamlessly moves among model development, ecological theory, and ecological applications. Too frequently, the latter two (theory and application) are viewed as antithetical by young scientists—a misconception for which gap models are a prominent counterexample.

7.2.2 Forest Landscape Models

Forest landscape models (FLMs) have been developed approximately over the past 40 years. Since the 1980s, new developments in landscape ecology, quantitative ecology, computer sciences, geospatial statistics, geographic information systems (GIS), remote sensing technologies, and decision-support systems have provided opportunities for developing FLMs that are increasingly valuable for forest ecological management and long-term research in macroecology, such as studying the equilibrium between landscape-scale biodiversity and human activities under global climate change (Mladenoff and Baker 1999; Scheller and Mladenoff 2007; He 2008; Xi et al. 2009a, 2009b; Dai et al. 2015).

7.2.2.1 Concepts

Compared with IBMs (e.g., gap models and hybrid forest succession models), FLMs (the term "raster" model is also used) simulate forest change through time across a broader spatial scale (i.e., landscape or regional scales) generally larger than a single forest stand. Spatial interactions among forest components are key for such models since spatially referenced data (e.g., maps and GIS layers) are often used. They consider that ecological processes occur at different scales, among forest ecosystems within a landscape or a forest region. FLMs can incorporate spatiotemporal processes such as succession, seed dispersal, natural disturbances (e.g., wildfires, hurricanes, outbreaks of native and exotic invasive pests and diseases), and human influences (e.g., harvesting and commercial thinning, tree planting, fire suppression, and global climate change). Thus, these models are increasingly used as tools for studying different scenarios of forest management, ecological assessment, restoration planning, and climate change. LANDIS is a good example of this type of model, originally designed to simulate forest landscapes varying from hundreds of ha to thousands of km^2 (Mladenoff 2004). Its structure is based on a raster GIS format that contains information and data on ecological attributes, species composition and natural disturbances, and forest management practices such as harvesting and tree planting (Figure 7.3).

FLMs are founded on concepts and theories of ecological succession, disturbance, and ecological equilibrium and non-equilibrium. Also, they incorporate forest landscape changes, biological characteristics of tree species within forest ecosystems, competition process among species, and

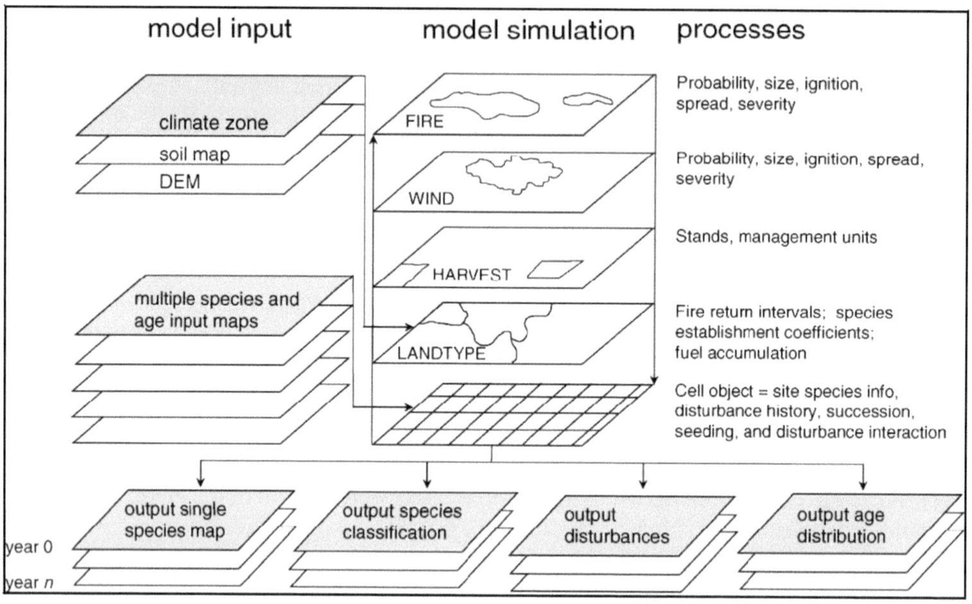

FIGURE 7.3 Computer operational design of LANDIS. (From Mladenoff 2004.)

synergistic effects with environmental disturbances. They can quantitatively describe the spatial distribution of trees, illustrate the relationship between dynamic variations in tree species composition and forest communities, as well as other influences. These models have been broadly applied in various fields, including forest sciences, ecosystem ecology, resource management, and wildlife habitat evaluation. The widely used FLMs listed in Table 7.2 are good examples that highlight the main forest types that were used in model development and application and their main contributions. Even though they share common functions, varied FLMs differ in several aspects, including representation of processes, methods of landscape change, and required data inputs and outputs (Table 7.2).

It is essential to understand the concept and scope of FLMs. According to the literature, there are diverse definitions of FLMs. For example, Mladenoff and Baker (1999) define FLMs as computer models that simulate the change in forest landscapes across broad spatiotemporal scales. Scheller and Mladenoff (2007) broadly define FLMs as computer programs or software packages for projecting landscape change over time. He (2008) provided both general and specific definitions and key terminology commonly used in classifying FLMs. These models can be used to predict variations in spatial characteristics (i.e., distribution, shape, abundance) for simulated objects. More specifically, an FLM is one that simulates the spatiotemporal characteristics of at least one recurrent spatial process in a spatially interactive manner. According to these definitions, He (2008) indicates that an FLM under this specific definition should have the following characteristics: (1) it is a simulation model; (2) it is able to run simulations of spatial processes repeatedly; and (3) it operates at a large spatial and temporal scale that is adequate to simulate the spatial process.

Xi et al. (2009a, 2009b) considered that FLMs are computing models that simulate and predict the spatiotemporal trends of forest changes at the landscape level based on mechanisms of forest dynamics and the interaction of disturbances. Common research questions include the interaction between diverse ecological processes and their implications for spatial patterns of forest landscapes, disturbance mechanisms (i.e., forest growth and succession, carbon, nitrogen and water cycles, forest wildfire, insect and disease dispersal, climate change and other factors), as well as the effects on animal and plant habitats. A typical FLM usually has the following characteristics: (1) projection of landscape-level spatiotemporal changes; (2) simulation of spatially interactive processes; and (3)

TABLE 7.2
Main Characteristics of Forest Landscape Models (FLMs)

Model Name	References	Main Processes and Model Type	Type of Landscape Change	Forest Type(s)/ Region(s) of the World	Main Drivers and Outputs	Spatial Scaling	Main Contributions
LANDIS	Mladenoff et al. (1996), He and Mladenoff (1999), Mladenoff (2004), Wang et al. 2013	Succession, seed dispersal, disturbances (windthrow, fire, insect), harvest management. A grid model derived from JABOWA-FORET gap models and LANDSIM model. Process-based simulation model	A spatially explicit and stochastic model	Second-growth forests in the northern Lake States region. Temperate, southern, and boreal forests of eastern North America. Montane and boreal western forests. Coastal California forests and shrub types. Boreal Finnish forests. Montane forests in Switzerland and northeastern China/ North America, Europe and East Asia	Species life history parameters (shade tolerance, seed dispersal, longevity). Map layers of ecological land and soil types, and climate zone. Outputs: maps based on landscape composition	Allow operations at different scales of resolution by modifying cell size.	Analysis of spatial dynamics and succession. Analysis of the impacts of fire, windthrows, and forest harvesting on landscape structure. Analysis of management consequences
FIRESCAPE	Cary (1997, 1998); Cary and Banks (1999)	Vegetation development/ dynamics. Spatial pattern and variation in fire regime (fire ignition, location of ignition sites, fireline intensity, head fire rate of spread, drought factors, fuel load) and spatial extent of ignition neighborhoods. Process-based simulation model	Spatial landscape	Eucalyptus-dominated forest landscapes/ South eastern Australia	Digital elevation maps. Weather data, fire danger, litter accumulation, fire regime parameters. Outputs: digital or GIS maps that represent the spatial variation of fire regime components (predicted frequency, predicted stand variation)	Variable complex landscape areas	Includes terrain-induced effects. Identifies the spatial extent of fire ignition neighborhoods as important process in topographically complex landscapes

(Continued)

TABLE 7.2 (*Continued*)
Main Characteristics of Forest Landscape Models (FLMs)

Model Name	References	Main Processes and Model Type	Type of Landscape Change	Forest Type(s)/ Region(s) of the World	Main Drivers and Outputs	Spatial Scaling	Main Contributions
LANDIS-II	Scheller et al. (2007)	Same as LANDIS	Same as LANDIS	Boreal forests, mixed hardwood forests, pine forests, and Mediterranean woodlands/North America and Europe	Mostly the same as LANDIS, additional biomass and carbon succession	Same as LANDIS, but allows spatial and temporal flexibility for ecological processes that differ in their temporal resolution	Same as LANDIS
Fire-BGC	Keane et al. (1996a, 1996b)	Simulation of biogeochemically driven forest succession (carbon, nitrogen, and water cycles). Seed dispersal and fire modeled in a spatial domain using raster data layers. Individual-tree succession model with stochastic properties implemented in a spatial domain. Use of FOREST-BGC routines Process-based simulation model	Landscape-level using raster data layers (greater than 10,000 ha).	Glacier National Park, Montana. The McDonald and St. Mary drainages of Glacier National Park/ North America	Ecophysiological properties of species. Carbon and nitrogen contents in ecosystem pools. Outputs: carbon and nitrogen contents, tree data, fire occurrence, seed dispersal	Tree to landscape levels	Analysis of the impacts of climate change and the role of fire on long-term (~200 years) landscape dynamics

Fire-BGCv2	Keane et al. (2011)	Second version of the original FIRE-BGC model Mechanistic, individual-tree, gap model that is implemented in a spatial domain. Vegetation development/dynamics. Integration of the latest ecophysiological simulations of BIOME-BGC. Complex, spatially explicit landscape fire and vegetation model. Simulation of the initiation, spread and effects of mountain pine beetle and white pine blister rust epidemics. Process-based simulation model	Same as Fire-BGC	Same as Fire-BGC	Same as Fire-BGC, plus a wildlife suitability model	Same as Fire-BGC	Simulation of wildland fire ignition and spread, along with cone crops and seed dispersal. Species phenology are simulated at the site level at a daily time step. Better cross-scale integration of processes

(*Continued*)

TABLE 7.2 *(Continued)*
Main Characteristics of Forest Landscape Models (FLMs)

Model Name	References	Main Processes and Model Type	Type of Landscape Change	Forest Type(s)/ Region(s) of the World	Main Drivers and Outputs	Spatial Scaling	Main Contributions
ALFRESCO	Rupp et al. (2000, 2007)	Transient landscape-level model of vegetation dynamics, changing climate and disturbance regime. Changes in vegetation flammability. Stochastic simulation of fire regime driven by climate, vegetation type, and time since last fire. Simulation of fire spread across the landscape using a cellular automation approach. Process-based simulation model	Landscape scale using a cellular automation approach. Pixel resolution: 1 km × 1 km	Upland tundra, black spruce forest, white spruce forest, deciduous vegetation, and coastal forest/ cold regions of North America	Inputs: maps of climate, initial vegetation type, and topography. Outputs: maps of vegetation dynamics, generation of growing season climate	Frame-based spatially explicit model. Use of remote sensing data	Representation of spatial processes of disturbance propagation and seed dispersal. Effect of moose browse pressure on the rate of succession
LANDSUM	Keane et al. (2002, 2006)	Simulation model, including the effects of disturbances (e.g., fire, insects and disease) Process-based simulation model	Spatially explicit, vegetation dynamics. PVT, cover type, and structural stage. Landscapes were defined by watershed Boundaries	Lodgepole pine and Douglas fir forests/ Northwestern United States	Input: data on vegetation and disturbance, management plan. Outputs: vegetation and disturbance maps	Square landscapes of 100,000 ha. Derivation of potential vegetation type over landscapes	Use of a stochastic approach to model disturbances. U.S. mapping development project based on the integration of remote sensing biophysical modeling and landscape simulation to produce maps of historical fire regimes

LandClim	Schumaker et al. (2004, 2006)	Same as LANDIS, but the following model components were added: (1) succession submodel that models the effects of competition, climatic conditions and soil characteristics on the dynamics of tree populations and (2) disturbance submodel of forest fire dynamics	Spatially explicit and stochastic	Montane forests, subalpine forests/ USA	Species' life history parameters (shade tolerance, seed dispersal, longevity). Map layers of ecological land and soil types, and climate zone. Fire parameters (e.g., ignition coefficient). Outputs: maps based on landscape composition	Same as LANDIS	Improvement of the representation of fire dynamics, quantitative forest structures (tree cohorts), and ecological processes

prediction of long-term interactions among multiple factors. FLMs are used increasingly often in studies of forest planning, forest management, resource conservation, ecological restoration, and global climate change.

The processes of forest succession and disturbance may take place over long time frames and broad spatial extents. Thus, there is a limitation to solving problems using traditional fine-scale ecological methods. FLMs provide a solution to overcome these issues. Hence, FLMs can be useful tools to assist long-term forest landscape planning and decision-making in resource management. Scale is important for FLM theory and applications. The object that is simulated using FLMs is the forest landscape. In general, the temporal scale of forest landscape simulations ranges approximately from decades to hundreds of years (i.e., ~50–500 years) while the spatial scale (e.g., spatial extent) is generally about 100–10,000 km^2.

In landscape ecology, extent and resolution are used to define scale. Extent, especially temporal extent, plays a key role in understanding forest ecosystem dynamics (e.g., physico-chemical reactions of wildfires) or long-term succession processes (e.g., soil development, organism decomposition, nutrient loss, or tree species competition). Landscape resolution, also known as grain, indicates the finest unit that can be identified in landscape spatial and temporal extents. Generally, considering the constraints of computational power and interpretation of the results, studying broader landscapes will necessarily increase the grain of a study while small-extent landscapes allow the simulation and integration of more detailed information.

The scale of FLMs refers to the suitable spatiotemporal scale for the study area. It includes the spatial (i.e., spatial extent, grain size) and temporal scales, which are defined by the temporal resolution (i.e., length of a time step) and the duration of the simulation. Temporal resolution for the FLMs, which is not as obvious as spatial resolution, is usually specified as a model parameter. An FLM can have multiple spatial and temporal resolutions when multiple processes are considered. The spatiotemporal scale is an indicator of the degree of complexity of a model. "The challenge of FLMs is to find the equilibrium between resolution level, amount of details and practical considerations" (Mladenoff et al. 1996).

A key issue for the development and application of FLMs is choosing a suitable scale. Due to the complexity and variability of spatial extents, no individual model can ever predict all of the forest ecological information and attributes precisely. On the one hand, landscape ecologists may wish to understand fine-scale ecosystem processes; on the other hand, forest landowners and forest managers need to determine their management plans across broad scales. Forest landscape modelers need to consider both these factors and make compromises based on specific study areas, research interests, and computer hardware and software capabilities. How to define the "best" or the most reasonable scales for FLMs is an important and outstanding research question for theoretical and practical landscape ecologists (e.g., scaling and across-scale landscape modeling).

7.2.2.2 Development History

FLMs have a younger history than gap models. Major developments in FLMs took place in the late 1980s and coincided with the development of landscape ecology (Mladenoff 2004). The need was identified to address large-scale environmental concerns and land management issues, which were becoming more complex. The development of FLMs represents the advance and integration of the disciplines of forest ecology and landscape ecology.

Landscape ecologists, particularly those who work in North America, have made significant contributions to the advancement and evolution of FLMs. With the goal of finding "the best" policy for both forest management and ecological sustainability, North American forest ecologists and forest industry managers have a long history of developing forest models and applying them to important ecological problems. Early in this development, North American forest ecologist focused on forest management problems occurring within forest stands (i.e., forest areas with similar environmental conditions, tree species compositions, and disturbance history) or at the scale of small watersheds.

In the 1980s, driven by the wider use of satellite images (30-m resolution Landsat Thematic Mapper (TM) data) and GIS software, larger-scale spatial analysis became easier. Computational speed and storage capacity achieved significant breakthroughs. At the same time, various new programming languages and software made model coding and data input and output more efficient. Those advances in computer science greatly improved the capabilities for developing multi-scale simulation models. Forest ecologists then began to focus on increasingly important, large-scale landscape management issues, and the development of spatially explicit FLMs became a rapidly developing research area. These new spatially explicit models began to be used to simulate several decades or even hundreds of years of landscape change.

Since the late 1980s, more diverse spatial data are available to forest ecologists and managers. Besides the continuous use of stand-scale, in situ data, other data sources include land cover remote sensing images, large-scale spatial database of soil types, vegetation plots (e.g., Forest Inventory and Analysis National Program), land-use survey data, as well as historical landscape information.

In the 1990s, initial data collection, parameter design, and data validation techniques for FLMs were greatly improved. During this era, parameter design of FLMs mainly used two methods: the physical method and the empirical method. The physical method uses mathematical equations to link physical variables to the resulting phenomena deterministically while the empirical method synthesizes the modeled processes using aggregated parameters generated from physical variables.

During the 1990s, many FLMs that can simulate many processes at multiple scales were developed rapidly, including LANDIS, LANDSIM, FORMOSAIC, and DELTA. He (2008) considers that FLMs developed during this period have similar characteristics to ecosystem process simulation models; not only they can track the spatial variation of individual trees, but they can also use integrated physical simulation methods to model key ecological processes, material cycles, and energy flow. FLMs of this period have adapted stochastic methods and can simulate the long-term effects of ecosystem processes, such as forest harvesting, wind, pests, and diseases. They were no longer limited to a single stand and can be used to represent larger areas including entire forests or ecosystems. Moreover, some smaller-scale FLMs became the integrated components for these large-scale models.

7.2.2.3 Methods

Early forest spatial dynamic models combined cellular automata methods with principles of forest dynamics. Cellular automatons are grid-based, space–time discrete models. It is a classic method for the study of the complexity of a system and particularly well suited to simulate multi-scale spatial and temporal dynamics. The cellular automata approach has become increasingly complex and has been widely used in ecological research. Cellular automata were also incorporated into certain rule-based methods for studying relationships between fire disturbance and forest changes.

The continuous development of forest succession theories and knowledge on ecosystem disturbance and non-equilibrium hypothesis have established the foundation of ecological landscape models. Ecological theories have experienced a paradigm transition from the classic concepts of equilibrium, ecological succession, and climax community to dynamic structure representing non-equilibrium dynamics and the spatial heterogeneity of ecosystems. This transition of core ecological concept and theory is one of the fundamental motivations for the expansion of forest landscape modeling. The advance of FLMs in the past two or three decades can be viewed as an integrated and quantitative process linking knowledge of forest landscape succession, influencing factors, and the non-equilibrium nature of ecosystems.

Hierarchy theory is a theory of the complexity of the system structure, function, and dynamics that form the theoretical basis for building a coherent modeling approach across multiple-scale and complex systems. From a scaling perspective, hierarchy theory acts as a tool to partition complex systems to minimize model error. It can help ecologists to understand landscape patterns and scale-dependent properties of ecosystems and to simplify their complexity and analytical features. Hierarchy theory provides a theoretical basis for scaling and for examining the cross-correlation between spatial and temporal scale functions. Together with other theories (including systems theory, information theory, modern philosophy and mathematics, and cross-scaling research), it provides a systematic and scientific foundation for FLMs.

The development of scaling methodologies allows multiple-scale more realistic modeling. Scaling refers to the translation of information between or across spatial and temporal scales or organizational levels, and two general scaling approaches can be distinguished: similarity-based and dynamic model-based methods. In the context of scaling, landscape ecologists are able to seek the spatial process in the vertical integration of scales. For spatially explicit FLMs, the scaling process is also the key to integrating social and spatial data. For example, Geoghegan et al. (1998) suggested that spatial scaling is a useful tool for the integration of spatial data for the society using social data.

At the dawn of the 21st century, with advances in remote sensing (RS) technology, landscape ecologists are able to quickly obtain time-series remote sensing images. Also, remote sensing imagery and other related GIS data can be directly used as the initial input data for FLMs, allowing pixel-based simulations for the entire target area. The emergence of supercomputers and the progress of computer graphics technology enhanced hardware capabilities to perform large-scale and more complicated imagery or quantitative analyses. Hence, more and more researchers around the world have begun to use remote sensing imagery and spatial analysis software to conduct landscape change modeling and prediction.

7.2.2.4 Classification

Unlike gap models and many ecological models, FLMs include a spatial dimension, i.e., they can simultaneously examine the dynamics of landscape change and interactions in both time and space. Generally, FLMs can be divided into two categories: stochastic landscape models and process-based landscape models.

Stochastic models evolved from the Markov process theory and are based on the transition probability, which combines probability distribution and spatial information. This type of model has been widely used in studies of forest ecology and forest management planning. Process-based landscape models simulate spatiotemporal interactions through the establishment of more realistic computer models. Because this type of model focuses on the study of landscape composition and ecosystem spatial structure, it is also known as the true structural model. For example, Costanza and Voinov (2001) developed a spatial dynamic model for predicting the transitions of coastal landscapes through the use of integrated water cycle nutrient dynamics and the response of biological factors.

More recently, rule-based landscape models have been developed that have adopted artificial intelligence technology. Although these types of models are still in the developmental stage, the wide application of the artificial intelligence theory in other scientific fields suggests that they will soon become an effective tool to address regional resources, ecosystem management, and complex landscape questions.

Horn et al. (1989) classified ecological models that simulate landscape changes into two categories: analytical models and simulation models. Analytical models derive equations using mathematical analysis, analyze the mechanism of forest dynamics, and focus on long-term integrated ecosystem dynamics, which is also known as a strategic model or a generic model, and is usually used for long-term landscape planning. Simulation models are usually based around more intuitive ecological principles and nonlinear equations, and often incorporate more details about the simulation system. Simulation models are often known as tactical models that can be used to develop specific and short-term management plans.

Perry and Enright (2006) classified simulation models into spatially explicit models and forest gap dynamic models based on whether or not a model simulates forest vegetation dynamics. Spatially explicit models are a tool for simulating and studying landscape dynamics mechanisms. They are also considered as "the essence of landscape ecological methods." This type of model assumes that landscape spatial composition and structure change over time, and these dynamic changes can be expressed by mathematical relationships. Such models are generally applicable to larger spatial- and temporal-scale landscape questions. Among them, the LANDIS model is a successful example. Grid-based spatially explicit models can be used to simulate landscape-scale forest changes and natural (e.g., wildfires and hurricanes) and anthropogenic (e.g., logging) disturbance impacts. Compared with the earlier spatially explicit models, LANDIS is much closer to the basic principles of landscape processes.

Baker (1989) suggested that landscape models could be categorized as "whole," "distributional," or "spatial" landscape models based on the details emphasized by the model. Gardner et al. (1999) developed an FLM for simulating the landscape-scale patterns of fire effects. At the same time, they classified landscape models into six categories: theoretical models, exploratory models, spatially explicit models, physical models, probabilistic models, shape models, and statistical models. Although they did not provide specific criteria for each sub-category of the classification, this classification scheme did offer a useful summary of the relationship between forest succession and fire disturbance and also provided a technical framework for further development of forest fire models. Perry and Millington (2008) divided FLMs into predictive and exploratory models based on the motivations for their development and their use. The first category (i.e., predictive models) includes empirical–statistical models, transition models, and forest gap models, which are mainly used to predict future changes in an ecosystem; the second category of models (i.e., exploratory models, also known as heuristic models) are mainly used to explore the relationship between spatial and temporal ecological process.

Scheller and Mladenoff (2005) classified FLMs into eight types from the perspective of ecological function. They mainly considered three ecological processes: reciprocal spatial interaction, tree species community dynamics, and ecosystem processes. This classification emphasized the ecosystem process for the subset of FLMs. Based on Scheller and Mladenoff's (2005) classification, He (2008) further developed quantitative criteria for FLM classification; he also applied similar criteria as those proposed by Scheller and Mladenoff (i.e., whether an FLM simulates a spatial process) to make the first distinction and then used temporal resolution and the forest succession simulation method as secondary criteria to separate physical fire growth models (e.g., FARSITE, Finney 1998) from the rest of FLMs. The criteria He (2008) used emphasize the choice of model resolution, the amount of spatial process, the methods of simulating forest succession in the model design, modeling approaches, and the scope of model applications.

7.2.2.5 Applications

The application of FLMs refers not only to the synthesis of practical issues in forestry and ecology but also to the specific interpretation of model results, including refinement of the models and the exchange of knowledge with experts and practitioners in the field. Although FLMs are often viewed as a research tool, they do provide new ideas and an effective way to examine the relationship among environmental factors, forest landscape spatial distribution, and forest landscape response to climate change.

FLMs have been increasingly applied to study and solve practical management issues in forestry. He (2008) suggested that the application of FLMs generally falls into one of three categories: (1) spatiotemporal patterns of model outputs, (2) sensitivity of model outputs to input parameters, and (3) comparison of different simulation scenarios. Currently, FLMs are used in forestry management, watershed planning and management, post-damage forest landscape restoration, as well as forest land-use development and planning. The application of these models to different forestry disciplines continues to expand (Table 7.2).

In North America, FLMs have most often been designed for forest landscape management and resource assessment. A typical example is the landscape model DISPATCH (Baker et al. 1991). The model uses GIS-based spatial data to simulate the environmental impacts of climate change on the landscape in the Minnesota Boundary Waters Canoe Area. The LANDMAN model represents another good example of a landscape management model. The model suggests that initial landscape structure and harvesting patterns are the major factors affecting the future landscape patterns and that forest managers need to develop planning strategies around the existing landscape structure and harvesting patterns. The LANDMAN model uses quantitative methods to measure spatial structure and includes logging disturbance patterns and a management efficiency evaluation index to develop optimal solutions for landscape management. The LANDFIRE project started in 2004. Between 2004 and 2009, the project developed over 1,200 unique state and transition models to predict the pre-European settlement conditions for all of the major ecosystems of the United States and served as a solid foundation for model development efforts related to land management and planning (Kerns et al. 2012).

LANDIS-II and LANDIS Pro, which were developed from the founder model LANDIS, are among the most widely used FLMs and have been used by many research groups and agencies throughout the world. In addition to simulate the changes in forest succession, structure pattern, and the cycle accumulation of biomass and carbon, several extensions have been developed with LANDIS-II and LANDIS Pro for incorporating timber harvesting, land use and other natural disturbances, such as wildfire, wind, insects, pathogen, fuels, browse, road and habitat suitability. For example, Swanson (2009) used the LANDIS-II model to model the effects of alternative management strategies on forest carbon. Gustafson et al. (2011) used the LANDIS-II model to study how the forest management strategy mitigates the climate change in south-central Siberia. Thompson et al. (2011) simulated the effects of climate change and land change on the biomass in the forest in the Massachusetts by using LANDIS-II model.

The application of existing models to investigate Chinese forest dynamics and management practices is underway. In the past few years especially, these applications have grown considerably. For example, Chinese ecologists have used LANDIS to examine forest succession and management in northeastern China. Hu et al. (2004) used LANDIS to assess the long-term forest landscape changes under various harvesting and logging scenarios in the Daxinganling region and also quantitatively evaluated the effects of harvesting on forest landscape changes. He et al. (2005) used LANDIS to simulate the long-term forest landscape dynamics in the Changbai Mountain National Nature Reserve. In general, current applications of FLMs in China are mainly limited to the northeastern region, and the current focus of applied studies is to simulate forest succession, landscape-scale disturbance effects and mechanisms, and resource management strategies in various forest ecosystems of China. LANDIS family models have been used to simulate the forest biomass and carbon changes under warming climate in recent years. For example, Scheller et al. (2007) used LANDIS-II to simulate the effects of various disturbances, including forest fires, harvesting strategies, and wind damage on northern broad-leaved forests of the United States. They also investigated the effects of spatial relationships of disturbance and re-introduction of natural forest fire on long-term forest landscape changes under various climate change scenarios. Dai et al. (2016) used LANDIS-II, coupled with a forest ecosystem process model (PnET-II), to simulate spatial interactions and ecological succession processes under RCP scenarios, RCP2.6, RCP4.5, and RCP8.5, respectively. Moreover, they modeled a control scenario of extrapolating current climate conditions to examine changes in distribution and aboveground biomass (AGB) among five different forest types for the period of 2010–2100 in Taihe County in southern China. The results of the simulation show that climate change will significantly influence forest distribution and AGB. Molina et al. (2022) used the LANDIS-II spatially explicit landscape model to evaluate the effects of forest management scenarios and projected fire regimes under climate change in northeastern Canadian boreal forests (Molina et al. 2022).

Visualization tools have been developed in the past years for forest managers and LANDIS users. LandViz is a visualization tool that allows researchers or managers to quickly and easily view LANDIS-II output (Gustafson et al. 2016), and LANDISVIEW was developed as an open-sourced geospatial image viewer designed to provide the most common functions of a LANDIS output viewer and geospatial data converter, to integrate key software features, and to enhance the data processing ability (Xi et al. 2020).

7.2.2.6 Interactions between Gap Models and FLMs

Forest resource monitoring and operational planning are other major fields for which FLMs have been developed and used. For example, LANDIS has been widely used for the simulation of broad-scale (>10^5 ha) landscape dynamics, including succession, disturbance, seed dispersal, forest management, carbon dynamics, and climate change effects. At the landscape scale, LANDIS uses age-defined cohorts to represent the age structure of tree species and a ten-year time-step to determine changes in these age classes driven by key ecological processes such as establishment, competition, and dispersal. In addition, it allows the user to quantitatively define disturbances such as forest fires and wind damage. He et al. (1999) provided a method for linking the forest gap model LINKAGES with LANDIS to study species' responses to large spatial scale climate warming in northern Wisconsin, United States.

Xi et al. (2008) recently applied LANDIS and provided a more integrated and automatic method for integrating LINKAGES with LANDIS. The LANDIS modeling environment has been used to examine forest landscape restoration strategies and forest management following southern pine beetle (*Dendroctonus frontalis*) outbreaks in the southern Appalachian Mountains of the United States, including pine-oak forest succession, the reciprocal interaction of forest landscape structure and southern pine beetle herbivory, the relationship between natural disturbances, such as wildfires and changes in forest composition and spatial patterns.

Other applications of FLMs include forest fire management and forestry practice policies. For example, LANDIS has been used to investigate the effects of climate change and forest fire dynamics on the California coastal shrub landscape in the United States. He et al. (2004) and Yang et al. (2007) used LANDIS to simulate the effects of different forest harvesting methods and fire disturbances caused by human activities on forest composition and productivity. LANDIS-II is a new generation of LANDIS and is distinguished by the inclusion of variable time steps for different ecological processes. The model builds upon and preserves the functionality of previous LANDIS forest landscape simulation models. Recent LANDIS-II developments have expanded the cohort definition to include other relevant data, including aboveground biomass and density and tree diameter data. These additions expand the range of ecosystem processes that can be represented in the model and provide additional quantitative output. FARSITE, a fire behavior and growth model, uses grid-cell input data, but models the spread of fire using a vector format and has exogenous climate drivers that control fire spread. FARSITE incorporates existing models of surface fire, crown fire, point-source fire acceleration, spotting, and fuel moisture to provide a comprehensive fire prediction model. This is particularly useful to forest managers for exploring the connections between different fire behavior models, for revealing the implications of their assumptions to fire growth, and for identifying missing components among the various models. SAFE FORESTS has been used to simulate and analyze the effects of fire dynamics and timber harvesting in the forests of Sierra Nevada, California, USA, and to support forest management decisions on wildfire and harvesting. LINKNZ, an extended version of the LINKAGES model, was used to simulate the effects of disturbance on forest succession in native evergreen New Zealand forests.

7.2.2.7 Advantages and Limitations of FLMs

Currently, the increasing availability of FLMs provides forest ecologists and managers with opportunities to conduct spatial simulation research. However, all computer models have limitations. The customized, user-friendly interfaces still cannot overcome the inherent limitations in each model. Current FLMs mainly focus on the simulations of certain species or communities within a relatively small spatial area. Therefore, simulating overall species diversity on larger-scale landscapes may be more difficult. The development of landscape models requires certain trade-offs among landscape extent, data resolution, prediction accuracy, parameterization, and validation. Costanza and Maxwell (1994) indicate that the conflicts between model resolution and model predictability exist in spatially explicit landscape models. Although increasing resolution provides more descriptive information about the patterns in the data, it also increases the difficulty of accurately simulating those patterns. In addition, Wennergren et al. (1995) indicated that there is uncertainty in processing spatial data (e.g., the phenomenon of seed mortality during seedling dispersal). Hence, landscape spatial models are more appropriate to use as a management planning tool rather than as a tool to predict variations in species composition.

The existing limitations of FLMs are largely the result of the incomplete understanding of forest ecological processes and patterns. Simulating landscape dynamics is essential not only to understand evolution patterns and processes of landscapes from their current status to the future, but also to understand the causes of landscape change. One of the key factors for the current limitation of the FLMs is the lack of understanding of landscape changes and development. While interest in the development of FLMs grows over time, these models are still difficult to apply on a greater range of time and space. Running these models requires a lot of parameter settings and advanced computing power. In addition, if current model users ignore inherent model limitations, the risk of model misuse increases.

The development of forest landscape modeling still faces critical issues in model result verifications. First, independent temporal or spatial series data, which may be necessary for model validation, are not always available. According to the conventional method, the results require specific time and spatial datasets to verify model prediction results. Normally, if the results confirm the effective phase, the following simulations will be considered valid. However, it is often impossible to perform overall time series verification on the FLMs using traditional methods; doing so would make the FLM no longer significant. Each landscape is unique in nature and cannot be replicated. In fact, the data used to conduct validation exercises are often difficult to obtain or cannot be easily collected. Second, when the effects of biological or non-biological factors on model simulations are discussed, these factors are actually included in the model parameters and have already been expected to show in simulation results.

Inadequate post-simulation data analysis techniques may also limit the capability of FLMs. To effectively analyze complex spatial patterns, the post-simulation analysis tool, data management methods, output visualization technique, and model validation still need to be further improved. LANDISVIEW, and a more recent upgraded version LANDISVIEW 2.0, partly contributed to solving this problem (Birt et al. 2009; Xi et al. 2020).

FLMs have become a useful and, to some extent, indispensable tool in modern forest ecosystem studies and ecological forest management. Improving the design and performance of the models has become the focus of forest landscape ecologists in this research field. FLMs can simulate landscape processes and patterns in time and space. The models are able to store information on past and existing vegetation and on the status of disturbance and management, and, more importantly, they can also be used to predict changing trends and contribute to a more effective study of forest responses to various disturbances and forest landscape management. Because landscape models can simulate the complicated process of spatial patterns and reflect the spatial and temporal characteristics of information, FLMs will be further developed and have good prospects for use in both theoretical and practical application research. In the coming decades, ecologists and foresters will have a better understanding of forest landscape patterns and processes. The technologies and methods used for the development of FLMs will become more comprehensive, and the types of FLMs will be more diversified in accordance with different research questions.

7.3 APPLICATION OF FOREST SUCCESSION MODELS AND LINKAGES WITH ECOLOGICAL INDICATORS

7.3.1 Introduction to Ecological Indicators and Definitions

Ecological indicators, defined here as an index of measures that characterize an ecosystem or one of its critical components, are typically used to assess the state of a specific environment and to evaluate the trajectory of that system under current and future conditions. The future trajectories of an ecosystem are contingent on its ecological memory (Padisak 1992; Sun et al. 2013), i.e., the state of the environment, which has been influenced by historical events. Ecological memory is fundamental in influencing future ecosystem response, spatial resilience, and the direction of ecosystem change. Ecological indicators should represent key information about the composition, structure, function, and vigor of an ecosystem. Ideally, there should be a wide range of indicators that represent a suite of ecosystem processes and that can be measured repeatedly. Examples of critical ecological indicators are species composition or diversity (i.e., "composition"), vegetation structure, related light environments and spatial heterogeneity (i.e., "structure"), and ecosystem productivity, nutrient cycling, and life history strategies (i.e., "function") (Noss 1990; Dale and Beyeler 2001). Within these three main categories are many processes that occur at different scales, such as (but not limited to) species distribution, abundance and richness, soil conditions, canopy openness, fragmentation, growth rates, phenology, regeneration metrics, response to abiotic stress, resilience, and resistance (Table 7.3). While conducting this review of forest succession models and linkages with ecological indicators we found that the definition of ecological indicators may vary between studies or that an indicator might fall into one or several of the broad categories used here (i.e., composition, structure, function). Therefore, we are keeping each author's definition and rationale for categorizing an indicator. Ecological indicators are explained in more detail in Chapters 11 to 17.

TABLE 7.3
Examples of Ecological Indicators Separated into Three Main Indicator Categories (i.e., Composition, Structure and Function) and Repeated into Two Ecosystem Scales (i.e., Community and Landscape)*

Ecosystem Hierarchy	Indicator Category	Indicator Example	Examples of Gap Models and FLMs (Not Exhaustive)
Community	Composition[§]	Abundance, distribution, diversity	FORCLIM (Bugmann 1996), FORMIX3 (Huth and Ditzer 2001), PICUS (Lexer et al. 2002), FAREAST (Shuman et al. 2011; Lutz et al. 2013)
	Structure	Age, size, canopy structure, canopy openness, light conditions, spatially explicit	SORTIE (Pacala et al. 1993, 1996), ZELIG (Urban 1990), FORSKA (Linder et al. 1996, 1997), Mountain (Cordonnier et al. 2008), ZELIG-CTF (Larocque et al. 2011)
	Function	Productivity, growth and mortality rates, phenology, regeneration metrics, nutrient cycling, and nutrient limitations	LINKAGES (Pastor and Post 1985), PICUS (Lexer and Honninder 1998), ZELIG-TROP (Holm et al. 2012, 2014), ForestMAS (Kolmanič et al. 2014)
Landscape	Composition	Richness, diversity	LANDSUM (Keane et al. 2008), 4C (Kint et al. 2009), SELES (Hope et al. 2006)
	Structure	Spatial heterogeneity, soil conditions, fragmentation	LandClim (Schumaker et al. (2004, 2006), LANDSUM (Keane et al. 2006)
	Function	Response to abiotic stress, nutrient cycling resilience, resistance, disturbance	FIRESCAPE (Cary and Bank 1999), TreeMIG (Lischke et al. 2006), FireBGCv2 (Keane et al. 2011), LANDIS-II (Kretchun et al. 2014; Loudermilk et al. 2013)

Notes:
* Derived from Dale and Beyeler 2001.
§ By default, all species-specific gap models simulate community composition behavior. A limited list of gap models that focus mainly on community distribution, diversity, or compositional changes are listed here.

Forest gap models and landscape models can simulate the state and trajectory of many of the ecological indicators mentioned previously and in Table 7.3, thus allowing managers and scientists to alleviate some of the challenges pertaining to on-the-ground diagnosis and indicators monitoring programs. Ecological systems are complex, and measuring only a few indicators can fail to capture this complexity. Long-term monitoring programs are costly, and they usually do not cover a large spatial extent. However, they are needed to evaluate sensitivity to stress and climate change. Gap and landscape models are able to overcome many of these barriers. Gap models take into account environmental constraints and, as a result, predict resource competition among trees (Shugart and West 1980), thus achieving the capability to quantify many key ecological indicators over time. For example, a timber harvest forest simulator was able to link ecosystem indicators at the landscape scale to measure forest sustainability and evaluate the impacts of multiple forest management strategies (Gustafson et al. 2007). It has also been demonstrated that multiple indicators, e.g., structural complexity and indicators of stand- and landscape-level heterogeneity, are more successful at assessing biodiversity and sustainable forest management (Lindenmayer et al. 2000). Fortunately, these ecosystem characteristics can be described in forest succession models, and it can be done at fine resolutions when using gap models. While these results help to evaluate the cumulative effects of management and landowner-decision-making, models need to be improved to be able to predict a wider range of indicators, and more modeling efforts are needed (Makela et al. 2012).

7.3.2 Review of Forest Succession Models Linked to Ecological Indicators

7.3.2.1 Composition: Biodiversity and Species Composition

Biodiversity can be an important ecosystem service (Mace et al. 2012) and an indicator of forest health. For example, increased diversity in species composition leads to increased genetic diversity and resistance to disease, invading species, and to the ecological impacts of climate change (Duffy 2009; Thompson et al. 2009), and thus to potentially more stable ecosystems. However, it might have limited validity in increasing forest function (Nadrowski et al. 2010; Grossiord et al. 2014). The relationship between species diversity and ecosystem functioning is more complex and at times could be uncorrelated. For example, functional diversity may play a stronger role than species diversity in ecosystem functioning (Naeem et al. 2000). However, species composition does play a major role as an indicator of forest health and resistance to disturbances (Thompson et al. 2009; Montréal Process Working Group 2009). Representing species diversity in models can be a challenge. However, forest succession models, in particular gap models, have the capability of predicting change in species composition over time and simulating the trajectories of species as influenced by site conditions, seasonal shifts, disturbances, or meteorological changes.

While the majority of individual-based forest gap models represent separate species and their attributes, the following models are useful for investigating ecological indicator species composition. FORCLIM (Bugmann 1996) is a simplified forest gap model that uses a minimum number of ecological assumptions to assess the effect of climate on plant population dynamics. The model is robust when it comes to matching tree species composition to empirical records, as well as aboveground biomass patterns and dominating tree species niches across climate gradients. The FORMIX family of models (Bossel and Krieger 1994), specifically FORMIX3 (Huth and Ditzer 2001), investigates the change in four variables: total yield, yield per cut, species composition, and canopy opening as indicators of long-term impacts caused by logging in rainforests. They found that longer logging cycles using low impact methods were needed to conserve species composition and minimize the erosion risk. The shorter logging cycles (<40 years) resulted in low yields and highly disturbed species composition. Shuman et al. (2011) utilized the gap model FAREAST to evaluate species composition shifts in response to a changing climate (i.e., shifts in temperature and precipitation) in boreal forests and found that low diversity sites exhibited an abrupt vegetation shift from deciduous larch forest to evergreen conifer forest in response to increased temperatures. Due to albedo changes, the shift in species composition enhanced the climate-vegetation feedback process and created a positive radiative forcing effect. Using the same model, Lutz et al. (2013) looked further into interactions and feedback between species composition shifts, vegetation cover changes, and a changing climate.

FLMs have also been successful tools for predicting changes in species composition, which is an indicator of diversity. Using the landscape fire succession model LANDSUM, Keane et al. (2008) assessed landscape composition over time in response to different scenarios of climate and fire regimes. Keane et al. (2008) used the area occupied by the dominant vegetation class and the Sorenson index for measuring similarity as indicators of change in landscape composition. The landscape model 4C has been used as a tool to describe forest biodiversity after a dozen forest conversion regimes (Kint et al. 2009), a useful practical application. The goal of conversion management was to convert Scots pine stands into a mixed oak-birch stand, promoting biodiversity and stand productivity. As another example, the FLM SELES evaluated forest-woodland recovery and woodland expansion in Scotland as a means to offset habitat loss and fragmentation and promote biodiversity conservation (Hope et al. 2006).

7.3.2.2 Structure: Vegetation and Canopy Structure

Canopy and stand structure parameters, such as canopy coverage, range in age of canopy trees, and canopy stratification, can be used as ecological indicators of forest vitality and tools for forest management (Gao et al. 2014). For examples of vegetation structure used as a modeling tool, we first look at the FORSKA-based family of models (Leemans and Prentice 1989; Desanker 1996; Lindner et al. 1996, 1997), which have adopted a cylindrical crown model as opposed to the original disk crown geometry to represent more realistic vertical canopy structures. FORSKA-based models take into account the length of the crown and the amount of leaf area determined from the pipe model theory (Shinozaki et al. 1964). Lindner et al. (1997) improved the capabilities to model stand structure by improving the height growth functions, which included accounting for indicators such as the competition status and shade tolerance of the tree. They argued that the updated height growth function was flexible, simple, and could be applied to forest management simulations. Improvements to vertical canopy structure can increase the accuracy of light competition estimates, thereby influencing growth and survival. The SORTIE model (Pacala et al. 1993, 1996) then added width of tree crowns to canopy depth and varying crown transmissivity, dropping the horizontal homogeneity assumption. Urban (1990) developed a spatially explicit model (i.e., ZELIG) that computes 3D canopy interactions among trees, which was then further developed into later versions by Weishampel and Urban (1996) and Larocque et al. (2011). ZELIG-CFS (Larocque et al. 2011; Larocque and Bell 2021; Elzein et al. 2020; Searle et al. 2021) updated the application of the ZELIG family of gap models by developing algorithms to improve the realism of the representation of crown interactive effects between relative light availability, crown recession rate, foliage distribution, and shade tolerance and thus predicted crown structure characteristics that were consistent with species' life history. Cordonnier et al. (2008) used the individually based, spatially explicit model called "Mountain" as a management tool and modeled structural indicators (i.e., diversity of diameter distribution, sapling cover, and basal area) to evaluate stand resilience and forest stability.

7.3.2.3 Functions: Competition, Forest Resistance, and Resilience

Competition can influence the composition, structure, and function of all forest types around the globe. In gap models, competition is modeled as the performance of trees under varying environmental conditions (Shugart 1998). Individual trees can influence environmental conditions (e.g., by modifying light conditions and soil characteristics), and environmental conditions can influence each tree (e.g., temperature, disturbance). Accounting for resource competition among species and for the related forest resilience and resistance within gap models and FLMs makes it possible to track different aspects of forest function, which is a critical ecological indicator.

For example, further modifications to ZELIG (i.e., ZELIG-TROP, Holm et al. 2012) led to successful simulations of tropical forest dynamics and evaluated the response of tropical forests to disturbance. ZELIG-TROP was applied to look at the resilience of secondary forest succession and found that forest recovery on abandoned agricultural farmland with degraded soils had a delayed response prior to full recovery. Ecological indicators related to early resource acquisition, such as canopy height, canopy coverage, and high stem density, were the fastest to recover, but structural indicators, such as biomass acquisition and basal area, took longer to recover and were less resilient. Holm et al. (2014) used ZELIG-TROP to investigate the impacts of increased natural disturbance in the Central Amazon and evaluated the extent to which variations in disturbance regimes in the Amazon contributed to existing forest gradients across the Amazon Basin. Application of the gap model found that aboveground biomass stocks in the Central Amazon forest significantly decreased but were otherwise resistant to continual, elevated disturbance. Additionally, after controlling for disturbance effects, the forest structure and functional variability across the Amazon Basin was not entirely explained by the variability in disturbance regimes, suggesting that forests have the capacity to adapt to varying degrees of disturbance or other intrinsic environmental factors that contribute to the regional gradients.

The "Mountain" model (Cordonnier et al. 2008) was used as a management tool in Norway spruce stands using structural indicators to evaluate stand resilience. This study determined that forest resilience decreased with thinning intensity and that applying different management scenarios led to the best protection against different natural disturbance events, suggesting that a compromise strategy was found to achieve the best results for all natural disturbance events. The succession gap model ForestMAS (Kolmanič et al. 2014) utilizes Ellenberg indicator values (also used in ForClim) to look at secondary forest succession on abandoned cropland in Slovenia and to forecast regeneration processes after clearcutting or natural disasters. The spatially explicit FLM TreeMig, which is based on a forest gap model (Lischke et al. 1998), incorporates spatial interactions and competition to address tree migration from the stand to the continental scale (Lischke et al. 2006).

As stated in the Montréal Process: Criteria and Indicators (Montréal Process Working Group 2009), two of the indicators used to quantify "maintenance of ecosystem health and vitality" are (1) the area and percentage of forest affected by disturbance, processes, or agents beyond the range of historic variations, and (2) the area and percentage of forest land with diminished biological components. Multiple FLMs have evaluated the effect on forest functioning resulting from multiple disturbances, with a strong emphasis on modeling fires (Baker et al. 1991 using DISPATCH; Cary and Banks 1999 using FIRESCAPE; Keane et al. 2011 using Fire-BGCv2), but also a changing climate (Loudermilk et al. 2013 using LANDIS-II), or land-use change (Voinov et al. 1999 using PLM), as well as biotic processes and agents such as insects (Kretchun et al. 2014 using LANDIS-II), and exotic species invasions (Waldron et al. 2010 using LANDIS). Up to 44 landscape-fire succession models have been listed and classified based on the approach of modeling fire-vegetation interactions (Keane et al. 2004).

7.3.2.4 Global Geographic Coverage of Forest Succession Models

The application of gap and landscape models to address ecological indicators has covered a wide range of ecosystem types and many regions of the world (Tables 7.1 and 7.2). However, there are still some ecosystems that are lacking representation. For instance, the evaluation of compositional, structural, and functional changes in all tropical regions could benefit from future model development. Current gap models that simulate tropical forest dynamics are KIAMBRAM (Shugart et al. 1980), FORICO (Doyle 1981), OUTENQUA (Van Daalen and Shugart 1989), FORMOSAIC (Liu and Ashton 1998), FORMIND (Kohler and Huth 1998), FORMIX3 (Huth and Ditzer 2001), SORTIE (Uriarte et al. 2009), and ZELIG-TROP (Holm et al. 2012, 2014). Tropical forests cycle more water and carbon than any other biome on the planet and are high priority ecosystems for predictive modeling. The overwhelming majority of successional models have been developed for temperate deciduous and mixedwood forests of North America and Europe (Shugart and Smith 1996). A growing group of gap models and FLMs have been developed for boreal forests. In addition to tropical forests, boreal forests exhibit strong feedback in response to climatic change and are critical regions to study. These models include BOFORS (Shugart et al. 1992), EXE (Martin 1992), ALFRESCO (Rupp et al. 2000), SORTIE-ND (Coates et al. 2009), FAREAST (Zhao et al. 2009), ZELIG-CFS (Larocque et al. 2011; Larocque and Bell 2021), LANDIS and LANDIS II (Scheller et al. 2007), and UVAFME (Shuman et al. 2014). While some successional models cover only a specific geographical location and ecosystem type, these models have a strong potential to be adapted and transferred to other regions. For example, models such as SORTIE, ZELIG, and LANDIS have been used in multiple ecosystems.

7.4 DYNAMICS OF REGENERATION AND SUCCESSION MODELS

Regeneration dynamics is a very complex process that remains poorly understood. For most species, many regeneration surveys have been conducted, at least in European and North American forests. However, regeneration surveys consist merely in estimating the number and distribution of seedlings or saplings. Even though regeneration surveys provide basic information that can be used in the decision-making process in forest management, they have limited usefulness for the study of forest succession since they consist of only a picture at one point in time. Regeneration dynamics includes several processes, including seed production, dispersal and germination, and seedling survival and development (Price et al. 2001). When established, the survival and growth of seedlings can be affected by competition from neighboring species or other random factors such as wildlife browsing. All these processes vary considerably among species and from years to years, and they all have an impact on the long-term development of forest ecosystems (Bugmann 2001; Price et al. 2001). Several studies have been conducted to better understand regeneration dynamics, including the growth response of different species to the availability of light and soil resources (e.g., Dai 1996; Finzi and Canham 2000; Kincaid 2012; Salazar et al. 2012; Major et al. 2013). However, most of them took place within short periods of time. They provide rich information on the relationships between site factors and seedling growth, but the duration of these experiments was too short to answer questions on long-term dynamics to better understand interactions between changes in site conditions, seedling development, and survival over time. Being able to answer these questions would be important to improve the modeling of regeneration dynamics in succession models (Larocque et al. 2006; Pabst et al. 2008).

Despite the lack of empirical information, regeneration dynamics in most succession models has been modeled using biologically consistent algorithms based on species-specific seedling ecological characteristics, including shade, water, and nutrient tolerance factors. For instance, JABOWA, FORET, and ZELIG account for the effect of understory light conditions, degree-days, water availability, and fertility level. For model initialization, users provide species-specific potential seedling numbers that may appear during each year of the simulation. Then, this potential number is modified by understory light conditions, water and nutrient factors, and a random number to account for annual variability. As a consequence, when there are gaps in forest ecosystems due to tree mortality, gap models are sensitive enough to model the change in regeneration dynamics.

Most of the studies on the simulation of regeneration dynamics with succession models were conducted using gap models. There are studies that focused on the simulation of the impacts of different regeneration scenarios or disturbances on potential successional pathways. Using the model ZELIG, Pabst et al. (2008) modified the regeneration probabilities for different ecoregions and forest types. Historical datasets were used to compare predictions and observations. Kolmanič et al. (2014) used the single-tree succession model ForestMAS to predict regeneration dynamics over landscapes of farmland regions abandoned in the 1940s in Slovenia. In other studies, the prediction of regeneration dynamics between models was compared. Garman (2004) compared the predictions of regeneration dynamics between the landscape model LandMod and the gap model PNWGap in three forest types of western Oregon Cascades, United States. Papaik et al. (2010) performed a similar exercise with the landscape model LANDIS and the gap model SORTIE-ND in an eastern boreal mixedwood forest type in northwestern Quebec, Canada. Simulations of the modifications of site factors were also performed. Beaudet et al. (2011) compared predicted and observed light levels using the SORTIE-ND gap model in boreal mixedwood forests and discussed the effects on regeneration dynamics. Working with the original version of ZELIG, Larocque et al. (2006) simulated the effect of the variation in "plot size," which represents the zone within which trees use site resources, on regeneration dynamics in two southeastern Canadian mixedwood forest types.

These studies contributed to evaluating the performance of gap models on regeneration dynamics, since they provided biologically consistent results. In a way, these efforts may be considered as a form of model validation (see Rykiel 1996). However, there is still a need to determine how well existing regeneration model components in succession models represent reality with a reasonable level of accuracy or realism. There are some studies that compared simulated results with long-term historical observations (e.g., Larocque et al. 2006, 2011; Larocque and Bell 2021; Pabst et al. 2008), but they were limited to mature trees. The comparison of simulated regeneration dynamics with observations still remains to be performed. One study that would come close to this type of comparison was conducted by Chen and Twilley (1998). They simulated the dynamics of mangrove forests along gradients of soil salinity and nutrient availability using the model FORMAN. The number of saplings added in each year of a simulation is a function of available light below the canopy, random effect, and effect of flooding and seed consumption. Their equation was based on Botkin (1993). Using different sapling recruitment scenarios, they compared the predicted patterns of mangrove development with observations made 34 years after a hurricane. This exercise allowed them to estimate initial values of sapling density in their model that would match the observed mangrove pattern 34 years after the hurricane. This procedure is an indirect way to compare predictions and observations, because the adjustment of initial sapling density to approximate observations made 34 years later does not guarantee that the new initial values were even close to reality. In addition to these issues, additional research in the modeling of regeneration dynamics is needed to improve the capacity of succession models to predict how forests may respond to climate change (Price et al. 2001).

7.5 APPLICATION OF REMOTE SENSING

The use of remote sensing technology with succession models is relatively new. It can be used to obtain information at the individual-tree or landscape levels to initialize or validate models. Using remotely sensed aerial images or digitized aerial photographs at high spatial resolution, it is possible to depict the crowns of individual trees and identify species using semi-automated recognition software (Gougeon et al. 1998; Gougeon 2010). It is even possible to obtain estimates of stand height (Mora et al. 2013). There are not many study cases on the use of this methodology jointly with succession models, particularly gap models. However, this method has a lot of potential for the use of gap models at the landscape levels. The delineation of individual-tree crowns over large areas offers the potential to provide individual-tree information in the input file of a gap model to run simulations at the landscape level. Thus, it is possible to run simulations at the single-tree level over a landscape.

However, for large-scale models, including landscape or global vegetation models, it is generally difficult to initialize or validate these models. Large-scale datasets are rarely available. Zhao et al. (2009) mentioned that the examination of land cover change over large areas is difficult to estimate using conventional forest inventory and, thus, remote sensing facilities are very useful. Remote sensing also facilitates data integration with ecological models (Xu and Chen 2013). For Song and Woodcock (2002; Song et al. 2002, 2007), remote sensing is the only realistic way to examine succession over large areas in an efficient manner. Their project consisted of linking the GORT canopy reflectance model with ZELIG, and they acknowledged the necessity for further developments.

However, relatively few studies on linkages between remote sensing and ecological models have been conducted. Among good examples, the study by Wallentin et al. (2008) included remote sensing analysis based on aerial photos and LiDAR data. Photos were good enough to recognize individual trees. Their individual-tree model is based on the prediction of seedling germination, tree height growth, and mortality. Zhao et al. (2009) combined the use of remote sensing data with the gap model FAREAST to model spatial variation and uncertainties of East Siberian boreal forest biomass change over time. Remote sensing was also used in conjunction with simple models. Both Chuang et al. (2011) and Petit et al. (2001) use Markov chains to simulate succession in forests using data from SPOT images.

REFERENCES

Aber, J.D. and J.M. Melillo. 1982. FORTNITE: A computer model of organic matter and nitrogen dynamics in forest ecosystems. Madison: University of Wisconsin Research Bulletin R3130.

Ager, A.A., A.M. Barros, R. Houtman, R. Seli, and M.A. Day. 2020. Modelling the effect of accelerated forest management on long-term wildfire activity. *Ecol. Modell.* 421:108962.

Acevedo, M.F., D.L. Urban, and H.H. Shugart. 1996. Models of forest dynamics based on roles of tree species. *Ecol. Modell.* 87:267–284.

Baker, W.L. 1989. A review of models of landscape change. *Landsc. Ecol.* 2:111–133.

Baker, W.L., S.L. Egvert, and G.F. Frazier. 1991. A spatial model for studying the effects of climate change on the structure of landscapes subject to large disturbances. *Ecol. Modell.* 56:109–125.

Beaudet, M., B.D. Harvey, C. Messier, K.D. Coates, J. Poulin, D.D. Kneeshaw, S. Brais, and Y. Bergeron, 2011. Managing understory light conditions in boreal mixedwoods through variation in the intensity and spatial pattern of harvest: A modelling approach. *For. Ecol. Manage.* 261:84–94.

Birt, A.G., W. Xi, and N.R. Coulson. 2009. LANDISVIEW: A visualization tool for landscape modeling. *Environ. Model. Softw.* 24:1339–1341.

Bose, A.K., B.D. Harvey, K.D. Coates, S. Brais, and Y. Bergeron. 2015. Modelling stand development after partial harvesting in boreal mixedwoods of eastern Canada. *Ecol. Modell.* 300:123–136. https://doi.org/10.1016/j.ecolmodel.2015.01.002

Bossel, H. and H. Krieger. 1994. Simulation of multi-species tropical forest dynamics using a vertically and horizontally structured model. *For. Ecol. Manage.* 69:123–144.

Botkin, D.B. 1993. *Forest dynamics—An ecological model.* New York: Oxford University Press.

Botkin, D.B., J.F. Janak, and J.R. Wallis. 1972. Some ecological consequences of a computer model of forest growth. *J. Ecol.* 60:849–872.

Brazhnik, K., H.H. Shugart. 2016. SIBBORK: A new spatially-explicit gap model for boreal forest. *Ecol. Modell.* 182–196. https://doi.org/10.1016/j.ecolmodel.2015.09.016

Bugmann, H. 1994. *On the ecology of mountainous forests in a changing climate: A simulation study.* PhD Thesis No. 10638. Swiss Federal Institute of Technology, Zurich, Switzerland.

Bugmann, H. 1996. A simplified forest model to study species composition along climate gradients. *Ecology* 77:2055–2074.

Bugmann, H. 2001. A review of forest gap models. *Clim. Change* 51:259–305.

Bugmann, H.K.M., S.D. Wullschleger, D.T. Price, K. Ogle, D.F. Clark, and A.M. Solomon. 2001. Comparing the performance of forest gap models in North America. *Clim. Change* 51:349–388. https://doi.org/10.1023/A:1012537914881

Caplat, P. and M. Anand. 2009. Effects of disturbance frequency, species traits and resprouting on directional succession in an individual-based model of forest dynamics. *J. Ecol.* 97:1028–1036.

Cary, G.J., 1997. FIRESCAPE—a model for simulation theoretical long-term fire regimes in topographically complex landscapes. In *Australian Bushfire Conference: Bushfire '97*, ed B.J. McKaige, R.J. Williams, and W.M. Waggitt, 45–67. Darwin: Australian Bushfire Association.

Cary, G.J. 1998. Predicting Fire Regimes and their Ecological Effects in Spatially Complex Landscapes. Ph.D. diss., The Australian National University, Canberra, Australia.

Cary, G.J. and J.C.G. Banks. 1999. Fire regime sensitivity to global climate change: An Australian perspective. *In Advances in Global Change Research: Biomass burning and its inter-relationships with the climate system*, eds. J.L. Innes, M. Beniston, and M.M. Verstraete, 233–246. London: Kluwer Academic Publishers.

Chen, R. and R.R. Twilley, 1998. A gap dynamic model of mangrove forest development along gradients of soil salinity and nutrient sources. *J. Ecol.* 86:37–51.

Chivoiu, B., G. Shao, J. Moser, and W.L. Mills. 2006. ArcFVS: An integration of a geographic information system and forest vegetation simulator. *Sci China Serie E: Technol. Sci.* 49:35–44. https://doi.org/10.1007/s11431-006-8105-1

Chuang, C.-W., C.-Y. Lin, C.-H. Chien, and W.-C. Chou. 2011. Application of Markov-chain model for vegetation restoration assessment at landslide areas caused by a catastrophic earthquake in Central Taiwan. *Ecol. Modell.* 222:835–845.

Coates, K.D., M. Boldor, E. Hall and R. Astrup. 2009. Evaluation of the Complex Stand Simulation Model SORTIE-ND for Timber Supply Review in Sub-Boreal Forests of Northern BC. Executive summary: Year 2. FIA–FSP Technical Report for Forest Science Program Project Y093187. Smithers, BC.

Cordonnier, T., B. Courbaud, F. Berger, and A. Franc. 2008. Permanence of resilience and protection efficiency in mountain Norway spruce forest stands: A simulation study. *For. Ecol. Manage.* 256:347–354.

Costanza, R. and T. Maxwell. 1994. Resolution and predictability—an approach to the scaling problem. *Landsc. Ecol.* 9:47–57.

Costanza, R. and A. Voinov. 2001. Modeling ecological and economic systems with STELLA: Part III. *Ecol. Modell.* 143(1–2):1–7.

Crookston, N.L. and G.E. Dixon. 2005. The forest vegetation simulator: A review of its structure, content, and applications. *Comput. Electron. Agric.* 49(1):60–80.

Dai, E., Z. Wu, Q. Ge, W. Xi, X. Wang. 2016. Predicting the responses of forest distribution and aboveground biomass to climate change under RCPs scenarios in southern China. *Global Change Biology* 22(11):3642–3661.

Dai, E., Z. Wu, X. Wang, H. Fu, W. Xi, and T. Pan. 2015. Progress and prospect of research on forest landscape model. *J. Geogr. Sci.* 25:113–128.

Dai, X. 1996. Influence of light conditions in canopy gaps on forest regeneration: A new gap light index and its application in a boreal forest in East-Central Sweden. *For. Ecol. Manage.* 84:187–197.

Dale, V.H. and S.C. Beyeler. 2001. Challenges in the development and use of ecological indicators. *Ecol. Indic.* 1:3–10.

Dale, V.H., M.L. Tharp, K.O. Lannom, and D.G. Hodges. 2010. Modeling transient response of forests to climate change. *Sci. Total Environ.* 408:1888–1901.

Desanker, P.V. 1996. Development of a MIOMBO Woodland dynamics model in Zambezian Africa using Malawi as a case study. *Clim. Change* 34:279–288.

Didion, M., A.D. Kupferschmid, M.J. Lexer, W. Rammer, R. Seidl, and H. Bugmann. 2009. Potentials and limitations of using large-scale forest inventory data for evaluating forest succession models. *Ecol. Modell.* 220:133–147. https://doi.org/10.1016/j.ecolmodel.2008.09.021

Dixon, G.E. 2002. *Essential FVS: A user's guide to the Forest Vegetation Simulator*, pp. 204, 226. Fort Collins, CO: USDA Forest Service Internal Report, Forest Management Service Center.

Doyle, T.W. 1981. The role of disturbance in the gap dynamics of a montane rain forest: An application of a tropical forest succession model. In *Forest Succession: Concepts and Application*, eds. D.C. West, H.H. Shugart, and D.B. Botkin, pp. 56–73. New York: Springer-Verlag.

Duffy, J.E. 2009. Why biodiversity is important to the functioning of real-world ecosystems. *Front. Ecol. Environ.* 7:437–444.

Ek, A.R. and R.A. Monserud. 1974. Trials with program FOREST: Growth and reproduction simulation for mixed species forest stands. In *Growth models for tree and stand simulation*, ed. J. Fries. Stockholm: Royal College of Forestry, Research Notes No. 30.

Elzein, T., G.R. Larocque, L. Sirois, and D. Arseneault. 2020. Comparing the predictions of gap model with vegetation and disturbance data in south-eastern Canadian mixed forests. *For. Ecol. Manage.* 455:117649. https://doi.org/10.1016/j.foreco.2019.

Finney, M.A. 1998. *FARSITE: Fire area simulator-model development and evaluation*, pp. 1–47. Fort Collins, CO: USDA Forest Service, Research Paper RMRS-RP-4.

Finzi, A.C. and C.D. Canham. 2000. Sapling growth in response to light and nitrogen availability in a Southern New England forest. *For. Ecol. Manage.* 131:153–165.

Fischlin, A., H. Lischke, and H. Bugmann. 1995. *The fate of forests in a changing climate: Model validation and simulation results from the Alps*. Zurich: Systems Ecology ETHZ, Bericht/Report Nr. 25.

Foster, A.C., J.K. Shuman, H.H. Shugart, K.W. Dwire, P.J. Ffornwalt, J. Sibold, and J. Negron. 2017. Validation and application of a forest gap model to the Southern Rocky Mountains. *Ecol. Modell.* 351:109–128.

Foster, A.C., J.K. Shuman, H.H. Shugart, and J. Negron. 2018. Modeling the interactive effects of spruce beetle infestation and climate on subalpine vegetation. *Ecosphere* 9:1–21.

Gao, T., M. Hedblom, T. Emilsson, and A.B. Nielsen. 2014. The role of forest stand structure as biodiversity indicator. *For. Ecol. Manage.* 330:82–93.

Gardner, R H., W.H. Romme, and M.G. Turner. 1999. Predicting forest fires effects at landscape scales, In *Spatial modeling of forest landscapes: Approaches and applications*, ed D.J. Mladenoff and W.L. Baker, 163–185. Cambridge: Cambridge University Press.

Garman, S.L. 2004. Design and evaluation of a forest landscape change model for western Oregon. *Ecol. Modell.* 175:319–337.

Gause, G.F. and A.A. Witt. 1935. Behavior of mixed populations and the problem of natural selection. *Am. Nat.* 69:596–609.

Geoghegan, J., L. Pritchard jr., Y. Ogneva-Himmelberger, R.R. Chowdhury, S. Sanderson and B.L. Turnere II. 1998. Socializing the pixel and pixelizing the social in land-use and land-cover change. In *People and pixels: Linking remote sensing and social science*, ed. D. Liverman, E.F. Moran, R.R. Rindfuss and P.C. Stern, 51–69. Washington, DC: National Academy Press.

Gougeon, F.A., 2010. *The ITC suite manual: A semi-automatic individual tree crown (ITC) approach to forest inventories*, p. 92. Victoria, BC, Canada: Pacific Forestry Centre, Canadian Forest Service, Natural Resources Canada.

Gougeon, F.A., D.G. Leckie, D. Paradine and I. Scott, 1998. Individual tree crown species recognition: The Nahmint study. In *Automated interpretation of high spatial resolution digital imagery for forestry*. Ed. D.A. Hill and D.G. Leckie, pp. 209–233. Victoria, BC: Natural Resources Canada, Canadian Forest Service, Pacific Forestry Centre, Canada.

Grimm, V., E. Revilla, U. Berger, H. Jeltsch, W.M. Mooij, S.F. Railsback, H.-H. Thulke, J. Weiner, T. Wiegand, and D.L. DeAngelis. 2005. Pattern-oriented modeling of agent-based complex systems: Lessons from ecology. *Science* 310:987–991.

Grossiord, C., A. Granier, A. Gessier, T. Jucker, and C. Bonal. 2014. Does drought influence the relationship between biodiversity and ecosystem functioning in boreal forests? *Ecosystems* 17:394–404.

Gustafson, E.J. et al. 2016. *Seeing the future impacts of climate change and forest management: A landscape visualization system for forest managers*. Gen. Tech. Rep. NRS-164. Newtown Square, PA: U.S. Department of Agriculture, Forest Service, Northern Research Station. 18 p.

Gustafson, E.J., D.E. Lytle, R. Swaty, and G. Loehle. 2007. Simulating the cumulative effects of multiple forest management strategies on landscape measures of forest sustainability. *Landsc. Ecol.* 22:141–156

Gustafson, E. J., A.Z. Shvidenko, and R.M. Scheller, 2011. Effectiveness of forest management strategies to mitigate effects of global change in south-central Siberia. *Can. J. For. Res.* 41(7):1405–1421. doi:10.1139/x11-065.

He, H.S. 2008. Forest landscape models: Definitions, characterization, and classification. *For. Ecol. Manage.* 254:484–498.

He, H.S., Z. Hao, D.J. Mladenoff, G. Shao, Y. Hu, and Y. Chang, 2005. Simulating forest ecosystem response to climate warming incorporating spatial effects in North-Eastern China. *J. Biogeogr.* 32:2043–2056.

He, H.S., D.J. Mladenoff, and T.R. Crow. 1999. Linking an ecosystem model and a landscape model to study forest species response to climate warming. *Ecol. Modell.* 114:213–233.

He, H.S., B.Z. Shang, T.R. Crow, E.J. Gustafson, and S.R. Shifley. 2004. Simulating forest fuel and fire risk dynamics across landscapes—LANDIS fuel module design. *Ecol. Modell.* 180:135–151.

Holm, J.A., J.Q. Chambers, W.D. Collins, and N. Higuchi. 2014. Forest response to increased disturbance in the central Amazon and comparison to western Amazonian forests. *Biogeosci.* 11:5773–5794.

Holm, J.A., H.H. Shugart, S.J. Van Bloem, and G.R. Larocque. 2012. Gap model development, validation, and application to succession of secondary subtropical dry forests of Puerto Rico. *Ecol. Modell.* 233:70–82.

Hope, J.C.E., J.W. Humphrey, and D. Stone. 2006. Modelling the effects of forest landscape on local species in Glen Africa, Northern Scotland. *Forestry* 79:294–302.

Horn, H.S., H.H. Shugart, and D.L. Urban. 1989. Simulators as models of forest dynamics. In *Perspectives in ecological theory*, eds. J. Roughgarden, R.M. May, and S.A. Levin, pp. 256–267. Princeton, NJ: Princeton University Press.

Hu, Y.M., C.G. Xu, and Y. Chang. 2004. Application of spatially explicit landscape model (LANDIS): A case research in Huzhong area, Mt. Daxing'anling. *Acta Ecologica Sinica* 24:1846–1856.

Huo, C., G. Cheng, X. Lu, and J. Fan. 2010. Simulating the effects of climate change on forest dynamics on Gongga Mountain, Southwest China. *J. For. Res.* 15:176–185.

Huston, M., D.L. DeAngelis, and W.M. Post. 1988. New computer models unify ecological theory. *Biosci.* 38:682–691.

Huth, A. and T. Ditzer. 2001. Long-term impacts of logging in a tropical rain forests—a simulation study. *For. Ecol. Manage.* 142:33–51.

Johnsen, K., L. Samuelson, R. Teskey, S. McNulty, and T. Fox. 2001. Process models as tools in forestry and management. *For. Sci.* 47:2–8.

Keane, R.E., G.J. Cary, I.D. Davies, M.D. Flannigan, R.H. Gardner, S. Lavorel, J.M. Lenihan, C. Li, and S. Rupp. 2004. A classification of landscape fires succession models: Spatial simulations of fire and vegetation dynamics. *Ecol. Modell.* 179:3–27.

Keane, R.E., L.M. Holsinger, R.A. Parsons, and K. Gray. 2008. Climate change effects on historical range and variability of two large landscapes in western Montana, USA. *For. Ecol. Manage.* 254:375–389.

Keane, R.E., L.M. Holsinger and S.D. Pratt. 2006. Simulating historical landscape dynamics using the landscape fire succession model LANDSUM version 4.0. Fort Collins: USDA Forest Service, Rocky Mountain Research Station, Gen. Tech. rep. RMRS-GTR-171CD.

Keane, R.E., R.A. Loehman and L.M. Holsinger. 2011. The FireBGCv2 landscape fire and succession model: A research simulation platform for exploring fire and vegetation dynamics. Gen. Tech. Rep. RMRS-GTR-255. Fort Collins, CO: U.S. Department of Agriculture, Forest Service, Rocky Mountain Research Station. 137 p.

Keane, R.E., P. Morgan and S.W. Running. 1996a. Fire-BGC: A mechanistic ecological process model for simulating fire succession on coniferous forest landscapes of the northern Rocky Mountains. Ogden Ut: USDA Forest Service, Intermountain Research Station.

Keane, R.E., R.A. Parsons, and P.F. Hessburg. 2002. Estimating historical range and variation of landscape patch dynamics: Limitations of the simulation approach. *Ecol. Modell.* 151:29–49.

Keane, R.E., K.C. Ryan, and S.W. Running. 1996b. Simulating effects of fire on northern Rocky Mountain landscapes with the ecological process model FIRE-BGC. *Tree Physiol.* 16:319–331.

Kerns, B.K., M. Hemstrom, D. Conklin, G. Yospin, B. Johnson and S. Brigham. 2012. Approaches to incorporating climate change effects in vegetation state and transition models. In *The First State- and-Transition Landscape Simulation Modeling Conference Proceedings*, ed Kerns, B.K., A.J. Shlisky and C.J. Daniel, 161–171. Portland: U.S. Department of Agriculture, Forest Service, Pacific Northwest Research Station. PNW-GTR-869.

Kienast, F. and N. Kuhn. 1989. Simulating forest succession along ecological gradients in Southern Central Europe. *Vegetatio.* 79:7–20.

Kincaid, J.A. 2012. Regeneration dynamics of *Tsuga Canadensis* in the southern Appalachian Mountains, USA. *Trees* 26:1045–1056.

Kint, V., P. Lasch, M. Lindner, and B. Muys. 2009. Multipurpose conversion management of Scots pine towards mixed oak-birch stands—a long-term simulation approach. *For. Ecol. Manage.* 257:199–214.

Köhler, P. and A. Huth. 1998. The effects of tree species grouping in tropical rainforest modelling: Simulations with the individual-based model FORMIND. *Ecol. Modell.* 109:301–321.

Kohyama, T. 1993. Size-structured tree populations in gap-dynamic forest—the forest architecture hypothesis for stable coexistence of species. *J. Ecol.* 81:131–143.

Kolmanič, S., N. Guid, and J. Diaci, 2014. ForestMAS—A single tree based secondary succession model employing Elleberg indicator values. *Ecol. Modell.* 279:100–113.

Kretchun, A.M., R.M. Scheller, M.S. Lucash, K.L. Clark, J. Hom, and S. van Tuyl. 2014. Predicted effects of gypsy moth defoliation and climate change on forest carbon dynamics in the New Jersey pine barrens. *PLOS One* 9:e102531. https://doi.org/10.1371/journal.pone.0102531

Lacerte, V., G.R. Larocque, M. Woods, W.J. Parton, and M. Penner. 2004. Testing the Lake States variant of FVS (Forest Vegetation Simulator) for the main forest types of Northern Ontario. *For. Chron.* 80:495–506.

Lacerte, V., G.R. Larocque, M. Woods, W.J. Parton, and M. Penner. 2006a. Calibration of the Forest Vegetation Simulator (FVS) for the major forest species of Ontario, Canada. *Ecol. Modell.* 199:336–349.

Lacerte, V., G.R. Larocque, M. Woods, W.J. Parton, and M. Penner. 2006b. Forest Vegetation Simulator model validation for Ontario (FVSOntario). Ont. Min. Nat. Resour., Southern Sci. Inf. Sect., North Bay, SSI ON. Tech. Rep. 120.398 p.

Larocque, G.R. 2008. Forest models. Vol. [2] of Encyclopedia of Ecology. In: Jøgensen SE, Fath BD (eds) Ecological models. Elsevier, Oxford, pp. 1663–1673.

Larocque, G.R., L. Archambault, and C. Delisle. 2006. Modelling forest succession in two southeastern Canadian mixedwood ecosystem types using the ZELIG model. *Ecol. Modell.* 199:350–362.

Larocque, G.R., L. Archambault, and C. Delisle. 2011. Development of the gap model ZELIG-CFS to predict the dynamics of North American mixed forest types with complex structures. *Ecol. Modell.* 222:2570–2583.

Larocque, G.R. and F.W. Bell. 2021. Evaluating the performance of a forest succession model to predict the long-term dynamics of tree species in mixed boreal forests using historical data in northern Ontario, Canada. *Forests* 21. www.mdpi.com/1999-4907/12/9/1181

Leemans, R. and I.C. Prentice. 1989. *FORSKA, a General Forest Succession Model.* Uppsala: Institute of Ecological Botany.

Lexer, M.J. and K. Hönninger. 1998. Simulated Effects of Bark Beetle Infestations on Stand Dynamics in *Picea abies* Stands: Coupling a Patch Model and a Stand Risk Model. In *The Impacts of Climate Variability on Forests*, ed. M. Beniston and J.L. Innes, 289–308. Berlin: Springer-Verlag.

Lexer, M.J., K. Hönninger, H. Scheifinger, C. Matulla, N. Groll, H. Kromp-Kolb, K. Schadauer, F. Starlinger, and M. Englisch. 2002. The sensitivity of Austrian forests to scenarios of climatic change: A large-scale risk assessment based on a modified gap model and forest inventory data. *For. Ecol. Manage.* 162:53–72.

Lindenmayer, D.B., C.R. Margules, and D.B. Botkin. 2000. Indicators of biodiversity for ecologically sustainable forest management. *Conserv. Biol.* 14:941–950.

Lindner, M., P. Lasch, and W. Cramer. 1996. Application of a forest succession model to a continentality gradient through Central Europe. *Clim. Change* 34:179–190.

Lindner, M., R. Sievänen, and H. Pretzsch. 1997. Improving the simulation of stand structure in a forest gap model through a flexible height growth function. *For. Ecol. Manage.* 95:183–195.

Lischke, H., T.J. Loffler, and A. Fischlin. 1998. Aggregation of individual trees and patches in forest succession models: Capturing variability with height structured, random, spatial distributions. *Theor. Popul. Biol.* 54:213–226.

Lischke, H., N.R. Zimmermann, J. Bolliger, S. Rickebusch, and T.J. Loffler. 2006. TreeMig: A forest-landscape model for simulating spatio-temporal patterns from stand to landscape scale. *Ecol. Modell.* 199:409–420.

Liu, J. and P.S. Ashton. 1995. Individual-based simulation models for forest succession and management. *For. Ecol. Manag.* 73:157–175.

Liu, J. and P.S. Ashton. 1998. FORMOSAIC: An individual-based spatially explicit model for simulating forest dynamics in landscape mosaics. *Ecol. Modell.* 106:177–200.

Loudermilk, E.L., R.M. Scheller, P.J. Weisberg, J. Yang, T. Dilts, S.L. Karam, and C.N. Skinner. 2013. Carbon dynamics in the future forest: The importance of climate-fire interactions and long-term successional legacy. *Glob. Chang. Biol.* 19:3502–3515.

Lutz, D.A., H.H. Shugart, and M.A. White. 2013. Sensitivity of Russian forest timber harvest and carbon storage to temperature increase. *Forestry* 86:283–293.

Ma, J., H.H. Shugart, X. Yan, C. Cao, S. Wu, and J. Fang. 2017. Evaluating carbon fluxes of global forest ecosystems by using an individual tree-based model FORCCHN. *Sci. Total Environ.* 586:939–951. http://dx.doi.org/10.1016/j.scitotenv.2017.02.073

Mace, G.M., K. Norris, and A.H. Fitter. 2012. Biodiversity and ecosystem services: A multilayered relationship. *Trends Ecol. Evol.* 27:19–26.

Mailly, D., J.P. Kimmins, and R.T. Busing. 2000. Disturbance and succession in a coniferous forest of northwestern North America: Simulations with DRYADES, a spatial gap model. *Ecol. Modell.* 127:183–205.

Major, K.C., P. Nosko, C. Kuehne, D. Campbell, and J. Bauhus. 2013. Regeneration dynamics of non-native northern red oak (*Quercus rubra* L.) populations as influenced by environmental factors: A case study in managed hardwood forests of southwestern Germany. *For. Ecol. Manage.* 291:144–153.

Mäkelä, A., M. del Rĭo, J. Hynynen, M.J. Hawkins, C. Reyer, P. Soares, M. van Oijen, and M. Tome. 2012. Using stand-scale forest models for estimating indicators of sustainable forest management. *For. Ecol. Manage.* 285:164–178.

Maleki, K., M.A. Gueye, B. Lafleur, A. Leduc, et Y. Bergeron. 2020. Modeling post-disturbance successional dynamics of the Canadian boreal mixedwoods. *Forests* 11:1–283. https://doi.org/10.3390/f11010003

Martin, P. 1992. EXE: A climatically sensitive model to study climate change and CO_2 enrichment effects on forests. *Austr. J. Bot.* 40:717–735.

McGaughey, R.J. 1997. Visualizing forest stand dynamics using the stand visualization system. In: Proceedings of the 1997 ACSM/ASPRS Annual Convention and Exposition; April 7–10, 1997. Seattle, WA. In: American Society for Photogrammetry and Remote Sensing. Bethesda, MD Vol. 4, pp. 248–257.

Mladenoff, D.J. 2004. LANDIS and forest landscape models. *Ecol. Modell.* 180:7–19.

Mladenoff, D.J., G.E. Host, J. Boeder, and T.R. Crow. 1996. A spatial model of forest landscape disturbance, succession, and management. In GIS and Environmental Modeling: Progress and research issues, ed. M.F. Goodchild, L.T. Steyaert, B.O. Parks, C. Johnston, D. Maidment, M. Crane and S. Glendinning, 175–180, Fort Collins: GIS World Books.

Mladenoff, D.L. and W.B. Baker, eds. 1999. *Spatial modeling of forest landscape change: Approaches and application.* Cambridge: Cambridge University Press.

Molina, E., O. Valeria, M. Martin, M. Montoro Girona, and J.A. Ramirez. 2022. Long-term impacts of forest management practices under climate change on structure, composition, and fragmentation of the Canadian Boreal landscape. *Forests* 13:1292. https://doi.org/10.3390/f13081292.

Monserud, R. 2003. Evaluating forest models in a sustainable forest management context. *FBMIS.* 1:35–47.

Montréal Process Working Group. 2009. *Criteria and indicators for the conservation and sustainable management of temperate and boreal forests*, 4th edn. Beijing: The Montreal Process.

Moorcroft, P.R., G.C. Hurtt, and S.W. Pacala. 2001. A method for scaling vegetation dynamics: The ecosystem demography model (ED). *Ecol. Monogr.* 71:557–586.

Moore, A.D. 1989. On the maximum growth equation used in gap simulation models. *Ecol. Modell.* 45:63–67.

Moore, A.D. and I.R. Noble. 1993. Automatic model simplification: The generation of replacement sequences and their use in vegetation modeling. *Ecol. Modell.* 70:137–157.

Mora, B., M.A. Wulder, G.W. Hobart, J.C. White, C.W. Bater, F.A. Gougeon, A. Varhola, and N.C. Coops. 2013. Forest inventory stand height estimates from very high spatial resolution satellite imagery calibrated with lidar plots. *Int. J. Remote Sens.* 34:4406–4424.

Morin, X., H. Bugmann, F. de Coligny, N. Martin-StPaul, M. Cailleret, J.-M. Limousin, J.-M. Ourcival, B. Prevosto, G. Simioni, M. Toigo, M. Vennetier, E. Catteau, and H. Geuillemot. 2021. Beyond forest succession: A gap model to study ecosystem functioning and tree community composition under climate change. *Funct. Ecol.* 35:955–975.

Nadrowski, K., C. Wirth, and M. Scherer-Lorenzen. 2010. Is forest diversity driving ecosystem function and service? *Curr. Opin. Environ. Sust.* 2:75–79.

Naeem, S., D.R. Hahn, and G. Schuurman. 2000. Producer-decomposer co-dependency influences biodiversity effects. *Nature* 403:762–764.

Newnham, R.M. 1964. The Development of a Stand Model for Douglas-Fir. Ph.D. diss., University of British Columbia, Vancouver.

Ngugi, M.R. and D.B. Botkin. 2011. Validation of a multispecies forest dynamics model using 50-year growth from *Eucalyptus forests* in eastern Australia. *Ecol. Modell.* 222:3261–3270. https://doi.org/10.1016/j.ecolmodel.2011.06.016

Noss, R.F. 1990. Indicators for monitoring biodiversity: A hierarchical approach. *Conserv. Biol.* 4:355–364.

Pabst, R.J., M.N. Goslin, S.L. Garman, and T.A. Spies. 2008. Calibrating and testing a gap model for simulating forest management in the Oregon Coast Range. *For. Ecol. Manage.* 256:958–972.

Pacala, S.W., C.D. Canham, J. Saponara, J.A. Silandeer Jr., R.K. Kobe, and E. Ribbens. 1996. Forest models defined by field measurements: Estimation error analysis and dynamics. *Ecol. Monogr.* 66:1–43.

Pacala, S.W., C.D. Canham, and J.A. Silander Jr. 1993. Forest models defined by field measurements: I. The design of a northeastern forest simulator. *Can. J. For. Res.* 23:1980–1988.

Padisák, J. 1992. Seasonal succession of phytoplankton in a large shallow lake (Balaton, Hungary)—a dynamic approach to ecological memory, its possible role and mechanisms. *J. Ecol.* 80:217–230.

Papaik, M.J., A. Fall, B. Sturtevant, D. Kneeshaw, C. Messier, M.-J. Fortin, and N. Simon. 2010. Forest processes from stands to landscapes: Exploring model forecast uncertainties using cross-scale model comparison. *Can. J. For. Res.* 40:2345–2359.

Pastor, J. and W.M. Post. 1985. *Development of a linked forest productivity-soil process model.* Oak Ridge: Environmental Sciences Division, Publication No. 2455.

Perry, G.L.W. and N.J. Enright. 2006. Spatial modelling of vegetation change in dynamic landscapes, a review of methods and applications. *Progr. Phys. Geogr.* 30:47–72.

Perry, G.L.W. and J.D.A. Millington. 2008. Spatial modeling of succession-disturbance dynamics in forest ecosystems: Concepts and examples. *Perspect. Plant Ecol. Evol. Syst.* 9:191–210.

Petit, C., T. Scudder, and E. Lambin. 2001. Quantifying processes of land-cover change by remote sensing: Resettlement and rapid land-cover changes in south-eastern Zambia. *Int. J. Remote Sens.* 22:3435–3456.

Post, W.M. and J. Pastor. 1996. LINKAGES—An individual-based forest ecosystem model. *Clim. Change* 34:253–261.

Price, D.T., N.E. Zimmermann, P.J. van der Meer, M.J. Lexer, P. leadly, I.T.M. Jorritsma, J. Schaber, D.D.F. Clark, P. Lasch, S. McNulty, J. Wu, and B. Smith. 2001. Regeneration in gap models: Priority issues for studying forest responses to climate change. *Clim. Change* 51:475–508.

Rasche, L., L. Fahse, A. Zingg, and H. Bugmann. 2012. Enhancing gap model accuracy by modeling dynamic height growth and dynamic maximum tree height. *Ecol. Modell.* 232:133–143. https://doi.org/10.1007/s13280-017-0919-5

Roth, P. and J.C. Finley. 2007. Visualize your forest—using forest simulation software to communicate forest management concepts to private forestland owners. *J. For.* 105:15–19. https://doi.org/10.1093/jof/105.1.15

Rupp, T.S., A.M. Starfield, and F.S. Chapin III. 2000. A frame-based spatially explicit model of subarctic vegetation response to climatic change: A comparison with a point model. *Landsc. Ecol.* 15:383–400.

Rupp, T.S., X. Chen, M. Olson, and A.D. McGuire. 2007. Sensitivity of simulated land cover dynamics to uncertainties in climate drivers. *Earth Interactions* 11:1–21.

Rykiel, E.J., Jr. 1996. Testing ecological models: The meaning of validation. *Ecol. Modell.* 90:229–244.

Salazar, A., G. Goldstein, A.C. Franco, and F. Miralles-Wilhelm. 2012. Differential seedling establishment of woody plants along a tree density gradient in Neotropical savannas. *J. Ecol.* 2012:1411–1421.

Scheller, R.M., J.B. Domingo, B.R. Sturtevant, J.S. Williams, A. Rudy, E.J. Gustafson, and D.J. Mladenoff. 2007. Design, development, and application of LANDIS-II, a spatial landscape simulation model with flexible temporal and spatial resolution. *Ecol. Modell.* 201:409–419.

Scheller, R.M. and D.J. Mladenoff. 2005. A spatially interactive simulation of climate change, harvesting, wind, and tree species migration and projected changes to forest composition and biomass in northern Wisconsin, USA. *Glob. Chang. Biol.* 11:307–321.

Scheller, R.M. and D.J. Mladenoff. 2007. An ecological classification of forest landscape simulation models, tools and strategies for understanding broad-scale forested ecosystems. *Landsc. Ecol.* 22:491–505.

Schumacher, S., H. Bugmann, and D.J. Mladenoff. 2004. Improving the formulation of tree growth and succession in a spatially explicit landscape model. *Ecol. Modell.* 2004:175–194.

Schumacher, S., B. Reineking, J. Sibold, and H. Bugmann. 2006. Modeling the impact of climate and vegetation on fire regimes in mountain Landscapes. *Landsc. Ecol.* 21:539–554.

Schuman, J.K., H.H. Shugart, and O.N. Krankina. 2014. Testing individual-based models of forest dynamics: Issues and an example from the boreal forests of Russia. *Ecol. Modell.* 293:102–110. https://doi.org/10.1016/j.ecolmodel.2013.10.028

Searle, E.B., F.W. Bell, G.R. Larocque, M. Fortin, J. Dacosta, R. Sousa-Silva, M. Mina, and H.D. Deighton. 2021. Simulating the effects of intensifying silviculture on desired species yields across a board environmental gradient. *Forests* 12. www.mdpi.com/1999-4907/12/6/755

Shao, G., H.H. Shugart, and T.M. Smith. 1996. A role-type model (ROPE) and its application in assessing climate change impacts on forest landscapes. *Vegetatio* 121:135–146.

Shinozaki, K., K. Yoda, K. Hozumi, and T. Kira, 1964. A quantitative analysis of plant form—the pipe model theory. I. Basic analyses. *Jap. J. Ecol.* 14:97–105.

Shugart, H.H. 1984. *Theory of forest dynamics: The ecological implications of forest succession models.* New York: Springer-Verlag.

Shugart, H.H. 1998. *Terrestrial ecosystems in changing environments.* Cambridge: Cambridge University Press.

Shugart, H.H., B. Wang, R. Fischer, J. Ma, J. Fang, X. Yan, A. Huth, and A.H. Armstrong. 2018. Gap models and their individual-based relatives in the assessment of the consequences of global change. *Environ. Res. Lett.* 13:033001. https://doi.org/10.1088/1748-9326//aaaacc

Shugart, H.H., M.S. Hopkins, I.P. Burgess, and A.T. Mortlock. 1980. The development of a succession model for subtropical rain forest and its application to assess the effects of timber harvest at Wiangaree State Forest, New South Wales. *J. Environ. Manage.* 11:243–265.

Shugart, H.H., S. Saatchi, and F.G. Hall. 2010. Importance of structure and its measurement in quantifying function of forest ecosystems. *J. Geophys. Res. Biogeosci.*115: G00E13. https://doi.org/10.1029/2009JG000993

Shugart, H.H. and T.M. Smith. 1996. A review of forest patch models and their application to global change research. *Clim. Chang.* 34:131–153.

Shugart, H.H., T.M. Smith, and W.M. Post. 1992. The potential for application of individual-based simulation models for assessing the effects of global change. *Annu. Rev. Ecol. Syst.* 23:15–38.

Shugart, H.H. and D.C. West. 1977. Development of an Appalachian deciduous forest succession model and its application to assessment of the impact of the chestnut blight. *J. Environ. Manage.* 5:161–179.

Shugart, H.H. and D.C. West. 1980. Forest succession models. *Biosci.* 30:308–313.

Shugart, H.H. and F.I. Woodward. 2011. Global Change and the Terrestrial Biosphere: Achievements and Challenges. Wiley-Blackwell Press, Oxford. 242 pp.

Shuman, J.K., H.H. Shugart, and O.N. Krankina. 2014. Testing individual-based models of forest dynamics: Issues and an example from the boreal forests of Russia. *Ecol. Modell.* 293:102–110.

Shuman, J.K., H.H. Shugart, and T.L. O'Halloran. 2011. Sensitivity of Siberian larch forests to climate change. *Glob. Chang. Biol.* 17:2370–2384.

Song, C. and C.E. Woodcock. 2002. The spatial manifestation of forest succession in optical imagery: The potential of multiresolution imagery. *Remote Sens. Environ.* 82:271–284.

Song, C., C.E. Woodcock, and X. Li. 2002. The spectral/temporal manifestation of forest succession in optical imagery: The potential of multitemporal imagery. *Remote Sens. Environ.* 82:285–302.

Song, C., T.A. Schroeder, and W.B. Cohen, 2007. Predicting temperate conifer forest successional stage distributions with multitemporal landsat thematic mapper imagery. *Remote Sens. Environ.* 106:228–237.

Stadt, K.J., C. Huston, K.D. Coates, Z. Feng, M.R.T. Date, V.J. Lieffers. 2007. Evaluation of competition and light estimation indices for predicting diameter growth in mature boreal mixed forests. *Ann. For. Sci.* 64:477–490. https://doi.org/10.1051/forest:2007025

Stage, A.R. 1973. Prognosis Model for Stand Development. USDA Forest Service Research Paper INT-173. Intermountain Forest & Range Experiment Station, Forest Service, U.S. Department of Agriculture, Ogden, Utah.

Sun, Z., H. Ren, V. Schaefer, H. Lu, J. Wang, L. Li, and N. Liu. 2013. Quantifying ecological memory during forest succession: A case study from lower subtropical forest ecosystems in South China. *Ecol. Indic.* 34:192–203.

Swanson, M. 2009. Modeling the effects of alternative management strategies on forest carbon in the Nothofagus forests of Tierra del Fuego, *Chile. For. Ecol. and Manage.* 257:1740–1750. 10.1016/j.foreco.2009.01.045

Taylor, A.R., H.Y.H. Chen, and L. VanDamme. 2009. A review of forest succession models and their suitability for forest management planning. *For. Sci.* 55(1):23–36, February. https://doi.org/10.1093/forestscience/55.1.23

Thompson, I., B. Mackey, S. McNulty, and A. Mosseler. 2009. *Forest resilience, biodiversity, and climate change. A synthesis of the biodiversity/resilience/stability relationship in forest ecosystems.* Montreal: Secretariat of the Convention on Biological Diversity, CBD Technical Series Number No. 43.

Thompson, J.R., D.R. Foster, R. Scheller, and D. Kittredge, 2011. The Influence of Land Use and Climate Change on Forest Biomass and Composition in Massachusetts, USA. *Ecol. Appl.* 21:2425–2444.

Thorpe, H.C., M.C. Vanderwel, M.M. Fuler, S.C. Thomas, and J.P. Caspersen. 2010. Modelling stand development after partial harvests: An empirically based, spatially explicit analysis for lowland black spruce. *Ecol. Model.* 221:256–267. https://doi.org/10.1016/j.ecolmodel.2009.10.005

Urban, D.L. 1990. A versatile model to simulate forest pattern: A user's guide to ZELIG Version 1.0. University of Virginia, Charlottesville, Virginia.

Uriarte, M., C.D. Canham, J. Thompson, J.K. Zimmerman, L. Murphy, A.M. Sabat, N. Fetcher, and B.L. Haines. 2009. Natural disturbance and human land use as determinants of tropical forest dynamics: Results from a forest simulator. *Ecol. Monogr.* 79:423–443.

Vaillant, N.M. and A.A. Ager. 2014. ArcFuels: An ArcMap toolbar for fuel treatment planning and wildfire risk assessment. Fire Management Today 74:21.23.

Van Daalen, J.C. and H.H. Shugart. 1989. OUTENIQUA—a computer model to simulate succession in the mixed evergreen forests of the Southern Cape, South Africa. *Landsc. Ecol.* 24:255–267.

Voinov, A., R. Costanza, L. Wainger, R. Boumans, F. Villa, T. Maxwell, and H. Voinov. 1999. Patuxent landscape model: Integrated ecological economic modeling of a watershed. *Environ. Model. Softw.* 14:473–491.

Waldron, J.D., R.N. Coulson, D.M. Cairns, C.W. Lafon, M.D. Tchakerian, W. Xi, K.D. Klepzig, and A.G. Birt. 2010. Evaluating the impact of invasive species in forest landscapes: The Southern Pine Beetle and the Hemlock Woolly Adelgid. In Advances in threat assessment and their application to forest and rangeland management, ed. J.M. Pye, M.H. Rauscher, Y. Sands, D.C. Lee, J.S. Beatty, 687–698. Portland: USDA Forest Service, Technical Report PNW-GTR-802.

Wallentin, G., U. Tappeiner, J. Strobl, and E. Tasser. 2008. Understanding alpine tree line dynamics: An individual-based model. Testing individual-based models of forest dynamics. *Ecol. Modell.* 218:235–246.

Wang, W.J., H.S. He, M.A. Spetich, S.R. Shifley, F.R. Thompson III, D.R. Larsen, J.S. Fraser, and J. Yang. 2013. A large-scale forest landscape model incorporating multi-scale processes and utilizing forest inventory data. *Ecosphere* 4:106.

Weishampel, J.F. and D.L. Urban. 1996. Coupling a spatially-explicit forest gap model with a 3-D solar routine to simulate latitudinal effects. *Ecol. Modell.* 86:101–111.

Wennergren, U., M. Ruckelshaus, and P.M. Kareiva. 1995. The promise and limitations of spatial models in conservation biology. *Oikos* 74:349–356.

Xi, W., R.N. Coulson, A.G. Birt, Z. Shang, J.D. Waldron, C.W. Lafon, D.M. Cairns, M.D. Tchakerian, and K.D. Klepzig. 2009a. Review of forest landscape models: Types, methods, development and applications. *Acta Ecologica Sinica* 29(1):69–78.

Xi, W., M, R.N. Coulson, J.D. Waldron, M.D. Tchakerian, C.W. Lafon, D.M. Cairns, A.G. Birt, and K.D. Klepzig. 2008. Landscape modeling for forest restoration planning and assessment: Lessons from the southern Appalachian Mountains. *J. For.* 106:191–197.

Xi, W, J. Waldron, C. Lafon C W, D. Cairns, A. Birt, M. Tchakerian, R. Coulson, and K. Klepzig. 2009b. Modeling long-term effects of altered fire regimes following southern pine beetle outbreaks (North Carolina). *Ecol. Restor.* 27:24–26.

Xi, W., X. Zhou and Jianwei Zhang. 2020. LANDISVIEW 2.0: An upgraded visualization and analytical tool for landscape modeling. *Environ. Model. Softw.* 34:104849. https://doi.org/10.1016/j.envsoft.2020.104849

Xiaodong, Y. and H.H. Shugart. 2005. FAREAST: A forest gap model to simulate dynamics and patterns of eastern Eurasian forests. *J. Biogeogr.* 32:1641–1658.

Xu, C. and M. Chen, 2013. Application of remote sensing in ecosystem and landscape modelling. In *Remote sensing of natural resources*, ed. G. Wang and Q. Weng, 173–190. New York: CRC Press.

Yang, J., H.S. He, S.R. Shifley, and E.J. Gustafson. 2007. Spatial patterns of modern period human-caused fire occurrence in the Missouri Ozark Highlands. *For. Sci.* 53:1–15.

Zhao, T., K.M. Bergen, D.G. Brown, and H.H. Shugart. 2009. Scale dependence in quantification of land-cover and biomass change over Siberian boreal forest landscapes. *Landsc. Ecol.* 24:1299–1313.

8 Process-Based Models

A Synthesis of Models and Applications to Address Environmental and Management Issues[1]

Guy R. Larocque, Alexander Komarov, Oleg Chertov, Vladimir Shanin, Jinxun Liu, Jagtar S. Bhatti, Weifeng Wang, Changhui Peng, Herman H. Shugart, Weimin Xi, and Jennifer A. Holm

8.1 INTRODUCTION

The development of process-based models, also known as mechanistic or ecosystem models, has a shorter history than empirical or gap models (Chapters 6 and 7). Numerous empirical models have been developed since the early 1960s to simulate forest growth using forest inventory data. For forest managers, empirical models became more useful tools than the traditional growth and yield tables, as their implementation on computers facilitated the management of large datasets of predicted stand growth attributes and the simulation of silvicultural treatments. The development of gap models, which simulate forest succession using semi-empirical relationships, began in the early 1970s with the JABOWA model (Botkin et al. 1972) (see Chapter 7). It is only in the 1980s that the development of process-based models really began when the need to simulate the effects of environmental disturbances, such as acid rain, air pollution, or climate change, on the functioning of forest ecosystems was identified.

Process-based models represent and simulate physiological and biogeochemical processes and their interactions with the abiotic environment (water, climate, and nutrients) in forest ecosystems by using functional relationships (Johnsen et al. 2001; Landsberg and Sands 2011). These functional cause–effect relationships are structured within a complex hierarchy of processes of different levels of importance within a system. As forest ecosystems are characterized by many interactions and feedbacks between vegetation, soil, and the atmosphere, the capacity of process-based models to make realistic predictions depends on a logical integration of matter and energy fluxes (Friend et al. 1997). Key physiological processes that may be considered for tree growth modeling include photosynthesis; mineral metabolism; respiration; partitioning and assimilation of carbon; absorption and accumulation of water; minerals, and gases; translocation; growth regulation, and litter decomposition (Pastor and Post 1985; Dixon 1990; Landsberg 2003a). Other processes that may be found in process-based models include tree phenology, plant functional type competition, mortality, responses to fire or pest disturbances, and effects of human and land management. However, few process-based models include representations of all these processes, as the amount of data to initialize them is enormous. Also, for the prediction of some ecosystem processes, such as growth or turnover rate of some ecosystem pools, it is not necessary to include model components for all possible ecophysiological processes. The challenge that modelers face when they develop process-based

DOI: 10.1201/9781003431084-10

models is to determine the extent to which the complexity in the representation of processes is sufficient to meet specific objectives and make realistic predictions. As there are many processes and interactions in forest ecosystems, the relationships can have a high degree of complexity that cannot necessarily be solved using analytical mathematical techniques, which explains why the majority of process-based models have been developed by applying the basic principles of systems analysis. Process-based models are seen as powerful tools that allow scientists to better understand ecosystem functioning and give a realistic prognosis if some factors will change in the future (Landsberg 2003b).

The basic framework of the majority of process-based models includes model components that interact to simulate productivity and nutrient, water, and carbon cycles. There are different levels of complexity in these models. Also, some models focus on the simulation of only some of the ecosystem components or fluxes, such as nutrient or water cycles. These are abiotic interactions, but there are also biotic interactions, such as competition. Typical models include the representation of the effects of environmental factors, including temperature, precipitation, photosynthetic active radiation (PAR), or soil properties, on the ecosystem processes: photosynthesis, respiration, organic matter decomposition, etc. Concerning productivity, two basic levels are to be considered. For trees and understory woody and herbaceous vegetation, productivity is the allocation of the net production of carbohydrates in the different compartments: stems, branches, foliage, and roots. Productivity at the ecosystem (or stand) level is the result of biotic processes, such as competition, that affect carbon allocation among individual trees. As stand density increases, each individual tree sees its available growing space reduced by the presence of neighboring trees. As a result, crown development is inhibited because the potential amount of PAR they can absorb for photosynthesis is reduced by neighboring crowns, and the amount of nutrients they can potentially uptake is captured by neighboring root systems. Carbon, water, and nutrients are transferred among ecosystem compartments, from trees to the forest floor and to the soil. The soil system is a biological reactor that is responsible for the biomass decomposition and elements' return in trees and forest vegetation. The rate of soil processes influences tree and forest productivity as well as photosynthesis. These cycles influence tree and ecosystem productivity by regulating the movement of water from the canopy to the soil, transferring and synthesizing carbon from vegetation to the soil and recycling nutrients contained in the litters of vegetation and roots.

There are two main benefits of process-based models. First, their development, validation, and application may lessen the need for expensive experimental studies. For instance, the analysis of the dynamics of a forest ecosystem type to all possible combinations of climatic conditions, site properties, and management regimes that would require the establishment of a statistically replicated large-scale experimental design is theoretically possible but very expensive and impractical (Korzukhin et al. 1996). The development of models based on the representation of the rules of changes in the main processes may be used as substitutes for experimentation, as long as the rules of change are adequately represented. Second, process-based models have the potential to address issues related to the basic principles of ecosystem management and to provide sound predictions of the dynamics of ecosystems when changes in environmental conditions are unprecedented (Korzukhin et al. 1996). This is the case of climate change or any external impacts such as pollution. Many short-term studies have examined the effects of increase in temperature and atmospheric CO_2 concentration either in growth chambers or in FACE experiments. However, the contributions of these types of experiments have limits, as they cannot consider all ecosystem components and interactions and capture the longevity factor (Ceulemans and Mousseau 1994; Luo and Reynolds 1999).

The objective of this chapter is to review different types of process-based models and describe how they can be used in a context of the application of the basic principles of ecological forest management. Relatively few process-based models have been used to address forest management questions (Battaglia and Sands 1998; Johnsen et al. 2001; Shanin et al. 2011; Bond-Lamberty et al. 2014). However, as the international community increasingly demands that forests be managed sustainably, which requires that forest managers evaluate the impacts of environmental disturbances on ecosystem dynamics, it is likely that process-based models will be used more frequently in the decision-making process for forest management. The next section gives a short overview of physiological models that focus on the modeling of single processes: light-use and photosynthesis models. Section 8.3 describes the basics of different ecosystem process-based models. The term ecosystem model is used here to include different types of model formulations to simulate ecological processes. Thus, hybrid models may combine empirical as well as process-based formulations and may facilitate the linkage with traditional forest inventory data. Process-based models are commonly viewed as focusing on ecosystem-level predictions, but there are global-scale process-based models that were developed to simulate ecological processes over large regions. Finally, linkages between process-based models and ecological indicators are discussed in the context of biodiversity and ecosystem management. In Section 8.4, model evaluation and uncertainty analyses will be discussed to highlight their importance for model development and usefulness. Different case studies that used process-based models to predict the effects of climate change, fire and management will be reviewed in Section 8.5.

8.2 PHYSIOLOGICAL SINGLE PROCESS MODELS

8.2.1 Light-Use Models

Light-use models are among the simplest forms of process-based models to predict forest growth. They have the following general form (Landsberg and Sands 2011):

$$P_n = \varepsilon(\theta,N,T)\phi_{abs} - R \tag{8.1}$$

where P_n is net primary production (NPP), ε the light-use efficiency factor that converts the absorbed photosynthetic radiation, ϕ_{abs}, to carbohydrates, θ the soil water effect, N the nutrient effect, T the temperature effect, and R total respiration. According to Landsberg and Sands (2011), the model developed by Byrne et al. (1986) is an early application example of forest growth prediction of the monthly biomass production of *Pinus radiata* grown in three sites differing in climatic conditions. Light-use models are relatively simple models that can easily be linked to remote sensing data to predict NPP or global terrestrial balance (Medlyn 1998). However, there are examples to estimate light interception at the individual crown level (see Duursma and Mäkelä 2007).

8.2.2 Photosynthesis Models

The modeling of photosynthesis has been the subject of many papers since the 1970s and 1980s. This situation can be attributed to the increased availability of gas exchange instruments (Landsberg 2003a). Several types of models were developed to predict photosynthetic rate. There are different levels of complexity with respect to the details included in the representation of biochemical processes and scaling methods, which may vary from individual leaves to canopy levels. Only the models that have been studied and used most often will be reviewed in this subsection.

One of the simplest forms of photosynthesis model was derived by McMurtrie et al. (1990) using a Blackman function:

$$A_n = \min\left(\delta\varphi(1-\rho_l-\tau_l), P_{max}\right) - R_d \tag{8.2}$$

where, A_n is the net photosynthetic rate, δ quantum yield, φ the photosynthetic photon flux density on a leaf, ρ_l and τ_l the leaf reflectance and transmittance to photosynthetic active radiation, respectively, P_{max} the light saturated photosynthetic rate, and R_d the dark respiration rate.

Photosynthesis models based on the non-rectangular hyperbola model have been used very often to model canopy photosynthesis for different species (Johnson et al. 2010). It is defined as follows (Johnson and Thornley 1984; Thornley 1998):

$$P_{leaf} = \frac{\alpha I_{leaf} + P_{max} - \sqrt{\left[\left(\alpha I_{leaf} + P_{max}\right)^2 - 4\theta\alpha I_{leaf} P_{max}\right]}}{2\theta} \tag{8.3}$$

where P_{leaf} is leaf gross photosynthetic rate, α the initial slope value of the photosynthetic light response curve (PLR) (also known as photochemical efficiency), I_{leaf} the incident light on the upper leaf surface, P_{max} the maximum photosynthetic rate that can be reached at saturated light conditions, and θ the convexity parameter of the PRL curve. The non-rectangular hyperbola model was used in several studies in which the effect of environmental factors was integrated. For instance, Johnson et al. (2010) incorporated equations to account for the effect of temperature, nitrogen, protein, and CO_2 atmospheric concentrations on the parameters of the non-rectangular hyperbola model. Calama et al. (2013) examined the effect of leaf temperature and soil water content on P_{max} and of leaf temperature on α. Xu et al. (2014) incorporated the influence of chlorophyll and leaf nitrogen.

The photosynthesis model that has had the greatest influence in terms of conceptual development and application was developed by Farquhar et al. (1980) and Farquhar and von Caemmerer (1982) (Landsberg 2003a). This model is universally recognized and is used to simulate photosynthesis in many ecosystem process-based models. This model assumes that net photosynthetic rate is regulated by the ribulose-bisphosphate (RUBP) concentration, the activity of RUBP carboxylase/oxygenase (Ribusco) at saturating RUBP, or by triose phosphate utilization. However, most applications of the model do not consider the triose phosphate utilization. Only the basic model components will be presented here. For more details, readers are encouraged to browse the abundant literature on this model.

The model component regulated by the ribulose-bisphosphate (RUBP) concentration is as follows:

$$A_n = J\frac{C_i - \Gamma}{4(C_i + 2\Gamma)} - R_d \tag{8.4}$$

where A_n is the net photosynthetic rate, J the potential electron transport rate, C_i the CO_2 intercellular concentration, Γ the CO_2 compensation point, and R_d the dark respiration rate.

The model component when A_n is regulated by the activity of RUBP carboxylase/oxygenase is:

$$A_n = V_{cmax}\frac{C_i - \Gamma}{C_i + K_c\left(1 + \frac{O}{K_o}\right)} - R_d \tag{8.5}$$

where V_{cmax} is the Rubisco potential capacity for CO_2 fixation per unit leaf area, K_c the Michaelis-Menten coefficient for oxygenation, O the intercellular O_2 concentration, and K_o the Michaelis-Menten coefficient for oxygenation.

Since this model was developed, several studies have been conducted to improve its potential applicability for different species. Wullschleger (1993) conducted an analysis of A_n/C_i curves for 109 species that resulted in a dataset of coefficients for V_{cmax} and maximum potential electron transport rate (J_{max}). Parameters were related to site conditions, including the influence of temperature or concentrations of nitrogen and proteins in the foliage (e.g., McMurtrie and Wang 1993; Leuning et al. 1995; Niinemets and Tenhunen 1997; Evans and Poorter 2001; Larocque 2002; Sharkey et al. 2007). The spatial variation within crowns through change in specific leaf area in different crown sections was modeled by Niinemets and Tenhunen (1997), Larocque (2002), and Evans and Poorter (2001).

8.3 ECOSYSTEM PROCESS-BASED MODELS

8.3.1 Biogeochemical Cycles

Process-based models that simulate the functioning of forest ecosystems usually include mathematical representations of several ecophysiological processes and carbon, water, and nutrient cycles in a hierarchy of cause–effect relationships. They are characterized by a sufficient degree of flexibility to predict the impacts of external environmental influences (Landsberg and Sands 2011). The modeling of the allocation of carbohydrates synthesized in the different physiological processes and carbon partitioning in the different pools in forest ecosystems has been the focus of the great majority of ecosystem process-based models. Table 8.1 includes examples of biogeochemical models, with a summary of the main processes modeled, domains of application, and main contributions. As indicated by the examples of Table 8.1, these models simulate several processes; some of them are common to nearly all models, such as photosynthesis, but some models focus on the detailed modeling of specific processes, such as the cycling of soil organic matter (e.g., ROMUL, EFIMOD).

The majority of models that have been developed in the last three decades are descendants of FOREST-BGC (Running and Coughlan 1988; Running and Gower 1991). FOREST-BGC was one of the early models with detailed mechanistic model components that established a benchmark for the modeling of carbon, water, and nitrogen cycles (Landsberg and Sands 2011). The design of its framework facilitates the use of remote sensing data of surface climate and vegetation structure and topographic and physical site characteristics obtained from geographic information systems to make predictions at regional scales. For processes at the canopy level, such as photosynthesis or respiration, FOREST-BGC is based on the big-leaf approach in which the canopy is viewed as a "giant leaf," using leaf area index (LAI) values that can be obtained from remote sensing data. Thus, the predictions of FOREST-BGC are very sensitive to LAI. FOREST-BGC simulates the main processes of the carbon, nitrogen, and water cycles, including photosynthesis, maintenance and growth respiration, evapotranspiration, root turnover, litterfall, litter decomposition, and carbon allocation in the foliage, stems, and roots (Figure 8.1). There are 20 state variables and 41 parameters, and daily temperature and precipitation data must be provided as input. While photosynthesis, evaporation, maintenance respiration, evaporation, evapotranspiration, and litter mineralization are simulated daily, the allocation of carbon and nitrogen in trees, foliage, roots, and soil is simulated on an annual time cycle. Good examples of the application of FOREST-BGC for different forest ecosystem types can be found in Running (1994), Chiesi et al. (2002), Hoff et al. (2002), Luckai and Larocque (2002), Rodriges et al. (2010), or Veganzones et al. (2010). Subsequently, FOREST-BGC evolved toward BIOME-BGC (Running and Hunt 1993), which retained the basic framework of FOREST-BGC, but with modifications in some model components, such as photosynthesis and the dynamics of soil fluxes. Also, the interface with remote sensing data was improved. Recent examples of the application of BIOME-BGC can be found in Tatarinov and Cienciala (2006), Chiesi et al. (2007), Jung et al. (2007), Pietsch et al. (2005), Petritsch et al. (2007), Wang et al. (2009).

The "giant leaf" approach used in FOREST-BGC to model canopy CO_2 fluxes is one of the methods that have been considered in different models. The modeling of the scaling of physiological processes from leaf to canopy is very complex, as there are nonlinear relationships involved (Leuning et al. 1995; Friend 2001). In addition to the "giant leaf" approach, other methods were proposed, such as the stratification of the canopy into foliage under sunlight and foliage in shade conditions, the establishment of scale factors based on photochemical efficiency and PAR absorption, average illumination or foliage energy balance related to environmental gradients (Norman 1993). Examples of the application of sunlit and shaded leaf portions of the canopy can be found in Chen et al. (1999) and Wang (2000). Further developments were inspired by these approaches. For instance, Larocque (2002) used the Weibull function to represent the change in foliage area and biomass with canopy depth and a linear relationship between foliage nitrogen content per unit leaf area and leaf dry mass area from the top to the bottom of crowns of sugar maple (*Acer saccharum* Marsh.).

TABLE 8.1
Examples of Process-Based Models for Forest Ecosystems

Model Name	References	Cycles	Main Processes[1]	Forest type(s)/ Scale(s)/ Region(s)	Main Contributions
FOREST-BGC	Running and Coughlan (1988), Running and Gower (1991)	Carbon, nitrogen, and water	PSN, IPAR, RM, RG, LF, LD, CF, CS, CR, NF, NS, NR, EVP, SR, NM, RT	*Picea glauca*, *Pinus elliotti*, *Pinus resinosa*, *Tsuga heterophylla*, *Alnus rubra*, *Pseudotsuga menziesii*, *Pinus Ponderosa*, *Quercus pubescens*, *Quercus cerris*, *Ostrya*, *Acer campestre*, *Picea mariana*/ Ecosystems and regions/South to north transects in Oregon, United States Northwestern Ontario	Major advances in the modeling of biogeochemical cycles and linkages with remote sensing data. Study of the impacts of climate change scenarios and comparison of predictions with CENTURY
BIOME-BGC	Running and Hunt (1993)	Carbon, nitrogen, and water	PSN, IPAR, RM, RG, LF, LD, CF, CS, CR, NF, NS, NR, EVP, SR, NM, RT	Mediterranean forest types (*Quercus ilex*, *Q. cerris*, *Pinus Pinaster*), *Picea abies*, *Pinus sylvestris*, *Fagus sylvatica*, *Quercus robur*, *Quercus petraea*, *Larix dedidua*, *Pinus cembra*/Ecosystem and regions/ Northwestern Italy, Central Europe, United States	Major advances in the modeling of biogeochemical cycles and linkages with remote sensing data
TREE-BGC	Korol et al. (1995a, 1995b)	Carbon, nitrogen, and water	PSN, IPAR, RM, RG, LF, LD, CF, CS, CR, NF, NS, NR, EVP, SR, NM, RT, TD, TH, COM	*Pseudotsuga menziesii*/Tree and ecosystem/British Columbia (Canada)	Application of a biogeochemical process-based model at the individual-tree level
Hybrid	Friend et al. (1997), Friend and White (2000)	Carbon, nitrogen, and water	PSN, RM, RG, TR, NU, PN, LF, CF, CS, CR, NF, NS, NR, TB, TD, TH, SR, SOMM, NM, COM, EVP, RT	Winter deciduous Oak-Hickory forest/ Tree and stand level/South-western Pennsylvania, USA and C3 and C4 grasses, broadleaf evergreen, broadleaf cold deciduous, broadleaf dry deciduous, needleleaf evergreen, needleleaf cold deciduous, needleleaf cold deciduous/global scale/world	Prediction of the global distribution of vegetation primary productivity, carbon content in vegetation and soil under preindustrial conditions for different plant functional types

TREEDYN & TREEDYN3	Bossel and Schafer (1989),Bossel (1996)	Carbon and nitrogen	PSN, IPAR, RM, RG, LF, LD, LH, LM, CF, CS, CR, CFU, CL, NF, NS, NR, EVP, SR, NM, HD, HM, NIF, ND, NI, AND, AV, AF, NIM, RT	Tropical acacia, *Picea abies*/tree and ecosystem/South China, Central Europe	Parameterized for several European species and applied several sites across Europe to examine the long-term effects of climate change on carbon dynamics and tree growth
3-PG	Landsberg and Waring (1997)	Carbon and water	PSN, IPAR, EVP, CF, CS, CR, LF, SMOR, RT	*Eucalyptus* spp., *Pinus contorta*/ ecosystem/Australia. *Eucalyptus nitens*/Ecosystem/Spain. *Betula platyphylla*/ecosystem/Japan. *Pinus sylvestris*/ecosystem/United Kingdom	Improvement of linkages with remote sensing and geographic information system data
ROMUL	Chertov et al. (2001), Nadporozhskaya et al. (2006), Yurova et al. (2010), Kalinina et al. (2013)	Carbon, nitrogen	LM, SOMM, LH, LM, SOMH, NRE, NM, CE	*Pinus sylvestris*, *Picea abies*, *Betula pendula*/Soil: stand, local, regional, global/Russia	Simulation of primary succession and effects of climate change
EFIMOD	Komarov et al. (2003), Bhatti et al. (2009), Chertov et al. (2009), Shanin et al. (2011)	NPP, H, DBH, BA, growing stock, soil carbon and nitrogen	IPAR, LF, LD, NF, NS, NR, SR, NM, NU, CA, TB, TD, TH, SOMM, COM, CW, CL, LH, LM, HD, HM, SM, RT	*Pinus sylvestris*, *Picea abies*, *Betula pendula*/Ecosystem: stand, local, regional/Russia	Prediction of the impact of management regimes, climate changes, N deposition, forest fires

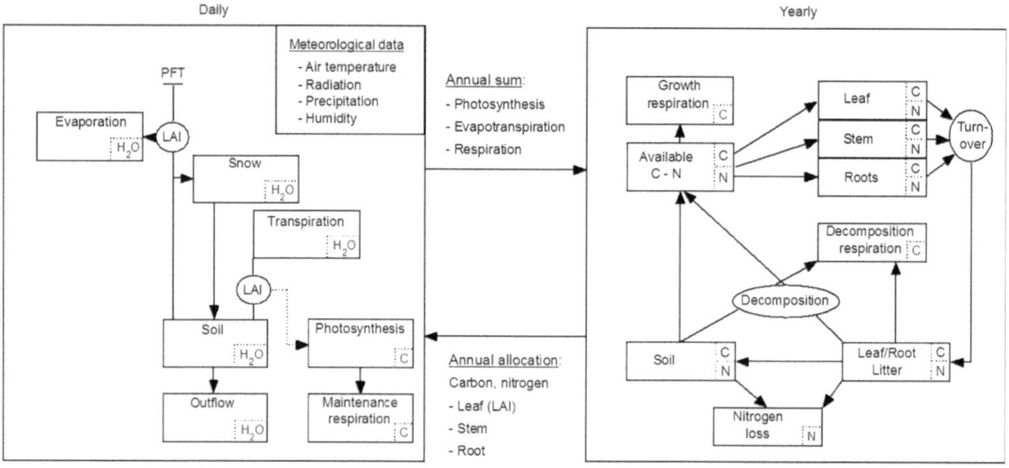

FIGURE 8.1 Basic diagram illustrating the main pools and daily and annual fluxes of FOREST-BGC. (C: carbon; N: nitrogen; LAI: leaf area index.) (Adapted from Running, S.W. and J.C. Coughlan. 1988. A general model of forest ecosystem processes for regional applications. Part I: Hydrologic balance, canopy gas exchange and primary production processes. *Ecol. Model.* 42:125–154, Copyright 1988, with permission from Elsevier.)

For the last three decades, the literature on process-based models has considerably increased. A complete review of all the models is beyond the scope of this chapter, but some models will be discussed to illustrate the evolution in the concepts or modeling approaches that took place. The model hybrid is a biogeochemical model that is similar in concept to FOREST-BGC (Friend et al. 1997). In fact, its development was based on both FOREST-BGC and ZELIG (Urban 1990), a gap model, and the soil decomposition model component shares similarities with CENTURY (Parton et al. 1993). Hybrid is an individual-tree model that simulates 14 processes in two time cycles, daily and annual time steps (Table 8.1). It also includes a grass layer that is subject to the same processes as the trees and the values of 40 parameters and daily temperature and precipitation data must be provided as input. Hybrid was subsequently modified to allow global-scale simulations of different plant functional types under preindustrial conditions (Friend and White 2000). In particular, it was parameterized for two grass and six tree plant functional types by accounting for differences in physiological and phenological characteristics. The model TREEDYN is also similar in concept to FOREST-BGC (Bossel and Schäfer 1989) (Table 8.1). However, compared to FOREST-BGC, the modeling of the physiological processes is simplified, and there are more state variables and fluxes among the ecosystem pools. Good examples of other models with similar concepts that have been developed and applied include BIOMASS (McMurtrie et al. 1990, 1992; Bergh et al. 1998, 2003) or G'DAY (Comins and McMurtrie 1993; Marsden et al. 2013).

Note: [1]AF: Ammonia fixation; AND: Atmospheric nitrogen deposition; CA: Carbon allocation; CE: CO_2 emission; CF: Carbon allocation in foliage; CFU: Carbon allocation in fruits; Cl: carbon allocation in litter; COM: Effect of competition; CR: carbon allocation in roots; CS: carbon allocation in stems; CW: carbon allocation in stems, branches, twigs, coarse and small roots; EVP: Evapotranspiration; HD: Decomposition of humus; HM: Mineralization of humus; IPAR: Intercepted photosynthetic active radiation; LD: Litter decomposition; LF: Litterfall; LH: Litter humification; LM: Litter mineralization; ND: Denitrification; NF: Nitrogen allocation in foliage; NI: Nitrification; NIF: Nitrogen fixation; NIM: Nitrogen immobilization; NM: Nitrogen mineralization; NR: Nitrogen allocation in roots; NRE: Nitrogen release; NS: Nitrogen allocation in stems; NU: Nitrogen uptake; PN: Phenology; PSN: Photosynthesis; RM: Maintenance respiration; RG: Growth respiration; SMOR: Stem mortality based on the −3/2 power law; SOMH: Soil organic matter humification; SOMM: Soil organic matter mineralization; SR: Soil respiration; RT:

Fine roots turnover; TB: Tree biomass; TD: Tree diameter growth; TH: Tree height growth; TR: Transpiration

PnET-CN is another process-based model that simulates the carbon, nitrogen, and water dynamics with a monthly time step. Temperature, precipitation, PAR conditions, wet and dry nitrogen deposition, and atmospheric CO_2 concentration drive the model (Aber and Driscoll 1997; Aber et al., 1997; Ollinger et al. 2002). The developers of PnET-CN-modeled complex interactions using measurable parameters or state variables (e.g., foliar nitrogen concentration) to simulate key carbon, nitrogen, and water dynamics. For example, the model simulates potential photosynthesis as a function of foliar nitrogen concentration and light-use efficiency with a multilayered canopy approach (Aber and Federer 1992). CO_2 fertilization is considered through constant ratios of leaf internal to ambient CO_2 concentrations (Ollinger et al. 2002). More details about the model structure and functions were previously reported (Aber et al. 1997; Ollinger et al. 2002). PnET-CN was used to simulate the effects of disturbances. Good examples can be found in Peters et al. (2013) and Wang et al. (2014). Forest management operations in PnET-CN can be treated as prescribed disturbance events that can be simulated in the model through four parameters: disturbance year, disturbance intensity, biomass removal fraction, and the loss rate of soil organic matter. Regarding the tests performed by Peters et al. (2013), PnET-CN was used to investigate how disturbance types, intensity, and frequency influenced forest net primary productivity across 107 stands in Midwest region, USA. They suggested that soil disturbance had greater negative effects on net primary productivity than stand mortality and biomass removal. Wang et al. (2014) used PnET-CN to simulate the effects of clearcuts on forest carbon dynamics against observations from two northern temperate forest eddy covariance chronosequences, indicating that forest management practices should aim at decreasing soil disturbance caused by clearcut operations.

Recognizing the need for simplified ecophysiological models that could be accessible by forest practitioners for operational needs and easily linked to remote sensing and geographic information system data, Landsberg and Waring (1997) developed the generalized model 3-PG. This model, which drew much attention, is based on the use of general knowledge of physiological processes. Compared to the models described earlier, soil processes are more simplified. The allocation of photosynthate production to foliage, stems, and roots is based on empirically derived ratios obtained from field data, allometric equations, or the literature. Respiration rate is only indirectly considered by computing the ratio of net primary production to gross primary production. The value of this ratio, 0.45, is assumed to be constant for many forest types. The changes in stand density over time is based on the application of the –3/2 power law and the physiological effect of age on stand growth rate decrease is represented in the model. As mentioned earlier, 3-PG drew much attention and has been applied for many forest types around the world. Good examples can be found in Landsberg et al. (2003), Paul et al. (2003), Nightingale et al. (2008), Xenakis et al. (2008), Rodríguez-Suárez et al. (2010), Pérez-Cruzado et al. (2011), Potithep and Yasuoka (2011). In particular, Xenakis et al. (2008) combined 3-PG with a soil organic matter model. Paul et al. (2003) combined 3-PG with CAMFor, a carbon accounting model, the litter decomposition model GENDEC, and the soil carbon turnover model ROTHC.

The majority of biogeochemical models perform simulations at the stand, landscape, or regional levels. Relatively few models were developed to simulate the development of individual trees using mechanistic relationships. Korol et al. (1995a, 1995b) developed TREE-BGC using FOREST-BGC as a framework to better integrate the effect of stand density and structure on productivity. Photosynthate production is computed at the stand level, but its partitioning among individual trees is computed using a competition algorithm based on tree height, leaf area, and light saturation threshold. The amount of carbon allocated to each tree in a stand is converted into diameter and height growth using allometric relationships. Zavala and Bravo de la Parra (2005) used a similar approach for Mediterranean forests. However, compared to TREE-BGC, the allocation to individual trees was directly linked to physiological resource limitation factors, including light and water availability and transpiration. A review of different single-tree process-based models may be found in Le Roux et al. (2001).

8.3.2 Soil Processes and Tree and Stand Development

Soil is a destructive component of terrestrial ecosystems responsible for nutrients and water supply to ecosystem organisms (first of all for plants) due to (a) dead organic matter decomposition both with its full mineralization and accumulation in soil, and (b) weathering of minerals in geological substrate. These processes lead to the release of nutrients for plant growth from decomposing organic matter and soil minerals, accumulation of transformed organic matter, nutrients and water in a soil profile, and the formation of underground pathways of biological cycle of elements in terrestrial ecosystems.

There are abiotic and biotic factors responsible for soil functioning and especially for forest ecosystem productivity. Abiotic, environmental factors are as follows.

(1) Solar radiation and heat flow from the atmosphere responsible for energy input and the rate of all biochemical, chemical and physical processes.
(2) Atmospheric input (deposition) of water, elements (N, S, P etc.), mineral and organic dust, and pollutants. These substances seriously influence soil development. It is especially clear in relation to water and nitrogen. Nitrogen is absent in geological rocks and sediments. The nitrogen input from the atmosphere goes in a form of nitrogen oxides and ammonium that can be consumed by soil biota. In pure atmosphere, nitrogen input varies from 1 to 4 kg [N] ha^{-1} $year^{-1}$, but due to air pollution nitrogen deposition can reach 40–60 kg [N] ha^{-1} $year^{-1}$.
(3) Weathering of minerals in a soil leads, first, to element release for green plants and soil biota (all elements with the exception of nitrogen) and, second, to formation of friable loose soil material, which are favorable for plants and soil biota physical properties. It is the disintegration of solid geological rocks, clay formation in sandy soils, and loss of clay in heavy loamy soils.
(4) Lateral input and output of water and nutrients that can significantly modify soil development and ecosystem functioning.
(5) Heterogeneity of water regime and parent material that form complex spatial mosaics of micro-sites in forest ecosystems. This mosaic structure can increase ecosystem stability.

Biotic factors of soil functioning can be arranged as follows.

(1) Biological fixation of atmospheric nitrogen (symbiotic and non-symbiotic) is a very powerful process in all terrestrial and forest ecosystems, especially after their disturbance by various natural and anthropogenic factors. It is a natural process to cover nitrogen deficit, and value of fixed nitrogen can be 10–20 kg [N] ha^{-1} $year^{-1}$ in temperate forests and up to 150 kg [N] ha^{-1} $year^{-1}$ in humid tropics.
(2) Accumulation of soil organic matter (SOM), humus and elements in topsoil due to input of organic litter as a product of ecosystem functioning. This is a main process of soil formation from an ecological point of view.
(3) Spatial heterogeneity of soil organic matter and elements' accumulation as determined by trees' growth. This "biogenic" mosaics of soil conditions (that can interfere with abiotic mosaics) play an important role in ecosystem functioning.

Minimally, the soil functioning can be expressed by the following equations:

$$\begin{aligned} \mathrm{dH/dt} &= \mathrm{L} - \left(k_1 + - \mathrm{k}_2 + \ldots + \mathrm{k}_n\right) H \\ \mathrm{dM/dt} &= \mathrm{D} - \left(k_4 + \mathrm{k}_5 + \ldots + \mathrm{k}_m\right) M \\ k_i &= \mathrm{f}\left(\mathrm{C,R,P,W,B,A,t}\right) \end{aligned} \tag{8.6}$$

where H and M are soil organic matter and mineral matrix, respectively; L is input of organic matter to soil (litter fall); D is input of mineral particles to the soil due to various deposition processes; k_1,

. . . k_i, . . . k_m are kinetic parameters expressing the rate of various processes of soil organic matter (SOM) and soil minerals transformation and destruction; C is climate; R is landform (relief); P is parent rock; B is soil biota and vegetation, and t is time; W is soil water regime; A represents the effect of anthropogenic factors. In this conceptual model, consideration of H and M in definite temperature and moisture regimes allows for calculations of all other soil parameters (physical and chemical characteristics, cation exchange capacity (CEC), available nutrients etc.), depending on these two basic soil components.

The aim of SOM dynamic modeling is to solve a mathematical problem: if the initial values of SOM pools are known, then it is possible to predict the amounts and pools of available nutrients for plant growth and carbon dioxide evolution using data on (a) organic litter input and its chemical composition, (b) soil temperature and moisture, (c) impact of other environmental factors (for example, effect of frost and melting, soil drainage, erosion, etc.), and (d) land use regimes (various regimes of forest management, damages due to forest fires and insect attacks, etc.).

In recent SOM models, the rate of organic matter mineralization is mostly expressed by a negative exponential function as in early Olson's (1963) model for prediction of the rate of forest floor mineralization:

$$X/X_0 = e^{-kt} \tag{8.7}$$

where X is the residual mass of organic layer at time "t," X_0 is the initial mass of the layer; and k is the mineralization rate. The model by Pastor and Post (1985), a component of the ecosystem model LINKAGE, was applied in numerous forest ecosystem models, resulting in a good correlation between the rate of organic debris mineralization and vegetation evapotranspiration at yearly time step. It is a one-compartment model where the rate of decomposition is determined by the following equation:

$$\text{PWTLOS} = (0.9804 + 0.09352{*}\text{AET})(-0.4956 + 0.00193{*}\text{AET}{*}\text{R}) \tag{8.8}$$

where PWTLOS is the annual percent weight loss; AET is the actual evapotranspiration rate (mm $year^{-1}$); R is the lignin/nitrogen ratio of organic litter.

The ITE (Institute of Terrestrial Ecology) model of Thornley and Cannell (1992) is a two-compartment (SOM and soil bugs) model. It takes into account four types of organic debris and calculates the mineral nitrogen transformation from ammonium to nitrates.

The CENTURY model (Parton et al. 1988, 1993) for grassland ecosystems was successfully tested in different natural zones and forest ecosystem models as well. The litter is divided into two classes: metabolic (fast decomposing compounds) and structural (slow decomposing lignin and cellulose). The SOM compartments are represented by pools of active soil C (biomass of microorganisms), slow soil C, and passive soil C. The model also calculates soil nitrogen and other elements' dynamics with the formation of a mineral nutrient pool for plant growth. There are four other agricultural SOM models with forest versions as well. The ROTHC model (Coleman and Jenkinson 1995) has two inputs of litter and a cascade of repeated pairs of biomass (microorganisms) and humified material, distinguished by the rate of SOM mineralization. The additional specific feature of ROTHC is a pool of inert organic matter (IOM) that is not participating in SOM turnover. This pool is included in the DAISY model (Hansen et al. 1991; Bruun et al. 2003) and CANDY model (Franko et al. 1995) by taking into consideration litter and SOM C/N ratios and transformation of nitrogen with more detailed SOM compartmentalization. The Yasso forest soil model (Liski et al. 2005) is also related to these types of models.

The SOM modeling is a crucial issue for the development of forest ecosystems models. These models are widely implemented for the evaluation of the carbon balance dynamics in terrestrial ecosystems under the impact of changing environmental factors. The SOM criterion of ecosystem sustainability has been also proposed (Morris et al. 1997). This criterion postulates that the sustainable use of ecosystems can be reached only if SOM and soil elements' pools will not be decreased or unbalanced by ecosystem exploitation, modification, or transformation to other ecosystem types. The following is a description of the ROMUL model of SOM dynamics in forest soils.

ROMUL (Chertov et al. 2001) and its earlier version, SOMM (Chertov and Komarov 1997; Chertov et al. 2002), describe the biological turnover of organic matter and nitrogen. It simultaneously allows for the calculation of the rate of mineralization and humification of litter and soil organic matter with the corresponding CO_2 emission and nitrogen release for plant growth. The initialization of the model uses pools of organic matter and nitrogen in correspondence with usually discriminable horizons of forest floor in field studies: L, F, H, and top horizon of mineral soil Ah. This definition of main variables shows that the ROMUL model best describes the specificity of forest boreal soils, even though it is possible to apply it with corresponding average of the variables for arable and more differentiated soils. The experimental basis of the ROMUL model consists of laboratory experiments (Chertov and Komarov 1997; Chertov et al. 2001) in controlled temperature and moisture conditions with validation using field experimental data.

The model also has input pools of fractions of above- and below-ground litter that differ in nitrogen and ash contents (decomposition rates for belowground litter fractions are calculated with different temperature and moisture dependencies) and a cascade of SOM fractions: undecomposed litters, partly humified organic materials (forest floor, peat, and so-called labile humus in mineral topsoil) and relatively stable humus bonded with the mineral matrix of the topsoil. The litter input in the model can be represented by an unlimited number of fractions that have species-specific ash and nitrogen contents. There are three processes of SOM humification by three communities of microorganisms and soil fauna (without consideration of their biomass) and three processes of mineralization. The model is represented by a system of ordinary differential equations with coefficients that depend on the soil temperature and moisture, the nitrogen and ash content in litter, and the C/N ratio in the mineral topsoil. The model also considers nitrogen release and gross carbon dioxide emission from the soil.

There are several variables in ROMUL that represent different pools of organic matter (Figure 8.2). L_{above} (kg m^{-2}) is a pool of organic matter corresponding with the L horizon of the forest floor. It contains undecomposed and partly decomposed litter with big plant fragments. L_{below} (kg m^{-2}) is a pool of SOM that contains undecomposed and partly decomposed root litter. Part of this pool is metabolized by the decomposing organisms and the corresponding carbon is released as gaseous CO_2 (flows with rates R_i marked arrows in Figure 8.2). The majority of the organic matter ends up in a more slowly decomposing complex of partly decomposed organic matter F (kg m^{-2}), which is a pool of organic matter that contains partly decomposed litter with small plant fragments treated as humus substances and corresponds to the F horizon of the forest floor. This pool is transformed by several groups of organism-destructors into different forms. The community of bacteria, fungi, and arthropoda with rate dk_F transforms part of the SOM into a H pool and part of the SOM ($1-dK_F$) into stable humus of mineral horizons pool Sh (stable humus) that produces raw humus with a C/N ratio of 15. The parameter d is a ratio between both pathways that depends on the C/N and F/B (fungi to bacteria) ratios in F. The greater part of humified matter stays in the organic layer H as C/N is increased. Transformation (humification) of SOM by community of earthworms and soil invertebrates forms the mull humus Sh in mineral horizons with rate k_E and a C/N ratio ≤8. Lh (Labile humus) (kg m^{-2}) is a pool of "belowground forest floor." The transformation of this pool also goes by two groups of organism-destructors. k_{Lh} is the rate of transformation by bacteria and arthropoda. k_{Ef} is the rate of transformation by earthworms. The immobilized humus pool H decomposes at a rather slow rate modified by the soil temperature and moisture conditions. The decomposition flux has a range of a minimum 1–2% annually up to 15% annually, depending on the soil texture and the clay content. Maximal rate of H decomposition may be observed for arable soils. On the other hand, as roughly half of the soil organic matter in the boreal zone is in the compartment H, the value of the rate factor k_{Ef} has a significant effect on the total storage of organic matter in the soil.

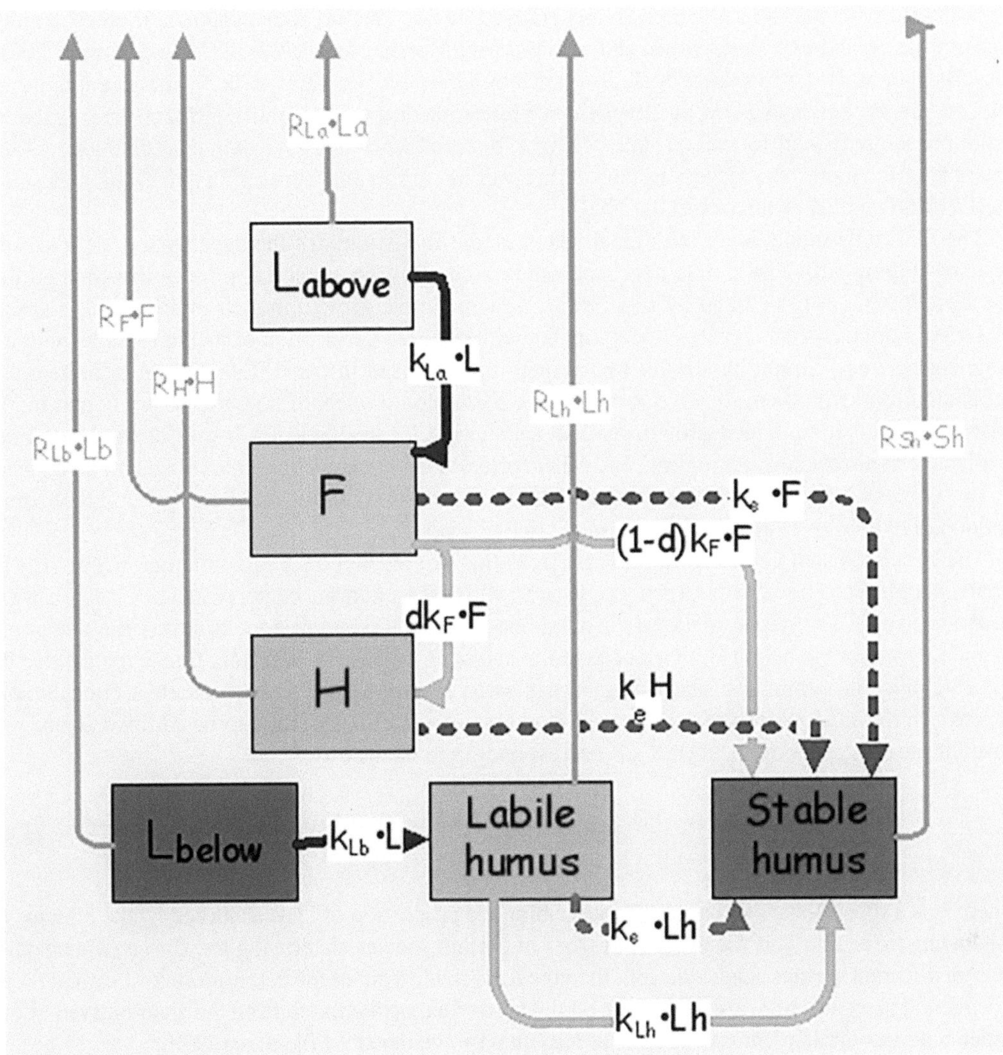

FIGURE 8.2 Flow chart of the ROMUL model. Explanations of symbols are in the text.

The temperature and soil moisture modifiers of the fluxes have various patterns given as stepwise defined functions. Typically, all these show an optimal range of values, where the decomposition takes place at a full rate and tapers off outside the optimal conditions. The optimal conditions for different fluxes are, however, somewhat different (Linkosalo et al. 2013). This simple scheme of SOM transformation allows for the description of elements contained in the pools of organic matter but using some modifiers in the main scheme of SOM decomposition, because carbon and nitrogen dynamics have different rates of transformation (Chertov et al. 2001). For instance, the transformation of nitrogen in the compartment H, i.e., two different mass flows (Figure 8.2), is based on the assumption that all the matter in the compartment H is mostly produced by the metabolism of the decomposing organisms (bacteria, protozoa arthropoda and lumbricidae), and the C/N ratio is used for these groups obtained from soil biology experiments. Therefore, nitrogen has a special role in this phase: first, the rates k_E and k_F of nitrogen moving from compartments F_i into the compartment H are calculated and, thereafter, a corresponding amount of organic matter, typical for the type of decomposers and depending on the C/N ratio of produced humus (24.0 for bacteria and arthropoda and 12.8 for lumbricidae), is moved from the compartments F_i to the compartment H.

Soil stability is an important additive criterion of sustainable forest management in the changing environmental and economic world (Kimmins 1996; Morris et al. 1997; Boyle and Powers 2001; Akselsson et al. 2007; Komarov et al. 2007). Many forest ecosystem models do not accentuate and analyze the dynamics of SOM and elements of forest nutrition, particularly nitrogen, in relation to stand productivity, soil formation, management, disturbances, and environmental changes. As an example of the model that describes closed turnover of carbon and nitrogen in a tree-soil system of the EFIMOD model (Komarov et al. 2003).

The EFIMOD model is an individual-based model that simulates the development of tree-soil systems (Figure 8.3). The simple approach used successfully in statistical physics was applied for the representation of the stand by separate trees with simple growth models and located in space as a sessile organism with precise locations but with detailed description of its interactions with its neighboring trees. Simple models of tree growth may be used in the EFIMOD system of models, depending on the problem to solve: from simple regression models of potential (usually maximal) growth to more complicated growth models accounting for available PAR with dependencies on available soil nitrogen, water regime, and other factors if necessary. Then, tree increment is reduced by two types of competition: shadowing that defines concrete light flux and belowground root competition that distributes available nitrogen for tree growth.

The simulated stand consists of separate trees that are located on a square lattice whose cell is small enough. Each tree creates a shadow zone S_i, which is a quadrate centered at a tree's cell x_i with quadrate size R_i. The tree x_i shadows the neighbor tree x_j if (1) shadow zone S_i covers the tree's cell x_j, and (2) ratio of the height H_i of a tree x_i to the height H_j of a tree x_j is greater than a fixed value β, i.e., $H_i/H_j > \beta$. The symmetric shadowing occurs when two trees shadow each other. A contribution of a tree in the shadow formed by all trees shadowing the cell is assumed to be additive. Thus, we have for the shadow coefficient $0 \le K_E \le 1$ corresponding to the tree *xi*,

$$K_E(x_i) = 1 - \sum \lambda_l k_l \tag{8.9}$$

where λ_l is a species-specific parameter describing the proportion of PAR absorbed by the crowns of shadowing trees, and k_l is the number of trees of certain species shadowing the x_i tree (when some trees of different species shade one cell, then the total shadowing of the cell represents a sum of their λ_l values). The values of λ_l are calculated on the base of experimental data on the absorbed radiation under canopies of different tree species at maximal crown density (Tselniker 1978).

After the calculation of a spatial mosaic of shadowing (available light in every cell) the light response multiplier, K_{SH}, is calculated for every tree. The value of K_{SH} as a function of shadowing has different shapes for shade-tolerant and shade-intolerant species, i.e., K_{SH} is linearly dependent

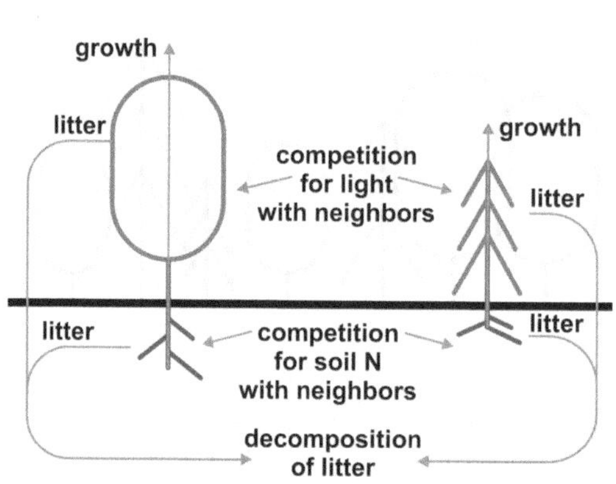

FIGURE 8.3 Flowchart of the EFIMOD system of models.

on K_E for Scots pine (*Pinus sylvestris* L.) and Silver birch (*Betula pendula* Roth) (shadow-intolerant trees):

$$K_{SH} = K_E \tag{8.10}$$

and is nonlinearly dependent for shadow-tolerant Norway spruce (*Picea abies* (L.) H. Karst.) (Tselniker 1978):

$$K_{SH} = 1 + 0.44 \log K_E \tag{8.11}$$

Thus, the light response multiplier K_{SH} reflects the local stand density in relation to a set of neighbors for each tree in the form of overlapping of shadowing zones and species-specific light response of trees.

Each tree has also a nutrition zone that is a second form of competition acting together with shadowing. The average and maximal distance of lateral root spreading are calculated on the basis of stem diameter derived from empirical data using a logistic equation. As maximum and average root spreading distances decrease with increasing soil fertility and moisture, these parameters have site-specific modifiers. The model calculates potential rooting area as the surface of a circle with a radius equal to the average root spreading distance. Then, for each cell, the probability of occupation is calculated, which is positively related to the amount of available nitrogen in the current cell and inversely related to the distance between this cell and stem base and the mass of roots of other plants. The distance from each cell to the rooting cell cannot be more than the maximal root spreading distance for the given tree. All cells in the rooting area of individual trees must be arranged in continuous connected contours. The vertical distribution of roots depends on species-specific features, soil characteristic, and strength of belowground competition. It is calculated as decreasing exponential function of depth. The model also calculates horizontal distribution of biomass of coarse and fine roots as a function of the distance from stem base. The uptake rate for water and nutrients depends on the biomass of fine roots in each cell, tree age, and species-specific modifier.

Areas of nutrition for neighboring trees can overlap, and available nitrogen is consumed proportionally to the mass of fine roots of neighboring trees in these overlapped areas. This definition of belowground competition is species-specific and depends on the spreading of roots and their density per square unit, which are different for different species. Thus, two possible types of tree increment can be calculated; that due to light or soil nitrogen (Komarov et al. 2003). The calculation requires species-specific estimates of foliage and fine roots biomass, maximal biological productivity of foliage, and the specific consumption rate of nitrogen. The joint impact of both types of competition can be evaluated either by the minimal value of two types of increment taken as the annual increment, following the Liebig's principle or by using weighting of particular reducing coefficients. Due to the two different types of competition, simple procedures were inserted to simulate adaptations of trees to the lack of some resources: (1) if tree increment is restricted by light, then its biomass gain will be allocated mainly to increases in stem height and foliage biomass; (2) conversely, if the growth of a tree is limited by nitrogen uptake, then the relative increase in biomass of fine roots and stem diameters will be higher than those of foliage and stem height. A deterministic procedure of tree death is based on the idea of lethal threshold, defined by the ratio of leaf mass B_l to total biomass B_T, below which the tree dies:

$$r = g_1 - g_2 \ln B_T,$$

where r is the lethal ratio B_l/B_T, and g_1 and g_2 are species-specific parameters evaluated from corresponding regional growth tables. In addition, we inserted a probabilistic mortality function that depends on tree age, which reaches 1.0 at a species-specific maximal tree age.

The sensitivity analysis and simulation experiments with the model showed that the spatial structure of plant community had a stronger influence on competition than species-specific features or environmental conditions.

8.3.3 Hybrid Models

Models that contain both empirical and process-based components are termed hybrid models (Mäkelä et al. 2000; Landsberg 2003a, 2003b). The hybrid approach explores methods that combine forest growth and yield equations, based on statistical relationships derived from empirical data (see Chapter 6) and mechanistic or theoretical relationships that represent physiological and biogeochemical processes that govern the dynamics of forest ecosystems (see Mäkelä et al. 2000). These models may be driven by both environmental variables (e.g., PAR, temperature, precipitation, or soil properties) and dendrometric variables obtained from forest inventories to simulate a set of tree and stand growth characteristics (e.g., diameter at the breast height, tree height, and stand density) and the effects of forest management (e.g., harvesting and thinning) on forest productivity (e.g., gross primary productivity and net primary productivity) for a given stand. In addition, many empirical relationships (e.g., diameter distribution and biometrical equations) have been incorporated into hybrid models to simulate some largely less known ecological processes (e.g., plant carbon allocation). For example, diameter distribution—a type of growth and yield model—has been demonstrated to be able to help ecosystem process-based models to predict stand structure (e.g., Lasch et al. 2005; Wang et al. 2011), since climate change has little effect on competition mechanisms for even-aged pure stands. Therefore, these models can be effective tools for exploring the effects of forest management and climate change on forest growth and carbon sequestration (Landsberg 2003).

Several hybrid models have been developed for this purpose like 3-PG (Landsberg and Waring 1997), CroBas (Mäkelä 1997), FORCYCTE-11 (Kimmins 1993), FORECASE (the successor to FORCYTE-11, Kimmins et al. 1999), FVS-BGC (Milner et al. 2003), CanSPBL (Pinjuv et al. 2006), TRIPLEX (Peng et al. 2002), and SECRETS-3-PG (Sampson et al. 2006). These models have been used to explore the best "practices" of forest management under climate change. For example, Wang et al. (2013) conducted a simulation experiment that combined three climate change scenarios and 36 forest management scenarios with different thinning intensities and occurrence and rotation length using the TRIPLEX model. They argue that pre-commercial thinning does not change mean annual net ecosystem productivity for boreal coniferous stands over their simulation period (2010–2100), while moderate warming with the rotation length of 50 years could enhance mean annual net ecosystem productivity. Simulations showed that intensive forest harvests may obstruct forest recovery for a very long time for Douglas fir stands in British Columbia, Canada (Blanco 2012). Using FORECAST, Wei and Blanco (2014) suggested that mixed plantations in subtropical regions with stand densities in the 2,000–3,000 stems ha^{-1} range could increase their amount of carbon by 68% compared with traditional pure plantations. As the understanding of physiological processes improves, the use of such hybrid models will likely continue to investigate the effects of forest management and climate on forest growth (Twery and Weiskittel 2013).

8.3.4 Global Scale Models

Local process-based models can be used in very detailed situations, such as being used against a forest stand with single tree species. At large regional scales, tree species usually have to change to loosely defined tree groups such as plant functional types in some Dynamic Global Vegetation Models (DGVMs) like MCI (Bachelet et al. 2001a) or IBIS (Foley et al. 1996). Also, there is a need to better represent biosphere–atmosphere interactions with vegetation dynamics (Olchev et al. 2008; Quillet et al. 2010). Many large-scale modeling exercises need to consider potential new processes because environmental conditions become different from that of the site scale. For example, the effect of lateral C redistribution across landscapes become noticeable at large-scale C accounting and has been considered to link with carbon models (e.g., the General Ensemble Biogeochemical Modeling System (GEMS), Liu et al. 2003). Also, some expected constant driving factor, such as atmospheric CO_2 concentration, can become variable across a large region, and therefore a regional model application may need additional treatment on using spatial varying CO_2 data instead of a fixed ppm value (e.g., Zhang et al. 2014). Broader model expansions can be seen when simulation considers wider simulation scope like including fire disturbance, land use change effect, and movement of C from land to aquatic systems (e.g., USGS LandCarbon Project, Zhu et al. 2010). The NCAR community earth system model (CESM) is a recent development trend for large-scale process models.[2] The biogeochemical models used in the system needs additional links with other models. In addition, models like CESM also need modern high-performance computing (HPC) resources to perform intensive simulations. This alone requires significant computational changes of the models.

When process-based models are applied to large regions, some key sources of uncertainty will become significant, such as model incompleteness, limited input and validation data, limited computing ability over large regions, and spatial scaling issues. When the study region is large enough, differences between sub-regions could be quite different, such as tropical region compared to temperate region. Ecoregions, provinces, states, and counties are all examples of regions that each may have their distinct geographical features, including climate, soil, vegetation, and land use. Process-based models may need to change parameter sets, driving variables, or some algorithms to deal with significant location changes. Without those significant model processes or component changes, spatial variability can sometimes be dealt with empirical spatial scalars. For example, The USGS LandCarbon project (Zhu et al. 2012) used county-level forest growth and grain yield data to calibrate the GEMS model (Liu et al. 2003) to generate spatial scalars to deal with the unknown causes of county-level differences. Another example is the GHG modeling in the Triplex-GHG model, in which a different Q_{10} parameter value is used to control GHG emission for different regions: boreal = 2.03, temperate = 2.66; sub-tropical and tropical = 3.75 (Zhu et al. 2014, 2015).

For large regions, simulation on higher spatial resolutions faces the issue of availability of high-resolution input data. Many models like IBIS (Liu et al. 2011) use internal data interpolation procedure to solve this problem. A simple nearest neighbor procedure was built in IBIS to do spatial interpolation so that all input layers are of the same resolution. On the other hand, spatial sampling simulation is useful for quick testing and calibration runs for large region with higher resolution input data. With a sampling interval of 5, which means selecting one land pixel every five rows and five columns on the land cover map, IBIS will first read in the original coarse resolution inputs, perform spatial interpolation to base resolution, conduct spatial data sampling by the interval, perform actual simulation on sample data, and produce outputs at the coarser resolution defined by the sampling interval.

Pure land cover at high resolution will change to mixed land pixel when aggregated to coarse resolution. IBIS is a DGVM that allows multiple plant functional types (PFT) to potentially coexist in a single land pixel. Also, it allows competition among the PFTs under the changing climate and land use conditions. In recent modified IBIS, unlike most biogeochemical models, IBIS takes in fractional land cover data and allows fractional vegetation changes following fire or land use change events. Although a typical land pixel in simulation can change relative leaf area index (LAI) of trees and grasses, the extent of change is limited by both physiological parameter and physical area fraction. For example, a land pixel with a small forest fraction may have healthy trees and high LAI per forest area, but the whole pixel may be dominated by grass or crop. Therefore, the land pixel could be classified as a grassland or cropland. The non-vegetation fraction is a major controlling factor in rocky areas, desert, water body, and urban development areas so that simulated vegetation will not take over a whole land pixel in the model. When logging happens, the forest area fraction will remain unchanged, and only the forest biomass level will be reset. When deforestation happens, the forest cover is converted to the target land cover (crop land, grassland, or urban), and the forest fraction decreases. When reforestation happens, other land covers convert to forest, hence forest fraction increases. Fractional land cover data products can be seen in a recent large global land cover dataset generated from the Global Change Assessment Model (GCAM[3]), which indicates the up-coming trend of ecosystem modeling with fractional input data. Other related data changes also exist. For example, biomass initialization for a single land pixel could include forest biomass, shrub biomass, and grass biomass separately. Fire disturbance on a land pixel could be containing percentages of low, medium, and high burn severity.

Process-based models can be calibrated at local scale with actual field observation data (e.g., flux tower data). When changed to large regions, local calibrated parameters may not always fit in other locations or broader extents. This is partly because regional input datasets are often interpolated data and may be at very coarse resolution. For example, the spatial resolution of interpolated climate data for the United States could be 4–50 km, which is potentially much different from the field data. So simulation with interpolated data will likely give a different output than simulation with local field observation data. This indicates that regional-level simulations need regional observations for calibration. There exist some regional level inventories and statistics that serve as validation data. For example, the global MODIS NPP maps (1-km MODIS NPP of 2001–2005, Zhao et al. 2005), the forest inventory and analysis (FIA) growth curves at county level and the crop grain yield statistics by county are used in IBIS for coterminous U.S. simulations. To validate the model, an automatic calibration procedure against county-level averages of forest biomass, NPP, and grain yield is developed as a post-processing step. NPP observations from remote-sensing products are first aggregated to county level and then compared with IBIS NPP outputs averaged at county level. Then IBIS calculates the county-level differences and the related adjustment scalars. The scalars are assumed to help dealing with unknown environmental factors. For algorithm, the scalars for each county will be used to modify the forest V_{max} parameter (Maximum Rubisco-limited rate of carboxylation) in a second simulation. On the other hand, forest biomass inventory data and crop grain yield statistics data are also summarized at county level for IBIS to compare with. The forest growth curves published by USDA Forest Service, the Carbon OnLine Estimator (COLE[4]) database, are the general forest growth references to be compared with IBIS growth curves. An indicator of growth rate, e.g., total biomass carbon increase from 1 to 100 years, can be calculated from the IBIS biomass pool. Comparing simulated biomass values with the COLE 100 year growth values for the same geolocations, a scalar can be generated to increase or decrease IBIS biomass pools in a renewed simulation. When iterated 3~4 times, a stable carbon scalar can be obtained. For forest ecosystems, the NPP and biomass scalars can be used at the same time. It may happen that one observation could be better than other observations. Or the IBIS model can be calibrated to match one type of observation well but fail with the other. Then more in-depth manual adjustments are used.

8.3.5 Process-Based Models and Ecological Indicators

Ecological indicators are defined here, and in Chapter 7, as an index of measures that characterize an ecosystem or one of its critical components. In addition, ecological indicators are typically used to assess the state of a specific environment and evaluate the trajectory of that system under current and future conditions. Ecological indicators should represent key information about the composition, structure, function, and/or vigor of an ecosystem. For example, biodiversity, including genetic diversity and functional diversity, can be an important ecosystem service (Mace et al. 2012) and an ecological indicator of forest health.

To be able to quantify the potential changes in biodiversity, it is necessary to understand the role and impact that climate has on species' ranges at the regional and continental scale. There are many ecological, evolutionary, and physiological processes that influence species' distribution and abundance and thus biodiversity (Kearney and Porter 2009). Similarly, limits and enhancements to plant functional processes, which can be simulated with process-based models, can be used to predict changes to biodiversity patterns, species abundances, tree distributions, and potential future migration patterns. As an example, the process-based model PHENOFIT was used to identify the climatic factors and biological processes that limit the ranges of 17 North American boreal and temperate tree species, ultimately with a goal of predicting potential species distributions (Morin et al. 2007).

The model PHENOFIT (Chuine and Beaubien 2001) simulates species' distributions based on a phenological submodel, a frost-injury submodel, a survival submodel, and a reproductive-success submodel. This study showed that species distributions were limited due to climatic constraints that impact phenological processes, such as the inability to achieve full fruit ripening or flowering. Later, the same model was used to reduce uncertainty in species range shifts under future climate change scenarios (Morin and Thuiller 2009). The need for understanding the implications of climate change on biodiversity and forecasting the shifts in biodiversity under climate change is immediate (Thuiller 2007).

Araujo and New (2007) argue that utilizing process-based models in conjunction with additional model classes and ensemble forecasting are an improved framework to diagnosing bioclimatic constraints on species distributions. An example of using multiple process-based perspectives was conducted in the study by Kleidon and Mooney (2000), which used components from the land-surface model ECHAM4 (Roeckner et al. 1996) and the terrestrial biosphere models CASA (Potter et al. 1993), SDBM (Knorr and Heimann 1995), and SILVAN (Kaduk and Heimann 1996) to create a "generic" plant model. This method was applied at a global scale with multiple climatic forcing constraints, finding that a major factor leading to distribution of diversity occurs during the early stage of a plant's life when a plant is more sensitive to changes in water availability due to variability in precipitation.

Ecological indicators linked with models can be used as a guide and/or enhance forest management (Lindenmayer et al. 2000) but can have challenges (Dale and Beyeler 2001; Failing and Gregory 2003). For example, indicators that express vulnerability to environmental disturbance and land use change can be useful to forest managers. For example, the Morales et al. (2005) study compared and evaluated the ability of four process-based models (i.e., RHESSys, GOTILWA+, LPJ-GUESS, and ORCHIDEE) to simulate fluxes of water and carbon within terrestrial ecosystems and can be a useful tool and methodology when evaluating ecosystem flux vulnerability to disturbance and/or climate change impacts. In addition quantifying the consequences of land use change on the terrestrial carbon budget, water resources, and nutrient availability can be achieved with process-based models (McGuire et al. 2001). For example, four process-based terrestrial biosphere models were used to evaluate the 20th-century carbon balance as a result of land use, in addition to CO_2 and climate effects (McGuire et al. 2001). Cramer et al. (2001) used six DGVMs as a tool to report the varying ranges and uncertainties in the terrestrial carbon sink as a function of rising CO_2 and climate change. It is likely that process-based models will be used more often in forest management and will be used to evaluate the impact on ecological indicators as a result of environmental disturbance, climatic changes, and land use change. However, variations among and between models and large uncertainties in simulating response variables and ecological indicators (Cramer et al. 2001; Morales et al. 2005; Balshi et al. 2007) warrant improvements to the models and our understanding of ecosystem processes.

8.4 MODEL EVALUATION AND UNCERTAINTY ANALYSIS

The majority of process-based models are deterministic: their predictions are provided without estimates of the errors. Regardless of their level of complexity or degree of realism, accuracy, or precision, there is always a degree of uncertainty in the predictions because models are imperfect representations of reality. For biogeochemical as well as biological models, uncertainty originates from different sources (O'Neil and Rust 1979; Medlyn et al. 2005; Larocque et al. 2008a, 2008b, 2011; Matott et al. 2009): (1) errors in measurements, estimates, or spatial or temporal scales; (2) lack of understanding of the mechanisms of the biological processes and interactions; (3) model structure and adequacy of the mathematical representations of processes; (4) parameter estimates; and (5) amplitude of natural variation within ecosystems under study.

Any modeling exercise should include two major steps: evaluation and uncertainty analysis. The model evaluation step contributes to the development of more reliable and accurate models to address the problems of the real world and is one of the important steps in model development and application. It can help to either build up confidence for model applications or allow model selections

(Tedeschi 2006). Uncertainty analysis, which consists of computing the variability in model outputs that may be attributed to errors associated with the sources mentioned earlier, allows model users to evaluate the level of confidence in the predictions and draw appropriate inferences. For instance, the uncertainty analysis that Larocque et al. (2008b) performed in a soil carbon cycle model to compare different scenarios of temperature increase allowed them to conclude that the gradual increase in temperature was likely to have a marginal effect over a short time period. Significant effects were evident only on some soil carbon pools and over a long time period.

To understand the model evaluation process, it is necessary to understand the following concepts (Rykiel 1996):

- Verification is defined as a technical step on the extent to which a mathematical model is logically consistent or represents well the mechanisms being modeled. In some cases, it is not easy to verify whether a complex computer program (model) is entirely error-free. For model developers, the model verification process needs to be addressed prior to any other processes in testing process-based models.
- Calibration is the estimation and adjustment of model parameters to improve the accuracy and precision of model simulations against a measured dataset.
- Validation is often defined as a process to evaluate the model output against another independent dataset using the calibrated parameters. This process indicates if the model is acceptable for use within a certain domain. It is important to distinguish the term of "verification" and "validation" in the modeling circumstance. As validation of ecological models is impossible to some extent (Oreskes et al. 1994), hereafter such a process is referred to as model evaluation.

Model evaluation/model testing could be generally categorized into two groups: (1) qualitative evaluations that compare the model behavior with hypothesized patterns and (2) quantitative methods that compare model simulations with observatory data.

Qualitative testing aims at evaluating the model behaviors, which is included in a standard development routine (Jakeman et al. 2006). The key point of qualitative testing is to find out if model behaviors match hypothesized patterns or observational patterns by using visualization techniques. If the comparison of simulated results between the model and observation data or theoretical patterns is unfavorable, then the model users or developers have to figure out why and how the model behaviors do not match as expected. A list of graphical display techniques is as follows:

- Comparison of matched estimated and measured time-series values and/or residuals
- Comparison of measurements and estimates (bivariate plots)
- Comparison of ranges, medians, and means (box plots)
- Cumulative/probability frequency distributions (histograms)

Many different statistical techniques can be used to test the performance of a model in the case when observatory data are available, and the model behaved as expected. It could be more useful to statistically test the model directly against observation data than to compare the time series (Kirchner et al. 1996). Previous review work has summarized these techniques for accessing model accuracy and precision, which are two important aspects of model testing (Tedeschi 2006). Accuracy measures the ability of a model to predict the right values, and precision measures the ability of a model to predict similar values consistently (Tedeschi 2006). Some quantitative methods are listed as follows:

- Linear regression and its analysis (Tedeschi 2006)
- Correlation coefficient (parametric and nonparametric)
- Nonparametric analysis: Kolmogorov-Simirnov test, rank test etc.

- Distribution comparison and its analysis: $\chi 2$ test, Kolmogorov-Smirnov's D test
- Residual analysis: mean bias (MB), mean square error (MSE), root mean square error (RMSE), mean absolute error (MAE), etc.
- Model efficiency: Willmott index (d) (Willmott 1982), Nash–Sutcliffe model efficiency coefficient (E) etc.

No single evaluation index is sufficient to adequately assess the model performance when comparing observed and predicted values. Therefore, we suggest that multiple statistics in evaluating precision and accuracy should be used in model evaluation. In the context of modeling forest management under changing climate conditions, multiple variables (e.g., dbh, height, biomass, and gross primary productivity) and multiple statistics should be used to compressively evaluate the performance of models.

Model evaluation is important and delivers important messages in model structure and/or sensitive model parameters to model developers and users. Such a feedback can contribute to developing useful models. Identifiability analysis is one of such a process that is defined as the identification of the model structure and its corresponding parameter set that is most representative of the study area by taking into account research questions (Wagener et al. 2003). In many cases, identifiability analysis was used to estimate sequential parameters or to identify the discrepancy of the model behavior from the expected ecosystem behavior (Wagener et al. 2003; Matott et al. 2009). Sensitivity analysis is performed by varying systematically the model parameters or inputs and quantifying their relative effect on model predictions. Thus, it contributes to identifying how the uncertainty in the output of an ecological model can be attributed to uncertainty in its inputs and parameters (Saltelli et al. 2008). Overall, model evaluation is not a straightforward process over the course of model development, as such analyses (e.g., sensitivity and identifiability analysis) play an important role in developing credible and useful ecological models for decision-making purpose.

Different approaches are available to conduct uncertainty analysis, but two methods have been used most often in process-based models: Taylor series technique and Monte Carlo technique. Both techniques are described in Chapter 6.

8.5 CASE STUDIES: PREDICTION OF THE EFFECTS OF CLIMATE CHANGE, FIRE, AND MANAGEMENT

It is in the 1980s that the development of process-based models really began when the need to simulate the effects of climate change and physical and human disturbances, such as acid rain, air pollution, harvesting or wildfire, on the functioning of forest ecosystems. This trend accentuated in the 1990s. An important advantage of process-based models is that they can simulate multiple "what-if"" scenarios. It is especially true when novel climates are predicted.

Many aspects of projected climate change and altered disturbances will likely affect forest growth and productivity. Climate change could alter the frequency and intensity of forest disturbances, such as insect outbreaks, invasive species, wildfires, or storms. These disturbances can reduce forest productivity and change the distribution of tree species. In some cases, forests can recover from a disturbance. In other cases, existing species may shift their range or die out. In these cases, the new species of vegetation that colonize the area create a new type of forest (U.S. Climate Change Science Program 2008).

8.5.1 Climate Change

There is increasing evidence that climate change has had and will continue to have a significant effect on forest ecosystems by affecting the ecophysiological processes that govern their dynamics (Dresner et al. 2009; Lucier et al. 2009; Fischlin et al. 2009; Medlyn et al. 2011; Seidl et al. 2014). Forest models are key tools that have the potential to contribute to better understanding the long-term changes in the complex interactions between changing environmental conditions and ecosystem processes, allowing the prediction of the likely impact of climate change on forest productivity (Medlyn et al. 2011). There are good examples in the literature of studies that examined different scenarios of climate change on forest ecosystem dynamics.

The EFIMOD model was used to predict the long-term effects of different scenarios of climate change, fire, and harvesting on the carbon cycle of black spruce (*Picea mariana* (Mill.) B.S.P.) and jack pine (*Pinus banksiana* Lamb.) forests in Central Canada (Bhatti et al. 2009; Chertov et al. 2009). The scenarios of change in temperature and precipitations were estimated from three global circulation models: CGM2, HadCM3, and CSIRO. For black spruce, 150-year simulation results indicated that the disturbances resulted in a decrease in net primary productivity (NPP), stand productivity, soil organic matter content, and an increase in CO_2 emission to the atmosphere. Similar results were obtained for jack pine but differed among the three climate change scenarios. Bergh et al. (2003) used the model BIOMASS (McMurtrie et al. 1990) to predict the effect of both increase in temperature and atmospheric CO_2 concentration on NPP of Scots pine, Norway spruce, European beech, and black cottonwood (*Populus trichocarpa* Torr. & A. Gray) forests located in Denmark, Finland, Iceland, Norway, and Sweden. Simulation results indicated that both temperature and atmospheric CO_2 increase resulted in increasing NPP, but there was substantial variation among forest types.

Existing studies using process-based models to examine the effect of climate change have provided realistic results but have also generated many questions and uncertainties. For instance, Luckai and Larocque (2002) compared the long-term simulation results between CENTURY and FOREST-BGC by using identical scenarios of climate change in black spruce in Northwestern Ontario, Canada. Predictions between both models differed in the amount of carbon content in large wood (stems, coarse roots and branches) in the scenario of both temperature and CO_2 increase. While CENTURY predicted a decrease, FOREST-BGC predicted an increase. They attributed the discrepancies in the predictions to differences in the structures, algorithmic approaches, and assumptions between both models and a lack of empirical and scientific knowledge. Medlyn et al. (2011) recently highlighted these issues by suggesting specific questions to consider in modeling studies to improve the capacity of models to predict climatic impacts on forest productivity: the type of model under study, the representation of the effects of increase in atmospheric CO_2 and temperature, the representation of water availability and nutrient cycling, the potential influence of other global change effects, and the climate change scenarios used in the simulations.

Lousteau et al. (2005) used three process-based models, CASTANEA, GRAECO, and ORCHIDEE, to model the effects of climate change and two forest management scenarios on wood production and forest carbon balance in French forests. All models predicted a slight increase in potential forest yield until 2030–2050, followed by a plateau or a decline around 2070–2100, with overall, a greater increase in yield in northern France than in the south.

PnET is a suite of nested process-based models (PnET-Day, PnET-II, PnET-CN) that provide a modular approach to simulate the carbon, water, and nitrogen dynamics of forest ecosystems. They have been used to predict forest productivity under climate change in various forests. Aber et al. (1995) described an improved version of a forest carbon and water balance model (PnET-11) and the application of the model to predict stand- and regional-level effects of changes in temperature, precipitation, and atmospheric CO_2 concentration. They found that a combination of increased temperature (+6 C), decreased precipitation (−15%), and increased water-use efficiency (2x, due to doubling of CO_2) generally resulted in increase in NPP and decrease in water yield over regional levels.

Process-based models also are better able to deal with the enhanced productivity possible via elevated CO_2 and increased water-use efficiency. Peters et al. (2013) used PnET-CN to compare the long-term effects of changing climate and atmospheric CO_2 on productivity, evapotranspiration, runoff, and net nitrogen mineralization in current Great Lakes forest types. The results suggested that ecosystem responses varied among geographic locations and forest types. With rising CO_2 and without changes in forest type, average regional productivity could increase from 67% to 142%, changes in evapotranspiration could range from –3% to +6%, runoff could increase from 2% to 22%, and net nitrogen mineralization could increase by 10–12%.

Process-based models are used to predict the effects of climate change on various aspects of forest ecosystems. Bachelet et al. (2001b) used the dynamic global vegetation model (DVGM) MC1 to predict continental-scale changes in tree life forms and carbon in a warming climate. The warmer and drier scenarios in these experiments showed some potential for the habitats of the eastern deciduous and mixed forests to shift to a more open canopy woodland or savanna type while the boreal forests disappeared almost entirely from the Great Lakes area by the end of the 21st century (Bachelet et al. 2001b). These changes would obviously markedly modify the character of these biomes and greatly reduce carbon storage in the eastern United States. Other studies using DGVMs have shown potential large impacts of climate change in high latitudes, over the tropics on the African continent and globally.

Another process-based model that incorporates climate is the Regional Hydro-Ecologic Simulation System (RHESSys) (Tague and Band 2004). In a Sierra Nevada mountain system, Tague et al. (2009) found significant elevational differences in vegetation water use and sensitivity to climate, which will probably be critical for the vulnerability of similar ecosystems under climate change. In these model results, transpiration was consistent across years at the lowest elevations because of topographically controlled high moistures. The middle-elevation transpiration rates were controlled primarily by precipitation, and the high-elevation transpiration rates were controlled primarily by temperature. These dynamics, along with evidence for reduced snow accumulation and earlier melt of seasonal snowpacks, are expected to influence future species composition in mountainous systems. These studies generally provide evidence, as with the SDMs, that forest composition and productivity are likely to change, often substantially, as the climate changes during this century and beyond.

8.5.2 Impact of Forest Fires on Coniferous Forest Ecosystems

The EFIMOD model was also used to simulate the dynamics of coniferous forest ecosystem dynamics in West European taiga and North American boreal forests with different forest fire scenarios (Chertov et al. 2013). The simulation scenarios comprise canopy and ground fires in combination with clearcutting (Table 8.2). The simulation results showed that forest fires led to a decrease in growing stock and soil organic matter in both regions in comparison with a clearcutting scenario without fires. The long-term reduction of net biological productivity was also found with a negative carbon balance (net ecosystem exchange, NEE). A maximal simulated loss of carbon took place after crown and ground fires and after clearcutting. In Europe, the negative carbon balance with a loss of carbon from forest ecosystems was found only in a case of crown fires with burned wood cutting. In other cases, there was a slight amount of carbon sequestered even after fires. In North America, all fires led to negative carbon balances with the maximal carbon losses at crown fires with cutting burned wood. This effect was determined by the soil reaction on any forest fires: a carbon loss from burned forest floor and a decrease in soil carbon in young stands due to imbalance between carbon input with litter and carbon output with soil respiration. These differences reflect the influence of a rather mild climate of European sites in comparison with extra-continental cold climate of modeled sites in Central Canada.

8.5.3 Forest Management

Ecosystem models have played key roles by providing guidance on uncertain future or unexpected problems to decision-makers and information for public debate in the management of terrestrial ecosystems (Harris et al. 2003). As they focus on the modeling of processes and flows of nutrients, water, and energy in ecosystems, process-based models are well adapted to predict the impacts of environmental disturbances (see Mäkelä et al. 2000). For instance, Driscoll et al. (2003) used two

TABLE 8.2
Changes in Soil Carbon Pools and Growing Stock at the End of a Second 70-Year Rotation with One Thinning and Clearcutting Management Scenario and Average Net Ecosystem Exchange (NEE) over Two Rotations (From Chertov et al. 2013)

Forest Stands	Disturbance Regimes[a]	Changes of Soil C Pools, %	Changes of Growing Stock, %	Average NEE, kg[C] m^{-2} yr^{-1}
East European taiga				
Scots pine	1	0	0	0.053
Pinus sylvestris	2	**−31.80**	**−3.1**	0.054
	3	**−11.20**	**−13.8**	**−0.008**
	4	**−40.20**	**−9.9**	0.015
Norway spruce	1	0	0	0.024
Picea abies	2	**−23.30**	**−6.1**	0.022
	3	**−7.30**	**−8.5**	**−0.018**
	4	**−17.70**	**−6.7**	0.015
North American boreal forest				
Jack pine	1	0	0	0[b]
Pinus banksiana	2	**−34.40**	**−25.9**	**−0.034**
	3	**−18.70**	**−23.4**	**−0.053**
	4	**−27.80**	**−16.9**	**−0.004**
Black spruce	1	0	0	0.003
Picea mariana	2	**−43.30**	**−19.5**	**−0.002**
	3	**−24.60**	**−21.3**	**−0.034**
	4	**−35.60**	**−14.6**	**−0.011**

Negative values are in bold.
Notes:
[a] 1—two 70-year rotations and clearcutting without fires; 2—two canopy fires at the end of every 70-year rotations before cutting; 3—two ground fires after thinning at 30 year, 4—two ground fires after final clearcutting.
[b] NPP = Rh, that is: carbon of net primary productivity is equal to one of soil respiration.

process-based models, PnET-BGC and WATERSN, to examine the effects of different management options on the reduction of nitrogen deposition to forests and watersheds in the northeastern United States. Process-based models can also contribute to improving the modeling of the productivity of mixed-species or heterogeneous tropical forests and planning for the analysis of the impact of management scenarios on water balance and nutrient cycling (Mäkelä et al. 2000) or ecosystem services. Ooba et al. (2010) used the biogeochemical model BGC-ES to evaluate the impacts of different management practices on ecosystem services, including water supply, carbon sequestration, timber production, and biodiversity conservation. The predictions of the different scenarios were associated with economic values. The next two subsections are case studies that illustrate the application of process-based modeling to address management issues.

8.5.3.1 Modeling the Effects of Rotation Length on Forest Productivity and Soil

The aim of this simulation using EFIMOD was to demonstrate the effect of various silvicultural regimes on carbon balance and wood production on forest ecosystems. The experimental data on soil and climate for a very productive Norway spruce ecosystem on Distric Cambisols in Höglwald, Germany, were used as initial parameters for stand, soil, and climate (Kreutzer and Göttlein 1991). The site had mild temperate climate with mean annual temperature of +8.2°C and total annual precipitation of 811 mm. The simulations were conducted between 1961 and 2050 with scenarios of

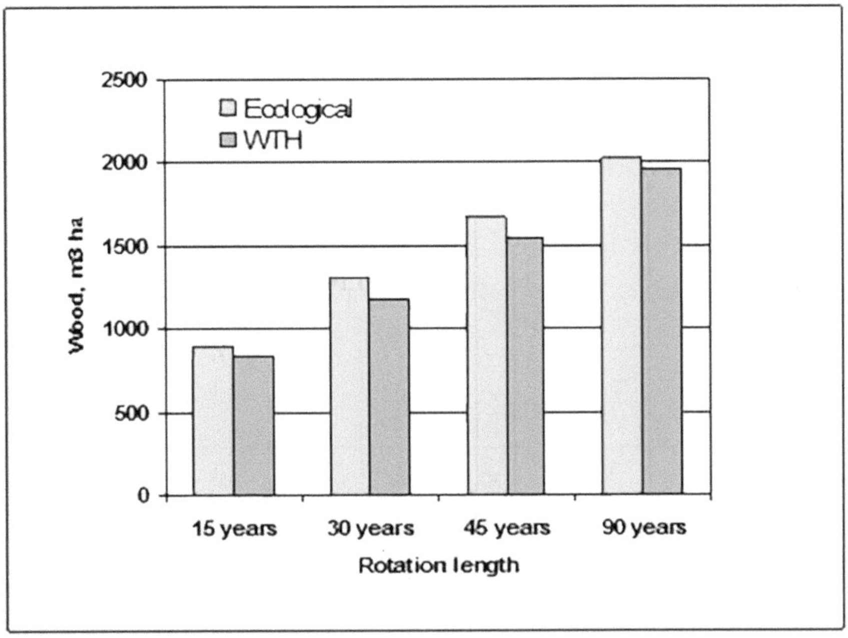

FIGURE 8.4 Accumulated harvested wood over a 90-year time span with different rotation lengths and two different clearcutting scenarios: ecological with the exposure of cutting residues for decomposition in the forest and whole tree harvesting (WTH) with extraction of all aboveground tree biomass from ecosystems.

increasing temperature and high atmospheric nitrogen deposition. Initial stand density was 10,000 3-year-old seedlings per hectare. The rotation lengths simulated were 15, 30, 45, and 90 years. There was regular thinning at 35, 45, 55, and 65 years in the 90-year rotation scenario, and one thinning in the 45-year one. Two types of cuttings were simulated: ecological (EC) with the remaining cutting residues on the harvested area for decomposition and whole tree harvesting (WTH) with the removing of all the aboveground biomass from the ecosystem.

The results demonstrated the productivity potential of the clearcutting regimes with different rotation lengths (Figure 8.4). They indicated that the shortest rotation had twice lower accumulated wood productivity over a long-term time interval. One can also see that the WTH had a little bit lower productivity in comparison with the "ecological" wood harvesting. These results can be seen in relation to soil carbon dynamics as well (Figure 8.5). There was no negative soil organic carbon effect following the ecological cutting. While the soil carbon slightly decreased at 15-years rotation with WTH, there is no soil carbon loss at 45- and 90-year rotations. These results clearly showed that the short rotation system was not an ecologically and economically optimal forestry practice without site amendments. It can be concluded that the methodology of this simulation can be used to evaluate in more depth carbon sequestration potentials of different wood harvesting methods.

8.5.3.2 Combination of Silvicultural Regimes

EFIMOD can be used to demonstrate the possibilities of forestry regime combination at the forest enterprise/regional level to reach s satisfactory balance between carbon budget and forest productivity from an environmental and economic point of view. The case study is an area of forest unit in central Eastern Europe 100 km south from Moscow with Scots pine and mixed stands of average ages of 30–70 years that had high productivity on sandy mesic forest sites on the river Oka terraces. The modeling of four management regimes was performed for every stand in the selected forest lots to compare ecological and silvicultural consequences of different management regimes (Chertov et al. 2005; Komarov et al. 2007, Figure 8.6). It was assumed that the initial structure of

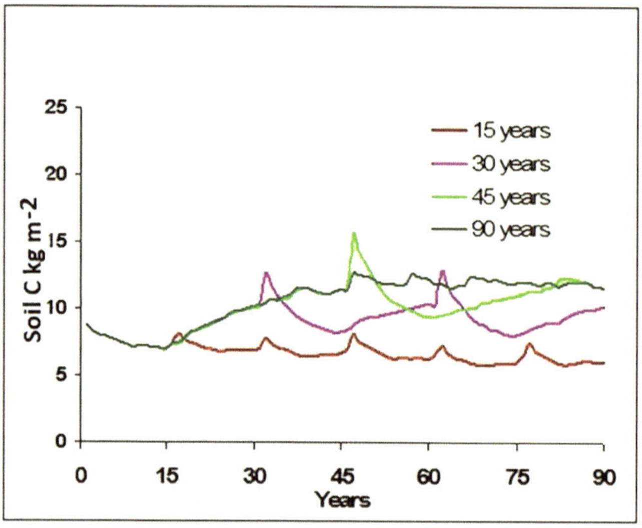

FIGURE 8.5 Predicted soil carbon dynamics for different rotation lengths and a scenario of whole tree harvesting (WTH).

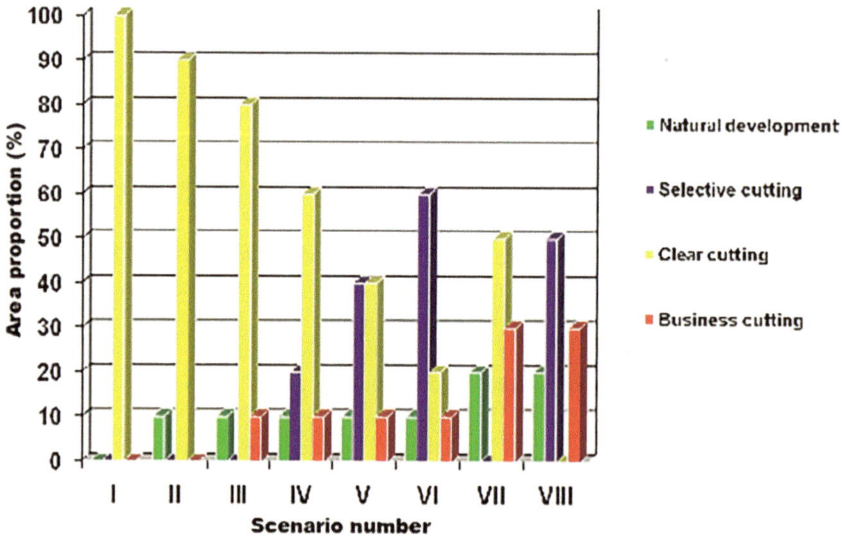

FIGURE 8.6 Scenarios based on the combinations of proportions of silvicultural regimes (natural development, selective cutting, clearcutting, and business cutting) based on the percentages of total areas in forest units.

forest sites, stand composition, and age structure were identical for all simulated forest regimes. The total simulated area corresponded to the area of the Dankovsky management unit (7,000 ha). It was also assumed that the scenario of "business cutting" corresponded to "legislation breach" with the following principle "take the best just now." Some preliminary model runs showed that this regime was very close to the short rotation forest plantations demonstrating the negative carbon balance on the territory. The scenarios represented a consecutive series from "Russian legislative clear cut system" (scenario I) to scenarios with decreased ratio of clearcutting and finally to combinations of natural development and "business cutting" with clear and selective cutting (scenarios VII and VIII). Relations between different silvicultural operations are presented in Figure 8.6.

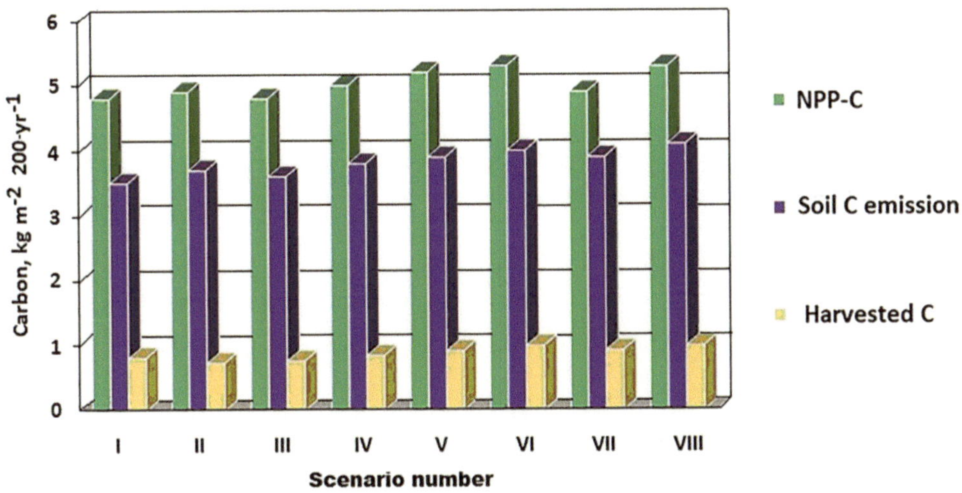

FIGURE 8.7 Components of carbon balance for different combinations of silvicultural scenarios (see Figure 8.6) averaged for the entire territory over 200-year simulations. Carbon emission from soil includes also carbon loss with burning cutting residues.

The parameters of carbon balance from the basic simulations were recalculated for every scenario as weighted averages proportionally to the percentage of every regime area:

$$P_{comb} = P_1A_1 + P_2A_2 + P_3A_3 + P_4A_4/100, \tag{8.12}$$

where P_{comb} is a calculation parameter of carbon balance or productivity at the combination of forestry regimes, P_1–P_4 are these parameters for every separate regime and A_1–A_4 are the areas of every regime in the modeled scenario.

The results of this exercise (Figures 8.7 and 8.8) showed a consistent increase in net primary productivity and soil carbon dioxide emissions as soon as the proportion of clearcutting decreased on the territory (Figure 8.7). A small admixture of naturally developed forests and "business cutting" did not change this trend. The maximal values of NPP and soil carbon dioxide emission occurred for scenarios with a domination of selective cutting with 20% clearcutting and 10% natural forest and "business cutting" (scenario VI). This scenario also demonstrated a maximum carbon production of harvested wood: 1.05 ton [carbon] ha^{-1} $year^{-1}$ in average for 200-year simulations that correspond approximately to 4.5 m^3 ha^{-1} of wood annually. It was found unexpectedly that two regimes with 30% of "business cutting" and 20% of naturally developed forest (scenarios VII and VIII) demonstrated rather good parameters of carbon balance and forest productivity. Moreover, scenario VIII with combination of selective and "business cutting" had just the same capacity of harvested wood carbon as scenario VI with domination of selective cuttings.

These results were confirmed by the patterns of net ecosystem exchange (NEE) for different combinations of silvicultural regimes (Table 8.3). The table shows that all combinations with some ratios of naturally developing forests had a positive carbon budget. Again, the maximum carbon sequestration in the territory for this 200-year simulation was observed for the combination of natural development with selective and "business cutting" (scenario VIII).

Figure 8.8 shows carbon pools in forest biomass and soil organic matter at the end of the simulations and exhibits similar dynamic trends as previous ones: the accumulated carbon in stands and soils increased as soon as clearcutting reduction took place. However, the largest pools were found for scenario VIII and the lowest pools for clearcutting scenario I. In addition, similar work was

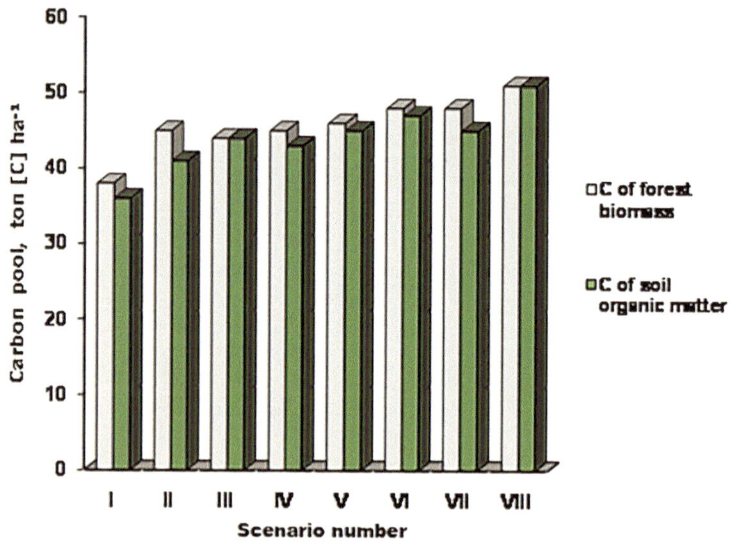

FIGURE 8.8 Amounts of carbon in forest biomass and soil organic matter averaged for the entire territory over 200-year simulations.

TABLE 8.3

Components of Carbon Budget (Ton [C]) Averaged over 200-Year Simulations for an Entire Forest Unit in Central Eastern Europe Composed of Scots Pine and Mixed Stands with Combinations of Different Silvicultural Proportions of Management Regimes

Proportion of the Area of Natural Development/Selective Cutting/Clearcutting/Plantation Forests, %	NPP[a]	Soil C Respiration (Emission)	C Removed with Cuttings	NEE[b]	Tree Biomass C[c]	Soil C[c]
	ton [C] ha^{-1} year^{-1}				ton [C] ha^{-1}	
0/0/100/0	4.75	3.48	0.81	0.000	38	36
10/0/90/0	4.91	3.67	0.73	0.099	45	41
10/0/80/10	4.85	3.61	0.76	0.074	44	44
10/20/60/10	5.01	3.75	0.84	0.112	45	43
10/40/40/10	5.17	3.89	0.91	0.150	46	45
10/60/20/10	5.33	4.03	0.99	0.188	48	47

Notes:

[a] NPP, C of a net primary productivity, increment of total biomass (leaves, branches, bole, roots);

[b] NEE, net ecosystem exchange, a measure of carbon balance:

NEE_C = NPP_C − (Soil_respiration_C + Cutting_C + Deadwood_decomposition_C + Disturbance_loss_C); deadwood decomposition and disturbance losses did not reflect in the table;

[c] Average values of stand, soil (with deadwood) C for the entire forest area at the end of 200-year simulation.

conducted for suburban forests in dense populated areas and recreational management (Table 8.4). Recreational scenario means a modest selective cutting to increase a scenic beauty of landscapes in combination with creation of recreational infrastructure. In this case, the results give also a possibility for the selection of optimal combination of regimes taking into account both recreational and productive functions of suburban forests. The fourth and fifth scenarios seem to be optimal from a carbon balance point of view, allowing also to get some wood from forest plantations. Though, it is disputable that plantations represent a good option for this area.

TABLE 8.4
Components of Carbon Budget (Ton [C] year^{-1} ha^{-1}) Averaged over 200-Year Simulations for an Entire Forest Unit in Central Eastern Europe Composed of Scots Pine and Mixed Stands with Combinations of Different Silvicultural Regimes in a Dense Populated and Industrial Region

Proportion of the Area of Natural Development/ Recreational/ Plantation Forests, %	NPP[a]	Soil C Respiration (Emission)	C Removed with Cuttings	NEE[b]	Tree Biomass C[c]	Soil C[c]
	ton [C] ha^{-1} year^{-1}				ton [C] ha^{-1}	
0/100/0	5.54	4.17	1.18	0.190	45	47
0/90/10	5.40	4.04	1.20	0.146	43	46
10/80/10	5.48	4.16	1.06	0.226	49	49
30/60/10	5.65	4.41	0.84	0.386	62	57
50/40/10	5.82	4.65	0.61	0.546	75	64
100/0/0	6.37	5.38	0.00	0.990	110	83

Note:

[a] NPP, C of a net primary productivity, increment of total biomass (leaves, branches, bole, roots);

[b] NEE, net ecosystem exchange, a measure of carbon balance:
NEE_C = NPP_C – (Soil_respiration_C + Cutting_C + Deadwood_decomposition_C + Disturbance_loss_C); deadwood decomposition and disturbance losses did not reflect in the table;

[c] Average values of stand, soil (with deadwood) C for the entire forest area at the end of 200-year simulation.

We can conclude that this is an idealized picture because the productivity and stand characteristics can be in reality different combinations of forestry practices. For example, the best productive sites can be used for "business cutting" and forests of lower productivity for other regimes.

These case studies demonstrate an exploratory potential of the forest ecosystems simulation for the investigation of carbon budget, wood productivity, and even biodiversity of different silvicultural regimes. These simulations demonstrate an effectiveness of the "offset areas" idea (Knaus et al. 2006): making of some areas with a positive effect of any management action to compensate negative effects of the other ones. In our case, it is the regimes of "Natural development" versus "Business cutting" in relation to carbon balance of forest territory at a long-term prospect. The other conclusion can be as follows: the larger the diversity of management regimes, the more optimal are parameters of carbon balance and forest productivity, but it needs additional verification. Nevertheless, it is clear that forest model could be an effective science-based forest planning instrument for an optimization of forest area splitting by management regimes to harmonize conflicting environmental and economic criteria of sustainable forest management.

There are other examples in the literature on the application of process-based models to predict the effects of human disturbances, such as harvesting forest ecosystems. Thompson et al. (2006) used the Landscape Management Policy Simulator (LAMPS) to project the effects of several forest policies, including disturbance-based policies. Simulated disturbance-based policies produced age-class distributions closer to the estimated historical range than those created by the current policy. Within 100 years, the proportions of younger forests were within the historical range while older forests moved closer to, but remained below, historical conditions. In the near term, disturbance-based policies produced annual harvest volumes 20– 60% lower than those produced by the current policy. However, relative costs of disturbance-based policies diminished over time. The results suggest that if expediting a return to historical age-class distributions at a provincial-scale was a goal, then public lands would be needed to provide large patches of old forests. In addition, this experiment illustrated that distributing costs and benefits of conservation policies equitably across multiple private landowners is a significant challenge.

8.5.4 Limitations and Challenges for Practical Implementation

Process-based models have limitations in that they require (1) detailed parameterization of life histories and physiologies, (2) capacity to capture the complexity of many interacting disturbance factors, and (3) high-resolution modeling over very large areas. A limitation that affects the precision and realism of many process-based models is that they do not cover the full range of forest dynamics processes. Some of the important aspects that models often lack include the effects of climatic conditions, disturbance, and site quality. In addition to model complexity, difficulties associated with their implementation, accessibility to user guidance, and degree of relevancy to answer questions important for decision-makers are other factors that explain their marginal application to forest management planning (Sands et al. 2000). Besides these logistical limitations, there are intrinsic problems with controlling cumulative errors associated with the sheer number and diverse spatiotemporal scales of parameters, most of which ultimately being derived from a limited number of empirical studies.

Continued research is needed to move the science forward on detecting and modeling and potential assisting of actual and potential forest change in a changing climate and disturbance. Understanding the effect of individual assumptions allows the identification of major uncertainties in the models and working toward improving them. Process-based models also are better able to deal with the enhanced productivity possible via elevated CO_2 and increased water-use efficiency. An assumption-centered approach to modeling could improve our understanding of the effects of rising CO_2, temperature, and water availability on forest function.

Much of this uncertainty can be incorporated in reporting projections. Medlyn et al. (2011) reviewed several process-based models that simulate forest productivity under climate change and provided a checklist for evaluating modeling studies and argued that models and experimental research should be closely integrated. Long-term, intensively studied experiments are needed to generate sufficient data to test alternative model hypotheses. Many additional advances will be possible by making continued improvements in the integration among the avenues here. The merging of process-based models, demography models, and species distribution models allows for some of the best attributes of each.

In summary, process-based models of forest ecosystems are widely used to predict the likely future impacts of environmental change and disturbance. These models are valuable tools for predicting the effect of climate change and disturbances on forest ecosystems and have the potential to address issues related to the basic principles of ecosystem management and provide sound predictions of the dynamics of ecosystems when changes in environmental conditions are unprecedented. This is the case of climate change or any external impacts such as large disturbance events.

NOTES

1 Note: This chapter was not modified. It is the original version published in the first edition of the handbook.
2 http://journals.ametsoc.org/doi/abs/10.1175/JCLI-D-12-00565.1
3 https://gcims.pnnl.gov/modeling/gcam-global-change-analysis-model
4 https://www.ncasi.org/resource/carbon-on-line-estimator-cole/

REFERENCES

Aber, J.D. and C.T. Driscoll. 1997. Effects of land use, climate variation, and N deposition on N cycling and C storage in northern hardwood forests. *Global Biogeochem. Cycles* 11:639–648.

Aber, J.D. and C.A. Federer. 1992. A generalized, lumped-parameter model of photosynthesis, evapotranspiration and net primary production in temperate and boreal forest ecosystems. *Oecologia* 92:463–474.

Aber, J.D., S.V. Ollinger, and C.T. Driscoll. 1997. Modeling nitrogen saturation in forest ecosystems in response to land use and atmospheric deposition. *Ecol. Modell.* 101:61–78.

Aber, J.D., S.V. Ollinger, C.A. Federer, P.B. Reich, M.L. Goulden, D.W. Kicklighter, J.M. Melillo, and R.G. Lathrop, Jr., 1995. Predicting the effects of climate change on water yield and forest production in Northeastern U.S. *Clim. Res.* 5:207–222.

Akselsson, C., O. Westling, H. Sverdrup, and P. Gundersen. 2007. Nutrient and carbon budgets in forest soils as decision support in sustainable forest management. *For. Ecol. Manage.* 238:167–174.

Araujo, M.B. and M. New. 2007. Ensemble forecasting of species distributions. *Trends Ecol. Evol.* 22:42–47.

Bachelet, D., J.M. Lenihan, C. Daly, R.P. Neilson, D.S. Ojima and W.J. Parton. 2001a. *MC1. A dynamic vegetation model for estimating the distribution of vegetation and associated ecosystem fluxes of carbon, nutrients and water.* Fort Collins: USDA Forest Service, Gen. Tech. Rep. PNW-GTR-508.

Bachelet, D., R.P. Neilson, J.M. Lenihan, and R.J. Drapek. 2001b. Climate change effects on vegetation distribution and carbon budget in the U.S. *Ecosystems* 4:164–185.

Balshi MS, A.D. McGuire, Q. Zhuang, J. Melillo, D.W. Kicklighter, E. Kasischke, C. Wirth, M. Flannigan, J. Harden, J.S. Clein, T.J. Burnside, J. McAllister, W.A. Kurz, M. Apps, and A. Shvidenko. 2007. The role of historical fire disturbance in the carbon dynamics of the pan-boreal region: A process-based analysis. *J. Geophys. Res.* 112:G02029.

Battaglia, M. and P.J. Sands. 1998. Process-based forest productivity models and their application in forest management. *For. Ecol. Manage.* 102:13–32.

Bergh, J., M. Freeman, B. Sigurdsson, S. Kellomäki, K. Laitinen, S. Niinistö, H. Peltola, and S. Linder. 2003. Modelling the short-term effects of climate change on the productivity of selected tree species in Nordic countries. *For. Ecol. Manage.* 183:327–340.

Bergh, J., R.E. McMurtrie, and S. Linder. 1998. Climatic factors controlling the productivity of Norway spruce: A model-based analysis. *For. Ecol. Manage.* 110:127–139.

Bhatti, J., O. Chertov, and A. Komarov. 2009. Influence of climate change, fire, insect and harvest on C dynamics for jack pine in central Canada: Simulation approach with the EFIMOD model. *Int. J. Clim. Change: Impacts and Responses* 1:43–61.

Blanco, J.A. 2012. Forests may need centuries to recover their original productivity after continuous intensive management: An example from Douglas-fir stands. *Sci. Total Environ.* 437:91–103.

Bond-Lamberty, B., J. Fisk, J.A. Holm, V. Bailey, and C.M. Gough. 2014. Moderate forest disturbance as a stringent test for gap and big-leaf models. *Biogeosci. Discuss.* 11:11217–11248.

Bossel, H. 1996. TREEDYN3 forest simulation model. *Ecol. Modell.* 90:187–227.

Bossel, H. and H. Schäfer. 1989. Generic simulation model of forest growth, carbon and nitrogen dynamics, and application to tropical acacia and European spruce. *Ecol. Modell.* 48:221–265.

Botkin, D.B., J.F. Janak, and J.R. Wallis. 1972. Some ecological consequences of a computer model of forest growth. *J. Ecol.* 60:849–872.

Boyle, J.R. and R.F. Powers, eds. 2001. *Forest Soils and Ecosystem Sustainability.* Amsterdam: Elsevier.

Bruun, S., B.T. Christensen, E.M. Hansen, J. Magid, and L.S. Jensen. 2003. Calibration and validation of the soil organic matter dynamics of the Daisy model with data from the Askov long-term experiments. *Soil Biol. Biochem.* 35:67–76.

Byrne, G.F., J.J. Landsberg, and M.L. Benson. 1986. The relationship of aboveground dry matter accumulation by *Pinus radiata* to intercepted solar radiation and soil water status. *Agric. For. Meteor.* 37:63–73.

Calama, R., J. Puértolas, G. Madrigal, and M. Pardos. 2013. Modeling the environmental response of leaf net photosynthesis in *Pinus pinea* L. natural regeneration. *Ecol. Modell.* 251:9–21.

Ceulemans, R. and M. Mousseau. 1994. Effects of elevated atmospheric CO_2 on woody plants. *New Phytol.* 127:435–446.

Chen, J.M., J. Liu, J. Cihlar, and M.L. Goulden. 1999. Daily canopy photosynthesis model through temporal and spatial scaling for remote sensing applications. *Ecol. Modell.* 1124:99–119.

Chertov, O.G., J. Bhatti, A. Komarov, A. Mikhailov, and S. Bykhovets. 2009. Influence of climate change, fire and harvest on the carbon dynamics of black spruce in Central Canada. *For. Ecol. Manage.* 257:941–950.

Chertov, O.G. and A.S. Komarov. 1997. SOMM: A model of soil organic matter dynamics. *Ecol. Modell.* 94:177–189.

Chertov, O.G., A.S. Komarov, S.S. Bykhovets, and K.I. Kobak. 2002. Simulated soil organic matter dynamics in forests of the Leningrad administrative area, Northwestern Russia. *For. Ecol. Manage.* 169:29–44.

Chertov, O.G., A.S. Komarov, A.V. Gryazkin, A.P. Smirnov, and J.S. Bhatti. 2013. Simulation modeling of the impact of forest fire on the carbon pools in coniferous forests of European Russia and Central Canada. *Contemp. Probl. Ecol.* 6:747–753.

Chertov, O.G., A. Komarov, A. Mikhailov, G. Andrienko, N. Andrienko, and P. Gatalsky. 2005. Geovisualisation of forest simulation modelling results: A case study of carbon sequestration and biodiversity. *Comp. Electr. Agric.* 49:175–191.

Chertov, O.G., A.S. Komarov, M. Nadporozhskaya, S.S. Bykhovets, and S.L. Zudin. 2001. ROMUL—a model of forest organic matter dynamics as a substantial tool for forest ecosystem modeling. *Ecol. Modell.* 138:289–308.

Chiesi, M., R. Maselli, M.M. Bindi, L. Fibbi, L. Bonora, A. Raschi, R. Tognetti, J. Cermak, and N. Nadezhdina. 2002. Calibration and application of FOREST-BGC in a Mediterranean area by the use of conventional and remote sensing data. *Ecol. Modell.* 154:251–262.

Chiesi, M., F. Maselli, M. Moriondo, L. Fibbi, M. Bindi, and S.W. Running. 2007. Application of BIOME-BGC to simulate Mediterranean forest processes. *Ecol. Modell.* 206:179–190.

Chuine, I. and E. Beaubien. 2001. Phenology is a major determinant of temperate tree range. *Ecol. Lett.* 4:500–510.

Coleman, K. and D.S. Jenkinson. 1995. ROTHC-26.3: A Model for the Turnover of Carbon in Soil. Harpenden: IACR Rothamsted.

Comins, H.N. and R.E. McMurtrie. 1993. Long-term response of nutrient-limited forests to CO_2 enrichment—equilibrium behavior of plant-soil models. *Ecol. Appl.* 3:666–681.

Cramer, W., A. Bondeau, F.I. Woodward, I.C. Prentice, R.A. Betts, V. Brovkin, P.M. Cox, V. Fisher, J.A. Foley, A.D. Friend, C. Kucharik, M.R. Lomas, N. Ramankutty, S. Sitch, B. Smith, A. White, and C. Young-Molling. 2001. Global response of terrestrial ecosystem structure and function to CO2 and cli mate change: Results from six dynamic global vegetation models, *Glob. Change Biol.* 7:357–373.

Dale, V.H. and S.C. Beyeler. 2001. Challenges in the development and use of ecological indicators. *Ecol. Indic.* 1:3–10.

Dixon, R.K. 1990. Physiological processes and tree growth. *In Process modeling of forest growth responses to environmental stress*, ed. R.K. Nixon, R.S. Meldahl, G.A. Ruark, and W.G. Warren, 21–32. Portland: Timber Press.

Dresner, S., P. Ekins, K. McGeevor, and J. Tomei. 2009. Forests and climate change: Global understanding and possible responses, In *Forestry and climate change*, eds. P.H. Freer-Smith, M.S.J. Broadmeadow and J.M. Lynch, pp. 38–48. Wallingford: CABI.

Driscoll, C.T., D. Whitall, J. Aber, E. Boyer, M. Castro, C. Cronan, C.L. Goodale, P. Groffman, C. Hopkinson, K. Lambert, G. Lawrence, and S. Ollinger. 2003. Nitrogen pollution in the northeastern United States: Sources, effects, and management options. *Biosci.* 53:357–374.

Duusrma, R.A. and A. Mäkelä. 2007. Summary models for light interception and light-use efficiency of non-homogeneous canopies. *Tree Physiol.* 27:859–870.

Evans, J.R. and H. Poorter. 2001. Photosynthetic acclimation of plants to growth irradiance: The relative importance of specific leaf area and nitrogen partitioning in maximizing carbon gain. *Plant, Cell Environ.* 24:755–767.

Failing, L. and R. Gregory. 2003. Ten common mistakes in designing biodiversity indicators for forest policy. *J. Environ. Manage.* 68:121–132.

Farquhar, G.D. and S. von Caemmerer, 1982. Modelling of photosynthetic response to environmental conditions. In *Physiological Plant Ecology II. Water relations and carbon assimilation*, eds. O.L. Lange, P.S. Nobel, C.B. Osmond and H. Ziegler, 549–599. Berlin: Springer-Verlag.

Farquhar, G.D., S. von Caemmerer, and J.A. Berry. 1980. A biochemical model of photosynthetic CO_2 assimilation in leaves of C3 species. *Planta* 149:78–90.

Fischlin, A., M. Ayres, D. Karnosky, S. Kellomaki, B. Louman, C. Ong, G.-K. Plattner, H. Santoso and I. Thompson. 2009. Future environmental impacts and vulnerabilities. In *Forest responses and vulnerabilities to recent climate change: A global assessment report*, ed. R. Seppälä, A. Buck, P. Katila, 53–100. Helsinki: IUFRO World Series.

Foley, J.A., I.C. Prentice, N. Ramankutty, S. Levis, D. Pollard, S. Sitch, and A. Haxeltine. 1996. An integrated biosphere model of land surface process, terrestrial carbon balance and vegetation dynamics. *Global Biogeoche. Cycles* 10:603–628.

Franko, U., B. Oelschlael, and S. Schenk. 1995. Simulation of temperature, water and nitrogen dynamics using the model CANDY. *Ecol. Modell.* 81:213–222.

Friend, A.D. 2001. Modelling canopy CO2 fluxes: Are 'big-leaf' simplifications justified? *Glob. Ecol. Biogeogr.* 10:603–619.

Friend, A.D., A.K. Stevens, R.G. Knox, and M.G.R. Cannell. 1997. A process-based, terrestrial biosphere model of ecosystem dynamics (Hybrid v3.0). *Ecol. Modell.* 95:249–287.

Friend, A.D. and A. White. 2000. Evaluation and analysis of a dynamic terrestrial ecosystem model under pre-industrial conditions at the global scale. *Global Biogeochem. Cycles* 14:1173–1190.

Hansen, S., H.E. Jensen, N.E. Nielsen, and H. Svendsen. 1991. Simulation of nitrogen dynamics and biomass production in winter wheat using the Danish simulation model DAISY. *Fertil. Res.* 27:245–259.

Harris, G.P., S.W. Bigelow, J.J. Cole, H. Cyr, L.L. Janus, A.P. Kinzig, J.f. Kitchell, g.E. Likens, K.H. Reckhow, D. Scavia, D. Soto, L.M. Talbot and P.M. Templer. 2003. The role of models in ecosystem management. In *Models in ecosystem science*, ed. C.D. Graham, J.J. Cole and W.K. Lauenroth, 299–307. Princeton: Princeton University Press.

Hoff, C., S. Rambal, and R. Joffre. 2002. Simulating carbon and water flows and growth in a Mediterranean evergreen *Quercus ilex* coppice using the FOREST-BGC model. *For. Ecol. Manage.* 164: 121–136.

Jakeman, A.J., R.A. Letcher, and J. Norton. 2006. Ten iterative steps in development and evaluation of environmental models. *Environ. Model. Softw.* 21:602–614.

Johnsen, K., L. Samuelson, R. Teskey, S. McNulty, and T. Fox. 2001. Process models as tools in forestry research and management. *For. Sci.* 47:2–8.

Johnson, I.R. and J.H.M. Thornley. 1984. A model of instantaneous and daily canopy photosynthesis. *J. Theor. Biol.* 107:531–545.

Johnson, I.R., J.H.M. Thornley, J.M. Frantz, and B. Bugbee, 2010. A model of canopy photosynthesis incorporating protein distribution through the canopy and its acclimation to light, temperature and CO_2. *Ann. Bot.* 106:735–749.

Jung, M., M. Vetter, M. Herold, G. Churkina, M. Reichstein, S. Zaehle, P. Ciais, N. Viovy, A. Bondeau, Y. Chen, K. Trusilova, F. Feser, and M. Heimann. 2007. Uncertainties of modeling gross primary productivity over Europe: A systematic study on the effects of using different drivers and terrestrial biosphere models. *Global Biogeochem. Cycles* 21:GB4021, 1–23.

Kaduk, J. and M. Heimann. 1996. A prognostic phenology scheme for global terrestrial carbon cycle models. *Climate Research* 6:1–19.

Kalinina, O., O. Chertov, A.V. Dogikh, S.V. Goryachkin, D.I. Lyuri, S. Vormstein, and L. Giani. 2013. Self-restoration of post-agrogenic Albeluvisols: Soil development, carbon stocks and dynamics of carbon pools. *Geoderma* 207–208:221–233.

Kearney, M. and W.P. Porter. 2009. Mechanistic niche modelling: Combining physiological and spatial data to predict species' ranges. *Ecol. Lett.* 12:334–350.

Kimmins, J.P. 1993. *Scientific foundations for the simulation of ecosystem function and management in FORCYTE11*. Edmonton: FCNFC. Inf. Rep. NOR-X-328.

Kimmins, J.P. 1996. Importance of soil and role of ecosystem disturbance for sustained productivity of cool temperate and boreal forest. *Soil Sci. Soc. Am. J.* 60:1643–1654.

Kimmins, J.P., D. Mailly, and B. Seely. 1999. Modelling forest ecosystem net primary production: The hybrid simulation approach used in forecast. *Ecol. Modell.* 122:195–224.

Kirchner, J.W., R.P. Hooper, C. Kendall, C. Neal, and G. Leavesley. 1996. Testing and validating environmental models. *Sci. Total. Environ.* 183:33–47.

Kleidon, A. and H.A. Mooney. 2000. A global distribution of biodiversity inferred from climatic constraints: Results from a process-based modelling study. *Glob. Chang. Biol.* 6:507–523.

Knaus, M., D. Löhr, and B. O'Regan. 2006. Valuation of ecological impacts—a regional approach using the ecological footprint concept. *Environ. Impact Assess. Rev.* 26:156–169.

Knorr, W. and M. Heimann. 1995. Impact of drought stress and other factors on seasonal land biosphere CO_2 exchange studied through an atmospheric tracer transport model. *Tellus* 47B:471 489.

Komarov, A., O. Chertov, S. Zudin, M. Nadporozhskaya, A. Mikhailov, S. Bykhovets, E. Zudina. and E. Zoubkova. 2003. EFIMOD 2—a model of growth and elements cycling in boreal forest ecosystems. *Ecol. Modell.* 170:373–392.

Komarov, A.S., O.G. Chertov, M.A. Nadporozhskaya, A.V. Mikhailov, S.S. Bykhovets, S.L. Zudin, I.V. Priputina, A.A. Larionova, P.Ya. Grabarnik, E.V. Zubkova, E.V. Zudina, V.N. Shanin, G. Andrienko, N. Andrienko, A.V. Martynkin, G.M.J. Mohren, E.V. Abakumov, A.M. Lukjanov, T.S. Kubasova, J. Bhatti, S. Shaw, M. Apps, M.V. Bobrovsky, L.G. Khanina, V.E. Smirnov, E.M. Glukhova. 2007. *Modelling of Organic Matter Dynamics in Forest Ecosystems (in Russian)*. Moscow: Nauka Publ.

Korol, R.L., K.S. Miller, and S.W. Running. 1995a. Testing a mechanistic model for predicting stand and tree growth. *For. Sci.* 42:139–153.

Korol, R.L., S.W. Running, and K.S. Miller. 1995b. Incorporating intertree competition into an ecosystem model. *Can. J. For. Sci.* 25:413–424.

Korzukhin, M.D., M.-T. Ter-Mikaelian, and R.G. Wagner. 1996. Process versus empirical models: Which approach for forest ecosystem management? *Can. J. For. Res.* 26:879–887.

Kreutzer, K. and A. Göttlein. 1991. *Ökosystemforschung Höglwald*. Forstwissenschaftliche Forschungen 39. Berlin: Parey.

Landsberg, J. 2003a. Physiology in forest models: History and tree future. *FBMIS* 1:49–63.

Landsberg, J. 2003b. Modelling forest ecosystems: State of the art, challenges, and future directions. *Can. J. For. Res.* 33:385–397.

Landsberg, J. and P. Sands, 2011. Modelling tree growth: Concepts and review. In *Physiological ecology of forest production*, eds. J. Landsberg and P. Sands, 221–240. London: Elsevier Inc.

Landsberg, J.J. and R.H. Waring. 1997. A generalised model of forest productivity using simplified concepts of radiation-use efficiency, carbon balance and partitioning. *For. Ecol. Manage.* 95:209–228.

Landsberg, J., R.H. Waring, and N.C. Coops. 2003. Performance of the forest productivity model 3-PG applied to a wide range of forest types. *For. Ecol. Manage.* 172:199–214.

Larocque, G.R. 2002. Coupling a detailed photosynthetic model with foliage distribution and light attenuation functions to compute daily gross photosynthesis in sugar maple (*Acer saccharum* Marsh.) stands. *Ecol. Modell.* 148:213–232.

Larocque, G.R., J.S. Bhatti, J.C. Ascough II, J. Liu, N. Luckai, D. Mailly, L. Archambault, and A.M. Gordon. 2011. An analytical framework to assist decision makers in the use of forest ecosystem model predictions. *Environ. Model. Softw.* 26:280–288.

Larocque, G.R., J.S. Bhatti, R. Boutin, and O. Chertov. 2008b. Uncertainty analysis in carbon cycle models of forest ecosystems: Research needs and development of a theoretical framework to estimate error propagation. *Ecol. Modell.* 219:400–412.

Larocque, G.R., J.S. Bhatti, A.M. Gordon, N. Luckai, M. Wattenbach, J. Liu, C. Peng, P.A. Arp, S. Liu, C.-F Zhang, A. Komarov, P. Grabarnik, J. and T. White, 2008a. Uncertainty and sensitivity issues in process-based models of carbon and nitrogen cycles in terrestrial ecosystems. In *Environmental Modelling, Software and Decision Support. State of the art and new perspectives. Developments in Integrated Environmental Assessment, vol. 3*, ed. A.J. Jakeman, A.A. Voinov, A.E. Rizzoli and S. Chen, 307–327. Amsterdam, Elsevier IDEA Series.

Lasch, P., F.-W. Badeck, F. Suckow, M. Lindner, and P. Mohr. 2005. Model-based analysis of management alternatives at stand and regional level in Brandenburg (Germany). *For. Ecol. Manage.* 207:59–74.

Le Roux, X., A. Lacointe, A. Escobar-Gutiérrez, and S. Le Dizès. 2001. Carbon-based models of individual tree growth: A critical appraisal. *Ann. For. Sci.* 58:469–506.

Leuning, R., R.M. Kelliher, D.G.G. DePury, and E.-D. Schulze. 1995. Leaf nitrogen, photosynthesis, conductance and transpiration: Scaling from leaves to canopies. *Plant Cell Environ.* 18:1183–1200.

Lindenmayer, D.B., C.R. Margules, and D.B. Botkin. 2000. Indicators of biodiversity for ecologically sustainable forest management. *Conserv. Biol.* 14:941–950.

Linkosalo, T., P. Kolari, and J. Pumpanen. 2013. New decomposition rate functions based on volumetric soil water content for the ROMUL soil organic matter dynamics model. *Ecol. Modell.* 263:109–118.

Liski, J., T. Palosuo, M. Peltoniemi, and R. Sievänen. 2005. Carbon and decomposition model Yasso for forest soils. *Ecol. Modell.* 189:168–182.

Liu, J., J.E. Vogelmann, Z. Zhu, C.H. Key, B.M. Sleeter, D.T. Price, J.M. Chen, M.A. Cochrane, J.C. Eidenshink, S.M. Howard, N.B. Bliss, and H. Jiang. 2011. Estimating California ecosystem carbon change using process model and land cover disturbance data: 1951–2000. *Ecol. Modell.* 222:2333–2341.

Liu, S., N. Bliss, E. Sundquist, and T.G. Huntington. 2003. Modeling carbon dynamics in vegetation and soil under the impact of soil erosion and deposition. *Global Biogeochem. Cycles* 17:1074–1097.

Loustau, D., A. Bosc, A. Colin, J. Ogée, H. Davi, C. François, E. Dufrêne, M. Déqué, E. Cloppet, E., D. Arrouays, C. Le Bas, N. Saby, G. Pignard, N. Hamza, A. Granier, N. Bréda, P. Ciais, N. Viovy, and F. Delage. 2005. Modeling climate change effects on the potential production of French plains forests at the sub-regional level. *Tree Physiol.* 2:813–823.

Lucier, A., N, M. Ayrers, D. Karnosky, and I. Thompson. 2009. Forest responses and vulnerabilities to recent climate change. In *Forest responses and vulnerabilities to recent climate change: A global assessment report*, ed. R. Seppälä, A. Buck, P. Katila, 29–52. Helsinki: IUFRO World Series.

Luckai, N. and G.R. Larocque. 2002. Challenges in the application of existing process-based models to predict the effect of climate change on pools in forest ecosystems. *Clim. Change* 55:39–60.

Luo, Y. and J.F. Reynolds. 1999. Validity of extrapolating field CO_2 experiments to predict carbon sequestration in natural ecosystems. *Ecology* 80:1568–1583.

Mace. G.M., K. Norris and A.H. Fitter. Biodiversity and ecosystem services: a multilayered relationship. *Trends Ecol. Evol.* 27:19–26.

Mäkelä, A. 1997. A carbon balance model of growth and self-pruning in trees based on structural relationships. *For. Sci.* 43:7–24.

Mäkelä, A, J. Landsberg, A.R. Ek, T.E. Burk, M. Ter-Mikaelian, G.I. Ågren, C.D. Oliver, and P. Puttonen. 2000. Process-based models for forest ecosystem management: Current state of the art and challenges for practical implementation. *Tree Physiol.* 20:289–298.

Marsden, C., Y. Nouvellon, J.-P. Lactau, M. Corbeels, R.E. McMurtrie, J.L. Stape, D. Epron, and G. Le Maire. 2013. Modifying the G'DAY process-based model to simulate the spatial variability of Eucalyptus plantation growth on deep tropical soils. *For. Ecol. Manage.* 301:112–128.

Matott, L.S., J.E. Babendreier, and S.T. Purucker. 2009. Evaluating uncertainty in integrated environmental models: A review of concepts and tools. *Water Resour. Res.* 45:W06421.

McGuire, A.D., S. Sitch, J.S. Clein, R. Dargaville, G. Esser, J. Foley, M. Heimann, F. Joos, J. Kaplan, D.W. Kicklighter et al. 2001. Carbon balance of the terrestrial biosphere in the Twentieth Century: Analyses of CO_2, climate and land use effects with four process-based ecosystem models. *Global Biogeochem. Cycles* 15:183–206.

McMurtrie, R.E., R. Leuning, W.A. Thompson, and A.M. Wheeler. 1992. A model of canopy photosynthesis and water use incorporating a mechanistic formulation of leaf CO_2 exchange. *For. Ecol. Manage.* 52:261–278.

McMurtrie, R.E., D.A. Rook, and F.M. Kellher. 1990. Modelling the yield of *Pinus radiata* on a site limited by water and nitrogen. *For. Ecol. Manage.* 30:381–413.

McMurtrie, R.W. and Y.-P. Wang. 1993. Mathematical models of the photosynthetic response of tree stands to rising CO_2 concentrations and temperatures. *Plant Cell Environ.* 16:1–13.

Medlyn, B.E. 1998. Physiological basis of the light use efficiency model. *Tree Physiol.* 18:167–176.

Medlyn, B.E., R.A. Duursma, and M.J.B. Zeppel. 2011. Forest productivity under climate change. A checklist for evaluating model studies. *Wiley Interdisciplinary Rev. Climat Change* 2:332–355.

Medlyn, B.E., A.P. Robinson, R. Clement, and R.W. McMurtrie. 2005. On the validation of models of forest CO_2 exchange using eddy covariance data: Some perils and pitfalls. *Tree Physiol.* 25:839–857.

Milner, K.S., D.W. Coble, A.J. McMahan, and E.L. Smith. 2003. FVSBGC: A hybrid of the physiological model STAND-BGC and the forest vegetation simulator. *Can. J. For. Res.* 33:466–479.

Morales, P., M.T. Sykes, I.C. Prentice, P. Smith, B. Smith, H. Bugmann, B. Zierl, P. Friedlingstein, N. Viovy, S. Sabaté, A. Sánchez, E. Pla, C.A. Gracia, S. Sitch, A. Arneth, and J. Ogee. 2005. Comparing and evaluating process-based ecosystem model predictions of carbon and water fluxes in major European forest biomes. *Global Change Biol.* 11:2211–2233.

Morin, X., C. Augspurger, and I. Chuine. 2007. Process-based modeling of tree species' distributions: What limits temperate tree species' range boundaries. *Ecology* 88:2280–2291.

Morin, X. and W. Thuiller. 2009. Comparing miche- and process-based models to reduce prediction uncertainty in species range shifts under climate change. *Ecology* 90:1301–1313.

Morris, D.M., J.P. (Hammish) Kimmins, and D.R. Duckert. 1997. The use of soil organic matter as a criterion of the relative sustainability of forest management alternatives: A modelling approach using Forecast. *For. Ecol. Manage.* 94:61–78.

Nadporozhskaya, M.A., G.M.J. Mohren, O.G. Chertov, A.S. Komarov, and A.V. Mikhailov. 2006. Soil organic matter dynamics at primary and secondary forest succession on sandy soils in The Netherlands: An application of soil organic matter model ROMUL. *Ecol. Modell.* 190:399–418.

Niinemets, U. and J.D. Tenhunen. 1997. A model separating leaf structural and physiological effects on carbon gain along light gradients for the shade-tolerant species *Acer saccharum*. *Plant Cell Environ.* 20:845–866.

Nightingale, J.M., M.J. Hill, S.R. Phinn, I.D. Davies, and A.A. Held. 2008. Use of 3-PG and 3-PGS to simulate forest growth dynamics of Australian tropical rainforests II. An integrated system for modeling forest growth and scenario assessment within the wet tropics bioregion. *For. Ecol. Manage.* 254:122–133.

Norman, J.M. 1993. Scaling processes between leaf and canopy photosynthesis. In *Scaling physiological processes: Leaf to globe*, ed. J.R. Ehleringer and C.B. Field, 41–74.New York: Academic Press, Inc.

Olchev, A., A. Ibrom, J. Priess, S. Erasmi, C. Leemhuis, A. Twele, K. Radler, H. Kreilein, O. Panferov, and G. Gravenhorst. 2008. Effects of land-use changes on evapotranspiration of tropical rain forest margin area in Central Sulawesi (Indonesia): Modelling study with a reginal SVAT model. *Ecol. Modell.* 212:131–137.

Ollinger, S.V., J.D. Aber, P.B. Reich, and R.J. Freuder. 2002. Interactive effects of nitrogen deposition, tropospheric ozone, elevated CO_2 and land use history on the carbon dynamics of northern hardwood forests. *Glob. Chang. Biol.* 8:545–562.

Olson, J.S. 1963. Energy storage and balance of producers and decomposers in ecological systems. *Ecol.* 44:322–331.

O'Neil, R.V. and B. Rust. 1979. Aggregation error in ecological models. *Ecol. Modell.* 7:91–105.

Ooba, M., Q. Wang, S. Murakami, and K. Kohata. 2010. Biogeochemical model (BGC-ES) and its basin-level application for evaluating ecosystem services under forest management practices. *Ecol. Modell.* 221:1979–1994.

Oreskes, N., K. Shrader-Frechette, and K. Belitz. 1994. Verification, validation, and confirmation of numerical models in the earth sciences. *Science* 263:641–646.

Parton, W.J., J.M.O. Scurlock, D.S. Ojima, T.G. Gilmanov, R.J. Scholes, D.S. Schimel, T. Kirchner, H.-C. Menaut, T. Seastedt, E. Garcia Moya, A. Kalmarut, and J.I. Kinyamario. 1993. Observations and modelling of biomass and soil organic matter dynamics for the grassland biome worldwide. *Global Biogeochem. Cycles* 7:785–809.

Parton, W.J., J.W.B. Stewart, and C.V. Cole. 1988. Dynamics of C, N, P and S in grasslands soils: A model. *Biogeochemistry* 5:109–131.

Pastor, J. and W.M. Post. 1985. Development of a Linked Forest Productivity—Soil Process Model. Oak Ridge: Oak Ridge National Laboratory ORNL/TM-9519.

Paul, K.I., P.J. Polgase, and G.P. Richards. 2003. Predicted change in soil carbon following afforestation or reforestation, and analysis of controlling factors by linking a C accounting model (CAMFor) to models of forest growth (3PG), litter decomposition (GENDEC) and soil C turnover (RothC). *For. Ecol. Manage.* 177:485–501.

Peng, C., J. Liu, Q. Dang, M.J. Apps, and H. Jiang. 2002. TRIPLEX: A generic hybrid model for predicting forest growth and carbon and nitrogen dynamics. *Ecol. Modell.* 153:109–130.

Pérez-Crusado, C.F. Muñoz-Sáez, F. Basurco, G. Riesco, and R. Rodriguez-Soalleiro. 2011. Combining empirical models and the process-based model 3-PG to predict *Eucalyptus nitens* plantations growth in Spain. *For. Ecol. Manage.* 262:1067–1077.

Peters, E.B., K.R. Wythers, J.B. Bradford, and P.B. Reich. 2013. Influence of disturbance on temperate forest productivity. *Ecosystems* 16:95–110.

Petritsch, R., H. Hasenauer, and S.A. Pietsch. 2007. Incorporating forest growth response to thinning within biome-BGC. *For. Ecol. Manage.* 242:324–336.

Pietsch, S.A., H. Hasenauer, and P.E. Thornton. 2005. BGC-model parameters for tree species growing in central European forests. *For. Ecol. Manage.* 211:264–295.

Pinjuv, G., E.G. Mason, and M. Watt. 2006. Quantitative validation and comparison of a range of forest growth model types. *For. Ecol. Manage.* 236:37–46.

Potithep, S. and Y. Yasuoka. 2011. Application of the 3-PG model for gross primary productivity estimation in deciduous broadleaf forests: A study area in Japan. *Forests* 2:590–609.

Potter, C.S., J.T. Randerson, C.B. Field et al. 1993. Terrestrial ecosystem production: A process model based on global satellite and surface data. *Global Biogeo. Cycles* 7:811–841.

Quillet, A., C. Peng, and M. Garneau. 2010. Toward dynamic global vegetation models for simulating vegetation-climate interactions and feedbacks: Recent developments, limitations, and future challenges. *Can. J. For. Res.* 18:333–353.

Rodrigues, M.A., D.M. Lopes, S.M. Leite, and V.M. Tabuada, 2010. Analyzing the carbon dynamics in North Western Portugal: Calibration and application of Forest-BGC. *Earth Syst. Dynam. Discuss.* 1:41–62.

Rodriguez-Suárez, J.A., B. Soto, M.L. Iglesias, and F. Diaz-Pierros. 2010. Application of 3PG forest growth model to a *Eucalyptus globulus* plantation in Northwest Spain. *Eur. J. Forest Res.* 129:573–583.

Roeckner, E, K. Arpe, L. Bengtsson et al. 1996. The Atmospheric General Circulation Model ECHAM-4: Model Description and Simulation of Present-Day Climate. Report 218. Max-Planck-Institut für Meteorologie, Hamburg, Germany.

Running, S.W. 1994. Testing forest-BGC ecosystem process simulations across a climatic gradient in Oregon. *Ecol. Appl.* 4:238–247.

Running, S.W. and J.C. Coughlan. 1988. A general model of forest ecosystem processes for regional applications. I. Hydrologic balance, canopy gas exchange and primary production processes. *Ecol. Modell.* 42:125–154.

Running, S.W. and S.T. Gower. 1991. FOREST-BGC, a general model of forest ecosystem processes for regional applications. II: Dynamic carbon allocation and nitrogen budgets. *Tree Physiol.* 9:147–160.

Running, S.W. and E.R. Hunt jr., 1993. Generalization of a forest ecosystem process model for other biomes, BIOME-BGC, and an application for global scale models. In *Scaling physiological processes: Leaf to globe*, ed. J.R. Ehleringer and C. Field, 141–158, San Diego: Academic Press.

Rykiel, E.J., Jr. 1996. Testing ecological models: The meaning of validation. *Ecol. Modell.* 90:229–244.

Saltelli, A., M. Ratto, T. Andres, F. Campolongo, J. Cariboni, D. Gatelli, M. Saisana, and S. Tarantola, 2008. Introduction to sensitivity analysis. In *Global sensitivity analysis. The primer*, ed. A. Saltelli, M. Ratto, T. Andres, F. Campolongo, J. Cariboni, D. Gatelli, M. Saisana and S. Tarantola, 1–51. New York: John Wiley & Sons, Ltd.

Sampson, D.A., R.H. Waring, C.A. Maier, C.M. Gough, M.J. Ducey, and K.H. Johnsen. 2006. Fertilization effects on forest carbon storage and exchange, and net primary production: A new hybrid process model for stand management. *For. Ecol. Manage.* 221:91–109.

Sands, P.J., M. Battaglia, and D. Mummery. 2000. Application of process-based models to forest management: Experience with PROMOD, a simple plantation productivity model. *Tree Physiol.* 20:383–392.

Seidl, R., M.-J. Schelhass, W. Rammer, and P.J. Verkerk. 2014. Increasing forest disturbances in Europe and their impact on carbon storage. *Nat. Clim. Chang.* 4:806–810.

Shanin, V.N., A.S. Komarov, A.V. Mikhailov, and S.S. Bykhovets. 2011. Modelling carbon and nitrogen dynamics in forest ecosystems of Central Russia under different climate change scenarios and forest management regimes. *Ecol. Modell.* 222:2262–2275.

Sharkey, T.D., C.J. Bernacchi, G.D. Farquhar, and E.L. Singsaas. 2007. Fitting photosynthetic carbon dioxide response curves for C3 leaves. *Plant Cell Environ.* 30:1035–1040.

Tague, C.L. and L.E. Band. 2004: RHES sys: Regional hydro-ecologic simulation system—an object-oriented approach to spatially distributed modeling of carbon, water, and nutrient cycling. *Earth Interact.* 8:1–42.

Tague, C.L., K. Heyn, and L. Christensen. 2009. Topographic controls on spatial patterns of conifer transpiration and net primary productivity under climate warming in mountain ecosystems. *Ecohydrol.* 2:541–554.

Tatarinov, F.A. and E. Cienciala. 2006. Application of BIOME-BGC model to managed forests! Sensitivity analysis. *For. Ecol. Manage.* 237:267–279.

Tedeschi, L.O. 2006. Assessment of the adequacy of mathematical models. *Agric. Sys.* 89:225–247.

Thompson, J.R., K.N. Johnson, M. Lennette, T.A. Spies, and P. Bettinger. 2006. Historical disturbance regimes as a reference for forest policy in a multiowner province: A simulation experiment. *Can. J. For. Res.* 36:401–417.

Thornley, J.H.M. 1998. Dynamic model of leaf photosynthesis with acclimation to light and nitrogen. *Ann. Bot.* 81:421–430.

Thornley, J.H.M. and M.G.R. Cannell. 1992. Nitrogen relations in a forest plantation-soil organic matter ecosystem model. *Ann. Bot.* 70:137–151.

Thuillier, W. 2007. Climate change and the ecologist. *Nature* 448:550–552.

Tselniker, J.L. 1978. *The physiological basis of trees shade-tolerance* (in Russian). Moscow: Nauka.

Twery, M.J. and A.R. Weiskittel. 2013. Forest-management modelling, in environmental modelling: Finding simplicity in complexity, 2nd edn, eds. J. Wainwright and M. Mulligan, 379–398. Chichester: John Wiley & Sons, Ltd.

Urban, D.L. 1990. A versatile model to simulate forest pattern: A user's guide to ZELIG Version 1.0. University of Virginia, Charlottesville, Virginia.

U.S. Climate Change Science Program. 2008. *The Effects of Climate Change on Agriculture, Land Resources, Water Resources, and Biodiversity in the United States.* A Report by the U.S. Climate Change Science Program and the Subcommittee on Global Change Research. Backlund, P., A. Janetos, D. Schimel, J. Hatfield, K. Boote, P. Fay, L. Hahn, C. Izaurralde, B.A. Kimball, T. Mader, J. Morgan, D. Ort, W. Polley, A. Thomson, D. Wolfe, M. Ryan, S. Archer, R. Birdsey, C. Dahm, L. Heath, J. Hicke, D. Hollinger, T. Huxman, G. Okin, R. Oren, J. Randerson, W. Schlesinger, D. Lettenmaier, D. Major, L. Poff, S. Running, L. Hansen, D. Inouye, B.P. Kelly, L Meyerson, B. Peterson, and R. Shaw. U.S. Environmental Protection Agency, Washington, DC, USA.

Veganzones, M.A., L.C.S. Nunes, D.M.M. Lopes, C. Hernandez, and M. Graña, 2010. Implementation of an user-friendly interface of the Forest-BGC model for regional/global scale researches. *Silva Lusitana* N° special:13–25.

Wagener, T., N. McIntyre, M.J. Lees, H.S. Wheater, and H.V. Gupta. 2003. Towards reduced uncertainty in conceptual rainfall-runoff modelling: Dynamic identifiability analysis. *Hydrol. Process.* 17:455–476.

Wang, W., K. Ichii, H. Hashimoto, A.R. Michaelis, P.E. Thornton, B.E. Law, and R.R. Nemani. 2009. A hierarchical analysis of terrestrial ecosystem model Biome-BGC: Equilibrium analysis and model calibration. *Ecol. Modell.* 220:2009–2023.

Wang, W., C. Peng, D.D. Kneeshaw, G.R. Larocque, X. Lei, Q. Zhu, X. Song, and Q. Tong. 2013. Modeling the effects of varied forest management regimes on carbon dynamics in jack pine stands under climate change. *Can. J. For. Res.* 43:469–479.

Wang, W., C. Peng, S.Y. Zhang, X. Zhou, G.R. Larocque, D.D. Kneeshaw, and X. Lei. 2011. Development of TRIPLEX-management model for simulating the response of forest growth to pre-commercial thinning. *Ecol. Modell.* 222:2249–2261.

Wang, W., J. Xiao, S.V. Ollinger, A.R. Desai, J. Chen, and A. Noormets. 2014. Quantifying the effects of harvesting on carbon fluxes and stocks in Northern temperate forests. *Biogeosci.* 11:6667–6682.

Wang, Y.P. 2000. A refinement of the two-leaf model for calculating canopy photosynthesis. *Agric. For. Meteorol.* 101:143–150.

Wei, X. and J.A. Blanco. 2014. Significant increase in ecosystem C can be achieved with sustainable forest management in subtropical plantation forests. *PLoS ONE* 9:

Willmott, C.J. 1982. Some comments on the evaluation of model performance. *Bull. Am. Meteor. Soc.* 63:1309–1313.

Wullschleger, S.D. 1993. Biochemical limitations to carbon assimilation in C3 plants—a retrospective analysis of the A/C_i curves from 109 species. *J. Exp. Bot.* 44:907–920.

Xenakis, G., D. Ray, and M. Mencuccini. 2008. Sensitivity and uncertainty analysis from a coupled 3-PG and soil organic matter decomposition model. *Ecol. Modell.* 219:1–16.

Xu, J.Z., Y.M. Yu, S.Z. Peng, S.H. Yang, and L.X. Liao. 2014. A modified nonrectangular hyperbola equation for photosynthetic light-response curves of leaves with different nitrogen status. *Photosynthetica* 52:117–123.

Yurova, A.Y., E.M. Volodin, G.I. Ågren, O.G. Chertov, and A.S. Komarov. 2010. Effects of variations in simulated changes in soil carbon contents and dynamics on future climate projections. *Glob. Change Biol.* 16:823–835.

Zavala, M.A. and R. Bravo de la Parra. 2005. A mechanistic model of tree competition and facilitation for Mediterranean forests: Scaling from leaf physiology to stand dynamics. *Ecol. Modell.* 188:76–92.

Zhang, Z., H. Jiang, J. Liu, X. Zhang, C. Huang, X. Lu, J. Jin, and G. Zhou, 2014. An analysis of the global spatial variability of column-averaged CO_2 from SCIAMACHY and its implications for CO2 sources and sinks. *Int. J. Remote Sens.* 35:2047–2066.

Zhao, M., F.A. Heinsch, R.R. Nemani, and S.W. Running. 2005. Improvements of the MODIS terrestrial gross and net primary production global data set. *Remote Sens. Environ.* 95:164–176.

Zhu, Q., J. Liu, C. Peng, H. Chen, X. Fang, H. Jiang, G. Yang, D. Zhu, W. Wang, and X. Zhou. 2014. Modelling methane emissions from natural wetlands by development and application of the TRIPLEX-GHG model. *Geoscien. Mod. Devel.* 7:981–999.

Zhu, Q., C. Peng, H. Chen, X. Fang, J. Liu, H. Jiang, Y.Yang, and Y. Yang. 2015. Estimating global natural wetland methane emissions using process modeling: Spatiotemporal patterns and contributions to atmospheric methane fluctuations. *Glob. Ecol. Biogeogr.* (Accepted for publication).

Zhu, Z., B. Bergamaschi, R. Bernknopf, D. Clow, D. Dye, S. Faulkner, W. Forney, R. Gleason, T. Hawbaker, J. Liu, S. Liu, S. Prisley, B. Reed, M. Reeves, M. Rollins, B. Sleeter, T. Sohl, S. Stackpoole, S. Stehman, R. Striegl, A. Wein and Z. Zhu. 2010. A method for assessing carbon stocks, carbon sequestration, and greenhouse-gas fluxes in ecosystems of the United States under present conditions and future scenarios. U.S. Geological Survey Scientific Investigations Report 2010–5233.

Zhu, Z. and B.C. Reed, eds. 2012. Baseline and projected future carbon storage and greenhouse-gas fluxes in ecosystems of the Western United States: U.S. Geological Survey Professional Paper 1797, 192 p. (Also available at http://pubs.usgs.gov/pp/1797/.).

9 Modeling Forest Carbon Budgets toward Ecological Forest Management

Challenges and Future Directions

Weifeng Wang, Changhui Peng, and Guy R. Larocque

9.1 INTRODUCTION

Forests play an important role in the global carbon (C) cycle, as they occupy about 30% of the continental areas of the earth and contain from 80% to 90% of terrestrial C (Thornton et al. 2002). The forest C budget varies among forest types and is affected by development stages, disturbance history (e.g., fires, insects, and harvests), management practices (e.g., pre-commercial thinning), climate variability, and nutrient availability (e.g., nitrogen). The scientific community, forest managers, and policymakers require information on the amount of C stored in forest ecosystems (Kurz et al. 2009). In particular, the Intergovernmental Panel on Climate Change requires countries to provide integrated assessments of the impact of forests and forestry activities on the mitigation of climate change (Karjalainen et al. 2002). To fulfill this requirement, it is essential to quantify forest C stocks and their rate of change and evaluate the effects of different scenarios of forest management practices and land use policy options (Kurz et al. 2009). This information is essential to determine if forests are acting as C sources or sinks over long time periods.

C accounting generally refers to the practice of making scientifically robust and verifiable measurements of greenhouse gas (GHG) emissions (Watson 2009). This information is used to create the C credit commodity for the C market. Policy demands for the ability to project future C stocks and C stock changes are also increasing. C budget models based on forest inventory are tools to estimate forest C stocks and/or exchanges between the atmosphere and forest ecosystems. Many process-based models on C cycles can also simulate the C exchange in forest ecosystems, which allow them to predict C budgets (see Chapter 8). However, C accounting models are a special type of C cycle models. Some of their model components are based on the use of mechanistic relationships, but empirical relationships obtained from large forest inventory datasets are also used. This chapter focuses on the use of forest inventory-based models that can be applied to provide estimates of C stocks and evaluate the impacts of disturbance regimes on C budget at landscape and national levels.

The objective of this chapter is to describe the structure and characteristics of this type of C budget models and discuss their limitations and future development. Following the introduction, the overview of forest C accounting models will be given. Three models will be introduced, and their applications will be discussed. Then, the limitations and possible improvements under changing climatic conditions will be addressed as well.

9.2 FRAMEWORK OF FOREST CARBON ACCOUNTING MODELS

C accounting models are basically used to report national-level GHG exchanges in forests and the forestry sector. The forest growth rate of stand volume is estimated from growth and yield tables or models that foresters use in volume estimation and forest management planning. Volume

DOI: 10.1201/9781003431084-11

is then converted into C biomass using volume–biomass equations that represent the relationships between C stock and net merchantable volume based on forest inventory data (Figure 9.1a). Allometric relationships are used to estimate biomass for each component (e.g., foliage, stems, and roots). Tree mortality in these models is usually calculated based on the intensity of competition and the effect of disturbances (e.g., fires, harvests, and insects). Dead organic matter and soil organic C decomposition and mineralization rates have also been estimated using first-order decay models (e.g., CENTURY, Parton et al. 1993, and YASSO, Liski et al. 2005) to fully represent the forest C cycle. More importantly, the effects of natural and anthropogenic disturbances on the forest C cycle can be represented and calculated in these models. In addition, these C accounting models include the accounting of harvested wood product C and/or bioenergy (Figure 9.1b).

Some of the characteristics and differences in key processes among the existing inventory-based C accounting models are presented in Table 9.1. Most of these models have a time-step ranging from one to five years, except for the Full Carbon Accounting Model (FullCAM, Richards and Evans 2004; Waterworth et al. 2007) that was developed for Australia's national GHG emissions for the land sector. FullCAM is a monthly time-step model that comprises the physiological forest growth model 3-PG (Landsberg and Waring 1997) and the Rothamsted soil carbon model (www.rothamsted.ac.uk/rothamsted-carbon-model-rothc) and the carbon accounting model CAM for for forests (Mohren and Goldewijk 1990). This makes the FullCAM model more like a hybrid model (see Hybrid Models in Chapter 8). These models have been used in different countries and regions to predict the effects of management scenarios and natural disturbance regimes (e.g., fires and insects) for different forest types.

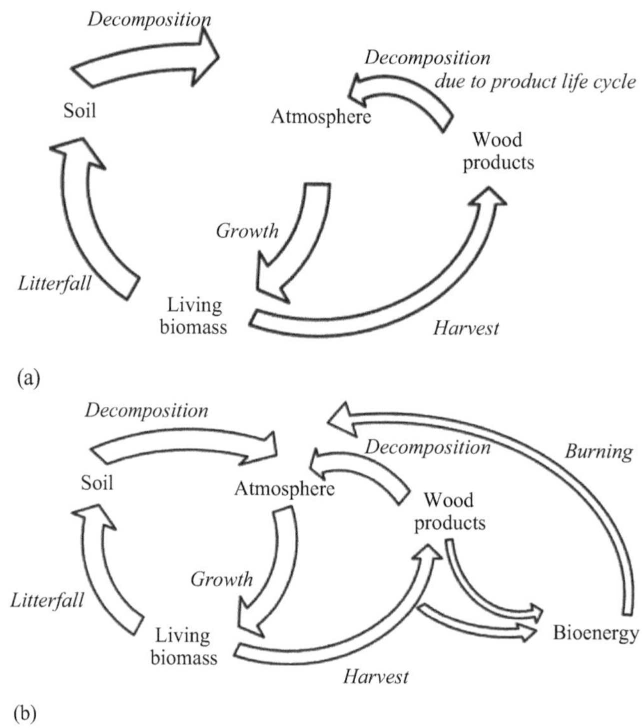

FIGURE 9.1 Basic frameworks of carbon accounting models representing the carbon cycle in the forest and forestry sector without (a) and with (b) bioenergy consideration.

TABLE 9.1
Comparison of Inventory-Based Carbon Accounting Models

Model Name	Time-Step	Forest Growth and Carbon Stocks	Soil Organic Matter	Forest Disturbance	Forestry Carbon Sector	Bioenergy	Nation	Forest Type/ System
CBM-CFS (Kurz et al. 1992, 2009; Kurz and Apps, 1999)	1 year	Growth and yield curves	FODM[a]	Fire, insects, and forest harvesting	Yes	No	Canada	Boreal and temperate rain forests
CO2FIX (Masera et al. 2003; Schelhaas et al. 2004)	1 year	Growth and yield curves	FODM[a]	Fire, insects, wind, and forest harvesting	Yes	Yes	Europe	Boreal, temperate, and tropical forests
EFISCEN (Nabuurs et al. 2000)	5 years	Growth and yield curves	FODM[a]	Forest harvesting	No	No	Europe	Temperate forests
FORCARB (Heath and Birdsey 1993; Heath et al. 2002, 2010)	1 year	Growth and yield curves	FODM[a]	Forest harvesting and fire	Yes	No	United States and Canada	Boreal and temperate forests
FullCAM (Richards and Evans 2004; Waterworth et al. 2007)	1 month	Photosynthesis	FODM[a]	Forest harvesting	Yes	No	Australia	Temperate forest and agricultural systems

Note:
[a] FODM, first-order decay model

9.3 MODELING APPROACH AND CASE STUDIES

We have briefly introduced the basic framework of inventory-based C accounting models. In this section, we consider three examples of inventory-based C accounting models—Carbon Budget Model of the Canadian Forest Sector (CBM-CFS) (Kurz et al. 1998, 2002, 2009), FORest CARBon Budget Model (FORCARB) (Heath and Birdsey 1993; Heath et al. 2002, 2010), and CO2FIX (Masera et al. 2003; Schelhaas et al. 2004). For each of these models, the modeling approach and its potential applications toward sustainable forest management and C reporting across the stand, landscape, and national scales are described as follows.

9.3.1 CBM-CFS

CBM-CFS is the key model of the Government of Canada's National Forest Carbon Monitoring, Accounting, and Reporting System. Its development began in the early 1990s by scientists of the Canadian Forest Service. This model was originally designed to estimate C fluxes and stocks at both the landscape and national levels. Thus, the landscape is designed as a collection of spatial units in this model (Figure 9.2). The early versions of the model, CBM-CFS (Kurz et al. 1992) and CBM-CFS2 (Kurz and Apps 1999), estimated the contribution of Canada's 404 Mha forest area and the forest product sector to the global C budget. With the increasing concerns that forest management activities can contribute to national C budget, the updated version CBM-CFS3 (Kurz et al. 2009) included the development/improvement of model components for forest C dynamics (i.e., growth, pools, biomass turnover, litterfall transfers, and dead organic matter decay) and forest management

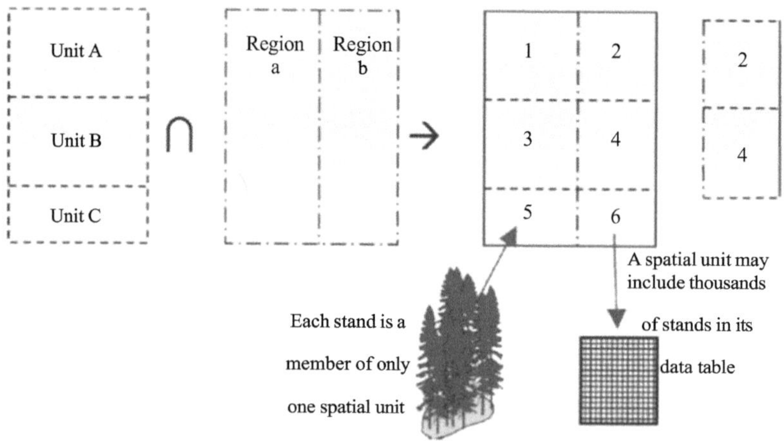

FIGURE 9.2 A heterogenetic forest landscape comprising administrative regions (A–C) and ecological regions (a and b), spatial units (1–6), and a disturbance group. (After Kurz, W.A. et al., *Ecol. Model.*, 220, 480, 2009. With permission.)

activities at the stand level to meet forest industry's needs. CBM-CFS3 has the potential to be used for reporting the C balance of Canada's managed forests.

CBM-CFS3 includes five C biomass pools (merchantable stemwood, branches, foliage, fine roots, and coarse roots) and eight dead organic matter C pools (dead standing stemwood, dead branches, and above- and belowground fast, very fast, and slow dead organic matter). The model is used for simulating the effects of different forest management operations on forest C balance. It can be used to some extent to assess potential soil organic C and dead organic matter response to climate change for managed forests. The first-order kinetic model is used to simulate organic matter decomposition (Kurz et al. 2009). The decay rates are modified by temperature (Kurz et al. 2009) as well as moisture (Smyth et al. 2011) in the most recent version. White et al. (2008) assessed the sensitivity of dead organic matter model of the CBM-CFS3 to variation in model parameters (e.g., biomass turnover rates and decay rates). The parameters of foliage and fine root turnover might be the most sensitive parameters in the model on their study sites. Many studies suggested to include additional processes in dead organic matter decomposition process such as a lignin effect and a moss effect on dead organic matter decay (Hararuk et al. 2017). Therefore, a new moss-derived C submodule was added in CBM-CFS3 to determine whether moss contributes to boreal black spruce forest C budgets (Bona et al. 2013, 2016).

At the stand level, Taylor et al. (2008) estimated the effects of harvesting intensity on C stocks in a chronosequence of 24 red spruce (*Picea rubens* Sarg.) dominated stands in eastern Canada using the CBM-CFS3 model. They found that clearcutting might lead to larger C storage than partial cutting over the whole simulation period in their study. At the landscape level, Trofymow et al. (2008) applied the model in a 2,500 ha landscape in British Columbia to simulate forest dynamics and C budgets for an 86-year period (1920–2005). They concluded that forest disturbances (e.g., fire and logging) resulted in large losses of forest C biomass in the area. Net biome productivity varied from −15 to 6 Mg C ha^{-1} year^{-1} depending on disturbance type and year. In addition, Bernier et al. (2010) calibrated the model for a 62 km^2 landscape of boreal forest in eastern Canada using a reconstructed dataset of forest growth and disturbances during a 71-year period (1928–1998). Their study found that the parameterization of yield curves, especially for the scaling between regional/national yield curves and local yield curves, caused the largest bias of net primary productivity. Luckai et al. (2012) conducted a similar study for selected forest types of the Canadian boreal shield by examining the effects of harvest and fire on C dynamics. Their results indicated that longer rotations increased total ecosystem C. From 2000 to 2012, the net biome productivity for European forests was estimated on average to 122 Tg C yr^{-1} with the consideration of carbon loss induced by forest harvesting and natural disturbance such as fire using the CBM-CFS3 model (Pilli et al. 2017). Toward stem-only harvesting, Sage et al. (2019) predicted boreal forest carbon pools using the CBM-CFS3 model, suggesting that model initialization of dead organic matter pools might not adequately simulate the long-term history of carbon dynamics following harvesting. Relatively stable pools of cumulative ecosystem C and decrease in C emissions were also predicted following fire disturbances. Kurz et al. (2008a) used a Monte Carlo design for simulating future net biome production from 2000 to 2020, using the model with the consideration of the impacts of the beetle, forest fires, and harvests on forest productivity and C balance for the south-central region (374,000 km^2) of British Columbia. They identified that insect outbreaks converted the forest from a net C sink to a large net C source in the area. At the national level, Kurz et al. (2008b) reported that the managed forests of Canada could emit 30–245 Mt CO_2 e $year^{-1}$ during the 2008–2012 period. The authors concluded that natural disturbances largely contributed to the total emissions of the Canadian forest sector.

9.3.2 FORCARB

The FORest CARBon Budget Model (FORCARB), which was developed by the U.S. Forest Service, is an empirical model used to estimate C budgets of U.S. forest ecosystems and the forest product sector (Heath and Birdsey 1993; Heath et al. 2002). The model has been used to provide estimates of C stocks and stock change for policy-related needs (e.g., the Resources Planning Act Timber Resource Assessment and Greenhouse Gas Inventories of the United States, land use, land use change, and forestry to the United Nations Framework Convention on Climate Change). Recently, FORCARB2, an updated version of the model, used a new methodology for C in harvested wood products (Heath et al. 2002, 2010). Also, it can now be used for public and reserved forestlands and forestlands with low productivity. They updated the methodology for soil C estimates according to forest vegetation maps and soil databases (State Soil Geographic dataset). They presented estimates of soil organic C for certain forest types and soil organic C transition with forest type change and discussed the effects of harvesting and land use change on soil organic C. FORCARB2 was modified and adapted for analyzing forest and wood product C for Ontario, Canada (Chen et al. 2010). FORCARB-ON, the modified version for Ontario's forests, was modified using a timber supply model (strategic forest management model) (Kloss 2002) and a harvested wood product C model using Ontario data (Chen et al. 2008).

Forest growth is estimated using the growth and yield curves that were derived from the timber assessment market model and aggregated timberland assessment system (ATLAS, Mills and Kincaid 1992). Theoretically, the driven module that was used to estimate C stocks could be replaced by other forest productivity models or other timber supply models. Similar to other models (e.g., CBM-CFS), algometric equations were used to estimate forest C stocks (Chen et al. 2010). The model has six C pools: living trees (i.e., above- and belowground), standing dead trees, down deadwood, understory vegetation, forest floor, and soil (Chen et al. 2010).

The FORCARB model has been seldom applied for simulating the effects of forest management on C dynamics at the stand and landscape levels. However, Colombo et al. (2012) investigated the effects of management strategies on trade-offs between forest C stocks and sustainability under five management scenarios in a 3.4 Mha forest area in northeastern Ontario using the FORCARB-ON model. Under the forest protection scenario, forests had the greatest C stocks but became C neural within 30 years. Harvesting scenarios resulted in less on-site forest C stock compared to that under the forest protection scenario, but total C storage in forests and harvested wood products was greater. These analyses provide a potential to use the model for evaluating the contribution of forests to climate change mitigation in the context of forestry.

9.3.3 CO2FIX

CO2FIX is also an empirical model that quantifies the C stocks and stock change in trees, soils, and wood products for a given stand (Schelhaas et al. 2004). The CO2FIX model has been developed since the first initiatives on the estimates of C sequestration by trees in the early 1990s. During the past decades, the model has been improved considerably as follows (Schelhaas et al. 2004):

- The cohort approach (multispecies and uneven-aged stands, Figure 9.3)
- Growth by stand density
- Intercohort competition
- Life cycle of harvested wood
- Soil organic C dynamics
- Selective logging systems
- A financial module to calculate costs and revenues (including C credits)
- Bioenergy module

The recent version of CO2FIX, CO2FIX V3.1, consists of five modules: biomass, soil, wood products, bioenergy, and finance (Masera et al. 2003; Schelhaas et al. 2004). In the model, C is accumulated in the biomass based on growth and yield curves and allocation pattern by species, region, size class, and owner type. The C biomass is then transferred to the soil organic C pool through litterfall. If harvests occur, the C biomass is mostly converted to wood products (Schelhaas et al. 2004). The part of C in the wood products is transferred into bioenergy through burning and/or by-products to generate energy. One of the unique features of the model is its consideration of reduced emissions from bioenergy when calculating the overall C budget.

The model has been applied toward forest management. For example, Alvarez et al. (2016) predicted the effects of forest management and climate change scenarios on carbon sequestration

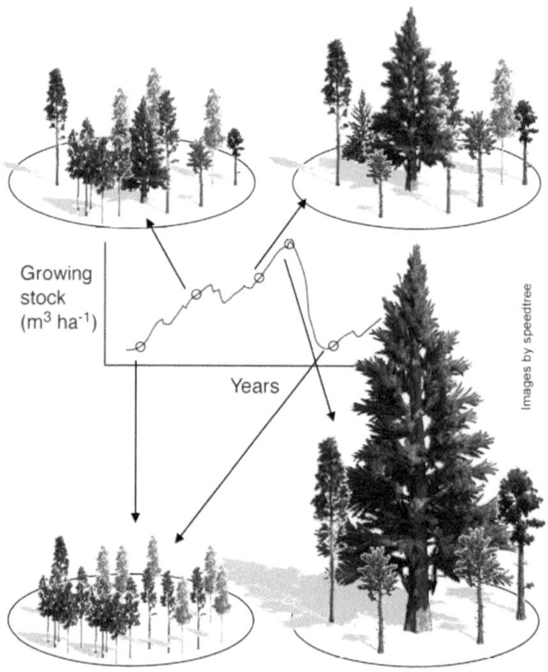

FIGURE 9.3 Cohort approach used in CO2FIX. (From ETH, Concept of forest gap models, www.fe.ethz.ch/research/standdynamics/forclim_model/index_EN, accessed April 10, 2015. With permission.)

in Mediterranean mountain forests. They discovered that mixed forests have the highest levels of carbon storage, implying that switching from pure to mixed forests would enhance carbon capture. Comic (2021) applied the CO2FIX V3.1 model to simulate C sequestration in poplar plantations in northern Serbia. Across the region, the average sequestration during the 30-year production cycle was 44.02 Mg C ha^{-1}. Gaboury et al. (2009) calculated the net C balance for boreal open woodland afforestation using the model. They concluded that afforestation could lead to a net C sink rate of 1 Mg C ha^{-1} $year^{-1}$ over 70 years if no disturbance occurs. Lemma et al. (2007) examined what factors control soil organic C sequestration under three tree plantations using the CO2FIX V3.1 model in southwestern Ethiopia. They suggested that the total litter input and the proportion of fine woody litter were the main factors that governed soil organic C accumulation. Therefore, choosing proper tree species is important in afforestation projects in terms of soil organic C accumulation. A new version, named CO2LAND, has been under development for the application of the model at the landscape level. Recently, the model was more often used in agroforestry system (Negash and Kanninen 2015; Ajit et al. 2017; Rizvi et al. 2019). Negash and Kanninen (2015) applied the CO2FIX V3.1 model to simulate C dynamics of three Indigenous agroforestry systems in southeastern Ethiopia highlighting that the tree cohort accounted for over 89% of the total biomass C stocks, and the soil organic C stocks accounted for over 60% of the total C in the studied agroforestry systems. At the national level, the existing agroforestry system in India was estimated to reduce 19.34 Tg CO_2 using the CO2FIX model (Ajit et al. 2017). Agroforestry could have significant contribution in mitigating atmospheric CO_2 if the number of trees on agricultural land are increased (Rizvi et al. 2019).

9.3.4 Combinations of Different Model Types

The vast majority of C budget models are based on the use of a single model, such as the models mentioned earlier. However, some studies combined the use of different types of models to simulate C fluxes. For instance, Song and Woodcock (2003) integrated elements of the forest succession model ZELIG (Urban 1990) and CENTURY (Parton et al. 1993) to simulate forest growth and soil respiration, respectively, at the ecosystem level. Then, they used RegCarb, a regional carbon budget model, to simulate carbon budget at the regional level based on the results from ecosystem-level simulations. This combined approach enhances the integration of widely different ecosystem structures and characteristics among forest types within a region, such as age, structure, and soil type. Polli et al. (2022) combined dynamic global vegetation models and process-based climate models to simulate the dynamics of C sinks in forests of the European Union by examining the effects of climate change, forest management scenarios, and disturbances. They aimed at establishing a methodological framework to evaluate the long-term capacity of forests to act as a C sink. Also, Smyth et al. (2017) combined CBM-CFS3 and a harvested wood product model to estimate bioenergy emissions and the substitution benefits of using bioenergy. Furthermore, stationary models, such as CBM-CFS3, in conjunction with remote sensing models, could be utilized to evaluate net biome productivity in nations with abundant forests (Krenke et al. 2021).

9.4 CHALLENGES AND FUTURE DIRECTIONS

These models can be useful but are generally applicable only to the nations or regions where they have been developed, limiting their broader application. For a better understanding of climate impacts to the C cycle in forest and forestry sectors, the best approach is to improve our current modeling techniques. Policy demands for climate change mitigations require models that are able to project C emissions from world forests and the global wood product market in the long term (centuries). We identify the following limitations for the current modeling approach and provide the insights of future directions.

9.4.1 Uncertainty Issue

Uncertainty usually implies a lack of knowledge or information that may or may not be obtainable. Uncertainties may result from errors in the sampling procedures of forest inventories, computation of wood volume estimates at regional scales, description of suitable C pools, and a lack of understanding of processes governing pool sizes and fluxes (Birdsey and Heath 1995). Better understanding of ecosystem components and processes that cause uncertainties may contribute to improving the development of possible research priorities and optimal management scenarios (Heath and Smith 2000; Böttcher et al. 2008). Undertaking uncertainty analysis of C budget models is a complex undertaking, as it is not common practice, although this need was identified. In practice, the implementation of procedures to estimate uncertainty in model outputs requires considerable modifications in the code of C budget models. However, there are relatively few examples in the literature. Good examples can be found in Heath and Smith (2000) and Böttcher et al. (2008). Heath and Smith (2000) used a Monte Carlo approach based on Latin hypercube sampling in the model FORCARB by randomly varying basic parameters related to C content in different pools. The probability distributions of C pool estimates approached normal distributions. Alternative methods can be used to simplify the undertaking of uncertainty analysis. For instance, White et al. (2008) used the Gaussian emulation machine for sensitivity analysis (GEM-SA), a statistical emulator based on Bayesian statistical methods, to conduct uncertainty analysis of the predictions of CBM-CFS for forest types in northwestern Ontario, Canada. The main benefit of using GEM-SA is to eliminate the need to modify the code of the model. Also, the application of common methods for complex models, such as Monte Carlo, requires many iterations, which increase the computation time. The use of a specialized application for sensitivity and uncertainty analysis, such as GEM-SA, reduces considerably the number of iterations required.

9.4.2 Climate Change

Disturbances and climate may jointly govern the forest C cycle (Williams et al. 2012; Bond-Lamberty et al. 2013; Kasischke et al. 2013). One of the advantages of inventory-based C accounting models is that they can simulate various forest disturbances that have serious effects on forest C dynamics (e.g., Kurz et al. 2008b). Because forest growth in the inventory-based C accounting models is driven by empirical growth and yield curves, they have limitations to simulate the effects of climate change on forest growth and tree mortality and their application in remote regions where few data are available to develop empirical growth and yield curves. Although much efforts have been made to investigate climate-induced tree mortality (Phillips et al. 2009; van Mantgem et al. 2009; Peng et al. 2011) and CO_2 fertilization (Friedlingstein et al. 1995; Hickler et al. 2008) and their interactions (e.g., Reich and Hobbie 2013; Feng et al. 2015), current forest inventory-based C accounting models have little ability to simulate the newly merged climate-induced processes (e.g., CO_2 fertilization and climate change-induced drought).

One of the possible solutions is to develop climate-driven growth and yield curves. For example, Albert and Schmidt (2010) developed a generalized additive site index model that is capable of predicting changes in productivity in changing climate. Crookston et al. (2010) modified the widely used Forest Vegetation Simulator to account for climate effects on growth rates, mortality, and regeneration. Overall, the climate-sensitive modeling approach could help fitting the C accounting models for predicting C budget under changing climate conditions and geological limitations without restructuring the model.

The hybrid modeling approach of integrating both empirical and mechanistic components could be used to simulate the effects of climate for C accounting models, bridging the gap between empirical forest growth and yield models and process-based C balance models. Some models have been developed for this purpose such as 3-PG (Landsberg and Waring 1997), TRIPLEX (Peng et al. 2002), and Full Carbon Accounting Model (FullCAM Waterworth et al. 2007; Waterworth and Richards 2008). Trait-based forest productivity models, such as 3-PG (Landsberg and Waring 1997), TRIPLEX (Peng et al. 2002), and PnET (Aber and Federer 1992), could assist the inventory-based C accounting models by being more sensitive to climate. For example, the driving forces (i.e., NPP) of FORCARB were modified by NPP change produced by the Terrestrial Ecosystem Model (Raich et al. 1991), which uses spatial information on climate, soil, and vegetation to predict C fluxes and stocks within ecosystems, within the ATLAS. In recent years, C budget predictions have been performed using strictly process-based models with ecophysiological components, such as photosynthesis, respiration, light interception, soil C, nitrogen, and water fluxes or climatic conditions. A good example is the FORCCHN model to simulate C fluxes in forest ecosystems in northern China (Zhao et al. 2012; Fang et al. 2022). In particular, the integration of temperature and precipitation variables facilitated the development of climate change scenarios. Process-based models (see Chapter 8) that include several ecophysiological components may be more difficult to parameterize and validate than empirical or hybrid models due to the large amount of data required, but they offer the possibility to test many hypotheses about relationships among the different complex processes in forest ecosystems. Also, as they are based on the representation of mechanistic processes, they offer more flexibility to simulate C budgets of forest ecosystems that were not necessary within the range of data used for their derivation, contrary to empirical models.

9.4.3 Harvested Wood Products Carbon and Bioenergy

Modeling efforts may be needed to account for bioenergy (forest biomass used for energy purposes) production in the forest and forestry sectors (Figure 9.1b). In the current policy framework (e.g., Europe), bioenergy is considered as a "C neutral" energy source. Ter-Mikaelian et al. (2015) argue that bioenergy should be calculated in C accounting system, because forest biomass offsets the consumption of fossil fuels. Therefore, the total GHG emissions caused by harvested wood product (HWP) and bioenergy are given by

$$\mathrm{GHG_T} = \Delta_{Mforest} + \Delta_{Mwp} - M_{bioenergy}$$

where,

GHG_T refers to the total GHG emissions;
ΔM_{forest} refers to the net change in forest C stocks;
ΔM_{wp} refers to the net changes in harvested wood product C; and
$M_{bioenergy}$ refers to GHG benefits of replacing fossil fuel with forest bioenergy.

The storage of C in harvested wood products is the supplementary of terrestrial carbon storage (Skog et al. 2004). The United Nations Framework Convention on Climate Change and the guidance of 1996, 2006, 2019 of the Intergovernmental Panel on Climate Change allow countries to estimate C storage in HWP. Many studies have estimated C storage in HWP in the national or the continental levels, such as New Zealand (Wakelin et al. 2020), China (Ji et al. 2013; Geng et al. 2019), and Europe (Pilli et al. 2015). However, few studies estimated the combined C storage between the forest and HWP sectors. This needs a combination method of different models which can simulate the forest C cycle and account for HWP C.

9.5 CONCLUDING REMARKS

In this chapter, we reviewed inventory-based C accounting models driven by empirical growth and yield curves and provided case studies of how to use these models toward sustainable forest management activities (e.g., harvests and afforestation) at stand, landscape, and national levels. Accurately estimating the C stock and stock change in the forest sector is a considerable challenge because they depend on climate, soil nutrient availability, disturbance, and management regimes.

We acknowledge that there is considerable uncertainty in simulating ecological processes and their effects on the forest C cycle under changing conditions of climate. In addition, there is a high degree of variability in wood product life cycle analysis.

Considering these limitations and challenges, it is clear that further research is needed in modeling to quantify the contributions of forest and forest management on the global C cycle under the context of climate change. Developing climate-sensitive forest growth models is necessary to improve our predictive capacity to C reporting at different spatial and temporal scales. Ultimately, C accounting models provide forest resource managers with tools for evaluating the effects of climate change on the C cycle in the forest and forestry sectors.

REFERENCES

Aber, J.D. and C.A. Federer. 1992. A generalized, lumped-parameter model of photosynthesis, evapotranspiration and net primary production in temperate and boreal forest ecosystems. *Oecologia* 92:463–474.

Ajit, D.S.K., A.K. Handa, R. Newaj, S.B. Chavan, B. Alam, R. Prasad, A. Ram, R.H. Rizvi, A.K. Jain et al. 2017. Estimating carbon sequestration potential of existing agroforestry systems in India. *Agrofor. Syst.* 91:1101–1118.

Albert, M. and M. Schmidt. 2010. Climate-sensitive modelling of site-productivity relationships for Norway spruce (*Picea abies* (L.) Karst.) and common beech (*Fagus sylvatica* L.). *For. Ecol. Manage.* 259:739–749.

Alvarez, S., C. Ortiz, E. Diaz-Pines, and A. Rubio. 2016. Influence of tree species composition, thinning intensity and climate change on carbon sequestration in Mediterranean mountain forests: A case study using the CO2 fix model. *Mitig. Adapt. Strateg. Glob. Chang.* 21:1045–1058.

Bernier, P.Y., L. Guindon, W.A. Kurz, and G. Stinson. 2010. Reconstructing and modelling 71 years of forest growth in a Canadian boreal landscape: A test of the CBM-CFS3 carbon accounting model. *Can. J. For. Res.* 40:109–118.

Birdsey, R.A. and L.S. Heath. 1995. Carbon changes in US forests. In: Joyce, L.A. (Ed.), Productivity of America's Forests and Climate Change, General Technical Report RM-271, USDA Forest Service, Rocky Mountain Forest and Range Experiment Station, Fort Collins, CO, pp. 56–70.

Bona, K.A., J.W. Fyles, C. Shaw, and W.A. Kurz. 2013. Are mosses required to accurately predict upland black spruce forest soil carbon in national-scale forest C accounting models? *Ecosystems* 16:1071–1086.

Bona, K.A., C.H. Shaw, J.W. Fyles, and W.A. Kurz. 2016. Modelling moss-derived carbon in upland black spruce forests. *Can. J. For. Res.* 46:520–534.

Bond-Lamberty, B., A.V. Rocha, K. Calvin, B. Holmes, C. Wang, and M.L. Goulden. 2013. Disturbance legacies and climate jointly drive tree growth and mortality in an intensively studied boreal forest. *Glob. Change Biol.* 20:216–227.

Böttcher, H., A. Freibauer, M. Obersteiner, and E.-D. Schulze. 2008. Uncertainty analysis of climate change mitigation options in the forestry sector using a generic carbon budget model. *Ecol. Model.* 213:45–62.

Chen, J., S.J. Colombo, M.T. Ter-Mikaelian, and L.S. Heath. 2008. Future carbon storage in harvested wood products from Ontario's Crown forests. *Can. J. For. Res.* 38:1947–1958.

Chen, J., S.J. Colombo, M.T. Ter-Mikaelian, and L.S. Heath. 2010. Carbon budget of Ontario's managed forests and harvested wood products, 2001–2100. *For. Ecol. Manage.* 259:1385–1398.

Colombo, S.J., J. Chen, M.T. Ter-Mikaelian et al. 2012. Forest protection and forest harvest as strategies for ecological sustainability and climate change mitigation. *For. Ecol. Manage.* 281:140–151.

Comic DR. 2021. Regression models of carbon and CO2 sequestration of hybrid poplar plantations in Northern Serbia. *Can. J. For. Res.* 51:1859–1874.

Crookston, N.L., G.E. Rehfeldt, G.E. Dixon, and A.R. Weiskittel. 2010. Addressing climate change in the forest vegetation simulator to assess impacts on landscape forest dynamics. *For. Ecol. Manage.* 260:1198–1211.

Fang, J., H.H. Shugart, F. Liu, X. Yan, Y. Song, and F. Lv. 2022. FORCCHN V2.0: An individual-based model for predicting multiscale forest carbon dynamics. *Geosci. Model Dev.* 15:6863–6872.

Feng, Z., T. Rütting, H. Pleijel et al. 2015. Constraints to nitrogen acquisition of terrestrial plants under elevated CO_2. *Glob. Change Biol.* 21:3152–3168.

Friedlingstein, P., I. Fung, E. Holland et al. 1995. On the contribution of CO_2 fertilization to the missing biospheric sink. *Glob. Biogeochem. Cycle* 9:541–556.

Gaboury, S., J.-F. Boucher, C. Villeneuve, D. Lord, and R. Gagnon. 2009. Estimating the net carbon balance of boreal open woodland afforestation: A case-study in Québec's closed-crown boreal forest. *For. Ecol. Manage.* 257:483–494.

Geng, A., J. Chen, and H. Yang. 2019. Assessing the greenhouse gas mitigation potential of harvested wood products substitution in China. *Environ. Sci. Technol.* 53:1732–1740.

Hararuk, O., C. Shaw, and W.A. Kurz. 2017. Constraining the organic matter decay parameters in the CBM-CFS3 using Canadian national forest inventory data and a Bayesian inversion technique. *Ecol. Model.* 364:1–12.

Heath, L.S. and R. Birdsey. 1993. Carbon trends of productive temperate forests of the coterminous United States. In *Terrestrial Biospheric Carbon Fluxes Quantification of Sinks and Sources of CO2*, eds. J. Wisniewski and R.N. Sampson. Dordrecht, the Netherlands: Springer.

Heath, L.S., R.A. Birdsey, and D.W. Williams. 2002. Methodology for estimating soil carbon for the forest carbon budget model of the United States, 2001. *Environ. Pollut.* 116:373–380.

Heath, L.S., M.C. Nichols, J.E. Smith, and J.R. Mills. 2010. FORCARB2: An updated version of the U.S. Forest Carbon Budget Model. General Technical Report NRS-67. Newtown Square, PA: U.S. Department of Agriculture, Forest Service, Northern Research Station.

Heath, L.S. and J.E. Smith. 2000. An assessment of uncertainty in forest carbon budget projections. *Environ Sci Policy* 3:73–82.

Hickler, T., B. Smith, I. Prentice et al. 2008. CO_2 fertilization in temperate FACE experiments not representative of boreal and tropical forests. *Glob. Change Biol.* 14:1531–1542.

Ji, C., H. Yang, Y. Nie, and Y. Hong. 2013. Carbon sequestration and carbon flow in harvested wood products for China. *Int. For. Rev.* 15:160–168.

Karjalainen, T., A. Pussinen, J. Liski et al. 2002. An approach towards an estimate of the impact of forest management and climate change on the European forest sector carbon budget: Germany as a case study. *For. Ecol. Manage.* 162:87–103.

Kasischke, E.S., B.D. Amiro, N.N. Barger et al. 2013. Impacts of disturbance on the terrestrial carbon budget of North America. *J. Geophys. Res. Biogeosci.* 118:303–316.

Kloss, D. 2002. *Strategic Forest Management Model, Version 2.0 User Guide.* Sault Ste. Marie, Ontario, Canada: Ontario Ministry of Natural Resources, Forest Management Planning Section.

Krenke, A.N., A.V. Ptichnikov, E.A. Shvarts, and I.K. Petrov. 2021. Assessments of the forest carbon balance in the national climate policies of Russia and Canada. *Doklady Earth Sci.* 501:1091–1095.

Kurz, W.A. and M.J. Apps. 1999. A 70-year retrospective analysis of carbon fluxes in the Canadian forest sector. *Ecol. Appl.* 9:526–547.

Kurz, W.A., M. Apps, E. Banfield, and G. Stinson. 2002. Forest carbon accounting at the operational scale. *For. Chron.* 78:672–679.

Kurz, W.A., M.J. Apps, T.M. Webb, and P.J. McNamee. 1992. *The carbon budget of the Canadian forest sector: Phase I*, p. 93. Forestry Canada, Northwest Region. Edmonton, Alberta, Canada: Northern Forestry Centre. Information Report NOR-X-326E.

Kurz, W.A., S.J. Beukema, and M.J. Apps. 1998. Carbon budget implications of the transition from natural to managed disturbance regimes in forest landscapes. *Mitig. Adapt. Strategies Glob. Change* 2:405–421.

Kurz, W.A., C.C. Dymond, G. Stinson et al. 2008a. Mountain pine beetle and forest carbon feedback to climate change. *Nature* 452:987–990.

Kurz, W.A., C.C. Dymond, T.M. White et al. 2009. CBM-CFS3: A model of carbon-dynamics in forestry and land-use change implementing IPCC standards. *Ecol. Model.* 220:480–504.

Kurz, W.A., G. Stinson, G.J. Rampley, C.C. Dymond, and E.T. Neilson. 2008b. Risk of natural disturbances makes future contribution of Canada's forests to the global carbon cycle highly uncertain. *Proc. Natl. Acad. Sci. USA* 105:1551–1555.

Landsberg, J.J. and R.H. Waring. 1997. A generalised model of forest productivity using simplified concepts of radiation-use efficiency, carbon balance and partitioning. *For. Ecol. Manage.* 95:209–228.

Lemma, B., D.B. Kleja, M. Olsson, and I. Nilsson. 2007. Factors controlling soil organic carbon sequestration under exotic tree plantations: A case study using the CO2Fix model in Southwestern Ethiopia. *For. Ecol. Manage.* 252:124–131.

Liski, J., T. Palosuo, M. Peltoniemi, and R. Sievänen. 2005. Carbon and decomposition model Yasso for forest soils. *Ecol. Model.* 189:168–182.

Luckai, N., G.R. Larocque, L. Archambault, D. Paré, R. Boutin, and A. Groot. 2012. Using the carbon budget model of the Canadian forest sector (CBM-CFS3) to examine the impact of harvest and fire on carbon dynamics in selected forest types of the Canadian Boreal shield. *For. Chron.* 88:426–438.

Masera, O.R., J.F. Garza-Caligaris, M. Kanninen et al. 2003. Modeling carbon sequestration in afforestation, agroforestry and forest management projects: The CO2FIX V.2 approach. *Ecol. Model.* 164:177–199.

Mills, J.R. and J.C. Kincaid. 1992. *The Aggregate Timberland Assessment System—ATLAS: A Comprehensive Timber Projection Model.* U.S. Department of Agriculture, Forest Service, Pacific Northwest Research Station, Portland, OR.

Mohren, G.M.J. and K.C.G.M. Goldewijk. 1990. *CO2Fix: A dynamics model of the CO2 fixation in forest stands*, p. 96. De Dorschkamp, Research Institute of Forestry and Urban Ecology, Wageningen. Rapport 624.

Nabuurs, G.-J., M.-J. Schelhaa, and A. Pussinen. 2000. Validation of the European forest information scenario model (EFISCEN) and a projection of Finnish forests. *Silva Fennica* 34(2):167–179.

Negash, M. and M. Kanninen. 2015. Modeling biomass and soil carbon sequestration of indigenous agroforestry systems using CO2FIX approach. *Agric. Ecosyst. Environ.* 203:147–155.

Parton, W., J. Scurlock, D. Ojima et al. 1993. Observations and modeling of biomass and soil organic matter dynamics for the grassland biome worldwide. *Glob. Biogeochem. Cycle* 7:785–809.

Peng, C., J. Liu, Q. Dang, M.J. Apps, and H. Jiang. 2002. TRIPLEX: A generic hybrid model for predicting forest growth and carbon and nitrogen dynamics. *Ecol. Model.* 153:109–130.

Peng, C., Z. Ma, X. Lei et al. 2011. A drought-induced pervasive increase in tree mortality across Canada's boreal forests. *Nature Clim. Change* 1:467–471.

Pilli, R., R. Alkama, A. Cescatti, W.A. Kurz, and G. Grassi. 2022. The European forest carbon budget under future climate conditions and current management practices. *Biogeosciences* 19:3263–3284.

Pilli, R., G. Fiorese, and G. Grassi. 2015. EU mitigation potential of harvested wood products. *Carbon Bal. Manage.* 10:6.

Pilli, R., G. Grassi, W.A. Kurz, G. Fiorese, and A. Cescatti. 2017. The European forest sector: Past and future carbon budget and fluxes under different management scenarios. *Biogeosciences* 14:2387–2405.

Phillips, O.L., L.E.O.C. Aragao, S.L. Lewis et al. 2009. Drought sensitivity of the Amazon rain forest. *Science* 323:1344–1347.

Raich, J.W., E.B. Rastetter, J.M. Melillo et al. 1991. Potential net primary productivity in South America: Application of a global model. *Ecol. Appl.* 1:399–429.

Reich, P.B. and S.E. Hobbie. 2013. Decade-long soil nitrogen constraint on the CO_2 fertilization of plant biomass. *Nature Clim. Change* 3:278–282.

Richards, G.P. and D.M.W. Evans. 2004. Development of a carbon accounting model (FullCAM Vers. 1.0) for the Australian continent. *Aust. For.* 67:277–283.

Rizvi, R.H., R. Newaj, O.P. Chaturvedi, R. Prasad, A.K. Handa, and B. Alam. 2019. Carbon sequestration and CO2 absorption by agroforestry systems: An assessment for Central Plateau and Hill region of India. *J. Earth Syst. Sci.* 128.

Sage, L.K., C.T. Smith, W. Kurz, E. Thiffault, D. Pare, and P. Bernier. 2019. Empirical and predicted boreal forest carbon pools following stem-only harvesting in Quebec, Canada. *Soil Sci. Soc. Am. J.* 83: S59–S81.

Skog, K.E., K. Pingoud, and J.E. Smith. 2004. A method countries can use to estimate changes in carbon stored in harvested wood products and the uncertainty of such estimates. *Environ. Manag.* 33: S65–S73.

Schelhaas, M., P. Van Esch, T. Groen et al. 2004. CO2FIX V 3.1-A modelling framework for quantifying carbon sequestration in forest ecosystems. Wageningen, the Netherlands: Alterra; rapport 1068.120 blz.

Smyth, C., W.A. Kurz, G. Rampley, T.C. Lempriere, and O. Schwab. 2017. Climate change mitigation potential of local use of harvest residues for bioenergy in Canada. *Glob. Change Biol.* 9:817–832.

Smyth, C.E., W.A. Kurz, and J.A. Trofymow. 2011. Including the effects of water stress on decomposition in the carbon budget model of the Canadian forest sector CBM-CFS3. *Ecol. Model.* 222:1080–1091.

Song, C. and C.E. Woodcock. 2003. A regional forest ecosystem carbon budget model: Impacts of forest age structure and landuse history. *Ecol. Model.* 164:33–47.

Taylor, A.R., J.R. Wang, and W.A. Kurz. 2008. Effects of harvesting intensity on carbon stocks in eastern Canadian red spruce (*Picea rubens*) forests: An exploratory analysis using the CBM-CFS3 simulation model. *For. Ecol. Manage.* 255:3632–3641.

Ter-Mikaelian, M.T., S.J. Colombo, and J. Chen. 2015. The burning question: Does forest bioenergy reduce carbon emissions? A review of common misconceptions about forest carbon accounting. *J. For.* 113: 57–68.

Thornton, P.E., B.E. Law, H.L. Gholz et al. 2002. Modeling and measuring the effects of disturbance history and climate on carbon and water budgets in evergreen needleleaf forests. *Agric. For. Meteorol.* 113:185–222.

Trofymow, J.A., G. Stinson, and W.A. Kurz. 2008. Derivation of a spatially explicit 86-year retrospective carbon budget for a landscape undergoing conversion from old-growth to managed forests on Vancouver Island, BC. *For. Ecol. Manage.* 256:1677–1691.

Urban, D.L. 1990. *A versatile model to simulate forest pattern: A user's guide to ZELIG Version 1.0*. Charlottesville, VA: University of Virginia.

van Mantgem, P.J., N.L. Stephenson, J.C. Byrne et al. 2009. Widespread increase of tree mortality rates in the western United States. *Science* 323:521.

Wakelin SJ, Searles N, Lawrence D, Paul TSH. 2020. Estimating New Zealand's harvested wood products carbon stocks and stock changes. *Car. Bal. Manag.* 15.

Waterworth, R.M. and G.P. Richards. 2008. Implementing Australian forest management practices into a full carbon accounting model. *For. Ecol. Manage.* 255:2434–2443.

Waterworth, R.M., G.P. Richards, C.L. Brack, and D.M.W. Evans. 2007. A generalised hybrid process-empirical model for predicting plantation forest growth. *For. Ecol. Manage.* 238:231–243.

Watson, C. 2009. *Forest carbon accounting: Overview and principles*, p. 39. United Nations Development Programme.

White, T., N. Luckai, G.R. Larocque, W.A. Kurz, and C. Smyth. 2008. A practical approach for assessing the sensitivity of the carbon budget model of the Canadian forest sector (CBM-CFS3). *Ecol. Model.* 219:373–382.

Williams, C.A., G.J. Collatz, J. Masek, and S.N. Goward. 2012. Carbon consequences of forest disturbance and recovery across the conterminous United States. *Glob. Biogeochem. Cycle* 26:GB1005.

Zhao, J., X. Yan, and G. Jia. 2012. Simulating net carbon budget of forest ecosystems and its response to climate change in Northeastern China using improved FORCCHN. *Chin. Geogra. Sci.* 22:29–41.

10 Modeling the Impacts of Pest Damage

Case Studies for Conifers in British Columbia, Canada

Michael G. Cruickshank, Rona N. Sturrock, Mario Di Lucca, and René I. Alfaro

10.1 INTRODUCTION

A pest, by definition, is an organism that causes harm to mankind, crops, or property (Dent 2000). Therefore, the concepts of pest and damage are closely linked. These are usually, although not exclusively, caused by insects and pathogens. Forested lands are usually managed on a multiple-use basis, that is, for the simultaneous production of several resources from the same unit of land. Resources include timber products and ecosystem services such as water for irrigation or human consumption, livestock and wildlife forage, wildlife protection, recreation, and fish habitat. Consequently, in multiple-use forestry, pest damage can be defined as the negative effects of an organism on the quantity and quality of the entire array of resources that are expected from a unit of land. The role of the silviculturist is to create a forest that will produce resources of the desired kind, in the right amount, and of the right value (Smith 1962) while maximizing all timber and nontimber values and minimizing the impacts on ecosystem function (Leech et al. 2009).

To quantify the impacts of pests on a particular resource, pest damage must be separated from other changes that normally occur as the stand matures and undergoes the process of succession, such as competition and natural attrition (Spurr and Barnes 1973). The stand is a dynamic entity where the complex of tree species, herbaceous plants, and other organisms may compete for light, air, water, heat, and nutrients. The consequences of this competition are a continuous change in the composition, structure, and general character of the stand as trees or other species die, others expand and take more space, and new species colonize the forest. Pests usually become a problem when they alter the expected quality and quantity of products provided by an ecosystem. The damage from pests is governed by one or more changes in host life history traits of growth, survival, and senescence.

Pest damage reduces forest productivity in four ways: some have selective mortality of certain tree species, such as the bark beetles in the genus *Dendroctonus* on conifers; some reduce tree vigor and growth rates by feeding on various parts of the tree, such as the fungal leaf blight (*Didymascella thujina* (Dur.)) infecting cedar needles and branches; some compete with trees for sunlight and nutrients, such as weed species; and some reduce tree quality through stem defects that destroy or degrade the wood, as is the case of the shoot weevil (*Pissodes strobi* Peck). These effects can interact; for example, changes in stem growth can also affect wood quality (Cruickshank 2010). Pests are often normal processes of ecosystems, periodically killing mature trees like the mountain pine beetle (MPB, *Dendroctonus ponderosae* Hopkins), which accelerates the succession of pure pine stands to multispecies stands (Alfaro et al. 2008) but often leads to economic losses to forest industry. The widespread impact of the 1999–2015 outbreak of MPB in British Columbia (BC) created short-term uplift in the harvest of large-scale salvage of merchantable dead timber as well as susceptible timber.

 DOI: 10.1201/9781003431084-12

This chapter (1) reviews the requirements that need to be considered when developing models that simulate tree and pest interactions, (2) evaluates pest damage to timber resources, and (3) describes case studies based on a suite of models that assist forest managers in BC, Canada, to determine growth and yield and the subroutines used to simulate pest damage. This chapter underscores how pests affect forest management decisions.

10.2 REQUIREMENTS FOR PEST DAMAGE FORECASTING

Forest pest damage forecasting has several requirements depending on the resource evaluated. It is important to understand the following.

1. *Effects of pests on stand structures*: Have a clear understanding of pest damage at the stand and tree levels. This is usually based on repeated measurements in permanent plots of damaged and undamaged sites that compare stand structure, age classes, tree growth and reproduction, tree mortality, and product quantity and quality.
2. *Pest epidemiology*: Understanding what triggers an attack, how long and how many attacks are expected between the seedling stage and maturity, how do organisms spread to invade new stands and trees, and how do they survive from one year to the next.
3. *Pest interactions with management actions*: Certain forest practices can increase or mitigate the damage.
4. *Have reliable stand projection models capable of projecting individual tree growth from seedling to maturity*: Because some pests selectively damage certain trees leaving others undamaged, these models should be able to capture the interaction between trees, that is, they should be able to mimic the competition between trees and the release of competitive interference that occurs in surviving trees when their neighbors are killed. Ideally, these stand projection models could mimic the action of pest control alternatives and how they affect the level of damage at tree and stand levels.

10.3 EFFECTS OF PESTS ON STAND STRUCTURES

Forest stands are characterized by their structure: the proportion in which the various species occur, their age, and the distribution of tree sizes. Depending on the ecosystem, stand structure can be highly variable over time. Typically, forest pests affect tree size distribution in a stand by selectively killing trees of particular size classes or species. For example, the western spruce budworm (*Choristoneura occidentalis* Freeman) causes the heaviest mortality among the smaller, understory trees. Since the trees in these lower canopy layers are expected to form the next crop under uneven-aged management, this creates losses because of the reduced future crop and harvest delays. At the other extreme, MPB kills the largest diameter classes in the stand, creating immediate timber losses.

The volume grown during a time interval is the balance between that added by the existing trees plus new tree entering the stand (ingrowth), minus the mortality occurring in the stand. Mortality is a natural result of the complex interaction among ecosystem components and of processes such as competition, suppression, senescence, insects, and diseases or sometimes through abiotic physical factors such as moisture deficits. The consequence of tree death is that the woody structure of trees that die is returned to the ecosystem via various ecological cycles (Franklin et al. 1987) and more space and resources (light, nutrients, water, and energy) are usually, but not always, available to the trees that survive. Therefore, mortality up to a certain threshold level may be beneficial for the stand. Especially in overstocked stands, some level of mortality may cause the surviving dominant trees to attain larger sizes and increasing product value. However, mortality levels higher than a threshold reduce the stand yield because larger areas are taken out of production and surviving trees cannot provide sufficient compensatory growth. Thus, tree mortality triggers intrastand responses, which forest simulators of pest impact must take into account. In the end, losses at the stand level may not be the actual sum of the individual trees killed but the difference between volume of the mortality minus compensatory growth in the surviving trees. Compensatory growth does not always follow density reduction if the remaining trees also become affected by the pest. For example, Armillaria root disease (RD) kills trees and colonizes the root system, and then the fungus transfers via root contacts to infect neighbors; as a result, compensatory growth is not always coupled with density reduction through mortality (Cruickshank et al. 2009).

Armillaria root disease is an example of a polyphagous pest that is not specific to one host and actually infects most plants with woody roots.

The spatial and temporal patterns of tree mortality also affect the level of damage. Spatially, a clumped distribution of affected trees usually results in more severe losses, relative to an even distribution of mortality, as entire areas are taken out of production. On the temporal scale, damage is hard to estimate to maturity, and many factors may modify loss estimations such as ingrowth and other disturbances occurring in the later stand stage. A damage function describes the relationship over time between the intensity of tree attack (e.g., percent defoliation) and the resulting damage on product quantity and quality; therefore, it links the probability of damage with its consequences. When this is calculated on an area basis, it determines the level of risk in that area at any time in the future. Simulations to incorporate this pest risk are usually done at the stand level or higher.

10.4 PEST EPIDEMIOLOGY

The population of certain pest agents sporadically rises into large outbreaks that cause widespread tree mortality. This type of acute loss is characteristic of bark beetle outbreaks or pathogen molds such as certain types of Phytophthora species. For instance, in BC, the MPB kills the lodgepole pine overstory in mixed stands, which hastens succession to a different species, such as the Douglas fir or spruce (Heath and Alfaro 1990), but the massive killing of the pine reduces timber product availability. The pest population changes dramatically following such an acute attack as the beetle exhausts its food supply (large-diameter trees). Chronic losses occur from pests such as Armillaria or Phellinus root diseases, which occur more slowly, are much less visible, but have a similar effect of hastening the succession of the Douglas fir to hemlock or cedar (Morrison 2011). In contrast to many insects, the population of some of these root diseases is relatively stable and widespread and continues on the site by colonizing stumps after harvest and thinning operations, creating a threat for the next forest generation.

Many pests cause little or no tree mortality; however, they reduce stand yield by reducing the tree growth rate. Pests that cause growth loss may cause damage in distinct episodes. For example, the distinct outbreaks of defoliating insects denude the trees over a finite period. After the infestation collapses, the surviving tree growth may continue at a lower, or a higher, or a similar rate with respect to the period prior to the infestation. Other pests such as the dwarf mistletoes (DMs) cause chronic type of growth loss over the life of the tree and often go unrecognized.

The outcome of any pest attack on a tree varies with the tree host species, the pest, and the environment this occurs in. Thus, the environment such as climate or other organisms whose frequency changes over the range and lifespan of the tree will also affect the outcome. For example, Douglas fir stands in more productive areas are also associated with higher root disease impact because the roots cover the site belowground sooner and touch the fungal inoculum more frequently (Cruickshank et al. 2011). Thus, it is important to recognize the need to sample over a large part of the host range so that the host–pest interaction with other factors can be determined and applied over the appropriate area. Linked to climate and soils is the potential productivity of the stand, which in some cases determines the magnitude of attack by a pest and the growth response of the surviving trees.

Also linked to pest attack is the number of tree species that the agent can affect. Pests that attack multiple hosts must be taken into consideration when considering the complex interaction between several tree species in the case of mixed stands.

10.5 OVERVIEW OF CURRENTLY AVAILABLE PEST SIMULATORS AND DECISION SUPPORT SYSTEMS

Assessment of disease and insect impacts on forest development and knowledge of the relative consequences of alternative strategies are critical to the selection of effective forest management strategies (Erdle and MacLean 1999). Currently, there are several decision support systems (DSSs)/ models available to simulate the quantitative and/or qualitative effects of insect- and disease-caused disturbance on forested stands and landscapes. Table 10.1. provides summary information for several such impact models applicable to temperate and boreal forests in North America and to forests in Europe. There are many insect- and disease-specific epidemiological and population dynamics models and other types of models and likely thousands of datasets that, when combined, support pest DSSs, but the identification of these goes much beyond the scope of this chapter. Here, we review a few of the models available, but we provide detailed examples of work done in BC. Readers are encouraged to use the selected references and other sources provided in Table 10.1. if they want detailed information on specific pest impact models.

There are several impact models currently under development in North America, including the following diseases: Swiss Needle Cast caused by *Phaeocryptopus gaeumannii* (T. Rohde) Petrak and Hard Pine Stem Rust caused by *Cronartium comandrae* Peck, *Cronartium coleosporioides* Arthur, and *Endocronartium harknessii* (J.P. Moore) Y. Hiratsuka (Woods et al. 2000). In addition, there are examples of models which take a broad view of pest impacts and model disturbances in combination with forest succession across large (>1 million ha) landscapes (He et al. 2005) or, in the case of the paper by Dietze and Matthes (2014), where they propose and outline the application of an ecophysiological framework for modeling the impacts of insects and diseases on forest ecosystems.

The Forest Vegetation Simulator (FVS) is a nonspatial, statistical (regression based) model used by the USDA Forest Service (FS) and several other U.S. agencies for predicting forest stand dynamics (Dixon, 2002). FVS descended from the Prognosis Model, which was developed in the early 1970s for forest conditions in the U.S. Northern Rocky Mountains (Stage 1973). The FVS system actually represents a collection of models for a variety of forest characteristics, including tree growth, mortality, regeneration, understory vegetation, fire, fuels, wood volumes, carbon, biomass, insects, diseases, and economics. While FVS typically operates at the stand level, many stands can be included in a simulation. FVS is linked to USDA FS corporate software such as databases and geographic information systems. For example, FVS-EMAP is a tool for moving FVS simulation results into the ArcView® Geographic Information System (McMahan et al. 2002).

10.5.1 Description of Model

A range of information such as forest inventory data, budworm monitoring data and defoliation predictions, stand growth, and timber supply models; the SBW DSS model quantifies marginal timber supply (m^3/ha) benefits of protecting stands against SB defoliation and allows users to quantify volume benefit of protecting different areas and determine the effects on forest development and AAC. SBW DSS includes Protection Planning System, which models timber supply benefits of different management actions.

Simulates damage to individual regenerating trees, including leader destruction, crown recovery, and formation of stem defects.

Simulates increased mortality and reduced growth effects of four RDs on 40 tree species; the ability to simulate RD and four functional beetle types and windthrow is also provided in the model. The model is highly parameterized, but the user can alter many parameters through keywords.

Simulates infection and spread of *Heterobasidion* spp. in Norway spruce or Scots pine stands, estimates economic outcomes, and optimizes management options for *Heterobasidion* spp. for European conditions. The model consists of a stand dynamics module and a disease dynamics module.

TABLE 10.1
Forest Insect and Disease Impact Models

Forest Insect or Disease	Model Name	Model Description	Selected References
Douglas-fir beetle *Dendroctonus pseudotsugae* Hopk.	Douglas-fir Beetle Impact Model (DFBIM) extension to FVS	Main impact is accelerated mortality of large diameter, dominant and co-dominant trees; no height or diameter growth reductions imposed.	Marsden et al. 1994; http://www.fs.fed.us/foresthealth/technology/products.shtml
Douglas-fir Tussock Moth *Orgyia pseudotsugata* (McDunnough)	Douglas-fir Tussock Moth Model (DFTM) extension to FVS	Simulates reduced growth and increased mortality of host tree species due to defoliation by tussock moth larvae; some pest control methods and FVS management actions can be used in simulations.	Monserud and Crookston 1982; http://www.fs.fed.us/foresthealth/technology/products.shtml
Mountain Pine Beetle *Dendroctonus ponderosae* Hopk.	Mountain Pine Beetle (MPB) Impact Model extension to FVS	Simulates MPB mortality impacts on lodgepole pine (*Pinus contorta* Dougl.); is a single-stand model with two outbreak/epidemic simulation methods available: first method utilizes MPB Rate of Loss model while second method utilizes MPB Population Dynamics model.	Burnell 1977; Crookston et al. 1978; Cole and McGregor 1983; http://www.fs.fed.us/foresthealth/technology/products.shtml
Mountain Pine Beetle, *D. ponderosae*; western pine beetle, *D. brevicomis* Leconte & *Ips* spp.	Westwide Pine Beetle (WWPB) Model	Simulates impacts of three beetle species for which western pines are host; it is a multi-stand, landscape-scale model that recognizes the conditions and events occurring in all simulated stands.	Crookston and Stage 1991; http://www.fs.fed.us/foresthealth/technology/products.shtml
Mountain Pine Beetle *Dendroctonus ponderosae* Hopk.	Mountain Pine Beetle (SELES-MPB)	Uses two MPB simulation models, which operate at distinct spatial and temporal scales, to produce a third model: a spatially explicit landscape-scale simulation of mountain pine beetle outbreak activity. SELES-MPB is a stochastic, process-based simulation that can project MPB spread and impacts on the landscape and can be used to evaluate management strategies; SELES-MPB can also project stand level impacts.	Safranyik et al. 1999; Riel et al. 2004

Pine Weevil *Hylobius abietis* Linn.	Hylobius Management Support System	A decision support system for managing plantation establishment around pine weevil; system allows users to plan site monitoring exercise to determine insect density and examine management/control options to minimize transplant losses.	http://www.forestry.gov.uk/fr/INFD-6T6LP4
Western Spruce Budworm *Choristoneura freemani* Razowski Syn.: *Choristoneura occidentalis* Freeman	Western Spruce Budworm (WSB) Damage model extension to FVS	Predicts top-kill, height growth, diameter growth, and tree mortality impacts of western spruce budworm defoliation on its spruce and fir host tree species.	Sheehan et al. 1989; Crookston et al. 1990; Crookston1991; http://www.fs.fed.us/foresthealth/technology/products.shtml
Spruce Budworm *Choristoneura fumiferana* (Clem.)	Spruce Budworm Decision Support System (SBW DSS)	A range of information such as forest inventory data, budworm monitoring data and defoliation predictions, stand growth, and timber supply models, the SBW DSS model quantifies marginal timber supply (m^3/ha) benefits of protecting stands against SB defoliation and allows users to quantify volume benefit of protecting different areas and determines effects on forest development and AAC. SBW DSS includes Protection Planning System (PROPS), which models timber supply benefits of different management actions.	MacLean et al. 2001
Spruce Weevil/white pine weevil *Pissodes strobi* Peck	Spruce Weevil Attack (SWAT) Model module of TASS	Simulates damage to individual regenerating trees, including leader destruction, crown recovery, and formation of stem defects	Alfaro et al. 1997
Annosus Root Disease (RD), *Heterobasidion annosum* sensu lato; Armillaria RD, *Armillaria ostoyae* (Romagnesi) Herink; & Phellinus RD/Laminated root rot (LRR), *Coniferiporia weirii*	Western Root Disease Model extension to FVS	Simulates increased mortality and reduced growth effects of four RDs on 40 tree species; ability to simulate RD and four functional beetle types and windthrow also provided in model. Model highly parameterized but user can alter many parameters through keywords.	McNamee et al. 1991; Shaw et al. 1991; Frankel et al. 1992; Frankel 1998; http://www.fs.fed.us/foresthealth/technology/products.shtml

(Continued)

TABLE 10.1 (*Continued*)
Forest Insect and Disease Impact Models

Forest Insect or Disease	Model Name	Model Description	Selected References
Annosus Root Disease (RD) *H. annosum* sensu lato	RotStand Model	Simulates infection and spread of *Heterobasidion* spp. in Norway Spruce or Scots Pine stands, estimates economic outcomes and optimizes management options for *Heterobasidion* spp. for European conditions. Model consists of a stand dynamics module and a disease dynamics module.	Pukkala et al. 2005; Wang et al. 2014
Armillaria Root Disease (RD) *A. ostoyae*	ROTSIM module of TASS	Biological model of Armillaria RD that assesses disease impacts, disease-management interactions, and potential control strategies in managed stands of interior BC Douglas-fir. ROTSIM mimics key processes in disease initiation and development such as distribution of tree roots, root and inoculum contacts, height growth reduction in infected stems, and inoculum persistence.	Stearns-Smith et al. 2004; Cruickshank et al. 2011; Cruickshank and Filipescu 2012
Armillaria RD, *A. ostoyae*; Phellinus RD/LRR, *Coniferiporia weirii* & Tomentosus RD, *Onnia tomentosa* (Fr.) P. Karst	Root Rot Tracker	A newer model that differs from earlier models by incorporating basic processes of spread by individual root contact; model stochastically simulates individual root growth and disease development using known physiological variables; estimates for numbers of infected trees and mortality are provided in tabular and graphical form.	Peet and Hunt 2005
Damping off & root rot *Fusarium oxysporum* Schlecht. Emend. Snyder & Hansen	Douglas-fir Damping Off & Root Rot Model	Model simulates *Fusarium oxysporum* infection of Douglas-fir roots to predict seedling infection, mortality, root growth in various temperature regimes, root growth rates and sizes, and concentrations of inoculum. The most important factors in simulated root infection are those affecting size and number of infection sites (root tips), size and concentration of inoculum particles, and rate of lesion elongation.	Bloomberg 1979

Dwarf Mistletoe (DM) *Arceuthobium spp.*	Interim DM Impact Model (IDMIM) extension to FVS	Simulates reduced growth and increased mortality caused by DM dwarf mistletoes on several different host tree species throughout western USA; model based on Hawksworth six-class dwarf mistletoe rating (DMR); model outputs enable users to evaluate relative effectiveness of multiple management scenarios.	Hawksworth 1977; Hawksworth et al. 1992; http://www.fs.fed.us/foresthealth/technology/products.shtml
DM *Arceuthobium spp.*	Three Scale Model	Simulates DM reproduction, dispersal & spatial patterns to assess long-term, cumulative effects.	Robinson and Geils 2006
Phellinus RD/LRR *Coniferiporia weirii*	ROTSIM module of TASS	Biological model of Phellinus RD that assesses disease impacts, disease-management interactions, and potential control strategies in managed stands of interior BC Douglas-fir. ROTSIM mimics key processes in disease initiation and development such as distribution of tree roots, root and inoculum contacts, height growth reduction in infected stems, and inoculum persistence.	Bloomberg 1988
White Pine Blister Rust (WPBR) *Cronartium ribicola* J.C. Fisch.	Blister Rust Infection Model for the Greater Yellowstone Ecosystem (GYE)	Simulates rate of BR progression to predict future levels of BR infection/impact in GYE	Hatala et al. 2011
White Pine Blister Rust (WPBR) *C. ribicola*	White Pine Blister Rust Model extension to FVS	Simulates life cycle of *C. ribicola* and resulting mortality of white pine; blister rust index (measure of WPBR infection potential for site, area, or stand) drives the model; three principal alternate hosts (*Ribes* spp.) represented in the model and different management options can be evaluated.	McDonald et al. 1981; http://www.fs.fed.us/foresthealth/technology/products.shtml
Western gall rust, (WGR), *Cronartium harknessii*	Western gall rust model to TASS	Simulates stand-level rust incidence and tree-level rates of infection over time, survival rates, and the location of galls requiring removal during lumber manufacturing.	Sattler et al. 2019

Biological model of Armillaria root disease that assesses disease impacts, disease–management interactions, and potential control strategies in managed stands of interior BC Douglas fir. ROTSIM mimics key processes in disease initiation and development such as distribution of tree roots, root and inoculum contacts, height growth reduction in infected stems, and inoculum persistence.

10.6 ECOLOGICAL FOREST MANAGEMENT OF PESTS THROUGH MODELING: CASE STUDIES FOR BRITISH COLUMBIA

10.6.1 TASS/TIPSY Growth and Yield Model Overview

The Tree and Stand Simulator (TASS-BC Ministry of Forests 2010) is a decision support tool that synthesizes and integrates multidisciplinary research, with an emphasis on response to silviculture treatments mainly for single-species stands (Mitchell 1975). TASS is a spatially explicit, density-dependent stand growth model capable of simulating the interaction and competition of trees as they develop. When a tree dies, the crowns of the surrounding trees may take advantage of the available space, resulting in accelerated growth. TASS and its derivatives, including the Table Interpolation Program for Stand Yields (TIPSY) (TIPSY–BC Ministry of Forests 2010) generated from TASS, are used in BC to project the growth and value of managed stands. Silviculturists use TASS and TIPSY to evaluate the impact of silvicultural treatments, such as planting, thinning, fertilization, and genetics of the stock, on future tree size, volume, and product value. A recent version of TASS (TASS III) was linked to a light interception model to improve predictions for stands with complex vertical structure and multiple species. This will facilitate a link with other models that forecast a wide range of future nontimber forest values, including wildlife habitat, biodiversity, visual quality objectives, and carbon. TIPSY is primarily used for timber supply analyses to support economic and ecological decisions at the landscape level in a management unit such as a timber supply area where multiple-use decisions of that area are required. TIPSY is available online from the BC Ministry of Forests, Lands and Natural Resource Operations.

10.6.2 TASS Pest Submodel Overview

The structure of TASS is amenable to input pest impacts through submodels that mimic the impact of pests on tree growth (Figure 10.1, biophysical factors). Two insect submodels, the Spruce Weevil Attack Trial (SWAT) model and the Mountain Pine Beetle (BEETLE) model, and a Rust and Root Disease submodel (ROTSIM), are incorporated into TASS to predict disturbance effects on the stand growth attributes listed previously (Table 10.2).

In addition to inputting pest impacts through TASS, users can estimate pest losses using TIPSY. TIPSY interpolates table format information, that is, output from multiple TASS runs under various stand conditions such as regeneration method, site index, species, and initial destiny. Utilizing this information, TIPSY predicts future volumes for a variety of harvesting patterns ranging from clearcutting to variable retention, silvicultural regimes, and pest impact. Additional factors that affect growth and survival such as disturbance agents can be input into TIPSY as Operational Adjustment Factors (OAFs) that lower or raise the model outputs using defaults and/or user-defined adjustment proportions at a stand age or several ages. Usually, OAFs are first simulated by interaction of the TASS model with a pest submodel under various stand conditions (Figure 10.1) and then input into TIPSY. Investment analyses based on predicted log and lumber yields are also possible with Silviculture on Yield, Lumber Value, and Economic Return, an integrated value-chain decision support tool that incorporates TASS and linked submodels (Mitchell 1988) and whose outputs are also available in TIPSY (Figure 10.1). Three pest submodels are described here (ROTSIM, SWAT, and BEETLE).

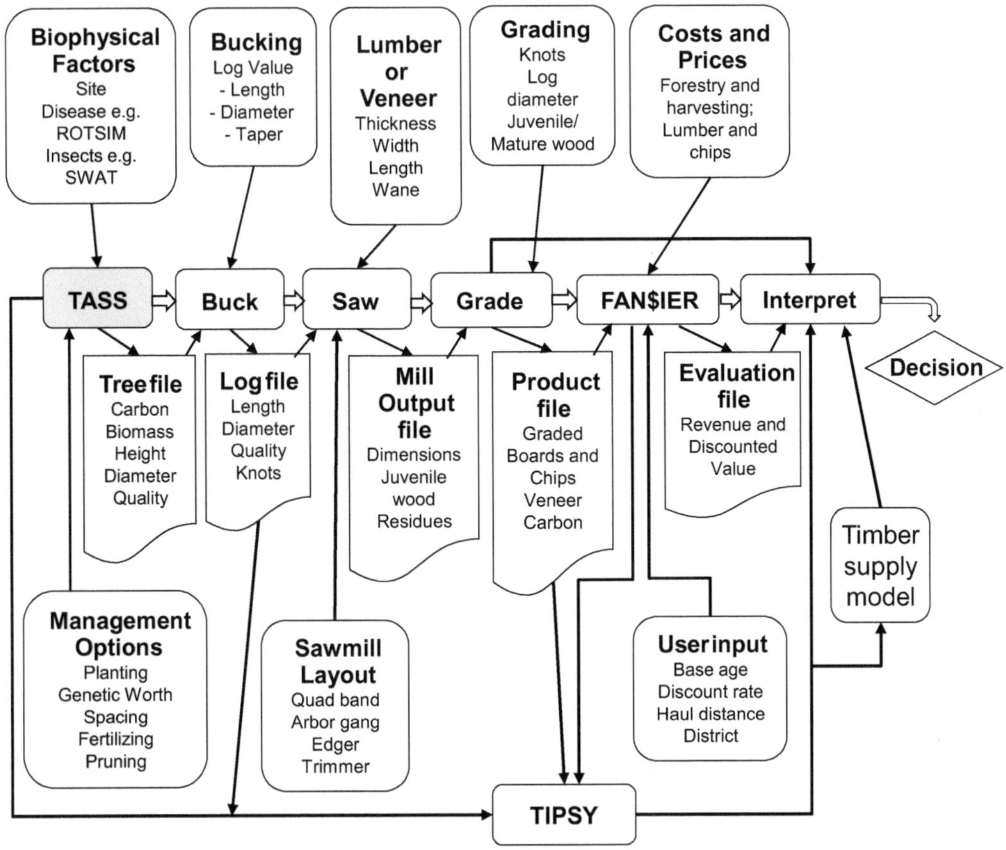

FIGURE 10.1 The integration of the Tree and Stand Simulator, Table Interpolation Program for Stand Yields, and Silviculture on Yield, Lumber Value, and Economic Return models and the inclusion of pest disturbance submodels (biophysical factors) for forest management decisions.

10.6.2.1 Phellinus Root Disease Simulation Using the TASS-ROTSIM and TIPSY Models: An Application with Operational Adjustment Factors

Phellinus root disease, also known as laminated root rot (LRR), is one of the most serious diseases affecting the growth and survival of Douglas fir in western North America (Thies and Sturrock 1995). LRR is caused by the fungus *Coniferiporia weirii* (Murrill) L.W. Zhou & Y.C. Dai, *comb. nov.*, previously known as *P. weirii* (Murr.) Gilb., a Douglas fir form. *Coniferiporia weirii* spreads among living trees and between colonized stumps and live roots via root contact. LRR persists on colonized roots and stumps (disease inoculum) in infected stands and can increase in incidence and impact where inoculum is not managed and stands are regenerated to susceptible coniferous species (Hansen 1979). LRR is estimated to cause annual landscape-level Douglas fir volume losses of less than 5% through mortality, windthrow, and growth reduction, except in heavily infected stands, where stand losses are much higher (Goheen and Hansen 1993; Thies 1983) (Figure 10.2). Productivity losses to LRR may be partially offset by infill of less susceptible species in root disease-created canopy gaps or increased growth of surviving trees. In stands not solely managed for wood production, tree mortality may provide wildlife habitat and increased biodiversity of species and other nontimber forest products (Thies and Sturrock 1995).

The ROTSIM model simulates root system development and disease epidemiology in communication with the TASS model. The model biological underpinnings follow empirical disease

TABLE 10.2
Attributes of Pests Impacts Simulated under the Tree and Stand Simulator and Table Interpolation Program for Stand Yields

Pest Impact Attribute	Mountain Pine Beetle	Spruce Weevil	Armillaria Root Disease	Laminated Root Rot	Western Gall Rust
Age affected	Mature trees 80 years old	Young stands 5-30 years old	All ages	All ages	<20 years
Mode of contagion	Widespread	Clumped	Random to clumped	Random to clumped	Random
Periodic on stand life?	Yes	No	Yes	Yes	Yes
Parts attacked	Trunk	Apical leader	Roots	Roots	Branches, Stem
Effect on tree	–Kills tree	–Reduce growth –Product downgrade	–Kills tree –Reduce growth –Product downgrade	–Kills tree –Reduce growth	–Kills tree –Product downgrade
Stand response	Triggers regeneration –single storied stand to single or multi-aged, multispecies	Eventually stands outgrow weevil damage –stem defects –longer rotation	Causes species replacement to less susceptible –reduced yields of tree most species	Causes replacement to less susceptible –reduced yield of Douglas-fir	Causes replacement to less susceptible –thinning agent –stem defects –reduced yield
Tree species affected/modelled	*Pinus contorta* var. *latifolia*	Attacks all species of Picea and most Pinus species.	Attacks all species, *Pseudotsuga menziesii* modelled	*Pseudotsuga menziesii*	*Pinus contorta* var. *latifolia*

FIGURE 10.2 Crown of a Douglas fir infected by *Coniferiporia weirii* (laminated root rot). Note the contrast between the rounded top and thinning foliage of this tree and the tops of the healthy Douglas firs in the immediate background.

epidemiology from sample plots related to pathogen inoculum quantity and quality and its spatial and temporal distribution. More specifically, these are root growth, the probability of root contacts between healthy and infected roots, fungal transfer to healthy roots, mycelial spread, tree growth reduction, mortality, and inoculum viability in infected roots and stumps.

The root system is simulated by dividing the ground horizontally around a tree base into 0.5 m concentric rings, then the rings into four quadrants (i.e., 90° each), and finally vertically into 25 cm deep soil zones (Bloomberg 1988). Within the smallest unit described earlier are given the estimated number of healthy and infected roots and their probability of contact, the probability of fungal transfer among roots, and the disease spread rate within root systems by means of regression equations (Reynolds and Bloomberg 1982). The growth of the root system is driven by the stem size governed by TASS and the rooting depth within each unit. The growth reduction in annual radial increment is applied to infected trees in TASS until a reduction level of 35% is reached, which then results in tree mortality. Mortality from the pest submodel is allowed to interact within the TASS model to account for changed spatial arrangement of stems. The model requires the size and distribution of infected stumps (primary inoculum), the planting density, and site index as initial factors. Infected roots and stump inocula are allowed to remain viably infective for a time period after initial colonization. ROTSIM is currently configured to simulate the effects of Phellinus and Armillaria root diseases in Douglas fir.

Using the Phellinus version of TASS–ROTSIM to investigate the impacts of LRR on interior Douglas fir yields, root disease experts in BC ran a simulation of three different levels of LRR on interior Douglas fir planted at uniform 1,200 stems/ha at a site index of 28 m. The initial levels of LRR were no root disease, low root disease (15 infected stumps/ha), and high root disease (40 infected stumps/ha). The volume–age curves generated from these TASS–ROTSIM simulations were then used to estimate proportional reduction values for infected stands compared to healthy stands given as custom OAFs that vary over time (Table 10.3). OAFs can be included in TIPSY to adjust for the impacts of root disease across all of the TIPSY outputs.

These OAFs were used in TIPSY to determine the net present lumber quantity and quality values (NPV) at varying stand ages and for two LRR levels (Figure 10.3) relative to the unaffected stand. Impacts were calculated for each tree and then summed per hectare. Tree stems were bucked into logs by the buck simulator, logs sawn into lumber by the sawmill simulator, and the resulting lumber graded and costed according to quantity, size, and grade by the economic module FAN$IER (Figure 10.1). FAN$IER also calculated the silviculture, manufacturing, and harvest costs that were subtracted from the lumber revenue. All revenue and costs were discounted at 4%. Currently, Douglas fir stands are harvested at about the age of 80–100 years depending on other nontimber constraints. For example, by the age of 80, in a monoculture Douglas fir stand, the simulated NPV of

TABLE 10.3
Custom Operational Adjustment Factors Derived from the Tree and Stand Simulator – ROTSIM for Use in the Table Interpolation Program for Stand Yields as a Proportion of Healthy Stands

Age (Years)	Low RD	High RD
20	1	1
50	0.97	0.95
80	0.92	0.73
100	0.73	0.49
120	0.56	0.26

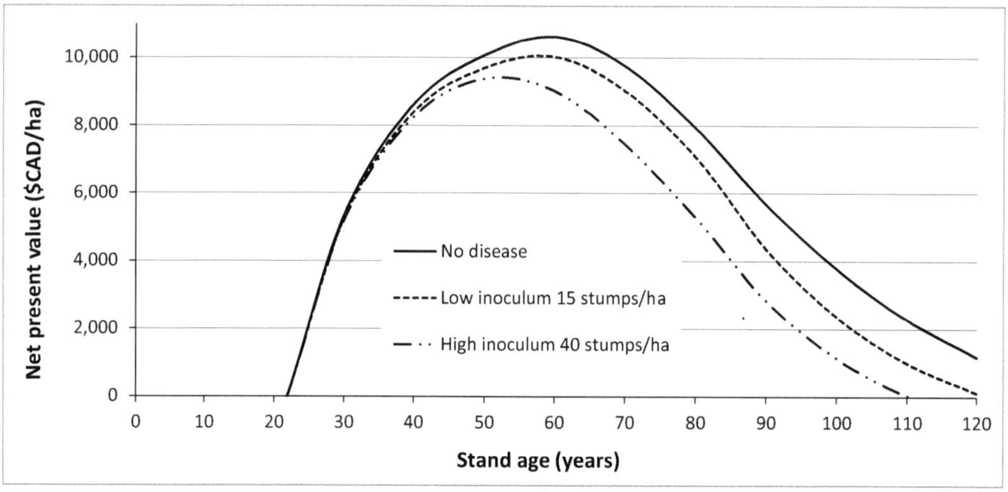

FIGURE 10.3 TIPSY outputs for net present lumber value (NPV) in an interior Douglas fir stand with the application of custom operational adjustment factors for none, low-, and high-inoculum levels of laminated root rot. NPV is the difference between the revenue and costs discounted at 4%. The lumber value is derived from all lumber grades and sizes, including mill residues, and include silviculture, manufacturing, and harvest costs.

all lumber grades and all piece sizes (2 × 4, 2 × 6, 2 × 8, and 2 × 10 nominal sizes) total about $8K/ha without root disease, while this value is reduced by about $1K/ha and $2.6K/ha for the same products with low and high fungal inoculum, respectively (Figure 10.3). The proportional value differences were approximately the same by the stand age of 100, although slightly greater for the low inoculum, but total NPVs were reduced for all stands because of the discounting forward in time. These data provided useful information for setting harvest dates. There was an economic advantage to harvest at the age of 60 years without root disease, but this was reduced in time by about ten years for the high inoculum starting conditions only. There was also low economic impact of LRR before the stand age of 40 years.

Figure 10.4 demonstrates TIPSY outputs of different levels of LRR on volume yields and related jobs (silviculture, harvesting, and manufacturing). Not surprisingly, volumes and jobs quickly begin to decline around the tree age of 80 whether disease levels are low or high. Employment yield curve measures total average person days per hectare (PD/ha) over time. By the age of 80, a reduction in person days occurred by about 8–27%, and by the age of 100 the same reduction was 27–51% for low- to high-inoculum levels compared to no disease. Harvesting stands before the stand age of 80 years predicts that the impact on forest-related employment can be reduced, and this can be very important in some areas.

Figure 10.5 demonstrates the degree to which live carbon stocks can be significantly reduced by LRR. Stands not affected by root disease increased steadily in cumulative carbon stocks while tree mortality and growth loss in stands with low or high root disease likely turn them into net sources and not sinks. Harvesting-infected stands about the stand ages of 60–70 years and replanting with fast-growing resistant or immune tree species would reduce the long-term disease impact on carbon removal.

The options for reducing root disease risk can also be investigated with the simulators described earlier and incorporated as management options. Usually, pest management treatments are applied

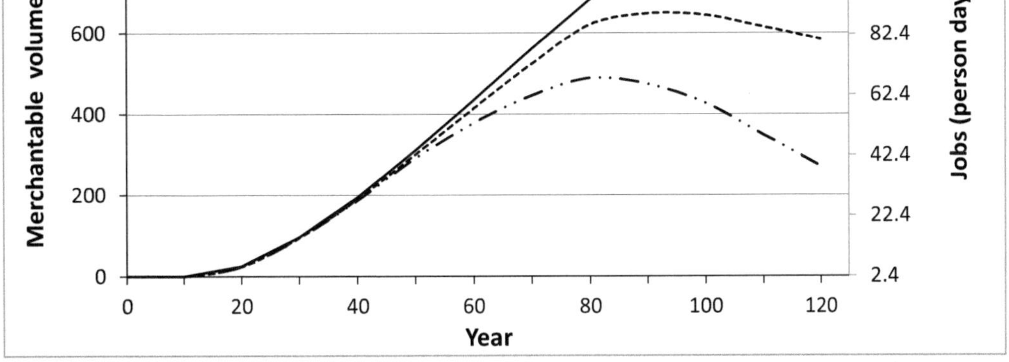

FIGURE 10.4 TIPSY outputs for yield volumes and employment yields/jobs in an interior Douglas fir stand with the application of custom operational adjustment factors for none, low-, and high-inoculum levels of LRR. Yield values were derived using merchantability limits of 12.5+ diameter at 1.3 m (excludes stumps and tops). TIPSY currently excludes any kind of thinning and assigns silviculture jobs a static value of at least 2.4 PD/ha over managed stand cycles.

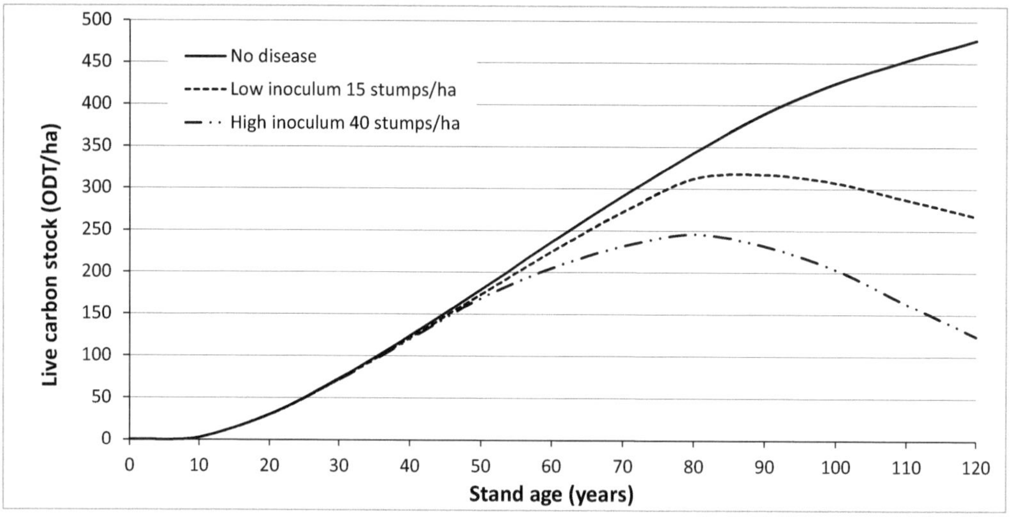

FIGURE 10.5 TIPSY output for live carbon stock values in an interior Douglas fir stand with the application of custom operational adjustment factors for none, low-, and high-inoculum levels of laminated root rot. Carbon values in oven-dried tons (ODT/ha) are calculated using belowground biomass of live roots and aboveground biomass of live stem, bark, foliage, and branches. Snags and coarse woody debris are excluded from the calculations.

using economic thresholds that relate pest damage costs to treatment costs and benefits. Treatment options are applied when pest impact rises above a break-even economic scenario. For example, greater tree density is known to increase LRR, and stump removal at harvest can reduce disease impact in some highly productive sites.

(Thies and Sturrock 1995). These options could be evaluated on low and high productive sites while also varying the level of fungal stump inoculum, the tree density, and stand age.

10.6.2.2 SWAT and Spruce Weevil Simulation with a TASS Overview

The white pine weevil is the most serious pest of interior and coastal spruce regeneration. By destroying the apical leader of the tree, this insect reduces growth and creates stem defects, which affect lumber quality (Alfaro 1994).

The SWAT model (Alfaro et al. 1997) was developed jointly by the Canadian Forest Service and British Columbia Ministry of Forests scientists over the last ten years to enable the forecasting of growth and yield of spruce plantations under weevil attack. The SWAT program works with TASS simulating the damage to individual trees, including destruction of the leader, crown recovery, and the formation of stem defects. The program reports tables representing stand development under different scenarios of weevil intensity and weevil management options. The model produces tabular output indicating the expected net yield after weevil-induced defects are removed from the stem.

SWAT allows the estimation of growth and yield of plantations infested to different degrees by weevil, with various tree-resistance configurations (Alfaro et al. 1997; Alfaro and Ying 1990). As operational planting of resistant stock is now a reality in BC, there is a need to forecast the potential yield gains to be realized by utilizing the resistant genotypes. Yield per hectare will vary depending on the growth characteristics of the resistant stock, on resistance level, and on the population pressure in the area of deployment. This information is crucial to providing the basis of a Genetic Worth system for orchard-deployed parent clones and for making informed decisions regarding the level of resistance to use (i.e., stock type) in different ecosystems. SWAT calculates the expected yield at a specified rotation age in spruce plantations (coastal and interior) initiated with mixtures of genotypes (user-defined) of known resistance or susceptibility (Figure 10.6). Users need to indicate the

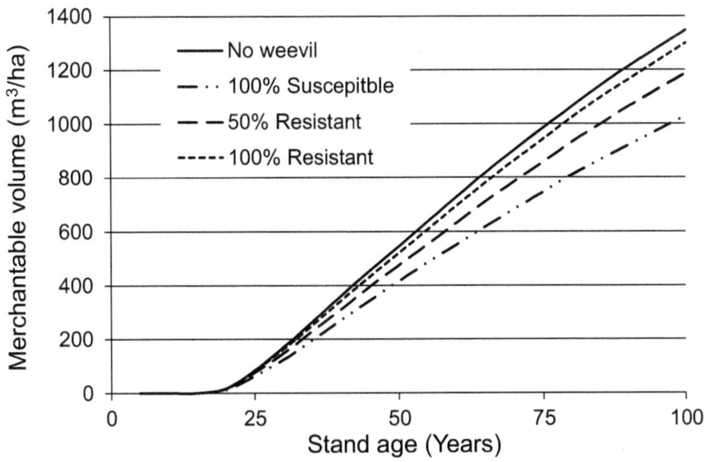

FIGURE 10.6 Results of runs of the Spruce Weevil Attack Trial model, a subroutine model of Tree and Stand Simulator. Merchantable volume in plantations initiated with Sitka spruce stock with different percent stand resistance mixes to the white pine weevil, *Pissodes strobi*.

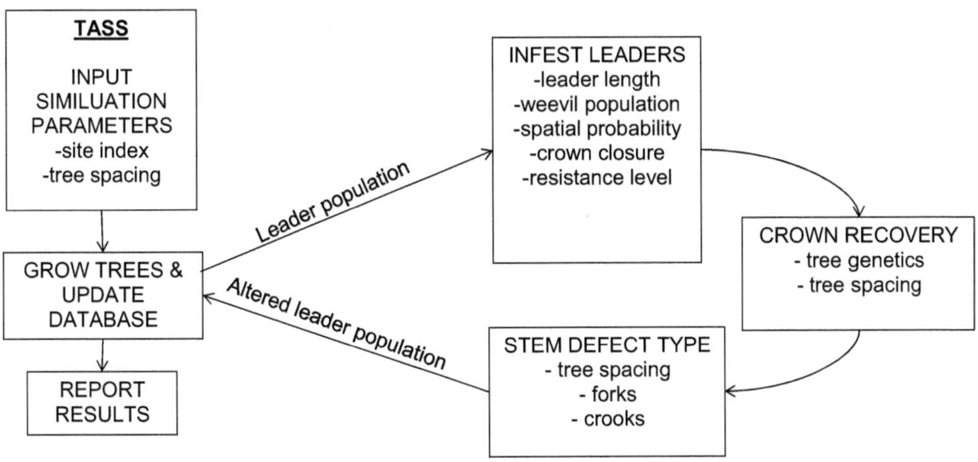

FIGURE 10.7 Interaction between the Tree and Stand Simulator model and the Spruce Weevil Attack Trial model. Small font indicates factors that are considered at each stage.

weevil hazard zone, the stock resistance ranking, and the proportion of the plantation in each stock type. Outputs include rotation age estimates of merchantable volume (to different utilization standards), stock tables, and yield/age curves. Merchantable volume is the volume of the stem between 30-cm-high stump and 7.5-cm-top diameter, after weevil-induced stem defects are removed.

The combined SWAT–TASS system runs on a one-year iteration. Every year, the TASS model passes onto SWAT a population of leaders grown by the plantation trees in the previous year (Figure 10.7). Leaders are assigned a probability of attack based on leader length, population level for the year, and plantation age. The resistance subroutine adds to every tree a probability of attack based on genotype resistance score. Trees are attacked using a stochastic, Monte Carlo simulation technique. Attacked trees can develop various stem defects at the base of each attacked leader (Figure 10.8), according to the probabilities observed in field studies (Alfaro 1989). A modified leader population is passed back

FIGURE 10.8 Sitka spruce tree with the leader killed by the leader weevil, *Pissodes strobi* Peck. Leader kill reduces height growth and creates stem defects that reduce lumber quality.

to TASS indicating whether the leader was attacked or not, the severity of the attack (number of internodes destroyed), and the defect type formed in the stem (Figure 10.7).

Two resistance mechanisms are represented in the system: leader repellency and leader toxicity. The weevil brood productivity on genotypes of different resistance levels is simulated based on data collected from genetic trials established by the BC Ministry of Forests and under screening by the Canadian Forest Service.

The user sets the hazard level for the plantation. Weevil hazard is represented in SWAT by different probability multipliers based on known epidemiology/ecosystem interactions. Hazard is maximum in fast-growing plantations, located on good sites, sustaining above-threshold heat accumulation. Attacks by the spruce weevil do not occur at random within a plantation. The dynamics of attack and spread within stands is determined by three characteristics of *P. strobi*: its preference for the most vigorous (longest and thickest) leaders, preference for a midrange of tree heights, and population aggregation. Research has indicated that aggregation is most important at low population levels.

10.6.2.3 Mountain Pine Beetle: An Example Simulation with TASS and BEETLE Models

The mountain pine beetle, a native insect of North America, periodically reaches population sizes that cause serious economic impact to the forest industry in western North America. During the last decade, the mountain pine beetle has had an unprecedented impact on pine forests of western North America, causing tree mortality in close to 20 million hectares (ha) of pine forests in Canada and the United States (Natural Resources Canada 2014; United States Department of Agriculture 2014).

All pine species in western North America are suitable hosts for this beetle (Furniss and Carolin 1977); however, lodgepole pine, *Pinus contorta* Douglas ex. Loudon, is one of the most susceptible and most widely distributed species. By volume, lodgepole pine comprises approximately 25% of BC's standing timber inventory, making the species critical to the BC timber supply (BC Ministry of Forests 1995).

The mountain pine beetle is a native insect of North America that typically exists at low, endemic levels feeding and reproducing in weakened trees. Outbreaks of the mountain pine beetle are natural disturbances that play a critical role in development, mortality, and rebirth of western North American pine stands. They are part of the gap dynamics cycle that helps to maintain biologically diverse and functionally healthy forest landscapes (Axelson et al. 2010). When conditions are favorable, outbreaks occur at landscape scales, killing mainly large-diameter trees and continue until the host has been depleted and only small-diameter pines and other unsuitable hosts remain (Safranyik and Carroll 2006).

The BEETLE subroutine of TASS was developed jointly by the Canadian Forest Service, BC Ministry of Forests, and Azura Formetrics to capture the dynamics of multiple-cohort lodgepole pine forests following outbreaks of the MPB. Stand reconstructions by Alfaro et al. (2008) and Axelson et al. (2009, 2010) demonstrated that the combined effects of low-intensity surface fires and attacks by the MPB can lead to the development of multiple cohort stands of lodgepole pine.

The BEETLE model, running with TASS, introduces beetle-induced mortality into multiple-cohort stands. These stands are characterized by having complex initial conditions containing different structures in terms of overstory, understory, and regeneration layers (Figure 10.9). These

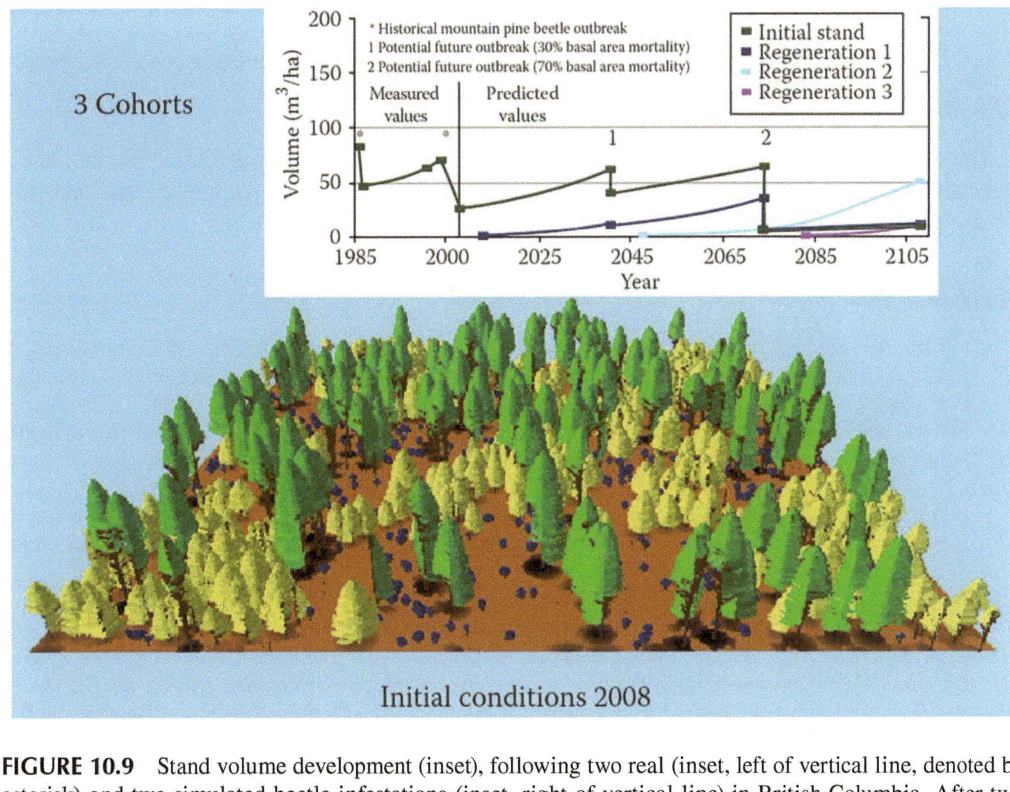

FIGURE 10.9 Stand volume development (inset), following two real (inset, left of vertical line, denoted by asterisk) and two simulated beetle infestations (inset, right of vertical line) in British Columbia. After two historic outbreaks, the forest had, in 2008, three cohorts (shown in thin, thick, and broken lines), each derived from a beetle outbreak.

initial conditions are grown for 100 years in TASS, applying beetle disturbance regimes that occur at user-defined intervals and have user-defined intensity. Disturbance regimes were defined in terms of year of outbreak initiation and proportion of basal area killed.

At the start of a simulated outbreak, pine trees are killed based on tree diameter, that is, tree mortality is concentrated in the overstory and intermediate canopy layers. The MPB has a preference for trees of larger diameter in the stand (Safranyik and Carroll 2006). In the BEETLE model, all trees are killed in one year, that is, outbreak duration was not considered important. Following an outbreak, a new-regeneration cohort of 5,000 trees, randomly spaced, is initiated. Based on stand reconstructions, a regeneration delay of five years was introduced.

The number of cohorts grown by TASS is four to six. The outputs of BEETLE include crown maps, graphical output of annual tree density volume per hectare by cohort, and the total of all cohorts combined (Figure 10.9).

10.7 CONCLUSION

Modeling pest impacts on forest stand dynamics requires information on the pest population dynamics and the response of the forest to the pest action, that is, a detailed knowledge of the pest–host interaction over different climactic regions. Significant advances have been made in pest management modeling in BC and other jurisdictions in North America. Several simulation tools have been developed to determine the effect of the pest over time on stand development, on the value chain from stand to mill level, and on a wide range of forest attributes affected by stand structure and spatial dynamics. To be of use to industrial and government decision-makers of pest management investments, the model outputs must be easily understood such as in terms of risk to standing volumes, stand structures, and potential product and job losses. The models described allow potential risk assessment and remediation outcomes to be developed and understood.

10.8 ACKNOWLEDGMENTS

The authors gratefully acknowledge the contributions of the BC Ministry of Forests Stand Development Modeling team and their extensive contributions to growth and yield and stand and pest simulation. Also, thank you to Mike Lipcsei for graphics creation.

REFERENCES

Alfaro, R.I. 1989. Stem defects in Sitka spruce induced by Sitka spruce weevil, Pissodes strobi (Peck). In *Insects affecting reforestation: Biology and damage*, eds. R.I. Alfaro and S.G. Glover, pp. 177–185. Proceedings of a Meeting of the IUFRO Working Group on Insects Affecting Reforestation, XVIII International Congress of Entomology, July 3–9, 1988. Vancouver, British Columbia, Canada: Forestry Canada. https://doi.org/10.13140/RG.2.1.3554.5122.

Alfaro, R.I. 1994. The white pine weevil in British Columbia: Biology and damage. In *The white pine weevil: Biology, damage and management*, eds. R.I. Alfaro, G. Kiss, and R.G. Fraser, pp. 7–22. Fraser Proceedings of a meeting held January 19–21, 1994. FRDA Report No. 226. Richmond, VA: Canadian Forest Service.

Alfaro, R.I., J. Axelson, and B. Hawkes. 2008. Mountain pine beetle increases the complexity of fire origin lodgepole pine stands in British Columbia, Canada. *BC J. Ecosyst. Manage.* 9(3):1–5.

Alfaro, R.I., R. Brown, K. Mitchell, K. Polsson, and R. McDonald. 1997. S.W.A.T.: A decision support system for spruce weevil management. In *Decision support systems for forest pest management*, eds. T.I. Shore and D.A. McLean, pp. 31–41. Victoria, British Columbia, Canada: Canadian Forest Service, FRDA Report No. 260. www.for.gov.bc.ca/hfd/pubs/Docs/Frr/Frr260-3.pdf. Accessed 25 March 2015.

Alfaro, R.I. and C.C. Ying. 1990. Levels of Sitka spruce weevil, *Pissodes strobi* (Peck), damage among Sitka spruce provenances and families near Sayward, British Columbia. *Can. Ent.* 122:607–615.

Axelson, J., R. Alfaro, and B. Hawkes. 2009. Influence of fire and mountain pine beetle on the dynamics of lodgepole pine stands in British Columbia, Canada. *For. Ecol. Manage.* 257(9):1874–1882.

Axelson, J., R. Alfaro, and B. Hawkes. 2010. Changes in stand structure in uneven-aged lodgepole pine stands impacted by mountain pine beetle outbreaks and fires in central British Columbia. *For. Chron.* 86(1):87–99.

Bloomberg, W.J. 1979. A model of damping-off and root rot of Douglas fir seedlings caused by *Fusarium oxysporum*. *Phytopathol.* 69:74–81.

Bloomberg, W.J. 1988. Modeling control strategies for laminated root rot in managed Douglas fir stands: Model development. *Phytopathol.* 78:403–409.

British Columbia Ministry of Forests. 1995. *Forest, Range and Recreation Resource Analysis*. Victoria, British Columbia, Canada: Crown Publications.

British Columbia Ministry of Forests and Range. 2010. *TASS: Predicting British Columbia's Future Forests*. Victoria, British Columbia, Canada: For. Sci. Prog. www.for.gov.bc.ca/hfd/pubs/Docs/Bro/Bro90.htm (accessed March 23, 2015).

Burnell, D.G. 1977. A dispersal-aggregation model for mountain pine beetle in lodgepole pine stands. *Res. Pop. Ecol.* 19(1):99–106.

Cole, W.E. and M.D. McGregor. 1983. *Estimating the rate and amount of tree loss from mountain pine beetle infestations*. Res. Pap. INT-318, p. 22. Ogden, UT: U.S. Department of Agriculture, Forest Service, Intermountain Forest and Range Experiment Station.

Crookston, N.L. 1991. *Foliage dynamics and tree damage components of the western spruce budworm modeling system*. Gen. Tech. Rep. INT-282, p. 40. Ogden, UT: U.S. Department of Agriculture, Forest Service, Intermountain Research Station.

Crookston, N.L., J.J. Colbert, P.W. Thomas, K.A. Sheehan, and W.P. Kemp. 1990. *User's guide to the western spruce budworm modeling system*. Gen. Tech. Rep. INT-274, p. 40. Ogden, UT: U.S. Department of Agriculture, Forest Service, Intermountain Research Station.

Crookston, N.L., R.C. Roelke, D.G. Burnell, and A.R. Stage. 1978. Evaluation of management alternatives for lodgepole pine stands using a stand projection model. In *Theory and Practice of Mountain Pine Beetle Management in Lodgepole Pine Forests*, eds. Berryman, A.A., Amman, G.D., and Stark, R.W. pp. 114–122. Symposium held at Washington State University, Pullman, Washington, April 25–27, 1978. Moscow, ID: University of Idaho, Forest, Wildlife and Range Experiment Station.

Crookston, N.L. and A.R. Stage. 1991. User's guide to the Parallel Processing extension of the Prognosis Model. USDA Forest Service, Intermountain Research Station, General Technical Report INT-281, p. 88. FHTET. Westwide Pine Beetle Model user's guide. Ft. Collins, CO: Forest Health Technology Enterprise Team, USDA Forest Service.

Cruickshank, M.G. 2010. Effect of Armillaria root disease on quality and value of green Douglas fir lumber. *For. Chron.* 86(2):263–266.

Cruickshank, M.G. and C.N. Filipescu. 2012. Allometries of coarse tree, stem, and crown measures in Douglas fir are altered by Armillaria root disease. *Botany* 90(8):711–721.

Cruickshank, M.G., D.J. Morrison, and A. Lalumière. 2009. The interaction between competition in Douglas fir plantations and disease caused by *Armillaria ostoyae*. *For. Ecol. Manage.* 257:443–452.

Cruickshank, M.G., D.J. Morrison, and A. Lalumière. 2011. Site, plot, and individual tree yield reduction of interior Douglas fir associated with non-lethal infection by Armillaria root disease in Southern British Columbia. *For. Ecol. Manage.* 261(2):297–307.

Dent, D. 2000. *Insect pest management*, p. 410. Wallingford, UK: CABI Publishing.

Dietze, M.C. and J.H. Matthes. 2014. A general ecophysiological framework for modelling the impacts of pests and pathogens on forest ecosystems. *Ecol. Lett.* 17:1418–1426.

Dixon, G.E. 2002. *Essential FVS: A user's guide to the Forest Vegetation Simulator*. Internal Rep., p. 226. Fort Collins, CO: U.S. Department of Agriculture, Forest Service, Forest Management Service Center. (Revised: December 24, 2014).

Erdle, T.A. and D.A MacLean. 1999. Stand growth model calibration for use in forest pest impact assessment. *For. Chron.* 75:141–152.

Frankel, S.J. 1998. User's guide to the Western Root Disease Model, version 3.0. Gen. Tech. Rep. PSWGTR-165, p. 166. Albany, CA: Pacific Southwest Research Station, Forest Service, U.S. Department of Agriculture.

Frankel, S., B. Eav, and G. Shubert. July 13–17, 1992. Workshop: Calibration of the annosus/bark beetle model with real data. In *Proceedings of the Fortieth Annual Western International Forest Disease Work Conference*, ed. S. Frankel, pp. 47–54. Fort Lewis College, Durango, CO: USDA Forest Service, Pacific Southwest Region, Forest Pest Management, San Francisco, CA.

Franklin, J.F., H.H. Shugart, and M.E. Harmon. 1987. Tree death as an ecological process. *BioSci.* 37:550–556.

Furniss, R.L. and V. Carolin. 1977. *Western forest insects*. Miscellaneous Publication No. 1339, 20402, p. 654. Washington, DC: United States Department of Agriculture.

Goheen, D.J. and E.M. Hansen. 1993. Effects of pathogens and bark beetles on forests. In *Beetle-Pathogen Interactions in Conifer Forests*, eds. T.B. Schowalter and G.M. Filip, pp. 175–196. London, UK: Academic Press.

Hansen, E.M. 1979. Survival of *Phellinus weirii* in Douglas fir stumps after logging. *Can. J. For. Res.* 9:484–488.

Hatala, J.A., M.C. Dietze, R.L. Crabtree, K. Kendall, D. Six, and P.R. Moorcroft. 2011. An ecosystem-scale model for the spread of a host-specific forest pathogen in the Greater Yellowstone Ecosystem. Ecosystem and Conservation Sciences Faculty Publications. Paper 40.

Hawksworth, F.G. 1977: The 6-class dwarf mistletoe rating system. USDA Forest Service Rocky Mt. Forest Range Experimental Station General Technical Report RM-48.

Hawksworth, F.G., J.C. Williams-Cipriani, B.B. Eav, B.W. Geils, R.R. Johnson, M.A. Marsden, J.S. Beatty, G.D. Schubert, and D.C.E. Robinson. 1992. Interim dwarf mistletoe impact modeling system: Users guide and reference manual. Report MAG-91-3. USDA, Forest Service, Forest Pest Management. Fort Collins, CO: Methods Application Group, 90pp (Revised: 1993)

He, H.S., W. Li, B.R. Sturtevan, J. Yang, B.Z Shang, E.J Gustafson, and D.J. Mladenoff. 2005. LANDIS 4.0 users guide. LANDIS: A spatially explicit model of forest landscape disturbance, management, and succession. Gen. Tech. Rep. NC-263. St. Paul, MN: U.S. Department of Agriculture, Forest Service, North Central Research Station, 93pp.

Heath, R. and R.I. Alfaro. 1990. Growth response in a Douglas fir/lodgepole pine stand after thinning of lodgepole pine by the mountain pine beetle. *J. Entomol. Soc. of B.C.* 87:16–21.

Leech, S., A. Wiensczyk, and J. Turner. 2009. Ecosystem management: A practitioners' guide. *BC J. Ecosyst. Manage.* 10(2):1–12. http://forrex.org/sites/default/files/publications/jem_archive/IS S51/vol10_no2_art1.pdf. Accessed 25 March 2015.

MacLean, D.A., T.A. Erdle, W.E. MacKinnon, K.B. Porter, K.P. Beaton, G. Cormier, S. Morehouse, and M. Budd. 2001. The spruce budworm decision support system: Forest protection planning to sustain long-term wood supply. *Can. J. For. Res.* 31:1742–1757.

Marsden, M.A., B.B. Eav, and M.K. Thompson. 1994. User's Guide to the Douglas-fir Beetle Impact Model. General Technical Report RM-250. Fort Collins, CO: U.S. Department of Agriculture, Forest Service.

McDonald, G.I., R.J. Hoff, and W.R. Wykoff. 1981. Computer simulation of white pine blister rust epidemics. I. Model formulation. Res. Pap. INT-258. Ogden, UT: U.S. Department of Agriculture, Forest Service, Intermountain Forest and Range Experiment Station, 136pp.

McMahan, A.J., A.W. Courter, and E.L. Smith. 2002. User's guide for FVS-EMAP: A simple tool for displaying FVS output in ArcView®. FHTET 02–01. Fort Collins, CO: USDA Forest Service, Forest Health Technology Enterprise Team, 30pp.

McNamee, P.J., W.A. Kurz, C.J. Daniel, D.C.E. Robinson, and M.G. Deering. 1991. Description of the annosus root disease model. Prepared by ESSA Ltd., Vancouver, British Columbia, Canada: U.S. Forest Service, 70pp.

Mitchell, K.J. 1975. *Dynamics and Simulated Yield of Douglas-fir. For. Sci. Monograph.* 17:39.

Mitchell, K.J. 1988. SYLVER: Modeling the impact of silviculture on yield, lumber value, and economic return. *For. Chron.* 64:127–131.

Monserud, RA and N.L. Crookston. 1982. A User's Guide to the Combined Stand Prognosis and Douglas fir Tussock Moth Outbreak Model. Gen. Tech. Rep. INT-GTR-127. Ogden, UT: Intermountain Forest and Range Experiment Station, Forest Service, U.S. Department of Agriculture, 56pp.

Morrison, D.J. 2011. Epidemiology of Armillaria root disease in Douglas fir plantations in the cedar-hemlock zone of the Southern interior of British Columbia. *For. Path.* 41:31–40.

Natural Resources Canada. 2014. Mountain pine beetle (factsheet). www.nrcan.gc.ca/forests/insects-diseases/13397 (last accessed August 18, 2014).

Peet, F.G. and R.S. Hunt. 2005. A new simulator for the spread of forest root diseases by individual root contacts. *For. Sci.* 51(5):425–437.

Pukkala, T., T. Möykkynen, M. Thor, J. Rönnberg, and J. Stenlid. 2005. Modeling infection and spread of *Heterobasidion annosum* in even-aged Fennoscandian conifer stands. *Can. J. For. Res.* 35:74–84.

Reynolds, K.M. and W.J. Bloomberg. 1982. Estimating the probability of intertree root contact in second-growth Douglas-fir. *Can. J. For. Res.* 12:493–498.

Riel, W.G., A. Fall, T.L. Shore, and L. Safranyik. 2004. A spatio-temporal simulation of mountain pine beetle impacts on the landscape. In *Mountain Pine Beetle Symposium: Challenges and Solutions* eds. T.L. Shore, J.E. Brooks, and J.E. Stone, pp. 106–113. October 30–31, 2003, Kelowna, British Columbia,

Canada: Natural Resources Canada, Canadian Forest Service, Pacific Forestry Centre, Victoria, British Columbia, Information Report BC-X-399, 298pp.

Robinson, D.C.E. and B.W. Geils. 2006. Modelling dwarf mistletoe at three scales: Life history, ballistics and contagion. *Ecol. Model.* 199:23–38.

Safranyik, L., H.J. Barclay, A.J. Thomson, and W.G. Riel. 1999. A population dynamics model for the mountain pine beetle, *Dendroctonus ponderosae* Hopk. (Coleoptera: Scolytidae). Victoria, British Columbia, Canada: Natural Resources Canada, Canadian Forest Service, Pacific Forestry Centre, Information Report BC-X-386, 35pp.

Safranyik, L. and A.L. Carroll. 2006. The biology and epidemiology of the mountain pine beetle in lodgepole pine forests. In *The Mountain Pine Beetle: A Synthesis of Its Biology, Management and Impacts on Lodgepole Pine*, eds. L. Safranyik and B. Wilson, pp. 3–66. Victoria, British Columbia, Canada: Pacific Forestry Centre, Canadian Forest Service, Natural Resources Canada.

Sattler, D.F., J.W. Goudie, and R.W. Reich. 2019. A module to simulate the impact of western gall rust (*Cronartium harknessii*) on merchantable volume and lumber yields for lodgepole pine (*Pinus contorta* var. *latifolia*) stands in British Columbia. *Can. J. For. Res* 49(11):1379–1391.

Shaw, C.G., III, A.R. Stage, and P. McNamee. 1991. Modeling the dynamics, behavior, and impact of Armillaria root disease. In *Armillaria Root Disease*, eds. C.G. Shaw, III and G.A. Kile, Agriculture Handbook No. 691, pp. 150–156. Washington, DC: U.S. Department of Agriculture, Forest Service.

Sheehan, K.A., W.P. Kemp, J.J. Colbert, and N.L. Crookston. 1989. The western spruce budworm model: Structure and content. Gen. Tech. Rep. PNW-Gtr-241. Portland, OR: U.S. Department of Agriculture, Forest Service, Pacific Northwest Research Station, 70pp.

Smith, M.S. 1962. *The Practice of Silviculture*. New York: John Wiley & Sons, Inc, 578pp.

Spurr, S.H. and B.V. Barnes. 1973. *Forest Ecology.* New York: Ronald Press, 570pp.

Stage, A.R. 1973. Prognosis model for stand development. Res. Pap. INT-137. Ogden, UT: U.S. Dept. of Agriculture, Forest Service, Intermountain Forest and Range Experiment Station, 32pp.

Stearns-Smith, S.G., G. Neinaber, M. Cruickshank, and A. Nussbaum. 2004. Demonstrating Growth and Yield Adjustments (TIPSY OAFs) for Armillaria root disease in a timber supply analysis. Victoria, British Columbia, Canada: Forestry Canada, Pacific Forestry Centre, 9pp.

Thies, W.G. 1983. Determination of growth reduction in Douglas fir infected by *Phellinus weirii. For. Sci.* 29:305–315.

Thies, W.G. and R.N. Sturrock. 1995. Laminated root rot in Western North America. USDA Forest Service, Pacific Northwest Research Station General Technical Report PNW-GTR-349, 32pp.

United States Department of Agriculture, Forest Service. 2014. Mountain Pine Beetle. www. fs.usda.gov/detail/mbr/home/?cid=stelprdb5139168 (last accessed August 18, 2014).

Wang, L., A. Gunulf, T. Pukkala, and J. Rönnberg. 2014. Simulated *Heterobasidion* disease development in *Picea abies* stands following precommercial thinning and the economic justification for control measures. *Scand. J. For. Res.* 30:174–185.

Woods, A.J., A. Nussbaum, and B. Golding. 2000. Predicted impacts of hard pine stem rusts on lodgepole pine dominated stands in central British Columbia. *Can. J. For. Res.* 30:476–481.

Section III

Ecological Indicators

11 Assessing Abundance, Biomass, and Complexity in the Context of Ecological Forest Management

Virginia Capmourteres and Madhur Anand

11.1 IMPORTANCE OF ABUNDANCE AND BIOMASS

11.1.1 Applications of Abundance and Biomass Assessments to Forest Management

Abundance usually refers to the relative numbers of individuals belonging to different species (Pagel et al. 1991), whereas biomass refers broadly to the material that is part of a living organism (Hames 2009) and is more related to productivity (Hui et al. 2012). Nevertheless, in plant communities, the definition and identification of individual plants are frequently difficult, and, thus, other measures of abundance such as plant density and percent cover (Chen et al. 2008) and even biomass (Chiarucci et al. 1999) are used. Currently, most ecologists refer to abundance as "some measure of the amount of species" that can be estimated through density, biomass, frequency, or cover (Chiarucci et al. 1999).

Although not exactly the same, vegetation cover and biomass have been reported to have a strong relationship (Jonasson 1988; Rottgermann et al. 2000). These relationships, however, might depend on the structure and vertical stratification of the vegetation community (Chiarucci et al. 1999), stand age, morphology, type of growth, dimensions of species, and species composition (Muukkonen et al. 2006). The assessment of biomass, as it will be explained in subsequent paragraphs, plays a major role in forest management. Abundance, as a proxy for biomass, can provide more information on the properties of the forests and lead to nondestructive determination methods of biomass (Muukkonen et al. 2006). Thus, the assessment of both contributes to forestry and scientific purposes (Parresol 1999).

The volume of wood from stems and branches has historically been of interest for foresters, who measured it with the aim of estimating "round wood" over or under the bark and cutoff points (Pardé 1980). Biomass estimations began later and were mainly meant to plan the extraction of forest resources (Zianis and Mencuccini 2004), including timber, roots, leaves, epiphytes, and seedlings and saplings (Tanner 1980). Inventories of biomass, and other tree metrics such as age, height, or basal area, were then inventoried for different species and countries as a reference for foresters, workers, and researchers (Cannell 1982).

In the past years, however, biomass estimations have gained importance for various aspects of forest management, and consequently, the development and use of indicators of biomass has become critical. In the context of climate change, for example, standing forest biomass represents carbon (C) stocks and the potential of the ecosystem for C sequestration (Jenkins et al. 2001). Also, standing biomass is seen as a sustainable alternative for fossil fuels, especially for subsistence communities. In addition, biomass is a fundamental state variable in several ecological and ecophysiological models and is linked to biodiversity and ecosystem services. Root biomass has also been implicated in regulating nutrient and C cycles (Cairns et al. 1997).

DOI: 10.1201/9781003431084-14

Forests are key ecosystems that have a high C density affecting the net balance of C worldwide (Lambert et al. 2005). They can act both as a sink and as a source of C, depending on the productivity and disturbance regime of the ecosystem, which determines the amount of plant C that is stored through photosynthesis. Approximately 80% and 40% of the above- and belowground terrestrial C are stored in forest biomass across the globe, respectively (Peichl and Arain 2007). Forest biomass has the potential of helping mitigate greenhouse gases (GHG), and, therefore, it has become a main research topic internationally (Brown 2002; Smith et al. 2003; Dhillon and von Wuehlisch 2013; Moroni 2013; Downie et al. 2014).

Currently, the quantification of biomass C densities and their distribution for major biomes on Earth can be used as inputs for climate–carbon models at large scales (Keith et al. 2009), although some estimations are based on poor databases, subjected to some level of error (Mooney 1991), or present different types and degrees of uncertainty (Chave et al. 2004; Houghton 2005). To improve such estimations, understanding and measuring how above- and belowground biomass vary over the life cycle of forests and how they respond to disturbances (pests, fire, pollutants, extreme weather events), stressors (variations in climate, site fertility, salinity), land uses (that modify the cover and stage of development of forests), or forest management practices (i.e., harvest or no harvest) may help quantify the biomass that remains stored in the ecosystem (Sherman et al. 2003; Peichl and Arain 2007; Pingoud et al. 2012). Of interest as well is to develop tools to identify forests with high biomass C stocks, such as those proposed by Keith et al. (2009).

Biomass estimates for C flux studies are also of relevance to comply with two international agreements: the United Nations Framework Convention on Climate Change (UNFCCC) and the Kyoto Protocol. In the former one, signatory countries have committed to developing national inventories of GHG and reporting them to the Conference of Parties. These should include C emissions and removal estimations under different land use scenarios (Brown 2002). In the UNFCCC, biomass was declared as an essential climate variable (ECV) to address climate change. Furthermore, the Operational Adjustment Factors (OAFs) reduction of emissions due to deforestation and forest degradation mechanisms was created with the aim of improving biomass estimations and monitoring C emissions produced by deforestation (Le Toan et al. 2011).

Another use of biomass valuations is the assessment of Net Primary Productivity (NPP), defined as "the balance between the light energy fixed through photosynthesis (gross primary productivity) and lost through respiration and mortality, representing the net C uptake from the atmosphere into vegetation" (Jenkins et al. 2001; Zhao and Zhou 2005). Both biomass and productivity are a function of similar ecological conditions and resources (moisture, temperature, and nutrients) (Keeling and Phillips 2007; Hui et al. 2012). Although it has been established that biomass responds positively to productivity with a consequent increase in C storage as well (Hui et al. 2012 and references therein), aboveground productivity and forest biomass are not linearly related (Keeling and Phillips 2007). This positive relationship is true at low-productivity sites, but it weakens at mid-productivity sites and in high-productivity areas, where biomass saturates and stops increasing. Even more, the relationship between biomass and NPP may indicate plant adaptation (Hui et al. 2012).

Another motif for the implementation of biomass measurements is the displacement and/or replacement of fossil fuel combustion through the use of renewable biomass fuels (Marland and Schlamadinger 1997). Biomass fuel is seen as sustainable since the C released during combustion can be taken up by vegetation (Freppaz et al. 2004; Stupak et al. 2007). This type of energy contributes to local employment and social development, allows silvicultural practices to be performed with a benefit, and makes waste recyclable (e.g., wood ash). Although the profits obtained from this source of energy are weak, it has been observed that with subsidies, fuel extraction from forests can be profitable (Stupak et al. 2007). Presently, forest biomass was a main energy source for subsistence economies, accounting for 9–13% of the world supply of energy (Parikka 2004; Le Toan et al. 2011). For community-based projects, biomass for direct combustion for heating or heat and power could be an option for offsetting GHG emissions (Paré et al. 2011). However, this use of biomass would not be recommended at large scales because it would imply an increase in human appropriation of NPP by 20%, resulting in younger forests and lower biomass pools (Schulze et al. 2012). Estimates of wood fuel vary greatly in the literature, indicating that data updates and a more accurate assessment of stocks were needed (Le Toan et al. 2011). This is especially important for regions where other sources of energy are difficult to exploit or where the use of biomass energy could reduce pollution and encourage welfare (Freppaz et al. 2004).

Biomass can also be an indicator of ecosystem services (Le Toan et al. 2011). Besides C and other GHG sequestration, which are the most valuable services for the planet (Foley et al. 2007), biomass is related with biodiversity conservation. For example, old-growth forests' biomass is habitat for many species that would be vulnerable if biomass decreased (Malhi and Phillips 2004). In turn, aboveground C biomass can be affected by species composition and hence affect biodiversity (Bunker et al. 2005). Biomass is a useful indicator for fibers, fuel, and other raw materials that humans can obtain from ecosystems, as well as the presence of species and abiotic resources with potential ornamental use (de Groot et al. 2010). Given these relationships between biomass and biodiversity, frequent and high-resolution mapping of forest biomass and biomass changes are crucial for conservation policies (Le Toan et al. 2011).

Biomass assessments can contribute to forest management by providing information about nutrient pools in the ecosystem that can be released through decomposition. When biomass is removed from forests through cleaning, thinning, or clearcutting, there is a significant export of nutrient from the ecosystem (Raulund-Rasmussen et al. 2008). The study of biogeochemical cycles and biomass in forest is thus tightly linked (Schroeder et al. 1997).

Biomass estimations of diverse groups of organisms can aid forest management because they reflect forest quality. For example, epiphytic lichens fix nitrogen (N) and take up other elements that are later released as leachates and litterfall and provide food resources and shelter for fauna (Esseen et al. 1996; Porada et al. 2014). Since they are sensitive to environmental disturbance and habitat destruction, they are good indicators of habitat quality and forest continuity. Plus, these organisms can represent between 6.2% and 8.6% of the amount of branch material, contributing to overall biomass in the lower canopy (Esseen et al. 1996). Also, liana abundance and biomass in tropical forests are noteworthy with regard to forest management because they compete with trees, altering community composition, C storage, nutrient cycling, and water fluxes (Schnitzer and Bongers 2011; van der Heijden et al. 2013; Laurance et al. 2014). The biomass of invertebrate communities can as well be related to show soil quality and food webs and therefore contribute to management goals (see subsequent sections). Understory communities, although sometimes undervalued, are important components of fuel loads, play a role in C and N fluxes, provide habitat, and protect soils against erosion (Carneiro et al. 2007; Suchar and Crookston 2010; Marinescu et al. 2011). In temperate and boreal forests, where canopies are open, understory layers are important contributors to aboveground biomass (Keeling and Phillips 2007).

The relevance of biomass estimations, the development of assessment tools, and consistent and updated datasets at the land use scale are imperative measures in forest ecology and management (Le Toan et al. 2011). In the following sections, we review some of the most used methodologies and approaches to assess forest biomass, including the use of cover and other forest attributes as a proxy for biomass, and we provide examples of their application in tropical, temperate, and boreal forests.

11.1.2 Methods for Measuring Abundance and Biomass

11.1.2.1 Trees

11.1.2.1.1 Direct Measurements

Obtaining exact biomass data requires tree harvesting and posterior fresh and dry weighing (Pardé 1980). However, this is hardly realistic, and therefore, measurements are taken from sampling standing trees (Neumann et al. 2012). For example, Bastien-Henri et al. (2010) studied the accumulation of biomass of native and exotic species in dry and humid environments of Panama. They collected subsamples, measured their wet biomass, and then converted it to dry biomass estimates for different compartments (stems, branches, leaves) using Eq. 11.1.

The dry biomass estimation for different tree compartments is given as follows (BastienHenri et al. 2010):

$$DW_c = \frac{DW_s}{FW_s} \times FW_c \qquad (11.1)$$

where,

DW_c is the dry weight of a section of a compartment;
FW_c represents the fresh weight of the same section;
DW_s is the dry weight of the subsample;
FW_s denotes the fresh weight of the same subsample.

Section biomasses are then summed to obtain estimates of the total dry biomass for the whole plant. In the case of belowground biomass, estimations are limited because of the difficulty associated with the sampling of woody roots of large trees (Drexhage and Colin 2001). Studying roots normally requires excavation and further harvesting and often presents inconveniences related to the depth of digging and limitations when harvesting small and fine roots (Castellanos et al. 1991).

11.1.2.1.2 Allometric Equations

For larger areas, obtaining subsamples and weighing wet or dry biomass can be tedious and time-consuming. However, taking into account the strong allometric relationships between tree features (e.g., diameter at breast height (DBH), height, form) and their aboveground biomass, the latter can be deducted (Case and Hall 2008). When using DBH, for example, calculations are performed with equations that have the following general form (Ter-Mikaelian and Korzukhin 1997) (Eq. 11.2).

$$M = aD^b \qquad (11.2)$$

where,

M represents the oven-dry weight of the biomass component of a tree (kg);
D is the DBH (cm);
a and b are the parameters.

This general equation has gained popularity because it provides a reasonable balance between accuracy and low data requirements, but it can still incorporate additional independent variables. Furthermore, an allometric relationship between biomass and volume exists, and thus, volume inventory data can be used to build equations (Ter-Mikaelian and Korzukhin 1997; Case and Hall 2008).

Allometric equations have been developed at several scales. For example, stand-specific volume equations are suited for those sites where ecological and environmental conditions are very similar. Conversely, generalized biomass equations are designed for landscape-scale studies and cover a range of environmental and site factors while maintaining their accuracy. Given that there are not many local equations available, general ones can provide relatively robust, average local predictions (Case and Hall 2008).

As an example of generalized equations, Lambert et al. (2005) proposed a set of tree biomass equations for Canada, using previously developed ones at the territorial and provincial scale (Eq. 11.3). Equations estimate compartment (foliage, branches, wood, bark) and total biomass. The diameter at breast height was used as the main predictor of biomass, but height can be included as well as a proxy for the competitive environment of the tree.

The following equations are for the estimation of tree compartment and total biomass (Lambert et al. 2005):

$$\begin{aligned} y_{wood} &= \beta_{wood1} D^{\beta wood2} H^{\beta wood3} + e_{wood} \\ y_{bark} &= \beta_{bark1} D^{\beta bark2} H^{\beta bark3} + e_{bark} \\ y_{stem} &= y'_{wood} + y'_{bark} + e_{stem} \\ y_{foliage} &= \beta_{foliage1} D^{\beta foliage2} H^{\beta foliage3} + e_{foliage} \\ y_{branches} &= \beta_{branches1} D^{\beta branches2} H^{\beta branches3} + e_{branches} \\ y_{crown} &= y'_{foliage} + y'_{branches} + e_{crown} \\ y_{total} &= y'_{wood} + y'_{bark} + y'_{foliage} + y'_{branches} + e_{total} \end{aligned} \tag{11.3}$$

where,

y_i is the dry biomass (kg) of compartment i;
i is the wood, bark, stem, foliage, branches, crown, and total
y' is the prediction of y_i
D is the DBH (cm)
H is the height (m)
β_{jk} are the model parameters with coefficient estimates b_{jk};
j is wood or bark or foliage or branches;
k is 1, 2, or 3;
e_i are the error terms.

These equations were produced for 33 hardwood and softwood species and are suited for C estimations in the framework of the Kyoto Protocol (Lambert et al. 2005).

In the United States, Schroeder et al. (1997) applied data from national forest inventories to estimate the aboveground biomass. Given that forest volume inventories emphasize commercially valuable wood, they developed biomass expansion factors (BEF) that convert volume to mass and include noncommercial compartments (branches, twigs, bark, stumps, foliage, seedlings, and saplings). They first developed biomass regression equations for hardwood and conifer species to estimate oven-dry biomass from DBH and applied them to the USDA Forest Service Forest Inventory and Analysis data. Afterward, the BEF was calculated as the aboveground biomass density divided by the growth stock volume.

Other studies focused on partitioning above- and belowground biomass. For example, Peichl and Arain (2007) used white pine (*Pinus strobus*) as a case study and came up with two sets of equations: one with DBH as a single input variable and another one with the option of adding height or age. Equations were built to include compartments belowground (roots) as well as aboveground (foliage, live branches, dead branches, stem wood, stem bark, total stem, total aboveground, and total tree). They estimated a BEF as stated in Eq. 11.4.

The BEF is given as follows (Peichl and Arain 2007):

$$\mathrm{BEF_i} = \frac{\mathrm{W_i}}{\mathrm{V}} \tag{11.4}$$

where,

W_i is the dry biomass (tons) of a tree compartment i (i.e., foliage, branches, aboveground, roots, total tree);
V is the stem volume (m^3).

Similarly, Návar (2009) developed allometric equations for compartments including belowground biomass for temperate and tropical forests in northwestern Mexico.

Other equations have been designed particularly for common tree species in Europe (Muukkonen 2007) or tropical commercial species of the *Dipterocarp* genus (Basuki et al. 2009).

Wang (2006) used allometric equations in temperate forests of China to help examine C, N, and water budgets of forest ecosystems in the northeast regions of the country. Also, Whittaker et al. (1974) estimated above- and belowground biomass and productivity as well as a mean climax biomass for a watershed in a cool-temperate forest region. Other applications are C sequestration estimates in secondary tropical forests in Brazil (Nelson et al. 1999), the analysis of old-growth forests biomass gains and losses in Borneo (Slik et al. 2010), and the observation of the relationship between gradients of aboveground biomass and patterns of species composition (Baker et al. 2004). More recently, allometry was employed to estimate the biomass in restoration projects of the Atlantic Forest in Brazil (Rodrigues Nogueira Junior et al. 2014) and to compare C stocks among different agroforestry systems (Jain and Ansari 2013). Combinations of some of the aforementioned methodologies and data sources, such as national inventories, systematic sampling, and BEFs, also have the potential to assess changes in forest biomass C storage over time (Fang et al. 2001).

11.1.2.1.3 Remote Sensing

The allometric equations explained in the previous section require consistency of inventory methods, but sampling remote areas, for instance, is highly labor demanding (Boudreau et al. 2008; Amini and Sumantyo 2011). In response to these limitations of biomass equations, Earth observation through remote sensing allows researchers to obtain information on a regular basis and across large areas.

Remote sensing can use passive or active optical imaging sensors. Passive sensing relies on vegetation spectral responses from the visible and near-infrared (NIR) spectral regions and provides measurements of horizontal distributions in vegetation canopies. Active sensing can obtain more accurate data in both the horizontal and vertical dimensions (Vazirabad and Karslioglu 2011). When using remote sensing, selecting variables that can potentially estimate the aboveground biomass (i.e., spectral signatures, vegetation indices, transformed images, and textures) is challenging (Amini and Sumantyo 2011).

Synthetic-aperture radar (SAR) is a technology that helps estimate forest type and aboveground biomass (Le Toan et al. 1992). Radar backscatters can accurately estimate biomass because of their sensitivity to water content in biomass (Austin et al. 2003), independent of daylight, and its capacity to penetrate through clouds (Englhart et al. 2012). However, the correlation between backscatter and biomass depends on wavelength, polarization, and incidence angle. Estimation methods from SAR data usually involve multiple linear regression, support vector machine regression, random forest regression (Maxim Neumann et al. 2012), and also logarithmic, exponential, and higher degree polynomial functional form (see Tanase et al. 2014 and references therein). These have been tested in coniferous, mixed deciduous, and coniferous, and tropical forests (Austin et al. 2003), although modifications are usually necessary to adjust estimations to forest type, particularly in tropical forests where vegetation density makes it difficult to assess some biophysical properties (Foody et al. 2001).

Kuplich et al. (2005) correlated texture measures obtained by SAR and the log of biomass with the aim of estimating biomass in regenerating forests of the Amazon and found a strong relationship between them (adjusted correlation coefficient >0.74). More recently, Sarker et al. (2013) assessed the biomass of complex subtropical forests of China using SAR and field data (DBH and height). Using stepwise multiple regressions, they developed a model between SAR parameters and field plot data, where SAR data were used as independent variables. Their results showed that texture parameters have a potential for biomass estimation. Alternatively, Amini and Sumantyo (2009) found that combining texture from SAR images and spectral responses from optical images helps improve biomass estimations.

For boreal forests, Rauste (2005) first developed regression analyses between radar backscatter data and stem volume obtained from forest inventories (Eq. 11.5) to later express dry biomass as a function of stem volume. SAR data from different seasonal configurations were used to account for the seasons' influence on biomass estimations. For their case study, the following regression for summer scenes was used.

$$V = 0.65 sqrt\left(10^{\left(\frac{\sigma^{o}+68.2}{10}\right)}\right) - 634 \tag{11.5}$$

where,

V is the stem volume (m^3/ha);

σ^o is a backscattering coefficient.

Expressed as dry biomass (B) from radar backscatter and stem volume data, the regression is as follows (Rauste 2005):

$$B = 0.39 sqrt\left(10^{\left(\frac{\sigma^{o}+68.2}{10}\right)}\right) - 380 \tag{11.6}$$

Among active sensors, light detection and ranging (LiDAR) allows users to estimate distances by emitting a laser toward a target object and measuring the time it takes the laser to reach that target (Lim et al. 2003a). This technology has been used to model forest heights and terrains; classify forest cover; estimate basal area, crown size, and canopy density (Lim et al. 2003b; van Aardt et al. 2008; Vazirabad and Karslioglu 2011); and assess biomass change (Meyer et al. 2013; Næsset et al. 2013).

LiDAR systems are useful for assessing vegetation characteristics and deriving forest biomass from small (tree) to large (landscape) spatial scales (Popescu 2007). Some applications of LiDAR data include that of Lefsky et al. (1999), who found the quadratic mean canopy height (QMCH) to be highly correlated (determination coefficient >0.80) to aboveground biomass. Relatedly, Hyde et al. (2007) employed linear regressions to create models of biomass as a function of LiDAR variables in ponderosa pine forests in the United States and showed that mean canopy height and QMCH explain over 80% of its variability. Lim et al. (2003a) derived three laser height metrics for hardwood forests near Sault Ste. Marie, Ontario. They assessed the relationship between these measures and biophysical properties of the forests, including aboveground biomass and total wood volume, DBH, canopy openness, leaf area index (LAI), and stem density. For that they used the following linear form (Eq. 11.7).

The relationship between laser height metrics and biophysical properties of forests is given as follows (Lim et al. 2003b):

$$\ln g = \ln \beta_0 + \beta_1 \ln h_L \tag{11.7}$$

where,

g is the forest metric;

h_L is one of the LiDAR height metrics; β_0 and β_1 are the regression coefficients.

Results showed that the maximum height measured by the laser pulse was the best estimator of aboveground biomass.

LiDAR data can aid with the estimation of spatial variability of forest biomass (Cao et al. 2014) and biomass and diversity (Hernández-Stefanoni et al. 2014). Maps of biomass of mixedwood forests in Canada (Thomas et al. 2006) and maps of C stocks of tropical forests across three continents (Saatchi et al. 2011) were also derived from this technology. When used with sampling data, LiDAR can be used to compare field and airborne biomass estimates in ecosystems with limited available allometry (Colgan et al. 2013). In view of the fact that aboveground biomass models derived from LiDAR metrics are significantly better than other passive optical methods or radar technologies (Zolkos et al. 2013), LiDAR stands as a promising tool for sustainable forest management (Wulder et al. 2008).

11.1.2.2 Other Biota

11.1.2.2.1 Lichen

Epiphytic lichen biomass estimations are time-consuming. Due to this, multistage sampling or subsampling techniques are used, and lichen abundance can be compared between tree branches. Subsequently, relationships between lichen mass and branch mass or age can be studied (Esseen et al. 1996). Other estimation methods include sampling late-summer litterfall lichen as a proxy to estimate their biomass, since a relationship between total biomass and litterfall (ratio of 100:1) was described for boreal (McCune 1994) and tropical forests (Li et al. 2011). Litterfall collection and subsequent drying at 70°C were also employed in temperate forests of the Himalayas to estimate dry biomass (Rawat et al. 2011).

Landscape approaches are also available for lichen biomass estimations. Lesmerises et al. (2011) developed an index that estimates terrestrial lichen biomass available for caribou foraging (Eq. 11.8). For this, they sampled lichens (species, mat thickness and cover, dry biomass, tree height) and estimated geographical variables of sampling sites (slope, altitude, and latitude). A linear relationship was found between mat thickness and biomass, whereby the biomass was estimated as follows:

$$\beta = \left[(ax+b)(c/100)\,160\right] \tag{11.8}$$

where,

a and b are the slope and intercept of the regression;
x is the mean mat thickness;
c is the mean cover;
160 is a constant to convert biomass units to kg/ha.

Through logistic regression, they found that lichen biomass is a function of altitude and tree density and thus built a map to show the areas that are more likely to sustain lichen biomass and therefore caribou populations. Satellite images can also be used for the classification of lichen-abundant and lichen-poor areas, as shown by Gilichinsky et al. (2011). Aligned with this study, lichen biomass mapping was applied to detected high-abundance areas where the use of chemicals that reduce the structural complexity of stands should be avoided (McMullin et al. 2013).

11.1.2.2.2 Liana

As with other woody species, field data on stem diameter of liana can be the input in allometric equations to estimate liana biomass (Alves et al. 2012; Laurance et al. 2014). However, allometric relationships between the aboveground biomass for lianas and stem diameter are not as good as with trees, and they also depend on forest type. Accounting for these difficulties, Schnitzer et al. (2006) developed a general equation that relates diameter (*D*) with aboveground dry biomass (*AGBM*) (Eq. 11.9).

The allometric equation for the estimation of liana's aboveground dry biomass is given as follows (Schnitzer et al. 2006):

$$AGBM = exp[-1.484 + 2.675 ln(D)] \tag{11.9}$$

Similarly, Addo-Fordjour and Rahmad (2013) developed allometric equations for liana stem biomass and total aboveground biomass. They obtained liana biomass data by destructive sampling and tested the equation for untransformed, weighted, and log-transformed (log10) data. As a result, they observed a significant linear relationship between biomass and the predictors (diameter, length, and both).

11.1.2.2.3 Understories

Understories of forests are good indicators of ecosystem health, and therefore, the measurement and prediction of their cover and biomass is relevant for management. To predict their cover and biomass, Suchar and Crookston (2010) suggested three possible analyses: (1) linear regressions between percent canopy and understory cover and biomass index; (2) regression of understory cover, biomass indices, and canopy cover against explanatory variables (elevation, slope, climatic variables, among others); and (3) maximum understory biomass indices by sampling the 99th percentile of shrub and herb biomass indices separately and then regressions against its constraints (competition, predation, disturbance, and soils quality).

Marinescu et al. (2011) assessed understory biomass in a forest in Florida, United States, with the aim of estimating its potential to meet energy needs. They measured the DBH of woody stems of noncommercial species, and some of them were harvested and weighed. Allometric regression equations for each species were developed for green biomass weight, and then dry biomass weight was calculated by multiplying green biomass weight values by (1−moisture content% [wet basis]) + 9.7%.

Not only it is possible to estimate the biomass of woody species, but grasses and forbs can also be included in the analysis (Reiner et al. 2010). For example, Porté et al. (2009) broke down the understory in groups: woody species, bracken, herbaceous species, and mosses. For all of them, relationships between biomass and phytovolume (m^3/ha) can be fitted with Eq. 11.10.

$$Vol_{species\ group} = \sum_{species} \frac{Cover_{species}}{100} 10{,}000\ Height_{species} \qquad (11.10)$$

where cover is expressed as percentage and height in meters. Subsequently, allometric models were fitted between biomass and volume and/or volume^2 using nonlinear procedures. Accordingly, Muukkonen et al. (2006) modeled the aboveground biomass of the understory vegetation of a boreal forest using the percent cover of different groups of species. For dwarf shrubs and herbs and grasses, they employed a linear model, but the biomass of bryophytes and lichens (y) was modeled as a function of percent cover (x) with a mixed linear model (Eq. 11.11).

The mixed linear model for the estimation of bryophytes and lichens' biomass is given as follows (Muukkonen et al. 2006):

$$y = \frac{u + x^2}{\left(\beta_0 + \beta_1 x\right)^2} \qquad (11.11)$$

where,

β_0 and β_1 are the population parameters;
u is a random parameter.

Other recent examples of understory assessment for forest management include the contribution of woody and herbaceous perennial understory species to total ecosystem biomass in pine temperate forests (Gonzalez et al. 2013), the examination of NPP and C allocation of understory saplings in subalpine coniferous forests (Takahashi and Obata 2014), and the quantification of carbon stocks at regional scales (Russell et al. 2014).

11.1.2.2.4 Invertebrates

Invertebrates' biomass and abundance can also be used to assess sustainable forest management practices. For example, Pettersson et al. (1995) analyzed the effects of commercial forestry on canopy-living invertebrates in the boreal forest. In this instance, abundance and biomass were measured via individual counts and fresh mass weight. Greenberg and McGrane (1996) measured both abundance and biomass of arthropods as indicators of the impact that forestry practices can have on ecosystem functions, since these taxa play a major role in food webs.

Bottom-up trophic linkages and forest productivity can be studied by assessing, among others, the dry mass of litterfall and soil invertebrates (Fittkau and Klinge 1973; Seagle and Sturtevant 2005). Furthermore, estimates of earthworms can serve as indicators of trends in soil pH, ectorganic horizon biomass, soil humidity, and humus quality (Muys and Granval 1997). Duguay et al. (2000), for example, used invertebrates' data to assess the impact of different silvicultural practices on those organisms as well as on breeding birds. Wilson et al. (2014) also compared early and late successional forest habitats to test which one supports a higher diversity and biomass of invertebrates and consequently insectivorous vertebrates, arriving at the conclusion that early successional forest habitats should be maintained in managed forest landscapes. Finally, other studies have used the abundance of certain sentinel species to examine the effect of different forest management practices (Zhao et al. 2013; Lange et al. 2014).

11.2 COMPLEXITY

11.2.1 Forests as Complex (Adaptive) Systems

The science of complex systems (CS) originated in early cybernetics and characterizes CS as process-dependent organic systems constituted by heterogeneous interacting units called agents. These agents interact and a collective behavior arises as an emergent property. Interactions include feedbacks and involve several scales. CS have a capacity for self-organization and are not deterministic, nor always predictable (Parrott and Kok 2000; Anand et al. 2010; Newman 2011).

Forests, as well as other ecosystems, can be seen through this CS perspective for they are composed by a heterogeneous collection of agents that interact locally (Folke 2006). Heterogeneity is also reflected in the nature of their components and behaviors, structural organization, spatial location, extent or history, and spatial and temporal scales at which they occur (Filotas et al. 2014). Interactions among components, usually affected by feedbacks, shape the way systems evolve in their genetics, distribution, and behaviors (Folke 2006; Parrott 2010). In addition, forests exhibit a hierarchical structure because their heterogeneous agents are organized over multiple scales. Cross-scale interactions and feedbacks take place and result in an interaction among subsystems where each level is subordinated to the one above it (Ulgiati and Brown 2009).

The properties of heterogeneity and hierarchy are associated with self-organization as well, because interactions among components can cause coherent patterns, entities, or behaviors to emerge at higher scales of the hierarchy (Filotas et al. 2014). Even more, positive feedback can act as organizing forces (Perry 1995; Green and Sadedin 2005). From a thermodynamics perspective, informational (e.g., biodiversity), energy (e.g., exergy), and material (e.g., nutrients, water) constraints shape the relationship between ecosystems' structure (species, biomass, soil etc.) and metabolism (functions) and thus promote or restrain self-organization (Kutsch et al. 2001).

Given that these natural systems typically display nonlinear dynamics in space and time (Anand 2000; Lansing 2003; Parrott 2010), their emergent properties or behaviors might be different from those expected by the single combination of its parts (Parrott and Kok 2000; Anand et al. 2010; Newman 2011). Similarly, forests are sensitive to initial conditions, and disproportionate and unexpected outputs can result from inputs: unexpectedly large responses to small disturbances or vice versa (Parrott and Lange 2013). Some of these emergent properties are evidenced in spatial canopy structure, trophic structures, diversity–productivity relationships, and patterns of nutrient flux (Levin 1998; Pagnutti et al. 2007).

Other properties of CS evidenced in forests are openness, memory, and uncertainty. Forest ecosystems are open as energy, matter, information, species, people, and capital can be exchanged with the external environment (Filotas et al. 2014). Moreover, they possess memory, meaning the past history of the system affects its present and future structure, composition, and behavior (Hendry and Mcglade 1995). This refers to both retrospective (reflecting past disturbances) and prospective (reflecting future trajectories) standpoints (Sun et al. 2013). Finally, uncertainty is derived from stochastic internal processes in these ecosystems, which make it difficult to predict behavior in the system. Forests have a certain degree of resilience as well, meaning the system can absorb change before losing its original function and structure. However, under fluctuations in external inputs, some forests may adapt by changing their behavior and attributes (Hartvigsen et al. 1998; Messier and Puettmann 2011; Filotas et al. 2014).

Anand et al. (2010) illustrated the components of ecological complexity in forests. In their example, forest structure was characterized by gap size distribution and dynamics were represented by succession. Previous treefalls and seed banks denoted the memory component, whereby cross-scale interactions could occur between the stand and landscape level. Self-reproducing elements, in this case tree individuals, originated interactions in a variable environment subjected to climatic fluctuations and stochastic events (e.g., windthrow). Gaps created by these stochastic events could be magnified through feedbacks between gap size, fetch, and boundary effects. Forest recovery would depend on memory-based spatiotemporal processes (e.g., seed banks) and growth patterns. In the case of fire, environmental heterogeneity can help predict their onset and propagation, and, in turn, fire can then affect weather conditions that drive its propagation, thereby generating a positive feedback.

11.2.2 Importance of the Complexity Approach for Forest Management

The science of CS can be applied to forest management because complexity is to some extent related to other holistic measures of ecosystem structure and functioning, such as resilience and integrity. For example, heterogeneity of components in a forest provides functional and structural redundancy, thus increasing both the complexity (in space, time, and structure) and resilience of the system (Parrott 2010). In addition, the complexity of a natural ecosystem increases with its maturity (Müller 2005), and the degree of maturity or organization of an ecosystem is complementary to the measures of self-organization as well (Parrott 2010).

Complexity measures could be used to assess the resilience and robustness of forests under different management scenarios. For example, aiming to preserve tree densities and arrangements that encourage diversity of plants as well as pollinators and insectivores contributes to the heterogeneity of these ecosystems and could protect them from temperature fluctuations, drought, frost, or insect damage (Filotas et al. 2014). Elements in the forests that can provide the system with memory, such as legacies left after disturbances (e.g., dead trees, snags), affect the ecosystem dynamics (e.g., succession, nutrient cycling) and structure over time and thus condition the timing and feasibility of management activities (James et al. 2007).

Additionally, considering and studying critical transitions between stability domains would shed some light on how forests respond to disturbances or different management practices (Messier and Puettmann 2011). The complexity associated with the disturbance–succession cycle of forests must also be embraced for management, since it reflects the maintenance of resilience in these ecosystems. Practices such as clearcutting and stand homogenization may fulfill short-term management goals but will lead to a decline in ecosystem resilience once ecological and evolutionary processes are affected (Kuuluvainen 2009). Furthermore, targeting the preservation of resilience and adaptive capacity of forests, instead of timber production alone, generates natural systems that in the long term will provide the desired ecosystem goods and services (Messier and Puettmann 2011).

Finally, uncertainty of forest systems should be taken into account for management, not only from a natural perspective but also from a social point of view. Reducing uncertainty requires collaborative processes among organizations and stakeholders at different scales (Armitage et al. 2009), an active adaptive management approach (Bell et al. 2008), and the monitoring of biotic and abiotic indicators of forests (Swift and Bell 2011). In the following sections, we review measures of complexity, many of them derived from other disciplines such as information theory and physics, and we provide examples of their application worldwide.

11.2.3 Complexity Measures Applied to Ecological Forest Management

11.2.3.1 Power Laws and Fractals

Statistical scaling relationships can be used to assess complex states. For example, to reflect the emergent property of scaling relationships that are fractal-like or self-similar over time and space, the following simple function can be used (Brown et al. 2002).

$$Y = Y_0\ X^b \tag{11.12}$$

where,

X and Y are the independent and dependent variables, respectively;
Y_0 is a normalization constant;
b is another constant.

This type of function can be applied, for instance, to phenomena that show Euclidean geometric scaling and have a scaling exponent b that is multiple of 1/3. In the cases where X is volume (V) or mass (M), some scaling relationships are surface areas as $V^{2/3}$ or linear scales as $V^{1/3}$. Likewise, for allometries where b is a multiple of 1/4, some power functions can be used, such as lifespan or development times expressed as $M^{1/4}$ or maximum rate of population growth as $M^{-1/4}$. Other applications for forest management are studying how population density changes with body size or mass (Enquist et al. 1998) or observe the scaling relationship between tree abundance and stem diameter and aboveground biomass (Enquist and Niklas 2001).

As for fractals, they are objects showing structures at all spatial scales, with a scale-dependent self-similarity. They can be applied in forest management to study the patchiness and heterogeneity of the forested landscape (see Li 2000 and references therein). Krummel et al. (1987), for instance, applied a perimeter–area method to determine the fractal dimension of deciduous forest patches in Mississippi, United States (Eq. 11.13).

The perimeter–area method for the estimation of fractal dimensions of forest patches is given as follows (Krummel et al. 1987):

$$A = \beta L^D \tag{11.13}$$

where,

A is the area of the patch;
L is its perimeter;
β is a constant;
D is the fractal dimension.

Comparing different scales, similar D values would designate no characteristic length scale (length determined with an order of magnitude) in the processes generating the shapes of the patches, whereas different D values would indicate the forest has structures of well-defined length scale. The exponent provides information on the complexity change of average patch shapes (Li 2000). Similarly, Rex and Malanson (1990) estimated the patch shape (C) as shown in Eq. 11.14.

$$C = P / A^{D/2} \tag{11.14}$$

where,

P is the patch perimeter;
A is the area;
D is the fractal dimension.

This shape index can then be regressed on hydrogeomorphic variables (e.g., total valley width, stream confluence, sinuosity) to test which one is more likely to control patch dimensions.

Furthermore, fractal geometry is suitable to study the crown surface of forests. Under the "the two-surface method" (Zeide 1991), it is assumed that a relationship exists between the leaf area of a tree (A) and the surface area of a convex hull that surrounds its crown (E). The fractal dimension, D, can be estimated as depicted in Eq. 11.15.

The fractal dimension of a forest crown surface is given as follows (Zeide 1991):

$$A = aE^{D/2} \tag{11.15}$$

where, a is a constant.

Other applications of fractal dimensions and power laws were employed to study fragmentation in tropical rainforests (Imbernon and Branthomme 2001) and landscape dynamics in post-fire and late-successional sites of the Canadian boreal forest (Walker and Kenkel 1998). More recently, Wei et al. (2013) used scaling laws to analyze biodiversity patterns along an altitudinal gradient. They found that richness and abundance have a fractal effect in the community despite the gradient, the several types of habitats, and the different vegetation strata.

11.2.3.2 Measures of Stand Structural Complexity

Several attributes can be described to account for stand complexity: canopy cover, foliar arrangement, understory vegetation, deadwood, biomass, and tree height, diameter, and spacing (McElhinny et al. 2005). However, indices that combine these and other attributes were developed to better represent the stand complexity of forests. Zenner (2000), for example, evaluated if overstory residual trees were associated with higher structural complexity of forests in the Willamette National Forest in Oregon, United States. He employed a structural complexity index (SCI), which was designed to associate the observed structure of the forest to other variables such as residual trees, tree density and size, species composition, and slope and elevation, among others. The observed structure was based on a 3D model of the structure of stands, where trees are 3D-spaced data points (x and y are coordinates, and z is DBH or height). In this x, y, z space, adjacent points (top of the trees) can be linked to form triangular surfaces that conform a triangulated irregular network (TIN). This can be seen as a canvas that covers the stand. Triangles are not overlapping, and points are connected to their nearest neighbor. The stand SCI is estimated as

$$SCI = \frac{SCI^*}{A_T} \tag{11.16}$$

where, SCI^* is the sum of the surface areas of the TINs for a stand; and A_T is the horizontal ground area covered by all triangles.

When SCI^* is large, the structural complexity of the stand increases. More recently, SCI was modified by Beckschäfer et al. (2013) to include stem density and the degree of mingling of tree attributes. They tested it over a gradient ranging from secondary to mature montane rainforests of China, proving that this modification provides a more complete description of forest complexity.

Forest canopies can exhibit complex vertical and horizontal structural patterns that can also be assessed. For example, Latham et al. (1998) developed TSTRAT, an algorithm that simulates the stratification that occurs during stand development. It employs basic data such as species, height, crown ratio, and DBH and assigns each tree to a stratum. According to the authors, some applications include health risk assessment and the assessment of aesthetic value and wildlife habitat.

Other measures of complexity that combine structural metrics include the complexity index (HC) by Holdridge (1967), defined by the top height (H), the basal area (BA), the stem number (n), and the number of species (N) found in the stand (Eq. 11.17). The forest stand complexity index is given as follows (Holdridge 1967):

$$HC = H \times BA \times n \times N \tag{11.17}$$

Metrics that account for the spatial distribution of trees were also developed, such as the index of aggregation (Eq. 11.18) of Clark and Evans (1954). These authors considered the nearest-neighbor distance (r_i) between trees (n) and their density per square meter (ρ). The index of aggregation is given as follows (Clark and Evans 1954):

$$IA = \frac{1}{n}\sum_{1}^{n} r_i 2\sqrt{\rho} \tag{11.18}$$

The Gini coefficient (G) of statistical dispersion of tree diameters is also used as a metric of live-tree structural complexity (Bradford and Kastendick 2010). For each stand, G is estimated according to Eq. 11.19.

$$G = \frac{\sum_{i=1}^{n} (2i - n - 1)x}{n^2 \mu} \tag{11.19}$$

where,

n is the number of trees in the stand;
x is the diameter of tree I;
μ is the mean diameter of the stand.

Indicators of complexity can also be directly applied to evaluate the probability of the presence of fauna. For instance, Watson et al. (2001) employed a habitat complexity score in temperate woodlands of Australia. Attributes considered were canopies, tall shrubs, low shrubs, ground herbage, logs and fallen branches, and litter. Each attribute was assigned a percent cover category (0–10, 10–20, 20–50, >50) and a respective score (0, 1, 2, 3) of complexity. Values of each attribute were then totaled for a final complexity score. Even more, a complex stand index (S-index) was applied by Pastorella and Paletto (2013). This was originally derived from other indices (diametric differentiation, mingling, and contagion indices), and it provides information on species composition, spatial distribution of trees, and horizontal structure that can be applied for biodiversity and stability maintenance in stands. Similar indices of diversity and complexity of forests were reviewed and compared by Neumann and Starlinger (2001).

11.2.3.3 Spatial Heterogeneity at the Landscape Scale

As proposed by Cadenasso et al. (2006), the biocomplexity of ecosystems can be measured through their heterogeneity (e.g., what and how patches are arranged), organization (e.g., how patches interact), and history (e.g., how they vary through time). Lindenmayer et al. (2002) also consider the connectivity between patches that describes how organisms move through elements in the landscape. Forests at the landscape scale are more complex when the number and type of patches increase and when their spatial arrangement and identity change shift through time. In addition, the structural heterogeneity of forests in the horizontal dimension includes the formation of gaps and the thinning of the overstory (Franklin and Van Pelt 2004).

Several metrics to measure heterogeneity exist and can be broadly classified as (1) areal (related to patch size, shape, and area), (2) linear (comprehends boundaries, distances, patch width, and connectivity), and (3) topological (refers to dispersion, spatial association, isolation between landscape elements) (Baskent and Jordan 1995). Haines-Young and Chopping (1996) made an extensive review on these metrics. Some of them could be classified as areal, such as mean density and size of patches and dominance of patches over the total area. Given that the whole area of the patch does not provide optimum conditions for the persistence of species, the core area of the patch is utilized as a more realistic metric, and it can be calculated by specifying edge buffer areas. Moreover, the number of core areas within each patch and the core area density (e.g., number of core areas/100 ha), the average core area per patch, or the proportion of the landscape occupied by core areas can be quantified as well. Some linear metrics that can be characterized for complexity purposes are, for example, the length and density of edges and the percent edge adjoining similar patch types. For topological measures that account for the spatial association between patches, the nearest neighbor is defined as the minimum distance between two patches of the same kind and can usually be estimated using centroids within each patch. This distance can be further used to estimate the dispersion of patches in the landscape. Other similar metrics of connectivity were reviewed and compared by Prugh (2009).

Indices that combine metrics have also been developed. For instance, the landscape shape index is calculated as the ratio of edge lengths to total area, and the mean shape index is estimated as the sum of the patch perimeter divided by the square root of the patch area, adjusted by a constant/number of patches (Haines-Young and Chopping 1996).

Some of these metrics can be applied, for example, to evaluate if small-scale disturbances within forests create different spatial patterns (Kane et al. 2011). Similarly, patch dynamics can be applied to the formation of gaps and community recovery following disturbances (Levin and Paine 1974). Hernández-Stefanoni et al. (2011) employed some patch metrics to depict the relationship between landscape configuration and community structure in tropical dry forests of Mexico. In the case of boreal forests, Janssen et al. (2009) looked into the relationship between landscape heterogeneity and the presence of dwelling invertebrates, and Kivinen et al. (2012) studied how changes in landscape structure are affecting sustainable reindeer husbandry. In addition, given that the composition or configuration of the forest landscape can affect the provision of ecosystem services, metrics of spatial heterogeneity can contribute to analyzing trade-offs between landscape and services in forest management (Turner et al. 2013).

11.2.3.4 Indicators for Quantifying Self-Organizing Open Systems

As it was explained in the previous sections, the self-organization of an ecosystem is associated with its metabolism. As such, simple measures of self-organization can be drawn from thermodynamic concepts. For example, when a system is exposed to a gradient of energy and matter and thus maintained far from a thermodynamic equilibrium, its complexity increases as well as its effectiveness in energy dissipation. Given that solar radiation is the main source of energy gradient in terrestrial ecosystems, comparing the way ecosystems dissipate that energy can describe their complexity. This can be estimated as R_n/K^*, where R_n is the net radiation flux (net radiation transformed into nonradiative energy at the surface) and K^* is the net flux of solar radiation (shortwave radiation balance) (Schneider and Kay 1994; Kay et al. 1999; Kutsch et al. 2001).

In addition, Kutsch et al. (2001) proposed other measures from thermodynamics that are of relevance to assess the system's capacity to self-organize and that indicate the integrity of ecosystems as well. These are, for instance, the ratio between transpiration and total evapotranspiration and the metabolic heat flux (e.g., photosynthesis as a measure of conversion of radiative to biochemical energy). To indicate energy efficiency, two suggested measures are the energy or C units per unit of biomass and time (qCO_2) and a ratio between gross ecosystem production and N availability in the soil.

While growing and self-organizing, systems show more complex structures and functions, which can be reflected in an increase of biomass, networks, species, and more complex life strategies. Hence, some structural properties of ecosystems such as biological diversity can be employed as measures of self-organization. Specifically, richness, biomass, and intrabiotic nutrients have been proposed as indicators (Kutsch et al. 2001).

As for applications to management, Norris et al. (2012) applied thermodynamic indicators to a chronosequence of forests and found that old-growth woodlands are more effective than plantations in attenuating surface temperature, suggesting that the conservation of complex old-growth forests contributes to climate regulation services. Relatedly, Wagendorp et al. (2006) suggested that these indicators could be used to assess the impact of anthropic land uses, which simplify biodiversity and result in loss of internal complexity of ecosystems. They observed that more complex undisturbed forests were more efficient in dissipating energy than other land covers. The measures presented, among others, can also play a role in the modeling of forest fires (Séro-Guillaume and Margerit 2002).

11.2.3.5 Information Theoretical Measures of Complexity

Complexity can be manifested by diversity, that is, the number of elements in a system and how they are proportioned. To measure this, Shannon's (Shannon 2001) entropy function (H) is commonly used (Eq. 11.20).

Shannon's entropy function is given as follows (Shannon 2001):

$$H = -\sum_{j=1}^{q} p_j \log_2 p_j \text{ bits} \tag{11.20}$$

where,

q is the number of community elements;
p_j indicates the proportion of community element j.

However, some authors make a distinction between measures of diversity and complexity, claiming that the former is a disorder-based complexity measure that lacks the entropy component (Anand and Orlóci 1996). Thus, for plant communities, they propose to transform Shannon's index with a coding system, where S is the species list, q is the number of species, s_j represents each species, p_j is their relative presence, and l_j is a code word (sequence of symbols). Low p_j is given long code words and vice versa to represent the notion that rare species convey more information and are more unpredictable. The code length (L) specific of S is calculated as in Eq. 11.21.

The transformation of Shannon's index with a coding system is given as follows (Anand and Orlóci 1996):

$$L(S) = \sum_{j=1}^{q} p_j l_j \text{ bits} \tag{11.21}$$

When tested with real data, Anand and Orlóci (1996) proved that $L(S)$ values are equal to or larger than Shannon's index and can be seen as a measure of "total complexity." L is statistically linked to vertical layering and nonlinear species responses and would therefore reflect not only entropy but also emergent properties. Moreover, Anand and Orlóci (2000) showed that there is variation in complexity when considering different vegetation strata and propose two partitioning schemes to assess that using an example in the boreal forests of Ontario.

Measures of statistical complexity were also proposed by Feldman and Crutchfield (1998), where large correlations between system components indicate more structured systems.

For instance, a measure can consider that the information gain between two probability distributions is defined as a one-way divergence. This can be applied to compare how an observed distribution, such as a community assemblage, differs from a theoretical one (Carlo Ricotta and Anand 2006). Later, a statistical complexity was defined as the product of the divergence and Shannon's index value.

Ricotta and Anand (2006) also showed that Juhász-Nagy's information theoretical model combines fuzzy analogs of the Shannon entropy and the traditional Shannon entropy for assessing spatial complexity in plant communities. Finally, taxonomic entropy (see Ricotta and Avena 2003) has been found to be a useful measure of biocomplexity to quantify perturbation in plant communities (Desrochers and Anand 2005).

11.2.3.6 Contribution of Digital Imaging and Remote Sensing Technologies

As it was mentioned in previous sections, remote sensing technologies are playing a major role in gathering data for forest research and management. Not only they aid with biomass estimations, but they are also used for obtaining information on forest heterogeneity, which then has further applications. Using remote sensing, one of the most used measures of vegetation biophysical properties is the normalized difference vegetation index (NDVI), which as developed by NASA in the 1980s. The NDVI uses reflectance measures in the red and NIR portion of the spectrum, and it is estimated using Eq. 11.22.

The following equation is the normalized difference vegetation index:

$$NDVI = \frac{R_{NIR} - R_{Red}}{R_{NIR} - R_{Red}} \tag{11.22}$$

where,

R_{NIR} is the reflectance of NIR radiation;
R_{Red} is the reflectance of visible red radiation.

More recently, the wide dynamic range vegetation index (WDRVI) developed by Gitelson (2004) was proposed as an improvement of the NDVI because it allows a better characterization of vegetation biophysical properties and surface conditions in cases of medium-to-high biomass.

The wide dynamic range vegetation index is given as follows (Gitelson 2004):

$$WDRVI = \frac{(a * R_{NIR} + R_{Red})}{(a * R_{NIR} + R_{Red})} \tag{11.23}$$

As it can be observed, Eq. 11.23 adds a weighting coefficient $a < 1$ to decrease the disparity between the contributions of the R_{NIR} and the R_{Red} to the NDVI. These indices are related to biomass estimations and variables of forest structure such as successional and phenological stage, chlorophyll content, NPP, LAI, and the fraction of absorbed photosynthetically active radiation. Some examples of the use of WDRVI include Aguilar-Amuchastegui and Henebry (2007), who employed it as a variable of structural heterogeneity when looking for relationships between forest structure, dung beetle communities, and logging intensity in tropical forests of Costa Rica. In addition, Lee et al. (2014) used these indices in combination with patch metrics to study how forest regeneration is affected by prefire heterogeneity and burn severity in Samcheok, Korea.

As for laser technology, it can be used to measure the structure of the outer forest canopy. Parker and Russ (2004) employed it in forests of the mid-Atlantic region to observe how stands of different developmental stage have dissimilar canopy surface and complexity. Measures assessed in this case included canopy height, fractal dimension, and shape of the canopy, among others. Data obtained from LiDAR have been used to improve habitat assessments in the Northern Prealps and the Central Eastern Alps (Zellweger et al. 2014) and to describe successional stages in conifer forests (Kane et al. 2010).

Lamonaca et al. (2008) showed the usefulness of satellite imagery with a case study on temperate forest of Italy. Through multiresolution segmentation of very-high-geometric resolution images, they obtained data on tree spatial distribution and sizes, tree species mingling, and occurrence of canopy gaps. Furthermore, they were able to identify patterns of structural heterogeneity at different spatial scales. Also, digital camera pictures can be analyzed through information theoretic indices, as shown by Proulx and Parrott (2008) in old-growth forests of Québec. Using pixel values, they estimated an index of aspatial heterogeneity via Shannon's formula for entropy (H_γ) and another one of spatial heterogeneity (H_χ). These were then utilized in a mean information gain (MIG) index that quantifies structural complexity (Eq. 11.24).

The MIG for structural complexity is the following (Proulx and Parrott 2008):

$$MIG = \frac{H_{[x]} - H_{[y]}}{\log N^4 - \log N^1} = \frac{JE - ME}{\log N^4 - \log N^1} \tag{11.24}$$

where,

JE and *ME* are joint and marginal entropy values, respectively;

N^4 is the number of theoretical 2 × 2 colored combinations that can be found in the picture.

The equation was then fourth-ordered to calculate the mean mutual information (MMI) index. Finally, the product of MIG and MMI was used as a convex complexity index (Γ). This photographic approach can be used for the monitoring of structural complexity (Proulx and Parrott 2009).

11.2.3.7 Indicators and Simple Models to Assess Forest Memory

The property of forest memory has been assessed in many ways using different phenomena, variables, and indicators to represent it. For example, it can be measured as a self-reinforcing process between trees and forest soil in which trees influence pedologic characteristics that make them more likely to establish on those same sites in the future (Phillips and Marion 2004). This works as a two-way process, where trees are preferentially established on nutrient-rich and rock fragment poor sites. As a feedback, the biomechanical effect of trees along with decomposition maintains rock fragmentation and nutrient availability in the soil, increasing the chances of subsequent tree establishment in the area.

Sun et al. (2013) expanded on this by selecting indicators of past vegetation and events (soil age, pollen, and mineral distribution), community estimates (plant species, soil microbes and animals, and light environment), potential ecosystem development (seed bank), and a mobile link species (birds). Using data from a subtropical forest in China, they proposed the quantification of memory (*EM*) between different successional stages.

The following equation is the quantification of forest memory (Sun et al. 2013):

$$EM = \sum_{i=1}^{n} \frac{\omega_i R_i}{n} \tag{11.25}$$

where,

R_i is the resource overlap ratio of the *i*th proxy indicator between the research and reference ecosystem;

ω_i is the weight of R_i;

n is the number of indicators.

A stage IV successional climax forest was used as a reference. The authors suggest that this framework could be applied in restoration to assess how far the restored ecosystem is from the reference or from natural successional trajectories.

Another approach to analyze forest memory is the study of individual spatial patterns as a result of previous abundance and dispersal processes. For example, Flügge et al. (2012) developed an individual-based simulation model with local dispersal and density-dependent competition for a tropical forest. They showed that recently growing populations of canopy trees present more aggregation than shrinking populations of the same abundance. Also related to spatial patterns of trees, Wild et al. (2014) examined memory as recruitment patterns around parental trees following disturbances. After an initial self-thinning of recruits, trees would emulate the spatial pattern of the stand before the disturbance.

Memory has been examined from a gap pattern perspective as well. Pagnutti et al. (2011) described a spatiotemporal correlation function that can be seen as a measure of gap pattern memory. Specifically, it gives the proportion of neighboring cells around a certain distance of a focal gap cell that presented a gap at a specified time in the past. Interestingly, they found a correlation time of 160 years, indicating that present gap patterns can affect the forest gap pattern for that long. Furthermore, long-term memory patterns were also observed in western Europe, where plots of recovering forests with different past agricultural use were compared in terms of community and soil variables. It was found that historic agricultural activities that modified soil properties were reflected in nutrient availability and plant diversity differences for more than two millennia after land abandonment (Dupouey et al. 2002). As shown by the aforementioned examples, memory is a dimension of forest complexity that should be considered by managers, as past events or scenarios are able to influence the future behavior of forests.

11.3 SUMMARY AND CONCLUDING REMARKS

In this chapter, we begin by examining the importance of abundance and biomass in forest management. We discuss trees, the most obvious variable in this process, and also other biota, including lichens, liana, understory vegetation, and invertebrates. These are some of the most commonly studied ecosystem components with respect to forest management. A large amount of research has also linked forest management to wildlife habitat and thus focused on many other taxa (e.g., birds and mammals). An increasing realization that forest management practices have implications for biodiversity at many levels is helping to understand the complex relationships between these systems. This has been the motivator for the conceptualization of forest management systems as complex adaptive systems and in the development of indicators of complexity, which we review here. There still remains many opportunities however to link concepts and measures of abundance and biomass with those of complexity. We need to ensure not only comprehensive biodiversity assessments in managed systems but also measures of ecosystem processes to better integrate information to link structures to functions.

REFERENCES

Addo-Fordjour, P. and Z.B. Rahmad. 2013. Development of allometric equations for estimating above-ground Liana biomass in tropical primary and secondary forests, Malaysia. *Int. J. Ecol.* 2013:1–9.

Aguilar-Amuchastegui, N. and G.M. Henebry. 2007. Assessing sustainability indicators for tropical forests: Spatio-temporal heterogeneity, logging intensity, and dung beetle communities. *For. Ecol. Manage.* 253(1–3):56–67.

Alves, L.F., M.A. Assis, J. Melis, A.L.S. Barros, S.A. Vieira, F.R. Martins, L.A. Martinelli, and C.A. Joly. 2012. Variation in Liana abundance and biomass along an elevational gradient in the tropical atlantic forest (Brazil). *Ecol. Res.* 27(2):323–332.

Amini, J. and J.T.S. Sumantyo. 2009. Employing a method on sar and optical images for forest biomass estimation. *IEEE Trans. Geosci. Remote Sens.* 47(12):4020–4026.

Amini, J. and J.T.S. Sumantyo. 2011. SAR and optical images for forest biomass estimation. In *Biomass- detection production and usage*, ed. D. Matovic, 508pp. Rijeka, Croatia: InTech.

Anand, M. 2000. The fundamentals of vegetation change-complexity rules. *Acta Biotheor.* 48(1):1–14.

Anand, M., A. Gonzalez, F. Guichard, J. Kolasa, and L. Parrott. 2010. Ecological systems as complex systems: Challenges for an emerging science. *Diversity* 2(3):395–410.

Anand, M. and L. Orlóci. 1996. Complexity in plant communities: The notion and quantification. *J. Theor. Biol.* 179(2):179–186.

Anand, M. and L. Orlóci. 2000. On hierarchical partitioning of an ecological complexity function. *Ecol. Model.* 132(1–2):51–62.

Armitage, D.R., R. Plummer, F. Berkes, R.I. Arthur, A.T. Charles, I.J. Davidson-Hunt, A.P. Diduck et al. 2009. Adaptive co-management for social-ecological complexity. *Front. Ecol. Environ.* 7(2):95–102.

Austin, J.M., B.G. Mackey, and K.P. Van Niel. 2003. Estimating forest biomass using satellite radar: An exploratory study in a temperate Australian Eucalyptus forest. *For. Ecol. Manage.* 176(1–3):575–583.

Baker, T., O. Phillips, Y. Malhi, S. Almeida, L. Arroyo, A. Di Fiore, T. Erwin et al. 2004. Variation in wood density determines spatial patterns in Amazonian forest biomass. *Glob. Change Biol.* 10(5):545–562.

Baskent, E.Z. and G.A. Jordan. 1995. Characterizing spatial structure of forest landscapes. *Can. J. For. Res.* 25:1830–1849.

Bastien-Henri, S., A. Park, M. Ashton, and C. Messier. 2010. Biomass distribution among tropical tree species grown under differing regional climates. *For. Ecol. Manage.* 260(3):403–410.

Basuki, T.M., P.E. van Laake, A.K. Skidmore, and Y.A. Hussin. 2009. Allometric equations for estimating the above-ground biomass in tropical lowland dipterocarp forests. *For. Ecol. Manage.* 257(8):1684–1694.

Beckschäfer, P., P. Mundhenk, C. Kleinn, Y. Ji, D.W. Yu, and R.D. Harrison. 2013. Enhanced structural complexity index: An improved index for describing forest structural complexity. *Open J. For.* 03(01):23–29.

Bell, F.W., J. Parton, N. Stocker, D. Joyce, D. Reid, M. Wester, A. Stinson, G. Kayahara, and B. Towill. 2008. Developing a silvicultural framework and definitions for use in forest management planning and practice. *For. Chron.* 84(5):678–693.

Boudreau, J., R.F. Nelson, H.A. Margolis, A. Beaudoin, L. Guindon, and D.S. Kimes. 2008. Regional aboveground forest biomass using airborne and spaceborne LiDAR in Québec. *Remote Sens. Environ.* 112(10):3876–3890.

Bradford, J.B. and D.N. Kastendick. 2010. Age-related patterns of forest complexity and carbon storage in Pine and Aspen–birch ecosystems of Northern Minnesota, USA. *Can. J. For. Res.* 40(3):401–409.

Brown, J.H., V.K. Gupta, B.-L. Li, B.T. Milne, C. Restrepo, and G.B. West. 2002. The Fractal nature of nature: Power laws, ecological complexity and biodiversity. *Philos. Trans. R. Soc. Lond. B. Biol. Sci.* 357(1421):619–626.

Brown, S. 2002. Measuring carbon in forests: Current status and future challenges. *Environ. Pollut.* 116(3):363–372.

Bunker, D.E., F. Declerck, J.C. Bradford, R.K. Colwell, I. Perfecto, O.L. Phillips, M. Sankaran, and S. Naeem. 2005. Species loss and aboveground carbon storage in a tropical forest. *Science* 310(2005):1029–1031.

Cadenasso, M.L., S.T.A. Pickett, and J.M. Grove. 2006. Dimensions of ecosystem complexity: Heterogeneity, connectivity, and history. *Ecol. Complex.* 3(1):1–12.

Cairns, M.A., E.H. Helmer, and G.A. Baumgardner. 1997. Root biomass allocation in the World' S upland forests. *Oecologia* 111:1–11.

Cannell, M.G.R. 1982. *World forest biomass and primary production data.* London, UK: Academic Press.

Cao, L., N.C. Coops, J. Innes, J. Dai, and G. She. 2014. Mapping above- and below-ground biomass components in subtropical forests using small-footprint LiDAR. *Forests* 5(6):1356–1373.

Carneiro, M., A. Fabião, M.C. Martins, C. Cerveira, C. Santos, C. Nogueira, M. Lousã et al. 2007. Species richness and biomass of understory vegetation in a Eucalyptus globulus Labill. Coppice as affected by slash management. *Eur. J. For. Res.* 126(4):475–480.

Case, B.S. and R.J. Hall. 2008. Assessing prediction errors of generalized tree biomass and volume equations for the boreal forest region of West-Central Canada. *Can. J. For. Res.* 38(4):878–889.

Castellanos, J., M. Maass, and J. Kummerow. 1991. Root biomass of a dry deciduous tropical forest in Mexico. *Plant Soil* 131(2):225–228.

Chave, J., R. Condit, S. Aguilar, A. Hernandez, S. Lao, and R. Perez. 2004. Error propagation and scaling for tropical forest biomass estimates. *Philos. Trans. R. Soc. Lond. B. Biol. Sci.* 359(1443):409–420.

Chen, J., M. Shiyomi, Y. Hori, and Y. Yamamura. 2008. Frequency distribution models for spatial patterns of vegetation abundance. *Ecol. Modell.* 211(3–4):403–410.

Chiarucci, A., J.B. Wilson, B.J. Anderson, and V. De Dominicis. 1999. Cover versus biomass as an estimate of species abundance: Does it make a difference to the conclusions? *J. Veg. Sci.* 10:35–42.

Clark, P.J. and F.C. Evans. 1954. Distance to nearest neighbor as a measure of spatial relationships in populations. *Ecology* 35(4):445–453.

Colgan, M.S., G.P. Asner, and T. Swemmer. 2013. Harvesting tree biomass at the stand level to assess the accuracy of field and airborne biomass estimation in Savannas. *Ecol. Appl.* 23(5):1170–1184.

Desrochers, R.E. and M. Anand. 2005. Quantifying the components of biocomplexity along ecological Perturbation gradients. *Biodivers. Conserv.* 14(14):3437–3455.

Dhillon, R.S. and G. von Wuehlisch. 2013. Mitigation of global warming through renewable biomass. *Biomass Bioenergy* 48:75–89.

Downie, A., D. Lau, A. Cowie, and P. Munroe. 2014. Approaches to greenhouse gas accounting methods for biomass carbon. *Biomass Bioenergy* 60:18–31.

Drexhage, M. and F. Colin. 2001. Estimating root system biomass from breast-height diameters. *Forestry* 74(5):491–497.

Duguay, J.P., P.B. Wood, and G.W. Miller. 2000. Effects of timber harvests on invertebrate biomass and avian nest success. *Wildl. Soc. Bull.* 28(4):1123–1131.

Dupouey, J.L., E. Dambrine, J.D. Laffite, and C. Moares. 2002. Irreversible impact of past land use on forest soils and biodiversity. *Ecology* 83(11):978–2984.

Englhart, S., J. Franke, V. Keuck, and F. Siegert. 2012. Aboveground biomass estimation of tropical peat swamp forests using SAR and optical data. *Geosci. Remote* 1:6577–6580.

Enquist, B.J., J.H. Brown, and G.B. West. 1998. Allometric scaling of plant energetics and population density. *Nature* 395(September):163–165.

Enquist, B.J. and K.J. Niklas. 2001. Invariant scaling relations across tree-dominated communities. *Nature* 410(6829):655–660.

Esseen, P.-A., K.-E. Renhorn, and R.B. Pettersson. 1996. Epiphytic Lichen biomass in managed and old-growth boreal forests: Effect of branch quality. *Ecol. Appl.* 6(1):228–238.

Fang, J., A. Chen, C. Peng, S. Zhao, and L. Ci. 2001. Changes in forest biomass carbon storage in China between 1949 and 1998. *Science* 292(5525):2320–2322.

Feldman, D.P. and J.P. Crutchfield. 1998. Measures of statistical complexity: Why? *Phys. Lett. A* 238(97):244–252.

Filotas, E., L. Parrott, P.J. Burton, R.L. Chazdon, K.D. Coates, L. Coll, S. Haeussler et al. 2014. Viewing forests through the lens of complex systems science. *Ecosphere* 5(January):1–23.

Fittkau, E.J. and H. Klinge. 1973. On biomass and trophic structure of the Central Amazonian rain forest ecosystem on biomass and trophic structure of the Central Amazonian1 rain forest ecosystem. *Biotropica* 5(1): 2–14.

Flügge, A.J., S.C. Olhede, and D.J. Murrell. 2012. The memory of spatial patterns-changes in local abundance and aggregation in a tropical forest. *Ecology* 93(7):1540–1549.

Foley, J.A., G.P. Asner, M.H. Costa, M.T. Coe, R. DeFries, H.K. Gibbs, E.A. Howard et al. 2007. Amazonia revealed: Forest degradation and loss of ecosystem goods and services in the Amazon basin. *Front. Ecol. Environ.* 5:25–32.

Folke, C. 2006. Resilience: The emergence of a perspective for social-ecological systems analyses. *Glob. Environ. Change* 16(3):253–267.

Foody, G.M., M.E. Cutler, J. Mcmorrow, D. Pelz, H. Tangki, D.S. Boyd, and I.A.N. Douglas. 2001. Mapping the biomass of Bornean tropical rain forest from remotely sensed data estimation and mapping. *Glob. Ecol. Biogeogr.* 10:379–387.

Franklin, J.F. and R. Van Pelt. May 2004. Spatial aspects of structural complexity in old-growth forests. *J. For.* 102(3):22–28.

Freppaz, D., R. Minciardi, M. Robba, M. Rovatti, R. Sacile, and A. Taramasso. 2004. Optimizing forest biomass exploitation for energy supply at a regional level. *Biomass Bioenergy* 26:15–25.

Gilichinsky, M., P. Sandström, H. Reese, S. Kivinen, J. Moen, and M. Nilsson. 2011. Mapping ground Lichens using forest inventory and optical satellite data. *Int. J. Remote Sens.* 32(July 2014):455–472.

Gitelson, A.A. 2004. Wide dynamic range vegetation index for remote quantification of biophysical characteristics of vegetation. *J. Plant Physiol.* 161:165–173.

Gonzalez, M., L. Augusto, A. Gallet-Budynek, J. Xue, N. Yauschew-Raguenes, D. Guyon, P. Trichet et al. 2013. Contribution of understory species to total ecosystem aboveground and belowground biomass in temperate Pinus Pinaster Ait. Forests. *For. Ecol. Manage.* 289:38–47.

Green, D.G. and S. Sadedin. 2005. Interactions matter—Complexity in landscapes and ecosystems. *Ecol. Complex.* 2(2 SPEC. ISS.):117–130.

Greenberg, C.H. and A. McGrane. 1996. A comparison of relative abundance and biomass of ground-dwelling arthropods under different forest management practices. *For. Ecol. Manage.* 89:31–41.

De Groot, R.S., R. Alkemade, L. Braat, L. Hein, and L. Willemen. 2010. Challenges in integrating the concept of ecosystem services and values in landscape planning, management and decision making. *Ecol. Complex.* 7(3):260–272.

Haines-Young, R. and M. Chopping. 1996. Quantifying landscape structure: A review of landscape indices and their application to forested landscapes. *Prog. Phys. Geogr.* 20:418–445.

Hames, B.R. 2009. Biomass compositional analysis for energy applications. In *Methods in molecular biology*, ed. J. Mielenz, Vol. 581, pp. 145–167. Clifton, N.J., USA: Humana Press.

Hartvigsen, G., A. Kinzig, and G. Peterson. 1998. Use and analysis of complex adaptive systems in ecosystem science: Overview of special section. *Ecosystem* (June):427–430.

Hendry, R.J. and J.M. Mcglade. 1995. The role of memory in ecological systems. *Proc. R. Soc. B Biol. Sci.* 259(1355):153–159.

Hernández-Stefanoni, J., J. Dupuy, K. Johnson, R. Birdsey, F. Tun-Dzul, A. Peduzzi, J. Caamal-Sosa, G. Sánchez-Santos, and D. López-Merlín. 2014. Improving species diversity and biomass estimates of tropical dry forests using airborne LiDAR. *Remote Sens.* 6:4741–4763.

Hernández-Stefanoni, J.L., J.M. Dupuy, F. Tun-Dzul, and F. May-Pat. 2011. Influence of landscape structure and stand age on species density and biomass of a tropical dry forest across spatial scales. *Landsc. Ecol.* 26:355–370.

Holdridge, L.R. 1967. *Life Zone Ecology*, 2nd edn. San José, Costa Rica: Tropical Science Centre.

Houghton, R.A. 2005. Aboveground forest biomass and the global carbon balance. *Glob. Chang. Biol.* 11:945–958.

Hui, D., J. Wang, X. Le, W. Shen, and H. Ren. 2012. Influences of biotic and abiotic factors on the relationship between tree productivity and biomass in China. *For. Ecol. Manage.* 264:72–80.

Hyde, P., R. Nelson, D. Kimes, and E. Levine. 2007. Exploring LiDAR-RaDAR synergy-predicting aboveground biomass in a Southwestern Ponderosa Pine forest using LiDAR, SAR and InSAR. *Remote Sens. Environ.* 106:28–38.

Imbernon, J. and A. Branthomme. 2001. Characterization of landscape patterns of deforestation in tropical rain forests. *Int. J. Remote Sens.* 22(9):1753–1765.

Jain, A. and S.A. Ansari. 2013. Quantification by allometric equations of carbon sequestered by Tectona Grandis in different agroforestry systems. *J. For. Res.* 24:699–702.

James, P.M.A, M.J. Fortin, A. Fall, D. Kneeshaw, and C. Messier. 2007. The effects of spatial legacies following shifting management practices and fire on Boreal forest age structure. *Ecosystems* 10:1261–1277.

Janssen, P., D. Fortin, and C. Hébert. 2009. Beetle diversity in a matrix of old-growth Boreal forest: Influence of habitat heterogeneity at multiple scales. *Ecography (Cop.).* 32(October 2008):423–432.

Jenkins, J.C., R.A. Birdsey, and Y. Pan. 2001. Biomass and NPP estimation for the mid-Atlantic region (USA) using plot-level forest inventory data. *Ecol. Appl.* 11(4):1174–1193.

Jonasson, S. 1988. Evaluation of the point intercept method for the estimation of plant biomass. *Oikos* 52(1):101–106.

Kane, V.R., R.F. Gersonde, J.A. Lutz, R.J. McGaughey, J.D. Bakker, and J.F. Franklin. 2011. Patch dynamics and the development of structural and spatial heterogeneity in Pacific Northwest forests. *Can. J. For. Res.* 41:2276–2291.

Kane, V.R., R.J. McGaughey, J.D. Bakker, R.F. Gersonde, J.A. Lutz, and J.F. Franklin. 2010. Comparisons between field- and LiDAR-based measures of stand structural complexity. *Can. J. For. Res.* 40:761–773.

Kay, J.J., H.A. Regier, M. Boyle, and G. Francis. 1999. An ecosystem approach for sustainability: Addressing the challenge of complexity. *Futures* 31:721–742.

Keeling, H.C, and O.L. Phillips. 2007. The global relationship between forest productivity and biomass. *Glob. Ecol. Biogeogr.* 16:618–631.

Keith, H., B.G. Mackey, and D.B. Lindenmayer. 2009. Re-evaluation of forest biomass carbon stocks and lessons from the world's most carbon-dense forests. *Proc. Natl. Acad. Sci. USA.* 106(28):11635–11640.

Kivinen, S., A. Berg, J. Moen, L. Östlund, and J. Olofsson. 2012. Forest fragmentation and landscape transformation in a reindeer husbandry area in Sweden. *Environ. Manage.* 49:295–304.

Krummel, J.R., R.H. Gardner, G. Sugihara, R.V. O'Neill, and P.R. Coleman. 1987. Landscape patterns in a disturbed environment. *Oikos* 48(3):321–324.

Kuplich, T.M., P.J. Curran, and P.M. Atkinson. 2005. Relating SAR image texture to the biomass of regenerating tropical forests. *Int. J. Remote Sens.* 26:4829–4854.

Kutsch, W.L., W. Steinborn, M. Herbst, R. Baumann, J. Barkmann, and L. Kappen. 2001. Environmental indication: A field test of an ecosystem approach to quantify biological self-organization. *Ecosystems* 4:49–66.

Kuuluvainen, T. 2009. Forest management and biodiversity conservation based on natural ecosystem dynamics in Northern Europe: The complexity challenge. *Ambio* 38(6):309–315.

Lambert, M.-C., C.-H. Ung, and F. Raulier. 2005. Canadian national tree aboveground biomass equations. *Can. J. For. Res.* 35:1996–2018.

Lamonaca, A., P. Corona, and A. Barbati. 2008. Exploring forest structural complexity by multi-scale segmentation of VHR imagery. *Remote Sens. Environ.* 112:2839–2849.

Lange, M., M. Türke, E. Pašalić, S. Boch, D. Hessenmöller, J. Müller, D. Prati et al. 2014. Effects of forest management on ground-dwelling beetles (Coleoptera; Carabidae, Staphylinidae) in Central Europe are mainly mediated by changes in forest structure. *For. Ecol. Manage.* 329:166–176.

Lansing, J.S. 2003. Complex adaptive systems. *Annu. Rev. Anthropol.* 32(1):183–204.

Latham, P.A., H.R. Zuuring, and D.W. Coble. 1998. A method for quantifying vertical forest structure. *For. Ecol. Manage.* 104(March 1997):157–170.

Laurance, W.F., A.S. Andrade, A. Magrach, J.L.C. Camargo, J.J. Valsko, M. Campbell, P.M. Fearnside, W. Edwards, T.E. Lovejoy, and S.G. Laurance. 2014. Long-term changes in liana abundance and forest dynamics in undisturbed amazonian forests. *Ecology* 95(6):1604–1611.

Le Toan, T., A. Beaudoin, J. Riom, and D. Guyon. 1992. Relating forest biomass to SAR data. *IEEE Trans. Geosci. Remote Sens.* 30(2):403–411.

Le Toan, T., S. Quegan, M.W.J. Davidson, H. Balzter, P. Paillou, K. Papathanassiou, S. Plummer et al. 2011. The BIOMASS mission: Mapping global forest biomass to better understand the terrestrial carbon cycle. *Remote Sens. Environ.* 115(11):2850–2860.

Lee, J.M., S.W. Lee, J.H. Lim, M.S. Won, and H.S. Lee. 2014. Effects of heterogeneity of pre-fire forests and vegetation burn severity on short-term post-fire vegetation density and regeneration in Samcheok, Korea. *Landsc. Ecol. Eng.* 10:215–228.

Lefsky, M.A., D. Harding, W.B. Cohen, G. Parker, and H.H. Shugart. 1999. Surface LiDAR remote sensing of basal area and biomass in deciduous forests of Eastern Maryland, USA. *Remote Sens. Environ.* 67(1):83–98.

Lesmerises, R., J.-P. Ouellet, and M.H. St-Laurent. 2011. Assessing terrestrial lichen biomass using ecoforest maps: A suitable approach to plan conservation areas for forest-dwelling Caribou. *Can. J. For. Res.* 41:632–642.

Levin, S.A. 1998. Ecosystems and the biosphere as complex adaptive system. *Ecosystems* 1(5):431–436.

Levin, S.A. and R.T. Paine. 1974. Disturbance, patch formation, and community structure. *Proc. Natl. Acad. Sci. USA* 71(7):2744–2747.

Li, B.L. 2000. Fractal geometry applications in description and analysis of patch patterns and patch dynamics. *Ecol. Model.* 132:33–50.

Li, S., W. Liu, L. Wang, W. Ma, and L. Song. 2011. Biomass, diversity and composition of epiphytic macrolichens in primary and secondary forests in the subtropical Ailao Mountains, SW China. *For. Ecol. Manage.* 261:1760–1770.

Lim, K., P. Treitz, K. Baldwin, I. Morrison, and J. Green. 2003a. LiDAR remote sensing of biophysical properties of tolerant Northern Hardwood forests. *Can. J. Remote Sens.* 29(5):658–678.

Lim, K., P. Treitz, M. Wulder, B. St-Onge, and M. Flood. 2003b. LiDAR remote sensing of forest structure. *Prog. Phys. Geogr.* 27:88–106.

Lindenmayer, D.B., R.B. Cunningham, C.F. Donnelly, and R. Lesslie. 2002. On the use of landscape surrogates as ecological indicators in fragmented forests. *For. Ecol. Manage.* 159:203–216.

Malhi, Y. and O.L. Phillips. 2004. Tropical forests and global atmospheric change: A synthesis. *Philos. Trans. R. Soc. Lond. B. Biol. Sci.* 359(March):549–555.

Marinescu, M., P. Dwivedi, and J. Nowak. 2011. Assessing forest understory biomass in Northwest Florida (USA) and its potential to meet the state energy needs. *For. Stud. China* 13(4):303–310.

Marland, G. and B. Schlamadinger. 1997. Forests for carbon sequestration or fossil fuel substitution? A sensitivity analysis. *Biomass Bioenergy* 13(6):389–397.

McCune, B. 1994. Using epiphyte litter to estimate epiphyte biomass. *Bryologist* 97(4):396–401.

McElhinny, C., P. Gibbons, C. Brack, and J. Bauhus. 2005. Forest and woodland stand structural complexity: Its definition and measurement. *For. Ecol. Manage.* 218:1–24.

Mcmullin, R.T., I.D. Thompson, and S.G. Newmaster. 2013. Lichen conservation in heavily managed boreal forests. *Conserv. Biol.* 27(5):1020–1030.

Messier, C. and K.J. Puettmann. 2011. Forests as complex adaptive systems: Implications for forest management and modelling. *L'Italia For. E Mont.* 66(3):249–258.

Meyer, V., S.S. Saatchi, J. Chave, J.W. Dalling, S. Bohlman, G.A. Fricker, C. Robinson, M. Neumann, and S. Hubbell. 2013. Detecting tropical forest biomass dynamics from repeated airborne LiDAR measurements. *Biogeosciences* 10:5421–5438.

Mooney, H.A. 1991. Biological response to climate change: An agenda for research. *Ecol. Appl.* 1(2):112–117.

Moroni, M.T. 2013. Simple models of the role of forests and wood products in greenhouse gas mitigation. *Aust. For.* 76(September 2014):50–57.

Müller, F. 2005. Indicating ecosystem and landscape organisation. *Ecol. Indic.* 5:280–294.

Muukkonen, P. 2007. Generalized allometric volume and biomass equations for some tree species in Europe. *Eur. J. For. Res.* 126:157–166.

Muukkonen, P., R. Mäkipää, R. Laiho, K. Minkkinen, H. Vasander, and L. Finér. 2006. Relationship between biomass and percentage cover in understorey vegetation of boreal coniferous forests. *Silva Fenn.* 40(August 2005):231–245.

Muys, B. and P. Granval. 1997. Earthworms as bio-indicators of forest site quality. *Soil Biol. Biochem.* 29(314):323–328.

Næsset, E., O.M. Bollandsås, T. Gobakken, T.G. Gregoire, and G. Ståhl. 2013. Model-assisted estimation of change in forest biomass over an 11 year period in a sample survey supported by airborne LiDAR: A case study with post-stratification to provide "Activity Data." *Remote Sens. Environ.* 128:299–314.

Návar, J. 2009. Allometric equations for tree species and carbon stocks for forests of Northwestern Mexico. *For. Ecol. Manage.* 257:427–434.

Nelson, B.W., R. Mesquita, J.L.G. Pereira, S. Garcia Aquino De Souza, G. Teixeira Batista, and L. Bovino Couto. 1999. Allometric regressions for improved estimate of secondary forest biomass in the Central Amazon. *For. Ecol. Manage.* 117:149–167.

Neumann, M., S.S. Saatchi, L.M.H. Ulander, and J.E.S. Fransson. 2012. Assessing performance of L- and P-band polarimetric interferometric SAR data in estimating Boreal forest above-ground biomass. *IEEE Trans. Geosci. Remote Sens.* 50(3):714–726.

Neumann, M. and F. Starlinger. 2001. The significance of different indices for stand structure and diversity in forests. *For. Ecol. Manage.* 145:91–106.

Newman, M.E.J. 2011. Resource letter CS–1: Complex systems. *Am. J. Phys.* 79(8):800.

Norris, C., P. Hobson, and P.L. Ibisch. 2012. Microclimate and vegetation function as indicators of forest thermodynamic efficiency. *J. Appl. Ecol.* 49(Dale 1997):562–570.

Pagel, M.D., P.H. Harvey, and H.C.T. Godfray. 1991. Species-abundance, biomass, and resource-use distributions. *Am. Nat.* 138(4):836–850.

Pagnutti, C., M. Anand, and M. Azzouz. 2011. Estimating gap lifetime and memory from a simple model of forest Canopy dynamics. *J. Theor. Biol.* 274(1):154–160.

Pagnutti, C., M. Azzouz, and M. Anand. 2007. Propagation of local interactions create global gap structure and dynamics in a tropical rain forest. *J. Theor. Biol.* 247:168–181.

Pardé, J. 1980. Forest biomass. *For. Abstr.* 41(8):343–359.

Paré, D., P. Bernier, E. Thiffault, and B.D. Titus. 2011. The potential of forest biomass as an energy supply for Canada. *For. Chron.* 87:71–76.

Parikka, M. 2004. Global biomass fuel resources. *Biomass Bioenergy* 27:613–620.

Parker, G.G. and M.E. Russ. 2004. The Canopy surface and stand development: Assessing forest Canopy structure and complexity with near-surface altimetry. *For. Ecol. Manage.* 189:307–315.

Parresol, B.R. 1999. Assessing tree and stand biomass: A review with examples and critical comparisons. *For. Sci.* 45(4):573–593.

Parrott, L. 2010. Measuring ecological complexity. *Ecol. Indic.* 10(6):1069–1076.

Parrott, L. and R. Kok. 2000. Incorporating complexity in ecosystem modelling. *Complex. Int.* 7:1–19.

Parrott, L. and H. Lange. 2013. Chapter 2: An introduction to complexity science. In *Managing forest as complex adaptive system: Building resilience to the challenge of global change*, eds. C. Messier, K. Puettmann, and D. Coates. New York: Routledge.

Pastorella, F. and A. Paletto. 2013. Stand structure indices as tools to support forest management: An application in Trentino Forests (Italy). *J. For. Sci.* 59(4):159–168.

Peichl, M. and M.A. Arain. 2007. Allometry and partitioning of above- and belowground tree biomass in an age-sequence of White Pine Forests. *For. Ecol. Manage.* 253:68–80.

Perry, D.A. 1995. Self-organizing systems across scales. *TREE* 10(6):241–244.

Pettersson, R.B., J.P. Ball, K.-E. Renhorn, P.-A. Esseen, and K. Sjöberg. 1995. Invertebrate communities in Boreal forest Canopies as influenced by forestry and lichens with implications for Passerine birds. *Biol. Conserv.* 74:57–63.

Phillips, J.D. and D.A. Marion. 2004. Pedological memory in forest soil development. *For. Ecol. Manage.* 188:363–380.

Pingoud, K., T. Ekholm, and I. Savolainen. 2012. Global warming potential factors and warming payback time as climate indicators of forest biomass use. *Mitig. Adapt. Strateg. Glob. Chang.* 17:369–386.

Popescu, S.C. 2007. Estimating biomass of individual pine trees using airborne LiDAR. *Biomass Bioenergy* 31:646–655.

Porada, P., B. Weber, W. Elbert, U. Pöschl, and A. Kleidon. 2014. Estimating impacts of lichens and bryophytes on global biogeochemical cycles. *Global Biogeochem. Cycles* 28:71–85.

Porté, A.J., J.-C. Samalens, R. Dulhoste, R. Teissier Du Cros, A. Bosc, and C. Meredieu. 2009. Using cover measurements to estimate aboveground understorey biomass in maritime Pine stands. *Ann. For. Sci.* 66:307.

Proulx, R. and L. Parrott. 2008. Measures of structural complexity in digital images for monitoring the ecological signature of an old-growth forest ecosystem. *Ecol. Indic.* 8:270–284.

Proulx, R. and L. Parrott. 2009. Structural complexity in digital images as an ecological indicator for monitoring forest dynamics across scale, space and time. *Ecol. Indic.* 9:1248–1256.

Prugh, L.R. 2009. An evaluation of patch connectivity measures. *Ecol. Appl.* 19(5):1300–1310.

Raulund-Rasmussen, K., I. Stupak, N. Clarke, I. Callesen, H.S. Helmisaari, E. Karltun, and I. Varnagiryte-Kabasinskiene. 2008. Effects of very intensive forest biomass harvesting on short and long term site productivity. In *Sustainable use of forest biomass for energy*, eds. D. Röser, I. Stupak, A. Asikainen, and K. Raulund-Rasmussen. Dordrecht, The Netherlands: Springer.

Rauste, Y. 2005. Multi-temporal JERS SAR data in Boreal forest biomass mapping. *Remote Sens. Environ.* 97:263–275.

Rawat, S., D.K. Upreti, and R.P. Singh. 2011. Estimation of epiphytic Lichen litterfall biomass in three temperate forests of Chamoli District, Uttarakhand, India. *Trop. Ecol.* 52(2):193–200.

Reiner, A.L., R.J. Tausch, and R.F. Walker. 2010. Estimation procedures for understory biomass and fuel loads in Sagebrush Steppe invaded by woodlands. *West. North Am. Nat.* 70(3):312–322.

Rex, K.D. and G.P. Malanson. 1990. The fractal shape of Riparian forest patches. *Landsc. Ecol.* 4(4):249–258.

Ricotta, C. and M. Anand. 2006. Spatial complexity of ecological communities: Bridging the gap between probabilistic and non-probabilistic uncertainty measures. *Ecol. Modell.* 197:59–66.

Ricotta, C. and G.C. Avena. 2003. An information-theoretical measure of taxonomic diversity. *Acta Biotheor.* 51:35–41.

Rodrigues Nogueira Junior, L., V.L. Engel, J. a Parrotta, A.C. Galvao de Melo, and D. Scorzoni Re. 2014. Allometric equations for estimating tree biomass in restored mixed-species Atlantic. *Biota Neotrop.* 14(2):1–9.

Rottgermann, M., T. Steinlein, W. Beyschlag, and H. Dietz. 2000. Linear relationships between aboveground biomass and plant cover in low open herbaceous vegetation. *J. Veg. Sci.* 11:145–148.

Russell, M.B., A.W. D'Amato, B.K. Schulz, C.W. Woodall, G.M. Domke, and J.B. Bradford. 2014. Quantifying understorey vegetation in the U.S. lake states: A proposed framework to inform regional forest carbon stocks. *Forestry* 2009:1–10.

Saatchi, S.S., N.L. Harris, S. Brown, M. Lefsky, E.T. Mitchard, W. Salas, B.R. Zuttaa, W. Buermann et al. 2011. Benchmark map of forest carbon stocks in tropical regions across three continents. *Proc. Natl. Acad. Sci. USA* 108(24):9899–9904.

Sarker, L.R., J. Nichol, H.B. Iz, B. Bin Ahmad, and A.A. Rahman. 2013. Forest biomass estimation using texture measurements of high-resolution dual-Polarization C-Band SAR data. *IEEE Trans. Geosci. Remote Sens.* 51(6):3371–3384.

Schneider, E.D. and J.J. Kay. 1994. Life as a manifestation of the second law of thermodynamics. *Math. Comput. Model.* 19(6):25–48.

Schnitzer, S.A. and F. Bongers. 2011. Increasing Liana abundance and biomass in tropical forests: Emerging patterns and putative mechanisms. *Ecol. Lett.* 14:397–406.

Schnitzer, S.A., S.J. Dewalt, and J. Chave. 2006. Comparison Lianas: Censusing and measuring of the common a quantitative methods. *Biotropica* 38(5):581–591.

Schroeder, P., S. Brown, J. Mo, R. Birdsey, and C. Cieszewski. 1997. Biomass estimation for temperate broadleaf forests of the United States using inventory data. *For. Sci.* 43(3):424–434.

Schulze, E.D., C. Körner, B.E. Law, H. Haberl, and S. Luyssaert. 2012. Large-scale bioenergy from additional harvest of forest biomass is neither sustainable nor greenhouse gas neutral. *GCB Bioenergy* 4:611–616.

Seagle, S.W. and B.R. Sturtevant. 2005. Forest productivity predicts invertebrate biomass and ovenbird (*Seiurus Aurocapillus*) reproduction in Appalachian landscapes. *Ecology* 86(6):1531–1539.

Séro-Guillaume, O. and J. Margerit. 2002. Modelling forest fires. Part I: A complete set of equations derived by extended irreversible thermodynamics. *Int. J. Heat Mass Transf.* 45:1705–1722.

Shannon, C.E. 2001. A mathematical theory of communication. *ACM Sigmobile Mob. Comput. Commun. Rev.* 5(1):3–55.

Sherman, R.E., T.J. Fahey, and P. Martinez. 2003. Spatial patterns of biomass and aboveground net primary productivity in a Mangrove ecosystem in the Dominican republic. *Ecosystems* 6(September 2001):384–398.

Slik, J.W.F., S.I. Aiba, F.Q. Brearley, C.H. Cannon, O. Forshed, K. Kitayama, H. Nagamasu, et al. 2010. Environmental correlates of tree biomass, basal area, wood specific gravity and stem density gradients in Borneo's tropical forests. *Glob. Ecol. Biogeogr.* 19:50–60.

Smith, J.E., L.S. Heath, and J.C. Jenkins. 2003. *Forest volume-to-biomass models and estimates of mass for live and standing dead trees of U.S. forests.* General Technical Report NE-298. United States Department of Agriculture, Washington, DC.

Stupak, I., A. Asikainen, M. Jonsell, E. Karltun, A. Lunnan, D. Mizaraite, K. Pasanen et al. 2007. Sustainable utilisation of forest biomass for energy-possibilities and problems: Policy, legislation, certification, and recommendations and guidelines in the Nordic, Baltic, and other European countries. *Biomass Bioenergy* 31:666–684.

Suchar, V.A. and N.L. Crookston. 2010. Understory cover and biomass indices predictions for forest ecosystems of the Northwestern United States. *Ecol. Indic.* 10:602–609.

Sun, Z., H. Ren, V. Schaefer, H. Lu, J. Wang, L. Li, and N. Liu. 2013. Quantifying ecological memory during forest succession: A case study from lower subtropical forest ecosystems in South China. *Ecol. Indic.* 34:192–203.

Swift, K.I. and F.W. Bell. 2011. What are the environmental consequences of using silviculturally effective forest vegetation management treatments? *For. Chron.* 87(2):201–216.

Takahashi, K. and Y. Obata. 2014. Growth, allometry and shade tolerance of understory saplings of four subalpine conifers in Central Japan. *J. Plant Res.* 127:329–338.

Tanase, M.A., R. Panciera, K. Lowell, S. Tian, J.M. Hacker, and J.P. Walker. 2014. Airborne multitemporal L-band polarimetric SAR data for biomass estimation in semi-arid forests. *Remote Sens. Environ.* 145:93–104.

Tanner, E.V.J. 1980. Studies on the biomass and productivity in a series of montane rain forests in Jamaica. *J. Ecol.* 68(2):573–588.

Ter-Mikaelian, M.T. and M.D. Korzukhin. 1997. Biomass equations for sixty-five North American tree species. *For. Ecol. Manage.* 97:1–24.

Thomas, V., P. Treitz, J.H. McCaughey, and I. Morrison. 2006. Mapping stand-level forest biophysical variables for a mixedwood boreal forest using LiDAR: An examination of scanning density. *Can. J. For. Res.* 36:34–47.

Turner, M.G., D.C. Donato, and W.H. Romme. 2013. Consequences of spatial heterogeneity for ecosystem services in changing forest landscapes: Priorities for future research. *Landsc. Ecol.* 28:1081–1097.

Ulgiati, S. and M.T. Brown. 2009. Energy and ecosystem complexity. *Commun. Nonlinear Sci. Numer. Simul.* 14:310–321.

Van der Heijden, G.M., S.A. Schnitzer, J.S. Powers, and O.L. Phillips. 2013. Liana impacts on carbon cycling, storage and sequestration in tropical forests. *Biotropica* 45(6):682–692.

Van-Aardt, J.A.N., R.H. Wynne, and J.A. Scrivani. 2008. LiDAR-based mapping of forest volume and biomass by taxonomic group using structurally homogenous segments. *Photogramm. Eng. Remote Sens.* 74(8):1033–1044.

Vazirabad, Y., and M. Karslioglu. 2011. LiDAR for biomass estimation. In *Biomass detection, production usage*, ed. D. Matovic, p. 508. Rijeka, Croatia: InTech.

Wagendorp, T., H. Gulinck, P. Coppin, and B. Muys. 2006. Land use impact evaluation in life cycle assessment based on ecosystem thermodynamics. *Energy* 31:112–125.

Walker, D.J. and N.C. Kenkel. 1998. Fractal analysis of spatio-temporal dynamics in boreal forest landscapes. *Abstr. Bot.* 22(Walker 1994):13–28.

Wang, C. 2006. Biomass allometric equations for 10 co-occurring tree species in Chinese temperate forests. *For. Ecol. Manage.* 222:9–16.

Watson, J., D. Freudenberger, and D. Paull. 2001. An assessment of the focal-species approach for conserving birds in variegated landscapes in Southeastern Australia\rEvaluación de La Aproximación de Especie Focal Para La Conservación de Aves En Paisajes Variegados En El Sureste de Australia. *Conserv. Biol.* 15(5):1364–1373.

Wei, S.G., L. Li, Z.L. Huang, W.H. Ye, G.Q. Gong, X.Y. Zhou, and J.Y. Lian. 2013. Multifractal analysis of diversity scaling laws in a subtropical forest. *Ecol. Complex.* 13:1–7.

Whittaker, R.H., F.H. Bormann, G.E. Likens, and T.G. Siccama. 1974. The Hubbard Brook ecosystem study: Forest biomass and production. *Ecol. Monogr.* 44(2):233–254.

Wild, J., M. Kopecký, M. Svoboda, J. Zenáhlíková, M. Edwards-Jonášová, and T. Herben. 2014. Spatial patterns with memory: Tree regeneration after stand-replacing disturbance in Picea Abies mountain forests. *J. Veg. Sci.* 25(6):1–14.

Wilson, M.K., W.H. Lowe, and K.H. Nislow. 2014. Family richness and biomass of understory invertebrates in early and late successional habitats of Northern New Hampshire 20. *J. For.* 112(4):337–345.

Wulder, M.A., C.W. Bater, N.C. Coops, T. Hilker, and J.C. White. 2008. The role of LiDAR in sustainable forest management. *For. Chron.* 84(6):807–826.

Zeide, B. 1991. Fractal geometry in forestry applications. *For. Ecol. Manage.* 46:179–188.

Zellweger, F., F. Morsdorf, R.S. Purves, V. Braunisch, and K. Bollmann. 2014. Improved methods for measuring forest landscape structure: LiDAR complements field-based habitat assessment. *Biodivers. Conserv.* 23:289–307.

Zenner, E.K. 2000. Do residual trees increase structural complexity in Pacific Northwest coniferous forests? *Ecol. Appl.* 10(3):800–810.

Zhao, J., Y. Shao, X. Wang, D.A. Neher, G. Xu, Z. Li, and S. Fu. 2013. Sentinel soil invertebrate Taxa as bioindicators for forest management practices. *Ecol. Indic.* 24:236–239.

Zhao, M. and G.S. Zhou. 2005. Estimation of biomass and net primary productivity of major planted forests in China based on forest inventory data. *For. Ecol. Manage.* 207:295–313.

Zianis, D. and M. Mencuccini. 2004. On simplifying allometric analyses of forest biomass. *For. Ecol. Manage.* 187(April 2003):311–332.

Zolkos, S.G., S.J. Goetz, and R. Dubayah. 2013. A meta-analysis of terrestrial aboveground biomass estimation using LiDAR remote sensing. *Remote Sens. Environ.* 128:289–298.

12 Considering Forest Biodiversity Indicators within the Pressure, State, Benefits, and Response Framework

F. Wayne Bell, Jennifer Dacosta, and Guy R. Larocque

12.1 INTRODUCTION

Forest biodiversity refers to the variability among all forest life forms (trees, plants, animals, fungi, and microorganisms) from genetics to species and ecosystems and their associated ecological processes. Conserving forest biodiversity is a fundamental goal of sustainable forestry. Therefore, governments, non-government organizations (NGOs), academic institutions, corporations, partnerships, and others tasked with managing and conserving the world's forests require mechanisms to monitor their progress toward this goal. One of the mechanisms to accomplish this is using a suite of biodiversity indicators. These indicators are data-based measures that convey information beyond what is directly measured or assessed (BIP 2011). Given the multitude of possible indicators, frameworks are used to organize them to understand better and show their relationships and interconnectedness. Indicators and indicator frameworks can be used to educate, support policy, and inform management decisions.

In this chapter, we do not provide an extensive list of forest biodiversity indicators; instead, we intend to discuss concepts associated with biodiversity conservation about indicator frameworks. Through this approach, we aim to provide the reader with sufficient understanding to review and select among the many contemporary indicators most relevant to their needs/questions. For readers seeking information about selecting forest biodiversity indicators, we suggest Ferris and Humphrey (1999), Jonsson and Jonsell (1999), Noss (1999), Lindenmayer et al. (2000), Puumalainen et al. (2003), Marchetti (2004), Hagen and Whitman (2006), and Levrel et al. (2007).

Following a brief background related to biodiversity conservation and indicator frameworks, our discussion of the broader concepts is primarily embedded in the State-Response-based, as well as planetary boundaries and Kunming Montreal Global biodiversity frameworks; the benefit, state, driving force, pressure, impacts, and policy response indicators; as well as active adaptive management approaches and means to visualize changes in indicators over time. This chapter is one of several in this handbook that focuses on ecological indicators and concepts that are important for sustainable forest management. The environmental indicators presented in other chapters, including complexity, integrity, abundance biomass, productivity, carbon balance, turnover, and ecosystem health, are intimately linked to biodiversity.

12.2 BACKGROUND

12.2.1 Biodiversity Conservation

Internationally, there is a commitment to conserve biodiversity and to practice sustainable forest management (SFM). In 1992, the Convention on Biological Diversity (CBD) was put forward,

DOI: 10.1201/9781003431084-15

and the Forest Principles document was established (UNCED 1992). By 1993, the CBD was in place with its primary objectives stated as (i) conservation of biodiversity, (ii) sustained use of the components of biodiversity, and (iii) sharing of benefits arising from the use of genetic resources. The Convention has parties from around the globe, representing most of the world's governments. Biodiversity conservation also includes many international, national, and local groups and initiatives, such as the Alliance for Zero Extinction, the Ramsar Convention on Wetlands, the Convention on Migratory Species, Millennium Development Goals, and national and regional biodiversity strategies. Complementary efforts to specifically conserve biodiversity are initiatives for sustainable forest management (see Chapter 14). These include certification schemes such as those produced by the Forest Stewardship Council (FSC), the Programme for the Endorsement of Forest Certification (PEFC), and the Sustainable Forestry Initiative (SFI). Publication of the Forest Principles in 1992 sparked work on SFM criteria and indicator frameworks such as those developed under the "Conservation and Sustainable Management of Temperate and Boreal Forests" (also called the Montreal Process), Forests Europe, and the International Tropical Timber Organization; each incorporates biodiversity conservation as a criterion (von Gadow et al. 2000).

12.2.1.1 Biodiversity-Related Conventions

The five signatories to the statement on behalf of biodiversity-related conventions are the Convention concerning the Protection of the World Cultural and Natural Heritage (World Heritage Convention), the Convention on Biological Diversity (CBD), the Convention on International Trade in Endangered Species of Wild Fauna and Flora (CITES); the Convention on the Conservation of Migratory Species of Wild Animals (CMS); and the Convention on Wetlands of International Importance especially as Waterfowl Habitat (Ramsar Convention on Wetlands).

12.2.1.2 Targets

In 2002, the signing parties of the CBD committed to a target to significantly reduce the current rate of biodiversity loss by 2010 (UNEP 2002). Although the agreement stimulated action to safeguard biodiversity in some areas, the consensus was that the target still needed to be met. In setting the strategic direction for 2011–2020, the parties of the CBD adopted 20 targets referred to as the Aichi Biodiversity Targets. These targets are organized under five strategic goals:

- Addressing underlying causes of biodiversity loss by mainstreaming biodiversity across government and society
- Reducing the many pressures on biodiversity and promoting sustainable use
- Improving the status of biodiversity by safeguarding ecosystems, species, and genetic diversity
- Enhancing the many benefits to all from biodiversity and ecosystem services
- Enhancing implementation through participatory planning, knowledge management, and capacity building

The Aichi Targets rely on indicators to report progress toward (i) reducing pressure on biodiversity, (ii) maintaining and improving the state of biodiversity, (iii) implementing conservation actions to ameliorate biodiversity loss, and (iv) providing benefits to human well-being (UNEP 2011). Parties to the Convention are required to produce national reports based on these targets, positioning the Aichi Targets to be the basis for biodiversity reporting worldwide (SCBD 2014; Tittensor et al. 2014).

12.2.2 Biodiversity Monitoring and Assessment Programs

Biodiversity monitoring and assessment programs (BMAP) provide a process that collects and reports scientific information for natural resource management (Dallmeier et al. 2013). The first task in developing the BMAP program is to define the objectives. Then, a baseline biological assessment is often required to determine the kinds of species and habitats that need to be considered as part of the program. After the evaluation, monitoring the selected indicators to answer questions using the scientific method will provide biodiversity trends in the ecosystem. Monitoring activities leads to insights on issues such as those requiring management strategies, for example, protecting threatened or endangered species or eradicating non-native invasive species. Monitoring is also used to assess progress toward meeting a management objective and is essential to a successful adaptive management program.

12.3 INDICATOR FRAMEWORKS

An indicator framework can provide organization and structure and facilitate awareness of connections among indicators. Indicators can be organized into a framework based on various overarching categories, including broad targets, goals, criteria, or casual and linked factors.

12.3.1 State-Response (S-R)-Based frameworks

Since 1979, state-response (S-R) based frameworks have evolved to represent causal relationships (i.e., one component causes a change in another) or linked relationships (one component can influence one or many of the other components). The components commonly part of these frameworks include benefits, states, driving forces, pressures, uses, capacity, effects/impacts, and policy responses. Multiple combinations of these components occur within these frameworks.

Statistics Canada developed the Stress-Response (S-R) framework in 1979, with the response category comprising both environmental and societal responses (Friend and Rapport 1991). The S-R framework evolved into the Pressure-State-Response (PSR) framework developed in 1993 by the Organization for Economic Co-Operation and Development (OECD 1993; Rigby et al. 2000; Levrel et al. 2009). The "PSR framework" reduced environmental problems and solutions into variables that stressed cause–effect relationships between human activities that exerted pressures on environments, the state of environments, and society's response to conditions. Since the PSR framework focused on anthropocentric pressures and responses, it poorly factored natural variability into the "pressure" category. This led to the development of the expanded Driving Force-State-Response (DSR) framework by the United Nations Commission on Sustainable Development (CSD) in 1997 and, subsequently, the Driving force-pressure-state-impact-response (DPSIR) framework (OECD 1993; Svarstad et al. 2008).

The Driving force-Pressure-State-Impact-Response (DPSIR) framework is at present a commonly used framework (see Smeets and Weterings 1999; Gabrielsen and Bosch 2003; Odermatt 2004; Wei et al. 2007, Benini et al. 2010, Martins et al. 2012, Kim et al. 2021). The DPSIR framework assumes a chain of causal links starting with "driving forces," through "pressures" to "states" and "impacts" on ecosystems, human health, and functions, eventually leading to political "responses" (Figure 12.1).

Since its conception by the European Environment Agency (EEA) in 1999, the DPSIR framework has evolved into modified frameworks such as Driver-Pressure-Chemical State-Ecological State-Response (DPCER; Rekolainen et al. 2003), Driver-Pressure-State-Ecosystem-Response (DPSER; Kelbel et al. 2013; Gari et al. 2015), and Driver-Pressure-State-Welfare-Response (DPSWR, O'Higgins et al. 2014). Another combination includes the Pressure-State-Benefits-Response (PSBR) framework (Sparks et al. 2011; Werner and Gallo-Orsi 2016; Hu et al. 2022).

12.3.2 Planetary Boundaries

"Planetary boundaries" is a framework designed to describe the limits to the effects of human activities on the Earth system (Rockström et al. 2009a, 2009b, 2023; Steffen et al. 2015; Steffen et al. 2020). Beyond pre-chosen limits, see Figure 12.2, the Earth's systems may not be able to self-regulate. This would imply that the Earth's systems would leave the period of stability of the

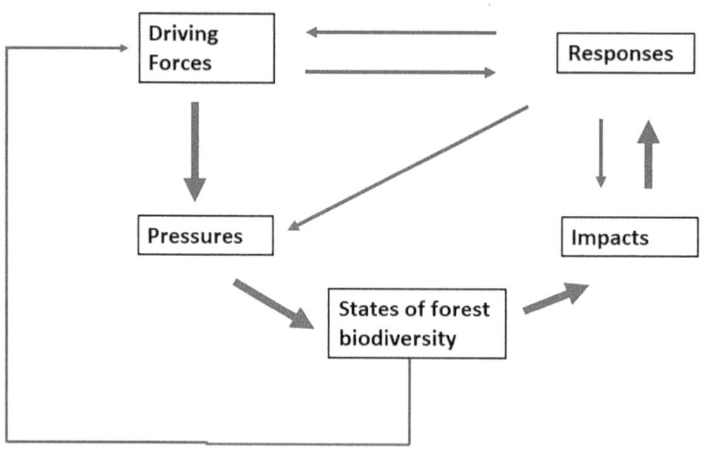

FIGURE 12.1 An example of a plausible Driver-Pressure-State-Impact-Response Framework. (Adapted from https://en.wikipedia.org/wiki/DPSIR#/media/File:DPSIR_Framework_.png.)

Holocene, in which human society developed (Rockström et al. 2009a, 2009b, 2023). The framework is based on the presumption that human actions, especially those of industrialized societies, have become the primary driver of global environmental change. According to the framework, "transgressing one or more planetary boundaries may be deleterious or even catastrophic due to the risk of crossing thresholds that will trigger nonlinear, abrupt environmental change within continental-scale to planetary-scale systems" (Rockström et al. 2009a, 2009b).

In 2015, an update was published, bringing in new co-authors and a new model-based analysis (Figure 12.2). At that time, four boundaries were crossed: climate change, loss of biosphere integrity, land-system change, and altered biogeochemical cycles (phosphorus and nitrogen) (Steffen et al. 2015). The authors also changed the name of the boundary "Loss of biodiversity" to "Change in biosphere integrity" to emphasize that the number of species and the functioning of the biosphere are vital for Earth system stability. In addition, the "Chemical pollution" boundary was renamed "Introduction of novel entities," thus widening the scope to consider different kinds of human-generated materials that disrupt the Earth system's processes.

In 2022, based on the available literature, the introduction of novel entities was concluded to be the fifth transgressed planetary boundary (Persson et al. 2022). Freshwater change was supposed to be the sixth transgressed planetary boundary in 2023 (Figure 12.2, Richardson et al. 2023).

12.3.3 Kunming Montreal Global Biodiversity Framework

The Kunming-Montreal Global Biodiversity Framework (GBF), adopted during the 15th meeting of the Conference of the Parties (COP 15), committed all parties to set national targets to implement it. At the same time, all other actors were invited to develop and communicate their commitments. The vision of the Kunming-Montreal Global Biodiversity Framework was a world of humans living in harmony with nature where "by 2050, biodiversity is valued, conserved, restored and wisely used, maintaining ecosystem services, sustaining a healthy planet and delivering benefits essential for all people" (COP 15). The mission of the Framework for the period up to 2030, toward the 2050 vision, is:

> To take urgent action to halt and reverse biodiversity loss to put nature on a path to recovery for the benefit of people and planet by conserving and sustainably using biodiversity and by ensuring the fair and equitable sharing of benefits from the use of genetic resources, while providing the necessary means of implementation.
>
> **(COP 15)**

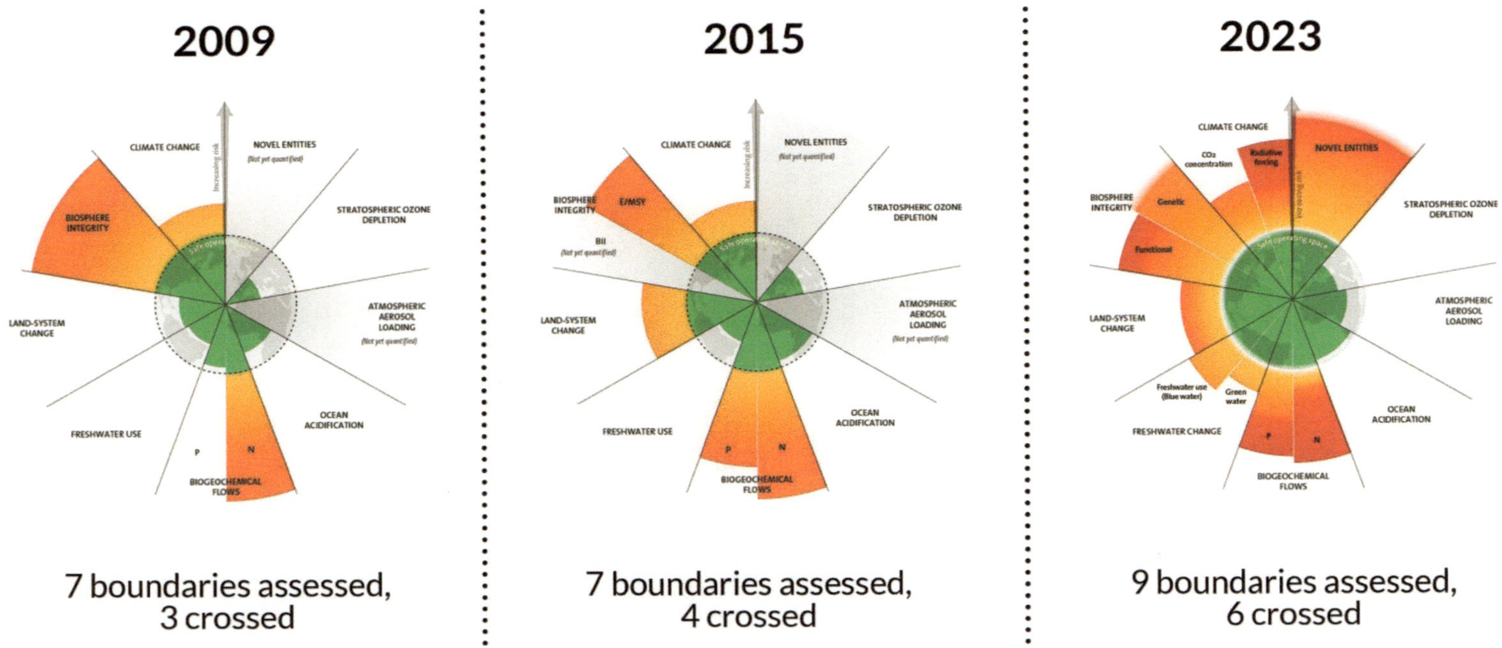

FIGURE 12.2 Impacts of transgression on planetary boundaries. (From the Stockholm Resilience Centre web site www.stockholmresilience.org/research/planetary-boundaries.html adapted from Rockström et al. 2009a, 2009b; Steffen et al. 2015; Persson et al. 2022; Richardson et al. 2023.)

The Kunming-Montreal Global Biodiversity Framework has four long-term goals for the 2050 Vision for Biodiversity (COP 15). The following goals are from COP (2015, page 8):

GOAL A: The integrity, connectivity, and resilience of all ecosystems are maintained, enhanced, or restored, substantially increasing the area of natural ecosystems by 2050. Human-induced extinction of known threatened species is halted, and by 2050, the extinction rate and risk of all species will be reduced tenfold, and the abundance of native wild species will be increased to healthy and resilient levels. The genetic diversity within wild and domesticated species populations is maintained, safeguarding their adaptive potential.

GOAL B: Biodiversity is sustainably used and managed, and nature's contributions to people, including ecosystem functions and services, are valued, maintained, and enhanced, with those currently in decline being restored, supporting the achievement of sustainable development for the benefit of present and future generations by 2050.

GOAL C: The monetary and non-monetary benefits from the utilization of genetic resources and digital sequence information on genetic resources, and of traditional knowledge associated with genetic resources, as applicable, are shared fairly and equitably, including, as appropriate with Indigenous peoples and local communities, and substantially increased by 2050, while ensuring traditional knowledge associated with genetic resources is appropriately protected, thereby contributing to the conservation and sustainable use of biodiversity, under internationally agreed access and benefit-sharing instruments.

GOAL D: Adequate means of implementation, including financial resources, capacity building, technical and scientific cooperation, and access to and transfer of technology to fully implement the Kunming-Montreal Global Biodiversity Framework, are secured and equitably accessible to all Parties, especially developing country Parties.

The Kunming-Montreal Global Biodiversity Framework has 23 action-oriented global targets for urgent action over the decade to 2030.

12.4 INDICATORS—BENEFITS, STATES, DRIVING FORCES, PRESSURES, IMPACTS, AND POLICY RESPONSES

An indicator is defined as "a measure or metric based on verifiable data that conveys information about more than itself" (BIP 2011). Indicators can provide information about the same issue from several perspectives and may convey contrasting messages. The characteristics of various indicators have been documented (BIP 2011) and include factors such as relevance, measurability, and utility. Variations of the SMART acronym (specific, measurable, assignable, realistic, and time-related (see Doran 1981) have also been applied to indicator characteristics to convey the importance of considering an essential set of attributes in defining an indicator. A well-thought-out development process ensures the ultimate utility and relevance of resulting indicators. Guidelines are available for developing relevant indicators at multiple scales (e.g., BIP 2011).

Here, we present in alphabetical order and briefly discuss each S-R-based framework indicator.

12.4.1 Benefits of Forest Biodiversity

Benefits describe the various ecosystem services that are provided by biodiversity to humans and ecosystems. The Millennium Ecosystem Assessment project (MA 2003) led to the development of a link between nature and human well-being using ecosystem services. Four categories of ecosystem services were defined: (i) supporting services such as nutrient cycling, soil formation and primary productivity, and crop pollination; (ii) provisioning services such as food, fuel, water, timber and medicine, and wildlife habitats; (iii) regulating services such as controlling erosion, regulating climate, sequestering CO_2, mitigating floods, purifying water and air, pollinating, and controlling pests and diseases; and (iv) cultural services such as recreation, ecotourism, educational, and spiritual values (MA 2003, 2005). Since these are discussed extensively in Chapter 5—Ecosystem Services, we limit our discussion to specific points relevant to biodiversity indicators.

The relationship between biodiversity and most forest ecosystem services is quite clear (MA 2005; Thompson et al. 2011, 2013; Hicks et al. 2014). Throughout history, forests have provided people with various ecosystem services, especially timber. However, scientists now understand that forest biodiversity underpins the full range of local and global ecosystem services (Thompson et al. 2011, 2013; Hicks et al. 2014) and that any losses of biodiversity will also reduce the associated ecosystem services (Hooper et al. 2005).

Although most of the focus is on the benefits that forests provide, forests can also result in significant costs to humans. They can invade other ecosystems and reduce the food-producing capacity of grazing and cultivated land (Scholes and Archer 1997), reduce the diversity of non-forest species (Parr et al. 2012, Ratajczak et al. 2012), reduce water available for humans and wildlife (Scott 1999, Wilcox et al. 2005, Zou et al. 2008), harbor dangerous and destructive wildlife (Nasi et al. 2002, Davidson et al. 2005, Belluscio 2009), act as reservoirs of human and livestock disease (Margaletic 2003, Wilcox and Ellis 2006), and diminish the enjoyment of natural areas (Fensham and Fairfax 1996, Gray and Bond 2013).

12.4.1.1 Supporting Services

Supporting services are natural processes "necessary for the production of all other ecosystem services" (MA 2005). They include (i) maintaining biogeochemical and nutrient cycles; (ii) soil formation, soil fertility, and primary production; (iii) natural forest dynamics; and (iv) ecosystem, species, and genetic diversity (see Chapter 5). These services indirectly benefit people by influencing provisioning, regulating, and cultural services.

Forest ecosystems are dynamic; they are generally recovering from some past disturbance (Chapin et al. 2002, Drever et al. 2006; Thompson et al. 2009). Ecosystem stability does not refer to a constant state but rather its ability to withstand environmental change, maintain its structure and species composition, and provide services consistently over time. Measures of stability refer to a system's capacity to remain unchanged in the face of ongoing disturbance (i.e., resistance) and its ability to return to its original state after being disturbed (i.e., resilience) (Chapin et al. 2002, Ives and Carpenter 2007, Reiners et al. 2012 as cited in Thompson et al. 2012). The rates of these processes can vary.

There is evidence that biodiversity confers resilience to some types of terrestrial ecosystems (van Ruijven and Berendse 2010, Hicks et al. 2014) and is essential for maintaining higher levels of ecosystem service provision (Gamfeldt et al. 2013). In forests, stability varies among types, especially over space, but usually refers to the recognizable mix of dominant tree species (e.g., Drever et al. 2006, Thompson et al. 2009). However, in a review of evidence from forest ecosystems, Thompson et al. (2013) concluded that any relationship between species richness and resistance is likely to be highly specific to the type and context of the ecosystem in question.

12.4.1.2 Provisioning Services

Provisioning services are goods or products obtained from ecosystems. These include, but are not limited to, food and medicines; freshwater; raw materials for buildings, clothing, and energy; and ornamentals and horticulture resources (MA 2005; Thompson et al. 2009, 2011, 2013). Provisions from forests are commonly categorized as either "primary timber products" or "non-timber forest products (NTFPs)" (Mohammed 1999, Belcher and Schreckenberg 2007, Hicks et al. 2014). Primary timber products include industrial board composites and paper products. The NTFPs are "botanical products harvested or originating from forest-based species" and "include foods, health and personal care products, materials and manufacturing products, environmental products, landscape and garden products, and decorative and aesthetic products" (Mohammed 1999).

Provisioning services are strongly related to biodiversity at both landscape and species scales (Quijas et al. 2010, Hicks et al. 2014). For example, some evidence indicates that species-rich forests are more important for timber production (Hicks et al. 2014). Since biodiversity is essential for provisioning services, and it can be negatively affected by deforestation (Shanley and Luz 2003), overharvesting (Thompson et al. 2013), and other pressures, to predict future losses in provisioning services it is essential to monitor both pressures and state of biodiversity.

12.4.1.3 Regulating Services

Regulating services are the benefits obtained from an ecosystem's control of natural processes such as (i) soil formation, stabilization, and erosion; (ii) water seasonal-flow regulation and filtration/purification, (iii) air quality regulation; (iv) carbon sequestration and storage, (v) climate regulation, (vi) pollination, (vii) pest and disease regulation, (viii) waste decomposition and detoxification, and (ix) natural hazard regulation (MA 2003, 2005; TEEB 2009; Thompson et al. 2009, 2011, 2013; Egoh et al. 2012).

Most regulating services are a function of biodiversity in that different ecotopes have different capacities to hold services. Within ecotopes, other species have various capacities to regulate services. A loss of species biodiversity could, therefore, lead to a reduction or loss of ecosystem services.

12.4.1.4 Cultural Services

Cultural services are people's nonmaterial benefits from ecosystems through spiritual enrichment, cognitive development, reflection, recreation, and aesthetic experiences (MA 2005). Health is a fundamental human right and one of the most important indicators of sustainable development. For most people, health means not only a lack of illness but an overall state of social, emotional, physical, spiritual, and cultural well-being, and therefore is the primary "cultural service" of interest. Ultimately, human health is a function of the state of the environment. Species richness and human population density are correlated, likely because both are positively influenced by increases in primary productivity (Chown et al. 2003). Without a healthy environment that can support us, we cannot exist.

12.4.2 States of Forest Biodiversity

The state is the current status, or condition, of biotic features at all scales (e.g., genetics to ecosystems) and abiotic features (e.g., water, air, and soil) that contribute to biodiversity (Hagen and Whitman 2006). State indicators may, for instance, describe the forest and wildlife resources present (Smeets and Weterings 1999).

In the context of biodiversity, the state is the condition of both the abiotic and biotic components of ecosystems. Those components are often described within a hierarchy of scales: landscape, ecosystem, species, and genetic. An alternative approach is to characterize the aspects of biodiversity that contribute to its conservation (Noss 1990), which are compositional, structural, and functional diversity. Each of these components can then be assessed at the appropriate scale. In this chapter, we focus on indicators related to the state of the biotic components of (i) compositional, (ii) structural, and (iii) functional diversity and (iv) indices related to the extent, remoteness, and naturalness of forests.

12.4.2.1 Compositional Diversity

Composition is the identity of and variety within an ecological system (Noss 1990). It refers to the number of elements within a geographic area, including landscape composition, species richness, and genetic diversity.

Landscape composition or ecosystem diversity refers to the diversity of ecologically homogeneous spatial units within a geographical area. These units, referred to as ecosystems, ecotopes, patches, or stands, are often defined using ecosystem classification systems. Ecosystem classification is commonly based on a combination of biotic and abiotic factors, including vegetation cover, plant indicator species, soil types, geomorphology, hydrology, and other factors, attributes or features such as land use history associated with anthropogenic disturbance (see Whittaker et al. 1973; Bailey 2009; Gilmore et al. 2010; Campos et al. 2012). Once identified, these landscape units can be characterized based on their patch composition, spatial relations and dynamics, patch richness, abundance, entropy, and evenness.

Species diversity refers to the number of species represented in a given community (richness) and their distribution (evenness). Indicators of compositional diversity in forests include the number of species and how they change over time. Some habitats have an exceptionally high number of species, which are referred to as "biodiversity hotspots" (Myers et al. 1988, 2000; Mittrmeier et al. 2000). Although species richness is commonly used as an indicator, its effectiveness is limited because it is not necessarily linked to conservation values or sustainability (Bunnell and Chan-McLeod 1998). In addition, species richness increases with the size of the area surveyed; thus, it is not very informative without an appropriate spatial reference (Brockerhoff et al. 2001). Indicators of change in species diversity also include changes in the status of specific species or species groups, the number and abundance of threatened or invasive species, and homogenization of species diversity.

The IUCN Red List of Threatened Species™ (IUCN 2008, Vié et al. 2008) provides taxonomic, conservation status, and distribution information on plants, animals, and fungi globally evaluated as facing a higher risk of global extinction. A growing number of species are candidates for red-listed and should be monitored. Candidate species may include those limited by area, dispersal, resources, or changes in ecological processes (Lambeck 1997, Noss 1999, Thompson 2006). Other species of interest include narrow endemics and evolutionarily distinct and globally endangered (EDGE) species (Isaac et al. 2007, Safi et al. 2013).

Although threatened species are commonly suggested as a biodiversity indicator, many other species should be regarded as having the potential to be surrogate species for conservation and monitoring actions (Siqueira et al. 2012). The use of surrogate species or taxa is based on the assumption that protection of the surrogate will also protect other species, taxa, or overall biodiversity (Lambeck 1997, Noss et al. 1997, Margules and Pressey 2000). Surrogates can be subdivided into indicator, flagship, umbrella, and keystone species (Simberloff 1998, Caro and O'Doherty 1999, Andelman and Fagan 2000).

Genetic diversity refers to the genetic variation (total number of genetic characteristics) in the genetic makeup of a population or a species (Carnus et al. 2006). It includes the intra-specific variation between individuals, e.g., genes, within and between populations (races, ecotypes, and provenances) (Carnus et al. 2006). Genetic diversity distinguishes breeds or races, thus providing the basis for locally well-adapted cultivars and a gene pool for further developing commercial crops and livestock.

12.4.2.2 Structural Diversity

Structure is a system's physical organization or pattern, from horizontal and vertical complexity within forest plant communities to the pattern of patches and other elements at the landscape scale (Noss 1990, Carnus et al. 2006). Structural measures refer to an assemblage's spatial and temporal (e.g., age structure) organization. Structure includes the heights and diameters of live and dead trees and the arrangement of foliage, crown layers, herbaceous plants, and downed woody material (Hummel 2003).

Evidence is increasing that, for some taxonomic groups, species numbers are correlated with habitat heterogeneity (Moser et al. 2002), but this is only sometimes the case (Rykken et al. 1997). Tews et al. (2004) show that ecological effects of habitat heterogeneity may vary considerably among species groups depending on several factors, including whether structural attributes are perceived as heterogeneity or fragmentation, the structural variable measured, the spatial scale, and the presence of structures necessary to that species.

Structural diversity can include landscape- and stand-scale structures. Landscape structure means the pattern of a landscape, which is determined by the size, shape, arrangement, and distribution of individual landscape elements (Walz 2011). Landscape structure parameters have been recognized as valuable surrogates for species diversity and are used in various stages of conservation planning (Williams et al. 2002, Lindenmayer et al. 2008, Walz 2011). Stand-scale structure, particularly the complexity and dynamics of forest canopies, is of fundamental importance to forest dynamics, e.g., spatial heterogeneity and temporal change in understory vegetation, patterns in regeneration mosaics and microclimate variation (Norman and Campbell 1989, Song et al. 1997, Perera and Buse 2014). In general, forests are rich in structures such as live trees and downed woody material (Franklin and Spies 1991, Spies 1997, Franklin et al. 2002) and correspondingly in ecological niches; hence, they can harbor plentiful biological diversity (Kanowski et al. 2003).

12.4.2.3 Functional Diversity

Functional diversity of forests includes the variation in ecological and evolutionary processes from the variety of species' active or life history traits to processes at a landscape scale (Noss 1990, Carnus et al. 2006). Although less intuitive and typically less common in biodiversity indicators than compositional diversity, functional diversity should be considered at all scales when developing biodiversity indicators (Vandewalle et al. 2010). Data for indicators of functional diversity can often be linked to that for structural diversity; for example, efforts to maintain canopy structure can also contribute to water interception and diversity of species traits (Franklin 1988).

Advances in our understanding and knowledge of functional diversity have the potential to be a proxy for quantifying the niche space of a given assemblage and niche differentiation among member species in functional space (Mouchet et al. 2010). Thus, functional diversity acts as an interface for improved understanding of ecological processes (e.g., environmental filter effects and dispersal limitation) that govern species sorting in a community (Pakeman 2011) and ecosystem functioning in terms of the complementarity of resource use and functional redundancy among component species (Naeem 1998, Díaz et al. 2007).

Examples of functions on the benefits of forest biodiversity are also discussed in Section 12.4.1, specifically in supporting and regulating service subsections.

Landscape function incorporates regional processes such as changes in disturbance frequency, land use trends, and biogeochemical cycling rates. Landscape patterns quantified in compositional state indicators must go beyond quantification to define relationships with ecological processes (Li and Wu 2004). Indicators of landscape function should emphasize those ecological processes, e.g., equating patch sizes to nutrient conservation (Ludwig et al. 2000).

Species functional traits are a species' morphological, physiological, or phenological characteristics that affect fitness and have demonstrable links to organism function (Violle et al. 2007; de Bello et al. 2010). Generally, a plant community with five species whose functional traits differ is likely to have a higher functional diversity than a community with species with similar traits.

The use of functional traits often has two relatively distinct goals, one of which is to investigate the effects of species on ecosystem properties (effect traits) and another which is to examine the response of species to changes in the environment, such as disturbance, resource availability, or climate (response traits) (Walker et al. 1999). For example, response traits such as seed weight and dispersal mechanisms are now commonly used to assess forest ecosystem responses to different management practices at species (Graae and Sunde 2000; Aubin et al. 2009, 2014; Biswas and Mallik 2010; Pidgen and Mallik 2013; Bell 2015) and genetic scales (Myking 2002).

Genetic and evolutionary functions consider the changes at a genetic or allele level that cause changes in a population or species. In the case of threatened forest species, reduced population sizes can affect their evolutionary processes.

12.4.2.4 Extent, Remoteness, and Naturalness of Forests

The extent of forest cover and the degree of remoteness and naturalness serve as proxies for biodiversity. Since species richness is strongly associated with the area, larger areas tend to contain higher numbers of species (Preston 1962), and the extent of forest cover influences the biodiversity that can be supported. Thus, an increase or decrease in forest area will typically indicate an increase or decrease in biodiversity.

Forest ecosystems' extent, remoteness, and naturalness can be monitored individually or through aggregate indices, such as the Hemeroby index, wilderness quality index, and biophysical naturalness index. The Hemeroby index measures the anthropogenic influence on landscapes and habitats (Wrbka et al. 2004, Fu et al. 2006, Walz and Stein 2014). The wilderness index (Lesslie and Taylor 1985, Lesslie and Maslen 1995) aggregates remoteness from settlements, access, and biophysical primitiveness. The biophysical naturalness index is rated according to land use intensity in areas where the primary vegetation structure is essentially intact.

12.4.2.5 Measuring Forest Biodiversity

The traditional method of measuring biodiversity consists of establishing inventory sample plots in the field and tallying the species in the different vegetation layers. Then, depending on the objectives of the biodiversity study, data are analyzed using various analytical and statistical methods. This approach may seem straightforward for boreal or temperate forests. However, tropical forests are characterized by more species and complex structures than boreal or temperate forests. For these reasons, Baraloto et al. (2013) tested different methods to estimate tree diversity and aboveground biomass in tropical forests by establishing inventory plots varying in shape and size. Even though this study aimed to improve the efficiency of biodiversity measurement in tropical forests, the methodology developed can be applied to boreal or temperate forests if future objectives of biodiversity evaluation increase in complexity. Future developments that will improve efficiency in biodiversity measurement and allow its evaluation at large scales will be the use of remote sensing technology. For instance, Bae et al. (2019) used the radar system Sentinel-1 to portray the large-scale biodiversity of five temperate European forest regions.

12.4.3 Driving Forces

Driving forces describe the social, demographic, and economic developments in societies and the corresponding changes in lifestyles, overall levels of consumption, and production patterns. Primary driving forces are population growth and developments in the needs and activities of individuals (Smeets and Weterings 1999).

12.4.3.1 Pressures on Forest Biodiversity

Monitoring changes in pressures on forests may provide an early warning of changes in the state and benefits of biodiversity. Evidence suggests that humans have altered biodiversity in most ecosystems on Earth, including tropical, temperate, and boreal forests. Threats to biodiversity in these forests have stemmed from the combined effects of habitat loss and degradation, mainly resulting from unsustainable human consumption (McLellan 2014). For example, Hung et al. (2014), citing the example of the passenger pigeon (*Ectopistes migratorius*), illustrated that even abundant species can be vulnerable to human pressures, such as deforestation and overhunting if they are also subject to dramatic population fluctuations.

In the foreseeable future, the cumulative effects of multiple pressures are expected to affect forest biodiversity. These pressures include (i) changing land use through deforestation, fragmentation, and degradation; (ii) overexploiting resources through unsustainable logging, hunting, gathering, and water use; (iii) changing climate; (iv) introducing invasive, genetically modified, or synthesized species and organisms; and (v) emitting and introducing fertilizers, pesticides, and other chemicals into the environment (Wilcove et al. 1998, Sala et al. 2000, Butchart et al. 2010, Rands et al. 2010).

12.4.3.2 Changing Land Use

Land use change is one of the most critical global drivers of changes in biodiversity (Sala et al. 2000; Foley et al. 2005; Lindenmayer and Fischer 2006; Laliberté et al. 2010). Land use change alters the quality and quantity of habitat that sustains many plant and animal species, impairs ecosystem function, and results in associated declines in ecosystem services (Krauss et al. 2010, Carreño-Rocabado et al. 2012; Newbold et al. 2015). It can cause declines in species range, reduced abundance, extinction, the spread of invasive species, changes in growing conditions such as temperatures, and disruption of ecosystem services such as water and nutrient availability (Pimm and Raven 2000, Brooks et al. 2002, Winter et al. 2009, Newbold et al. 2015). In forest ecosystems, indicators of land use change include deforestation, fragmentation, and degradation. Since the 1980s, these changes have been monitored through satellite imagery (Woodwell et al. 1987, Skole and Tucker 1993). Land uses such as roads (Laurance et al. 2001, Heilman et al. 2002, Lee and Cheng 2014), agriculture (Kemper et al. 2000, Cochrane and Laurance 2002, Broadbent et al. 2008, Gibbs et al. 2010), forest logging (Turner et al. 2003, Asner et al. 2005), oil and gas developments (Finer et al. 2008), and urbanization (Zhu et al. 2006) contribute to deforestation, fragmentation, and degradation.

12.4.3.3 Overexploiting Resources

Overexploitation is threatening global biodiversity (Wilcove et al. 1998). Also referred to as overharvesting, overexploitation means harvesting a renewable resource to the point of diminishing returns. If it occurs beyond the natural capacity for reproduction, it can lead to the extirpation of populations and even the extinction of a species. Overexploitation is associated with unsustainable logging, hunting, gathering, and water use.

Unsustainable logging threatens some of the world's most famous and valuable forests, including remote areas of the Amazon, Central Africa, Borneo (Kammesheidt et al. 2001, Laporte et al. 2007, Hall 2008, Schulze et al. 2008, Asner et al. 2009, Bryan et al. 2010), and the Russian Far East. Unsustainable logging occurs when tree species are logged at a faster rate than they are replanted. It can also include logging in unmanaged forests and illegal logging (Nawir and Rumboko 2007, Traffic 2012, Lawson 2014). In 2012, The World Bank estimated the market value of global annual losses from the illegal cutting of forests on public lands at over $10 billion—more than eight times the total official development assistance provided for the sustainable management of forests.

Unsustainable hunting includes game species hunted for meat, skins, or medicinal/chemical properties beyond their natural reproduction capacity. Hunting pressure and the indirect effects of hunting may be important indicators for conserving biodiversity (Levi et al. 2011). Overhunting leads to the immediate loss of species (Struhsaker 1975) and the loss or impairment of functional roles that species provide (Struhsaker 1975, Nasi et al. 2008, Harrison 2011, Thompson et al. 2013).

12.4.3.4 Changing Climate

Contemporary climate (i.e., temperature and precipitation) has long been recognized as a factor that accounts for most of the variation in species richness (see O'Brien 1998, Francis and Currie 2003, Hawkins et al. 2003, Field et al. 2005, Kreft and Jetz 2007). Climate is expected to change (first increasing then decreasing) in all forest regions over the next 50–100 years (Smith et al. 2015), and these changes are anticipated to affect ecosystems worldwide (Bergengren et al. 2001, Dale et al. 2001, Meehl et al. 2007, Maclean et al. 2008). It is necessary to increase knowledge about linkages between forest biodiversity, ecosystem properties, and climate change effects to derive optimal solutions integrating biodiversity evaluation and alleviation of climate change effects (Hisano et al. 2018).

Changes in temperature and precipitation patterns and increased frequency and magnitude of extreme events will likely force species' ranges to shift (Maclean et al. 2008) and disrupt ecological communities (Lemoine and Bohning-Gaese 2003; Hole et al. 2009). Bergengren et al. (2001) predict that, by 2100, global climate change will modify plant communities on almost half of Earth's land surface and drive the conversion of nearly 40% of land-based ecosystems from one primary ecological community type to another. Forest ecosystems will change because species' physiological tolerances may be exceeded, and the rates of biophysical forest processes will be altered (Olesen et al. 2007; Kellomaki et al. 2008; Malhi et al. 2008). In a recent summary of the impacts of climate change on different forest types worldwide, Hisano et al. (2018) indicated that forest growth will either increase or decrease, the mortality rate will increase, and the net rate of change in biomass will decrease. These effects can be associated with an increase in temperature and a reduction in water availability. These changes will lead to differential species-specific adaptations, creating favorable conditions for some species and unfavorable for others (Hisano et al. 2018). For example, in the boreal forest, there is a possibility that, under climate change, shade-intolerant species, such as trembling aspen or jack pine, which colonize sites in early succession, will have their presence gradually increase. In contrast, shade-tolerant late succession species, such as black spruce, will decrease in importance (Huang et al. 2009, 2013; Searle and Chen 2017).

12.4.3.5 Introducing Species and Organisms

Introduced invasive alien, genetically modified, or synthesized species and organisms can have complex effects on global forest biodiversity. Although the vast majority will do little more than change species diversity, others can compete with native species for resources, change disturbance regimes, and, in worst case scenarios, contribute to the final collapse of an ecosystem (Davis 2006, Lockwood et al. 2007).

12.4.3.6 Emitting and Introducing Chemicals

Introducing fertilizers, pesticides, and other chemicals into the biosphere can substantially negatively affect forest biodiversity.

Anthropogenic production of nitrogen (N) has increased rapidly since 1950, mainly from emissions of ammonia (NH_3) and nitrogen oxides (NOx). By 2050, the N deposition caused by increasing nitrogen emissions will be so intense and widespread that it will influence nearly all parts of the world (Galloway et al. 2004). According to Xiankai et al. (2008), increased atmospheric N deposition will probably affect at least three aspects of biodiversity: (i) plant diversity, including trees, understory plants, and cryptogams; (ii) soil microorganism diversity; and (iii) animal diversity, including underground soil fauna and aboveground herbivores (van Herk et al. 2003).

Pesticides used in forests include insecticides, herbicides, and fungicides. Each chemical has a unique mode of action and thus uniquely affects forest biodiversity. Although they are applied to control specific pests, they can have broader effects. For example, herbicide applications have short-term (i.e., <10 years) effects on plant species (NCASI 2009, Swift and Bell 2011) but may have longer-term effects on biodiversity as they affect succession trajectories, altering the resulting species mix.

Other chemicals, such as ozone, fluoride, sulfur dioxide, and soil pollutants, have been shown to affect the genetic diversity of forest plants and animals.

12.4.4 Impacts

Impact refers to how changes in the system's state affect human well-being. The effect can also be ecological (e.g., reduction of wetlands, biodiversity loss), socioeconomic (e.g., reduced tourism), or a combination of both (Smeets and Weterings 1999; Martins et al. 2012).

Impact indicators are used to describe these impacts. Impacts occur in a specific sequence: air pollution may cause global warming (primary effect), which may in turn cause an increase in temperature (secondary effect), which may provoke a rise of sea level (tertiary impact), which could result in the loss of biodiversity (Smeets and Weterings 1999).

12.4.5 Pressures

Pressures are the threats to biodiversity or the forces or actions causing conditions to change (Hagen and Whitman 2006, Sparks et al. 2011). Pressure indicators describe developments in the release of substances (emissions), physical and biological agents, the use of resources, and the use of land (Smeets and Weterings 1999).

Pressure represents the consequence of the driving force, which in turn affects the state of the environment. The pressures on our ecosystems include climate change, food production, chemicals, mining, energy production, waste, trade, and transport (UNEP 2013), each having an equity dimension. They are usually depicted as unwanted and harmful, based on the concept that any environmental change caused by human activities is damaging and degrading (Martins et al. 2012).

Pressures can affect the short run (e.g., deforestation) or the long run (e.g., climate change), which, if known with sufficient certainty, can be expressed as a probability. They can be both human-induced, like emissions, fuel extraction, and solid waste generation, and natural processes damaging and degrading (Martins et al. 2012). Pressures can also be sub-categorized as endogenic managed pressures when they stem from within the system and can be controlled (e.g., land claim, power generation) and as exogenic unmanaged pressures when they arise from outside the system and cannot be handled (e.g., climate change, geomorphic activities).

12.4.6 Responses

Responses are policies or actions to address changes in pressures, states, and benefits to prevent or reduce biodiversity loss (Sparks et al. 2011). Responses can be associated uniquely with policy action or to different levels of society, including groups and individuals from the private, government, or non-governmental sectors. Under the PSBR framework, policy response indicators are used to monitor the presence and, more importantly, the effectiveness of policies, laws, and activities related to the management and conservation of biodiversity. Such policy, laws, and actions may include (i) normative policies such as strategic documents and policy objectives, high-level recommendations, and guidance for voluntary implementation; (ii) procedural policies, including binding legislation; and (iii) implementation policies that result in action and change (Harley and Van Minnen 2010; Lenschow and Zito 1998 as cited in Meller et al. 2011). These may be implemented through adaptive policy responses or mitigation intended to reduce or offset significant adverse effects on biodiversity (Newton and Humphrey 1997).

12.5 ACTIVE ADAPTIVE MANAGEMENT APPROACH

12.5.1 Adaptive Management

Adaptive management is a six-step cycle of assessing, designing, implementing, monitoring, evaluating, and adjusting that allows managers to work iteratively toward improved understanding and management over time (Williams et al. 2007). Adaptive management was envisioned as a single loop (Walters and Hilborn 1978) incorporating technical learning. Later, a second loop (Argyris and Schön 1978) was added to integrate institutional learning and, together with the single loop, composes the six-step cycle that is most familiar to natural resource managers (Williams and Brown 2018). A third loop (Pahl-Wostl 2009) was later suggested to allow updates to underlying values and beliefs (see Williams and Brown 2018).

Within this adaptive management cycle(s), every step is intended to refine and improve movement toward a predefined management target. If the identified objectives continue to be feasible, continued targeted monitoring (designed around testable hypotheses over defined areas; Sparrow et al. 2020, p. 1706), experimentation, and pilot studies can be used incrementally to improve management action without questioning the underlying assumptions (Lynch et al. 2022). If not, the RAD adaptive management approach could be considered.

12.5.2 RAD Adaptive Management for Transforming Ecosystems

Climate-smart conservation has effectively drawn adaptive management into the climate change arena (Stein et al. 2014, Lynch et al. 2022). Anthropogenic ecological trajectories and ecosystem transformations are now recognized to be occurring at rates that render the historical range of variability less and less relevant as a management target (Walters and Holling 1990; Millar et al. 2007; Wiens et al. 2012; Schuurman et al. 2021). However, a dominant assumption is that future system behavior will mimic past behavior in management approaches (Nichols et al. 2011; Beever et al. 2013; Schuurman et al. 2021). To facilitate a transition to managing ecosystems in which past experiences no longer suffice, Lynch et al. (2022) placed adaptive management within the resist–accept–direct (RAD) conceptual framework (Figure 12.3).

The RAD framework lays out three approaches for making management decisions for systems undergoing ecosystem transformation: (1) resist, where managers work to maintain or restore ecosystem composition, structure, processes, or function based on historical or acceptable current conditions; (2) accept, where managers allow ecosystem composition, structure, process, or function to change autonomously, and (3) direct, where managers actively shape change in ecosystem composition, structure, processes, or function toward preferred new conditions.

12.6 VISUALIZING CHANGES IN INDICATORS OVER TIME

Organizations increasingly use indicator frameworks paired with innovative graphics to convey complex messages of indicators in an easy-to-understand format to policymakers, students, and the public (e.g., Biodiversity Indicators Dashboard; Han et al. 2014). Common approaches to displaying biological diversity data include bar and pie charts; other techniques, such as radar and polar area diagrams, have also been used.

12.6.1 Radar Diagrams

We recommend using radar diagrams to visualize the full suite of biodiversity indicators. A radar chart displays multivariate data as a two-dimensional chart of 3+ quantitative variables represented on axes starting from the same point.

Radar charts are a valuable way to display multivariate observations with an arbitrary number of variables (Chambers et al. 1983). Radar diagrams were used by Huber-Sannwald et al. (2012) to study ecosystem services, Thompson et al. (2013) to study forest degradation, Houdet et al. (2014) to study differences in the effects of industrial sectors on various indicators (including biodiversity) and Deighton et al. (2021) to study the impact of alternative silvicultural treatments on floral diversity. The radar diagram approach has been applied to the full suite of State-Response (S-R)-based indicators (see Section 12.4).

12.6.2 Polar Area Diagrams

The polar area diagrams have equal angles and differ from pie charts in how far each sector extends from the circle's center. The "planetary boundaries" authors use polar area diagrams (see Figure 12.2).

12.7 SUMMARY

In this chapter, we discuss the State-Response (S-R)-based framework, the planetary boundaries framework, and the Kunming Montreal Global Biodiversity framework. Also, we discuss a variety of indicators, an active adaptive management approach, and means to visualize changes in indicators over time.

Biodiversity indicators have been developed for many purposes, including monitoring (i) benefits of biodiversity, (ii) changes in the state(s) of biodiversity, (iii) forces driving changes, (iv) cumulative impacts of forces on biodiversity, (v) effects of pressures on biodiversity, and (iv) effectiveness of policy responses. The State-Response (S-R)-based framework of indicators is one of a suite of tools

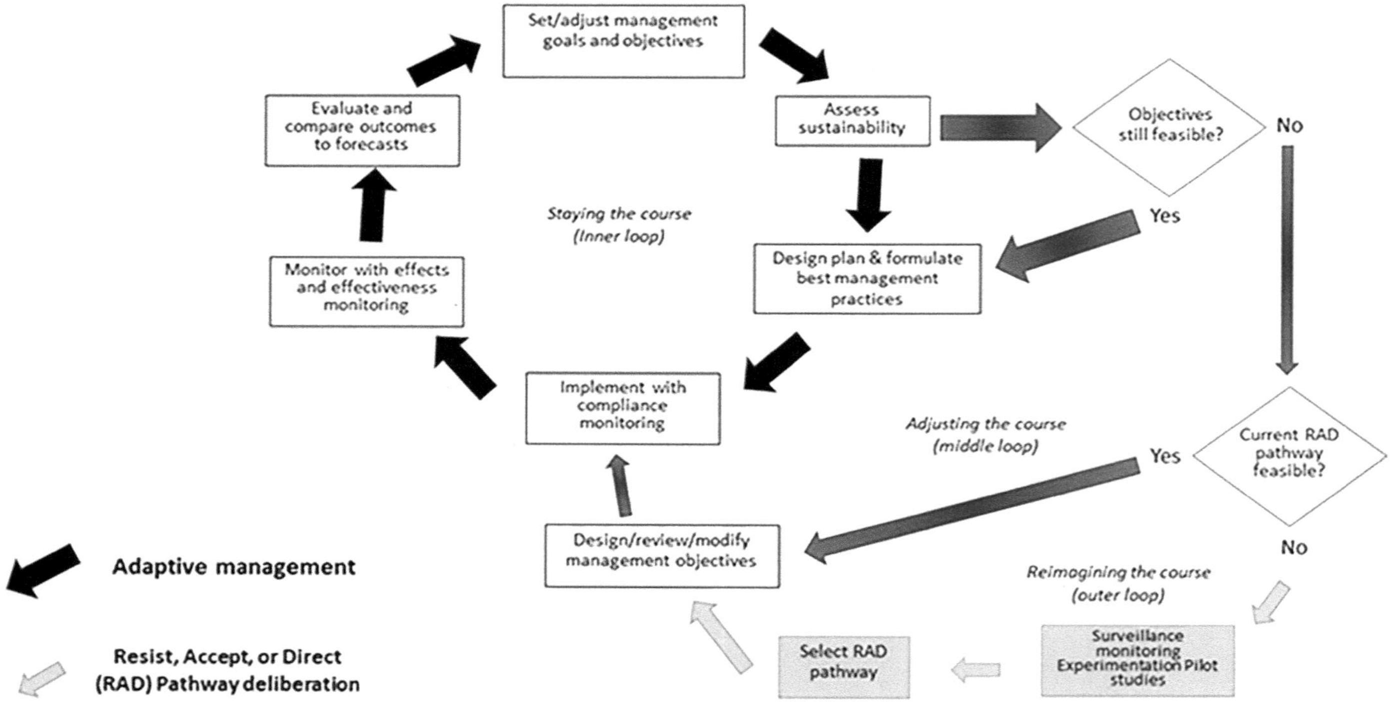

FIGURE 12.3 Six-step cycle involved in adaptive management in the resist–accept–direct (RAD) framework. (Adapted from Bell and Baker 2006 and Lynch et al. 2022.)

that should be considered to conserve biodiversity in boreal, temperate, and tropical forests. The S-R-based framework is presumed to provide a more holistic view of the effects of changing forest policies. It will help forest managers remain aware of the many aspects they are not monitoring and be cognizant that there are things they need to know they are not monitoring.

To use these frameworks rigorously, agencies should choose indicators and set related objectives *a priori*. By doing so, they can determine whether pressures on forests, the state of biodiversity within forests, or benefits derived from forests are improving and subsequently revise their policies as appropriate. An active adaptive management approach is recommended.

Producing a radar or polar diagram for each element will provide a means to visually distinguish among indicator levels, track changes through time, and reconcile among multiple perspectives.

REFERENCES

Abigail J.L., L.M. Thompson, J.M. Morton, et al. 2022. RAD Adaptive management for transforming ecosystems. *BioScience* 72:45–56. https://doi.org/10.1093/biosci/biab091.

Argyris, C. and D.A. Schön. 1978. *Organizational learning: A theory of action perspective*. Reading, MA: Addison-Wesley.

Anand, M., S. Laurence, and B. Rayfield. 2005. Diversity relationships among taxonomic groups in recovering and restored forests. *Conserv. Biol.* 19:955–962.

Andelman, S.J. and W.F. Fagan. 2000. Umbrella and flagships: efficient conservation surrogates or expensive mistakes? *Proc. Natl. Acad. Sci. U.S.A.* 97:5954–5959.

Asner, G.P., D.E. Knapp, E.N. Broadbent, et al. 2005. Selective logging in the Amazon. *Science* 310:480–482.

Asner, G., T. Rudel, T. Aide, et al. 2009. A contemporary assessment of change in humid tropical forests. *Conserv. Biol.* 23:1386–1395.

Aubin, I., O. Deshaies, F. Cardou, et al. 2014. Management legacy in the understory of North American mixed boreal regenerating stands. *For. Ecol. Manage.* 320:129–137.

Aubin, I., M.-H. Ouellette, P. Legendre, et al. 2009. Comparison of two plant functional approaches to evaluate natural restoration along an old-field—deciduous forest chronosequence. *J. Veg. Sci.* 20:185–198.

Bader, P., S. Jansson, and B. Jonsson. 1995. Wood-inhabiting fungi and substratum decline in selectively logged boreal spruce forests. *Biol. Conserv.* 72:355–362.

Bae, S., S.R. Levick, L. Heidrich, et al. 2019. Radar vision in the mapping of forest biodiversity from space. *Nat. Commun.* 10:4757.

Bailey, R.G., 2009. *Ecosystem geography* (2nd. ed.). New York, NY: Springer.

Baird, D., M. Rubach, and P. Van den Brink. 2008. Trait-based ecological risk assessment (TERA): the new frontier? *Integr. Environ. Assess. Manage.* 4:2–3.

Baraloto, C., Q. Molto, S. Rabaud, et al. 2013. Rapid simultaneous estimation of aboveground biomass and tree diversity across neotropical forests: A comparison of field inventory methods. *Biotropica* 45:288–298. 10.1111/btp.12006.

Baroloto, C., Q. Molto, S. Rabaud, et al. 2012. Rapid simultaneous estimation of aboveground biomass and tree diversity across neotropical forests: a comparison of field inventory methods. *Biotropica* 45:288–298.

Belcher, B., and K. Schreckenberg. 2007. Commercialization of non-timber forest products: A reality check. *Dev. Policy Rev.* 25(3):355–377.

Bell, F. 2015. *Effects of intensification of silviculture on plant diversity in northern temperate and boreal forests of Ontario, Canada*. Guelph, ON: Sch. Environ. Sci., Univ. Guelph, PhD diss.

Bell, F.W. and J.A. Baker. 2006. Adaptive forest management. *Can. Silv. Mag.* Nov 2006:20–23.

Belluscio, A. 2009. Planting trees can shift water flow. *Nature News*, online 1–2.

Benini, L., V. Bandin, D. Marazza, et al. 2010. Assessment of land use changes through an indicator-based approach: A case study from the Lamone river basin in Northern Italy. *Ecol. Indicators* 10:4–14.

Bergengren, J., D. Waliser, and Y. Yung. 2001. Ecological sensitivity: A biospheric view of climate change. *Climatic Change* 107(3):433–457.

Bergmann, F., and B. Hosius. 1996. Effects of heavy-metal polluted soils on the genetic structure of Norway spruce seedling populations. *Water Air Soil Pollut.* 89:363–373.

Bergmann, F., and F. Scholz. 1987. The impact of air pollution on the genetic structure of Norway spruce. *Silvae Genet.* 36:80–83.

Berrang, P., D. Karnosky, and J. Bennett. 1989. Natural selection for ozone tolerance in *Populus tremuloides*: field verification. *Can. J. For. Res.* 19:519–522.

Beever E.A., S.Z. Dobrowski, J. Long, et al. 2013. Understanding relationships among abundance, extirpation, and climate at ecoregional scales. *Ecology* 94:1563–1571.

[BIP] Biodiversity Indicators Partnership. 2010. *Biodiversity indicators and the 2010 target: Experiences and lessons learned indicators and the 2010 Target: Experiences and lessons learned.* Montréal, Canada: Secretariat of the Convention on Biological Diversity.

BIP (Biodiversity Indicators Partnership). 2011. *Guidance for national biodiversity indicator development and use.* Cambridge, UK: UNEP World Conservation Monitoring Centre.

Biswas, S.R., and A.U. Mallik. 2010. Disturbance affects species diversity and functional diversity in riparian and upland plant communities. *Ecology* 91:28–35.

Broadbent, E.N., G.P. Asner, M. Keller, et al. 2008. Forest fragmentation and edge effects from deforestation and selective logging in the Brazilian Amazon. *Biol. Conserv.* 141:1745–1757.

Brockerhoff, E.G., C.E. Ecroyd, and E.R. Langer. 2001. Biodiversity in New Zealand plantation forests: Policy trends, incentives, and the state of our knowledge. *N. Z. J. For.* 46:31–37.

Brockerhoff, E., C. Ecroyd, and E. Langer. 2001. Biodiversity in New Zealand plantation forests: Policy trends, incentives, and the state of our knowledge. *N. Z. J. For.* 46:31–37.

Brooks, T.M., R.A. Mittermeier, C.G. Mittermeier, et al. 2002. Habitat loss and extinction in the hotspots of biodiversity. *Conserv. Biol.* 16(4):909–923.

Bryan, J., P. Shearman, J. Ash, et al. 2010. Estimating rainforest biomass stocks and carbon loss from deforestation and degradation in Papua New Guinea 1972–2002: Best estimates, uncertainties, and research needs. *J. Environ. Manage.* 91:995–1001.

Bunnell, E.L., and A. Chan-McLeod. 1998. Forestry and biological diversity: Elements of the problem. In *Policy and practices for biodiversity in managed forests: The living dance,* eds. E. Bunnell and J.E. Johnson, pp. 3–18. Vancouver, Canada: UBC Press.

Butchart, S.H., M. Walpole, B. Collen, et al. 2010. Global biodiversity: Indicators of recent declines. *Science* 328:1164–1168.

Campos, M., A. Velázquez, G. Bocco, et al. 2012. The potential role of local knowledge and perception of landscape in land-use planning: Lessons from a rural area of the Mexican Pacific coast. *Soc. Nat. Resour.* 25:759–774.

Carnus, J.-M., J. Parrotta, E. Brockerhoff, et al. 2006. They planted forests and biodiversity. *J. For.* 104(2):65–77.

Caro, T. 2010. *Conservation by proxy. Indicator, umbrella, keystone, flagship, and other surrogate species.* Washington, DC: Island Press.

Caro, T., and G. O'Doherty. 1999. On the use of surrogate species in conservation biology. *Conserv. Biol.* 13:805–814.

Carreño-Rocabado, G., M. Peña-Claros, F. Bongers, et al. 2012. Effects of disturbance intensity on species and functional diversity in a tropical forest. *J. Ecol.* 100(6):1453–1463.

Catford, J., R. Jansson, and C. Nilsson. 2009. Reducing redundancy in invasion ecology by integrating hypotheses into a single theoretical framework. *Divers. Distrib.* 15:22–40.

Chambers, J., W. Cleveland, B. Kleiner, et al. 1983. *Graphical methods for data analysis*, pp. 158–162. Belmont, CA: Wadsworth.

Chapin, F., P. Matson, and H. Mooney. 2002. *Principles of terrestrial ecosystem ecology.* New York, NY: Springer.

Chown, S., B. van Rensburg, K. Gaston, et al. 2003. Energy, species richness, and human population size: Conservation implications at a national scale. *Ecol. Appl.* 13(5):1233–1241.

Cochrane, M., and W. Laurance. 2002. Fire as a large-scale edge effect in Amazonian forests. *J. Tropical Ecol.* 18:311–325.

[CBD] Convention on Biological Diversity. 2011. *Conference of the Parties Decision X/2: Strategic plan for biodiversity 2011–2020.* www.cbd.int/decision/cop/?id=12268.

[COP] Convention on Biological Diversity. 2015. *Decision adopted by the conference of the parties to the convention on biological diversity.* Conference of the Parties to the Convention on biological diversity. Fifteenth meeting – Part II Montreal, Canada, December 7–19, 2022 Agenda item 24.

Cornelissen, J.H., S. Lavorel, E. Granier, et al. 2003. A handbook of protocols for standardized and easy measurement of plant functional traits worldwide. *Austral. J. Bot.* 51:335–380.

Dale, V., L. Joyce, S. McNulty, et al. 2001. Climate change and forest disturbances: Climate change can affect forests by altering the frequency, intensity, duration, and timing of fire, drought, introduced species, insect and pathogen outbreaks, hurricanes, windstorms, ice storms, or landslides. *BioScience* 51(9):723–734.

Dallmeier, F., R.C. Szaro, A. Alonso, et al. 2013. Framework for assessment and monitoring of biodiversity. *Encycl. Biodiver.* 3:545–559. http://dx.doi.org/10.1016/B978-0-12-384719-5.00316-6

Davidson, A., L. Elliston, P. Kokic, et al. 2005. Native vegetation: cost of preservation in Australia. *Austral. Commodities* 12(3):543–548.

Davis, M. 2006. Invasion biology 1958–2005: The pursuit of science and conservation. In *Conceptual ecology and invasion biology: Reciprocal approaches to nature*, eds. M. Cadotte, S. McMahon, and T. Fukami, pp. 35–64. Berlin: Springer.

De Bello, F., S. Lavorel, S. Diaz, et al. 2010 Towards an assessment of multiple ecosystem processes and services via functional traits. *Biodiv. Conserv.* 19:2873–2893.

Deighton, H.D., F.W Bell, N. Thiffault, et al. 2021. Trade-offs among release treatments in jack pine plantations: Twenty-five-year responses. *Forests* 12(370). https://doi.org/10.3390/f12030370

Díaz, S., S. Lavorel, F. de Bello, et al. 2007. Incorporating plant functional diversity effects in ecosystem service assessments. *Proc. Natl. Acad. Sci. U.S.A.* 104:20684–20689.

Doran, G. 1981. There's a S.M.A.R.T. way to write management's goals and objectives. *Manage. Rev.* 70(11) (AMA FORUM):35–36.

Drever, C., G. Peterson, C. Messier, et al. 2006. Can forests management based on natural disturbances maintain ecological resilience? *Can. J. For. Res.* 36:2285–2299.

Egoh, B., E.G. Drakou, M.B. Dunbar, et al. 2012. *Indicators for mapping ecosystem services: A review.* JRC Scientific and Policy Reports. European Commission, Joint Research Centre, Ispra.

Faith, D.P. 1992. Conservation evaluation and phylogenetic diversity. *Biol. Conserv.* 61:1–10.

Fensham, R., and R. Fairfax. 1996. The disappearing grassy balds of the Bunya Mountains, south-eastern Queensland. *Austral. J. Bot.* 44:543–548.

Ferris, R., and J.W. Humphrey. 1999. A review of potential biodiversity indicators for application in British forests. *Forestry* 72(4):313–328.

Field, R., E. O'Brien, and R. Whittaker. 2005. Global models for predicting woody plant richness from climate: development and evaluation. *Ecology* 86:2263–2277.

Finer, M., C.N. Jenkins, S.L. Pimm, et al. 2008. Oil and gas projects in the western Amazon: Threats to wilderness, biodiversity, and indigenous peoples. *PLoS ONE* 3(8):e2932

Flynn, D., M. Gogol-Prokurat, T. Nogeire, et al. 2009. Loss of functional diversity under land use intensification across multiple taxa. *Ecol. Lett.* 12:22–33.

Foley, J.A., R. DeFries, G.P. Asner, et al. 2005. Global consequences of land-use. *Science* 309:570–574.

Francis, A., and D. Currie. 2003. A globally consistent richness-climate relationship for angiosperms. *Am. Nat.* 161:523–536.

Franklin, J.F. 1988. Structural and functional diversity in temperate forests. In *Biodiversity,* ed. E.O. Wilson, pp. 166–175. Washington, DC: National Academy Press.

Franklin, J., and T.A. Spies. 1991. Composition, function, and structure of old-growth Douglas-fir forests. In *Wildlife and vegetation of unmanaged Douglas-fir forests*, L.F. Ruggiero, K.B. Aubry, A.B. Carey, and M.H. Huff, pp. 71–80. Portland, OR: US Dept. Agric., For. Serv. Pacific Northwest Res. Stn., Gen. Tech. Rep. PNW-GTR-285.

Franklin, J.F., T.A. Spies, R. Van Pelt, et al. 2002. Disturbances and structural development of natural forest ecosystems with silvicultural implications, using Douglas-fir forests as an example. *For. Ecol. Manage.* 155:399–423.

Friend, A.M., and D.J. Rapport. 1991. Evolution of macro-information systems for sustainable development. *Ecol. Econ.* 3(1):59–76. https://doi.org/10.1016/0921-8009(91)90048-J.

Fu, B.J., C.X. Hu, L.D. Chen, et al. 2006. Evaluating change in agricultural landscape pattern between 1980 and 2000 in the Loess hilly region of Ansai County, China. *Agric. Ecosys. Environ.* 114:387–396.

Gabrielsen, P., and P. Bosch. 2003. *Internal working paper environmental indicators: Typology and use in reporting*. Copenhagen: European Environment Agency.

Galloway, J., F. Dentener, D. Capogne, et al. 2004. Nitrogen cycles: Past, present, and future. *Biochemistry* 70:53–226.

Gamfeldt, L., T. Snäll, R. Bagchi, et al. 2013 Higher levels of multiple ecosystem services are found in forests with more tree species. *Nat. Comm.* 4, Art. 1340.

Gari, S.R., A. Newton, and J.D. Icely. 2015. A review of the application and evolution of the DPSIR framework with an emphasis on coastal social-ecological systems. *Ocean Coast. Manag.* 103:63–77. https://doi.org/10.1016/j.ocecoaman.2014.11.013.

Gibbs, H.K., A.S. Ruesch, M.K. Achard, et al. 2010. Tropical forests were the primary sources of new agricultural land in the 1980s and 1990s. *Proc. Natl. Acad. Sci. U.S.A.* 107(38):16732–16737.

Gilmore, M.P., S. Ros Ochoa, and S. Rios-Flores. 2010. The cultural significance of the habitat Mañaco Taco to the Maijuna of the Peruvian Amazon. In *Landscape ethnoecology: Concepts of biotic and physical space*, eds. L.M. Johnson and E.S. Hunn, pp. 141–157. New York and Oxford: Berghahn.

Graae, B.J., and P.B. Sunde. 2000. The impact of forest continuity and management on forest floor vegetation evaluated by species traits. *Ecography* 23:720–731.

Gray, E., and W. Bond. 2013. Will woody plant encroachment impact the visitor experience and economy of conservation areas? *Koedoe* 55(1) Art. #1106, 9 pages.

Hagen, J., and A. Whitman. 2006. Biodiversity indicators for sustainable forestry: Simplifying complexity. *J. For.* 203–2010.

Hall, J. 2008. Seed and seedling survival of African mahogany (*Entandrophragma* spp.) in the Central African Republic: Implications for forest management. *For. Ecol. Manage.* 255:292–299.

Han, X., R. Smyth, B. Young, et al. 2014. A biodiversity indicators dashboard: Addressing challenges to monitoring progress towards the Aichi biodiversity targets using disaggregated global data. *PLoS ONE* 9(11):e112046.

Harley, M., and J.G. van Minnen. 2010. Adaptation indicators for biodiversity. *ETC/ACC Technical Paper 2010/15*; December 2010, 14p.

Harrison, R. 2011. Emptying the forest: hunting and the extirpation of wildlife from tropical nature reserves. *BioScience* 61:919–924.

Hawkins, B., R. Field, H. Cornell, et al. 2003. Energy, water, and broad-scale geographic patterns of species richness. *Ecology* 84:3105–3117.

Heilman, G., J. Strittholt, N. Slosser, et al. 2002. Forest fragmentation of the conterminous United States: assessing forest intactness through road density and spatial characteristics. *BioScience* 52:411–422.

Hicks, C., S. Woroniecki, M. Fancourt, et al. 2014. *The relationship between biodiversity, carbon storage and the provision of other ecosystem services: Critical Review for the Forestry Component of the International Climate Fund.* Cambridge, UK: United Nations Environment Programme.

Hisano, M., E.B. Searle, and H.Y.H. Chen. 2018. Biodiversity as a solution to mitigate climate change impacts on the functioning of forest ecosystems. *Biol. Rev.* 93:439–456.

Hole, D., S.G. Willis, D. Pain, et al. 2009. Projected impacts of climate change on a continentwide protected area network. *Ecol. Lett.* 12:420–431.

Hooper, D., F. Chapin, J. Ewel, et al. 2005. Effects of biodiversity on ecosystem functioning: A consensus of current knowledge. *Ecol. Monogr.* 75:3–35.

Houdet, J., B. Bellini, and M. Barra. 2014. La biodiversité comme enjeu stratégique pour l'entreprise. L'indicateur d'interdépendance de l'entreprise à la biodiversité. In *Les indicateurs de biodiversité forestière,* eds C. Nivet, I. Bonhême, and J.-L. Jean-Luc Peyron, pp. 115–122. Paris: Gip Ecofor.

Howard, P., P. Viskanic, R. Davenport, et al. 1998. Complementarity and the use of indicator groups for reserve selection in Uganda. *Nature* 394:472–475.

Hu, Y., M. Wang, T. Ma, et al. 2022. Integrated index-based assessment reveals long-term conservation progress in the implementation of the Convention on Biological Diversity. *Sci Adv.* 8(1):eabj8093. January 7. https://doi.org/10.1126/sciadv.abj8093. Epub 2022 Jan 5. PMID: 34985949; PMCID: PMC8730628.

Huang, J.-G., J.C. Tardif, Y. Bergeron, et al. 2009. Radial growth response of four dominant boreal tree species to climate along a latitudinal gradient in the eastern Canadian boreal forest. *Glob. Chang. Biol.* 16(2):711–731. https://doi.org/10.1111/J.1365-2486.

Huang, J.-G, Y. Bergeron, F. Berninger, et al. 2013. Impact of future climate on radial growth of four major boreal tree species in the eastern Canadian boreal forest. *PloS ONE* 8(2):e56758. https://doi.org/10.1371/journal.pone.0056758.

Huber-Sannwald, E., M.R. Palacios, J.T.A. Moreno, et al. 2012. Navigating challenges and opportunities of land degradation and sustainable livelihood development in dryland social–ecological systems: a case study from Mexico. *Phil. Trans. R. Soc. B* 367:3158–3177.

Hummel, S.S. 2003. Managing structural and compositional diversity with silviculture. Chap. 4. In *Compatible forest management*, eds. R. Monserud, R. Haynes, and A.C. Johnson, pp. 85–119. Dordrecht, The Netherlands: Kluwer Academic Publishers.

Hung, C., P. Shaner, R. Zink, et al. 2014. Drastic population fluctuations explain the rapid extinction of the passenger pigeon. *Proc. Natl. Acad. Sci. U.S.A.* 111(29):10636–10641.

IPCC. 2014. *Climate change 2014: Synthesis report. Contribution of working groups I, II and III to the fifth assessment report of the intergovernmental panel on climate change.* Geneva, Switzerland: IPCC.

Isaac, N.J., S.T., Turvey, B. Collen, et al. 2007. Mammals on the EDGE: Conservation priorities based on threat and phylogeny. *PLoS ONE* 2(3):e296.

IUCN. 2008. *Guidelines for using the IUCN Red List categories and criteria.* www.iucnredlist.org/static/categories_criteria

Ives, A., and S. Carpenter. 2007. Stability and diversity of ecosystems. *Science* 317:58–62.

Jonsson, B., and M. Jonsell. 1999. Exploring potential biodiversity indicators in boreal forests. *Biodiv. Conserv.* 8:1417–1433.

Kammesheidt, L., P. Köhler, and A. Huth. 2001. Sustainable timber harvesting in Venezuela: A modelling approach. *J. Appl. Ecol.* 38:756–770.

Kanowski, J., C.P. Catterall, G.W. Wardell-Johnson, et al. 2003. Development of forest structure on cleared rainforest land in eastern Australia under different styles of reforestation. *For. Ecol. Manage.* 183:265–280.

Kelble, C.R.; D.K. Loomis, S. Lovelace, et al. 2013. The EBM-DPSER conceptual model: Integrating ecosystem services into the DPSIR framework. *PLOS ONE* 8(8):e70766. https://doi.org/10.1371/journal.pone.0070766.

Kellomaki, S., H. Peltola, T. Nuutinen, et al. 2008. Sensitivity of managed boreal forests in Finland to climate change, with implications for adaptive management. *Phil. Trans. Roy. Soc. B* 363:2341–2351.

Kemper, J., R. Cowling, D. Richardson, et al. 2000. Landscape fragmentation in South Coast Renosterveld, South Africa, in relation to rainfall and topography. *Austral. Ecol.* 25:179–186.

Kim, Y.-S.; M. Rodrigues, F.-N. Robinne. 2021. Economic drivers of global fire activity: A critical review using the DPSIR framework. *For. Policy Econ.* 131:102563. https://doi.org/10.1016/j.forpol.2021.102563.

Krauss, J., R. Bommarco, M. Guardiola, et al. 2010. Habitat fragmentation causes immediate and time-delayed biodiversity loss at different trophic levels. *Ecol. Lett.*13:597–605.

Kreft, H., and W. Jetz. 2007. Global patterns and determinants of vascular plant diversity. *Proc. Natl. Acad. Sci. U.S.A.* 104(14):5925–5930.

Laliberté, E., J. Wells, F. DeClerck, et al. 2010. Land-use intensification reduces functional redundancy and response diversity in plant communities. *Ecol. Lett.* 13:76–86.

Lambeck, R. 1997. Focal species: A multi-species umbrella for nature conservation. *Conserv. Biol.* 11:849–856.

Landres, P., J. Verner, and J. Thomas. 1988. Ecological uses of vertebrate indicator species—a critique. *Conserv. Biol.* 2:316–328.

Laporte, N., J. Stabach, R. Grosch, et al. 2007. Expansion of industrial logging in Central Africa. *Science* 316:1451.

Laurance, W.F., M. Goosem, and S.G.W. Laurance. 2001. Impacts of roads and linear clearings on tropical forests. *Trends Ecol. Evol.* 24(12):659–669.

Lavorel, S., S. Diaz, J.H. Cornelissen, et al. 2007. Terrestrial ecosystems in a changing world. In *Plant functional types: Are we getting any closer to the holy grail?* eds. J. Canadell, I.F. Pitelka, and D. Pataki. New York, NY: The IGBP series. Springer-Verlag.

Lawson, S., 2014. Methodology for import-source estimates of illegally sourced wood imports: Thailand, South Korea and India. *Energy, Environment and Resources EER PP* 2014/05.

Laycock, W.A. 1991. Stable states and thresholds of range condition on North American rangelands: A viewpoint. *J. Range Manag.* 44:427–433.

Lee, P., and R. Cheng. 2014. Human access in Canada's landscapes. *Global Forest Watch Canada Bulletin.* 7 p.

Lemoine, N., and K. Böhning-Gaese. 2003. Potential impact of global climate change on species richness of long-distance migrants. *Conserv. Biol.* 17(2):577–586.

Lenschow, A., and A. Zito. 1998. Blurring or shifting of policy frames? Institutionalization of the economic-environmental policy linkage in the European Community. Governance 11:415–441.

Lepš, J., F. de Bello, S. Lavorel, et al. 2006. Quantifying and interpreting functional diversity of natural communities: practical considerations matter. *Preslia* 78:481–501.

Lesslie, R.G., and S.G. Taylor. 1985. The wilderness continuum concept and its implications for wilderness preservation policy. *Biol. Conserv.* 32:309–333.

Lesslie, R., and M. Maslen. 1995. *National wilderness inventory handbook.* 2nd edn. Canberra, Australia: Australian Heritage Comm., Australian Govern. Publ. Serv.

Levi, T., J. Shepard, J. Ohl-Schacherer, et al. 2011. Spatial tools for modeling the sustainability of subsistence hunting in tropical forests. *Ecol. Appl.* 21(5):1802–1818.

Levrel, H., D. Couvet, and G. Loïs. 2007. Indicateurs de biodiversité pour les forêts francaises. État des lieux et perspectives. *Revue Forestière Française* 59:45–56.

Levrel, H., C. Kerbiriou, D. Couvet, et al. 2009. OECD Pressure-State-Response indicators for managing biodiversity: A realistic perspective for a French biosphere reserve. *Biodiv. Conserv.* 18(7):1719–1732.

Li, H., and J. Wu. 2004. Use and misuse of landscape indices. *Landscape Ecol.* 19:389–399.

Lindenmayer, D. 1999. Future direction for biodiversity conservation in managed forests: indicator species, impact studies and monitoring programs. *For. Ecol. Manage.* 115:277–287.

Lindenmayer, D. and J. Fischer. 2006. *Habitat fragmentation and landscape change an ecological and conservation synthesis.* Washington, DC: Island Press.

Lindenmayer, D.B., C.R. Margules, and D.B. Botkin. 2000. Indicators of biodiversity for ecological sustainable forest management. *Conserv. Biol.* 14(4):941–950.

Lindenmayer, D., A. Manning, P. Smith, et al. 2002. The focal-species approach and landscape restoration: A critique. *Conserv. Biol.* 16(2):338–345.

Lindenmayer, D., R.J. Hobbs, R. Montague-Drake, et al. 2008. A checklist for ecological management of landscape for conservation. *Ecol. Lett.* 11:78–91.

Lockwood, J., M. Hoopes, and M. Marchetti. 2007. *Invasion ecology*. Oxford: Blackwell.

Longauer, R., D. Gömöry, L. Paule, et al. 2001. Selection effects of air pollution on gene pools of Norway spruce, European silver fir and European beech. *Environ. Pollut.* 115:405–411.

Ludwig J.A., J.A. Wiens, and D.J. Tongway. 2000. A scaling rule for landscape patches and how it applies to conserving soil resources in Savannas. *Ecosystems* 3:84–97.

Lynch, A.J., L.M. Thompson, J.M. Morton, et al. 2022. RAD Adaptive Management for Transforming Ecosystems. *BioScience* 72(1):45–56. January. https://doi.org/10.1093/biosci/biab091.

[MA] Millennium Ecosystem Assessment. 2003. Ecosystems and their services. Chap. 2 in *Ecosystems and human well-being: a framework for assessment*. Washington, DC, USA: Island Press.

[MA] Millennium Ecosystem Assessment. 2005. *Ecosystems and human well-being: Synthesis*. Washington: Synthesis Island Press, 155pp.

Maclean, I.M.D., G.E. Austin, M.M. Rehficsh, et al. 2008. Climate change causes rapid changes in the distribution and site abundance of birds in winter. *Glob. Change Biol.* 14(11):2489–2500.

Malhi, Y., J. Roberts, R. Betts, et al. 2008. Climate change, deforestation, and the fate of the Amazon. *Science* 319:169–172.

Marchetti, M. 2004. *Monitoring and indicators of forest biodiversity in Europe—From ideas to operationality*. Saarijåvi, Finland: Gummerus Printing.

Margaletic, J. 2003. Small rodents in the forest ecosystem as infectious disease reservoirs. *Acta Med Croatica (in Croatian)* 57(5):421–426.

Margules, C.R., and R. Pressey. 2000. Systematic conservation planning. *Nature* 405:243–253.

Martins, J.H., A.S. Camanho, M.B. Gaspar. 2012. A review of the application of driving forces—Pressure—State—Impact—Response framework to fisheries management. *Ocean Coast. Manag.* 69:273–281. https://doi.org/10.1016/j.ocecoaman.2012.07.029.

McLellan, R., L. Iyengar, B. Jeffries, et al. 2014. *Living planet report 2014: Species and spaces, people and places*. Gland: WWF International,

Meehl, G., R. Stocker, W.D. Collins, et al. 2007. Climate change 2007: The physical science basis. In *Contribution of working group i to the fourth assessment report of the intergovernmental panel on climate change*. Cambridge, United Kingdom and New York, NY, USA: Cambridge University Press.

Meller, L., A. van Teeffelen, J. van Minnen, et al. 2011. A matrix of biodiversity indicators. European responses to climate change: Deep emissions reductions and mainstreaming of mitigation and adaptation. (Responses project).

Millar C.I., N.L. Stephenson, and S.L. Stephens. 2007. Climate change and forests of the future: Managing in the face of uncertainty. *Ecol. Appl.* 17:2145–2151.

Mills, L., M. Soulé, and D. Doak. 1993. The key-stone species concept in ecology and conservation. *BioScience* 43:219–224.

Mittermeier, R.A., N. Myers, and C.Q. Mittermeier. 2000. *Hotspots: Earth's biologically richest and most endangered terrestrial ecoregions*. Cemex: Conservation International. ISBN: 978-968-6397-58-1

Mohammed, G. 1999. *Non-timber forest products in Ontario: An overview*. For. Res. Inf. Pap. No. 145. Sault Ste Marie, ON: Ontario Forest Research Institute.

Moser, D., H.G. Zechmeister, C. Plutzar, et al. 2002. Landscape patch shape complexity as an effective measure for plant species richness in rural landscapes. *Landscape Ecol.* 17:657–669.

Mouchet, M.A., S. Villéger, N. Mason, et al. 2010. Functional diversity measures: An overview of their redundancy and their ability to discriminate community assembly rules. *Func. Ecol.* 24:867–876.

Myers, N. 1988. Threatened biotas: 'Hotspots' in tropical forests. *Environmentalist* 8:187–208.

Myers, N., A.R., Mittermeier, C.G. Mittermeier, et al. 2000. Biodiversity hotspots for conservation priorities. *Nature* 403:853–858.

Myking, T. 2002. Evaluating genetic resources of forest trees by means of life history traits a Norwegian example. *Biodiv. Conserv.* 11:1681–1696.

[NCASI] National Council for Air and Stream Improvement, Inc. 2009. *Forest herbicide effects on Pacific Northwest ecosystems: A literature review*. Research Triangle Park, NC. National Council for Air and Stream Improvement, Inc., Tech. Bull. No. 970.

Naeem, S. 1998. Species redundancy and ecosystem reliability. *Conserv. Biol.* 12:39–45.

Nasi, R., D. Brown, D. Wilkie, et al. 2008. *Conservation and use of wildlife-based resources: The bushmeat crisis*. CBD Technical Series 33. Secretariat of the Convention on Biological Diversity, Montreal and Center for International Forestry Research (CIFOR), Bogor.

Nasi, R., S. Wunder, and J.J. Campos. 2002. Forest ecosystem services: Can they pay our way out of deforestation? A discussion paper prepared for the GEF for the Forestry Roundtable held in conjunction with the UNFF II, 11 March 2002. New York, NY. In *Landscape ecology: Theory and application,* 2nd edn, eds. Z. Naveh and A. Lieberman. 1994. New York, NY: Springer-Verlag.

Nawir, A.A., and L. Rumboko. 2007. History and state of deforestation and land degradation. In *Forest rehabilitation in Indonesia: Where to after more than three decades?* eds. A.A. Nawir, Murniati, and L. Rumboko, pp. 11–32. Jakarta: Center for International Forestry Research (CIFOR).

Newbold, T., L.N. Hudson, S.L.L. Hill, et al. 2015. Global effects of land use on local terrestrial biodiversity. *Nature* 520:45–69.

Newton, A., and J. Humphrey. 1997. Forest management for biodiversity: perspectives on policy, context and current initiatives. In *Biodiversity in Scotland: status, trends and initiatives,* eds. L. Fleming, A. Newton, J. Vickery, and M. Usher, pp. 179–198. Edinburgh: Scottish Natural Heritage/University of Edinburgh, The Stationary Office.

Nichols J.D., M.D. Koneff, P.J. Heglund, et al. 2011. Climate change, uncertainty, and natural resource management. *J. Wildl. Manag.* 75:6–18.

Nitare, J., and M. Noren. 1992. Woodland key-habitats of rare and endangered species will be mapped in a new project of the Swedish National Board of Forestry. *Sven. Bot. Tidskr.* 86:219–226.

Norman, J.M., and G.S. Campbell. 1989. Canopy structure. In *Plant physiological ecology: Field methods and instrumentation,* eds. R.W. Pearcy, J. Ehleringer, H.A. Mooney, and P.W. Rundel, pp. 301–325. London and New York, NY: Chapman and Hall.

Noss, F.R. 1990. Indicators for monitoring biodiversity: A hierarchical approach. *Conserv. Biol.* 4(4):355–364.

Noss, R. 1999. Assessing and monitoring forest biodiversity: A suggested framework and indicators. *For. Ecol. Manage.* 115:135–146.

Noss, R., M.A. O'Connell, and D.D. Murphy. 1997. *The science of conservation planning. habitat conservation under the Endangered Species Act.* Covelo, CA: Island Press.

[OECD] Organisation for Economic Co-operation and Development. 1993. *OECD core set of indicators for environmental performance reviews.* Paris, France: OECD Environment Monographs No. 83.

O'Brien, E. 1998. Water–energy dynamics, climate, and prediction of woody plant species richness: As interim general model. *J. Biogeogr.* 25:379–398.

Odermatt, S. 2004. Evaluation of mountain case studies by means of sustainability variables: A DPSIR model as an evaluation tool in the context of the North-South discussion. *Mountain Res. Devel.* 24:336–341.

OECD [Organisation for Economic Co-operation and Development]. 1993. *OECD core set of indicators for environmental performance reviews.* Paris, France: OECD Environment Monographs No. 83.

Oertli, S., A. Müller, D. Steiner, et al. 2005. Cross-taxon congruence of species diversity and community similarity among three insect taxa in a mosaic landscape. *Biol. Conserv.* 126:195–205.

O'Higgins, T., A. Farmer, G. Daskalov, et al. 2014. Achieving good environmental status in the Black Sea: scale mismatches in environmental management. *Ecol. Soc.* 19(3). https://doi.org/10.5751/ES-06707-190354.

Olesen, J., T. Carter, C. Az-Ambrona, et al. 2007. Uncertainties in projected impacts of climate change on European agriculture and terrestrial ecosystems. *Clim. Change* 81:123–143.

Pahl-Wostl, C., 2009. A conceptual framework for analysing adaptive capacity and multi-level learning processes in resource governance regimes. *Global Environmental Change* 19:354–365.

Pakeman, R.J. 2011. Functional diversity indices reveal the impacts of land use intensification on plant community assembly. *J. Ecol.* 99:1143–1151.

Parr, C., E. Gray, and W. Bond. 2012. Cascading biodiversity and functional consequences of a global change–induced biome switch. *Divers. Distrib.* 18:493–503.

Pausas, J., and S. Lavorel. 2003. A hierarchical deductive approach for functional types in disturbed ecosystems. *J. Veg. Sci.* 14:409–416.

Perera, A.H., and L. Buse. 2014. *Ecology of wildfire residuals in Boreal Forests.* West Sussex, UK: Wiley Blackwell.

Pérez-Harguindeguy, N., S. Díaz, E. Garnier, et al. 2013. New handbook for standardised measurement of plant functional traits worldwide. *Austral. J. Bot.* 61:167–234.

Persson, L., B.M. Carney Almroth, C.D. Collins, et al. 2022. Outside the safe operating space of the planetary boundary for novel entities. *Environ. Sci. Tech.* 56(3):1510–1521. https://doi.org/10.1021/acs.est.1c04158. hdl:20.500.11850/532277.

Pidgen, K., and A. Mallik. 2013. Ecology of compounding disturbances: The effects of prescribed burning after clearcutting. *Ecosystems* 16:170–181.

Pimm, S., and P. Raven. 2000. Biodiversity. Extinction by numbers. *Nature* 403:843–845.

Preston, F.W. 1962. The canonical distribution of commonness and rarity: Part I. *Ecology* 43:185–215 and 410–432.

Puumalainen, J., P. Kennedy, and S. Folving. 2003. Monitoring forest biodiversity: A European perspective with reference to temperate and boreal forest zone. *J. Environ. Manage.* 67:5–14.

Quijas, S., B. Schmid, and P. Balvanera. 2010. Plant diversity enhances provision of ecosystem services: A new synthesis. *Basic Appl. Ecol.* 11(7):582–593.

Rands, M.R., W.M. Adams, L. Bennun, et al. 2010. Biodiversity conservation: Challenges beyond 2010. *Science* 329:1298–1303.

Ratajczak, Z., J. Nippert, and S. Collins. 2012. Woody encroachment decreases diversity across North American grasslands and savannas. *Ecology* 93(4):697–703.

Reiners, W., K. Driese, T. Fahey, et al. 2012. Effects of three years of regrowth inhibition on the resilience of a clear-cut northern hardwood forest. *Ecosystems* 15(8):1351–1362.

Rekolainen, S., J. Kämäri, M. Hiltunen, T.M. Saloranta. 2003. A conceptual framework for identifying the need and role of models in the implementation of the water framework directive. *Int. J. River Basin Manag.* 1(4):347–352. https://doi.org/10.1080/15715124.2003.9635217.

Richardson, K., W. Steffen, W. Lucht, et al. 2023. Earth beyond six of nine planetary boundaries. *Sci. Adv.* 9(37):eadh2458. https://doi.org/10.1126/sciadv.adh2458.

Rigby, D., D. Howlett, and H. Woodhouse. 2000. *Sustainability indicators for natural resource management and policy: A review of indicators of agricultural and rural livelihood sustainability*. Manchester, UK: University of Manchester.

Rockström, J., J. Gupta, D. Qin, et al. 2023. Safe and just earth system boundaries. *Nature*. 619(7968):102–111. Bibcode:2023Natur.619.102R. https://doi.org/10.1038/s41586-023-06083-8. PMC 10322705. PMID 37258676

Rockström, J., W. Steffen, K. Noone, et al. 2009a. Planetary boundaries: Exploring the safe operating space for humanity. *Ecol. Soc.* 14(2):32.

Rockström, J., W. Steffen, K. Noone, et al. 2009b. A safe operating space for humanity. *Nature* 461:472–475.

Rodrigues, A., and T. Brooks. 2007. Shortcuts for biodiversity conservation planning: the effectiveness of surrogates. *Ann. Rev. Ecol. Evol. Syst*. 38:713–737.

Ruggiero, L.R.S. Holthausen, B.G. Marcot, et al. 1998. Ecological dependency: the concept and its implications for research and management. *Trans. North Am. Wildlife and Natural Resour. Conf.* 53:115–126.

Rykken, J.J., D.E. Capen, and S.P. Mahabir. 1997. Ground beetles as indicators of land type diversity in the Green Mountains of Vermont. *Conserv. Biol.* 11:522–530.

Sætersda, L.M., I. Gjerde, H. Blom, et al. 2003. Vascular plants as a surrogate species group in complementary site selection for bryophytes, macrolichens, spiders, carabids, staphylinids, snails, and wood living polypore fungi in a northern forest. *Biol. Conserv.* 115:21–31.

Safi, K., K. Armour-Marshall, J. Baillie, et al. 2013. Global patterns of evolutionary distinct and globally endangered amphibians and mammals. *PLoS ONE* 8(5): e63582.

Sala, O., F. Chapin III, J. Armesto, et al. 2000. Global biodiversity scenarios for the year 2100. *Science, New Series* 287(5459):1770–1774.

[SCBD] Secretariat of the Convention on Biological Diversity. 2010. *Global biodiversity* Outlook 3. Montréal, PQ. 94 pages.

[SCBD] Secretariat of the Convention on Biological Diversity. 2014. *Global biodiversity* Outlook 4. Montréal, PQ. 155 pages.

Scholes, R.J., and S.R. Archer. 1997. Tree-grass interactions in savannas. *Ann. Rev. Ecol. Syst.* 28:517–544.

Schulze, M., J. Grogan, C. Uhl, et al. 2008. Evaluating ipê (*Tabebuia, Bignoniaceae*) logging in Amazonia: Sustainable management or catalyst for forest degradation? *Biol. Conserv.* 141:2071–2085.

Schuurman, G.W., D.N. Cole, A.E. Cravens, et al. 2021. Navigating ecological transformation: Resist-accept–direct as a path to a new resource management paradigm. *BioScience* 72:16–29. https://doi.org/10.1093/biosci/biab067.

Scott, D. 1999. Managing riparian vegetation to sustain streamflow: Results of paired catchment experiments in South Africa. *Can. J. For. Res.* 29:1149–115.

Searle, E.B. and H.Y.H. Chen. 2017. Persistent and pervasive compositional shifts of western boreal forest plots in Canada. *Glob. Chang. Biol.* 23:857–866. https://doi.org/10.1111/gcb.13420.

Shanley, P. and L. Luz. 2003. The impacts of forest degradation on medicinal plant use and implications for health care in eastern Amazonia. *BioScience* 53(6):573–584.

Simberloff, D. 1998. Flagships, umbrellas, and keystones: Is single-species management passe in the landscape era? *Biol. Conserv.* 83:247–257.

Siqueira, T., L.M. Bini, F.O. Roque, et al. 2012. A metacommunity framework for enhancing the effectiveness of biological monitoring strategies. *PLOS ONE* 7(8):e43626.

Skole, D., and C. Tucker. 1993. Tropical deforestation and habitat fragmentation in the Amazon: Satellite data from 1978 to 1988. *Science* 260:1905–1910.

Smeets, E. and R. Weterings. 1999. Environmental indicators: Typology and overview. Copenhagen, Denmark. *Eur. Environ. Agen., Tech. Rep.* 25.

Smith, S.J., J. Edmonds, C.A. Hartin, et al. 2015. Near-term acceleration in the rate of temperature change. *Nature Climate Change.* https://doi.org/10.1038/nclimate2552.

Song, B., J. Chen, P. Desanker, et al. 1997. Modeling canopy structure and herterogeneity across scales: Fron crowns to canopy. *For. Ecol. Manage.* 96:217–229.

Sparks, T.H., S.H. Butchart, A. Balmford, et al. 2011. Linked indicator sets for addressing biodiversity loss. Fauna Flora Int. *Oryx* 45(3):411–419.

Sparrow, B.D., W. Edwards, S.E.M. Munroe, et al. 2020. Effective ecosystem monitoring requires a multi-scaled approach. *Biol. Rev.* 95:1706–1719.

Spies, T. 1997. Forest stand structure, composition, and function. In *Creating a forestry for the 21st century. The science of forest management*, eds. K.A. Kohm and J.F. Franklin, pp. 11–30. Washington, DC: Island Press.

Steffen, W., K. Richardson, J. Rockström, et al. 2015. Planetary boundaries: Guiding human development on a changing planet. *Science* 347:1259855.

Steffen, W., K. Richardson, J. Rockström, et al. 2020.The emergence and evolution of Earth system science. *Nat. Rev. Earth Environ.* 1:54–63.

Stein BA, P. Glick, N. Edelson, and A. Staudt. 2014. *Climate-smart conservation: Putting adaption principles into practice.* Washington, DC: National Wildlife Federation.

Struhsaker, T.T. 1975. *The red colobus monkey*, 311 pp. Chicago: University of Chicago Press.

Svarstad, H, L.K. Petersen, D. Rothman, H. Siepel, and F. Wätzold, 2008. Discursive biases of the environmental research framework DPSIR. *Land Use Policy* 25(1):116–125.

Swift, K.I. and F.W. Bell. 2011. What are the environmental consequences of using silviculturally effective vegetation management alternatives? *For. Chron.* 87(2):201–216.

TEEB. 2009. *TEEB climate issues update.* Nairobi: UNEP.

Tews, J., U. Brose, V. Grimm, et al. 2004. Animal species diversity driven by habitat heterogeneity/diversity: the importance of keystone structures. *J. Biogeogr.* 31:79–92.

Thompson, I. 2006. Monitoring of biodiversity indicators in boreal forests: A need for improved focus. *Environ. Mon. Assess.* 121:263–273.

Thompson, I., B. Mackey, S. McNulty, et al. 2009. *Forest resilience, biodiversity, and climate change. A synthesis of the biodiversity/resilience/stability relationship in forest ecosystems.* Montreal, PQ: Secretariat of the Convention on Biological Diversity.

Thompson, I.D., K. Okabe, J.M. Tylianakis, et al. 2011. The role of forest biodiversity in delivery of ecosystem goods and services: translating science into policy. *BioScience* 61:972–981.

Thompson, I.D., J. Ferreira, T. Gardner, et al. 2012. Forest biodiversity, carbon and other ecosystem services: relationships and impacts of deforestation and forest degradation. In. *Understanding relationships between biodiversity, carbon, forests and people: The key to achieving Redd+ objectives. A global assessment report,* eds. J.A. Parrotta, C. Wildburger, and S. Mansourian, 161 p. Prepared by the Global Forest Expert Panel on Biodiversity, Forest Management, and REDD+. IUFRO World Series Volume 31. Vienna.

Thompson, I.D., M.R. Guariguata, K. Okabe, et al. 2013. An operational framework for defining and monitoring forest degradation. *Ecol. Soc.* 18(2):20.

Thompson, L.M., A.J. Lynch, E.A. Beever, et al. 2021. Responding to ecosystem transformation: Resist, accept, or direct? *Fisheries* 46:8–21.

Tittensor, D., M. Walpole, S. Hill, et al. 2014. A mid-term analysis of progress toward international biodiversity targets. *Science* 346(6206):241–244.

TRAFFIC. 2012. *Precious woods: Exploitation of the finest timber.* Chatham House Workshop: Tackling the trade in illegal precious wood. 23–24 April 2012. Background Paper 1 (pp. 1–62). Chatham House Workshop: Tackling the trade in illegal precious wood. 23–24 April 2012. Background Paper 1.

Turner, M., S. Pearson, P. Bolstad, et al. 2003. Effects of land-cover change on spatial pattern of forest communities in the Southern Appalachian Mountains (USA). *Landscape Ecol.* 18:449–464.

UNEP. 2002. *Report on the sixth meeting of the Conference of the Parties to the Convention on Biological Diversity.* (UNEP/CBD/COP/20/Part 2) Strategic Plan Decision VI/26.

UNEP. 2011. *Report on the tenth meeting of the Conference of the Parties to the Convention on Biological Diversity*. UNEP/CBD/COP/10/27.

UNEP. 2013. *United Nations environmental programme annual report 2012*. Nairobi: United Nations Environment Programme (UNEP).

United Nations Conference on Environment and Development. 1992. *Agenda 21, Rio declaration, forest principles.* New York, NY: United Nations.

Vanderwalle, M., F. de Bello, M.P. Berg, et al. 2010 Functional traits as indicators of biodiversity response to land use changes across ecosystems and organisms. *Biodiv. Conserv.* 19(10):2921–2947.

van Herk, C.M., E.A.M. Mathijssen-Spiekman, and D. Zwart. 2003. Long distance nitrogen air pollution effects on lichens in Europe. *Lichenologist* 35:347–359.

van Ruijven, J. and F. Berendse. 2010. Diversity enhances community recovery, but not resistance, after drought. *J. Ecol.* 98(1):81–86.

Ventor, O., N.N. Brodeur, L. Nemiroff, et al. 2006. Threats to endangered species in Canada. *BioScience* 56(11):903–910.

Vessby, K., B. Söderström, A. Glimskär, et al. 2002. Species-richness correlations of six different taxa in Swedish seminatural grasslands. *Conserv. Biol.* 16:430–439.

Vié, J.-C., C. Hilton-Taylor, C. Pollock, et al. 2008. The IUCN red list: A key conservation tool. In *The 2008 review of the IUCN Red List of Threatened Species*, eds. J.-C. Vié, C. Hilton-Taylor, and S.N. Stuart. Gland, Switzerland: IUCN.

Vié, J.-C., C. Hilton-Taylor, and S. Stuart. 2009. *Wildlife in a changing world—An analysis of the 2008 IUCN Red List of Threatened Species.* Gland, Switzerland: IUCN.

Violle, C., M.-L. Navas, D. Vile, et al. 2007. Let the concept of trait be functional! *Oikos* 116:882–892.

Vitousek, P., H. Mooney, J. Lubchenco, et al. 1997. Human domination of Earth's ecosystems. *Science* 277:494–499.

von Gadow, K., T. Pukkala, and M. Tomé. 2000. *Sustainable forest management: Managing forest ecosystems.* Dordrecht, The Netherlands: Kluwer Academic Publishers.

Walker, B., A. Kinzig, and J. Langridge. 1999. Plant attribute diversity, resilience and ecosystem function: the nature and significance of dominant and minor species. *Ecosystems* 2:95–113.

Walters, C.J. and R. Hilborn. 1978. Ecological optimization and adaptive management. *Ann. Rev. Ecol. Syst.* 9:157–188.

Walters C.J. and C.S. Holling. 1990. Large-scale management experiments and learning by doing. *Ecology* 71:2060–2068.

Walz, U. 2011. Landscape structure, landscape metrics and biodiversity. *Living Rev. Landscape Res.* 5:3. www.livingreviews.org/lrlr-2011-3.

Walz, C. and C. Stein. 2014. Indicators of hemeroby for the monitoring of landscapes in Germany. *J. Nat. Conserv.* 22:279–289.

Wei, J., Z. Yountao, X. Houqin, et al. 2007. A framework for selecting indicators to assess the sustainable development of the natural heritage site. *J. Mountain Sci.* 4:321–330.

Werner, F.A. and U. Gallo-Orsi. 2016. *Biodiversity monitoring for natural resource management. An introductory manual.* Eschborn and Bonn, Germany: GIZ.

Whittaker, R., S. Levin, and R. Root. 1973. Niche, habitat, and ecotope. *Am. Nat.* 107:321–338.

Wiens J.A., G.D. Hayward, H.D. Safford, et al., eds. 2012. *Historical environmental variation in conservation and natural resource management.* Chichester: Wiley-Blackwell.

Wilcove, D.S., D. Rothstein, J. Dubow, et al. 1998. Quantifying threats to imperiled species. *BioScience* 48:607–615.

Wilcox, B.A., and B. Ellis. 2006. Forests and emerging infectious diseases of humans. Food and Agriculture Organization of the United Nations. *Unasylva* 224(57): online.

Wilcox, B.P., M.K. Owens, R.W. Knight, et al. 2005. Do woody plants affect stream-flow on semiarid karst rangelands? *Ecol. Appl.* 15:127–136.

Williams, B.K. and E.D. Brown. 2018. Double-loop learning in adaptive management: The need, the challenge, and the opportunity. *Environ. Manag.* 62:995–1006.

Williams, B.K., R.C. Szaro, C.D. Shapiro, eds. 2007. *Adaptive management: The US department of the interior technical guide.* US Department of the Interior.

Williams, P., D. Faith, L. Manne, et al. 2006. Complementarity analysis: Mapping the performance of surrogates for biodiversity. *Biol. Conserv.* 128:253–264.

Williams, P.H., C.R. Margules, and D.W. Wilbert. 2002. Data requirements and data sources for biodiversity priority area selection. *J. Biosci.* 27:327–338.

Winter, M., O. Schweiger, S. Klotz, et al. 2009. Plant extinctions and introductions lead to phylogenetic and taxonomic homogenization of the European flora. *Proc. Natl. Acad. Sci. U.S.A.* 106:21721–21725.

Woodwell, G., R. Houghton, T. Stone, et al. 1987. Deforestation in the tropics: new measurements in the Amazon basin using Landsat and NOAA advanced very high resolution radiometer imagery. *J. Geophysical Res.* 92:2157–2163.

Wrbka, T., K.H. Erb, N.B. Schulz, et al. 2004. Linking pattern and process in cultural landscapes. An empirical study based on spatially explicit indicators. *Land Use Policy* 21:289–306.

Xiankai, L., M. Jiangming, and D. Shaofeng. 2008. Effects of nitrogen deposition on forest biodiversity. *Acta Ecologica Sinica* 28(11):5532–5548.

Zhu, M., N. Jiang, J. Li, et al. 2006. The effects of sensor spatial resolution and changing grain size on fragmentation indices in urban landscape. *Int. J. Remote Sens.* 27:4791–4805.

Zou, C.B., D. Breshears, B. Newman, et al. 2008. Soil water dynamics under low- versus high-ponderosa pine tree density: Ecohydrological functioning and restoration implications. *Ecohydrology* 1:309–315.

13 Indicators of Forest Ecosystem Integrity

André Arsenault and Guy R. Larocque

There is nothing wrong with change if it is in the right direction.

—Winston Churchill

13.1 INTRODUCTION

Many countries in the world have developed policies and criteria and indicators to promote sustainable forest management (SFM) (Brand 1997). Although this term will mean different things to different people, it is generally accepted that SFM can be achieved only by using ecosystem-based management (Grumbine 1994), with the ultimate goal of maintaining forest ecosystem integrity over time so that future generations can benefit from the same goods and services currently and historically provided by forest landscapes. In some cases, where ecosystems have lost their integrity, measures can be put in place for restoration (DellaSala et al. 2003), but this is a risky proposition associated with many myths (Hilderbrand et al. 2005). Forest ecosystem integrity can be defined as a state in which forest ecosystems have enough resilience to "maintain ecosystem structure and function" (Dorren et al. 2004), unimpaired and sound (De Leo and Levin 1997; Tierney et al. 2009). De Leo and Levin (1997) emphasize that the idea of integrity "must recognize a human perspective, the ability of ecosystems to continue to provide services that humans expect." Carignan and Villard (2002) use a definition from Karr and Dudley (1981): ecological integrity is "The capacity of an ecosystem to support and maintain a balanced, integrated, adaptive community of organisms having a species composition, diversity, and functional organization comparable to that of similar, undisturbed ecosystems in the region." Ecological integrity is considered useful to assess ecologically based management goals (Tierney et al. 2009). The concepts of resilience and stability are pivotal in understanding the dynamics of ecosystems and their ability to keep their ecological integrity in a changing world. As forest ecosystems are dynamic and their stability and integrity cannot be determined using a single point in time, a range of natural variability should be examined (Landres et al. 1999). Therefore, to assess if forest management is sustainable, we must develop and measure ecological indicators of ecosystem integrity over time. Although this seems a relatively effortless task, it is convoluted by the simple fact that ecosystems are complex. As in the famous line by Bunnell (1999), a forest ecologist at the University of British Columbia, "Forest ecology is not rocket science, it is much more complex."

The use of environmental indicators to track forest ecosystem changes over time is appealing because it has clear application potential, and the alternative to monitor every species and every aspect of ecosystems is not realistic. Environmental indicators have been used for a long time by different societies. First Nations used phenological indicators, such as the flowering of a certain species announcing the arrival of a valuable resource species. Lantz and Turner (2003) found examples of 140 phenological indicators in 20 different linguistic groups during a review of ethnographic works on British Columbia's First Nations. The authors gave many interesting examples like this one from the Nlaka'pamux people, who used the flowering of the wild rose as an indicator that soapberries were ready to harvest. Environmental indicators have also been used as early warning systems. One of the most famous examples is the use of canaries as animal sentinels in British coal

DOI: 10.1201/9781003431084-16

FIGURE 13.1 The Haldane canary cage developed by John Haldane. If the canary collapsed, the miners would ensure their personal safety and seal the door and fill the cage with oxygen from the above tank to revive the canary. This early warning system was used in British coal mines from 1911 to 1987. (From www.healeyhero.co.uk/rescue/individual/Bob_Bradley/Email/Email-E.html.)

mines from 1911 to 1987. The canary is 20 times more sensitive to toxic gases than humans. John Scott Haldane, described as the first environmental physiologist, thus came up with the idea of using canaries in cages in coal mines to warn miners if their breathing environment became hazardous (Figure 13.1) (Brubakk and Lang 2009). The purpose of this chapter is to review basic concepts of forest ecosystem integrity and identify challenges in using ecological indicators to assess it.

13.2 STABILITY AND RESILIENCE

Discussions about stability and resilience have been taking place for a long time in the scientific literature in ecology. Holling (1973) distinguished both terms and suggested that they should be used together in the characterization of dynamic pathways. Stability is the capacity of an ecosystem to recover following a temporary disturbance and come back to an equilibrium state, whereas resilience is a measure of the endurance (resistance) of an ecosystem to "absorb" a disturbance and maintain similar state characteristics (identity, structure, composition, and function). Resilience is considered an emergent property of complex adaptive systems (Puettmann et al. 2009). One of the best approaches to foster resilience in managed forest landscapes is to maintain complexity. Recent advances in complex system science provide a promising approach to analyze and manage forests as integrated social–ecological systems (Filotas et al. 2014).

Ecosystem resilience can be quantified by the magnitude of disturbance an ecosystem can absorb without undergoing significant transformation leading to state change (Walker et al. 2004; Drever et al. 2006). Ecosystems that have lost considerable resilience are more susceptible to changing into an alternative steady state following stress and disturbance (Scheffer et al. 2001). For example, a forest that has lost resilience due to overgrazing may be susceptible to being transformed into grassland following a major windrow event. The estimation of ecosystem resilience needs to take into consideration the complex organization of ecosystems at multiple scales. It is therefore critical to specify the ecosystem state and types of disturbances being considered, as well as spatial and temporal scales of interest (Peterson et al. 1998; Carpenter et al. 2001; Walker et al. 2004; Drever et al. 2006). For instance, species turnover rates are directly linked to the time factor, and the successional phase may or may not affect the state of ecosystems, depending on the type and importance of disturbances (Drever et al. 2006). Normal changes in species composition that are part of the natural successional pathway do not necessarily affect the state of a forest ecosystem. On the other hand, succession occurrences that result in the conversion of forests into grasslands or swamps clearly result in a new ecosystem state.

The concepts of stability and resilience are important for the application of the basic principles of ecological forest management. Forest management practices that enhance the degree of uniformity in the structure and species composition of landscapes may result in increasing the risks of disturbances, such as fire or insect infestations (Franklin and Forman 1987; Turner and Romme 1994; Bergeron et al. 1998; Urrego et al. 2013), and consequently affect the resilience potential of forest ecosystems (Proença et al. 2010; Seidl et al. 2014). It has long been recognized that the degree of stability and resilience of forest ecosystems increases with the degree of complexity in their structure and species composition or biodiversity (see Drever et al. 2006; Thompson et al. 2009; Mori et al. 2013; Potter and Woodall 2014). Forest managers have the option to consider guidelines that will assist them in the decision-making process to ensure that their management decisions preserve the resilience capacity of the forests under their responsibility. Drever et al. (2006) suggested several recommendations at the ecosystem, landscape, and stand scales that contribute to maintaining ecological resilience in the context of a management approach based on the analysis of natural disturbances. In summary, the recommendations focus on the importance of basic knowledge of ecosystem properties (species composition, structure, and heterogeneity), understanding and prediction of the potential effects of disturbances, and diversification of silvicultural practices.

To decide whether a system is stable or not, we must first specify what we mean by a change in configuration or loss of integrity (Ludwig et al. 1997). Here, two worldviews of ecosystem dynamics collide. The balance of nature worldview considers that an ecosystem has resilience if it can return to equilibrium following disturbance (Pimm 1991). A radically different perspective is that most forest ecosystems are open systems with nonequilibrium dynamics (Holling 1973; May 1977). This worldview depicts forests as complex adaptive systems that are likely to have multiple steady states resulting from self-organizing abilities to adapt to natural disturbances (Wallington et al. 2005). This illustrates the importance of understanding the natural disturbance regime for assessing ecological resilience and forest integrity (see Section 13.4). The natural disturbance regime of the boreal forest of eastern continental Canada includes two key disturbance types, fire and spruce budworm outbreaks, that lead to ecosystem state changes. In this case, although a specific natural disturbance event will lead to local state change, the landscape will retain resilience and stability, illustrated by a mosaic of different ecosystem states resulting from different disturbances and different times since disturbance, provided that the disturbance regime stays within the natural range of variability.

The most extreme loss of forest ecosystem resilience and integrity results from deforestation leading to ecosystem degradation. This should not be confused with logging because many logged forests will regenerate to form young forests either naturally or through artificial regeneration. The degree to which a logged forest has lost resilience depends on how ecological functions and habitat have been compromised. For example, if logging transforms a subalpine forest into a meadow permanently, the ecosystem has lost a considerable amount of resilience.

Synergistic effects of different disturbance agents acting together can have cumulative effects on ecological resilience. For example, the combination of logging and introduced herbivores can have devastating effects on forest regeneration (Barrette et al. 2014). From a biodiversity perspective, the loss of species resulting from a combination of overexploitation, introduced species, habitat loss, and chains of linked extinctions (Diamond 1984) would represent an extreme loss of ecological resilience. Ecological indicators of ecosystem integrity can help managers guide their action to prevent loss of resilience before the fact or assist them in identifying positive interventions to increase resilience after the fact. Although there are many successful examples of afforestation throughout the world (Laganière et al. 2010), we must remain cautious in raising expectations that all degraded forests can be "restored" (see Hilderbrand et al. 2005 on the myths of restoration ecology). Some unique features associated with old trees and old forests cannot be easily restored and may require many centuries to return to a predisturbance state and thus potentially result in temporary extinctions (Lindenmayer et al. 2014). The importance of selecting key ecological indicators is critical to assess the success of a restoration effort. Another important consideration is the temporal and spatial scales for the different indicators (see Section 13.5).

13.3 DRIVERS OF SUSTAINABLE FOREST MANAGEMENT

A knowledge of the history of SFM and its link with ecological indicators is critical to understand how and why these are used today.

13.3.1 External Drivers and Indicators

Although humans are clearly dependent on forests, concern over their impact on forests at an international level is a relatively recent phenomenon. From the light touch of primitive humans to the contemporary conversion of forests through agriculture and other developments and to the more permanent deforestation as a result of urbanization, the speed and spatial extent of change in the world's forest is impressive. The magnitude of the change in the world's population over the last 1,000 years and the forecasted rate of increase to 2100 (Figure 13.2), and the associated hunger for natural resources (Ehrlich 1968; Kimmins 1987) have created pressures on forest ecosystems that will probably increase until 2100. The world's population can be considered as an ecological indicator of stress on the world's forests. The tension between the development of natural resources and their conservation creates an opportunity for a dialog on sustainable forest development. One key step for

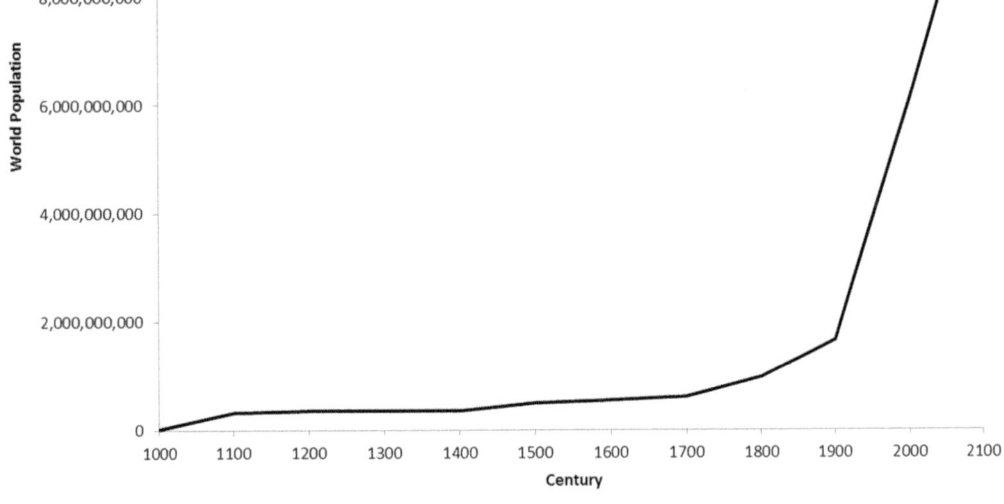

FIGURE 13.2 Estimated changes in world population between the beginning of the second millennium and 2100. (From multiple sources.)

a productive discussion is to develop tools and create an environment that leads toward an informed consensus. Ecological indicators represent a powerful way to communicate about our impact on forest ecosystems and find innovative ways to adapt management practices for optimal SFM.

13.3.2 Leadership

Another important driver of change from unregulated forest management to SFM is the result of the actions of strong leaders and critical thinkers. The last century has witnessed several generations of remarkable leaders who have contributed significantly toward the conservation of forests. One of the first leaders was Gifford Pinchot, the first chief of the USDA Forest Service. He is famous for the phrase "Without natural resources life itself is impossible" (Forest History Society 2015). Good leaders are driven individuals with an ability to see the big picture, solve problems, and seize opportunity when it presents itself. Pinchot certainly fits this description. In the late 1800s, there was no school of forestry in the United States. After graduating from Yale, he went to study in France at the L'École Nationale Forestière in Nancy. Upon his return to the United States, he made the following observation: "When I got home at the end of 1890 . . . the nation was obsessed by a fury of development. The American Colossus was fiercely intent on appropriating and exploiting the riches of the richest of all continents." Over the next 20 years, Pinchot led a nationwide movement to improve forest management and ecological conservation. He developed the USDA Forest Service and served as its first chief forester. The famous U.S. President, Theodore Roosevelt, remarked about Pinchot "among the many, many public officials who under my administration rendered literally invaluable service to the people of the United States, Gifford Pinchot on the whole, stood first" (USDA Forest Service 2015).

Another strong leader of environmental conservation is Aldo Leopold, who started his career with the USDA Forest Service in Arizona. He later became professor of game management at the University of Wisconsin. Leopold is famous for his concept of a land ethic published in his book *A Sand County Almanac* in 1949. "That land is a community is the basic concept of ecology, but that land is to be loved and respected is an extension of ethics" (Leopold 1949). His clear thinking still influences contemporary ideas about biological conservation. For example, panarchy and the nature-deficit disorder very much have their roots in the land ethic. More recently, the Brundtland Commission in 1987 achieved the first world-level conservation ethic "Our common future sustainable development," which was later followed by the Rio Earth Summit in 1992. This world summit was a turning point for how humanity defines its interaction with the planet (http://www.montrealprocess.org/The_Montreal_Process/About_Us/history. shtml). It became clear that to adapt management practices to conserve forest values, it was critical to monitor and understand changes in forests. The Government of Canada initiated the Montréal Process and hosted a key meeting in Montréal in 1993 to develop science-based criteria and indicators to assess forest management. European countries had a similar initiative referred to as the Helsinki Process. The Montréal Process was renamed the Working Group on Criteria and Indicators for the Conservation and Sustainable Management of Temperate and Boreal Forests and officially launched in 1994. It includes members of 12 countries. This was the most comprehensive list of criteria and indicators for SFM and included both biophysical values and socioeconomic values. The seven overarching criteria, under which are nested more than 50 indicators, are listed as follows and are discussed in detail in Chapter 14.

(1) Conservation of biological diversity
(2) Maintenance of productive capacity of forest ecosystems
(3) Maintenance of forest ecosystem health and vitality
(4) Conservation of soil and water resources
(5) Maintenance of forest contribution to global carbon cycles

(6) Maintenance and enhancement of long-term multiple socioeconomic benefits to meet the needs of society
(7) Legal, institutional, and economic framework for forest conservation and sustainable management

13.3.3 Environmental Movement, Markets, and Certification

Other important drivers of change also included actions from the environmental movement. In Canada, several protests like the road blockade over the coastal temperate rainforests on Meares Island and Clayoquot Sound (Figure 13.3) on Vancouver Island brought the issue of SFM to the forefront of public policy debates. A documentary film over the management of Québec's boreal forest, *L'erreur boréale* by singer/poet Richard Desjardins, transformed public perception about SFM in Québec overnight. All these actions transformed society and had influence on forest management legislation and policy. In the Pacific Northwest, protests and court challenges over old-growth forests and endangered species, such as the northern spotted owl (*Strix occidentalis caurina*), were also transformative. This led to the Northwest Forest Plan, which was the most significant shift in management focus of federal forests since their creation (Thomas et al. 2006). Similar protests and debates have occurred over the deforestation of tropical rainforests of the Amazon and Indonesia and old-growth forests of Russia. Some environmental groups targeted large companies, consumers of forest products, to have an effect on the forest sector markets, thereby hoping to exert an influence on forest practices (Luckert et al. 2011). To protect their markets, many forest companies started getting certified to different levels of SFM standards, which are audited by third parties. This has the advantage of selling wood products with different types of SFM labels, similar to the organic labels in agriculture. All these certification initiatives have criteria and indicators that are monitored by third parties to assess whether the companies are meeting the certification standards. Many of these criteria and indicators have evolved from the Montréal Process. The tension between environmental groups and the forest industry persists due to different values and visions about what SFM really means, leading to different agendas. However, there are interesting developments under the Canadian Boreal Forest Agreement, where both industry and ENGOS are working together toward a consensus for SFM in Canada's boreal forests. The Canadian Boreal Forest Agreement has six goals and also builds on the Montréal Process:

FIGURE 13.3 The Clayoquot Sound blockade in 1993 on Vancouver Island was one of the largest acts of civil disobedience in Canadian history. (From https://commons.wikimedia.org/wiki/File:Blocking_the_Logging_Road2.jpg.)

(1) Implement world-leading SFM practices
(2) Accelerate the completion of the protected spaces network for the boreal forest
(3) Fast-track plans to protect boreal forest species at risk, particularly woodland caribou
(4) Take action on climate change as it relates to forest conservation
(5) Improve the prosperity of the Canadian forest sector and communities that rely on it
(6) Promote and publicize the environmental performance of the participating companies

13.4 FOREST CHANGE AND NATURAL DISTURBANCE

Forest scenes (Figure 13.4) appear permanent and unchanging features, but the opposite is true. Forest ecosystems are very dynamic and change on a regular basis as a result of natural and anthropogenic disturbances, variation in climate, and interaction between species and between species and their environment. These changes operate at different spatial and temporal scales (Figure 13.5). Large fires in the boreal forests of Canada can extend over thousands of hectares, killing most trees in their path, whereas surface fires in dry forests can be relatively small and nonlethal (Figure 13.6). Understanding the natural disturbance regime and its effect on forest ecosystems is essential to assess the impact of anthropogenic disturbances on ecosystem integrity. In fact, metrics of disturbance regime can be used as indicators of ecosystem integrity (see the next section).

13.4.1 Characteristics of Natural Disturbances: Examples from British Columbia, Canada

13.4.1.1 Natural Range of Variation and Forest Ecosystem Integrity

The *Biodiversity Guidebook* (see BC Ministry of Forests and BC Environment 1995) has stratified the province into five coarse-level disturbance regimes (Figure 13.7). This broad classification has been very useful in providing a strategic direction for reducing the overall impact of timber harvesting on biodiversity and for determining strategic targets for maintaining old-growth forests. The habitats grouped as NDT4 are distinct from the other natural disturbance types (NDT) in the province, and fire has certainly been an essential process in shaping ecosystem characteristics. However, we argue that it does not necessarily follow that these ecosystems

FIGURE 13.4 A forest scene from the Canadian Rockies, Alberta. (Photo from André Arsenault.)

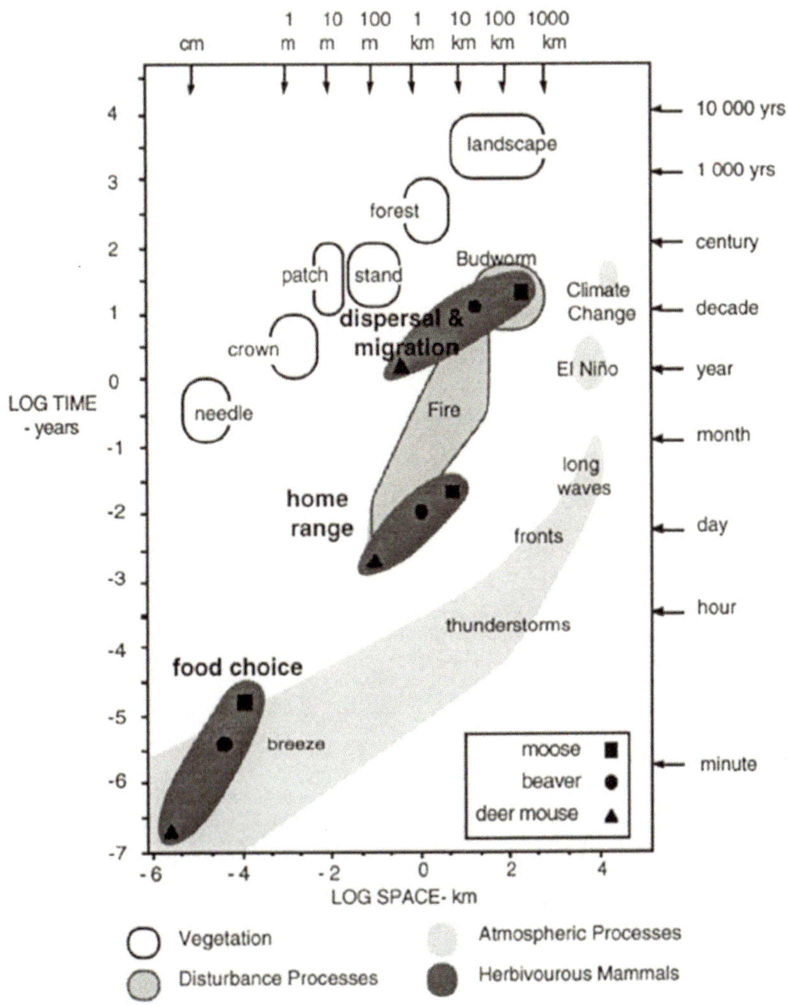

FIGURE 13.5 A stromel diagram showing natural disturbances and other ecological processes occurring over multiple temporal and spatial scales. (From Peterson et al., Ecosystems, 1, 6, 1998.)

(a) (b)

FIGURE 13.6 Examples of disturbances operating at different spatial scales. (a) A low severity fire in a Ponderosa pine forest in British Columbia's southern interior, less than 500 ha, and (b) a massive crown fire in the boreal forest (right). The large fires can exceed 100,000 ha.

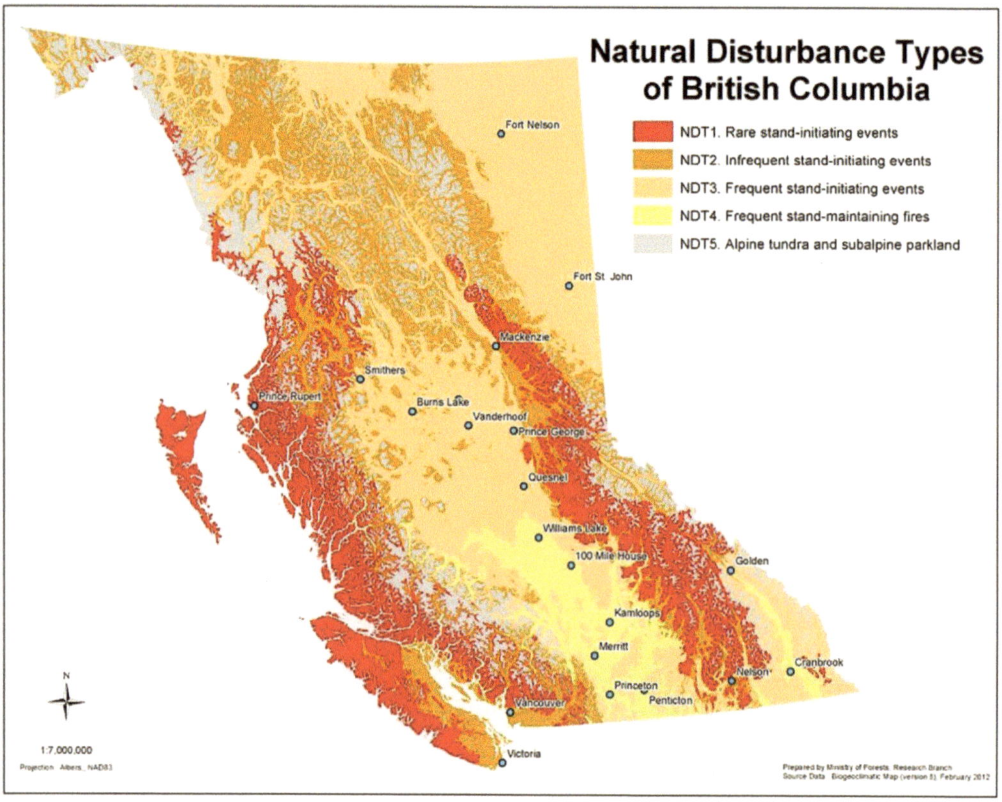

FIGURE 13.7 Map of NDTs in British Columbia. (From BC Ministry of Forests and BC Environment, Biodiversity guidebook—Forest practices code of British Columbia, 1995.)

are "fire maintained." A disturbance regime can be described as the general role disturbances play in a given ecosystem. Disturbances are often described using the characteristics outlined in Table 13.1.

13.4.1.2 Why Use "Natural" Disturbance Types for Conservation Planning?

The province of British Columbia developed many ecosystem management tools associated with the Forest Practices Code in the mid-1990s. One of these was the biodiversity guidebook, which used an NDT of classification for coarse- and fine-filter conservation planning (BC Ministry of Forests and BC Environment 1995). The ecological principle behind it is that the closer forest management resembles natural disturbances, the more chances that species that evolved under these "natural" conditions will persist in managed landscapes. It follows the same logic as to try to maintain managed landscapes within the range of "natural" variability (Figure 13.7).

13.4.1.3 Pros and Cons of Using Natural Disturbance Types

Although the intent of using natural disturbances and range of natural variability to guide natural resource management is good, challenging questions emerge during the implementation of the approach, such as the following: What is "naturalness" or "wilderness" (Sprugel 1991)? Or what is the ideal benchmark period to quantify the natural range of variability? 1650, 1851, 1890? This approach is also complicated by the fact that historical information is rarely complete and thus is subject to some level of interpretation (Arsenault and Klenner 2001). Furthermore, it is becoming very clear from most studies of disturbance and forest dynamics that landscapes are not in a steady

TABLE 13.1
Characteristics Used in Defining Disturbance Regimes

Descriptor	Definition
Type	The cause of disturbance (in our dry-belt forests, the major types are fire, defoliators, bark beetles, and pathogens)
Frequency	The level of recurrence of disturbance events
Severity	The amount of biomass removed by disturbance
Intensity	Physical strength of the disturbance event, for example, wind speed
Distribution	The spatial and temporal distribution of disturbance events (e.g., in dry-belt forests, the vast majority of fire events are generated in the mid-summer months)
Extent	Size of disturbance events
Synergism	The accrued effect of two or more different disturbance agent or events (e.g., a severe surface burn may damage roots and predisposes trees to windthrow)

Source: Adapted from Pickett, S.T.A. and White, P.S., eds., *The Ecology of Natural Disturbance and Patch Dynamics*, Academic Press, San Diego, CA, 1985; Agee, J.K., *Fire Ecology of Pacific Northwest Forests*, Island Press, Washington, DC, 1993.

state but are governed by nonequilibrium dynamics (Cumming et al. 1996; Mori 2011). This fact makes it difficult to use standard disturbance metrics, such as frequency and size, because they vary in time and space. Although the argument of using natural disturbance as a template to create an environment similar to what species that have evolved in is logical, it is not without flaws because a chainsaw and a skidder will never be exactly like a fire or an insect outbreak. However, it does make sense to have more old-growth forest retention in coastal temperate rainforests and inland rainforests that have very infrequent disturbances, compared with lodgepole pine (*Pinus contorta*) forests that are affected more regularly by severe disturbances (Figures 13.8 and 13.9). One of the challenges faced by the authors of the biodiversity guidebook was that very few peer-reviewed and reliable studies on natural disturbance regimes existed from British Columbia. So the guidebook relied heavily on papers from the United States, unpublished reports, and expert opinion. This turned out to be a serious problem, especially for dry forests that are at the northern extent of their range in British Columbia and have a very complex disturbance regime (Klenner et al. 2008). The other challenge the authors faced was to have to make a decision about when a forest attains old-growth status (Hunter 1989). This is not as simple as it may seem. Forests rarely follow the classic succession trajectories presented in first-year biology textbooks. Furthermore, the reconstruction of historical disturbance regimes is not a trivial task as there are many methodological constraints (Arsenault and Klenner 2001). For example, the majority of fire-history reconstruction studies in boreal and oro-boreal regions of Canada made a serious error by using reverse cumulative standing age distributions in fire-frequency analysis and often violated the assumption of stationarity (Huggard and Arsenault 1999, 2001; see also Bergeron and Fenton 2012).

13.5 SELECTING AND MONITORING INDICATORS TO ASSESS FOREST ECOSYSTEM INTEGRITY: SUCCESSES AND CHALLENGES

13.5.1 Characteristics of a Robust and Efficient Environmental Monitoring Program

The selection of indicators to assess forest ecosystem integrity in managed forest landscapes implies that an effective adaptive environmental monitoring program is required. However, Lovett et al. (2007) in their thought-provoking piece titled "Who Needs Environmental Monitoring?" noted that such programs have received a lot of criticisms, such as "they represent fishing expeditions not science" and "most monitoring data are never used," giving the discipline of environmental science a bad reputation

FIGURE 13.8 Examples of different disturbance types in Canada's dry forests.

(a) (b)

FIGURE 13.9 An example of an NDT3 (a) (lodgepole pine forest) and of NDT1 (b) (cedar–hemlock forests). The lodgepole pine forest originated from a severe wildfire in 1869 and suffered high mortality in 2006 following a mountain pine beetle outbreak in the Arrowstone watershed, southern interior of British Columbia. The cedar–hemlock forest in the Capilano watershed near Vancouver, British Columbia, has not seen a stand-replacing disturbance for over 1,000 years.

(Lovett et al. 2007). They present seven key characteristics of an efficient monitoring program: (1) center it on clear and compelling scientific questions; (2) include review, feedback, and adaptation in the design; (3) choose measurements carefully and with the future in mind; (4) maintain quality and consistency of the data; (5) plan for long-term data accessibility and sample archiving; (6) continually examine, interpret, and present the monitoring data; and (7) include monitoring within an integrated research program. Lindenmayer and Likens (2009b) further noted that many long-term monitoring and research programs are ineffective or fail completely due to a lack of focus and poor planning.

Two important problems with monitoring systems is that they are often driven by short-term funding opportunities and by an urgency to collect the data now and wait to develop a plan about how to use it (Roberts 1991). Lindenmayer and Likens (2009a, 2009b) state that two examples of this are the Alberta Biodiversity Monitoring Institute (ABMI) program and the Program for Planned Biodiversity and Ecosystem Research (PPBio) for biodiversity monitoring in southeastern Queensland, Australia. A description of the ABMI program and how it differed from traditional biodiversity monitoring and research was published by Boutin et al. (2009). The authors admitted that their approach was controversial but innovative. The strength of this approach is that "it is taxonomically broad and cumulative-effects oriented" (Haughland et al. 2010). However, Lindenmayer and Likens (2009a, 2009b) insist that a strong and efficient environmental monitoring program "must have well-developed and have readily tractable scientific questions, a conceptual model of the ecosystem(s) in question," such as the model used for the Hubbard Brooks study and a robust and statistically based experimental design and be driven by a human need to know about an ecosystem so that they pass the test of management relevance. Both camps in this academic dispute make excellent points and perhaps could be used to develop an optimal adaptive environmental monitoring system. Burton et al. (2014) make a strong case that both surveillance and targeted monitoring with different tasks are necessary to assess the cumulative effects across large spatial scales by ABMI (Figure 13.10). Although this makes good sense, it is not clear how the rigorous monitoring approach by ABMI is linked in an adaptive management system. We suggest that a good monitoring program is very similar to a good research project. The main difference is that the information from a monitoring program has immediate implications about ecosystem assessment and determining appropriate actions when necessary. Monitoring cannot answer all questions and should be closely linked with well-defined research projects to test assumptions. Klenner (2011) proposed that the main objective of a monitoring program should be to apply "corrective measures" when ecological indicators detect undesirable trends or conditions. He compared this with a medical analogy where the goal of monitoring a patient during surgery is to monitor physiological indicators to be ready to intervene if an undesirable situation develops. A critical ingredient for this to work is to identify acceptable ranges for the given indicators. This is much easier with a human patient when compared with complex forest landscapes. Klenner argues that monitoring ecological indicators without having clear predetermined thresholds that would trigger corrective actions is of "little use," which is in line with the fishing expedition described by Lovett et al. (2007; see Section 13.5.3).

13.5.2 Selecting the Right Indicators and Measurements

Many papers have been published on the selection of ecological indicators, and new international journals, such as *Ecological Indicator* or *Environmental Bioindicator*, are devoted to this topic. Typically, ecological indicators of forest biodiversity or ecosystem integrity are developed along three coarse components: composition, structure, and function (see Franklin et al. 1981; Noss 1990). Noss (1990) proposed a hierarchical approach for selecting ecological indicators to assess biodiversity using the three components listed across logical scales of biological/ecological organization: regional/landscape, community/ecosystem, population/species and genetic. The elaboration of a comprehensive monitoring program to monitor the biodiversity or ecosystem integrity of a region is a daunting task. Noss (1990) proposed ten steps for a hypothetical program to monitor biodiversity in the Pacific Northwest of the United States: (1) "What and why?" These refer to specific objectives

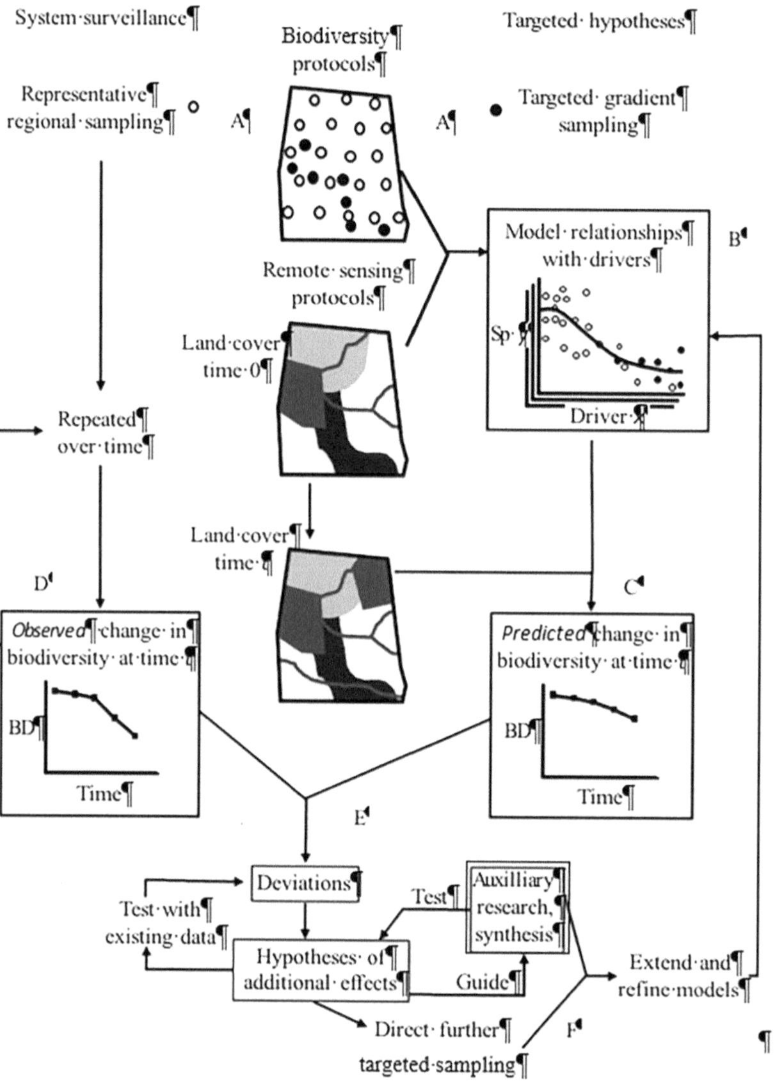

FIGURE 13.10 An illustration of the cumulative effects monitoring framework used by the ABMI. (From Burton, A.C. et al., *Environ. Monit. Assess. 186*, 3605, 2014.)

and sub-end points similar to what Klenner (2011) describes as a range of acceptable conditions. Here it is interesting that Noss acknowledges that this will result as much from policy as science. This is linked to the debate about defining ecological threshold developed in the next section. In the case of inevitable uncertainty that cannot be completely solved by science, some choices involving environmental ethics will be necessary. (2) Gather and integrate existing data. There are many datasets that are not fully exploited and digitized. New technology makes this task easier than before. However, it is important to be cautious about the limitations of various datasets. (3) Establish "baseline" conditions. This refers to the idea of natural range of variation. (4) Identify "hot spots" and ecosystems at high risk. (5) Formulate specific questions to be answered by monitoring. (6) Select indicators. (7) Identify control areas and treatments. (8) Design and implement a sampling scheme. (9) Validate relationships between indicators and sub-endpoints. (10) Analyze trends and recommend management actions. This framework is very similar to adaptive monitoring frameworks in more recent literature (Lindenmayer and Likens 2009a, 2009b; Lindenmayer et al. 2011).

The specific choice of an ecological indicator is potentially controversial and should be approached carefully (Carignan and Villard 2002; Niemi and McDonald 2004). They are typically used to assess ecological conditions, to investigate the cause of change, and to forecast future conditions under different management scenarios (Niemi and McDonald 2004). Some important attributes of ecological indicators would possess at least some of the following features: provide early warning on negative impact from anthropogenic actions, identify the cause(s) of change, provide continuous assessment of the effect of stressors, and be cost-effective (Carignan and Villard 2002). In addition, it is very important that the value of the indicators and their response can be easily and effectively communicated to stakeholders, including the public (Niemi and McDonald 2004). Choosing an indicator species is a daunting task because different groups of species will respond differently to disturbances. Noss (1999) proposes choosing from a suite of species groups, such as species limited by area, resources, or processes, that is, dispersal, keystone species, narrow endemic species, and species that are locally important but that do not fall within the groups described earlier.

A good example where indicators have been used to assess forest ecosystem integrity is an initiative by the National Park Service to monitor ecological integrity in parks of the Northeast Temperate Network. Their approach was to develop a conceptual model linking key ecosystem attributes with stressors in northeastern forest and to establish end points to differentiate desired from undesirable conditions. This allowed them to present easily communicable results about the state of the Acadia National Park by rating ecological indicators with good, caution, and significant concern.

Although society may never be able to attain a complete understanding of all species and all ecosystems everywhere, scientists and managers need to be very careful to test assumptions and inferences about species distributions and management effects. From a practical perspective in an environment of uncertain research and monitoring funding, another approach to developing ecological indicators is to focus on coarse-level filters. Klenner (2011) provides six landscape-level indicators to assess ecosystem trends that enable the assessment of habitat loss: "1. Seral condition and ecological representation, 2. Patch size of forest and openings, 3. Tree species composition, 4. Density and dispersion of roads, 5. Area disturbed and type of disturbance, and 6. Ownership and land use designation."

13.5.3 Concept of Thresholds

Elevators have security ratings indicating how many people and how much weight should not be exceeded for safety reasons. Engineers calculate weight thresholds beyond which the danger of the elevator breaking and becoming unsafe increases dramatically. Is it possible to do something analogous with species and ecosystems? This is very much a matter of perspectives and scales and goes back to our introductory section on ecosystem resilience and integrity. In the case of a balance of nature worldview, the concept of thresholds makes sense, but in the case of nonequilibrium worldview, it becomes much more complex. To reconcile the idea of ecological threshold with the open system with a nonequilibrium dynamics worldview, a new paradigm has been created—the natural range of variation, which we introduced in the section on natural disturbance. In this case, an ecosystem has not crossed a threshold unless the disturbance has created a pattern or influenced a process that was not observed historically for this ecosystem. There are two immediate complications with this paradigm: one is the question of spatial and temporal scale and the other is the complexity of reconstructing historical observations of ecological phenomena. Successful thresholds used to set targets for managed forest landscapes are presented in Villard and Jonsson (2009). Excellent examples of ecological thresholds include avi-fauna associated with standing dead trees and large trees (Guénette and Villard 2005) and the levels of deadwood associated with saproxylic insects (Work and Hibbert 2011). These examples are based on empirical studies of species and habitat relationships. In cases where these relationships are well developed and understood, these species or habitat characteristics can be valuable ecological indicators. However, it is important to acknowledge that these species–habitat relationships may not be good indicators for other taxa. In the case of species depending on deadwood, it is highly unlikely that single deadwood targets will be adequate for the conservation of saproxylic species (Jonsson and Ranius 2009). In the absence of ecological data, managers will sometimes develop targets, for example, the level of snags and coarse woody debris to be retained in logged areas. Although this may be desirable as a precautionary approach, it is critical to rigorously document and report the rationale for identifying these targets. In particular, if elements of forest structure are used as surrogates of biodiversity, it is important to test these assumptions with rigorous scientific studies.

13.6 CONCLUSION

The world is experiencing unprecedented environmental changes that are likely to increase if the current trends of anthropogenic disturbances continue. Ecological indicators of ecosystem integrity provide a useful tool to assess the direction and degree of change. There is a long history of using environmental indicators effectively, such as the phenological indicators used by First Nations and the canary used by miners in coal mines. The complexity of forest ecosystems and society should be taken into account when identifying ecological indicators for monitoring programs. In order to be effective, we suggest engaging stakeholders and peers throughout different stages of the process—probably one of the most important pieces of the puzzle (Klenner and Cameron 2013). An effective ecological monitoring program should be scientifically based, which means that clear and testable questions should be identified. We recommend integrating surveillance and targeted approaches combined with complementary research methods, such as space for time substitution, retrospective studies, or ecological modeling. Other key ingredients include the timely archiving and reporting of data and a clear feedback loop between monitoring and changes in policy and management. This will likely increase the level of engagement of all parties involved, give the program greater credibility, and thereby increase its chances of being successful and effective. Ecological monitoring should not be considered as different from science but should be integrated as part of a scientific process and subject to rigorous peer review. The critical importance of long-term and well-designed research projects at operational scales should become part of the modern ecological science and monitoring culture. Engaging the public in monitoring programs, for example, citizen science, could be a cost-effective way to collect reliable ecological information over large areas in a consistent fashion. Scientists and practitioners should consider increasing ecological literacy as a mission. This is the only way society can evolve toward informed consensus on wicked environmental issues.

13.7 ACKNOWLEDGMENTS

We would like to thank Robert LeBlanc, Bruce Pike, Candice Power, Kishan Sambaraju, and Caroline Simpson for providing useful comments on earlier versions.

REFERENCES

Agee, J.K. 1993. *Fire ecology of Pacific northwest forests.* Washington, DC: Island Press.

Arsenault, A. and W. Klenner. 2001. Smoke and mirrors? A critical view on the use of fire history paradigms in forest management. In *Proceedings of symposium on natural disturbance and forest management: What's happening and where it's going*, eds. K. McCleary, D. Andison, F. Ragan, et al., p. 45. Edmonton, Alberta, Canada: Co-sponsored by the Foothills Model Forest and the Sustainable Forest Management Network.

Barrette, M., L. Bélanger, L. De Grandpré, et al. 2014. Cumulative effects of chronic deer browsing and clear-cutting on regeneration processes in second-growth white spruce stands. *Forest Ecol. Manage.* 329:69–78.

BC Ministry of Forests and BC Environment. 1995. *Biodiversity guidebook—Forest practices code of British Columbia.* Victoria, British Columbia, Canada: Ministry of Forests and BC Environment.

Bergeron, Y. and N. Fenton. 2012. Boreal forests of eastern Canada revisited: Old growth, nonfire disturbances, forest succession, and biodiversity. *Botany* 90:509–523.

Bergeron, Y., P.J.H. Richard, C. Carcaillet, et al. 1998. Variability in fire frequency and forest composition in Canada's southeastern boreal forest: A challenge for sustainable forest management. *Conserv. Ecol.* 2. the Internet. http://www.ecologyandsociety.org/vol2/iss2/art6/. Accessed September 2015).

Boutin, S., L.D. Haughland, J. Schieck, et al. 2009. A new approach to forest biodiversity monitoring in Canada. *Forest Ecol. Manag.* 258:S168–S175.

Brand, D.G. 1997. Criteria and indicators for the conservation and sustainable management of forests: Progress to date and future directions. *Biomass Bioenergy* 13:247–253.

Brubakk, A.O. and M.A. Lang. 2009. JS Haldane, the first environmental physiologist. In *The future of diving: 100 years of haldane and beyond*, eds. M.A. Lang and A.O. Brubakk, pp. 5–9. Washington, DC: Smithsonian Institution Scholarly Press.

Bunnell, F. 1999. Forestry isn't rocket science—It's much more complex. *Forum* 6:7.

Burton, A.C., D. Huggard, E. Bayne, et al. 2014. A framework for adaptive monitoring of the cumulative effects of human footprint on biodiversity. *Environ. Monit. Assess.* 186:3605–3617.

Carignan, V. and M.A. Villard. 2002. Selecting indicator species to monitor ecological integrity: A review. *Environ. Monit. Assess.* 78:45–61.

Carpenter, S., B. Walker, M. Anderies, and N. Abel. 2001. From metaphor to measurement: Resilience of what to what? *Ecosystems* 4:765–781.

Cumming, S.G., P.J. Burton, and B. Klinkenberg. 1996. Boreal mixedwood forests may have no "representative" areas: Some implications for reserve design. *Ecography* 19:162–180.

De Leo, G.A. and S. Levin. 1997. The multifaceted aspects of ecosystem integrity. *Conserv. Ecol.* 1(1):3. URL: www.ecologyandsociety.org/vol1/iss1/art3/ (accessed April 15, 2015).

DellaSala, D., A. Martin, R. Spivak, et al. 2003. A citizen's call for ecological forest restoration: Forest restoration principles and criteria. *Ecol. Restor.* 21:15.

Diamond, J.M. 1984. 'Normal' extinction of isolated populations. In *Extinctions*, ed. M.H. Nitecki, pp. 191–246. Chicago, IL: Chicago University Press.

Dorren, L.K.A., F. Berger, A.C. Imeson, et al. 2004. Integrity, stability and management of protection forests in the European Alps. *For. Ecol. Manag.* 195:165–176.

Drever, C.R., G. Peterson, C. Messier, et al. 2006. Can forest management based on natural disturbances maintain ecological resilience? *Can. J. For. Res.* 36:2285–2299.

Ehrlich, P. 1968. *The population bomb*. New York: Ballantine.

Filotas, E., L. Parrott, P.J. Burton, et al. 2014. Viewing forests through the lens of complex systems science. *Ecosphere* 5:1–23.

Forest History Society. 2015. *In US forest service history-people-chiefs-Gilford Pinchot*. http://www.foresthistory.org/ASPNET/People/Pinchot/Pinchot.aspx. Accessed September 12, 2015.

Franklin, J.F., K. Cromack, W. Denison, et al. 1981. *Ecological characteristics of old-growth Douglas fir forests*. USDA Forest Service general technical report PNW-1 18. Portland, OR: Pacific Northwest Forest and Range Experiment Station.

Franklin, J.F. and R.T.T. Forman. 1987. Creating landscape patterns by forest cutting: Ecological consequences and principles. *Landsc. Ecol.* 1:5–18.

Grumbine, R.E. 1994. What is ecosystem management? *Conserv. Biol.* 8:27–38.

Guénette, J.S. and M.A. Villard. 2005. Thresholds in forest bird response to habitat alteration as quantitative targets for conservation. *Conserv. Biol.* 19:1168–1180.

Haughland D.L., J.-M. Hero, J. Schieck, et al. 2010. Planning forwards: biodiversity research and monitoring systems for better management. *Trends Ecol. Evol.* 25:199–200.

Hilderbrand, R.H., A.C. Watts, and A.M. Randle. 2005. The myths of restoration ecology. *Ecol. Soc.*10:19. www.ecologyandsociety.org/vol10/iss1/art19/. Accessed October 2, 2015.

Holling, C.S. 1973. Resilience and stability of ecological systems. *Annu. Rev. Ecol. Syst.* 4:1–23.

Huggard, D. and A. Arsenault. 1999. Reverse cumulative standing age distributions in fire-frequency analysis. *Can. J. For. Res.* 29:1449–1456.

Huggard, D. and A. Arsenault. 2001. Reply—Standing age distributions and fire frequency analysis. *Can. J. For. Res.* 31:369–371.

Hunter, M. 1989. What constitutes an old-growth stand? *J. For.* 87:33–35.

Jonsson, B.G. and T. Ranius. 2009. The temporal and spatial challenges of target setting for dynamic habitats: The case of dead wood and saproxylic species in boreal forests. In *Setting conservation targets for managed forest landscapes*, eds. A. Villard and B.G. Jonsson, pp. 207–226. Cambridge, UK: Cambridge University Press.

Karr, J.R. and D.R. Dudley. 1981. Ecological perspective on water quality goals. *Environ. Manage.* 5:55–68.

Kimmins, J.P. 1987. *Forest ecology.* London, UK: Macmillan.

Klenner, W. 2011. Changing landscapes of the Montane Cordillera Ecozone: Strategic indicators to monitor habitat change. In *Assessment of species diversity in the Montane Cordillera ecozone*, eds. G.G.E. Scudder and I.M. Smith, pp. 38–53. www.royalbcmuseum.bc.ca/assets/Montane-Cordillera-Ecozone.pdf

Klenner, W. and R. Cameron 2013. Perspectives on the use of long-term silvicultural research in British Columbia to inform sustainable forest management in an era of uncertainty. In *LongTerm silvicultural & ecological studies: Results for science and management*, eds. A.E. Camp, L.C. Irland, and C.J.W. Carroll, pp. 66–74. Yale: GISF research paper 013. New Haven, CT: Yale University, School of Forestry and Environmental Studies, Global Institute of Sustainable Forestry.

Klenner, W., R. Walton, A. Arsenault, and L. Kremsater. 2008. Dry forests in the Southern Interior of British Columbia: Historic disturbances and implications for restoration and management. *For. Ecol. Manag.* 256:1711–1722.

aganière, J., D.A. Angers, and D. Paré. 2010. Carbon accumulation in agricultural soils after afforestation: A meta-analysis. *Glob. Chang. Biol.* 16:439–453.

Landres, P.B., P. Morgan, and F.J. Swanson. 1999. Overview of the use of natural variability concepts in managing ecological systems. *Ecol. Appl.* 9:1179–1188.

Lantz, T.C. and N.J. Turner. 2003. Traditional phonological knowledge of aboriginal peoples in British Columbia. *J. Ethnobiol.* 23:263–286.

Leopold, A. 1949. *A sand county almanac, and sketches here and there*. New York: Oxford University Press.

Lindenmayer, D.B. and G.E. Likens. 2009a. Adaptive monitoring: A new paradigm for long-term research and monitoring. *Trends. Ecol. Evol.* 24:482–486.

Lindenmayer, D.B. and G.E. Likens. 2009b. Improving ecological monitoring. *Trends. Ecol. Evol.* 25:200–201.

Lindenmayer, D.B., G.E. Likens, A. Haywood, et al. 2011. Adaptive monitoring in the real world: Proof of concept. *Trends. Ecol. Evol.* 26:641–646.

Lindenmayer, D.B., W.F. Laurance, J.F. Franklin, et al. 2014. New policies for old trees: Averting a global crisis in a keystone ecological structure. *Conserv. Lett.* 7:61–69.

Lovett, G.M., D.A. Burns, C.T. Driscoll et al. 2007. Who needs environmental monitoring? *Front. Ecol. Environ.* 5:253–260.

Luckert, M.K., D. Haley, and G. Hoberg. 2011. *Policies for sustainably managing Canada's forests: Tenure, stumpage fees, and forest practices*. Vancouver, British Columbia, Canada: UBC Press.

Ludwig, D., B. Walker, and C.S. Holling. 1997. Sustainability, stability, and resilience. *Conserv. Ecol.* 1(1):7. Available from the Internet. www.ecologyandsociety.org/vol1/iss1/art7/. Accessed May 15, 2015.

May, R.M. 1977. Thresholds and breakpoints in ecosystems with a multiplicity of stable states. *Nature* 269:471–477.

Mori, A.S. 2011. Ecosystem management based on natural disturbances: Hierarchical context and non-equilibrium paradigm. *J. Appl. Ecol.* 48:280–292.

Mori, A.S., T. Furukawa, and T. Sasaki. 2013. Response diversity determines the resilience of ecosystems to environmental change. *Biol. Rev.* 88:349–364.

Niemi, G.J. and M.E. McDonald. 2004. Application of ecological indicators. *Ann. Rev. Ecol. Evol. Syst.* 35:89–111.

Noss, R.F. 1990. Indicators for monitoring biodiversity: A hierarchical approach. *Conserv. Biol.* 4:355–364.

Noss, R.F. 1999. Assessing and monitoring forest biodiversity: A suggested framework and indicators. *Forest Ecol. Manage.* 115:135–146.

Peterson, G., C.R. Allen, and C.S. Holling. 1998. Ecological resilience, biodiversity, and scale. *Ecosystems* 1:6–18.

Pickett, S.T.A. and P.S. White, eds. 1985. *The ecology of natural disturbance and patch dynamics*. San Diego, CA: Academic Press.

Pimm, S.L. 1991. *The balance of nature?* Chicago, IL: University of Chicago Press.

Potter, K.M. and C.W. Woodall. 2014. Does biodiversity make a difference? Relationships between species richness, evolutionary diversity, and aboveground live tree biomass across U.S. forests. *For. Ecol. Manag.* 321:117–129.

Proença, V., H.M. Pereira, and L. Vicente. 2010. Resistance to wildfire and early regeneration in natural broadleaved forest and pine plantation. *Acta Oecol.* 36:626–633.

Puettmann, K.J., K.D. Coates, and C. Messier. 2009. *A critique of silviculture: Managing for complexity*. Bota Raton, FL: Island Press.

Roberts, K.A. 1991. Field monitoring: Confessions of an addict. In *Monitoring for conservation and ecology*, ed. F.B. Goldsmith, pp. 179–212. New York, NY: Chapman & Hall.

Scheffer, M., S. Carpenter, J.A. Foley, C. Folke, et al. 2001. Catastrophic shifts in ecosystems. *Nature* 413:591–596.

Seidl, R., W. Rammer, and T.A. Spies. 2014. Disturbance legacies increase the resilience of forest ecosystem structure, composition, and functioning. *Ecol. Appl.* 24:2063–2077.

Sprugel, D.G. 1991. Disturbance, equilibrium, and environmental variability: What is 'natural' vegetation in a changing environment? *Biol. Conserv.* 58:1–18.

Thomas, J.W., J.F. Franklin, J. Gordon, et al. 2006. The Northwest Forest Plan: Origins, components, implementation experience, and suggestions for change. *Conserv. Biol.* 20:277–287.

Thompson, I., B. Mackey, S. McNulty, et al. 2009. *Forest resilience, biodiversity, and climate change. A synthesis of the biodiversity/resilience/stability relationship in forest ecosystems*, Technical series no. 43. Montréal, Québec, Canada: Secretariat of the Convention on Biological Diversity.

Tierney, G.L., D. Faber-Langendoen, B.R. Mitchell, et al. 2009. Monitoring and evaluating the ecological integrity of forest ecosystems. *Front. Ecol. Environ.* 7:308–316.

Turner, M.G. and W.H. Romme. 1994. Landscape dynamics in crown fire ecosystems. *Landsc. Ecol.* 9:59–77.

Urrego, D.H., M.B. Bush, M.R. Silman, et al. 2013. Holocene fires, forest stability and human occupation in south-western Amazonia. *J. Biogeogr.* 40:521–533.

USDA Forest Service. 2015. *Gifford Pinchot (1865–1946) in Grey Towers National historic site history*. www.fs.usda.gov/main/greytowers/aboutgreytowers/history. Accessed September 2015.

Villard, M.A. and B.G. Jonsson, eds. 2009. *Setting conservation targets for managed forest landscapes* (No. 16). Cambridge, UK: Cambridge University Press.

Walker, B., C.S. Holling, S.R. Carpenter, et al. 2004. Resilience, adaptability and transformability in social–ecological systems. *Ecol. Soc.* 9:5. www.ecologyandsociety.org/vol9/iss2/art5/. Accessed April 2015.

Wallington, T.J., R.J. Hobbs, and S.A. Moore. 2005. Implications of current ecological thinking for biodiversity conservation: A review of the salient issues. *Ecol. Soc.* 10(1):15. www. ecologyandsociety.org/vol10/iss1/art15/. Accessed March 15, 2015.

Work, T.T. and A. Hibbert. 2011. Estimating species loss of saproxylic insects under scenarios of reduced coarse woody material in eastern boreal forests. *Ecosphere* 2: art41. www. esajournals.org/doi/pdf/10.1890/ES10-00075.1. Accessed September 4, 2015.

14 Criteria and Indicators of Sustainable Forest Management

Michal Bosela, Guy R. Larocque, Tanya Baycheva, Ruben Valbuena, and Markus Lier

14.1 INTRODUCTION

Globally, over nine million hectares are deforested each year and considerably more are degraded by human intervention (FAO 2006). The loss of forests is usually closely related to demographic expansions leading to the conversion of forests to other uses. Therefore, increasing concerns from environmental nongovernmental organizations and the public about the impact of traditional forest management techniques on nontimber forest values have triggered national and international discussions to increase the sustainability of forest management and to develop standards for its definition and standards for monitoring, assessment (evaluation), and reporting. Sustainability is a human concept developed to recognize the interdependence of nature and human activities and the importance of the long-term effects of human activities on the environment (Burton et al. 2006). Although, for centuries, forest management has been based on the principle of sustained yield stating that the harvested volume should not exceed the volume increment gained over a given time period, there has been increasing recognition that forests may provide a wide range of nonwood production functions, such as biodiversity, contribution to hydrological cycles, and sociological functions (Wijewardana 2008). Sustainable forest management (SFM) can thus be generally thought of as the management of forests (public and private) to ensure that they continue to supply timber for future generations and also environmental and social values.

This chapter uncovers different issues on SFM. In the next section, controversies surrounding the definition and application of SFM that led to the development of processes on criteria and indicators (C&I) are presented. The basic elements of C&I that are common to all processes are reviewed and forest certification is briefly discussed. Then, the main processes that are most applied are reviewed. In Section 14.3, the pan-European and Montréal Processes for temperate and boreal forests are described, and some examples of application are provided. The processes for tropical forests are covered in Section 14.4, which include the International Tropical Timber Organization (ITTO), the African Timber Organization (ATO), and Tarapoto processes. In Section 14.5, the importance of collaboration and cooperation among processes is highlighted. In particular, efforts to promote collaborations among organizations involved in the development of processes and to harmonize the different processes are discussed. The application of C&I must be based on different sources of information and data to ensure that they report as exactly as possible the state of forests. These sources are discussed in Section 14.6. Regarding data sources, the importance of statistically based forest inventories and the use of remote sensing techniques are reviewed.

14.2 DEFINITION AND APPLICATION OF SFM LEADING TO PROCESSES ON C&I

The concept of sustainable development was first outlined in the Brundtland report, *Our Common Future* (WCED 1987). Since then, the principle of sustainable development has been widely

DOI: 10.1201/9781003431084-17

accepted and defined as *development that meets the needs of the present without compromising the ability of future generations to meet their own needs* (WCED 1987). The principle of sustainability can also be found in forestry since the 18th century when early forest managers in Europe developed an understanding of natural forestry productivity and how it might be enhanced through silviculture to maintain a continuous supply of wood, game, and other products for human use and consumption. Nowadays, SFM has become a key concept that underpins modern forestry and refers not just to the flow of goods and services but also to the *multiple uses* and *functions* of the forests (e.g., wood production, collecting nonwood forest products, recreation, protection of soil and water resources, biodiversity conservation, and carbon sequestration) and aims to maintain and enhance social, cultural, environmental, and economic values of forests for the benefit of present and future generations.

In a political context, the concept of SFM was first introduced at the United Nations Conference on Environment and Development (UNCED) in Rio de Janeiro, in 1992. The discussions at the conference were mainly focused on the management, conservation, and sustainable development of forests and whether or not to launch negotiations for an international legally binding instrument for forests. However, they did not reach a consensus, and this led to the adoption of the Forest Principles and Chapter 11 of Agenda 21 on "Combating Deforestation." Although there is no single globally agreed definition of SFM, the most widely intergovernmentally agreed definition is the one adopted by the General Assembly of the United Nations in December 2007:

> Sustainable forest management, as a dynamic and evolving concept, aims to maintain and enhance the economic, social and environmental value of all types of forests, for the benefit of present and future generations. It is characterized by seven elements, including: (i) extent of forest resources; (ii) forest biological diversity; (iii) forest health and vitality; (iv) productive functions of forest resources; (v) protective functions of forest resources; (vi) socio-economic functions of forests; and (vii) legal, policy and institutional frameworks.

In the field of SFM, C&I for a more detailed measurement of and reporting on progress toward SFM have taken a prominent role in the beginning of the 1990s. At UNCED, the world governments called for the formulation of scientifically sound criteria and guidelines for the management and sustainable development of all forest types. As a response, several international and regional processes and initiatives have developed C&I as a framework for reporting and evaluating progress toward SFM (The ITTO development of SFM policies and manuals for application in developing countries as the pioneer process. This was followed by the Montréal Process for the development of SFM principles for the temperate and boreal forests of non-European countries and FOREST EUROPE, the former Ministerial Conference on the Protection of Forests in Europe (MCPFE) (Castañeda 2000; McDonald and Lane 2004). C&I thus have an ambition to be the most important tools for forest management through which it is also possible to derive a global understanding of SFM (Wijewardana 2008). Nearly 150 countries that contain 97.5% of the world's forest area participate in at least one of the nine processes toward developing C&I of SFM (Castañeda 2000; Wijewardana 2008). Nevertheless, there have been many discussions, particularly in the context of the United Nations Economic Commission for Europe/Food and Agriculture Organization (UNECE/FAO), how to assess and evaluate progress toward SFM.

One of the main issues associated with negotiating a sustainable future is to define sustainability and then determine progress toward this goal (Hickey and Innes 2005). Following this, principles, criteria, indicators, and verifiers were developed. The aim of the efforts to create a set of C&I is to provide an objective and unambiguous way to evaluate environmentally SFM (Stork et al. 1997; Ghazoul and Hellier 2000).

Principle—A fundamental truth or law as the basis of reasoning or action. The principles, in the context of SFM, are considered as providing the primary framework for managing forests in a sustainable way. They provide the justification for criteria, indicators, and verifiers.

Criterion—A standard that a thing is judged by. A criterion can be seen as a "secondorder" principle, one that adds meaning to a principle without itself being a direct measure of performance.

Example: Maintenance or enhancement of forest resources, maintenance and conservation of biological diversity etc.

Indicator—An indicator is any variable or component of the forest ecosystem or the relevant management systems used to infer attributes of the sustainability of the resource and its utilization.

Example: Forest area, growing stock etc.

Verifier—Data or information that enhances the specificity or the ease of assessment of an indicator. It provides specific details that would indicate or reflect a desired condition of an indicator.

Example: Basal area, volume of dead trees, species richness.

14.2.1 Historical Background

Individual processes were established at different times in the past starting in 1991. By 2000, nine processes were developed (Baycheva et al. 2013) (see Figure 14.1).

In 1991, prior to UNCED, the ITTO developed C&I for SFM at the forest management unit (FMU) level to assess the conditions of natural tropical forests in producer member countries and help identify weaknesses in forest practices and improvements needed (ITTO 1992). In Europe, a set of national-level indicators was established in the 1990s to initialize and standardize pan-European reporting. The MCPFE C&I were developed in 1993 as "a basis for international reporting and the further development of national indicators" (Lisbon 2 Resolution, MCPFE, 1998). Only since 2003, the Forest Europe (former MCPFE process) has adopted a set of pan-European C&I that have become a guide for policymakers to monitor, evaluate, and report progress toward SFM (Barbati et al. 2014). C&I for West and Central Africa were developed at the same time by the ATO to promote the implementation of SFM at regional, national, and FMU levels. In 1995, following these initiatives, the Montréal Process was launched to develop an indicator set for temperate and boreal forests. In the same year, eight countries of the Amazon basin developed regional C&I. This process is known as the Tarapoto process. In addition to these processes, three other initiatives were established with the support of FAO and United Nations Environment Programme: the Dry-Zone Africa, the Near East Process, and the Dry Forests Asia.

14.2.2 Overview of Processes

The processes differ between each other by several attributes, such as forest type for which they were developed, level at which they are applied, and by C&I (its number and kind of forest-related attributes) to address SFM (Table 14.1). While the C&I sets differ in various aspects, they all reflect a holistic approach to forests as ecosystems with multiple values beyond wood and fiber production (ITTO 2012). Although the criteria address the same key components of SFM ("seven thematic elements of SFM"), the processes differ in the indicators by which they assess the status and changes of forest management and measures to reach SFM. The indicators are usually selected to address national-, regional-, or forest-type level, particularly to better assess SFM.

There is no globally agreed set of indicators, as they need to be adapted to the ecological, economic, social, and institutional conditions and needs of each country (Castañeda 2000). Governmental and nongovernmental stakeholders in the countries may, however, develop their own national- or regional-level sets of C&I to fit to international and regional C&I sets over the forest types, continents, or regions as much as possible, so that a common reporting would be possible at an adequate level. Despite differences in indicators among processes, there are globally agreed and applied thematic elements of SFM common to all processes drawn from the criteria:

- The extent of forest resources is an element that assesses forest cover and stocking, including other wooded land and trees outside forests, to support the social, economic, and environmental dimensions of forestry. It has an ambition to reduce deforestation and to restore or recover degraded forest areas. In addition, it includes the function of forests,

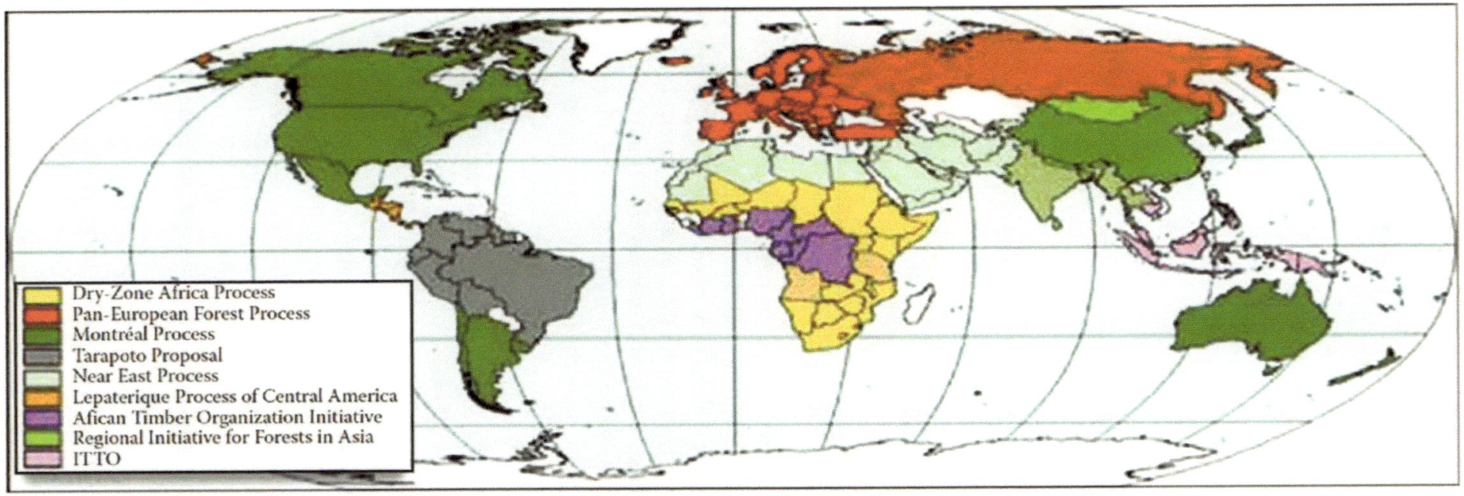

FIGURE 14.1 Distribution of the nine processes on criteria and indicators for sustainable forest management as of 2000/2001.

Note: Russia is part of both Montréal and Forest Europe (former MCPFE) processes. (From Food and Agriculture Organization of the United Nations, Membership of world countries in nine C&I processes, www.fao.org/forestry/ci/16609/en/, accessed September 22, 2014. Reproduced with permission.)

TABLE 14.1
List of Nine C&I Processes and Their Current Status

Process	Year Launched	Active?	Forest Type	Level	Number of Criteria/Indicators
ITTO	1992	Yes	Humid tropical forests	National, FMU	7/66
ITTO/ATO	1995	Yes	West and Central Africa	National, FMU	28/60
FOREST EUROPE (former MCPFE)	1993	Yes	European boreal, temperate, and Mediterranean forests	Regional, National	6/35
Montréal Process	1994	Yes	Temperate and boreal forests	National	7/67
The Tarapoto Proposal	1995	Yes	Amazon Forest	International, National, FMU	7/47
Lepaterique	1997	Yes	Central America	Regional, National	8/53 (national)
The Dry-Zone Africa	1995	No	North, East, and Southern Africa	Regional, National	7/47
The Near East Process	1996	No	Near East	National	7/65
Dry Forests Asia	1999	No	South and Central Asia	National	8/49

other wooded land, and trees outside forests to store carbon and thereby to moderate the global climate warming. This element is the first measure of SFM and has an ambition to maintain adequate forest resources of various forest types to future generations. The following six variables are used to assess the element: (1) area of forest and other wooded lands (if possible also of other lands with tree cover); (2) forest characteristics according to three classes: primary forests, other naturally regenerated forests, and planted forests; (3) area of selected forest types (mangroves, bamboo, and rubber plantations); (4) standing volume of wood (total growing stock in forests and other wooded land); (5) forest biomass; and (6) carbon stock in living woody biomass, deadwood, litter, and forest soil.

- Biological diversity assesses the conservation and management of biological diversity of the ecosystems, species, and genetic levels. Biological diversity encompasses the variety of existing life forms, the ecological roles they perform, and the genetic diversity they contain (FAO 1989). Monitoring of changes in biological diversity caused by forestry or other practices is essential for the evaluation of forest management effectiveness toward SFM. The assessment of biological diversity can be performed at different levels, such as ecosystems, landscapes, species, populations, individuals, or genes and requires the application of different methodologies. However, there is a high complexity behind the biological diversity. Therefore, assessment and monitoring are possible only for specific aspects. There is no single objective measure but only proxy by applying which one supposes to reach sustainability of the biological diversity.
- The variables used for forest resources assessment (FRA) to assess the biological diversity are as follows: (1) area of primary forests, (2) forest area designated primarily for the conservation of biological diversity, (3) area of forests in protected areas, and (4) tree species composition of forests.
- Forest health and vitality is directed toward determining whether forests are managed so as to minimize the risks and impacts of undesired disturbances, including wildfires, airborne pollution, storm felling, invasive species, pests, and insects, which might have had

an impact on social, economic, or environmental sphere of forestry. To assess the forest health and vitality, the following variables are used within FRA: (1) area of forests significantly affected by insects; (2) area of forests significantly affected by diseases; (3) area burned (separated into areas of forests, other wooded lands, and other lands); (4) number of wildfires (separated into affecting forests, other wooded lands, and other lands); (5) proportion of wildfires and planned fires; (6) area of forests significantly affected by other biotic factors such as wildlife browsing, grazing, and physical damage by animals; (7) area of forests significantly affected by abiotic factors such as air pollution, wind, snow, ice, floods, landslides, tropical storms, drought, and tsunami; and (8) area of forests significantly affected by invasive species (woody species only).

- Productive forest function is the element that aims to maintain an ample and valuable supply of primary forest products while at the same time ensuring that production and harvesting are sustainable and balanced. Earlier assessments included only timber supply. However, the concept has changed and widened to include all types of wood as well as nonwood forest products. For FRA, these attributes are assessed: (1) area of forests designated for production, (2) area of planted forests, (3) areas of afforestation and reforestation, (4) removal of wood products, and (5) removal of nonwood forest products.
- Protective forest functions of other wooded lands and trees outside forests (see Vidal et al. 2008 for reference definitions) include protection of soil, hydrological, and aquatic systems by continuous and adequate forest management and applying less-intensive and close-to-nature measures. The functions include the maintenance of clean water as well as to reduce risks of floods, avalanches, erosion, and droughts. Regarding FRA, there is only one variable to be assessed: the area of forests with protection of soil and water as the primary designated function. It means the areas with soil and water conservation as the main function or management objective.
- Socioeconomic benefits and needs address the contribution of forests to the overall economy (e.g., through employment, values generated through processing and marketing of forest products, trade and investments in the forestry sector). The economic benefits of forest management can be calculated as the quantity of products and services, each multiplied by an appropriate value and added together. However, it is more difficult to estimate values for subsistence uses of products or for outputs that are not sold in markets. To assess this element, the recent trends in forest ownership and management rights are analyzed. The fiscal measures that governments use to support forest management are also included. It then includes the current status and trends for an indicator of the economic benefits (the value of wood and nonwood forest products removals and social indicators such as employment and the area of forests designated for social services).
- The legal, policy, and institutional framework is the element that includes legal, policy, and institutional arrangements, which are necessary to support the previous six components of SFM. It includes participatory decision-making, governance, and monitoring of progress. In addition, it also addresses other aspects such as science research and education, infrastructure, transfer of technology and capacity building, and public information. Specifically, the following attributes are used for FRA: (1) the existence of national and/or subnational forest law, date of enactment, and latest amendment; (2) the existence of national and/or subnational forest policy and date of endorsement; (3) the existence of national forest program (NFP), date of commencement, and its current status; (4) the institutional structure related to forests and forestry; (5) human resources in public forestry institutions; (6) the number of graduates in forestry education; and (7) the number of professional staff in publicly funded forest research centers.

These key elements of SFM are based on the criteria of the nine processes on C&I of SFM, which are described in the following sections. They are used by the FAO for the purpose of global FRA as a reporting framework (it was used in 2005 for the first time), and the structure of the reports is arranged according to these key elements. These elements have also been identified at the International Conference on the Contribution of C&I for SFM (CICI) in Guatemala in 2003 as well as at the Expert Consultation on C&I for SFM (ECCI) in the Philippines in 2004.

However, the forestry experience shows that cross-sectoral collaborations are important for creating and ensuring SFM (De Montalbert and Schmithüsen 1993; De Montalbert 1995; Schmithüsen et al. 2001) since there are often many stakeholders, state administration, environmental agencies, and policymakers involved.

14.2.3 Forest Certification

Another way to save biodiversity and ensure SFM is through certification. It was introduced in the 1990s to address concerns of deforestation (Rametsteiner and Simula 2003). Certification is a tool whereby a certifier assesses forest management in relation to defined requirements (standards). However, its application differs among tropical, boreal, and temperate forests. According to Rametsteiner and Simula (2003), it is still evident that the original intention to save tropical biodiversity using certification has failed (only around 10% of all certified forests were located in tropics). The authors stated that "most of the certified areas are in the temperate and boreal zones, with Europe as the most important region." By 2013, the global area of certified forests (endorsed by Forest Stewardship Council (FSC), the Programme for the Endorsement of Forest Certification (PEFC)) was as much as 417 million hectares (UNECE/FAO 2013).

Some forest certification schemes (e.g., PEFC) have developed their principles and criteria based on the Pan-European Operational Level Guidelines for SFM (PEOLG), which are based on the pan-European C&I for SFM. Many national certification schemes are also using C&I sets (national or international) to develop, apply, and promote their national forest certification or standards.

14.3 THE USE OF C&I FOR TEMPERATE AND BOREAL FORESTS

14.3.1 Pan-European C&I for Sustainable Forest Management

Following the UNCED Rio Declaration (United Nations 1993), the pan-European countries agreed in the Second Ministerial Conference for the Protection of Forests in Europe (former MCPFE, now FOREST EUROPE*), in Helsinki in 1993, on the basic principles of SFM (Resolution H1, item D) with simultaneous respect to ecological, economic, and social functions of forests as follows:

> Agreeing that, for the purposes of this resolution, "sustainable management" means the stewardship and use of forests and forest lands in a way, and at a rate, that maintains their biodiversity, productivity, regeneration capacity, vitality and their potential to fulfil, now and in the future, relevant ecological, economic and social functions, at local, national, and global levels, and that does not cause damage to other ecosystems.
>
> **(Resolution H1)**

At the same time, in 1993, FOREST EUROPE member countries recognized the challenge of establishing an instrument for assessing and measuring the development of sustainable managed forests to produce balanced, politically relevant, scientifically sound, and practically feasible information, that is, information that is being used in the presentation of and reporting on forestry as well as in the monitoring and preparation of forest-related national policies. In this context, the pan-European set was first approved in 1998 and revised in 2003 as an internationally agreed policy instrument for promoting SFM by providing relevant information for forest policy development and evaluation, national forest policies, plans, and programs, and as a basis for cross-sectoral forest-related data collection. The current set consists of six criteria and 35 associated quantitative indicators describing

* www.foresteurope.org

the forest status and changes as well as a set of 17 qualitative indicators describing the national forest policies, institutions, and instruments toward SFM (Appendix 14A). By now, at pan-European level, the C&I set has served as the basis for State of Europe's Forests assessments in 2003, 2007, and 2011 and provided the basis for regional and national policy, analysis, and monitoring.

How these C&I are used by pan-European countries to assess SFM at the national level has been one of the research questions of a recent research project conducted by the European Forest Institute (see Baycheva et al. 2013). The study revealed that the pan-European C&I set is considered by FOREST EUROPE countries as a useful tool for monitoring, assessing, and reporting on the condition of forests and forest management; as a reference framework for the development, implementation, and evaluation of national forest policies and strategies (e.g., NFP); as a basis for guiding and setting the direction of forest management practices; as a framework for structuring and communicating information about forests and forestry intended for political decision-makers and other interested parties, identifying issues for research; and as a reference framework for developing national forest certification schemes or standards.

The revision of the pan-European indicator set is currently debated among specialists and decision-makers (Lier et al. 2013). Particular attention is given to the relevance of the C&I for SFM at the national level, lack of data for some indicators (e.g., expenditure for services [6.4] and landscape patterns [4.7]), identification of critical threshold values for indicators (e.g., minimum amount of deadwood left in the forest), and the development of new (e.g., *habitat trees*, *wood as building material and its effects on greenhouse gas emissions*, or *the share of urban forest of total forest area*) or combined indicators (e.g., *ecosystem services*). Recent developments in 2013–2014 take a step forward in this direction, as FOREST EUROPE has set up an expert group to analyze the use of the set of indicators for SFM and their linkages to other sectors and aims for a proposal for an updated set of the SFM indicators. In addition, there are ongoing initiatives by UNECE to improve the assessment of the status of forests by the current pan-European indicators.

In addition, two new experimental tools have been introduced within the "State of Forests and Sustainable Forest Management in Europe 2011" (listed in Barbati et al. 2014): (1) pilot reporting by 14 classes of European forest types (EFT), which included selected quantitative indicators and (2) key parameters for monitoring progress for a quantitative indicator.

As expected, these two new tools could be a step forward in the application of C&I set to a policy context. The use of EFT aims at stratifying a large forest territory into a smaller discrete number of more homogeneous ecological units to ease the analysis, interpretation, and reporting (Mc Roberts et al. 2011 in Barbati et al. 2014).

Barbati et al. (2014) tested the use of these new tools for monitoring progress on forest biodiversity conservation, and they concluded that it enables question-driven monitoring in many ways: (1) reporting by EFTs helps to interpret the variability in the values taken by the indicators explicitly considering ecological differences between EFTs, (2) temporal trends in forest area can be interpreted in terms of expansion of loss of forest habitats dominated by native and introduced tree species, and (3) progress in implementing biodiversity-friendly strategies can be quantitatively evaluated against baselines.

14.3.1.1 The Use of C&I to Assess SFM at a National Level: Measuring Changes in Forest Biodiversity in Finland

At the national level, Finland uses the pan-European C&I set for reporting and collecting comparable data on forest biodiversity and progress toward forest-related targets and international commitments. The Finnish example shows how a change in forest management policy has influenced the state of biodiversity, as assessed through repeated monitoring of red-listed species (FOREST EUROPE indicator on threatened forest species (indicator 4.8: Threatened forest species, see Appendix 14A)).

The application of the FOREST EUROPE indicators of SFM in Finland has produced a time series spanning more than 15 years. Although changes in forests are slow, the indicators have enabled certain trends to be distinguished and comparisons to be made. A national evaluation of threatened species has been conducted four times by the Ministry of the Environment: in 1983–1985, 1987–1991,

1997–2000, and 2007–2010. The last two evaluations are based on the IUCN criteria by the International Union for Conservation of Nature. The evaluations produce information about the number of threatened species, the causes of decline, risks, and proposals for improving their protection (indicator 4.8). The rate of decline of certain forest species has slowed in Finland or in some cases even stopped since the 1990s, although it has not been possible to halt the decline in forest species overall. The evaluation of threatened species (including vertebrates, invertebrates, vascular plants, cryptogams, and fungi and lichens) conducted in 2000 and 2010 showed that the decline slowed or stopped for 81 forest species but continued for 108 species. Retention trees at felling sites have been particularly important for curbing species decline (Parviainen and Västilä 2011; Lier et al. 2013).

Unlike in other European countries, strict forest protection is emphasized in Finland. Under various protection programs and decisions, the area of protected forests has been tripled over the past 35 years. The total area of protected forests is currently 2.2 million hectares or 9.6% of all forestland. The total area of protected forests and forests under restricted use is almost three million hectares or 13.0% of all forestland. The percentage of strictly protected forests in Finland (5.2% of forestland) is the highest in Europe. Most of the strictly protected forests are located in northern Finland, which is a consequence of its specific natural conditions. In southern Finland, where the percentage of strictly protected forests varies between 1.0% and 3.6%, biological diversity and protection is promoted through the Forest Biodiversity Programme for Southern Finland (METSO). The METSO program involves developing silvicultural methods, as well as voluntary measures by private forest owners to protect biodiversity and restoration management of protected areas in state ownership.

This example shows how a policy response to biodiversity decline (e.g., the plan created in 1997 for increasing the area of strictly protected forest area, the new biodiversity orientated forest law in 1997, and the program to increase the areas of protected forests by voluntary means supported with financial compensation in private forestland since 2003 (METSO)) has had an effect on the state of biodiversity and how it has been monitored according to a systematic and repeated data collection method and indicator.

14.3.1.2 The Use of C&I for Reporting on Sustainable Forest Management in Slovakia (Central Europe)

Slovakia joined the pan-European process for the development of C&I of SFM at the beginning in 1993. Slovakia, as a signatory state, obliged itself especially to (1) adopt C&I as a basis for international reporting on forest state and development; (2) support the development of national C&I taking specific conditions into account; (3) evaluate completeness, importance, and financial requirements of C&I, improve C&I when necessary—especially regarding forest functions that are not covered enough; and (4) evaluate improvement in SFM at a particular time interval using national C&I set.

In addition, a national system of C&I exists based on pan-European one for the purpose of forest certification. The system was developed by the Association of Forest Certification in Slovakia established under PEFC, which is the world's leading forest certification organization. PEFC Slovakia was established in 2002, and the Slovak Forest Certification System has been endorsed by PEFC Council since August 2005. The aim of the association is to promote SFM. Furthermore, it aims at the promotion of nature conservation principles and sustainable development in society. PEFC Slovakia currently includes 16 members from the forestry, wood processing, industry, and other stakeholders. The set of C&I was developed to be used at the regional and stakeholder level, but in compliance with pan-European set and PEOLG. For the assessment of the status of forests and for deciding whether the forests are to be certified, data and information from forest management plans are prevailingly used in order not to increase costs for the assessment. Besides the management plans, forest management evidence, National Forest Inventory (NFI), Green Report on forest status, reports from environmental, and other state administration are also often used.

Forests in Slovakia cover more than 40% of the country area, and almost 41% of that is managed by the state itself (under state enterprises) (Sarvašová et al. 2014). State forestry administration is separated from state enterprise and performs administration and the control of forest management. There are two levels of management planning in Slovakia: strategic national planning using NFP

SR and management unit level ensured by Forest Management Plans, which are created for forest owners (including state forest enterprise). SFM, considered as balanced yield meaning that annual cuttings should not exceed annual volume increment, has had a long tradition in Slovakia. However, SFM had been mostly thought of as sustainable yield management until recently when Slovakia was involved in several European activities toward increasing or sustaining not only forest yield but also biodiversity and socioeconomic values of forests (reporting for State of Europe's Forests (SoEF), forest certification activities (PEFC), NATURA 2000, etc.). Forest management at management unit is ensured by the authorized forest manager who is given a license and guarantee expert treatments to forest property in accordance with the law.

Nature conservation and protection have also a long history in Slovakia, and currently almost 30% of the forests are included in NATURA 2000 habitats, and around 60% of the forest area is under some level of nature protection (including NATURA 200) (Sarvašová et al. 2013). Over 70,000 ha of forests (3.6% out of the forests) are strictly protected, and no management is allowed there at all.

14.3.2 The Montréal Process C&I

The Montréal Process is essentially a voluntary binding agreement that different countries negotiated to layout the foundations of the application of a common set of C&I for the conservation and sustainable management of temperate and boreal forests in their respective countries. These C&I were developed by the Montréal Process working group. Originally, this group was formed in 1994 in Geneva, Switzerland, following the 1992 Earth Summit in Rio de Janeiro, Brazil, where the nonlegally binding Rio Forest Principles were adopted to acknowledge the importance of forests for the environment, the economic development, and the maintenance of all forms of life. There are 12 countries that are members of the Montréal Process working group: Argentina, Australia, Canada, Chile, China, Japan, Korea, Mexico, New Zealand, the Russian Federation, the Unites States, and Uruguay. The forests in these countries represent 83% of the temperate and boreal forests of the world, and 45% of the wood products manufactured in the world come from these countries (The Montréal Process 2009). Every country that participated in the Montréal Process was expected to use the C&I to report by 2012–2014 on their forest trends and progress to support SFM.

For the member countries, the Montréal Process C&I is a tool to share common basic concepts on the meaning of SFM and provides an international reference for the development of national policies. There are several indicators associated with each of the seven criteria (Appendix 14B). The indicators consist of qualitative or quantitative statements that are used to describe or support each criterion. In principle, C&I should be applied by considering linkages among them. This means that what indicates forest management sustainability is the combined use of different C&I. Indicators become most useful when measured periodically over time to evaluate if the evolvement of change and trends in forest management practices remains sustainable over time.

As indicated earlier, every country member is expected to use the C&I to report on forest sustainable management issues in their countries. There are more and more examples of forest national reports that use C&I of the Montréal Process.

In *Canada*, for example, the 2014 annual report on the state of Canada's forests is largely based on the indicators of the Montréal Process (Canadian Forest Service 2014). The intent was to lead discussions about the State of Canadian forests and improvements on forest management practices and policies and demonstrate the environmental credentials of Canada. The Government of Canada has presented sustainability or performance indicators in its annual reports for several years. A turning point on the use of indicators in Canada was the report published by the Canadian Council of Forest Ministers in 2006 to adapt the indicators of the Montréal Process to the Canadian context.

The contribution of the United States to report on the application of the Montréal Process indicators has also been important, with two reports produced in 2003 and 2010, respectively, to allow the public to evaluate the progress toward SFM practices in the forests of the United States.

For *Australia*, the most recent report on the state of the forests was produced in 2013 by the Montréal Process Implementation Group for Australia and the NFI Steering Committee. All the indicators of the Montréal Process were used to provide information on SFM on public, private, and leasehold forests for both native forests and plantations. The report is rich in information, as it covers all forest types, including the northern closed tropical rainforests, the tall forests of eastern and southwestern Australia, and the inland acacia and eucalypt woodland forests.

New Zealand also produced a report in recent years on the sustainable management of their forests using indicators of the Montréal Process (Novis et al. 2009). Their report focused on the presentation of indicators for both commercial plantation forests and the Indigenous forests that are largely protected. Even though data were missing for some of the indicators, the use of the Montréal Process was a priority for the Government of New Zealand to provide clear indications to the international community that all their forests are managed sustainably.

China, a country with a variety of forests, produced a national report on SFM using C&I of the Montréal Process (State Forestry Administration 2013). The indicators contributed to the development of policies to increase forest area and stemwood production on the basis of sustainable practices, which included plans to protect natural forests and endangered species and improve the livelihoods of forest farmers.

14.4 THE USE OF C&I FOR TROPICAL FORESTS

14.4.1 The ITTO Process

The ITTO is an intergovernmental organization promoting sustainable development through the sustainable management, use, and conservation of tropical forests. It has 56 member governments (and the European community), which collectively represent 90% of the world tropical timber trade and about 80% of the world's tropical forests. ITTO has developed a series of internationally agreed policy documents for achieving SFM and forest conservation and assists tropical member countries to adapt these to local circumstances and to implement them in the field. In addition, ITTO collects, analyzes, and disseminates data on the production and trade of tropical timber and facilitates the development of value-added, forest-based industries at both the community and industrial scales.

The ITTO was a pioneer in developing C&I for SFM and published the first set of C&I for the sustainable management of natural tropical forests before the summit in Rio in 1992. Since then, many sets have been developed, often with the assistance of international organizations, and almost every country is now covered by at least one set. ITTO revised and updated its 1992 set of C&I based on experiences acquired in the implementation of SFM in tropical countries and research advances in this field, publishing a new set in 1998.

ITTO has recently performed a global study to investigate the impact of C&I on SFM worldwide and to gain a better understanding of the experiences of countries across the world in using C&I (ITTO 2012). For the ITTO process, the study showed that a number of tropical producers are using C&I as a framework to monitor, assess, and report on SFM at the national level often with the support of ITTO. However, some countries have developed their own C&I to better reflect national particularities. These were usually based on or inspired by ITTO C&I set. Some ITTO producers use national or FMU C&I as a basis for monitoring, assessing, and reporting at subnational and FMU levels, also under ITTO support. Many countries reported that they use ITTO C&I as a framework for national-level forest management planning (FMP) and for the development of NFP. Some countries use C&I for FMP and strategies at subnational, local, and/or field levels. Several countries suggested the use of C&I as a basis for forest certification programs (FSC and PEFC). In addition, almost all respondents included in the study indicated that they involve stakeholders in the aforementioned C&I-related activities.

14.4.2 The ATO Process

The ATO is an intergovernmental organization created in 1976 for cooperation on forestry issues relating to its 14 member countries, which between them contain over 75% of the tropical natural forests on the African continent. One of the major objectives of the ATO is to promote the production and trade of African timber within the framework of SFM. Within the mandate of this objective, the ATO has undertaken several activities and also executed several projects on short-, medium-, and long-term bases. Most of the projects and activities have been executed with the technical and financial assistance of certain donor agencies and governments, which continue to be ATO's development partners. Such projects and activities include studies, training programs, workshops, and seminars designed to support the national efforts of the ATO member countries. The ATO developed, with the financial assistance of the European Union (EU) and the technical collaboration of the Center for International Forestry Research, a set of principles, criteria, and indicators (PCI) to promote the sustainable management of African forests.

14.4.3 The ATO/ITTO PCI

In 2001, the ATO/ITTO PCI were developed under the title ATO/ITTO Principles, Criteria and Indicators for the Sustainable Management of African Natural Tropical Forests as a result of the effective collaboration between ITTO and African tropical timber producers (Appendix 14C). The ATO/ITTO PCI set includes one principle, five criteria, 33 indicators, and 45 subindicators at the national level and three principles, 15 criteria, 57 indicators, and 140 subindicators at the FMU level.

ATO member countries are (only producers) Angola, Cameroon, Central African Republic, Congo, Côte d'Ivoire, Democratic Republic of Congo, Equatorial Guinea, Gabon, Ghana, Liberia, Nigeria, Sao Tome et Principe, and the United Republic of Tanzania.

ITTO includes the following member countries (only producers): Bolivia, Brazil, Cambodia, Cameroon, Central African Republic, Colombia, Congo, Côte d'Ivoire, Democratic Republic of the Congo, Ecuador, Fiji, Gabon, Ghana, Guatemala, Guyana, Honduras, India, Indonesia, Liberia, Malaysia, Mexico, Myanmar, Nigeria, Panama, Papua New Guinea, Peru, the Philippines, Suriname, Thailand, Togo, Trinidad and Tobago, Vanuatu, and Venezuela.

14.4.4 The Tarapoto Process

The Tarapoto process started in 1995, when member countries of the Amazon Cooperation Treaty drafted the proposal on the Amazon Forests Sustainability Criteria and Indicators at a regional workshop in Tarapoto—Peru. Member countries at that time were Bolivia, Brazil, Colombia, Ecuador, Guyana, Peru, Suriname, and Venezuela. The Amazon countries identified 12 criteria and as many as 77 indicators within three categories reflecting specific features of the ecosystems as well as social and cultural aspects. Evaluation and validation of the proposed indicators in each country were performed through a process of national consultations (NC) with the aim to assess the adequacy of the applicability of the criteria and the indicators to measure Amazon forest sustainability. During the period 1996–2000, the eight NC were carried out under the financial support of the government of Finland and the technical support from FAO, involving as many as 351 institutions and 830 participants from all sectors. In 2004, the Amazon Cooperation Treaty Organization (ACTO), with the support of the FAO, began validating the 15 priority indicators to assess the feasibility of their implementation (Appendix 14D). Recently, ITTO and Tarapoto C&I guided policy decisions in the Amazon countries and were also incorporated in national policies with various levels of influence. ITTO and ACTO signed a memorandum of understanding in 2010 to implement strategic policies related to the application of C&I for SFM. The ITTO global study mentioned in subsection 14.4.1 identified for the Tarapoto process resulted the same experience and challenges in using C&I of SFM for monitoring, assessing, and reporting at the national, subnational, or FMU levels as in the case of ITTO process (ITTO 2012).

14.5 COLLABORATION AND COOPERATION AMONG THE C&I PROCESSES

Repeated requests over the past years, especially from international expert conferences, for more and improved collaboration among C&I processes had led to collaborative work toward harmonization as well as improved communication and coordination among the C&I processes. Since 1995, a number of international seminars, conferences, and workshops have been launched upon the need on how to clarify terminology and facilitate comparison between countries, particularly in relation to monitoring and reporting. Although there was a general agreement on the need to ensure comparability between the different processes, it was also recognized that they should pursue their goals unimpeded and in a way that would be compatible with their particular environmental and socioeconomic contexts (FAO 1997). In recent years, further possibilities for improvements in harmonizing ongoing initiatives related to C&I (e.g., ITTO, Montréal Process, FOREST EUROPE, FAO FRA) have been discussed at the Inter-Criteria and Indicator Process Collaboration Workshop (Poland, 2006), the Forest Criteria and Indicators Analytical Framework and Report Workshop (Finland, 2008), and the International Seminar on Challenges of SFM—Integrating Environmental, Social and Economic Values of Forests (Japan, 2011). Subsequently, the joint workshop of the Montréal Process, ITTO, Forest Europe, and FAO in Canada, in October 2011, has led to a process to develop a Collaborative Forest Resources Questionnaire (CFRQ) to reduce the proliferation of monitoring requirements and associated reporting burdens imposed by the different processes and initiatives. The preparatory process for the collaborative reporting stimulated the cooperation among organizations, processes, and countries in the field of forest reporting. In this context, the CFRQ is foreseen to contain a set of shared variables from the FRA—and will be augmented by additional variables from the FRA, FOREST EUROPE/UNECE, ITTO, OFAC, and the Montréal Process. In addition, in 2012, the Technical Consultation on *Preparation* for *Global Forest Resources Assessments 2015* held in Ispra (Italy) and the Stream Lining Forest Based Reporting Workshop in Florida (United States) have further highlighted the benefits of collaborative work among the processes and continued relevance of C&I for SFM. However, in spite of these joint efforts to streamline and rationalize national reporting for SFM tropic, FRAs, and regional forest assessments, the need to continue to strengthen efforts to reach global consensus on key concepts and terms used in the international discussion on C&I has been underscored (Baycheva et al. 2013) (Appendices 14E and 14F).

14.6 INFORMATION SOURCES FOR REPORTING ON THE C&I OF SFM

14.6.1 Currently Used Information Sources for Reporting on SFM

Since C&I sets usually include various kinds of indicators from simple timber production to social and economic issues, no single data source can be found to cover all. Thus, various methods, approaches, and data sources are used to report on particular C&I often with low or unknown precision (Table 14.2).

In Europe, there have been several challenges identified in relation to the pan-European C&I (Baycheva et al. 2013): (1) no data available for certain indicators because they are not included in NFI, or national statistics do not exist to answer; (2) the need to convert, adapt, and extrapolate national data to fit to the SoEF reporting tables; (3) difficulties with the terminology and definitions and the structure of the SoEF reporting tables in general.

Although C&I were developed to account for sustainability of multipurpose forest functioning, it is often so that there are no data for many indicators at the national (or management unit) level, and they are reported only using expert assessment (judgment). Moreover, it is often not easy to fit national classifications to that provided in reporting tables. In Europe, most of the countries use some kind of forest management plans to manage their forests at the management unit and national-level sustainably. These management plans are often summarized to provide forest state and its change at the national level and are then often used for international reporting (as is the case of, e.g., Slovakia). Data from such management plans are able to cover only some indicators (mostly only yield-related indicators), and other sources are needed to fill in the other indicators. In most of the European countries, a comprehensive annually published report exists (e.g., Green Report in Slovakia), which are usually prepared so that it fit with international reporting on SFM at some level. However, there are still indicators (e.g., socioeconomic related) that require intensive

TABLE 14.2
Examples of Data Sources Used for the Purpose of Reporting on the C&I of SFM in Selected Countries

Country (Process)	Data Source
Finland (FOREST EUROPE)	NFI, National Statistical Agency
Slovakia (FOREST EUROPE)	Forest Management Plans, Forest Management Evidence, Green Report, Databases gathered by Ministry of the Environment, etc.
Canada (Montréal Process)	Annual reports of the State of Canada's forests published by the Canadian Forest Service (Natural Resources Canada)

cooperation with public and private owners, which is not always easy. Then, the worse is the cooperation between subjects, the less precise are the results.

In addition, to increase the plausibility of the outcome, more objective approaches would be desired. Without adequate scientific knowledge, the attempts to implement SFM have generally been prescriptive, ecologically conservative, and often disproportionate to the desired outcome (Burton et al. 2006).

14.6.2 Statistically Based Forest Inventories as a Challenge for Harmonized Reporting

Forest inventories and other ecosystem monitoring activities of EU member states are reliable sources of forest-related information. Nevertheless, due to the increasing demands by the forest and other related sectors, there is a clear need to make data collection and analysis more efficient and harmonized, to enhance interoperability, to provide information for the main parameters on complete coverage maps, and to provide up-to-date information. Better knowledge on forest resources, their availability, and their sustainable future supply can help to predict developments of forest ecosystem goods and services and thus can serve a broad range of affected stakeholders and create new opportunities for an innovative, sustainable, and inclusive bioeconomy.

There are often different methods over countries to gather information on forest state and changes at both national and management unit level. For instance, in Slovakia, there is a stand-wise inventory to develop management plans. The inventory is carried out annually only on 1/10 of the forest area so that the entire area is inventoried within ten years. Thus, there is a time lag using this source of information for reporting at the national level so that the status reported at present is 10 years old for 1/10 of the area and so on. Similar stand-wise inventory systems can be found in many other European countries, however, with different methodological approaches.

Sample-based NFIs provide a chance to develop more reliable data source with known precision of the results, which is important for objective decision-making at the national and regional levels. However, there are often big differences among countries especially in the following aspects: (1) some countries have very long history of such NFI (e.g., in Finland, the NFI has been carried out since 1920), but some countries have just launched their first NFI so the long-term trends are not possible to be assessed from now; (2) there is often a different time interval between two inventory cycles; (3) different thresholds for some important forest or tree attributes are frequent (e.g., DBH, definition of forest, and missing tree components). To reflect the challenge in using NFIs as comprehensive monitoring system and to address the mentioned shortcomings coming from different methodologies of NFIs over Europe, the European National Forest Inventory Network (ENFIN) was established as a platform for collaboration between NFI organizations. In 2004, the ENFIN was successful in competition and was granted a project to make a first step toward harmonization between European NFIs. The project named "COST E43 (Harmonization of NFI in Europe: Techniques for Common Reporting)" was supported by EU under the program called "European Cooperation in Science and Technology (COST)." Detailed information on NFI over Europe were gathered and summarized in a book as results from the project (Tomppo et al. 2010), and harmonized definitions of relevant tree and forest attributes were established. The next step toward harmonization of NFIs was a COST project (2010–2014) "COST FP1001 (Improving Data and Information on the Potential Supply of Wood Resources: A European Approach from Multisource National Forest Inventories [USEWOOD])." Nowadays, the group is still active and has an ambition to establish the basis for common reporting at the European level since there is a clear need for harmonized and relevant data and reporting obligations related to EU forest-related policies and international agreements (e.g., Millennium Development Goals on preventing deforestation, the Land Use Land Use Change and Forestry (LULUCF), the Convention on Biological Diversity, and United Nations Framework Convention on Climate Change) call for comparable data and information on Europe's forest ecosystems (Standing Forestry Committee 2012).

14.6.3 Remote Sensing Techniques to Assess Indicators of Sustainable Forest Management

For the assessment and monitoring of indicators for SFM, in addition to ground-based sampling data collected in NFI or other forest inventory data used, remote sensing techniques can provide important complementary information. Diverse remote sensors have potential for obtaining different types of information (Section 14.6.3.2 describes some of them more in detail), which can be employed to assist the assessment of SFM indicators:

- Passive optical sensors (e.g., multispectral cameras and hyperspectral scanners) acquire information on sunlight radiation reflected from a surface at selected bands along the spectrum, which in the case of forest canopies can be used to evaluate differences in photosynthetic activity. Therefore, they have potential for providing information about species, forest health, water content, and productivity. In the context of SFM indicator assessment, they can be beneficial for assisting the evaluation of these criteria: (1) Extent of forest resources—forest coverage and LULUCF activities; (2) Biological diversity—species-wise estimations and landscape patterns; (3) Forest health—water and chlorophyll content, leaf area index (LAI), and defoliation; (4) Productive functions—estimations for standing volume, growing stock, and so forth; and (5) Protective functions—assessment of forest evapotranspiration (ET) and its effect on hydraulic systems. The shortcoming of passive sensors in forest assessment is that they capture only information reflected or emitted by the dominant canopy, neglecting suppressed trees and understory.
- Active sensors (e.g., radar and LiDAR, radio and light detection and ranging, respectively) emit polarized electromagnetic radiation and detect the characteristics of their signals bouncing back from a surface. When directed to forest ecosystems, it is possible to achieve partial or total penetration of the canopy, depending on the wavelength used, providing details on the forest vertical profile and precise ground elevation. Thus, they are also employed to deliver accurate digital terrain models (DTM), which would otherwise be impractical to obtain in forested areas. This allows the derivation of predictors related to the height of canopy elements, and therefore, they have the best potential for biomass estimation and forest structure characterization. With regard to SFM indicators, their derived information can be advantageous for the assessment of these criteria: (1) Extent of forest resources—biomass and carbon budgets can be estimated more accurately when using active sensors, with potential to detect selective logging; (2) Biological diversity—forest structure and gap occurrence, deadwood, and tree size inequality; (3) Forest health—forest damage and LAI; (4) Productive functions—estimations can be carried out for entire diameter distributions, allowing users to even estimate separate timber assortments; and (5) Protective functions—DTMs can be employed for watershed analysis, and forest structure information may also be linked to erosion, runoff etc.

In the case of (6) socioeconomic functions, remote sensing data sources can be useful only in the sense of using forest cover estimates to aggregate/disaggregate statistics (see Section 14.6.3.1).

In remote sensing, the platform from which sensor's information is acquired makes an important difference affecting across all these types of sensors. Remote sensors may be mounted on board of satellites, aircraft, or they may as well be used to obtain field plot information. In this order, data are acquired with more detail at a higher cost. Consequently, there are many trade-offs involved in the characteristics and benefits of either type of remote sensing method since, for instance, lower spatial resolutions can usually be complemented with a higher coverage and/or temporal recurrence. The most cost-effective solutions integrate and combine various types of datasets, with the purpose of obtaining the final SFM indicators via multilevel estimation (see, e.g., Valbuena 2014).

14.6.3.1 Multilevel Estimation of SFM Indicators: Upscaling and Downscaling

Depending on the data available in each case, the use of remote sensing methods to assist the estimation of SFM indicators may pursue a number of different objectives. Various sensors can provide with a number of auxiliary variables that shall be used to assist statistical aggregation or disaggregation of data obtained on the ground at diverse scales (Figure 14.2). Therefore, the use of this auxiliary information allows users to obtain spatially explicit estimates. In other words, they enable the observation of spatial changes in a given indicator, as continuous information is obtained wall-to-wall throughout a target study area. For the same reason, they also allow users to estimate changes by comparing multitemporal data (e.g., Duveiller et al. 2008; FAO and JRC 2012). By means of calibrating remote sensing predictors with forest data obtained in the field, the accuracy of forest estimates can be improved in comparison to using the sample plots alone. The key to any such type of generalization is to know what are the key spatial determinants associated with a given SFM indicator (Eigenbrod et al. 2010) and whether or not they can be related to properties obtained with remote sensors. If this relation is only partial, ancillary sources of information may as well be employed to assist the remote sensing estimations, such as latitude, elevation, terrain ruggedness, and accessibility.

Downscaling may be carried out when national or regional statistics, or coarse-resolution estimates, are available, but the objectives require to estimate a given SFM indicator at a finer scale. Knowledge on the factors determining the spatial distribution of those statistics can be applied to disaggregating those statistics to the resolution provided by remote sensors. Such approaches have been used for estimating land use change (Dendoncker et al. 2007), forest biomass, carbon and growing stock (Kindermann et al. 2008; Verkerk et al. 2015), forest management systems (Hengeveld et al. 2012), and tree species distribution (Brus et al. 2012). As the auxiliary data are used as a proxy for the indicator, the resulting disaggregation is only to be used for the study of broadscale patterns in the target SFM indicator (Eigenbrod et al. 2010).

Upscaling consists of using the data available for estimating the target indicator in a stagewise manner, gradually lowering its resolution while increasing its spatial extent, until full coverage of the area of interest is achieved (McInerney et al. 2010). Traditionally, forest estimation has been carried out using information measured in sample forest plots, from which the desired variable may be derived and aggregated at the scale of management units with a certain level of uncertainty. Remote sensing provides auxiliary variables that can be related to the forest response, and they are acquired over larger extents although they provide less details than the sample plots (Valbuena 2014). Estimations for a target SFM indicator can be inferred throughout the area surveyed by the remote sensor, and the accuracy of these predictions will depend on the relation between the response and the auxiliary variables. Provided that the uncertainty of this relation is low, the overall accuracy of the estimation at the level of the area of interest can be more accurate than traditional plot-wise estimation (e.g., Patenaude et al. 2004; Pesonen et al. 2009). From forest plot information, these estimations can be upscaled to multiple scales of estimation, up to satellite imagery in the case of large-scale estimation, for a cost-effective use of the data available (Andersen et al. 2011).

Approaches can vary from two-stage procedures when using one remote sensing source such as satellite imagery (e.g., Tomppo et al. 2009) or airborne laser scanning (ALS, one type of airborne

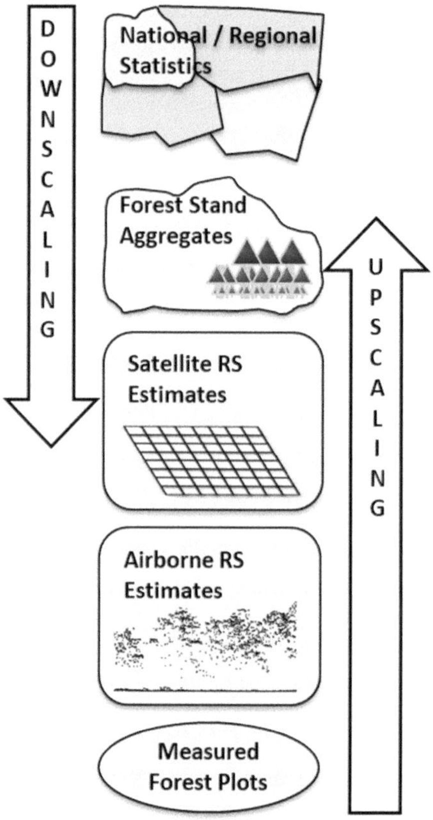

FIGURE 14.2 Upscaling and downscaling approaches for multilevel estimation of SFM indicators and their spatial distribution.

LiDAR which delivers discrete return information) (e.g., Næsset and Gobakken 2008), or multilevel approaches when combining various levels of auxiliary remote sensing data (McInerney et al. 2010; Andersen et al. 2011). SFM indicators can this way be estimated at separate phases: first, plot-level totals are computed from all the individual calculations obtained for each single tree in each plot. These values are then used to develop empirical relationships between these plot aggregates and various predictors obtained from remote sensing data. Examples of these predictors may be the normalized difference vegetation index for passive optical sensors or height percentiles for ALS. Upscaling approaches are widely employed for remote sensing-assisted estimation of forest variables traditionally used for forest management (Andersen et al. 2011). They can as well be employed to predict the spatial distribution of indicators that can be useful for SFM, for instance, in estimating LAI (Jensen et al. 2008), forest structure and tree size inequality (Valbuena et al. 2013), coarse woody debris (CWD) (Pesonen et al. 2009), tree defoliation (Solberg et al. 2006), or determining forest types and species dominance (Tomppo et al. 2009), or even species-specific tree diameter distributions (Packalén and Maltamo 2008).

14.6.3.2 Potential of Remote Sensing for Estimating SFM Indicators

14.6.3.2.1 Standing Volume, Growing Stock, and Carbon Budget

Estimates for carbon stock in living woody above (AGB) and below (BGB) ground biomass can be obtained from remote sensing-assisted estimation (Næsset and Gobakken 2008). Carbon stocks can be directly derived from biomass estimates, using allometric equations, which may include various AGB and BGB components (e.g., crown, foliage, stump, roots, etc.). This allometry is to be used for the remote sensing estimation of biomass components, which can be subsequently converted to CO_2 equivalents following the Guidelines for National Greenhouse Gas Inventories.* Furthermore, remote sensing allows users to obtain a more detailed information on the spatial variation in carbon content within woodlands, being this especially relevant in the case of structurally heterogeneous forest areas (Patenaude et al. 2004). Then, multitemporal analysis can be used for mapping deforestation (Duveiller et al. 2008; FAO and JRC 2012).

14.6.3.2.2 Deadwood

CWD may be estimated by remote sensing based on the relation of its dynamics with forest gap occurrence, as canopy gaps are originated by natural tree mortality. Biologically functioning prediction models have been constructed for the direct estimation of CWD volumes (Pesonen et al. 2009). Silvicultural operations, however, change this condition by generating gaps artificially. Therefore, CWD estimations by remote sensing are less reliable in managed forest areas with low CWD volumes.

14.6.3.2.3 Leaf Area Index, Defoliation, and Evapotranspiration

Remote sensing has been used to estimate the relative ET of forests, usually using estimations of the effective LAI or angular canopy closure as a proxy. Canopy evaporation is mostly governed by the biophysical properties of the forest canopy. The bigger the LAI is, the less amount of net solar radiation reaches the ground, reducing the amount of surface water that will be evaporated. Ge et al. (2011) analyzed the influence of LAI in the proportions of each one of the three water flux components of the ET: (1) canopy transpiration (Et), (2) canopy evaporation (Ec), and (3) ground evaporation (Eg). They observed that higher LAI influences both decreasing Eg and increasing Ec. As a result, although overall ET presents no significant correlation with LAI, its components do. These components can also be estimated by remote sensing (Jensen et al. 2008). From ALS, LAI, and defoliation may be estimated from the rate of penetration of laser pulses (Solberg 2010). Then, it can be further studied how LAI influences the ET of a certain forest area.

14.6.3.2.4 Tree Diameter Distributions and Inequality

To describe forest structure with an indicator independent from stand development, a statistic of relative dispersion (concentration) is needed, such as the coefficient of variation or the Gini coefficient (GC). The GC has been found as a reliable indicator of tree diameter inequality that can be directly applied to SFM (Duduman 2011). Valbuena et al. (2013) carried out estimations of GC for assisting management in continuous cover forestry systems and recommended to accompany relative dispersion with an indicator of asymmetry for a concise characterization of the complexity of diameter distributions. If more details are desired, ALS remote sensing may as well be used for estimating complete tree lists. One of the first approaches was to estimate quantiles for either diameter or basal area-weighted distributions. Alternatively, estimating Weibull model parameters allowed inferring diameter distributions presenting a wide range of simple shapes (Gobakken and Næsset 2004). Packalén and Maltamo (2008) retrieved these distributions by species, combining both active and passive sensors.

14.7 ACKNOWLEDGMENTS

The authors acknowledge FAO for providing permission to publish Figure 14.1. The authors also thank Matej Schwarz for sharing his experience in the implementation and use of Pan-European C&I in Slovakia. The work was supported by the Slovak Research and Development Agency via the project no. APVV-19-0183 and by the Ministry of Agriculture and Rural Development of the Slovak Republic (project EPRIBLES, No. 08V0301).

* IPCC. 2006. Guidelines for National Greenhouse Gas Inventories. https://www.ipcc-nggip.iges.or.jp/public/2006gl/index.html.

REFERENCES

Andersen, H., J. Strunk, H. Temesgen, et al. 2011. Using multilevel re-mote sensing and ground data to estimate forest biomass resources in remote regions: A case study in the boreal forests of interior Alaska. *Can. J. Remote Sens.* 37:596–611.

Barbati, A., M. Marhcetti, G. Chirici, et al. 2014. European Forest Types and Forest Europe SFM indicators: Tools for monitoring progress on forest biodiversity conservation. *For. Ecol. Manag.* 321:145–157.

Baycheva, T., H. Inhaizer, M. Lier, et al. 2013. *Implementing criteria and indicators for sustainable forest management in Europe*. Final Report. Joensuu, Finland: European Forest Institute.

Brus, D.J., G.M. Hengeveld, D.J.J. Walvoort, et al. 2012. Statistical mapping of tree species over Europe. *Eur. J. For. Res.* 131:145–157.

Burton, P.J., C. Messier, W.L. Adamowicz, et al. 2006. Sustainable management of Canada's boreal forests: Progress and prospects. *Ecoscience* 13:234–248.

Canadian Council of Forest Ministers. 2006. *Criteria and indicators of sustainable forest management in Canada: National Status 2005*. Ottawa, Ontario, Canada: Canadian Council of Forest Ministers.

Canadian Forest Service. 2014. *The state of Canada's forests: Annual report 2014*. Ottawa, Ontario, Canada: Natural Resources Canada, Canadian Forest Service.

Castaneda, F. 2000. Criteria and indicators for sustainable forest management: International processes. Current status and the way ahead. *Unasylva* 51(4):34–40.

De Montalbert, M.R. 1995. Cross-sectoral linkages and the influence of external policies on forest development. *Unasylva* 45:25–97.

De Montalbert, M.R. and F. Schmithüsen. 1993. Policy and legal aspects of sustainable forest management. *Unasylva* 44:3–9.

Dendoncker, N., P. Bogaert, M. Rounsevell, et al. 2007. Empirically derived probability maps to downscale aggregated land-use data. In *Modelling land-use change: Progress and applications*, eds. E. Koomen, J. Stillwell, A. Bakema, and H.J. Scholten, pp. 117–132. Dordrecht, The Netherlands: Springer.

Duduman, G. 2011. A forest management planning tool to create highly diverse uneven-aged stands. *Forestry* 84:301–314.

Duveiller, G., P. Defourny, B. Desclée, et al. 2008. Deforestation in Central Africa: Estimates at regional, national and landscape levels by advanced processing of systematically-distributed Landsat extracts. *Remote Sens. Environ.* 112:1969–1981.

Eigenbrod, F., P.R. Armsworth, B.J. Anderson, et al. 2010. The impact of proxy-based methods on mapping the distribution of ecosystem services. *J. Appl. Ecol.* 47:377–385.

FAO. 1989. *Plant genetic resources: Their conservation in situ for human use*. Prepared in collaboration with the United Nations Educational, Scientific and Cultural Organization (UNESCO), UNEP and IUCN, Rome, Italy.

FAO. 1997. The development of criteria and indicators for sustainable forest management. In *State of the world's forests*. Rome, Italy: Food and Agriculture Organization of the United Nations.

FAO. 2006. *Global forest resources assessment 2005—Progress towards sustainable forest management*. FAO Forestry paper 147. Rome, Italy. www.fao.org/forestry/fra2005/en/. Accessed September 22, 2014.

FAO. 2014. *Membership of world countries in nine C&I processes*. www.fao.org/forestry/ci/16609/en/. Accessed September 22, 2014.

FAO and JRC. 2012. *Global forestland-use change 1990–2005*, by E.J. Lindquist, R. D'Annunzio, A. Gerrand, et al. FAO Forestry Paper No. 169. Rome, Italy.

Ge, Z.M., X. Zhou, S. Kellomäki, et al. 2011. Climate, canopy conductance and leaf area development controls on evapotranspiration in a boreal coniferous forest over a 10-year period: A united model assessment. *Ecol. Modell.* 222:1626–1638.

Ghazoul, J. and A. Hellier. 2000. Setting critical limits to ecological indicators of sustainable tropical forestry. *Int. For. Rev.* 2:243–253.

Gobakken, T. and E. Næsset. 2004. Estimation of diameter and basal area distributions in coniferous forest by means of airborne laser scanner data. *Scand. J. Forest. Res.* 19:529–542.

Hengeveld, G.M., G.-J. Nabuurs, M. Didion, et al. 2012. A forest management map of European forests. *Ecol. Soc.* 17(4):53.

Hickey, G.M. and J.L. Innes. 2005. *Scientific review and gap analysis of sustainable forest management criteria and indicators initiatives*. FORREX Forest Research Extension Partnership, Kamloops, British Columbia, FORREX Series 17.55 p.

ITTO. 1992. *Criteria for the measurement of sustainable tropical forest management*. International Tropical Timber Organization Policy Development. Series No. 3. Yokohama, Japan.

ITTO. 2012. *Uses and impacts of criteria and indicators for sustainable forest management at the field/FMU level and other operational levels*. Forty-sixth Joint Session of the Committee. Yokohama, Japan.

Jensen, J.L.R., K.S. Humes, L.A. Vierling, et al. 2008. Discrete return LiDAR-based prediction of leaf area index in two conifer forests. *Remote. Sens. Environ.* 112:3947–3957.

Kindermann, G.E., I. McCallum, S. Fritz, et al. 2008. A global forest growing stock, biomass and carbon map based on FAO statistics. *Silva Fenn.* 42:387–396.

Lier, M., J. Parviainen, C. Nivet, et al. 2013. European initiatives on criteria and indicators for forest biodiversity. In *Integrative approaches as an opportunity for the conservation of forest biodiversity*, eds. D. Kraus and F. Krumm, pp. 32–43. Freiburg, Germany: European Forest Institute.

McDonald, G.T. and M.B. Lane. 2004. Converging global indicators for sustainable forest management. *For. Policy. Econ.* 6:63–70.

McInerney, D.O., J. Suárez-Mínguez, R. Valbuena, et al. 2010. Forest canopy height retrieval using LiDAR data, medium-resolution satellite imagery and kNN estimation in Aberfoyle, Scotland. *Forestry* 83:195–206.

McRoberts, R.E., G. Chirici, S. Winter, et al. 2011. Prospects for harmonized biodiversity assessments using national forest inventory data. In *National forest inventories: contributions to forest biodiversity assessments*, eds. G. Chirici, S. Winter, and R.E. McRoberts, pp. 41–97. Heidelberg, Germany: Springer.

Montréal Process Working Group. 2009. *Criteria and indicators for the conservation and sustainable management of temperate and boreal forests*, 4th edn. Montréal Process. www. montrealprocess.org/Resources/Publications/index.shtml. Accessed September 22, 2014.

Næsset, E. and T. Gobakken. 2008. Estimation of above- and below-ground biomass across regions of the Boreal forest zone using airborne laser. *Remote. Sens. Environ.* 112:3079–3090.

Novis, J., R.R. Palma, A. Reid, et al. 2009. *Sustainable management of New Zealand's forests. The 2008 New Zealand country report on the Montréal Process criteria and indicators*. Wellington, New Zealand: Ministry of Agriculture and Forestry.

Packalén, P. and M. Maltamo. 2008. Estimation of species-specific diameter distributions using airborne laser scanning and aerial photographs. *Can. J. For. Res.* 38:1750–1760.

Parviainen, J. and S. Västilä. 2011. *State of Finland´s Forests 2011. Based on the criteria and indicators of sustainable forest management*. Wellington, New Zealand: Ministry of Agriculture and Forestry and Finnish Forest Research Institute.

Patenaude, G., R.A. Hill, R. Milne, et al. 2004. Quantifying forest above ground carbon content using LiDAR remote sensing. *Remote Sens. Environ.* 93:368–380.

Pesonen, A., O. Leino, M. Maltamo, et al. 2009. Comparison of field sampling methods for assessing coarse woody debris and use of airborne laser scanning as auxiliary information. *For. Ecol. Manage.* 257:1532–1541.

Rametsteiner, E. and M. Simula. 2003. Forest certification—An instrument to promote sustainable forest management? *J. Environ. Manage.* 67:87–98.

Sarvašová, Z., J. Šálka, and Z. Dobšinská. 2013. Mechanism of cross-sectoral coordination between nature protection and forestry in the Natura 2000 formulation process in Slovakia. *J. Environ. Manage.* 127:S65–S72.

Sarvašová Z., Z. Dobšinská, and J. Šálka. 2014. Public participation in sustainable forestry: The case of forest planning in Slovakia. *iForest* 7:414–422. www.sisef.it/iforest/contents/?id=ifor1174-007. Accessed April 2015.

Schmithüsen, F., F. Bisang, and W. Zimmerman. 2001. *Cross-sectoral linkages in forestry. Review on available information and consideration on future research.* Rome, Italy: FAO. www. fao.org/docrep/003/AA002E/AA002E00.htm. Accessed October 4, 2014.

Solberg, S. 2010. Mapping gap fraction, LAI and defoliation using various ALS penetration variables. *Int. J. Remote Sens.* 31:1227–1244.

Solberg, S., E. Næsset, K.H. Hanssen, and E. Christiansen. 2006. Mapping defoliation during a severe insect attack on scots pine using airborne laser scanning. *Remote Sens. Environ.* 102:364–376.

Standing Forestry Committee. 2012. *Ad hoc working group on forest information and monitoring*. Final report "Forest information needs, required resources, ways and means". http://ec.europa.eu/agriculture/fore/publi/sfc-wg6–2012_en.pdf. Accessed October 8, 2014.

State Forestry Administration. 2013. *National report on sustainable forest management.* Beijing, China: Forestry Publishing House, People Republic of China.

Stork, N.E., T.J.B. Boyle, V. Dale, et al. 1997. *Criteria and indicators for assessing the sustainability of forest management: Conservation of biodiversity*. Bogor, Indonesia: Center for International Forestry Research, Working Paper No. 17.

Tomppo, E., C. Gagliano, F. De Natale, et al. 2009. Predicting categorical forest variables using an improved k-nearest neighbour estimator and Landsat imagery. *Remote Sens. Environ.* 113:500–517.

Tomppo, E., T. Gschwantner, M. Lawrence, et al. 2010. *National forest inventories – Pathways for common reporting*. Dordrecht, The Netherlands: Springer.

UNECE/FAO. 2013. *Forest products annual market review 2012–2013*. Geneva Timber and Forest Study Paper 33. Geneva: Forestry and Timber Section.

United Nations. 1993. *Report of the United Nations conference on environment and development.* Rio de Janeiro, June 3–14, 1992: Volume I: Resolution Adopted by the Conference. https://documents.un.org/conferences/environment/rio1992

Valbuena, R. 2014. Integrating airborne laser scanning with data from global navigation satellite systems and optical sensors. In *Forestry applications of airborne laser scanning. concepts and case studies: managing forest ecosystems series*, eds. M. Maltamo, E. Næsset, and J. Vauhkonen, pp. 63–88. Dordrecht, The Netherlands: Springer.

Valbuena, R., P. Packalén, A. García-Abril, et al. 2013. Characterizing forest structural types and shelterwood dynamics from lorenz-based indicators predicted by airborne laser scanning. *Can. J. For. Res.* 43:1063–1074.

Verkerk, P.J., C. Levers, T. Kuemmerle, et al. 2015. Mapping wood production in European forests. *For. Ecol. Manage.* 357:228–238.

Vidal, C., A. Lanz, E. Tomppo, et al. 2008. Establishing forest inventory reference definition for forest and growing stock: A study towards common reporting. *Silva. Fenn.* 42:247–266.

WCED. 1987. *Our common future.* Report of the World Commission on Environment and Development. United Nations.

Wijewardana, D. 2008. Criteria and indicators for sustainable forest management: The road travelled and the way ahead. *Ecol. Indic.* 8:115–122.

APPENDIX 14A: PAN-EUROPEAN CRITERIA AND INDICATORS FOR SUSTAINABLE FOREST MANAGEMENT, QUALITATIVE INDICATORS

A. Overall policies, institutions, and instruments for sustainable forest management
   - A.1 National forest programs or similar
   - A.2 Institutional frameworks
   - A.3 Legal/regulatory frameworks and international commitments
   - A.4 Financial instruments/economic policy
   - A.5 Informational means

B. Policies, institutions, and instruments by policy area
   - B1 Land use and forest area and OWL
   - B2 Carbon balance
   - B3 Health and vitality
   - B4 Production and use of wood
   - B5 Production and use of nonwood goods and services, provision of especially recreation
   - B6 Biodiversity
   - B7 Protective forests
   - B8 Economic viability
   - B9 Employment (including safety and health)
   - B10 Public awareness and participation
   - B11 Research, training, and education
   - B12 Cultural and spiritual values

C1 Maintenance and appropriate enhancement of forest resources and their contribution to global carbon cycles
   - 1.1 Forest area and other wooded land
   - 1.2 Growing stock
   - 1.3 Age structure and/or diameter distribution
   - 1.4 Carbon stock

C2 Maintenance of forest ecosystem health and vitality
- 2.1 Deposition of air pollutants
- 2.2 Soil condition
- 2.3 Defoliation
- 2.4 Forest damage

C3 Maintenance and encouragement of productive functions of forests
- 3.1 Increment and fellings
- 3.2 Round wood
- 3.3 Nonwood goods
- 3.4 Services
- 3.5 Forests under management plans

C4 Maintenance, conservation, and appropriate enhancement of biological diversity in forest ecosystems
- 4.1 Tree species composition
- 4.2 Regeneration
- 4.3 Naturalness
- 4.4 Introduced tree species
- 4.5 Deadwood
- 4.6 Genetic resources
- 4.7 Landscape pattern
- 4.8 Threatened forest species
- 4.9 Protected forests

C5 Maintenance and appropriate enhancement of protective functions in forest management
- 5.1 Protective forests—soil, water, and other ecosystem functions
- 5.2 Protective forests—infrastructure and managed natural resources

C6 Maintenance of other socioeconomic functions and conditions
- 6.1 Forest holdings
- 6.2 Contribution of forest sector to GDP
- 6.3 Net revenue
- 6.4 Expenditures for services
- 6.5 Forest sector workforce
- 6.6 Occupational safety and health
- 6.7 Wood consumption
- 6.8 Trade in wood
- 6.9 Energy from wood resources
- 6.10 Accessibility for recreation
- 6.11 Cultural and spiritual values

APPENDIX 14B: CRITERIA AND INDICATORS DEVELOPED BY THE MONTRÉAL PROCESS WORKING GROUP AS UPDATED IN 2009

Criteria	Indicators
1. Conservation of ecosystem diversity	1.1 Ecosystem diversity Area and percent of forest in protected areas by forest ecosystem type and by age class or successional stage Fragmentation of forests
	1.2 Species diversity Number of native forest-associated species Number and status of native forest-associated species at risk, as determined by legislation or scientific assessment Status of on-site and off-site efforts focused on the conservation of species diversity

(*Continued*)

(*Continued*)

Criteria	Indicators
	1.3 Genetic diversity Number and geographic distribution of forest-associated species at risk of losing genetic variation and locally adapted genotypes Population levels of selected representative forest-associated species to describe genetic diversity Status of on-site and off-site efforts focused on conservation of genetic diversity
2. Maintenance of productive capacity of forest ecosystems	Area and percent of forestland and net area of forestland available for wood production Total growing stock and annual increment of both merchantable and nonmerchantable tree species in forests available for wood production Area, percent, and growing stock of plantations of native and exotic species Annual harvest of wood products by volume and as a percentage of net growth or sustained yield Annual harvest of nonwood forest products
3. Maintenance of forest ecosystem health and vitality	Area and percent of forest affected by biotic processes and agents (e.g., disease, insects, invasive alien species) beyond reference conditions Area and percent of forest affected by abiotic agents (e.g., fire, storm, land clearance) beyond reference conditions
4. Conservation and maintenance of soil and water resources	4.1 Protective function Area in percent of forest whose designation or land management focus is the protection of soil or water resources 4.2 Soil Proportion of forest management activities that meet best management practices or other relevant legislation to protect soil resources Area and percent of forestland with significant soil degradation 4.3 Water Proportion of forest management activities that meet best management practices or other relevant legislation to protect water-related resources Area and percent of water bodies, or stream length, in forest areas with significant change in physical, chemical, or biological properties from reference conditions
5. Maintenance of forest contribution to global carbon cycles	Total forest ecosystem carbon pools and fluxes Total forest product carbon pools and fluxes Avoided fossil fuel carbon emissions by using biomass for energy
6. Maintenance and enhancement of long-term multiple socioeconomic benefits	6.1 Production and consumption Value and volume of wood and wood products production, including primary and secondary processing Value of nonwood forest products produced or collected Revenue from forest-based environmental services Total and *per capita* consumption of wood and wood products in round wood equivalents Total and *per capita* consumption of nonwood forest products Value and volume in round wood equivalents of exports and imports of wood products Value of exports and imports of nonwood forest products Exports as a share of wood and wood products production and imports as a share of wood and wood products consumption Recovery or recycling of forest products as a percent of total forest products consumption

(*Continued*)

Criteria	Indicators
	6.2 Investment in the forest sector
	Value of capital investment and annual expenditure in forest management, wood and nonwood forest product industries, forest-based environmental services, recreation, and tourism
	Annual investment and expenditure in forest-related research, extension and development, and education
	6.3 Employment and community needs
	Employment in the forest sector
	Average wage rates, annual average income, and annual injury rates in major employment categories
	Resilience of forest-dependent communities
	Area and percent of forests used for subsistence purposes
	Distribution of revenues derived from forest management
	6.4 Recreation and tourism
	Area and percent of forests available and/or managed for public recreation and tourism
	Number, type, and geographic distribution of visits attributed to recreation and tourism and related to facilities available
	6.5 Cultural, social, and spiritual needs and values
	Area and percent of forests managed primarily to protect the range of cultural, social, and spiritual needs and values
	The importance of forests to people
7. Legal, institutional, and economic framework for forest conservation and sustainable management	Legislation and policies supporting the sustainable management of forests
	Cross-sectoral policy and program coordination
	Taxation and other economic strategies that affect sustainable management of forests
	Clarity and security of land and resource tenure and property rights
	Enforcement of laws related to forests
	Programs, services, and other resources supporting the sustainable management of forests
	Development and application of research and technologies for the sustainable management of forests
	Partnerships to promote the sustainable management of forests
	Public participation and conflict resolution in forest-related decision-making
	Monitoring, assessing, and reporting on progress toward sustainable management of forests

APPENDIX 14C: CRITERIA AND INDICATORS OF SFM IN THE FRAMEWORK OF THE ATO/ITTO PROCESS COLLABORATION

Criteria	Indicators
1.1 The State has clear objectives for the sustainable utilization of the forest heritage and a realistic action program for their achievement	1.1.1 At the international level, the State has ratified or accepted the treaties, conventions, or recommendations related to the sustainable development of forests
	1.1.2 In the signatory countries, the clauses of all international agreements relating to the sustainable development of forests, such as the African Timber Organization, the International Tropical Timber Agreement, the Convention on International Trade in Endangered Species of Wild Fauna and Flora, the International Labour Organization, the Convention on Biological Diversity, and the Framework Convention on the Climate Change, are respected

(*Continued*)

(*Continued*)

Criteria		Indicators
	1.1.3	There is a forest policy and related forestry laws that are recognized by the various stakeholders. This forest policy and its laws prescribe the norms of forest management, resource allocation, and resource use in order to achieve sustainable management
	1.1.4	The forest laws and directives and all other national guidelines and norms of forest management are known by all stakeholders
	1.1.5	A permanent forest estate exists as a result of negotiation between different stakeholders within the framework of a procedure of coordinated planning for land use allocation. This land use allocation is consistent with the objectives of the forest laws
	1.1.6	The procedures for access to natural resources are clearly defined and respected
	1.1.7	The legal and customary rights of local populations related to the ownership, utilization, and management of the land and resources near the permanent forest estate are clearly defined, recognized, and respected
	1.1.8	At the national level, the definition of the underlying legislative framework for the sustainable management of the forest heritage is based on current information and knowledge of its ecology
	1.1.9	There is a functional industrial policy favorable to the sustainable management of the forestry sector
	1.1.10	The legislative texts relating to the use of nontimber forest products, especially the hunting of wildlife and the harvesting of wild produce, respect as well as ensure the sustainable management of these resources
	1.1.11	The State policy for sustainable forest management promotes the development of the local economy and entrepreneurship in the forest zone
	1.1.12	There is a policy for the protection and conservation of forest ecosystems and of sites of archeological interest and cultural value
	1.1.13	National forest policies take into account the potential value of the carbon storage functions of forests
1.2 An institutional framework exists to achieve the objectives of sustainable forest management	1.2.1	There is an established and functional forestry service/administration to ensure the direct or delegated management of all forests
	1.2.2	There is a research institution to develop the necessary scientific and technical knowledge for sustainable forest management
	1.2.3	Coordination and negotiation mechanisms exist at different stakeholder levels for discussing the objectives of the sustainable forest management policy, the implementation of modalities for conflict resolution, and the formulation of common rules between them
	1.2.4	Mechanisms exist to promote the participation of different stakeholders, particularly rural communities, in the management of forests
1.3 The State provides the necessary means for the sustainable management of forests	1.3.1	There is a forestry organizational structure responsible for the direct or delegated management of all forests that has adequate, well-trained, competent, and efficient human resources for carrying out its mission
	1.3.2	The research institution in charge of the development of the necessary scientific and technical knowledge for the sustainable management of forests has sufficient, well-trained, and competent human resources to achieve its mission
	1.3.3	The State ensures that its partners in the forestry sector have sufficient technical knowledge and capacities to sustainably manage the forest or participate in all activities relating to the management of the forest

(*Continued*)

Criteria		Indicators
	1.3.4	There is an adequate and sustained financing for forest management, including investment and reinvestment
	1.3.5	The State ensures that requisite information for the sensitization of stakeholders on their rights and obligations relating to the management of the forest is in conformity with forestry laws
1.4 The economic and fiscal policies of the State ensure the viability of forestry enterprises	1.4.1	There are financial incentives for the sustainable management of forests and the industrialization of forest products
	1.4.2	The functions of the forestry service/administration do not unnecessarily hamper the economic activities of enterprises
	1.4.3	The State provides incentives and implements effective measures to encourage the trade of forest products and facilitates the participation of economic operators in the promotion of such products
1.5 The forestry service/ administration implements effective measures to ensure the monitoring and evaluation of the implementation of its forestry policy in relation to production, the conservation of ecosystems, and social benefits	1.5.1	The State has a reliable and adequate system for updating forestry information and action plans and for adjusting the available resources as required to implement the action plans
	1.5.2	The forestry service/administration ensures that stakeholders are operating on a basis of a legal title deed, with a duration compatible with the objectives of sustainably managing the forest resource
	1.5.3	The forestry service/administration ensures that each forest concessionaire has a complete forest management plan and a "concession agreement," which sets out the obligations of all parties. Both documents are formally approved by the relevant authorities
	1.5.4	The forestry service/administration ensures that the forestry operations described in the forest management plan are implemented as soon as the plan is approved, and all the clauses in the concession agreement are respected
	1.5.5	The forestry service/administration implements mechanisms recognized by law to limit and control all non-forestry activities in the permanent forest estate
	1.5.6	The forestry service/administration implements mechanisms recognized by law to protect forest under management from illegal occupation and other encroachment activities
	1.5.7	The State ensures that all economic operators are treated equitably
	1.5.8	All forms of damage caused to the property of the local populations by forestry activities are compensated according to the norms in force or after negotiation

APPENDIX 14D: THE TARAPOTO SET OF SFM INDICATORS

National Level

1. Existence of policies and a legal framework for land use planning through ecological and economic zoning
2. Area, by forest type, categorized as conservation areas, in relation to total forest area
3. Rate of conversion of forest cover to other uses
4. Quantity and quality of appropriate techniques for forest management and sustainable production
5. Investment in research, education, and technology transfer
6. Quantity and quality of research and sustainable development projects in course of implementation

Management unit level

7. Forest management plan and other plans concerning the use of forest resources, approved by the authorities responsible

8. Frequency of the evaluation of management plan implementation and the average percentage of implementation
9. Degree of utilization of environmentally sound, appropriate, and compatible technologies
10. Proportion of environmental protection areas as against permanent production areas
11. Existence of prevention measures to protect water courses from the impact of forest extraction activities
12. Number of direct and indirect jobs and income levels

Global level

13. Contribution to the conservation of biological diversity
14. Contribution to the maintenance, restoration, and protection of the cultural values and diversity of the Indigenous and local population
15. Contribution to the economy, health, culture, science, and recreation

Source: Adapted from Food and Agriculture Organization of the United Nations, Membership of world countries in nine C&I processes, www.fao.org/forestry/ci/16609/en/, accessed September 22, 2014.

APPENDIX 14E: THE DRY FORESTS IN ASIA

Criteria	Indicators
1. Extent of forest and tree cover	1.1 Area of natural and planted forests
	1.2 Area of dense, open, and scrub forests
	1.3 Area under trees outside forests
	1.4 Forest areas diverted for nonforestry use
	1.5 Extent of encroachment in forest areas
2. Maintenance of ecosystem health and vitality	2.1 Extent of natural regeneration
	2.2 Extent of secondary forests
	2.3 Extent of forest areas affected by noxious weeds and pests/diseases
	2.4 Extent of forest area affected by human-induced and natural occurrences
3. Maintenance and enhancement of biodiversity	3.1 Extent of protected areas
	3.2 Number of threatened, keystone, flagship, and endemic species of plants and animals
	3.3 List of flora and fauna
	3.4 Extent of nondestructive harvest
	3.5 Percentage of cover by forest type
	3.6 Mechanisms for the conservation of genetic resources
4. Conservation and enhancement of soil and water resources and other environmental functions	4.1 Extent of watershed areas under management
	4.2 Area under shelter and green belts
	4.3 Duration of streamflow and water yield
	4.4 Extent of soil erosion
	4.5 Change in level of the water table
	4.6 Change in sediment load
5. Maintenance and enhancement of forest productivity	5.1 Extent of forest area under forest management plans
	5.2 Changes in growing stock of wood and NWFPs
	5.3 Difference between allowable and actual cuts
	5.4 Annual NWFP removable
	5.5 Area of afforestation and new plantations including agroforestry
	5.6 Degree of technological input
	5.7 Contribution of forest to GDP through total economic value
6. Extent of forest resource utilization	6.1 Per capita wood and NWFP consumption
	6.2 Import and export of wood and NWFPs
	6.3 Recorded and unrecorded removals of wood and NWFPs

(*Continued*)

Criteria		Indicators
7. Socioeconomic, cultural, and spiritual needs	7.1	Extent of contribution of forest management activities to food security and other livelihood needs
	7.2	Level of recreation, cultural, religious, and aesthetic needs
	7.3	Gender-related indices in forestry
	7.4	Extent of application of traditional knowledge
	7.5	Employment in forestry and forest industries
	7.6	Contribution of the forest to the income of forest dependent people
8. Policy, legal, and institutional framework	8.1	Existence of national forest policy and legal framework
	8.2	Extent of community, NGO, and private sector participation in forestry activities
	8.3	Investment in forestry research and development
	8.4	Human resource capacity building mechanisms
	8.5	Existence of forest resource accounting mechanisms
	8.6	Monitoring and evaluation mechanisms
	8.7	Existence of mechanisms for information dissemination
	8.8	Extent of transfer of technology
	8.9	Fiscal and monetary incentives for investing in forestry activities
	8.10	Benefit-sharing mechanism for stakeholders engaged in forest management activities
	8.11	Existence of conflict management mechanisms
	8.12	Changes in number of forest offenses

APPENDIX 14F: CRITERIA AND INDICATORS OF THE DRY-ZONE AFRICA PROCESSES

Criteria		Indicators
1. Maintenance and improvement of forest resources including their contribution to global carbon cycles	1.1	Total area and percentage of forests, plantations, and other wooded lands (and their changes over time)
	1.2	Biomass (and its changes over time)
	1.3	Available carbon stock
2. Conservation and enhancement of biological diversity in forest ecosystems		
Ecosystem indicators	2.1	Areas by types of vegetation (natural and man-made)
	2.2	Extent of protected areas
	2.3	Fragmentation of forests
	2.4	Area cleared annually of forest ecosystems containing endemic species
Species indicators	2.5	Number of forest-dependent species
	2.6	Number of forest-dependent species at risk
	2.7	Number of forest-dependent species that have disappeared
	2.8	Number of species that have reappeared
Genetic indicators (fauna, flora)	2.9	Average number of provenances
	2.10	Number of forest-dependent species with reduced range
	2.11	Population levels of key species across their range
	2.12	Management of genetic resources
3. Maintenance of forest ecosystem health, vitality, and integrity	3.1	Areas and percentages of forest (natural and man-made) modified, with the indication of severity of damage by such agents as fires (including frequency), storms (e.g., windthrow, flooded areas), insects and diseases, damage by wild and domesticated animals, competition from introduced plants, drought, damage by wind erosion
	3.2	Percentages of forest ecosystems with or without regeneration

(*Continued*)

(*Continued*)

Criteria		Indicators
	3.3	Changes in nutrient balance and soil acidity
	3.4	Bush encroachment
	3.5	Trends in crop yields
	3.6	Percentage of the population employed in crop and livestock farming
4. Maintenance and enhancement of productive functions of forests and other wooded lands	4.1	Percentage of forests and other wooded lands managed according to a management plan
	4.2	Growing stock
	4.3	Annual balance between growth and removal of wood products (and its change over time)
	4.4	Average annual consumption of wood for energy per capita
	4.5	Managed and sustainable extraction of nonwood forest products (and its change over time), including fodder (grass layer and fodder from trees/shrubs), consumptive wildlife utilization, honey, gum, misc. fruits, roots, edible leaves, medicinal substances, fibers for handicrafts
5. Maintenance and improvement of protective functions in forest management	5.1	Areas and percentages of forests and other wooded lands managed mainly for the protection and/or rehabilitation of agricultural or range lands and/or rehabilitation of degraded lands and relevant important infrastructure works
	5.2	Areas and percentages of forests and other wooded areas managed mainly for the production of water, protection of watersheds, riverine zones, and for flood control
	5.3	Areas of forests and other wooded lands managed for scenic and amenity purposes
6. Maintenance and enhancement of socioeconomic benefits Indicators of economic benefits	6.1	Value of wood products
	6.2	Value of nonwood forest products
	6.3	Ecotourism (including hunting, recreation)
	6.4	Share of forest sector in GNP
	6.5	Value from primary and secondary industries
	6.6	Value from biomass energy
	6.7	Forest sector trade balance
	6.8	Investment in forests and forest industries
Indicators of distribution of benefits	6.9	Employment generation and other social advantages
	6.10	Degree of satisfaction to which social, cultural, and spiritual needs are met
Indicators of sharing of benefits	6.10	Share of the forestry sector in the formation of local communities incomes (rural households, women, and youth)
	6.11	Level of contribution to food security
	6.12	Employment, opportunities, and other advantages created notably in rural areas
7. Adequacy of legal, institutional, and policies frameworks for sustainable forest management	7.1	Existence of a national forest policy that ensures the integration of forest management in rural land use planning and to the economic and social development strategic framework and to the Convention on Diversification Control and Biodiversity
	7.2	Existence of a comprehensive legislative and regulatory framework providing for equitable access to resources, alternative forms of conflict resolution, and consideration of land occupancy and cultural rights of local populations
	7.3	Institutional, human, and financial capacity to implement the national forestry policy and relevant national and international laws, instruments, and regulations

(*Continued*)

Criteria	Indicators
	7.4 Research and development capacity
	7.5 Existence of incentives for investments in the forestry sector
	7.6 Valorization of local expertise, knowledge, and technologies
	7.7 Existence of measures to facilitate the transfer and adaptation of appropriate technologies
	7.8 Existence of an administrative policy and legal framework for the effective participation of local communities, NGOs, and the private sector in forest policy formulation, implementation, and monitoring

15 Indicating Forest Ecosystem and Stand Productivity

From Deductive to Inductive Concepts

Hans Pretzsch and Thomas Rötzer

15.1 INTRODUCTION

The productivity of forest ecosystems and stands represents their capacity to synthesize dry matter and depends highly on external environmental conditions. The dry matter productivity is essential for ecological forest management as it determines most forest functions and services. Carbon sequestration, abundance of decomposers, financial income of forest owners, and most other forest functions and services increase with forest ecosystems and stand productivity. So, information on productivity in the narrower sense was pivot for the dominant wood use paradigm but became even more important for ecological forest management because of its strong effect on all achievements in the modern forest basket.

For the purpose of clarity, the chapter starts with definitions of different aspects of productivity, which are applied for characterizing the world's forests. Empirical knowledge on productivity comes mostly from long-term plots. However, forest practice cannot start long-term monitoring when information on productivity is required but needs easy-to-measure indicators. We introduce such indicators for the production that can be realized at a certain site with a given species or genotype. We further show how this potential productivity can be modified by specified management regimes.

Historic litter raking, acid rain pollutant, and climate change are just a few out of many disturbances which massively modify productivity. We stress long-term survey for quantifying such human footprints on forest and for updating the level of productivity of forest ecosystems and stands even in a rapidly changing environment. Beyond their application for monitoring and controlling, indicators of productivity play a crucial role in strategic planning and decision-making. Finally, it will be emphasized that earlier approaches deduced information about productivity from general models (e.g., yield tables, stand simulators), which were parameterized just by a small number of long-term experimental plots. Present-day inventories with repeated survey provide information on forest structure, growth, yield, and site conditions, which enable site-specific induction of productivity and replacement of the former inductive approaches.

Wood production and harvest were the exclusive purpose of productivity assessment in the past. This chapter will show that the productivity of forest ecosystems and stands determines nearly all sectors of ecological forest management covered in this handbook, reaching from forest inventory, to planning and decision-making, to monitoring, certification, conservation, regeneration, and even reclamation of devastated forest ecosystems.

 DOI: 10.1201/9781003431084-18

15.2 RELEVANCE OF PRODUCTIVITY TO ECOLOGICAL FOREST MANAGEMENT

15.2.1 Definitions and Base Processes of Forest Stand Productivity

Productivity of plants in an ecological sense means the fixation of solar energy and is referred to as gross primary productivity (GPP). In other words, GPP denotes the amount of carbon gained through photosynthesis for a given area over a given time. In forestry and ecology, GPP is often calculated as biomass production per hectare and year (Pretzsch 2009, Matyssek et al. 2010).

Regarding the energy (or the carbon) losses required to support plant respiration the remaining biomass increment is called net primary productivity (NPP). If only the stand productivity is considered, NPP can be calculated as

$$\mathrm{NPP} = \mathrm{GPP} - \mathrm{R_a} \tag{15.1}$$

with R_a as the sum of all autotrophic respiration shares, i.e., growth and maintenance respiration while heterotrophic respiration R_h is not regarded (Figure 15.1).

According to Ito (2011), who cited Monteith (1972), NPP can also be calculated using the light use efficiency of NPP, $\mathrm{lue_{NPP}}$, and the absorbed photosynthetically active radiation APAR:

$$\mathrm{NPP} = \mathrm{lue_{NPP}} * \mathrm{APAR} \tag{15.2}$$

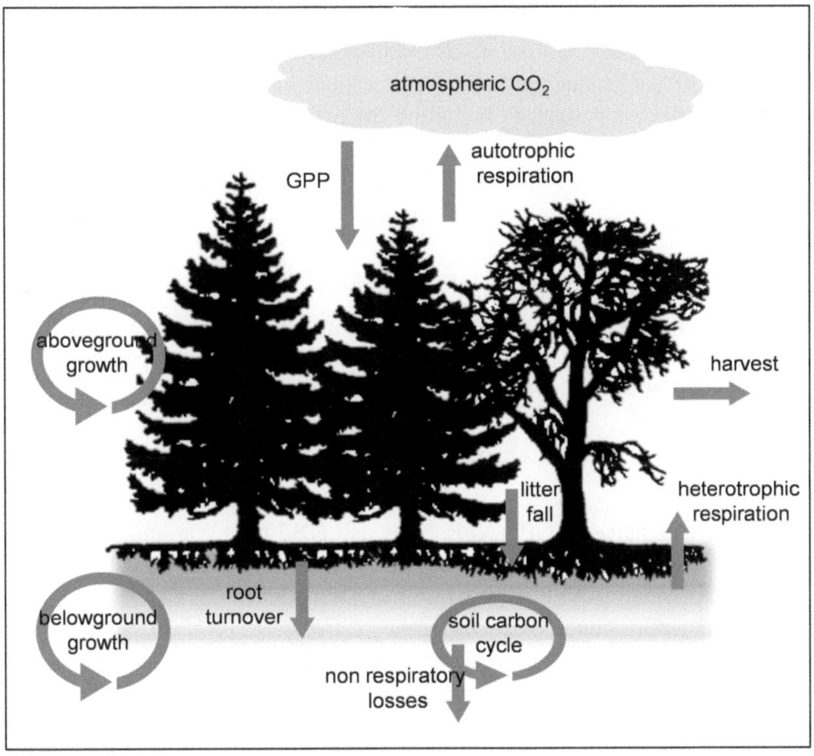

FIGURE 15.1 Carbon fluxes within forest stands.

Consequently, NPP can be described by the "production of new dry-matter by primary producers during a certain period" (Ito 2011). Biomass accumulation of a plant or an entire stand includes losses through plant organ turnover that consists of litterfall and branch, twig and root turnover, as well as losses through the mortality of the entire plants:

$$\mathrm{BA} = \mathrm{NPP} - \mathrm{losses} - \mathrm{mortality} \tag{15.3}$$

BA is often called net growth, which implies that gross growth is equal to NPP (Pretzsch 2009).

Following Luyssaert et al. (2007) net primary productivity for the individual compartments of a tree or a stand can be split up into

$$\mathrm{NPP} = \mathrm{NPP}_{\mathrm{foliage}} + \mathrm{NPP}_{\mathrm{wood}} + \mathrm{NPP}_{\mathrm{root}} + \mathrm{NPP}_{\mathrm{reproduction}} \tag{15.4}$$

where net primary productivity of foliage, wood (=stem and branches), roots (fine as well as coarse roots), and of reproduction (flowers, seeds, fruits) is summed up.

According to Ito (2011) and Luyssaert et al. (2007), the NPP-estimation can be extended for consumers:

$$\mathrm{NPP}^{+} = \mathrm{BA} + \mathrm{C}_{\mathrm{ha}} + \mathrm{C}_{\mathrm{he}} + \mathrm{l}_{\mathrm{em}} + \mathrm{l}_{\mathrm{ex}} + \mathrm{C}_{\mathrm{my}} \tag{15.5}$$

where NPP^{+} is the modified net primary productivity, C_{ha} and C_{he} are the consumption by human harvest and herbivores, respectively, l_{em}, l_{ex} denote losses by emissions such as biogenic volatile organic compounds (BVOCs) or CH_4, and losses by exudation of roots, and C_{my} is the consumption by mycorrhiza.

The described equations can be used at different scales: GPP, NPP etc. can be calculated for a single tree, for an entire forest stand but also for one species of a given area. In addition, they can be scaled up for an entire ecosystem by including the heterotrophic respiration (Smith and Smith 2009) (Figure 15.1):

$$\mathrm{NEP} = \mathrm{GPP} - \mathrm{R}_{\mathrm{a}} - \mathrm{R}_{\mathrm{h}} \tag{15.6}$$

where NEP is the net ecosystem productivity.

If further the import from surrounding ecosystems (im_{se}) and non-respiratory losses (l_{nr}) such as losses by forest fires, harvest, herbivores or erosion is known (Buchmann und Schulze 1999, Chapin et al. 2006) net biome productivity NBP can be calculated by

$$\mathrm{NBP} = \mathrm{NEP} - \mathrm{l}_{\mathrm{nr}} + \mathrm{im}_{\mathrm{se}} \tag{15.7}$$

The principal differences between GPP, NPP, NEP, and NBP based on temporal and spatial scale are summarized in Figure 15.2 according to Steffen et al. (1998).

Short-term carbon uptake (=NPP) is the result of fixed CO_2 by plants (=GPP) reduced by the plant respiration (=R_a) releasing CO_2 to the atmosphere. Medium-term carbon uptake (=NEP) denotes a further release of CO_2 to the atmosphere by decomposition of litter and soils (=R_h). Long-term carbon storage (=NBP) considers disturbances from both natural and anthropogenic sources (D = imse—lnr) and leads in most cases to further release of CO_2.

Consequently, the temporal scale of plant productivity is closely linked with the spatial scale (individual tree, stand, ecosystem) regardless of the climate zone. This is also crucial when analyzing forest functions and services.

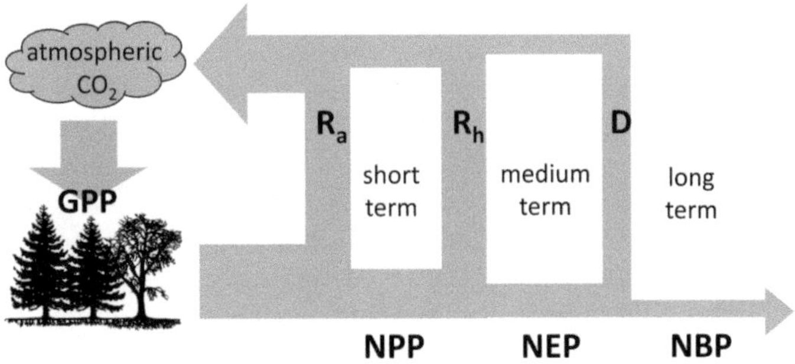

FIGURE 15.2 Temporal and spatial scaling of plant productivity. (Adapted from Steffen et al. 1998.)

15.2.2 Relationship between Productivity and Forest Functions and Services

Information on productivity in the narrower sense was pivot for the dominant wood use paradigm (Yaffee 1999) but became even more important for ecological forest management because of its strong effect on all achievements in the contemporary basket of forest functions and services. The effect of productivity on other forest functions and services than wood production can differ a lot. The carbon sequestration and fossil fuel substitution can increase with productivity (Marland and Schlamadinger 1997), but the groundwater supply often decreases (Dudley and Stolton 2003, Molden et al. 2003). The abundance of organisms rises (Vile et al. 2006), but the species richness subsides with increasing forest productivity (Waide et al. 1999). While high productivities mean favorable financial income for the forest owner, it may reduce the aesthetic value and recreation function due to high stocking densities and missing views on the landscape. So far, some theoretical pros and cons of high productivity levels in unmanaged forests.

Beyond the level of productivity, most forest functions and services are highly determined by the management intensity in terms of wood harvest and removal. Positive effect of high productivity can turn into negative effects by nutrient export, which can be extremely high especially on high productivity sites with high stocking and removal volume and biomass. The other way round negative effects of high productivity can be remedied, e.g., when thinning reduces high stocking densities and water consumption, species mixing and improve their aesthetic value, and creation of open spaces establishes lookout points for forest walkers.

Positive effects of low productivity can be spoiled when low productivity sites with high species richness are fertilized, drained, or transformed into plantations. Negative effects of low productivity, on the other hand, can be remedied by the establishment of species mixtures, among other with atmospheric nitrogen-fixing species, which may improve soils conditions and fertility.

As these examples illustrate, there is a trade-off between many productivity-determined forest functions and services already at the stand level and in a defined period. Including other spatial and temporal scales can further complicate the relationships and trade-offs. High productivity plantations exclusively for wood supply may appear negative in terms of biodiversity seen at the stand level. However, if they help to establish conservation areas elsewhere, the assessment may change. It finally depends on the legal frame and the objective of the forest owner how the various forest functions and services depending on and including productivity are weighted and realized.

15.3 FOREST ECOSYSTEM AND STAND PRODUCTIVITY WORLDWIDE

15.3.1 Global Values

Global terrestrial gross primary production is reported to be at 133 ± 15 Gt C per year (Piao et al. 2013). By using a light use efficiency model, Zheng et al. (2020) calculated a mean global GPP of

106.2 ± 2.9 Gt C per year for the period 1982–2017. Global terrestrial NPP, on the other hand, was estimated at a mean value of 56.2 ± 14.3 Gt C per year based on 251 estimates published in 178 studies between 1862 and 2011 (Ito 2011). Due to the estimation methods, NPP varies between 46.2 ± 22.3 Gt C yr^{-1} using inventory aggregation and 61.2 ± 9.4 Gt C yr^{1} using dynamic global vegetation models (Ito 2011). Carvalhais et al. (2014) estimate that the global gross primary production is 126.7 Gt yr^{1} on average, resulting in a net primary production of 54 Gt yr^{-1}.

Assuming a total land area of approximately 149 x 10^6 km^2 (Saugier et al. 2001 or www.spacenews.de) the area based GPP is 8.93 ± 1.01 t C ha^{-1} yr^{-1} while the global average of *NPP* is 3.77 ± 0.96 t C ha^{-1} yr^{-1}. Higher NPP values were given by Saugier et al. (2001) with 4.19 t C ha^{-1} yr^{-1}.

Based on model simulations, several studies reveal a GPP and NPP rise within the last decades. Zheng et al. (2020) found a trend for the GPP of 0.15 Gt C per year for the period 1982–2017, which was mainly due to a CO_2 fertilizing effect. Over the period 2000–2100, for the global GPP a trend between −0.15 and 1.09 Gt C per year was simulated by several light use efficiency models (Cai et al. 2014). Global GPP is positively affected by the change of land cover, by an increasing CO_2 concentration, by an increase temperature and changing water conditions while radiation changes have negative effects, the single contributions of the parameters on the GPP trend are 4.57%, 65.73%, 13.07%, 11.74% and –7.24, respectively (Sun et al. (2018).

An NPP increase from preindustrial time to the present which in most cases ranges between 10% and 30% was reported by Sitch et al. (2003), Ito (2011) and Piao et al. (2013). Piao et al. (2013) found that NPP rises on average by 16 % (simulations results) resp. 13 % (FACE experiments) per 100 ppm increase of atmospheric CO_2 concentration. However, future droughts are expected to enhance climate-induced NPP losses and mitigate CO_2 fertilization effects, meaning that future climate change will further threaten the stability of terrestrial carbon sinks due to droughts, especially in arid, semi-arid, and tropical rainforest areas (Cao et al. 2022).

The carbon use efficiency (cue) can be defined as the ratio between NPP and GPP. Because about 50% of the photosynthesis is spent for plant respiration, *cue* is commonly assumed to be 0.5. However, global cue values vary to a great extent between 0.33 and 0.88 (Ito 2011). Depending on forest type, DeLucia et al. (2007) found cue values ranging from 0.23 up to 0.83. Zhang et al. (2009) calculated an average NPP/GPP ratio of 0.52. They conclude that cue shows a pattern that depends on the key climate parameters, temperature and precipitation, as well as on the geographical parameters, altitude and latitude.

15.3.2 Productivity of Forest Ecological Zones

Along with the huge uncertainties when estimating global GPP and NPP (Pan et al. 2011), regional gross primary and net primary production differ to a great extent. The main reasons are the variations in environmental conditions like climate and vegetation physiognomy. Based on the large-scale changes of these parameters five different climate domains exist: tropical, subtropical, temperate, boreal, and polar (FAO 2012). Including mountain systems four domains of the global ecological zones can be derived for forests (FAO 2012, IPCC 2006) ranging from tropical rainforests to boreal coniferous forests (Table 15.1).

Land covered by the global forest ecological zones ranges from 1.37 × 10^9 ha for temperate forests up to 1.75 × 10^9 ha for tropical and subtropical forests (Table 15.2, IPCC 2001, Field and Raupach 2004, Bresinsky et al. 2008). In Table 15.2 Mediterranean xeromorphic forests are included, which occupy an area of 0.28 × 10^9 ha.

Plant biomass of the global forest ecological zones sums up to 57 Gt C in boreal forests while it amounts to 340 Gt C in tropical and subtropical forests. Based on these data, average stand biomass per hectare can be calculated. In tropical and subtropical forests, the mean biomass per hectare is 194 t of carbon. In these forest ecological zones, aboveground biomass is overall more abundant in old-growth forests than in secondary forests and is more abundant in older secondary

TABLE 15.1
Global Forest Ecological Zones (Dry Month: Months Are Defined as Months in Which the Precipitation Sum [mm] ≤ 2 mm. * Mean Temperature [°C]) (from FAO 2012)

Domain	Criteria (Number of Months with T > 10°C)	Dominant Zonal Forest Vegetation	Criteria (Number of Dry Months)
Tropical	No frost, coldest month > 18°C	Tropical rainforest	< 3 (winter)
		Tropical moist forest	3–5 (winter)
		Tropical dry forest	5–8 (winter)
		Tropical mountain systems	app. > 1000 m altitude
Subtropical	> 8	Subtropical humid forest	Humid, no dry season
		Subtropical dry forest	Winter rains, dry summer
		Subtropical mountain systems	App. > 800–1000 m altitude
Temperate	4–8	Temperate oceanic forest	Oceanic climate, coldest month > 0°C
		Temperate continental forest	Continental climate, coldest month < 0°C
		Temperate mountain systems	App. > 800 m altitude
Boreal	< 3	Boreal coniferous forest	Coniferous dense forest dominant
		Boreal mountain systems	App. > 600 m altitude

TABLE 15.2
Forest Area, Living Plant Biomass (IPCC 2001, Field and Raupach 2004, Bresinsky et al. 2008), and Total Biomass Density (Carvalhais et al. 2014) of the Global Forest Ecological Zones

Forest Ecological Zone	Area 10^9 ha	Plant Biomass Gt C	Plant Biomass Density t C ha^{-1}	Total Biomass Density t C ha^{-1}
Boreal	1.37	57	42	342
Temperate	1.04	139	134	234
Mediterranean (xeromorphic)	0.28	17	61	
Tropical/ subtropical	1.75	340	194	350
Total	**4.44**	**553**	**125**	**309**

forests, (i.e., in forest with growth of more than 20 years and degraded/deforested forests) than in young secondary forests (i.e., in forest with growth smaller or equal to 20 years (Roozendaal et al. 2022).

Average plant biomass density of boreal forests, on the other hand, is with 42 t C ha^{-1} less than a third of the plant biomass density in temperate forests (134 t C ha^{-1}). If, however, the total biomass of the forest ecological zones is compared, which also includes the carbon content of the soil (Carvalhais et al. 2014), boreal forests reveal approximately the same amounts as tropical forests.

Based on the studies of Luyssaert et al. (2007) and Fernandez-Martinez et al. (2014), the total plant biomass of the global forest ecological zones can be divided up in its above- and belowground parts (Figure 15.3). On average, boreal and tropical forests show the highest aboveground biomass

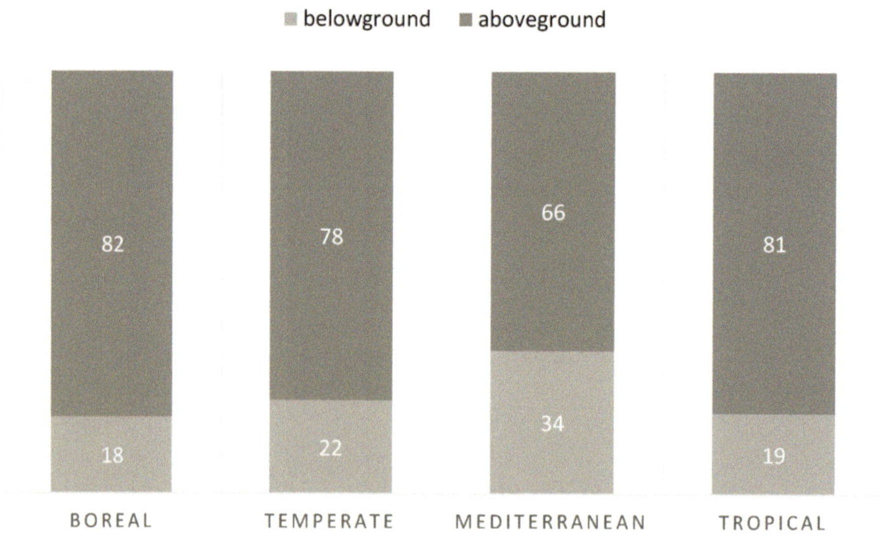

FIGURE 15.3 Partitioning of above- and belowground plant biomass of the global forest ecological zones. (Data based on Luyssaert et al. 2007, Fernandez-Martinez et al. 2014.)

amounts with 82% and 81%, respectively, which means only 18% and 19% of the total plant biomass can be found belowground. For Mediterranean forests, on the other hand, the aboveground share is at 66% and the belowground share at 34%.

Values for the productivity of the global forest ecological zones—independently whether gross primary production or net primary production—vary to a great extent depending on the scope and the sample size of the study, namely the methods used for estimation (Luyssaert et al. 2007, DeLucia et al. 2007, Beer et al. 2010, Fernandez-Martinez et al. 2014, Lyr et al. 1992, Roy et al. 2001, Whittaker and Likens 1973, Bresinsky et al. 2008, Field and Raupach 2004, Atjay et al. 1979, Saugier et al. 2001, Schultz 2005).

In Figure 15.4 estimated GPP and NPP values for the global forest ecological zones are given enlarged by Mediterranean forests and split in subtropical and tropical forests.

Despite high variations within the zones, a clear increase of the gross and the net primary production is obvious. Comparing the averages of the global forest ecological zones, GPP rises from 8.9 t C ha^{-1} yr^{-1} for boreal forests to 30.0 t C ha^{-1} yr^{-1} for tropical forests. NPP rises from 2.6 t C ha^{-1} yr^{-1} for boreal forests over 6.7 t C ha^{-1} yr^{-1} for temperate forests and 9.9 t C ha^{-1} yr^{-1} for subtropical forests to 10.8 t C ha^{-1} yr^{-1} for tropical forests.

As shown in Figure 15.3, the ratio between above- and belowground biomass clearly differs. In consequence also, the annual biomass production below- and aboveground varies among ecological zones (Figure 15.5).

Large differences in the GPP and NPP within each global forest ecological zones are obvious. However, the increase of the aboveground biomass production from boreal to tropical forest ecological zones is steeper than the respective belowground biomass production. The percentage of annual investment in the different compartments of trees, such as foliage, wood, and roots, vary clearly within the ecological zones (Figure 15.6).

The increase in the belowground biomass production from boreal forests to tropical forests, based on the studies of Roy et al. (2001) and Fernandez-Martinez et al. (2014) (Figure 15.5), can again be seen in the data of Luyssaert et al. (2007) (Figure 15.6). In tropical forests, approximately the same high share of NPP that is allocated to the roots (38%) is used for the construction of foliage (37%). Thus, only 25% of the NPP are left for the production of wood. All the

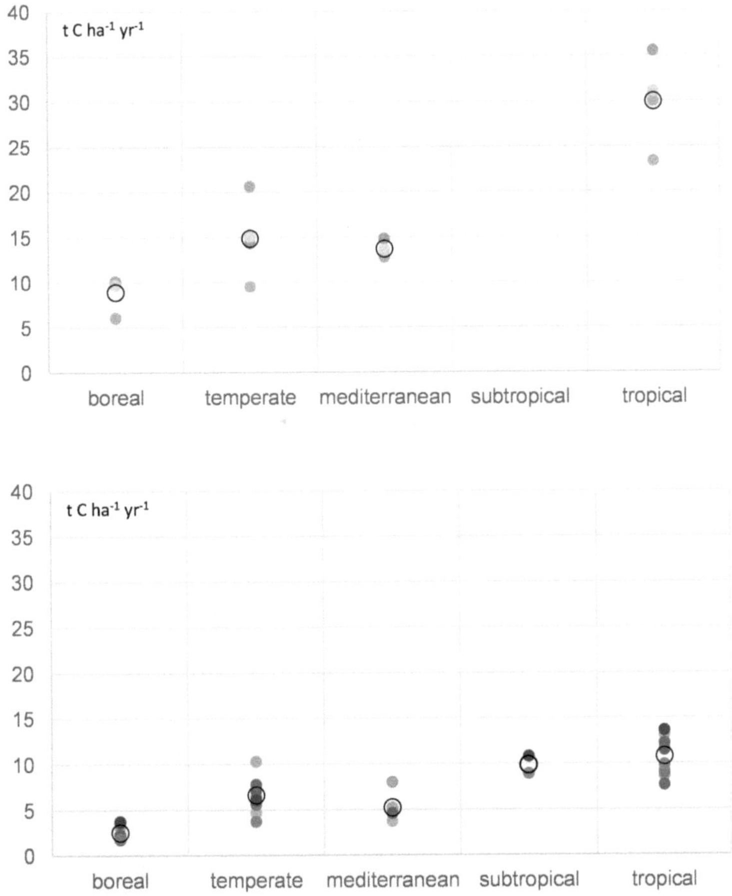

FIGURE 15.4 GPP (above) and NPP (below) of the global forest ecological zones (extended by the ecological zone "Mediterranean") based on several studies. (GPP: Luyssaert et al. 2007, DeLucia et al. 2007, Beer et al. 2010, Fernandez-Martinez et al. 2014: NPP: Luyssaert et al. 2007, DeLucia et al. 2007, Lyr et al. 1992, Roy et al. 2001, Whittaker and Likens 1973, Bresinsky et al. 2008, Field and Raupach 2004, Atjay et al. 1979, Saugier et al. 2001, Schultz 2005.)

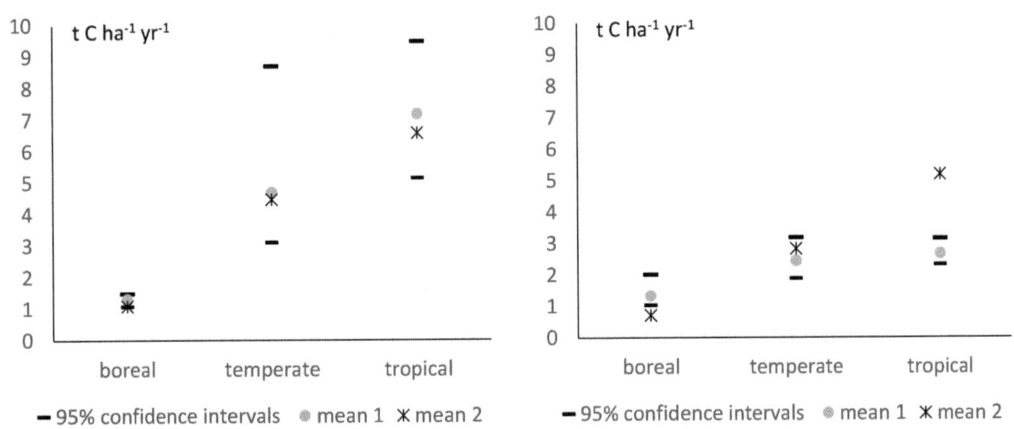

FIGURE 15.5 Above- and belowground biomass production for the global forest ecological zones. (Data based on Roy et al. 2001, Fernandez-Martinez et al. 2014.)

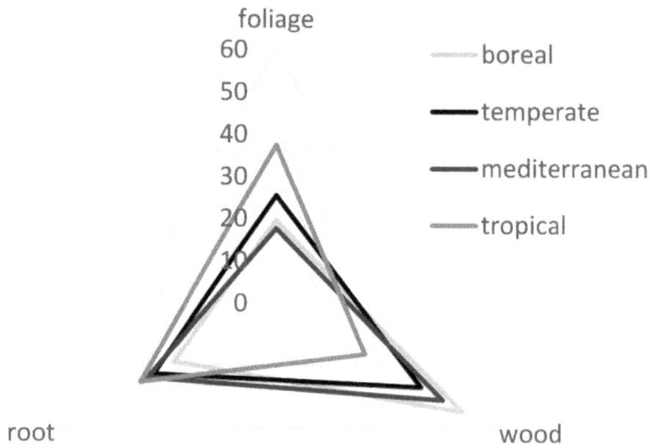

FIGURE 15.6 Average percentage of annual net primary production into foliage, wood, and roots for the forest ecological zones. (Database: Luyssaert et al. 2007.)

other forest ecological zones form distinctly higher wood biomass per year (52%, 41%, and 47%, respectively, for boreal, temperate, and Mediterranean forests). The average annual foliage NPP ranges from 17% in Mediterranean forests to 37% in tropical forests. But, of course, the percentages reveal a broad variation. For example, in temperate humid forests, the share of foliage NPP is 24% for evergreen forests, but 30% for deciduous, while the percentage for the root NPP is 35% and 27%, respectively (Luyssaert et al. 2007). And the differences between humid and semiarid evergreen forests of the boreal zone are −7% for the foliage and +30% for the roots.

The amount of harvested wood products (Pan et al. 2011), the quantity of BVOC emissions (Matyssek et al. 2010), and the type of forest management (e.g., species composition (DeLucia et al. 2007), stand age (Michaletz et al. 2014), or natural vs. plantation forests (IPCC 2006)) are additional factors that influence the quantity of the production in forest ecological zone. Along with climate-induced changes, the increase in atmospheric CO_2 concentration and N depositions (Pretzsch et al. 2014b) as well as the extension of the growing season (e.g., Chmielewski and Rötzer 2001) are causing significant variabilities in the productivity of ecological zones (e.g., Pan et al. 2011, le Quere et al. 2013). Nevertheless, climate controls forest productivity of an ecological zone to a great extent (e.g., Beer et al. 2010, Luyssaert et al. 2007, Michaletz et al. 2014), which enables to develop indicators to estimate site productivity.

15.4 INDICATORS FOR FOREST ECOSYSTEM AND STAND PRODUCTIVITY

15.4.1 Overall View of Indices for Estimating Site Productivity

The productivity at a certain site with a given tree species can be assessed directly by the synthesized plant biomass or produced stem volume or indirectly by quantifying the environmental condition and resource supply. A classic example of indirect assessment of forest stand productivity is the CVP-index “climate-vegetation-productivity index” by Paterson (1956, pp. 68–75)

$$CVP = \left(T_v * P * G * e\right) / \left(T_a * 12 * 100\right) \tag{15.8}$$

with T_v = temperature of warmest month, P = annual precipitation, G = length of the growing season, T_a = annual range of temperature and e: evapotranspiration reducer = function of the latitude dependent radiation.

The *Martonne-index* (Martonne 1926), which is a function of precipitation P and temperature T (=P/(T+10)), forms a simplified productivity indicator compared, for instance, to the CVP-index. Other productivity indicators that are more or less based on meteorological parameters are the *Bruscheck index* or the climatic water balance. The Bruscheck index is based on the annual precipitation sum and the number of summer days, i.e., the number of days with temperatures above 25°C (Bruscheck 1994). A more sophisticated index is the climatological water balance, i.e., difference between the precipitation sum and the potential evapotranspiration. Another drought index is the *ratio of actual to potential evapotranspiration* (Rötzer et al. 2012), whose values can be calculated using simple or complex water balance models (e.g., Rötzer et al. 2009, 2017, 2019). Other commonly used drought indicators are the Palmer Drought Severity Index (PDSI) (Palmer 1965) and the Standardized Precipitation Evapotranspiration Index (SPEI) (Vicente-Serrano et al. 2010). While PDSI is derived from a simplified soil water balance and provides relative soil moisture conditions, SPEI is based on the water balance, defined as the difference between precipitation and potential evapotranspiration.

Insertion of the climatic characteristics yield a CVP index that closely correlates with the productivity of forests stands in terms of mean annual volume growth per hectare. In addition to this geocentric type of indicator, lots of dendrometric indicators exist to describe the standing volume or the yield on stand scale on the base of growth and yield data with respect to stem volume or tree mass (Table 15.3). Biological and ecological indicators, such as forest or vegetation type, range from stand to regional scale while (eco-)physiological ones can also be used on continental scale.

In particular, ecophysiological indicators that assess forest site productivity like gross or net primary production are often used to estimate productivity on small- as well as on large-scale regions (Turner et al. 2005; Luyssaert et al. 2007; Weiskittel at al. 2011). They are either based on satellite (e.g., Turner et al. 2005, Weiskittel at al. 2011) or terrestrial data (Weiskittel at al. 2011). For the

TABLE 15.3
Overview of Indicators and Approaches for Assessment of Forest Site Productivity

Indicator	Scale	Type of Indicator	Reference
GPP, NPP	Stand, region, continent	Ecophysiological	Luyssaert et al. (2007) Weiskittel et al. (2011) Michaletz et al. (2014)
NDVI	Stand, region, continent	Physiological	Wang et al. (2004)
Standing volume = f (h)	Stand	Dendrometric direct	Eichhorn (1902)
Yield = f (h)	Stand	Dendrometric direct	Gehrhardt (1923)
Yield = f (h, SI)	Stand	Dendrometric, indirect	von Baur (1876)
Yield = f (h, SI, yield l)	Stand	Dendrometric, indirect	Assmann (1970)
CVP-index	Stand, region, continent	Geocentric	Paterson (1956)
Martonne-index	Stand, region, continent	Meteorological	Martonne (1926)
Bruschek	Stand, region, continent	Meteorological	Bruscheck (1994)
ETa/ETp	Stand, region, continent	Meteorological	Rötzer et al. (2012)
PDSI	Stand, region, continent	Meteorological	Palmer (1965)
SPEI	Stand, region, continent	Meteorological	Vicente-Serrano et al. 2010
Forest type	Stand, region	Biological	Cajander (1926)
Vegetation type	Stand, region	Ecological	Wykoff et al. (1982)

Note: Ecophysiological indicators based on synthesized total biomass, geographical, biological, or ecological indicators based on geobioecological characteristics of the site in question. Dendrometric indicators based on growth and yield in terms of stem volume or tree mass.

estimation of GPP and NPP, in most cases, process-based models form the base of the calculations. Weiskittel at al. (2011), for example, used the 3-PG model while Pretzsch et al. (2012) based their calculations on the model BALANCE, which is able to calculate productivity indicators also for structured forests like mixed stands (Rötzer et al. 2009). If the light use efficiency of a tree species is known, NPP can be derived from the APAR (Ito 2011). Eddy covariance measurements are another way to estimate ecophysiological indicators.

15.4.2 From Proxy Variables to "Primary" Factors for Explanations and Estimations of Stand and Tree Growth

The attempt to assess, classify, understand, and even estimate primary productivity, growth, or yield for a given forest stand from causal variables has been undertaken in forest research since the first experiments in the 18th century (von Gadow 2013, Assmann 1970). However, the approaches have become more and more mechanistic and more focused on the primary resource and environmental variables. Initially, in the 18th century, the standing stem volume of forest stands was used for the classification of a given stand into a site quality system (Pressler 1877).

$$\text{Site fertility class} = \text{f (standing volume, stand age)} \tag{15.9}$$

And then the site fertility class could be applied to estimate growth and yield, e.g.,

$$\text{Volume growth} = \text{f}\left(\text{site class, stand age}\right) \tag{15.10}$$

This classification was somehow circular; the standing volume must be estimated before determining the site fertility class, after which projections of the future volume productivity could be calculated. This approach only made sense as long as light and moderate thinnings were common. With the change to more intensive management concepts in the 19th century, the thinning component of total production increased so that standing volume became an increasingly poorer indicator of site fertility class.

As the relationship between stand age and stand height correlates closely with total stand production (Eichhorn 1902) and is less dependent on thinning intensity, it provides an alternative to the former approach. Thus, the use of age and mean height, introduced by von Baur (1876, 1881) and Perthuis de Laillevault (1803),

$$\text{Site fertility class} = \text{f}\left(\text{mean height, stand age}\right) \tag{15.11}$$

for the classification of stand growth became recognized despite some reluctance in the beginning (Heyer 1845). Estimates of stand growth and yield result from,

$$\text{Stand growth} = \text{f}\left(\text{site fertility class, stand age}\right) \tag{15.12}$$

With the intensification of thinning from below, which significantly influences calculations of mean height, another change was made in the mid-20th century toward the use of top height as an indicator of site fertility (Assmann 1970). The idea of using stand volume or height growth as a "phytometer" for the productivity of a site has continued to the present day. However, this approach is again being questioned as forest practice increasingly implements thinning from above in management regimes to encourage structurally diverse mixed stands. The more a stand deviates from an even-aged, mono-layered structure, the greater the influence of density and competition on the relationships between age and height, and the less suitable the age-height relationship becomes an indicator of site fertility. Especially in highly structured mixed stands, mean and top height are unsuitable as site fertility indicators.

Heyer (1845) demanded that yield studies not be directed exclusively toward the volume yield but toward the investigation and measurement of "primary site factors," such as temperature, nutrient supply, radiation, and so on. One step into this direction was made by Cajander (1926). He developed a classification system for boreal forests that facilitated the growth and yield estimation from the forest floor vegetation as the indicator (species lichens, mosses, grasses, herbs, shrubs). While this method became standard in the rather uniform and undisturbed boreal forests, the heterogeneity of forest types and human influences makes this approach inadequate for central European forests.

However, the core of Cajander's idea was to combine locally available indicator variables for classifying a particular stand with growth and yield information deduced from site-specific yield tables. The increasing availability of site information and growth and yield data from inventories led to an increased meshing of locally acquired information about site, growth, and yield with general growth and yield relationships deduced from models. Moosmayer and Schöpfer (1972), Wykoff et al. (1982), and Wykoff and Monserud (1988) developed relationships between site conditions and growth at the tree or stand level by regression analysis

$$\text{Volume growth} = f\left(\text{stand attributes, site characteristics}\right) \tag{15.13}$$

As independent variables, they used metric information (e.g., annual precipitation, mean temperature, slope, exposition) and nominal (e.g., levels of nutrition supply, levels of water supply) and ordinal (e.g., growth region, degree of disturbance of topsoil by machines) variables. To estimate the potential height growth, volume growth, and yield in the forest growth simulator SILVA, Kahn (1994) used a set of nine metric site variables.

Experimental field plots, monitoring plots, and chamber experiments enable the study of metabolic, physiologic, and growth process as well as the factors affecting growth (environmental conditions, resource supply) at increasingly more refined spatial and temporal resolution. For example, temperature measured per day, hour, or minute, and radiation recorded separately for different wavelengths can be used to estimate net primary production, according to the following approach

$$\text{NPP} = f\,(\text{leaf area, radiation, temperature, nutrient, water}) \tag{15.14}$$

Simultaneously recorded assimilation rate, respiration, height, and diameter enable a refined estimation also of gross production (gC min^{-1}), (cm day^{-1}), (mm day^{-1}) to be made. The development of inventories monitoring and innovative regionalization tools, which deliver all the relevant variables for such primary variable-based approaches Heyer (1845) already had in mind, is in progress. Knowledge about site-growth relationship forms the backbone of forest growth models, and hence the availability of site variables is therefore decisive for the applicability of these models. To overcome the lack of knowledge about the relationships between the primary causes affecting tree and stand development (environmental conditions, resource supply), forest research has gathered rich experience in the application of "surrogate variables" or "proxy variables" (Oliver and Larson 1996; Zeide 2003). Examples for the application of a "surrogate variable" or "proxy variable" include the use of age–height relations for estimations of stand growth, the area available for growth or growing space of a tree for estimations of its resource supply, or the application of competition indices for estimations of height and diameter increment of an individual tree in relation to resource supply. In all cases, the primary factors remain unsolved. However, they are replaced by easy to measure auxiliary variables although they merely reflect only the hidden relationships.

15.5 EFFECT OF FOREST MANAGEMENT ON FOREST ECOSYSTEM AND STAND PRODUCTIVITY

15.5.1 Competition-Based Mortality and Tree Losses: An Essential Component of Net Primary Production

Even-aged stands can start with many thousand seedlings per hectare before the density declines naturally to just a few hundred trees per hectare 100 years later. Without management the majority of individuals die naturally due to competition. Pretzsch et al. (2023a) used a unique set of 476 unmanaged, monospecific experimental plots of Norway spruce (*Picea abies* (L.) H. Karst.), silver fir (*Abies alba* Mill.), Scots pine (*Pinus sylvestris* L.), European larch (Larix decidua Mill.), Douglas fir (*Pseudotsuga menziesii* (Mirb.) Franco), European beech (*Fagus sylvatica* L.), and oak (*Quercus robur* L. and *Quercus petraea* (Matt.) Liebl.) throughout Europe to analyze the competition-based mortality of trees and its dependency on age and site conditions.

Figure 15.7 summarizes their main results. They found that up to an age of 100–150 years, the natural volume loss by mortality amounted to 500–1000 m^3 ha-1 which is 30–40% of the total volume production, an indicator of NPP. The loss corresponded to an average annual loss of 0.8–2.1 t ha-1 of biomass or 0.4–1.1 t ha-1 of carbon storage. The loss fraction of the gross stand growth increased continuously with stand age. Setting aside (previously managed) stands means that about one-third of the total production within a rotation period of 100–150 years would remain unutilized and become deadwood. This result indicates a trade-off between wood utilization (and related carbon storage and emission savings) and deadwood accumulation for facilitating biodiversity (Dieler et al. 2017, Pretzsch and Grote 2023).

15.5.2 Continuous versus Cyclic Stand Productivity

The worldwide applied silvicultural models for stand management lie in the continuum spanned and restricted by the borderline models of the age-class model and the selection forest. Figure 15.8 shows in schematic representation the current volume increment for four characteristic silvicultural model procedures.

In age-class forests (Figure 15.8a), the current volume increment follows a uni-modal curve and decreases to zero in the time of the transition to the new generation. In the selection forest (Figure 15.8d),

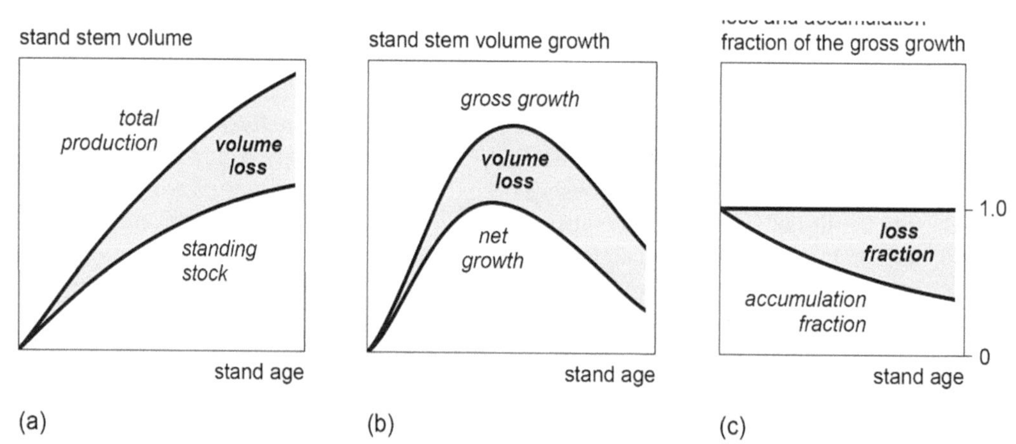

FIGURE 15.7 Competition-based mortality and tree losses. An essential component of net primary production. (After Pretzsch et al. 2023a.)

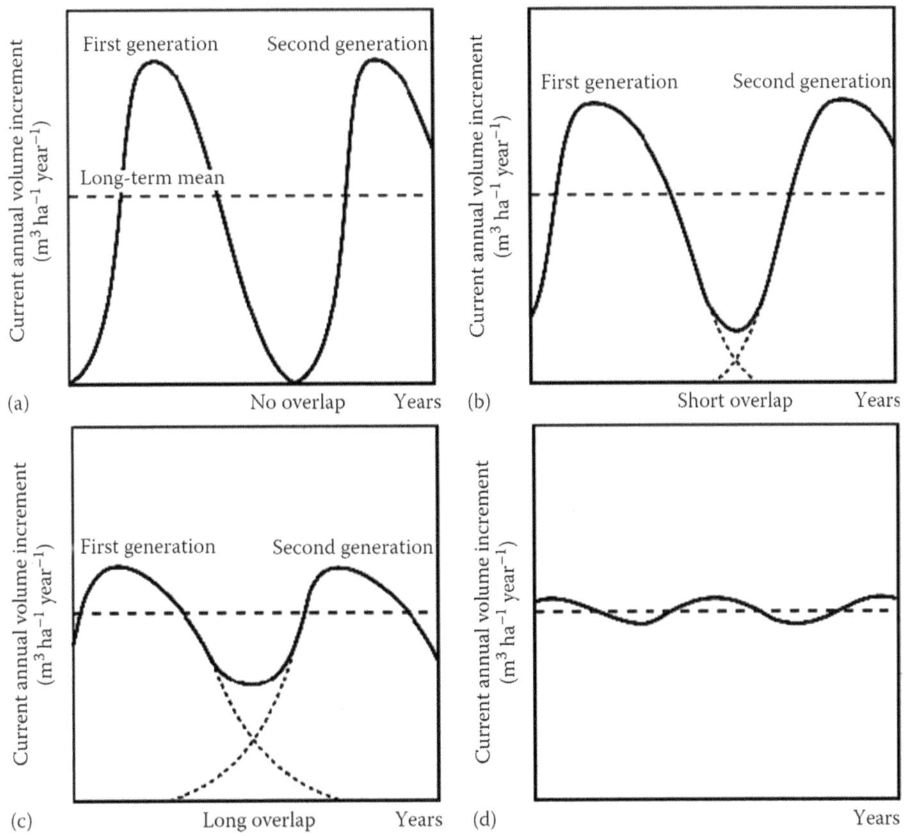

FIGURE 15.8 Continuous versus cyclic stand productivity. (a) clearcut system without generation overlap, (b) short overlap (b) and long overlap (c) between successive stand generations, (d) selection forest system. The longer the overlap between trees or cohorts of different ages and sizes the more steady the course of current volume increment. (Adapted from Assmann 1970, p 473.)

in contrast, the combination of trees of many age classes, sizes, and species results in a rather steady course of current volume increment, as the stand is continuously closed and equipped with trees in different ontogenetic stages (juvenile, mature, old), and therefore the stand density is permanently rather high and the growth at a continuous level. Non-productive phase caused by open field conditions after clearcut or low-growth phases in the early stand development phase before crown canopy closure are avoided by this model. Stands where the regeneration overlap temporarily, e.g., for short or long periods of 20 or 60 years, lie in between these borderline cases (Figure 15.8b and 15.8c, respectively). In the time of stand overlap, the stand volume increment of regeneration and overstory add up and result to a less strong oscillation of the total volume increment compared with the age-calls forest. The longer the overlap between trees or cohorts of different ages and sizes the steadier the course of current volume increment. The more multilayered the stand structure in terms of horizontal and vertical, the higher the growth resilience to natural and silvicultural interferences, such as regeneration cuts.

15.5.3 Species Mixing Can Modify Productivity

Recent empirical works on specific species mixtures (Griess and Knoke 2011, Pretzsch et al. 2010, 2013a), simulation model studies (Morin et al. 2011), as well as meta-analyses (Paquette and Messier

2011, Piotto 2008, Zhang et al. 2012) showed that the productivity of mixed stands can exceed the weighted mean productivity of pure stands by 50% in mixture with nitrogen-fixing species (Forrester et al. 2006) and by 20–30% in other mixtures (Pretzsch et al. 2013b). The benefit of species mixing can increase with the complementary of species' traits (Morin et al. 2011) and stand structure (Río et al. 2014a), be higher on pure compared with fertile sites (Río et al. 2014b), and stronger in low-growth compared with high-growth years (Pretzsch et al. 2013b). Further stability of forest can be improved by species mixing (del Rio et al. 2017, 2022).

For temperate forests in Central Europe, where the nutrient and water supply in the soil can be assumed as rather sufficient for all trees so that they are mainly competing above ground for light, Figure 15.9 shows mean overyielding of 19–24% of mixed versus neighboring pure stands. The data comes from long-term experimental plots. Closer analysis of the size distribution, structure, and individual tree behavior showed that, as a result of competition for light, tree size distribution, growth partitioning between the trees, and canopy density of mixed-species stands can change fundamentally and cause higher light interception light-use efficiency and overyielding compared with neighboring pure stands.

Whereas many studies show that mixed species stands can have higher gross growth, or so-called overyielding, compared with monocultures, much less is known about mortality in mixed stands. Pretzsch et al. (2023c) addressed this knowledge gap of mixed stand dynamics by evaluating 23 middle-aged, unthinned triplets of monospecific and mixed plots of Scots pine (*Pinus sylvestris* L.) and European beech (*Fagus sylvatica* L.) repeatedly surveyed over 6–8 years throughout Europe. They found that the gross growth of mixed stands was higher, but net growth was lower compared to monocultures (Figure 15.10). Mortality, natural thinning, and volume dropout were higher in mixed stands.

In the mixture, the dropout of pine was mainly responsible for the reduction from gross to net growth, and a demixing in terms of mixing proportion and spatial segregation. The stem volume dropout increased as water supply increased. Thinning may utilize the gross overyielding of mixed stands and counteract the demixing.

15.5.4 Relationship between Stand Density and Growth in Dependence on Forest Structuring

How silvicultural interference (thinning, regeneration cuts) and other structural disturbances (windthrow, snow-breakage, bark beetle damage) affect stand productivity depends mainly on the stand composition and stand structure. The vertically deeper and multilayered a canopy is, the higher can be the growth resilience to structural disturbances (Pretzsch 2003). Productivity losses due to tree removal by thinning or losses by damage can be better compensated by the presence of a second or third layer of trees that might immediately use the resources not any longer intercepted by removal trees.

The superior part of Figure 15.11 shows common stand structures. The deepness of the canopy increases from even-aged pure stands to uneven-aged mixed stands (from a to d). The respective kinds of silvicultural interference and removal trees are indicated by gray silhouetted trees. The bottom part of Figure 15.11 shows in schematic representation that the growth resilience of stands to tree removal by thinning or losses by damage increases with the deepness and layering of the canopy. All shown density growth relationships are uni-modal, i.e., with lowest productivity level at low densities, increasing productivity when stand density increases, and a slight decrease of productivity when approaching maximum stand density. The main causes for the growth resilience to density reductions are (i) the acceleration of highly efficient understory trees when light is flooding deeper canopy space, (ii) the activation of soil processes and improvement of nutrition due to higher soil moisture and temperature after density reduction, and (iii) the removal of low resource efficient trees and use of the released resources by remaining neighbors with higher resource use efficiency (Assmann 1970, Pretzsch 2005).

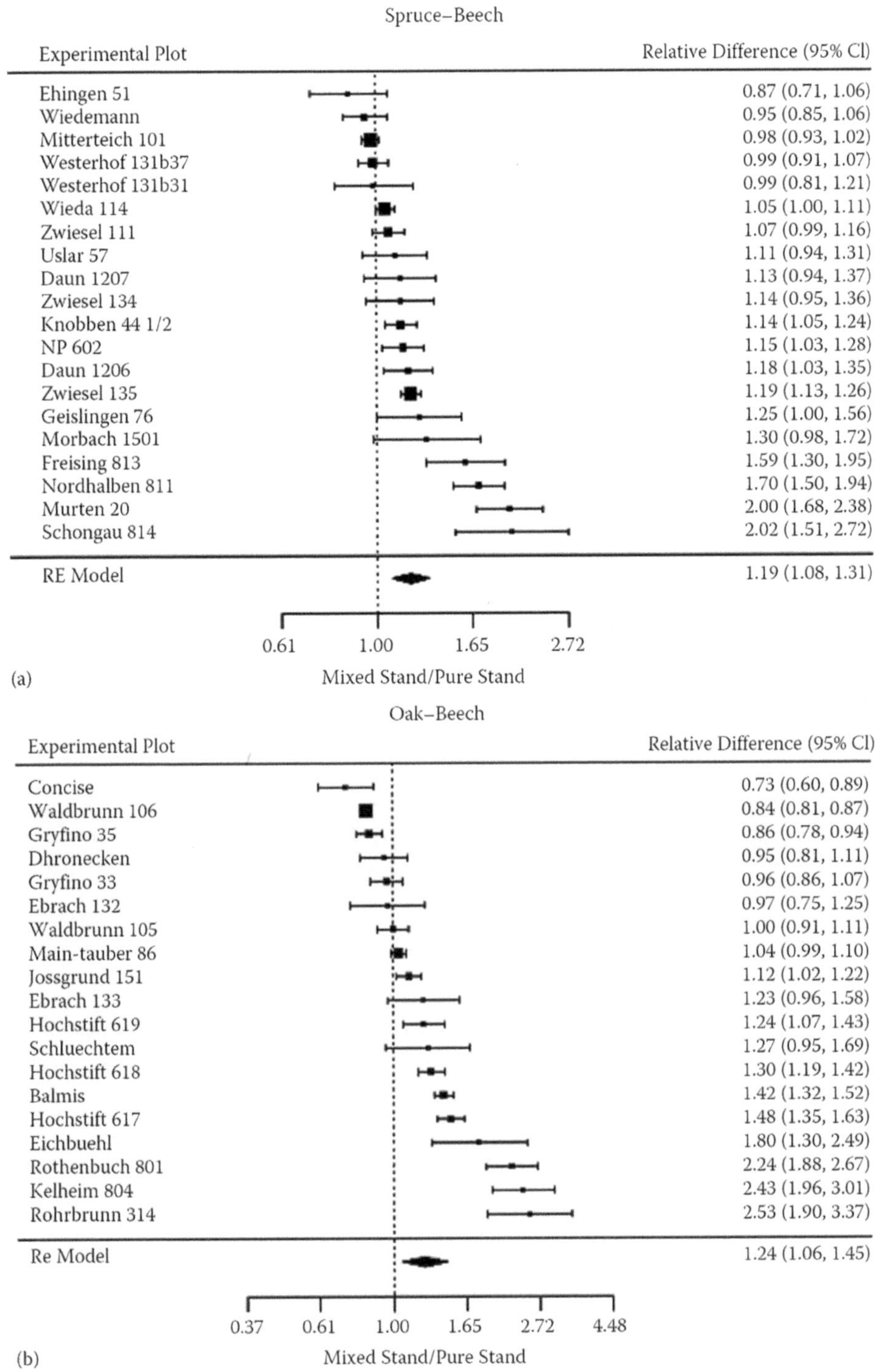

FIGURE 15.9 Comparison of the stand productivity of mixed versus pure forest stands bases on long-term experimental plots of (a) Norway spruce and European beech, (b) sessile oak and European beech, (c) Norway spruce, silver fir, and European beech in Germany. On average, the mixed stands produce 119%, 124%, and 120%, respectively, of the weighted mean of the neighboring pure stands. All three analyses revealed a broad variation of mixing effects including strong underyielding and overyielding. (According to Pretzsch et al. 2010, 2013a; Pretzsch 2014).

(*Continued*)

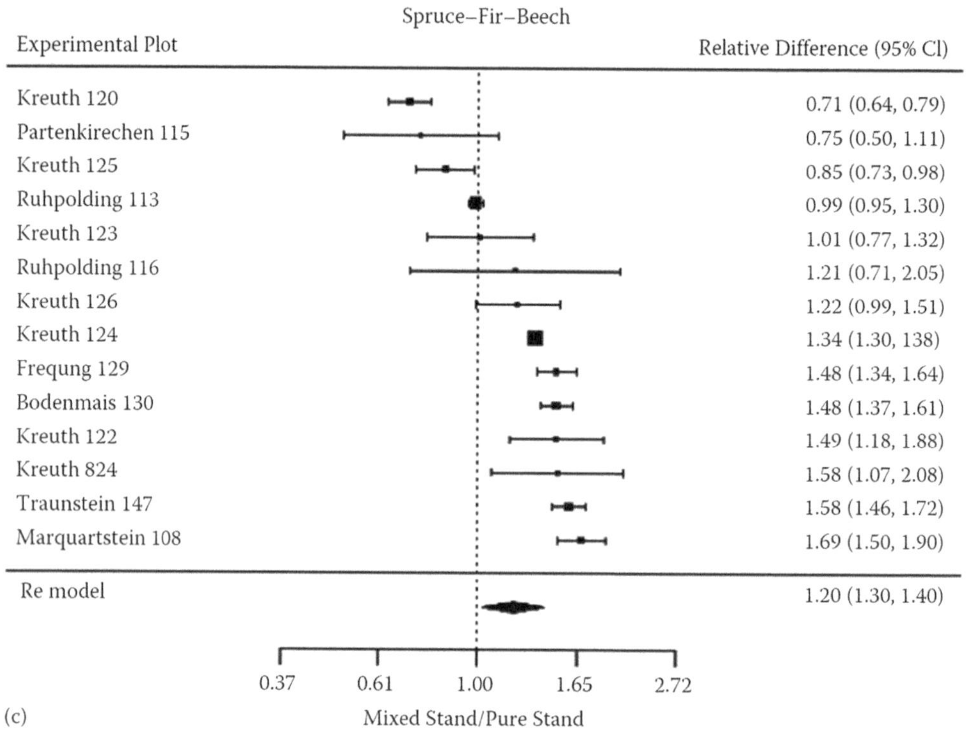

FIGURE 15.9 ***(Continued)***

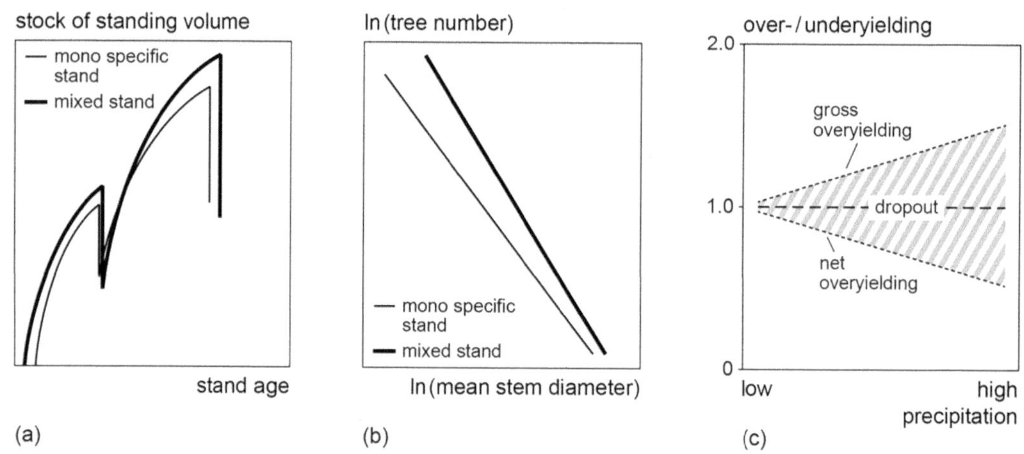

FIGURE 15.10 In the unthinned triplets of monospecific and mixed stands of Scots pine and European beech throughout Europe, we found that (a) gross stem volume growth was higher in mixed compared to monospecific stands. Nonetheless, net stem volume growth hardly differed, (b) stand density was higher in mixed stands, but the alien-thinning accelerated, (c) gross-overyielding increased in mixed stands with water supply, net-overyielding decreased. Thus, the stem volume dropout fraction (hatched) was higher on moist sites. (After Pretzsch et al. 2023c).

However, while the patterns are similar in principle, the position and extent of the curves' saddle is rather different. As tree removal can cause uncovered areas without wood production when thinning even-aged pure stands or femeling mixed stands with mono-layered canopies

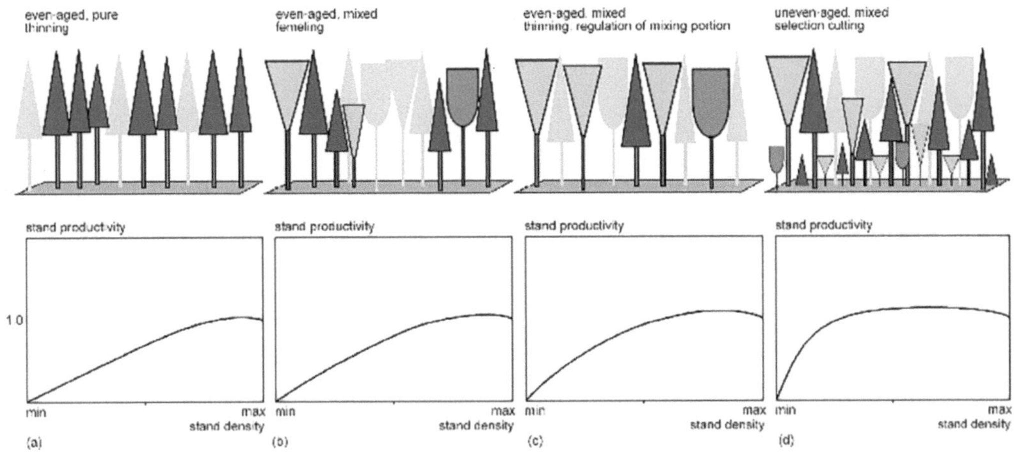

FIGURE 15.11 The growth resilience of stands to tree removal by thinning or losses by damage (grey silhouetted trees) increases with the deepness and layering of the canopy. Schematic representation of the strand structure and the density-productivity relationship of (a) an even-aged pure stand, (b) femel structured mixed stand, (c) even-aged mixed stand, (d) uneven-aged, multilayered mixed stand. The 1.0-line (broken horizontal line) represents the productivity at maximum stand density.

(Figure 15.11a and 15.11b), the productivity can decrease already after slight density reductions. Thinning in even-aged mixed stands or selection cutting in uneven aged mixed stands with deep canopy spaces (Figure 15.11c and 15.11d) causes productivity reduction not until heavy density reductions, as the trees in the medium and lower canopy layers can compensate for any losses in the upper layer. The range where productivity is rather stable, although density has been reduced, also called by Langsaeter (1941) the saddle of the density growth relationship, is particularly wide in selection forests, as their multilayered, uneven-aged structure buffers well most structural interferences. Notice that the response pattern in Figure 15.11 represents just the general dependency between structure and growth. The schematically shown density-growth relationships can be modified by site conditions, species, age, and the kind of silvicultural interference.

As stand structure determines many ecosystem services (e.g., resilience against disturbances, habitat provision, esthetics and recreational value), it received increased intention recently and turned out to also increase productivity. Recent studies address the relevance of stand structure in terms of tree size variation for growth of monospecific and mixed stands (Ali et al. 2016, Dănescu et al. 2016). Overyielding of mixed versus monospecific stands may be caused mainly by structural heterogeneity rather than by the species identify of mixtures. Especially in older stands structural heterogeneity can increase the growth, standing stock, and resistance and resilience against disturbances. Homogenization by thinning needs to be rethought; in young stands it may temporarily increase stand growth. However, it may sacrifice especially those trees which increase the structure and growth in older stands. Such findings need more consideration and integration in monitoring, planning, and silvicultural steering of forest stands.

Figure 15.12 shows the effect of tree size variation on growth and yield of even-aged Norway spruce stands (Pretzsch et al. 2024). The focus is on the effect on stand growth, maximum stand density, general and special yield level.

All four stand characteristics show a unimodal dependency on tree size variation with an optimum level at mean size variation.

The study stresses the relevance of tree size variation when analyzing, modeling, and silviculturally steering forest stands.

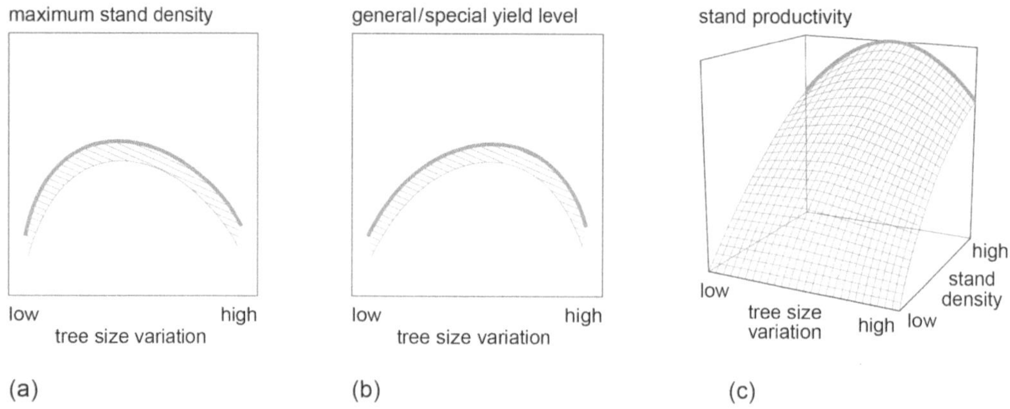

FIGURE 15.12 The stand characteristics (a) maximum stand density, (b) general and special yield level, and (c) stand growth show a unimodal dependency on tree size variation with an optimum level at mean size variation. (From ongoing investigations by Pretzsch et al. in preparation for publication.)

15.6 PRODUCTIVITY CHANGES AND CHANGE DETECTION: IMPACT OF DISTURBANCE EVENTS AND LONG-TERM TRENDS ON STAND PRODUCTIVITY

15.6.1 Long-Term Changes in Productivity

Forest ecosystems have been increasingly exposed to climate change since more than 100 years (Skeffington and Wilson 1988). While only two decades ago the "Waldsterben" caused by acid rain seemed to terminate forest ecosystems (Ulrich 1999), presently phenology observations (Chmielewski and Rötzer 2001), NDVI sensing (Myneni et al. 2001), and inventories (Spiecker et al. 1996) indicate rather rampant than collapsing forests in Europe. Global climate change will affect forest productivity in a very regional-specific way in the future (IPCC 2007). A crucial basis for understanding the ongoing changes in environmental conditions and their effect on forest stand dynamics and productivity, for avoidance of negative effects by environmental policy, or for mitigating stress effects by management measures, is a long-term biomonitoring. Three hundred years ago, von Carlowitz (1713) brought the idea of sustainability into forestry, and Hartig (1791, 1795), von Cotta (1828), and Pfeil (1860) developed concepts for establishing this idea in forest management. To procure growth and yield data as quantitative basis for sustainable forest management, farsighted researchers in the late 19th century started installing long-term observational and experimental plots (see von Ganghofer 1881; Verein Deutscher Forstlicher Versuchsanstalten 1873).

15.6.2 Long-Term Plots as Ultimate Arbiters

As the appropriate kind, severity, and intensity of thinnings were crucial information for sustainable management, most of the early experiments comprised plots in pure and mixed stands with different thinning grades as well as unthinned reference plots. In the intensively managed forests in Europe, the untreated plots represent the exceptional case of up to 140 years' unmanaged ecosystem development, where any growth trends can be attributed to changing environmental conditions as not superimposed by silvicultural treatment (Figure 15.13). Many of the observational plots dating back to these days are still an essential part of the forest observation network in Central Europe.

A growth comparison between previous and subsequent stands at the same site can diagnose long-term changes in the growth behavior over two stand generations (Kenk et al. 1991; Röhle 1994, 1997; Wiedemann 1923). It is clear that the analysis over such a large time span requires an

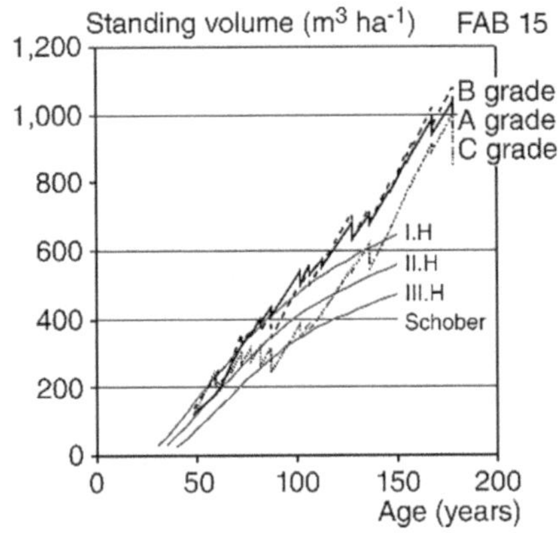

FIGURE 15.13 Long-living organisms require long-term research. Development of standing volumes (m^3 ha^1) on the European beech thinning experiment Fabrikschleichach 15 (A, B, and C grade) as the result of trans-generation measurements of the forest yield research. As one of more than 100 long-term experimental plots in Bavaria, the Fabrikschleichach plots have been surveyed regularly since 130 years.

exceptionally large database with consistent surveys of the present and previous generation, which might easily date back 200 years. Few long-term experimental plots provide such valuable databases. The comparison can be carried out for diameter, height, basal area, and volume, of which height is especially useful as it remains almost unaffected by stand silvicultural treatment. In an early study, Wiedemann (1923) detected an impeded height growth of Norway spruce in Saxony (Germany), which he attributed to silvicultural and climatic effects. Comparing the height growth behavior between stands established in 1700–1730 with the subsequent generation established in 1825–1845, he proved a degradation in the site fertility by two levels of Schwappach's (1890) yield tables for Norway spruce (Figure 15.14).

More recently, Kenk et al. (1991) detected a growth improvement of up to seven site index levels of the Assmann and Franz yield tables (1963) for Norway spruce for stands with poor-to-medium site fertility in Baden-Württemberg (Figure 15.14). The time span between the establishment of the previous stands (1820) and the subsequent stands (1950) was—similar to Wiedemann (1923)—approximately 130 years.

The growth comparison of two successive generations on the same site can prove the existence of a disturbance and the time and strength of its effect. Although the underlying causes are not determined this way, uncertainties about site comparability, present in the reference plot comparison, for instance, can be excluded. Furthermore, provided the stand treatment (e.g., A, B, or C grade thinning), and the genetic composition (establishment through natural regeneration or by sowing seeds of the previous stands) is similar, then effects due to treatment and genetics can be assumed negligible. In a strict sense, though, the seeds from the previous stand used for establishment of the subsequent stand originate from the collective of the individuals that survived self-differentiation, thinning, and calamities, and therefore the subsequent stands represent only a fraction of the previous stand's gene pool.

Long-term observation plots and experiments in forest stands deliver shorter time series than tree ring analysis and are probably less representative than inventories, but they can provide rather unique information about long-term performance on both tree and stand levels (Hilmers et al. 2019, Nagel et al. 2012). Compared with retrospective tree ring analysis and inventory data, long-term plots have the advantage that their stand history (including calamities etc.) is recorded, their experimental setup is standardized, and they often include completely unmanaged variants. Beginning in 1870, many long-term observational plots in Central Europe have been re-measured more than 20 times till the present. As this study is based on long-term observational plots and exploits their information for detection of growth trends, we briefly introduce the concept of these observation plots.

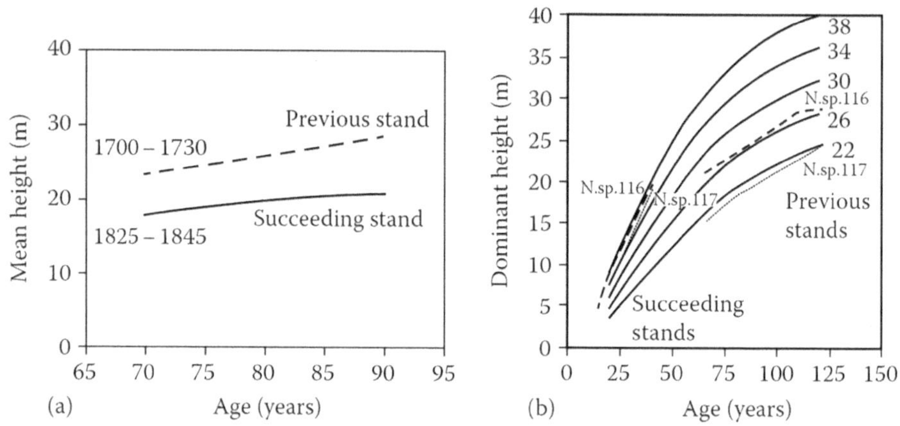

FIGURE 15.14 Long-term observational plots as basis for monitoring of productivity changes. (a) Detection of static growth of Norway spruce in the Forest near Tharandt (Eastern Germany) by comparison of previous and subsequent forest stands (Wiedemann 1923, p. 157, Tab. 1). Mean height of forest stands established between 1700 and 1730 (dashed line) and the subsequent stands between 1825 and 1845 (continuous line). (b) Comparison of height development of previous and subsequent generation on the Norway spruce (N. sp.) experimental trial plots N. sp. 116 and N. sp. 117 in the Black Forest (Kenk et al. 1991, p. 30, Figure 14). The comparison reveals a growth improvement of several site classes (from h_{100} = 22–26 m to h_{100} = 34–38 m) according to the yield table from Assmann and Franz (1963).

15.6.3 Current Growth Trends

Even 300 years after Carl von Carlowitz, long-term plots are ultimate arbiters of human footprints on forest ecosystems as they reflect whether and how forest growth is changing. The information potential of long-term experimental plots has been tapped by Pretzsch et al. (2014a) providing evidence of increased forest stand productivity in Central Europe since 1870 (Figure 15.15). Based on the oldest existing experimental forest plots in Central Europe, this study showed that currently, the dominant tree species, Norway spruce and European beech, exhibit significantly faster tree growth (+32–77%), stand volume growth (+10–30%), and standing stock accumulation (+6–7%) than in 1960. Stands still follow similar general allometric rules but proceed more rapidly through usual trajectories. As forest stands develop faster, tree numbers are currently 17–20% lower than in past same-aged stands. Self-thinning lines remain constant while growth rates increase, indicating the stock of resources have not changed while growth velocity and turnover have altered. Statistical analyses of the experimental plots, and application of an ecophysiological model, suggest that mainly the rise in temperature and extended growing seasons contribute to increased growth acceleration, particularly on fertile sites.

Especially untreated, so-called 0-plots or reference plots are indispensable, as they reflect just the effect of environmental changes on stand dynamics. They exclude silvicultural effects, e.g., spacing, thinning, fertilization, and breeding, which confound environmental effects when studied on the basis of inventories in managed stands. While long-term observational plots served mainly for quantifying growth and yield in the past, they address a much broader scope of forest functions and services at present and in future (Burkhart and Temesgen 2014). While countries with far back dating observational plot should keep and update them, countries with sustainable forestry just emerging should establish networks for long-term biomonitoring of forest components, structure, and functioning. As even 300 years after von Carlowitz (1645–1714) proposed the concept of sustainable forests by publishing *Sylvicultura Oeconomica* (von Carlowitz 1713), long-term plots remain the ultimate arbiters of human footprints on forest ecosystems.

Reports on how climate change affects forest growth are contradictory, even for same regions. Pretzsch et al. (2023b) used 415 unique long-term experiments including 642 plots across Europe

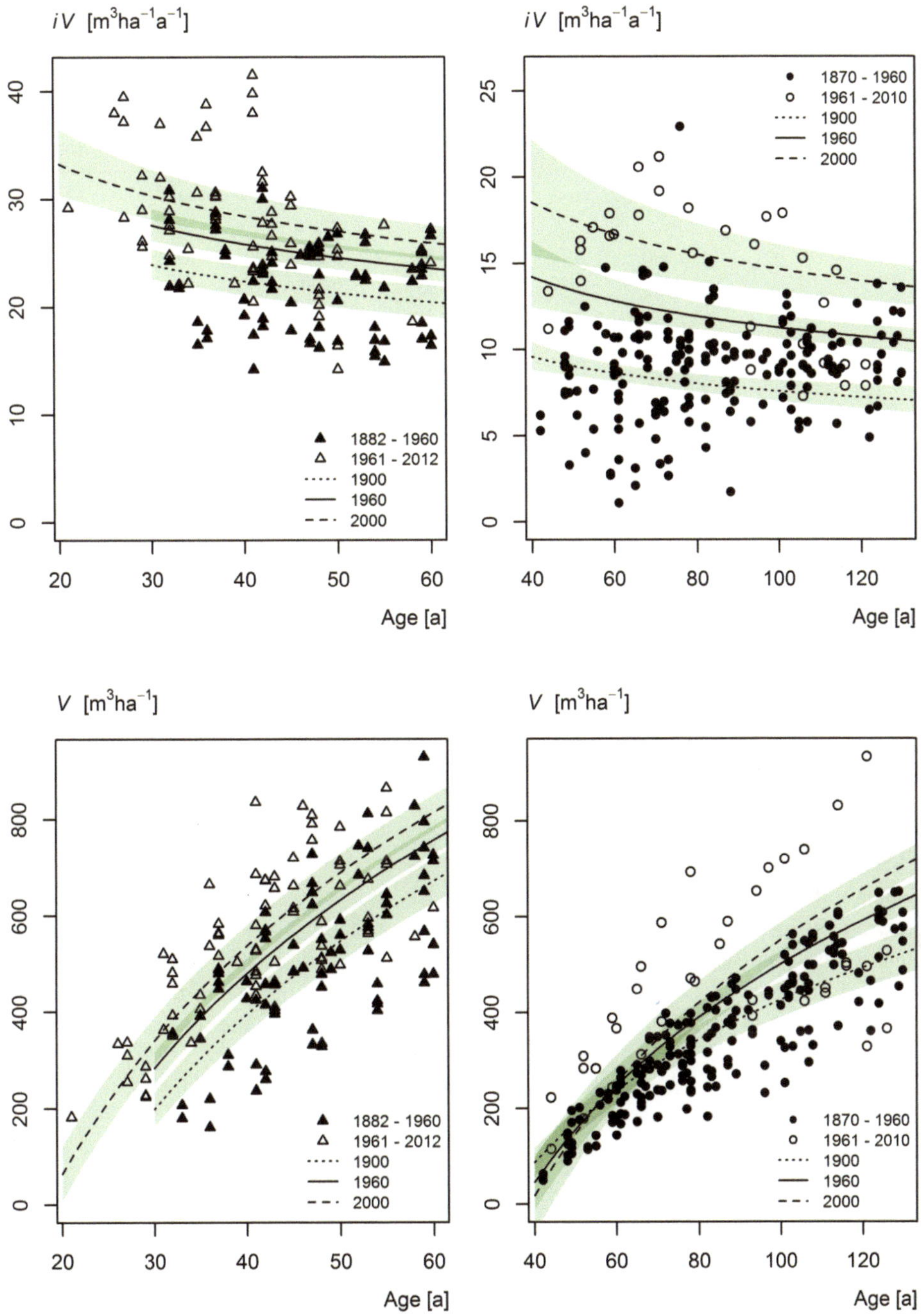

FIGURE 15.15 Statistical analysis of tree and stand dynamic changes since 1870 according to Pretzsch et al. (2014b). Trends in stand Periodic Annual Volume Increment (*PAIV*) and standing volume (V) for Norway spruce (above) and European beech (below) age ranges. Observations before 1960 (filled symbols); after 1960 (empty symbols); predictions with our fitted linear mixed models (LMM) for 2000 (dashed line); for 1960 (solid line); and as a reference for 1900 (dotted line). The gray-shaded areas illustrate the prediction standard error.

TABLE 15.4
Upscaling of the Expected Biomass Growth Trends for Scots Pine to the European Level* (After Pretzsch et al. 2023a)

Trend Class	Area Covered by Scots Pine	Biomass Growth		Change from 1975 to 2015	
		1975	2015	absolute	relative
	km^2	Gg a^{-1}	Gg a^{-1}	Gg a^{-1}	%
−1	8,990	4,081	3,3238	−843	−21
0	205,098	79,963	99,069	19,106	24
1	134,263	59,632	65,727	6,095	10
2	142,181	32,246	53,969	21,723	67
total	490,532	175,922	222,003	46,080	26

Notes:
−1 Spain, Italy, Southern France
0 Northern France, Italy, Balkan States
1 Central Europe, Baltic States
2 mostly Sweden and Norway

covering seven tree species and surveys from 1878 to 2016 and showed that on average forest growth strongly accelerated since the earliest surveys. Based on a subset of 189 plots in Scots pine (the most widespread tree species in Europe) and high-resolution climate data, they identified clear large-regional differences; growth is strongly increasing in Northern Europe and decreasing in the Southwest. A less pronounced increase, which is probably not mainly driven by climate, prevails on large areas of Western, Central, and Eastern Europe (Table 15.4).

15.6.4 Effects of Extreme Events

It is evident that extreme events such as droughts, frosts, storms or insect infestations have a strong influence on the growth of forests and thus on their productivity. In terms of climate change, droughts in particular will become increasingly common. Based on model simulations, Cao et al. (2022) report drought-induced losses of up to 8.7 Gt C in NPP by the end of the 21st century under RCP scenarios, with the most substantial NPP reductions occurring in the southern hemisphere and in many subtropical areas.

Decisive factors for the effects of droughts on growth and productivity of forest stands are intensity, duration, frequency, and occurrence of droughts in the course of the year (Rötzer et al. 2012). Given the complexity of quantifying drought and the different physiological responses to drought, the mechanisms by which plants in different climate zones cope with droughts on different timescales are still poorly understood (Liu et al. 2021).

On the top of that, there is limited understanding about the impacts of prolonged and recurrent droughts, specifically relating to the effects of consecutive drought years that are likely to occur in future (e.g., Seneviratne et al. 2012). The KROOF experiment set up ten years ago in the Kranzberg forest near Freising/Germany provides information on the decline in growth and possible adaptation and recovery reactions based on long-term droughts (Pretzsch et al. 2020, Grams et al. 2021). Here, six plots of 70-year-old Norway spruce and 90-year-old European beech were reduced in precipitation during the growing season from 2014 to 2018 by means of automatic throughfall exclusion roofs. Afterwards, in early summer 2019, the trees were watered and the roofs were removed. Six neighboring control plots were included for comparison.

Significant volume increment reductions can be seen for both tree species but clearly stronger for spruce in 2003 (Figure 15.16), a year with exceptionally low precipitation over entire Central Europe. The artificial period of drought from 2014 to 2018 resulted in significant losses for spruce,

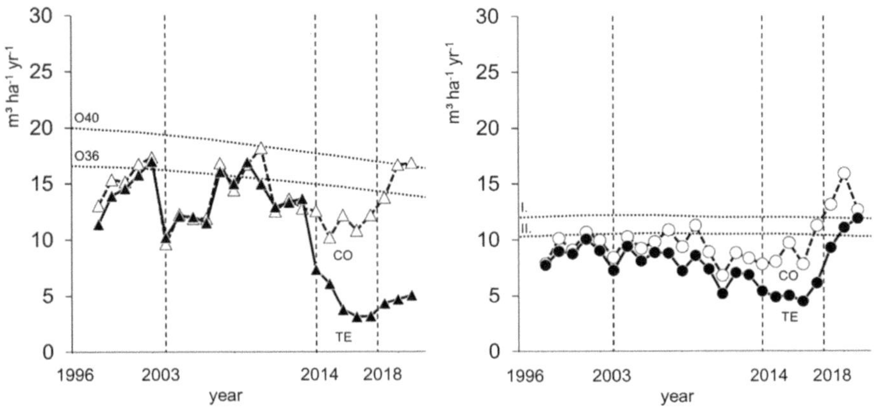

FIGURE 15.16 Annual volume increment at stand-level of spruce (left) and beech (right) on plots with drought treatments (solid line and filled marks) and on the control plots (dashed line and empty marks). (Adapted after Motte et al. 2023.)

with some trees dying and others remaining at low growth levels without significant recovery from drought stress. In the first drought years, beech also lost significant growth but recovered more strongly from the drought than trees in the control plots (Pretzsch et al. 2021).

The annual volume increment of spruce and beech decreased strongly under the long-term drought. At the stand level, the volume increment losses accumulated up to 34.4 m^3 ha^{-1} for spruce and 18.7 m^3 ha^{-1} for beech over the entire drought period. Compared to the trees on the control plots, spruce and beech trees experienced volume growth reductions of 59% and 42%, respectively due to the drought treatment (Motte et al. 2023). After irrigation, beech trees in particular showed a significant increase in volume growth and almost reached the level of the control trees after irrigation. By contrast, spruce trees did not display a clear upward trend in volume growth and did not approach the volume growth of trees in the control plots.

Anderegg et al. (2015) also found that severe droughts had legacy effects over a period of 1–4 years on tree growth and recovery in forests, which was most pronounced in dry ecosystems. The productivity, i.e., the NPP of forest ecosystems, may further be impaired by indirect effects of drought on ecosystems (Huang et al. 2016), such as forest fires or insect infestations (e.g., Reichstein et al. 2013, Anderegg et al. 2015). The combination of these factors points to drought as the main driver of the decline in global terrestrial NPP in the first decade of the 21st century (Zhao et al. 2010). Cao et al. (2022) concluded that future droughts will increase NPP losses due to climate change and weaken CO_2 fertilization effects, further threatening the stability of terrestrial carbon sinks.

15.7 INTEGRATION OF INDICATOR SYSTEMS INTO FOREST MANAGEMENT

15.7.1 Productivity Regulation as Basis for Many Other Forest Functions and Services

Because of their size, firm position, and longevity, trees are the founder species in ecosystems (Whitham et al. 2006). Their morphology and structure determine the living conditions of many ecosystem characteristics, their functions, and many services beyond the wood and timber quality (Hector and Bagchi 2007). By the shading effect of their crowns and the water uptake of their roots, trees influence the light and water supply and thus the habitats of flora and fauna in their surroundings.

Therefore, forest inventories, especially when based on fixed grid points, including spatial explicit tree and stand records, and when repeated over time, comprise valuable and hardly extracted aspects of functional, compositional, and structural diversity. The commonly measured annual growth rate not only provides information about the allowable cut but also correlates closely with leaf biomass turnover, which is essential for feeding all herbivores and decomposers. Tree size and structure not only determines standing stock and wood quality but also forms the habitat of many animals like birds, bats, and spiders.

15.7.2 Stand Productivity and Sustainable Productivity at Enterprise Level

The mean annual volume increment (MAI) and mean standing volume stock (V) that allows this MAI are the most fundamental indicators for any planning and management. They represent the potential mean productivity and mean standing volume within a time span from stand establishment to culmination of the MAI (rotation period) in age class forest and the long-term mean productivity and standing stock in uneven-aged forest, such as selection forest systems. Both indicators depend mainly on the prevailing site conditions, the chosen tree species, and occurring disturbances. MAI and V can be derived for both stand and management block level based on, e.g., observational plots, yield tables, or stand simulators. MAI indicates the level of the potential sustainable annual cut.

In contrast, the actually possible annual cut is induced in the course of silvicultural planning by assessing the appropriate treatment and resulting annual cut at the individual stand, stand type, or stratum level. However, this silviculturally derived possible annual cut reflects just the possible quantity of logging which might considerably differ from the long-term sustainable cut aimed at on overall enterprise level. A high stock of old stands may suggest a far above-average annual cut for the coming planning period, which might be distributed over a longer time span to avoid a long time period of under-average annual cut.

15.7.3 Productivity Regulation in Forest Practice

Based on the information about the potential and actual annual cut on the one hand and the long-term objective on the other hand, forest management deduces a long-term sustainable cut which might considerably differ from both potential and annual. The main reason for such deviations are, beside any legal, political, or ecological restrictions, that the planning normally aims at achieving a stable annual cut that can be realized continuously over a long or even infinite time span without exploiting, but only sustainably removing, the mean annual increment.

Especially when forest management planning aims at equalization of a so-far uneven age class distribution, replacement of pure by mixed-species stands, or transition from age class to uneven-aged management, simulation models can support the decision-making. They show the long-term effect of various silvicultural options and help to find out which set of silvicultural prescriptions (thinning, mixing, final cutting, regeneration) is necessary for achieving a target long-term annual cut and respective standing stock. The recursive planning process includes the provisional selection of silvicultural measures for all planning units → upscaling of the resulting long-term development and annual cut → comparison of the resulting long-term development and annual cut with the target state of the management block → modification of silvicultural measures for selected planning units to better arrive at the target state → upscaling of the modified treatments with respect to the long-term development at the overall management block level, and so on. It is much easier and more flexible when based on simulation models compared with classical approaches based on normal forest assumptions, yield tables, and formulas for indicating the sustainable annual cut.

A combined use of simulation models for strategic and recursive planning and long-term permanent inventory plots for model control is common in modern planning. Silvicultural measures, which strongly affect the standing stock, volume growth, and productivity at the individual stand level and the sustainable cut at the enterprise level on the long term, are in particular the amount of final cutting, thinning, and regeneration. Long-term effects have furthermore species and provenance choice, species mixing, mixing regulation, wild repellent, and mechanical stabilization by tending and thinning.

The long-term annual cut, as deduced by recursive planning at stand and overall enterprise level, is the most relevant indicator for sustainable forest management.

15.8 PERSPECTIVES: FROM DEDUCTIVE TO INDUCTIVE AND STATIC TO DYNAMIC INDICATION OF FOREST ECOSYSTEM AND STAND PRODUCTIVITY

Former approaches deduced forest ecosystem and stand productivity from general models, which were based on a rather small number of long-term observational plots. As examples for this approach, we introduced productivity indices, yield tables, and individual tree models which were based on long-term experiments.

Present-day inventories and in particular their repeated surveys of the forest state, increment, environmental conditions, and resource supply provide information about forest ecosystem to parameterize the relationship between site fertility and productivity or to calibrate models for the ecosystems in question. They even make it possible to initialize growth models site-specifically and constantly update the parameterization. The advantage over conventional model parameterization on the basis of long-term experimental plots is obvious: inventories cover a much broader range of site conditions and stand structures. Thus, even when plots with unknown history are left out, the remaining datasets are more extensive and representative than the experimental plot data (von Gadow 2005). The use of inventory data as start values for simulation runs and for the derivation of site-growth relations represents an inductive approach. In this case, information for indication, prognosis, and simulation are induced from the inventory data itself and replace the former deductive approaches of indication of forest ecosystem and stand productivity (Figure 15.17).

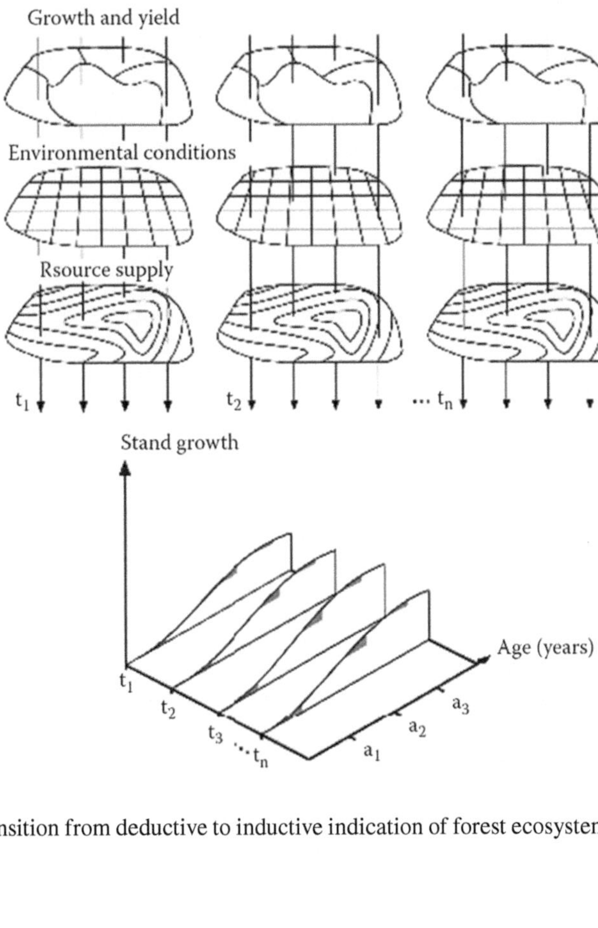

FIGURE 15.17 Transition from deductive to inductive indication of forest ecosystem and stand productivity.

REFERENCES

Ajtay, G.L., P. Ketner, and P. Duvigneaud. 1979. Terrestrial primary production and phytomass. *The Global Carbon Cycle Scope.* 13:129–182.

Ali, A., E.R. Yan, H.Y. Chen, et al. 2016. Stand structural diversity rather than species diversity enhances aboveground carbon storage in secondary subtropical forests in Eastern China. *Biogeosciences* 13(16):4627–4635.

Anderegg, W.R., C. Schwalm, F. Biondi, et al. 2015. Pervasive drought legacies in forest ecosystems and their implications for carbon cycle models. *Science* 349(6247):528–532.

Assmann, E. 1970. *The principles of forest yield study.* New York: Pergamon Press.

Assmann, E. and F. Franz. 1963. *Vorläufige Fichten-Ertragstafel für Bayern.* Forstl Forschungsanst München, Inst Ertragskd, 104 p.

Beer, C., Reichstein, M., E. Tomelleri, et al. 2010. Terrestrial gross carbon dioxide uptake: global distribution and covariation with climate. *Science* 329/5993:834–838.

Bresinsky, A., Körner, C., J.W. Kadereit, et al. 2008. *Strasburger Lehrbuch der Botanik.* Heidelberg: 36. Auflage. Spektrum Akademischer Verlag.

Brünig, E.F. 1971. *Forstliche Produktionslehre.* Europ. Hochschulschriften, Reihe XXV, Verlag Herbert Lang, Bern und Peter Lang, Frankfurt.

Bruscheck, G.J. 1994. Waldgebiete und Waldbrandgeschehen in Brandenburg im Trockensommer 1992. *PIK-Report* 2:245–264.

Buchmann, N. and E.D. Schulze. 1999. Net CO_2 and H_2O fluxes of terrestrial ecosystems. *Global Biogeochem Cycles* 13/3:751–760.

Burkhart, H. and H. Temesgen. 2014. Forest observational studies: Data sources for analysing forest structure and dynamics. *For. Ecol. Manage.* 316:147p.

Cajander, A.K. 1926. The theory of forest types. *Acta forestalia fennica* 29:108 p.

Cai, W., W. Yuan, S. Liang, et al. 2014. Improved estimations of gross primary production using satellite-derived photosynthetically active radiation. *J. Geophys. Res.-Biogeo.* 119:110–123. https://doi.org/10.1002/2013jg002456.

Cao, D., J. Zhang, J. Han, et al. 2022. Projected increases in global terrestrial net primary productivity loss caused by drought under climate change. *Earth's Future* 10(7):e2022EF002681.

Carvalhais, N., M. Forkel, M. Khomik, et al. 2014. Global co-variation of carbon turnover times in terrestrial ecosystems with climate. *Nature* 514:213–217.

Chapin III, F.S., G.M. Woodwell, J.T. Randerson, et al. 2006. Reconciling carbon-cycle concepts, terminology, and methods. *Ecosystems* 9/7:1041–1050.

Chmielewski, F.M. and Rötzer, T. 2001. Response of tree phenology to climate change across Europe. *Agr. For. Meteorol.* 108(2):101–112.

Crutzen, P.J. 2002: Geology of mankind. *Nature* 415:23.

Dale, V.H. and S.C. Beyeler. 2001. Challenges in the development and use of ecological indicators. *Ecol. Indic.* 1:3–10.

Dănescu, A., A.T. Albrecht, and J. Bauhus. 2016. Structural diversity promotes productivity of mixed, uneven-aged forests in southwestern Germany. *Oecologia* 182(2):319–333.

Davis, K.P. 1966. *Forest management.* New York, St. Louis, Sydney: McGraw-Hill Book Company.

del Río, M., H. Pretzsch, R. Ruiz-Peinado, et al. 2022. Emerging stability of forest productivity by mixing two species buffers temperature destabilizing effect. *J. Appl. Ecol.* 59(11):2730–2741.

del Río, M., H. Pretzsch, R. Ruíz-Peinado, et al. 2017. Species interactions increase the temporal stability of community productivity in Pinus sylvestris–Fagus sylvatica mixtures across Europe. *J. Ecol.* 105(4):1032–1043.

Delucia, E.H., Drake, J.E., R.B. Thomas, et al. 2007. Forest carbon use efficiency: Is respiration a constant fraction of gross primary production? *Glob. Chang. Biol.* 13:1157–1167.

Dieler, J., E. Uhl, P. Biber, J. Müller, T. Rötzer, and H. Pretzsch. 2017. Effect of forest stand management on species composition, structural diversity, and productivity in the temperate zone of Europe. *Eur. J. For. Res.* 136:739–766.

Dobbertin, M. 2002. Influence of stand structure and site factors on wind damage comparing the storms Vivian and Lothar. *For. Snow. Landsc. Res.* 77(1/2):187–205.

Dobbertin, M. 2005. Tree growth as indicator of tree vitality and of tree reaction to environmental stress: a review. *Eur. J. For. Res.*124/4:319–333.

Dudley, N., and S. Stolton. 2003. *Running pure: The importance of forest protected areas to drinking water.* Washington, DC: World Bank/WWF Alliance for Forest Conservation and Sustainable Use.

Eichhorn, F. 1902. *Ertragstafeln für die Weißtanne*. Berlin: Verlag Julius Springer.
Fang, J., Kato, T., Z. Guo, et al. 2014. Evidence for environmentally enhanced forest growth. *PNAS* 111/26:9527–9532.
FAO. 2012. *Global ecological zones for FAO Forest reporting: 2010 Update*. Forest resources assessment working paper 179. Food and Agriculture Organization of the United Nations, Rome, Italy.
Fernández-Martínez, M., Vicca, S., I.A. Janssens, et al. 2014. Spatial variability and controls over biomass stocks, carbon fluxes, and resource-use efficiencies across forest ecosystems. *Trees* 28/2:597–611.
Field, C.B., and M.R. Raupach. 2004. *The global carbon cycle: integrating humans, climate, and the natural world*. Washington, Covelo, London: Island Press.
Forrester, D.I., J. Bauhus, A.L. Cowie, and J.K. Vanclay. 2006. Mixed-species plantations of Eucalyptus with nitrogen-fixing trees: A review. *For. Ecol. Manag.* 233(2–3):211–230.
Gehrhardt, E. 1923. *Ertragstafeln für Eiche, Buche, Tanne, Fichte und Kiefer*. Berlin: Verlag Julius Springer.
Gough, A.D., J.L. Innes, and S.D. Allen. 2008. Development of common indicators of sustainable forest management. *Ecol. Indic.* 8:425–430.
Grams, T.E.E., B.D. Hesse, T. Gebhardt, et al. 2021. The Kroof experiment—realization and efficacy of a recurrent drought experiment plus recovery in a beech/spruce forest. *Ecosphere* 12(3):e03399. https://doi.org/10.1002/ecs2.3399.
Griess, V.C. and T. Knoke. 2011. Growth performance, windthrow, and insects: meta-analyses of parameters influencing performance of mixed-species stands in boreal and northern temperate biomes. *Can. J. For. Res* 41:1141–1158.
Hanewinkel, M. 2001. Neuausrichtung der Forsteinrichtung als strategisches Management Instrument. *AFJZ* 172:203–211.
Hartig, G.L. 1791. *Anweisung zur Holzzucht für Förster*. Marburg: Neue Akademische Buchhandlung.
Hartig, G.L. 1795. *Anweisung zur Taxation der Forste oder zur Bestimmung des Holzertrags der Wälder*. Wiesbaden: Georg-Ludwig-Hartig-Stiftung.
Hector, A. and R. Bagchi. 2007. Biodiversity and ecosystem multifunctionality. *Nature* 448:188–190.
Heyer, G. 1845. *Wedekinds Neue Jahrb*. 30:1–127.
Heyer, G. 1852. *Über die Ermittlung der Masse, des Alters und des Zuwachses der Holzbestände*. Dessau: Verlag Katz.
Hilmers, T., A. Avdagić, L. Bartkowicz, et al. 2019. The productivity of mixed mountain forests comprised of Fagus sylvatica, Picea abies, and Abies alba across Europe. *Int. J. For. Res.* 92:512–522.
Huang, L., B. He, A. Chen, H. Wang, J. Liu, A. Lű, and Z. Chen. 2016. Drought dominates the interannual variability in global terrestrial net primary production by controlling semi-arid ecosystems. *Sci. Rep.* 6(1):24639.
Ito, A. 2011. A historical meta-analysis of global terrestrial net primary productivity: are estimates converging? *Glob. Chang. Biol.* 17/10:3161–3175.
IPCC. 2001. www.ipcc.ch/ipccreports/tar/wg1/099.htm#tab32. Download November 4, 2014.
IPCC. 2006. *Guidelines for national greenhouse gas inventories. Vol. 4 agriculture, forestry and other land use, prepared by the national greenhouse gas inventories programme*. Eds. S. Eggleston, L. Buendia, K. Miwa et al. Kanagawa, Japan: Institute for Global Environmental Strategies.
IPCC. 2007. *Fourth assessment report: Climate change 2007*. Working Group I Report. Geneva, Switzerland: The Physical Science Basis, 104 p.
Jactel, H., J. Bauhus, J. Boberg, et al. 2017. Tree diversity drives forest stand resistance to natural disturbances. *Curr. For. Rep.* 3:223–243. https://doi.org/10.1007/s40725-017-0064-1
Jørgensen, S.E., B. Burkhard, and F. Müller. 2013. Twenty volumes of ecological indicators—An accounting short review. *Ecol. Indic.* 28:4–9.
Judeich, F. 1871: *Die Forsteinrichtung*. Dresden: Schönfelds Verlagsbuchhandlung.
Kahn, M. 1994. Modellierung der Höhenentwicklung ausgewählter Baumarten in Abhängigkeit vom Standort. *Forstliche Forschungsberichte München* 141:221 p.
Kenk, G., H. Spiecker, G. Diener. 1991. *Referenzdaten zum Waldwachstum*. Kernforschungszentrum Karlsruhe, KfK-PEF 82, 59 p.
Kramer, H. 1988. *Waldwachstumslehre*. Hamburg: Berlin: Paul Parey Verlag.
Knoke, T., T. Schneider, A. Hahn, et al. 2012. Forstbetriebsplanung als Entscheidungshilfe. Stuttgart: Ulmer.
Lal, P., Alavalapati, J., M. Marinescu, et al. 2011. Developing sustainability indicators for woody biomass harvesting in the United States. *J. Sust. For.* 30:736–755.
Landsberg, J.J. 1986. *Physiological ecology of forest production*. London: Academic Press.
Landsberg, J.J., and R.H. Waring. 1997. A generalised model of forest productivity using simplified concepts of radiation-use efficiency, carbon balance and partitioning. *Forest Ecol. Mod.* 95:209–228.

Langsaeter, A. 1941. Om tynning i enaldret granog furuskog Maddel. *Det Norske Skogforoksvesen* 8:131–216.
Larcher, W. 2003. *Physiological plant ecology: ecophysiology and stress physiology of functional groups.* Berlin: Springer Verlag.
Liu, Q., S. Piao, I.A. Janssens, et al. 2018. Extension of the growing season increases vegetation exposure to frost. *Nat. Commun.* 9(1):426.
Liu, Y., R. Zhou, Z. Wen, et al. 2021. Assessing the impacts of drought on net primary productivity of global land biomes in different climate zones. *Ecol. Indic.* 130:108146.
Luyssaert, S., Inglima, I., M. Jung, et al. 2007. CO_2 balance of boreal, temperate, and tropical forests derived from a global database. *Glob. Chang. Biol.* 13/12:2509–2537.
Lyr, H., H.J. Fiedler, and W. Tranquillini. 1992. *Physiologie und Ökologie der Gehölze.* Jena: G. Fischer Verlag.
Mantel, W. 1949. *Forsteinrichtungslehre.* Hamburg: Neumann-Neudamm.
Matyssek, R., J. Fromm, H. Rennenberg, et al. 2010. *Biologie der Bäume.* Stuttgart: Verlag Eugen Ulmer.
Marland, G., and B. Schlamadinger. 1997. Forests for carbon sequestration or fossil fuel substitution? A sensitivity analysis. *Biomass Bioenergy* 13/6:389–397.
Martonne de, E. 1926. Une novelle fonction climatologique: L'indice d'aridité. *La Météorologie* 21:449–458.
McMahon, S.M., G.G. Parker, and D.R. Miller. 2010. Evidence for a recent increase in forest growth. *PNAS* 107/8:3611–3615.
Michaletz, S.T., D. Cheng, A.J. Kerkhoff, et al. 2014. Convergence of terrestrial plant production across global climate gradients. *Nature* 512:39–43.
Myneni, R.B., J. Dong, C.J. Tucker, R.K. Kaufmann, P.E. Kauppi, J. Liski, . . . and Hughes, M.K. 2001. A large carbon sink in the woody biomass of Northern forests. *Proc. Natl. Acad. Sci.* 98(26):14784–14789.
Wald, Wachstum und Umwelt. 1. Band, Form und Wachstum von Baum und Bestand. Frankfurt am Main: JD Sauerländer's Verlag.
Molden, D., H. Murray-Rust, R. Sakthivadivel, et al. 2003. *A water-productivity framework for understanding and action.* In *Water productivity in agriculture: Limits and opportunities for improvement*, CAB International 2003. Water Productivity in Agriculture: Limits and Opportunities for Improvement (eds J.W. Kijne, R. Barker and D. Molden), No. 1, pp. 1–18.
Monteith, J.L. 1972. Solar radiation and productivity in tropical ecosystems. *J. Appl. Ecol.* 9:747–766.
Moosmayer, H.U. and W. Schöpfer. 1972. Beziehungen zwischen Standortsfaktoren und Wuchsleistung der Fichte. *AFJZ* 143/10:203–215.
Morin, X., L. Fahse, M. Scherer-Lorenzen, et al. 2011. Tree species richness promotes productivity in temperate forests through strong complementarity between species. *Ecol. Lett.* 14:1211–1219.
Motte F., T. Rötzer, P. Biber, E. Uhl, K. Pritsch, and H. Pretzsch. 2023. European beech recovered faster than Norway spruce after a five-year lasting artificial drought period within a mixed forest stand. *TREES* 37(6):1695–1715. https://doi.org/10.1007/s00468-023-02453-x.
Nagel, J., H. Spellmann, and H. Pretzsch. 2012. Zum Informationspotenzial langfristiger forstlicher Versuchsflächen und periodischer Waldinventuren für die waldwachstumskundliche Forschung. *AFJZ* 183:111–116.
Noss, R.F. 1990. Indicators for monitoring biodiversity: A hierarchical approach. *Conserv. Biol.* 4/4:355–364.
Oesten, G. and A. Roeder. 2012. *Management von Forstbetrieben.* Band I-III, Institut für Forstökonomie der Universität Freiburg, www.ife.uni-freiburg.de/lehre/lehrbuch-1 respectively lehrbuch-2 und -3, Volume I-III: 383 p, 365 p, and 291p.
Oliver, C.D. and B. Larson. 1996. *Forest stand dynamics.* New York, Chichester, Brisbane, Toronto, Singapore: John Wiley and Sons Inc.
Palmer, W.C. 1965. *Meteorological drought* (Vol. 30). US Weather Bureau Res. Paper, 45, 1-58.
Pan, Y., R.A. Birdsey, J. Fang, et al. 2011. A large and persistent carbon sink in the world's forests. *Science* 333/6045:988–993.
Paterson, S.S. 1956. *The forest area of the world and its potential productivity*, no. 51, p. 216. Goteborg: Meddelande fran Goteborgs Universitets Geografiska Institution.
Paquette, A. and C. Messier. 2011. The effect of biodiversity on tree productivity: from temperate to boreal forests. *Glob. Ecol. Biogeogr.* 20/1:170–180.
Perthuis de Laillevault, de R. 1803. *Traité de l'aménagement et de la restauration des bois et forêts de la France.* Paris: Madame Huzard.
Pfeil, W. 1860. *Die deutsche Holzzucht.* Leipzig: Verlag Baumgartner, 551 p.
Piao, S., S. Sitch, S., P. Ciais, et al. 2013. Evaluation of terrestrial carbon cycle models for their response to climate variability and to CO2 trends. *Glob. Chang. Biol.* 19:2117–2132.
Piotto, D. 2008. A meta-analysis comparing tree growth in monocultures and mixed plantations. *For. Ecol. Manage.* 255:781–786.
Pressler, M. 1877. *Forstliche Zuwachs-, Ertrags- und Bonitierungs-Tafeln mit Regeln und Beispielen.* Tharandt: 2nd edn. Selfpubl.

Pretzsch, H. 1995. Zum Einfluß des Baumverteilungsmusters auf den Bestandeszuwachs. *AFJZ* 166/9/10:190–201.

Pretzsch, H. 1996. Growth trends of forests in Southern Germany. In *Growth trends in European forests.* Berlin, Heidelberg: Springer Verlag.

Pretzsch, H. 2003. The elasticity of growth in pure and mixed stands of Norway spruce (*Picea abies* (L.) Karst.) and common beech (*Fagus sylvatica* L.). *J. For. Sci.* 49(11):491–501.

Pretzsch, H. 2005. Diversity and productivity in forests. In *Forest diversity and function*, eds. M. Scherer-Lorenzen, C. Körner, and E.D. Schulze. Ecol Studies 176. Berlin: Springer-Verlag.

Pretzsch, H. 2005. Stand density and growth of Norway spruce (*Picea abies* (L.) Karst.) and European beech (*Fagus sylvatica* L.). Evidence from long-term experimental plots. *Eur. J. For. Res.* 124:193–205.

Pretzsch, H. 2009. *Forest dynamics, growth and yield.* Berlin, Heidelberg: Springer Verlag.

Pretzsch, H. 2014. Canopy space filling and tree crown morphology in mixed-species stands compared with monocultures. *For. Ecol. Manag.* 327:251–264.

Pretzsch, H., P. Biber, and J. Ďurský. 2002. The single tree based stand simulator SILVA. Construction, application and evaluation. *For. Ecol. Manage.*162:3–21.

Pretzsch, H., P. Biber, G. Schütze, et al. 2014a. Changes of forest stand dynamics in Europe. Facts from long-term observational plots and their relevance for forest ecology and management. *For. Ecol. Manag.* 316:65–77.

Pretzsch, H., P. Biber, G. Schütze, et al. 2014b. Forest stand growth dynamics in Central Europe have accelerated since 1870. *Nat. Commun.* 5. https://doi.org/10.1038/ncomms5967.

Pretzsch, H., K. Bielak, J. Block, et al. 2013a. Productivity of pure versus mixed stands of oak (*Quercus petraea* (MATT.) LIEBL. and *Quercus robur* L.) and European beech (*Fagus sylvatica* L.) along an ecological gradient. *Eur. J. For. Res* 132/2:263–280.

Pretzsch, H., K. Bielak, A. Bruchwald, et al. 2013b. Species mixing and productivity in forest stands. Results from long-term experiments. *AFJZ* 184/7/8:177–196. [In German with English summary].

Pretzsch, H., J. Block, J. Dieler, et al. 2010. Comparison between the productivity of pure and mixed stands of Norway spruce and European beech along an ecological gradient. *Ann. For. Sci.* 67. https://doi.org/0.1051/forest/2010037.

Pretzsch, H., M. del Río, C. Arcangeli, et al. 2023a. Competition-based mortality and tree losses. An essential component of net primary productivity. *For. Ecol. Manag.* 544:121204.

Pretzsch, H., M. del Río, C. Arcangeli, et al. 2023b. Forest growth in Europe shows diverging large regional trends. *Sci. Rep.* 13(1):15373.

Pretzsch, H., J. Dieler, T. Seifert, et al. 2012. Climate effects on productivity and resource use efficiency of Norway spruce (*Picea abies* [L.] Karst.) and European beech (*Fagus sylvatica* [L.]) in stands with different spatial mixing patterns. *Trees* 26:1343–1360.

Pretzsch, H., T. Grams, K.H. Häberle, et al. 2020. Growth and mortality of Norway spruce and European beech in monospecific and mixed-species stands under natural episodic and experimentally extended drought. Results of the KROOF throughfall exclusion experiment. *Trees* 34:957–970.

Pretzsch, H., and R. Grote. 2023. Tree mortality: Revisited under changed climatic and silvicultural conditions. In *Progress in botany.* Heidelberg: Springer. https://doi.org/10.1007/124_2023_69.

Pretzsch, H., R. Grote, B. Reineking, et al. 2008. Models for forest ecosystem management: A European perspective. *Ann. Bot.* 101:1065–1087.

Pretzsch, H., M. Heym, T. Hilmers, et al. 2023c. Mortality reduces overyielding in mixed Scots pine and European beech stands along a precipitation gradient in Europe. *For. Ecol. Manag.* 539:121008.

Pretzsch, H., T. Hilmers, and M. del Río. 2024. The effect of structural diversity on the self-thinning line, yield level, and density-growth relationship in even-aged stands of Norway spruce. *For. Ecol. Manag.* 556:121736.

Pretzsch, H., F. Motte, T.E.E. Grams, et al. 2021. Fichten und Buchen bei mehrjähriger Trockenheit im Rein- und Mischbestand. *AFZ—Der Wald* 20:12–15.

Pretzsch, H., G. Schütze, and E. Uhl. 2013c. Resistance of European tree species to drought stress in mixed versus pure forests: Evidence of stress release by inter-specific facilitation. *Plant Biol.* 15:483–495.

Pretzsch, H., et al. In preparation. Diversity of tree size and growth in even-aged stands of Norway spruce. Effects on the density-growth relationship, self-thinning line, Eichhorn rule, and yield level.

Quéré, C.L., R.J. Andres, T. Boden, et al. 2013. The global carbon budget 1959–2011. *Earth Syst. Sci. Data* 5/1:165–185.

Reichstein, M., M. Bahn, P. Ciais, et al. 2013. Climate extremes and the carbon cycle. *Nature* 500(7462): 287–295.

Río, M. del, S. Condés, and H. Pretzsch. 2014a. Analyzing size-symmetric vs. size-asymmetric and intra-vs. inter-specific competition in beech (*Fagus sylvatica* L.) mixed stands. *For. Ecol. Manag.* 325:90–98.

Río, M. del, G. Schütze, and H. Pretzsch. 2014b. Temporal variation of competition and facilitation in mixed species forests in Central Europe. *Plant Biol.* 16:166–176.

Röhle, H. 1994. *Zum Wachstum der Fichte auf Hochleistungsstandorten in Südbayern. Ertragskundliche Auswertung langfristig beobachteter Versuchsreihen unter besonderer Berücksichtigung von Trendänderungen im Wuchsverhalten.* Habil Forstwiss Fak, LMU München, 249 p.

Röhle, H. 1997. Änderung von Bonität und Ertragsniveau in südbayerischen Fichtenbeständen. *Allg. Forst Jagdztg.* 168(6/7):110–114.

Rötzer, T., C. Dittmar, and W. Elling. 2004. A model for site specific estimation of the actual soil water content and the evapotranspiration in forest ecosystems. *J. Environ. Hydrol.* 12/7:1–14.

Rötzer, T., T. Seifert, and H. Pretzsch 2009. Modelling above and below ground carbon dynamics in a mixed beech and spruce stand influenced by climate. *Eur. J. For. Res*128:171–182.

Rötzer, T., T. Seifert, S. Gayler, et al. 2012. Effects of stress and defence allocation on tree growth—simulation results at the individual and stand level. In *Growth and defence in plants*, Ecology Studies 220. Berlin, Heidelberg: Springer-Verlag. https://doi.org/10.1007/978-3-642-30645-7_18

Rötzer, T., K.H. Häberle, C. Kallenbach, et al. 2017. Tree species and size drive water consumption of beech/spruce forests (Fagus sylvatica/Picea abies)—a simulation study highlighting growth under water limitation. *Plant Soil* 418(1):337–356.

Rötzer, T., M.A. Rahman, A. Moser-Reischl, et al. 2019. Process based simulation of tree growth and ecosystem services of urban trees under present and future climate conditions. *Sci. Total Environ.* 676:651–664.

Roy, J., H.A. Mooney, and B. Saugier. 2001. *Terrestrial global productivity.* London: Academic Press.

Rozendaal, D.M., D. Requena Suarez, V. De Sy, et al. 2022. Aboveground forest biomass varies across continents, ecological zones and successional stages: Refined IPCC default values for tropical and subtropical forests. *Environ. Res. Lett.* 17(1):014047.

Saugier, B., J. Roy, and H.A. Mooney. 2001. Estimations of global terrestrial productivity: Converging toward a single number? In *Terrestrial global productivity*, eds. J. Roy, B. Saugier, and H.A. Mooney. San Diego: Academic Press.

Schellnhuber, H.J. 1999. 'Earth system' analysis and the second Copernican revolution. *Nature* 402:19–23.

Schultz, J. 2005. *The ecozones of the world*, vol. 252. Berlin: Springer.

Schwappach, A. 1890. *Wachstum und Ertrag normaler Fichtenbestände.* Berlin: Verlag Julius Springer, 100 p.

Seneviratne, S., N. Nicholls, D. Easterling, et al. 2012. *Changes in climate extremes and their impacts on the natural physical environment.* Cambridge University Press. www.ipcc.ch/site/assets/uploads/2018/03/SREX-Chap3_FINAL-1.pdf

Sitch, S., B. Smith, I.C. Prentice, et al. 2003. Evaluation of ecosystem dynamics, plant geography and terrestrial carbon cycling in the LPJ dynamic global vegetation model. *Glob. Chang. Biol.* 9:161–185.

Skeffington, R.A. and E.J. Wilson. 1988. Excess nitrogen deposition: Issues for consideration. *Environ. Poll.* 54:159–184.

Smith, T.M. and R.L. Smith. 2009. *Ökologie.* München: Addison-Wesley Verlag; 6. Auflage.

Sohn, J.A., S. Saha, and J. Bauhus. 2016. Potential of forest thinning to mitigate drought stress: A meta-analysis. *Forest Ecology and Management, Special section: Drought and US Forests: Impacts and Potential Management Responses* 380:261–273. https://doi.org/10.1016/j.foreco.2016.07.046

Speidel, G. 1972. *Planung im Forstbetrieb.* Hamburg, Berlin: Verlag Paul Parey.

Spiecker, H., K. Mielikäinen, M. Köhl, et al. 1996. *Growth trends in European forests.* Europe Forest Institute, Res Rep 5. Heidelberg: Springer-Verlag.

Steffen, W., I. Noble, J. Canadell, et al. 1998. The terrestrial carbon cycle: Implications for the Kyoto Protocol. *Science* 280:1393–1394.

Sun, Z., X. Wang, H. Yamamoto, et al. 2018. Spatial pattern of GPP variations in terrestrial ecosystems and its drivers: Climatic factors, CO2 concentration and land-cover change, 1982–2015. *Ecol. Inform.* 46:156–165.

Turner, D.P., W.D. Ritts, W.B. Cohen, et al. 2005. Site-level evaluation of satellite-based global terrestrial gross primary production and net primary production monitoring. *Glob. Chang. Biol.* 11/4:666–684.

Ulrich, B. 1999. Entwicklungsprognosen für Waldökosysteme aus der Sicht der Hierarchietätstheorie. *Forstw. Cbl.* 118:118–126.

Verein Deutscher Forstlicher Versuchsanstalten. 1873. Anleitung für Durchforstungsversuche. In *Das Forstliche Versuchswesen*, vol 2, pp. 247–253, ed. von A. Ganghofer (1884). Augsburg: Schmid'sche Buchhandlung,

Vicente-Serrano, S.M., S. Beguería, J.I. López-Moreno. 2010. A multiscalar drought index sensitive to global warming: the standardized precipitation evapotranspiration index. *J. Clim.* 23:1696–1718.

von Baur, F. 1876. *Die Fichte in Bezug auf Ertrag, Zuwachs und Form.* Berlin: Springer-Verlag.

von Baur, F. 1881. *Die Rothbuche in Bezug auf Ertrag, Zuwachs und Form: unter Zugrundlegung der an der Kgl.* Württemberg'schen Forstlichen Versuchsanstalt angestellten Untersuchungen bearbeitet. Berlin: P. Parey.

von Carlowitz, H.C. 1713. *Sylvicultura Oekonomica oder Haußwirthliche Nachricht und Naturmäßige Anweisung zur wilden Baum-Zucht.* Leipzig: JF Braun.

von Cotta, H. 1828. *Anweisung zum Waldbau.* Dresden, Leipzig: Arnoldische Buchhandlung.

von Gadow, K. 2005. *Forsteinrichtung. Analyse und Entwurf der Waldentwicklung.* Göttingen: Univ-Verlag Göttingen, 342 p.

von Gadow, K. 2013. Messung und Modellforschung-Grundlagen der Forsteinrichtung. *AFJZ* 184/7/8:143–158.

von Ganghofer, A. 1881. *Das Forstliche Versuchswesen: Unter Mitwirkung forstlicher Autoritäten und tüchtiger Vertreter der Naturwissenschaften hrsg.* Band I, 505 p. Augsburg.

Waide, R.B., M.R. Willig, C.F. Steiner, et al. 1999. The relationship between productivity and species richness. *Ann. Rev. Ecol. Syst.* 257–300.

Wang, J., P.M. Rich, K.P. Price, et al. 2004. Relations between NDVI and tree productivity in the central Great Plains. *Int. J. Remote Sens.* 25/16:127–3138.

Weiskittel, A.R., N.L. Crookston, and P.J. Radtke. 2011. Linking climate, gross primary productivity, and site index across forests of the western United States. *Can. J. For. Res* 41:1710–1721.

Whitham, T.G., J.K. Bailey, J.A. Schweitzer, et al. 2006. A framework for community and ecosystem genetics: from genes to ecosystems. *Nat. Genet.* 7:510–523.

Whittaker, R.H. and G.E. Likens. 1973. Primary production: The biosphere and man. *Hum. Ecol.* 1/4:357–369.

Wiedemann, E. 1923. *Zuwachsrückgang und Wuchsstockungen der Fichte in den mittleren und unteren Höhenlagen der sächsischen Staatsforsten.* Tharandt: Kommissionsverlag W Laux, 181 p.

Wykoff, W.R., N.L. Crookston, and A.R. Stage. 1982. *User's guide to the stand prognosis model.* Ogden, UT: US Forest Serv, Gen Techn Rep INT-133.

Wykoff, W.R. and R.A. Monserud. 1988. Representing site quality in increment models: A comparison of methods. In *Proc IUFRO conference, Aug 1987*, eds. A.R. Ek, S.R. Shifley, and T.E. Burk. Minneapolis, MN: Gen Tech Rep NC-120.

Vanclay, J.K. 1994. *Modelling forest growth and yield. Applications to mixed tropical stands.* Wallingford, UK: CAB International.

Vile, D., B. Shipley, and E. Garnier. 2006. Ecosystem productivity can be predicted from potential relative growth rate and species abundance. *Ecol. Lett.* 9/9:1061–1067.

Yaffee, S.L. 1999. Three faces of ecosystem management. *Conser. Biol.* 13:713–725.

Zhang, Y., H.Y. Chen, and P.B. Reich. 2012. Forest productivity increases with evenness, species richness and trait variation: a global meta-analysis. *J. Ecol.* 100/3:742–749.

Zhang, Y., M. Xu, H. Chen, et al. 2009. Global pattern of NPP to GPP ratio derived from MODIS data: effects of ecosystem type, geographical location and climate. *Glob. Ecol. Biogeogr.* 18(3):280–290.

Zhao, M. and S.W. Running. 2010. Drought-induced reduction in global terrestrial net primary production from 2000 through 2009. *Science* 329:940–943.

Zheng, Y., R. Shen, Y. Wang, et al. 2020. Improved estimate of global gross primary production for reproducing its long-term variation, 1982–2017. *Earth Syst. Sci. Data* 12(4):2725–2746.

Zeide, B. 2001. Thinning and growth: A full turnaround. *J. For.* 99:20–25.

Zeide, B. 2002. Density and the growth of even-aged stands. *For. Sci.* 48:743–754.

Zeide, B. 2003. The U-approach to forest modeling. *Can. J. For. Res* 33:480–489.

Zohner, C.M., L. Mo, S.S. Renner, et al. 2020. Late-spring frost risk between 1959 and 2017 decreased in North America but increased in Europe and Asia. *Proc. Natl. Acad. Sci.* 117(22):12192–12200.

16 Modeling Forest Floor Biomass and N Accumulations and Related Turnover Rates

Paul A. Arp

16.1 INTRODUCTION

Knowing wood, litter, and root turnover rates is important for determining how organic matter accumulates over the years on and within mineral soils. For example, the annual forest litter and coarse woody debris accumulations determine the depth and the layering of the forest floor in terms of its litter, fermentation, and humification layers (LFH layers). Within mineral soils, roots are the dominant soil organic matter contributors by way of die-back, humification, and mineral-based organic-matter stabilization. As climate conditions vary, accumulations and turnover rates of forest organic matter (OM), including wood and roots, change as well (Bhatti et al. 2003; Deng et al. 2013). It is, however, difficult to know how extensive these changes will be and how they will accumulate over time with changing soil and site conditions across landscapes. This is because OM production and turnover times vary strongly by vegetation and by litter type composed of foliage, coarse woody debris, roots, and humus (Cou'teaux et al. 1995; Zhang et al. 2008; Jacob et al. 2010; Song et al. 2010; Ge et al. 2013).

Organic matter decomposition experiments generally aim to acquire specific information about the extent of OM loss according to prescribed experimental conditions under (i) carefully controlled temperature and moisture conditions, and/or (ii) under field conditions under changing weather and climate conditions. Such investigations vary from using bulked and unprocessed field samples to examining isolated organic matter fractions of interest. Fractioning involves physical and chemical processes, such as sample drying, grinding, and sieving, and using various solvents to isolate soluble from insoluble organic matter fractions (residues). The targets for isolation may refer to specific compound types (humin, cellulose, hemicellulose, lignin) and the extent to which these fractions are attached to mineral particles, especially clay particles. These determinations may also involve quantifying (i) the extent of organic matter mineralization, (ii) release of mineral N, Ca, Mg, K, etc. (McClaugherty 1985; Johnson and Lindberg 1992), (ii) the release of gases (NO_x, CO_2, CH_4), (iii) the change in chemical composition within the residue, and (iv) the effect of adding chemical substances or biological agents to enhance or retard the decomposition process.

Field-centered litter decomposition studies done across a wide range of field and climate conditions use single litter-filled mesh bags and spread these one by one across designated forest floor plots, and retrieve some of these each year over the course of ten years or so (Trofymow and CIDET Working Group 1998; Moore et al. 1999; Cusak et al. 2009). An alternative method uses 1 m^2 traps to collect and analyze multi-year forest litter accumulation using fiberglass screens as annual separators (Jorgensen et al. 1980; Piatek and Allen 2001). Upon retrieval, the samples are then analyzed for (i) mass remaining, (ii) nutrients remaining, and (iii) type extent of organic matter transformations.

Field-centered wood decomposition studies involve placing wood samples of same type, size, and shape into or onto the ground across many stand types. Alternatively, wood already lying on the ground can be classified by deadwood classes, and these classes are then evaluated in terms of mass loss composition over time and other parameters (e.g., Jurgensen et al. 1984; Lombardie et al. 2013).

DOI: 10.1201/9781003431084-19

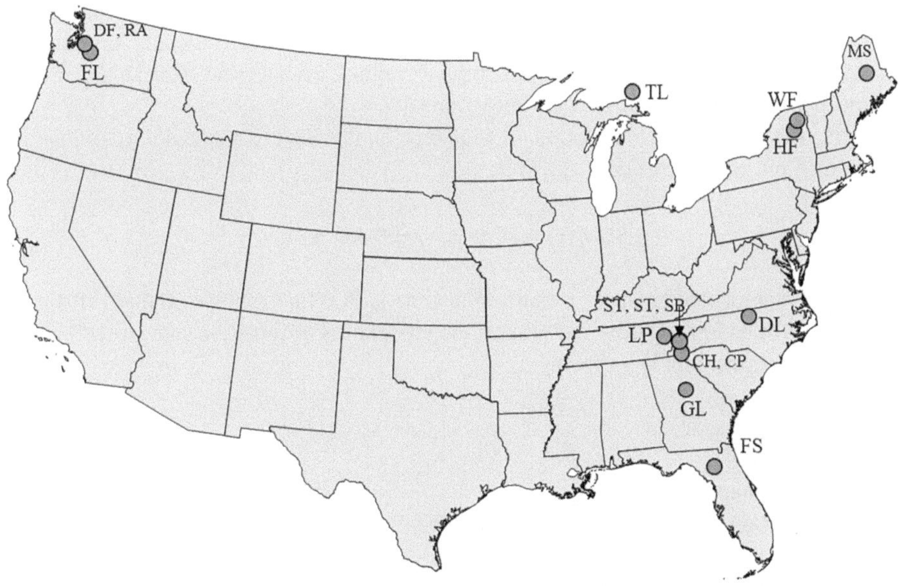

FIGURE 16.1 Locator points for the Integrated Forest Study Sites (IFS, Johnson and Lindberg 1992), overlaid on USA State Borders. There are 21 plots altogether, including two plots in Norway.

The wood decay experiment that was done across North America (LIDET 1995; Smith et al. 2011) involved wooden dowels with the lower part embedded in soil and the upper left surface exposed.

Alternative forest litter and wood studies involve (i) monitoring decomposition-generated CO_2 losses (Chiti 1999; Gaudinski et al. 2000; Song et al. 2010; Giardina et al. 2014) and (ii) radiocarbon dating (Pressenda et al. 2001). CO_2 monitoring varies from using soil-containing enclosures to whole-stand net flux determinations ranging from local to regional and global projects (Deng et al. 2013). Radiocarbon dating involves (i) determining the ^{14}C age of the decaying and humifying organic matter and (ii) tracking the decay-induced changes from the atmospheric ^{13}C as well ^{14}C concentrations (Aucour et al. 1999; Gaudinski et al. 2000; Matamala et al. 2003).

This chapter focuses on how forest floor biomass accumulation and turnover rates can be estimated across climate conditions by way of an empirical model formulation that is based on:

- Documented mass loss of CIDET and LIDET field-placed litter bags and wood blocks,
- Monitored soil respiration,
- Radiocarbon dating, and
- Mass balance modeling.

The objective of this chapter is to demonstrate how the annual accumulation of forest litter on the forest floor can be modeled to eventually reach a steady state. This demonstration is done for the Integrated Forest Site locations (IFS, Figure 16.1), as described and summarized in Johnson and Lindberg (1992). Preceding efforts to quantify some of these processes from an ecosystem modeling perspective were, e.g., summarized by Larocque et al. (2008).

16.2 MODEL FORMULATION, BACKGROUND

Modeling provides a means to integrate information generated from soil, plot, and stand-specific flux and age determinations into C-related stand- to forest-specific mass balance models (Oja and Arp 1996). This can be done by quantifying (i) the accumulating C or organic matter pools as represented by foliage, wood, roots, forest floor, mineral soil and (ii) addressing the C allocations to each pool. For this purpose, pool sizes, net annual primary productivity, and key photosynthate allocations to foliage, wood, root, forest floor compartments all need to be estimated, and these need to be balanced with realistic estimates for turnover rates.

To quantify the decomposition process, it is generally assumed that organic matter decays in an exponential fashion, as expressed by Eq. 16.1:

$$\mathrm{SOM}(t) = \mathrm{SOM}(t_0)\ \exp(-\mathrm{km}(S)\ t) \tag{16.1}$$

where SOM(t) is the organic matter remaining at time t, $\mathrm{SOM}(t_0)$ refers to the SOM(t) production at time = 0, and km(S) is the state- and climate-dependent decomposition rate (in yr^{-1}). Apart from local climate conditions, km(S) varies by

- OM producing species,
- OM type,
- OM size and shape,
- OM composition (e.g., cellulose, hemicellulose, lignin, humin),
- OM nutrient content, especially nitrogen (N),
- OM placement (air, soil, water, from aerobic to anaerobic conditions),
- OM concentration and extent of OM compaction,
- type, diversity, and succession of OM-consuming agents (bacteria, fungi, insects, earthworms, termites).

By focusing on organic mass (M) and N alone, Zhang et al. (2007, 2008) and Smith et al. (2011) developed and analyzed the following equations:

$$\frac{dM}{dt} = -\mathrm{km}(S)M\left(1 - \frac{[N]}{[N_f]}\left(1 - \frac{\mathrm{kn}}{\mathrm{km}}\right)\right) \tag{16.2}$$

and

$$\frac{dN}{dt} = -\mathrm{kn}(S)N \tag{16.3}$$

with $k_m(S)$ and kn(S) mass and N loss coefficients further symbolized by

$$\mathrm{km}(S) = \mathrm{km}\, f(\mathrm{climate}), \tag{16.4a}$$

$$\mathrm{kn}(S) = \mathrm{kn}\, f(\mathrm{climate}), \tag{16.4b}$$

and with [N] and $[N_f]$ referring to the N concentrations within the organic matter at time t and at steady state. The climate function is given by

$$f(\mathrm{climate}) = \left[\left(\frac{\mathrm{ppt}}{1000}\right)^{0.5} \min\left(1, \frac{T_{Jul}}{T_{Jul} - T_{Jan}}\right)\right] \exp\left[-\frac{Ea}{R}\left(\frac{1}{T_{Jul} + 273} - \frac{1}{288}\right)\right] \tag{16.5}$$

where ppt, T_{Jul}, and T_{Jan} refer to annual precipitation (in mm) and July and January temperatures (in °C). The activation energy Ea was determined to be 61,690 J $mole^{-1}$, and the universal gas constant R is given by 8.31 J $mole^{-1}\,{}^{\circ}C^{-1}$. In detail, Eq. 16.5 implies that organic matter composition and

N mineralization rates increase with increasing precipitation rates and with increasing July and January temperatures. In addition, f(climate) = 1 when ppt = 1,000 mm, T_{Jul} = 15°C, and T_{Jan} = 0°C.

Over the years, each year's litter contribution from leaves, wood, and root litter adds to the already existing albeit diminishing litter contributions from each of the preceding years. This realization can be summed for M and N over the years by litter type as follows:

$$M_j(\text{age}) = LP_j^0 \sum_{i=0}^{i=\text{age}} \exp\left[-km_j(S)i\right] \tag{16.6}$$

and

$$N_j = N_j^0 \sum_{i=0}^{i=\text{age}} \exp\left[-kn_j(S)i\right], \tag{16.7}$$

where i refers to litter age in years, j refers to litter type, and LP_j^0 refers to the annual litter production per litter type j. In addition, N_j^0 refers to the initial N content of LP_j^0, and $km_j(S)$ and $kn_j(S)$ are the rate of mass and N loss coefficients per litter type. The mean turnover rate for each of the accumulating litter types can then be determined from

$$MTT_j(\text{age}) = \frac{\sum_{i=0}^{i=\text{age}} i\exp\left[-km_j(S)i\right]}{\sum_{i=0}^{i=\text{soil age}} \exp\left[-km_j(S)\, i\right]}, \tag{16.8}$$

and the mean turnover rate for all organic matter accumulations for all the accumulating layers would be given by

$$MTT(\text{age}) = \frac{\sum_{j=1}^{j} \sum_{i=0}^{i=\text{age}} i\exp\left[-km_j(S)i\right]}{\sum_{j=1}^{j} \sum_{i=0}^{i=\text{soil age}} \exp\left[-km_j(S)i\right]} \tag{16.9}$$

Table 16.1 provides an overview for km and kn by forest litter type and whether placed above or within mineral soils by organic matter type and soil depth. In this, leaves, coarse woody debris, and roots all are part of the forest floor accumulations to the forest floor while root litter dominates the organic matter contributions to the mineral soil (Rasse et al. 2005). Note that litter-bag mass losses do not follow a single process over time but at least three climate- and location-specific exponential decay processes: fast, slow, and very slow (Minderman 1968). In contrast, mass losses from wooden dowels (no bark) follow a single exponential decay process. For root and wood litter, two fractions are considered, one which is generally comparable to the slow leaf litter component, and one that refers to the bark and suberized wood (bark) and root components (Matamala et al. 2003, Chen et al. 2001, Garrett et al. 2012), with the suberized portions amounting to about 10% of wood or root biomass. In general, coarse woody debris and root decay rates are quite similar, with, e.g., km (roots) = 1.4 km (coarse woody debris) according to Garrett et al. (2012). Due to increasing organic matter stabilization at increasing soil depth (Wang and Amundson 1996, Pressenda et al. 2001, Rasse et al. 2005, McFarlane et al. 2013), km needs to be adjusted by soil depth as follows:

$$km(\text{soil depth}) = km\left[1 + 0.335\,\text{soil depth}\left(\frac{\text{soil age}}{15{,}000}\right)^{\frac{1}{3}}\right] \tag{16.10}$$

TABLE 16.1
Parameters Useful for Estimating the Decay and Turnover Rates of Coarse Woody Debris, Forest Leaf Litter, and Roots in the Forest Floor and in the Mineral Soil

Forest Soil Layers	Litter Type		j	Mass Fractions	Decay Parameters		N Mineralization Parameters	Reference
Forest floor	Leaf litter	fast	1	e	km_1 =	$19.8\ km_2$	$kn_1 = kn_2\ km_1/km_2$	Zhang et al. (2007)
		slow	2	g(1-e)	km_2 =	0.377	$kn_2 = 0.555\ km_2\ [N]_{initial}$	
		very slow	3	(1-g) (1-e)	km_3 =	$0.292\ km_2$	$kn_3 = kn_2\ km_3/km_2$	
	Coarse woody debris	wood	4	1-f	km_4 =	0.11	$kn_4 = 0.35\ km_4$	Smith et al. (2011)
		bark	5	f	km_5 =	$0.292\ km_4$	$kn_5 = kn_4\ km_5/km_4$	
	Root litter	wood	6	1-f	km_6 =	0.175	$kn_6 = 0.35\ km_6$	Matamala et al. (2003)
		suberized	7	f	km_7 =	$0.292\ km_6$	$kn_7 = kn_6\ km_7/km_6$	
Mineral soil	Root litter	wood	8	1-f	km_8 =	0.175	$kn_8 = 0.35\ km_8$	Matamala et al. (2003)
		suberized	9	f	km_9 =	$0.292\ km_8/[1+0.33\ \text{depth}\ (\text{soil age}/15{,}000)^{0.33}]$	$kn_9 = kn_8\ km_9/km_8$	Wang et al. (1996), interpreted from graph

For leaf litter fractions and related e and g specifications, see Zhang et al. (2007); f refers to the suberized component of coarse woody debris (bark) and root litter. Also note: [N]initial = [N0], in %.

where the soil age component reflects cumulative retention and organo-metal stabilization of soil leached C. In addition, root litter production and accumulations decrease with increasing soil depth, as follows (Jackson et al. 1996):

$$\text{Cumulative fraction of } LP_{roots} = 1 - b^{\text{soil depth}}, \qquad (16.11)$$

with b varying from 0.914 (tundra) to, e.g., 0.943 (boreal forest) and 0.976 (temperate coniferous forest).

16.3 COMPUTATIONS

The computations, done for each IFS location and plot, were initially based on Eqs 16.1 to 16.9. The computational process was facilitated using the STELLA modeling software. The resulting calculations were done by year from time = 0 up to 2,000 years so that the steady-state condition for the forest floor biomass and N accumulations could be emulated for each location through calibration. The Runge-Kutta algorithm was used to ensure numerical stability. During the calibration process, it was realized that Eqs 16.2 and 16.3 needed to be reformulated to best capture the gradual conversion of fresh forest litter toward the steady-state forest floor forest biomass and N accumulations at each IFS location as follows:

$$\frac{dM}{dt} = -km(S)M\left(1 - \alpha\frac{[N]}{[N_f]}\left(1 - \frac{\beta kn}{km}\right)\right) \qquad (16.12)$$

and

$$\frac{dN}{dt} = -\beta kn(S)N, \qquad (16.13)$$

where α and β serve as plot-specific biomass and N mineralization adjustment parameters. For example:

(i) Reducing β toward 0 lowers the expected N mineralization rate but increases the expected biomass mineralization rate;

(ii) Setting β kn/km = 1 or α = 0 renders the biomass mineralization process independent of the N mineralization process;

(iii) At steady-state, the $1-\alpha\frac{[N]}{[N_f]}\left(1-\frac{\beta kn}{km}\right)$ expression simplifies to $1-\alpha\left(1-\frac{\beta kn}{km}\right)$.

The process of the changing N concentrations in the forest floor would occur as follows:

$$dN/dM = \frac{\beta kn}{km}[N]/\left[1-\pm\frac{[N]}{[N_f]}\left(1-\frac{\beta kn}{km}\right)\right]. \tag{16.14}$$

Consequently,

(i) when km = β kn, then dN/dM = $\frac{\beta kn}{km}[N]$;
(ii) when dN / dM < 1, M losses exceed N losses, so that [N] increases, and C/N decreases;
(iii) when dN / dM > 1, N losses exceed M losses, so that [N] decreases, and C/N increases.

At steady state, the M and N additions to the forest floor equal the M and N losses from the forest floor. For this condition—and noting from Table 16.1 that $kn/km = kn_1/km_1 = kn_2/km_2 = kn_3/km_3 = 0.555\,[N_0]$—it can be shown via Eqs 16.12 and 16.13 that

$$[N_f] = \frac{1-\alpha(1-0.555[N_0])}{0.555\beta} \tag{16.15}$$

and that

$$\alpha = \frac{1-0.555\ \beta[N_f]}{1-0.555\ \beta[N_0]} \tag{16.16}$$

Hence, α can be inferred from steady-state forest litter $[N_0]$ and forest floor $[N_f]$ concentrations, and from plot-based model-calibrated β values. In addition, β can be estimated from the steady-state N_0, N_f (t/ha) and $[N_0]$ (%) specifications by setting

$$\beta \approx 10\frac{N_0}{N_f\ f(clim)[N_0]}. \tag{16.17}$$

16.4 RESULTS AND DISCUSSION

The summary in Table 16.2 demonstrates that the IFS locations represent a wide range of forest type by tree age, climate condition, forest litter production, forest floor biomass and N accumulations, and related N concentrations and C/N ratios. Note that the C/N ratios generally decrease as the forest litter accumulations age, but exceptions of this occurred at the American beech (SB1, SB2) and red alder (RA) locations. At the CP pine location, the C/N ratio remained the same

TABLE 16.2

Leaf Area Index (LAI), Tree Age, Climate Conditions, and Forest Litter and Forest Floor Biomass Accumulation Including N Concentrations and C/N Ratios at the IFS Locations, with Information on Two Additional DL Plots (DL19, DL40, Jorgensen et al. 1980)

									Forest Litter				Forest Floor			
			LAI	Age	Tjan	Tjul	Precip.	f(climate)	Biomass	Litter N			Biomass	Forest Floor N		
Loc	Forest Type	Site	m²/m²	Years	°C	°C	mm	-	t/ha	t/ha	%	C/N	t/ha	t/ha	%	C/N
CP	White Pine	Plantation	17	30	1.3	20.9	1440	2.00	6.47	0.0431	0.666	75.1	28.1	0.282	1.003	49.9
DF	Douglas Fir	Plantation	5.3	55	3	16.5	1140	1.22	1.78	0.0148	0.831	60.1	36.7	0.379	1.033	48.4
DL	Loblolly pine	Plantation	6.5	23	3	25.2	1130	2.54	4.92	0.0187	0.380	131.4	63.0	0.450	0.714	70.0
FS	Slash Pine	Plantation	5.1	22	12.2	27.1	1120	2.96	4.45	0.0215	0.483	103.6	37.4	0.286	0.765	65.4
GL	Loblolly pine	Plantation	11	17	6.7	26.8	870	2.54	10.32	0.0760	0.737	67.9	32.7	0.310	0.948	52.8
LP1	Loblolly pine	Plantation	3.6	35	2	24.7	1140	2.45	2.60	0.0137	0.528	94.7	14.1	0.277	1.960	25.5
LP2	Loblolly pine	Plantation	3.6	35	2	24.7	1140	2.45	3.50	0.0194	0.554	90.2	14.1	0.277	1.960	25.5
NS1	Norway Spruce	Plantation	10.5	40	-6.9	16	1070	0.79	3.29	0.0287	0.873	57.2	63.3	0.842	1.330	37.6
NS2	Norway Spruce	Plantation	10.5	40	-6.9	16	1070	0.79	2.29	0.0203	0.886	56.4	57.8	0.605	1.048	47.7
CH	Tol. Hardwoods	Second growth	–	10	1.3	20.9	1380	1.96	4.98	0.0364	0.731	68.4	27.3	0.245	0.897	55.7
TL	Tol. Hardwoods	Second growth	5.3	135	-12.4	16	1210	0.68	3.73	0.0400	1.072	46.6	33.0	0.927	2.809	17.8
HF	Tol. Hardwoods	Old growth	6.5	100	-9.4	17.5	970	0.80	5.26	0.0596	1.132	44.2	44.9	0.531	1.182	42.3
SB1	Tol. Hardwoods	Old growth	–	200	-1.6	18.7	1510	1.56	2.24	0.0262	1.168	42.8	29.1	0.232	0.798	62.6
SB2	Tol. Hardwoods	Old growth	–	200	-1.6	18.7	1510	1.56	1.83	0.0234	1.280	39.1	28.8	0.247	0.857	58.3
RA	Red Alder	Old growth	4.5	55	3	16.5	1140	1.22	4.48	0.0800	1.785	28.0	98.5	1.629	1.654	30.2
FL	Silver Fir	Old growth	–	180	-3.9	12.7	1740	0.82	2.73	0.0165	0.604	82.7	54.8	0.575	1.049	47.7
WF	Mixed	Old growth	8.2	100	-12.6	14.3	1150	0.54	2.60	0.0237	0.913	54.8	141.4	2.642	1.868	26.8
MS	Red spruce	Old growth	5.7	85	-10.3	19.4	990	0.95	2.11	0.0122	0.580	86.3	150.7	1.121	0.744	67.2
SS	Red Spruce	Old growth	–	250	-3	17.1	1510	1.26	1.91	0.0112	0.587	85.2	166.1	1.916	1.154	43.3
ST1	Red Spruce	Old growth	9	250	-1.7	17.1	2440	1.71	1.77	0.0130	0.734	68.1	151.1	2.162	1.431	34.9
ST2	Red Spruce	Old growth	9	250	-1.7	17.1	2440	1.71	5.42	0.0323	0.596	83.8	108.9	1.517	1.393	35.9
DL19	Loblolly pine	Plantation	–	19	3	25.2	1130	2.54	7.71	0.0626	0.812	61.6	29.0	0.397	1.367	36.6
DL40	Loblolly pine	Plantation	–	40	3	25.2	1130	2.54	6.09	0.0409	0.671	74.5	32.9	0.416	1.266	39.5

while the lowest forest floor C/N ratio was recorded for the tolerant hardwood site at the Turkey Lakes (TL) location.

The numbers in Table 16.3 refer to:

(i) the best-fitted α and β values that account for the location-by-location variations pertaining to the model-emulated rates of litter decay and N mineralization,
(ii) the best-fitted forest floor biomass and N turnover times at (a) the mean turnover times and (b) reported tree age, and
(iii) the model-inferred forest flood decomposition index, given by:

$$\text{FFDI} = 1 - \alpha\,(1 - 0.555\,\beta\,[\text{N}_0]),\ \text{with}\,[\text{N}_0]\ \text{in}\ \%$$

An example of how the model-calculated forest floor biomass, N content, N concentration, and related C/N ratio all vary over time starting from time = 0 and by changed mean annual temperature (ΔT = −5, 0, 5 °C) and precipitation (ΔPPT = 0, 500 mm) is presented in Figure 16.2 for the TL location. As shown, the forest-floor biomass accumulations and related C/N ratios are modeled to decrease with increasing mean annual temperatures and precipitation rates while the N concentrations would increase. In biological terms, the rate of decay of the accumulating forest floor biomass is modeled to decrease as the availability of digestible C sources diminishes. This therefore leads to increasing N concentrations and decreasing C/N ratios. Decreasing C/N ratios, however, also lead to decreasing N retention. As a result, reported N leaching rates increase at locations with low forest floor C/N ratios, as detailed for the TL location by, e.g., Foster et al. (1992), Moayeri et al. (2001), and Murphy et al. (2009). In comparison, the reported N leaching rates were found to be low at the hardwood-covered Huntington Forest (HF) and the Great Smoky Mountains beech locations (SB1/SB2: Johnson and Lindberg 1992). The contrast between the TL and HF hardwood locations is striking: both accumulate forest biomass toward the steady-state conditions with about the same turnover times (15.5 versus 12.3 years, respectively, Table 16.3) but not so for N content (33.1 vs. 12.7 years, respectively; Table 16.3). Hence, forest floor N% increases slightly at HF but nearly triples at TL (Table 16.3, Figure 16.3). One reason for this difference could be related to the higher calcareousness of the TL soil parent material, which therefore implies higher biological activities and biomass turnover rates at TL compared to HF. In some cases, the forest litter to forest floor C/N ratio is about 1 (CP, white pine; RA, red alder) or increases to 1.6 (SB1, SB2, beech). This implies that biological activities at these locations consume forest litter N at the same or a higher rate than forest litter C.

The forest floor-to-forest litter C/N ratio appears to decrease with increasing forest litter C/N ratio from the IFS hardwood through the spruce-fir to the pine plantation locations as shown in Figure 16.4. In this, the forest floor-to-forest litter C/N ratio variations are largest for the hardwood locations (ranging from 0.4 to 1.5) and are least for the spruce-fir locations (ranging from 0.4 to 0.8). In contrast, the forest litter C/N ratio variations are largest for the pine plantation locations, spanning from 0.6 to 1.37. For the spruce-fir locations, the forest floor-to-forest litter C/N ratio appears to decrease with increasing tree age, but this is not the case for the hardwood and pine plantation locations (Table 16.2).

Figure 16.5 shows how the best-fitted N mineralization adjustment parameter α varies with the best-fitted biomass mineralization parameter β (left) and the Table 16.2 specified C/N ratio of the forest litter (right). The resulting patterns reveal again noticeable α-versus-β differences for the IFS hardwood, spruce-fir, and pine plantation locations. In addition, there is a significant decrease in the best-fitted N mineralization parameter (α) as the forest litter C/N ratio increases from the hardwood through the spruce-fir to the pine plantation locations, in which the TL and DL locations are recognizable outlier. The patterns displayed also reveal that the plot-to-plot α and β variations for the LP,

TABLE 16.3
Best-Fitted Calibration Parameters (α, β), Turnover Times, and Model-Assessed Forest Floor Biomass and N Accumulations at Turnover Time and IFS-Quoted Tree Age, Tree Age When 90% and 95% of Estimated Steady-State Forest Biomass Is to Occur, and Model-Generated Forest Floor Decomposition Index FFDI, for Each IFS Location

			Forest Floor Decomposition Adjustment Parameters	Forest Floor Turnover Times For:		Estimated Forest Floor Biomass and N at Turnover Time:				Best-Fitted Forest Floor Biomass at Tree Age	Tree Age When Reported Biomass Is Modeled to Occur at:	Forest Floor Decomposition Index (FFDI) = 1-α (1-β kn/km)	Biomass	N	Biomass
	Forest	Site		N		α	β	Years	Years				t/ha	% Reported	t/ha
Location	**Type**	**Condition**	**% Reported**	**t/ha**	**%**	**90%**	**95%**	**CP**	**White Pine**	**Plantation**	**0.610**	**1.15**	**7.8**	**9.2**	**19**
67.6	0.18	63.8	27.2	96.7	18	25	0.523	DF	Douglas Fir	Plantation	0.938	0.44	34	35.7	24.2
65.9	0.251	66.2	29.5	80.4	62	110	0.252	DL	Loblolly Pine	Plantation	0.912	0.43	30.7	33.3	41.7
66.2	0.328	72.9	36.2	57.5	77	104	0.171	FS	Slash Pine	Plantation	0.903	0.53	17.3	18.7	24.5
65.5	0.189	66.1	27.5	73.5	42	58	0.225	GL	Loblolly pine	Plantation	0.675	1.30	5.3	6.2	21.5
65.7	0.202	65.2	31.2	95.4	13	21	0.684	LP1	Loblolly pine	Plantation	0.660	0.38	20.1	28.1	9.8
69.3	0.182	65.7	11.6	82.1	44	75	0.413	LP2	Loblolly pine	Plantation	0.517	0.52	12.3	19.9	10.1
71.5	0.181	65.3	12.9	91.1	32	48	0.566	NS1	Norway Spruce	Plantation	0.835	0.495	35.6	40.5	42.5
67.2	0.552	65.6	44.4	70.2	113	120	0.366	NS2	Norway Spruce	Plantation	0.944	0.48	39.5	41.2	38
65.8	0.393	65.0	38.3	66.3	85	128	0.279	CH	Tol. Hardwoods	Second growth	0.844	1	9.2	10.2	18.3
67.0	0.165	67.3	19.1	70.0	20	24	0.498	TL	Tol. Hardwoods	Second growth	0.352	0.48	18.6	33.7	23.2
70.3	0.607	65.5	32.8	99.4	45	37	0.748	HF	Tol. Hardwoods	Old growth	0.839	1.25	12.3	12.7	29.5
65.7	0.352	66.3	44.9	99.9	29	39	0.820	SB1	Tol. Hardwoods	Old growth	1.224	0.64	13.1	12.4	18.1
62.3	0.15	64.7	29.1	100.1	33	42	0.283	SB2	Tol. Hardwoods	Old growth	1.166	0.47	15.6	15.3	18
62.5	0.165	66.8	28.8	100.0	36	51	0.224	RA	Red Alder	Old growth	1.022	0.23	28.2	28	61
61.9	1.059	65.0	83.4	84.7	69	91	0.211	FL	Silver Fir	Old growth	0.820	0.585	41.3	47.9	36.5
66.6	0.374	65.0	53.3	97.3	105	144	0.341	WF	Mixed	Old growth	0.895	0.18	135.5	149.9	93
65.8	1.712	64.8	80.5	56.9	345	466	0.187	MS	Red Spruce	Old growth	0.981	0.198	124.1	126.3	98.6
65.4	0.733	65.4	81.3	54.0	304	408	0.082	SS	Red Spruce	Old growth	0.974	0.0795	228.4	234	108
64.8	1.254	65.4	113.2	68.2	564	750	0.051	ST1	Red Spruce	Old growth	0.981	0.0482	222.2	227.1	99.3
65.7	1.411	65.3	104.9	69.4	535	708	0.038	ST2	Red Spruce	Old growth	0.900	0.21	59	64.8	72
66.1	0.994	65.5	106.0	97.3	149	202	0.162	DL19	Loblolly pine	Plantation	0.644	0.76	7.2	9.3	19.4
66.8	0.184	46.3	26.6	91.6	17	24	0.577	DL40	Loblolly pine	Plantation	0.653	0.755	11.8	14.4	22.2
67.6	0.242	58.2	31.1	94.6	30	41	0.530								

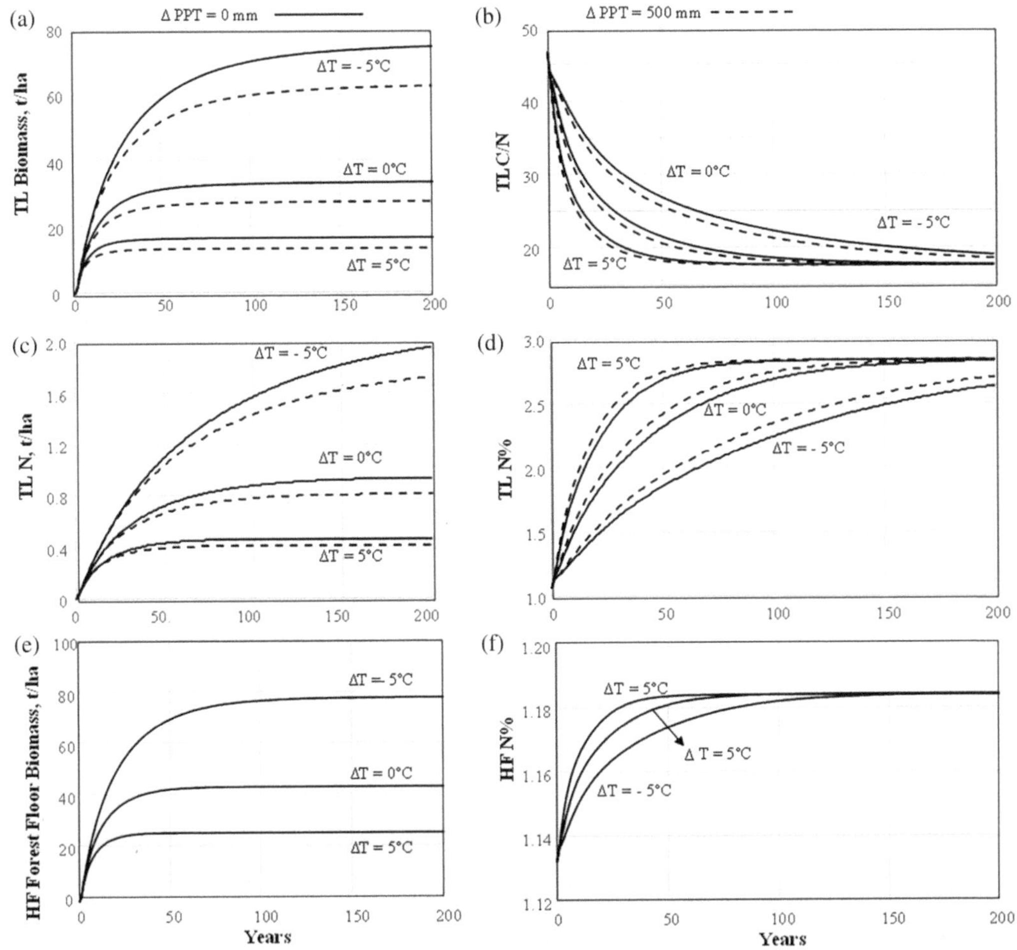

FIGURE 16.2 Modeled forest floor biomass (**A**) and nitrogen accumulations (**B**) along the y-axis in tonnes/ha for the Turkey Lakes (TL) location over 200 years (x-axis), assuming constant forest litter input from year to year as specified in Table 16.1. Also shown for the same location: related changes in C/N ratio (**C**) and N% (**D**) for the forest floor. For comparison, the corresponding forest floor biomass and N% concentrations are shown for the Huntington Forest location (HF) at bottom left (**E**) and right (**F**).

NS, and SB Plot 1 and Plot 2 locations are considerably smaller than the wider location-to-location variations. This is also true for the DL19 and DL40 locations of the Jorgensen et al. (1980) report but not so for the IFS-reported DL location (tree age = 23 years). For the latter, the forest floor C/N ratio is exceptionally high while the parameters for biomass and N mineralization are exceptionally low.

Figure 16.6 shows that the results so achieved are in general agreement with the sequential 8-year forest floor biomass and mineralization results reported by Jorgensen et al. (1980) for the DL19 and DL40 loblolly pine locations within the Duke Forest of Northern Carolina. Similar agreements were achieved for the loblolly pine forest floor biomass and mineralization results reported by Piatek and Allen (2001).

The scatterplot in Figure 16.7 shows that IFS-reported forest floor biomass at IFS-reported tree age generally exceeds modeled forest floor biomass by 36% and does not fall below the 1:1 line. Greater than modeled forest biomass accumulations are likely due to antecedent forest floor biomass prior to forest cover reestablishment. In contrast, the model calculations start at zero forest floor biomass at zero forest age.

The turnover times of the accumulating forest floor biomass and N are plotted in Figure 16.8 for each IFS location, grouped by forest type. These times are similar for the pine plantations and the tolerant hardwood sites while the reported tree age range is much wider for the latter. In contrast, the modeled turnover times are widest for the spruce-fir locations, likely due to the higher spruce-firlignin content.

Figure 16.9 shows how the forest biomass and N accumulate over time at the three DL locations overall and at modeled turnover times. In general, the biomass and N accumulations so derived amount to 66.2 ± 2.3% and 64.7 ± 4.6% of the biomass and N accumulations at steady state (see Table 16.3).

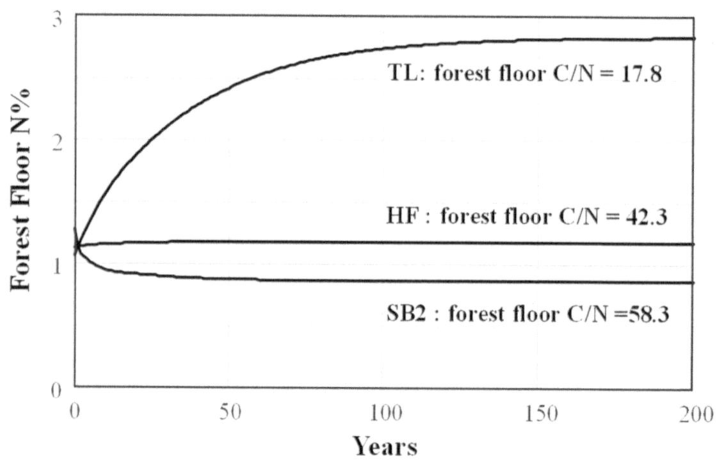

FIGURE 16.3 Modeled forest floor N% changes (y-axis) over time (x-axis) at the Turkey Lakes (TL), Huntington Forest (HF), and Great Smoky Mountains Beech (SB) locations, as implied by the low C/N = 17.8 ratio at TL (where N% increases), the CN = 42.3 ratio at HF (where N% stays the same), and the CN = 58.3 at SB (where N% decreases).

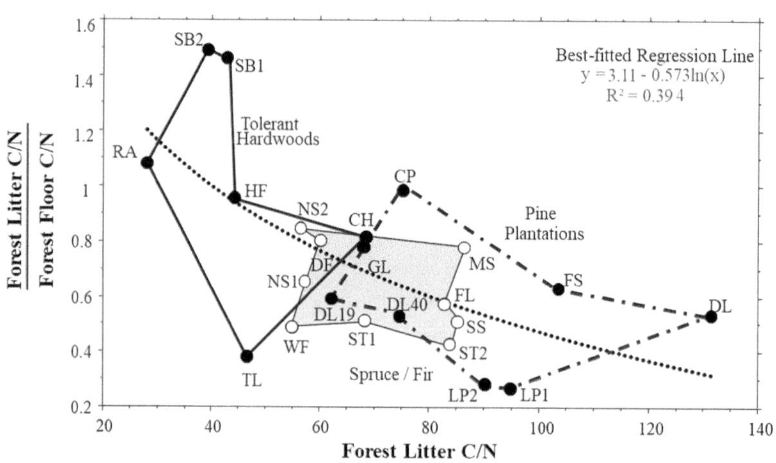

FIGURE 16.4 Dot plot showing the ratio of the forest litter C/N ratio over the forest floor C/N ratio (y-axis) against the forest floor C/N ratio (x-axis) at each IFS location. These dots are joined by polygons to reveal the dot-implied pattern by forest type, i.e., tolerant hardwoods (thick solid line, mostly at left), spruce-fir combinations (thin solid line, at center), and pine plantation (thick stippled line, mostly at right). Also overlaid is the best-fitted y-versus-x regression line (dotted points).

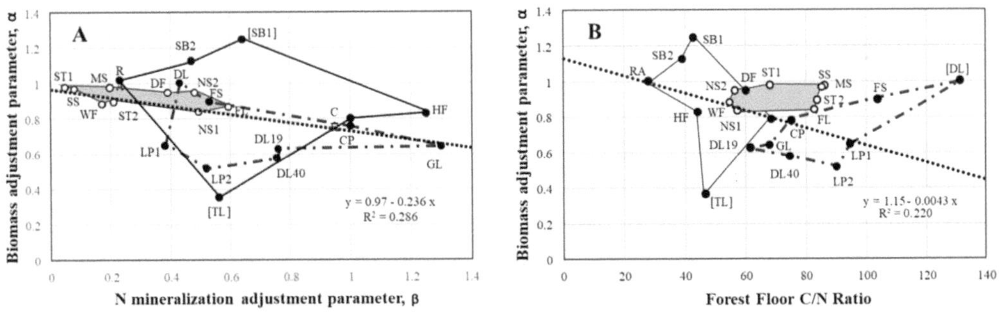

FIGURE 16.5 Dot plotting the biomass-adjustment parameter α versus (i) the N-adjustment parameter β (**Panel A**) and (ii) versus the forest litter C/N ratio (**Panel B**) for the IFS locations, polygon-joint by hardwood, spruce-fir, and pine plantation locations, as in Figure 16.4. Also shown are the best-fitted y-versus-x regression lines.

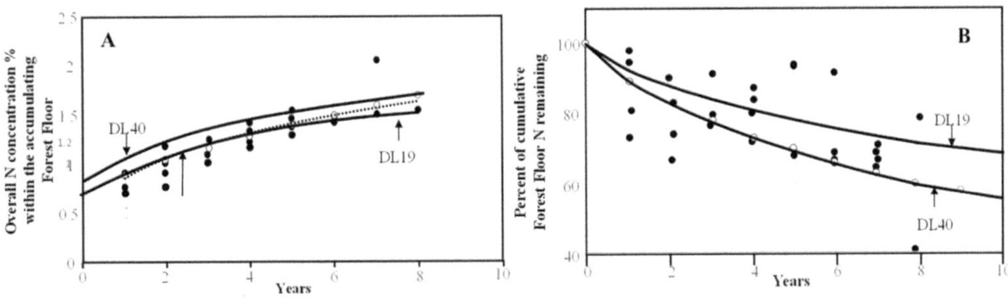

FIGURE 16.6 Actual (black dots) and modeled (heavy black lines) forest floor N concentrations (**Panel A**) and percent of N remaining (**Panel B**) within the sequentially accumulated forest litter at the Duke Forest DL19 and DL40 loblolly pine plantations (y-axes), plotted over the course of eight years (x-axis). Black dots: taken from Jorgensen et al. (1980); open dots with point-stippled lines: best-fitted y-versus-x regression model through the black dots.

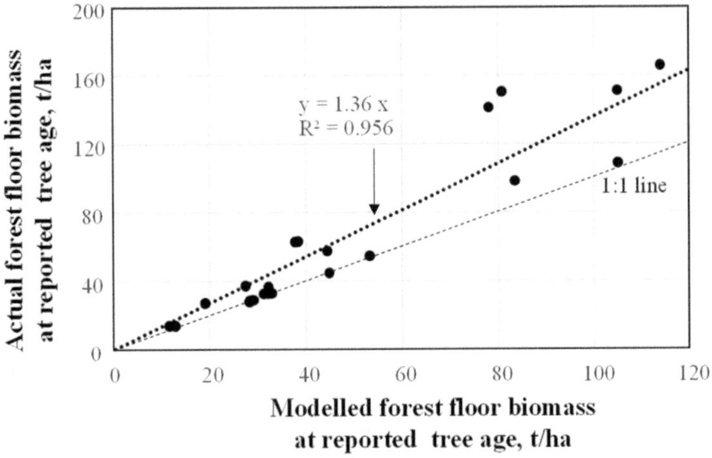

FIGURE 16.7 Dot plot of actual (y-axis) versus best-fitted (x-axis) forest floor biomass at reported tree age for each IFS location, with best-fitted regression equation and 1:1 line overlaid. On average, the reported biomass at tree age exceeds the modeled biomass by 36%.

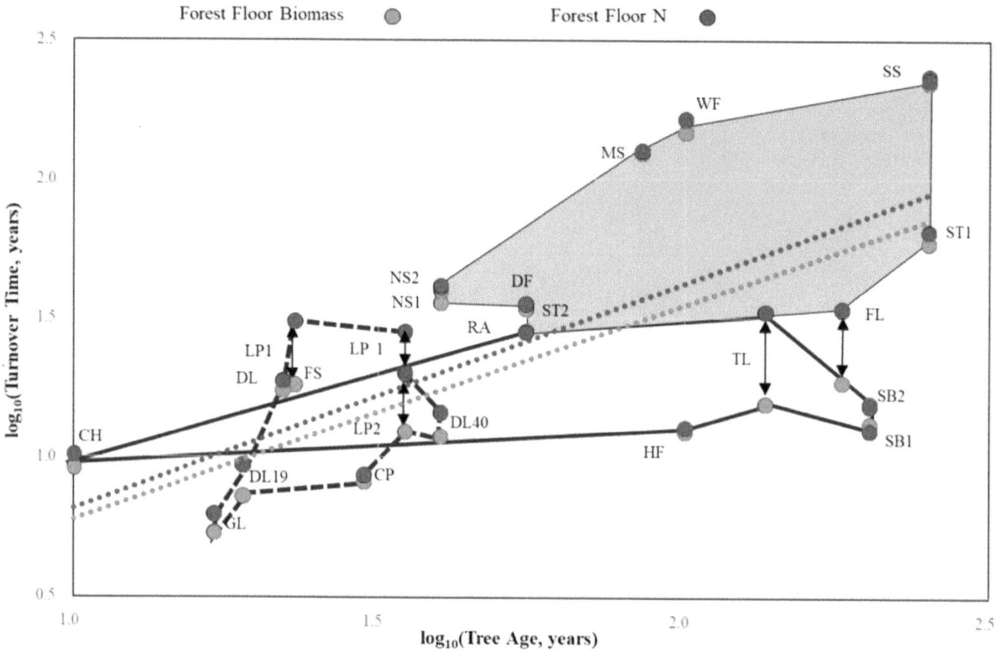

FIGURE 16.8 Log-log plot of forest-floor biomass (grey dots) and N turnover (black dots) times (y-axis) versus reported tree age (x-axis), with dots polygonized by forest type, i.e., tolerant hardwoods (thick solid line), spruce-fir combinations (thin solid line), and pine plantations (thick stippled line).

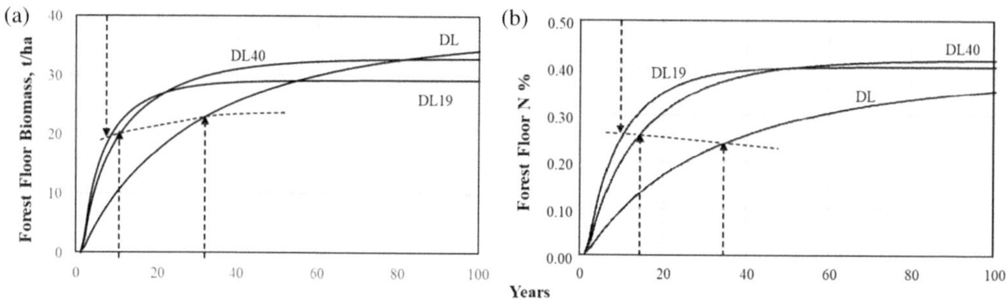

FIGURE 16.9 Line plots for modeled forest floor biomass (**Panel A**) and N% (**Panel B**) along the y-axes versus year along the x-axes at the Duke forest DL, DL19 and DL40 pine plantation plots, assuming constant litter production rates, initially falling on bare ground and at model-generated arrow-indicated turnover times.

Based on the report by Berg (2000) and the earlier results, there appears to be a general association between the forest floor decomposition index FFDI, litter lignin % and litter N % by forest type (Figure 16.10). While not in exact correspondence, the FFDI and lignin % versus forest plots have

(i) the widest variations across hardwood litter,
(ii) the least variations across spruce-fir litter, and
(iii) generally the lowest litter N% levels across the pine plantations.
(iv) a high 20–50% lignin range which, in part, refers to the final lignin content of decomposed forest litter, i.e., $[N_f]$.

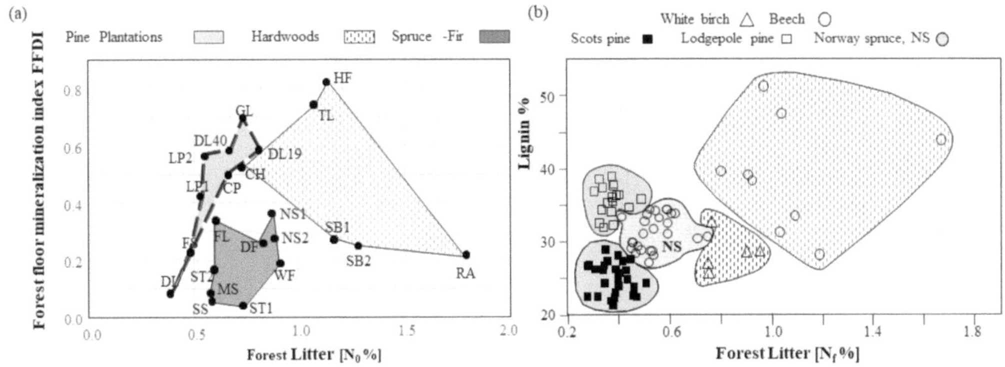

FIGURE 16.10 Polygonized dot plots of (i) the model-generated forest floor decomposition index FFDI = 1–α (1–0.555 β[N_0]) for the IFS locations (**Panel A**), and (ii) reported lignin% (Berg 2000; **Panel B**) versus the initial and final N concentrations of the decomposing forest litter along the x-axes, by forest type. [N_0] = initial forest litter N concentration; [Nf]= final forest litter N concentration.

Figure 16.10 therefore implies that the 0.2–1.8 N% range is in part due to increasing N retention with increasing lignin %, but with hardwood litter exhibiting wider initial and final N% ranges than spruce and pine litter (Berg 2000; Kreutzweiser et al. 2008; Wang et al. 2021).

16.5 CONCLUDING REMARKS

At the base of tracking changes in forest floor biomass and nitrogen accumulations over time is the CIDET-generated parametrization of fast, slow, and very slow forest litter decaying processes, as affected by litter type and climate (Table 16.1). While it is assumed that annual per hectare forest litter productions remain the same year after year, they differ substantially across the IFS locations (Table 16.2). The assumption of constant litter production per year is, however, not mandatory, because the model formulation can accommodate changing forest litter production rates if tracked or projected over time based on actual or projected climate and forest disturbance patterns.

To further generalize, the model can be applied to many more forest locations for which summaries regarding forest litter production, forest floor biomass (e.g., Reichle 1981) and N accumulations are also available. In addition, the formulation can be used to track other nutrients such as Ca, Mg, K, P, and S, and there is the need to address wood and root decay above and within the mineral soil as well. In this regard, Tables 16.1, 16.2, and 16.3 provide a starting point for quantifying long-term forest type and location-specific production and related above- and below-ground decay differences at the per hectare level.

As illustrated, the proposed model provides an efficient means to ascertain forest floor biomass and N turnover times: only one parameter (β) needs to be calibrated once reported or expected forest litter production and forest floor biomass and N accumulations are specified. The resulting model-estimated turnover rates are generally but not always somewhat slower for N than for biomass. Over time, the model consistently projects 62% to 64% of the tree-age reported biomass and N accumulations. This difference is in part due to setting modelled biomass and N content to 0 at time 0.

REFERENCES

Aucour, A.M., R. Bonnefille, and C. Hillaire-Marcel. 1999. Sources and accumulation rates of organic carbon in an equatorial peat bog (Burundi, East Africa) during the Holocene: carbon isotope constraints. *Palaeogeogr. Palaeoclimatol. Palaeoecol.* 150:179–189.

Berg, B. 2000. Litter decomposition and organic matter turnover in northern forest soils. *For. Ecol. Manag.* 133:13–22.

Bhatti, J.S., G.C. van Kooten, M.J. Apps, et al. 2003. Carbon balance and climate change in boreal forests. Chapter 20. In *Towards sustainable management of the boreal forest,* eds. P.J. Burton, C. Messier, D.W. Smith, et al., pp. 799–855. Ottawa, Ontario, Canada: NRC Research Press.

Chen, H., M.E. Harmon, and R.P. Griffiths. 2001. Decomposition and nitrogen release from decomposing woody roots in coniferous forests of the Pacific Northwest: A chronosequence approach. *Can. J. For. Res.* 31:246–260.

Chiti, T., R.E.M. Neubert, I.A. Janssens, et al. 1999. Radiocarbon dating reveals different past managements of adjacent forest soils in the Campine region, Belgium. *Chem. Geol.* 159:87–105.

Cou'teaux, M.-M., P. Bottner, and B. Berg. 1995. Litter decomposition, climate and litter quality. *Trends Ecol. Evol.* 10:63–66.

Cusack, D.A.F., W.W. Chou, W.H. Yang, et al. 2009. Controls on long-term root and leaf litter decomposition in neotropical forests. *Glob. Change Biol.* 15:1339–1355.

Deng, F., J.M. Chen, Y. Pan, et al. 2013. The use of forest stand age information in an atmospheric CO2 inversion applied to North America. *Biogeosciences* 10:5335–5348.

Foster, N., M. Mitchell, I. Morrison, et al. 1992. Cycling of acid and base cations in deciduous stands of Huntington Forest, New York, and Turkey Lakes, Ontario. *Can. J. For. Res.* 22:167–174.

Garrett, L.G., M.O. Kimberley, G.R. Oliver, et al. 2012. Decomposition of coarse woody roots and branches in managed Pinus radiata plantations in New Zealand—A time series approach. *For. Ecol. Manag.* 269:116–123.

Gaudinski, J.B., S.E. Trumbore, E.A. Davidson, et al. 2000. Soil carbon cycling in a temperate forest: radiocarbon-based estimates of residence times, sequestration rates and partitioning of fluxes. *Biogeochemistry* 51:33–69.

Ge, X., L. Zeng, W. Xiao, et al. 2013, Effect of litter substrate quality and soil nutrients on forest litter decomposition: A review. *Acta Ecol. Sin.* 33:102–108.

Giardina, C.P., C.M. Litton, S.E. Crow, et al. 2014. Warming-related increases in soil CO2 efflux are explained by increased below-ground carbon flux. *Nat. Clim. Chang.* 4:822–827.

Jackson, R.B., J. Canadell, J.R. Ehleringer, et al. 1996. A global analysis of root distributions for terrestrial biomes. *Oecologia* 108:389–411.

Jacob, M., K. Viedenz, A. Polle, et al. 2010. Leaf litter decomposition in temperate deciduous forest stands with a decreasing fraction of beech (*Fagus sylvatica*). *Oecologia* 164:1083–1094.

Johnson, D.W., and S.E. Lindberg, eds. 1992. *Atmospheric deposition and forest nutrient cycling: A synthesis of the integrated forest study*. Ecological Studies 91. New York, NY: Springer-Verlag.

Jorgensen, J.R., C.G. Wells, and L.J. Metz. 1980. Nutrient changes in decomposing loblolly pine forest floor. *Soil Sci. Soc. Am. J.* 44:1307–1314.

Jurgensen, M.F., M.J. Larsen, S.D. Spano, et al. 1984. Nitrogen fixation associated with increased wood decay in Douglas-fir residue. *For. Sci.* 30:1038–1044.

Kreutzweiser, D.P., K.P. Good, S.S. Capell, et al. 2008. Leaf-litter decomposition and macroinvertebrate communities in boreal forest streams linked to upland logging disturbance. *J. NorthAm. Benthol. Soc.* 27:1–15.

Larocque, G.R., J.S. Bhatti, A.M. Gordon, et al. 2008. Uncertainty and sensitivity issues in process-based models of carbon and nitrogen cycles in terrestrial ecosystems. In *Environmental modelling, software and decision support,* ed. A.J. Jakeman, pp. 307–368. Amsterdam: Elsevier.

LIDET. 1995. *Meeting the challenges of long-term, broad-scale ecological experiments.* Report No. 19. US LTER Network Office, Seattle.

Lombardi F., P. Cherubini, R. Tognetti, et al. 2013. Investigating biochemical processes to assess deadwood decay of beech and silver fir in Mediterranean mountain forests. *Ann. For. Sci.* 70:101–111.

Matamala, R., M.A. Gonzalez-Meler, J.D. Jastrow, et al. 2003. Impacts of fine root turnover on forest NPP and soil C sequestration potential. *Science* 302 (ANL/ER/JA-47577).

McClaugherty, C.A., J. Pastor, J.D. Aber, et al. 1985. Forest litter decomposition in relation to soil nitrogen dynamics and litter quality. *Ecology* 66:266–275.

McFarlane, K.J., M.S. Torn, P.J. Hanson, et al. 2013. Comparison of soil organic matter dynamics at five temperate deciduous forests with physical fractionation and radiocarbon measurements. *Biogeochemistry* 1:457–476.

Minderman, G. 1968. Addition, decomposition and accumulation of organic matter in forests. *J. Ecol.* 56:355–362.

Moayeri, M., F.-R. Meng, P.A. Arp, et al. 2001. Evaluating critical soil acidification loads and exceedances for a deciduous forest at the turkey lakes watershed. *Ecosystems* 4:555–567.

Moore, T.R., J.A. Trofymow, B. Taylor, et al. 1999. Litter decomposition rates in Canadian forests. *Glob. Change Biol.* 5:75–82.

Murphy, P.N.C., M. Castonguay, J. Ogilvie, et al. 2009. A geospatial and temporal framework for modeling gaseous N and other N losses from forest soils and basins, with application to the Turkey Lakes Watershed Project, in Ontario, Canada. *For. Ecol. Manag.* 258:2304–2317.

Oja, T., and P.A. Arp. 1996. Nutrient cycling and biomass growth at a North American hardwood site in relation to climate change: FORSVA assessments. *Clim. Change* 34:239–251.

Piatek, K.B., and H.E. Allen. 2001. Are forest floors in mid-rotation stands of loblolly pine (*Pinus taeda*) a sink for nitrogen and phosphorus? *Can. J. For. Res.* 31:1164–1174.

Pressenda, L.C.R., S.E.M. Gouveia, and R. Aravena. 2001. Radiocarbon dating of total soil organic matter and humin fraction and its comparison with ^{14}C ages of fossil charcoal. Proceedings of the 17th International 14 C Conference; I Carmi and E Boaretto eds. *Radiocarbon* 43:595–601.

Rasse, D.P., C. Rumpel, and M.F. Dignac. 2005. Is soil carbon mostly root carbon? Mechanisms for a specific stabilisation. *Plant Soil* 269:341–356.

Reichle, D.E. 1981. Dynamic properties of forest ecosystems. *Int. Biol. Programme 23.*

Smith, A.C., J.S. Bhatti, H. Chen, et al. 2011. Modelling above-and below-ground mass loss and N dynamics in wooden dowels (LIDET) placed across North and Central America biomes at the decadal time scale. *Ecol. Model.* 14:2276–2290.

Song, F., Z. Fan, and R. Song. 2010. Review of mixed forest litter decomposition researches. *Acta Ecol. Sin.* 30:221–225.

Trofymow, J.A. 1998. *The Canadian Intersite Decomposition Experiment (CIDET): Project and site establishment report.* CIDET Working Group. Natural Resources Canada, Canadian Forest Service, Pacific Forest Center, Victoria, B.C.; Inf. Rep. BC-X-378.eorest

Wang, Y., and R. Amundson. 1996. Radiocarbon dating of soil organic matter. *Quat. Res.* 45:282–288.

Wang, Q.W., M. Pieristè, C. Liu, et al. 2021. The contribution of photodegradation to litter decomposition in a temperate forest gap and understorey. *New Phytol.* 229:2625–2636.

Zhang, C.F., F.R. Meng, J.S. Bhatti, et al. 2008. Modeling forest leaf-litter decomposition and N mineralization in litterbags, placed across Canada: A 5-model comparison. *Ecol. Model.* 3:342–360.

Zhang, C.F., F.R. Meng, J.A. Trofymow, et al. 2007. Modeling mass and nitrogen remaining in litterbags for Canadian forest and climate conditions. *Can. J. Soil Sci.* 87:413–432.

17 Forest Ecosystem Health and Biotic Disturbances

Perspectives on Indicators and Management Approaches

Kishan Sambaraju, Simon Shamoun, Yan Boulanger, Véronique Martel, Pierre DesRochers, Danny Rioux, Nitin Kulkarni, Ram Keerti Verma, Marco Pautasso, Deepa Pureswaran, Michel Cusson, Johanne Delisle, Christian Hébert, Mohan Chinnarasu, and Darshan Kumara

17.1 INTRODUCTION

A properly functioning ecosystem that largely retains its attributes such as productivity, structure, and composition under various stresses, while contributing to human well-being and commerce, is a goal shared by multiple stakeholders ranging from environmentalists to the ordinary public (Rapport et al. 1998a). However, this worthy objective is being increasingly challenged by numerous threats, namely population expansions of native pest species, establishment and spread of aggressive non-native species, an increased frequency of extreme events (e.g., wildfires, floods, droughts, and hurricanes), habitat degradation, resource overexploitation by humans, and climate change (Gauthier et al. 2015; Millar and Stephenson 2015). It is essential to restore affected ecosystems to a more stable state and to maintain their health to realize long-term sustainability in land use. However, attempting to do this requires a satisfactory definition and broad-scale quantification of ecosystem health, which is challenging because what healthy ecosystems mean varies from one individual to another, by ecosystem type, and by perspectives under consideration (utilitarian or ecosystem-based) (Kolb et al. 1994; Rapport et al. 1998b; Costanza and Mageau 1999). Hence, many definitions of ecosystem health are available in the literature, from a normative concept (the desired end point of environmental management; Costanza (1992)) to an ecosystem engineering notion (the capacity of ecosystems to withstand threats; Rapport (1989)). The concept of resilience, i.e., the ability and speed with which an ecosystem can bounce back after a perturbation, has been shown to be a common thread in the health definitions for animals, ecosystems, humans, plants, and soils (Döring et al. 2015). While the philosophical debate defining the health of plants is underdeveloped in comparison with animal and human health (Döring et al. 2012), defining the health of ecosystems over regional levels is even more problematic, because of a complex spatiotemporal interplay of intrinsic and extrinsic processes and widespread human influences and stakes. Reflecting on ecosystem health concepts thus provides an opportunity for interactions between the health, natural, and social sciences (Rapport et al. 1998b).

The challenges inherent in striving for (and defining) forest ecosystem health are not simpler than for generic ecosystem health. Forests are generally resilient ecosystems capable of tolerating occasional, low-to-moderate intensity perturbances, and regenerating over time to different degrees and compositions (Oliver 1980; Clark and Covey 2012). However, when magnitudes of disturbances

DOI: 10.1201/9781003431084-20

exceed historically observed patterns or baselines, the impacts of these anomalies could portend undesirable or potentially irreversible impacts on forest ecosystem processes. Consequently, forest conditions are typically taken as a gauge to represent forest health. Government forestry departments therefore monitor and report statuses and trends of forest condition using indicators such as area affected by biotic (e.g., insects, diseases) and abiotic disturbances (wildfires, drought, snowstorms) (Rapport et al. 1998b; Tkacz et al. 2013; NRCAN 2022). Going a step further, however, a more holistic or systems-level approach would ideally include the recognition that forests are not just essential for the provision of ecosystem services such as clean air and water, carbon storage, recreation, biodiversity, food and wood supply (Boyd et al. 2013), but they are also critical for the sustenance of Indigenous communities that attach economic, cultural, and spiritual significances to natural forests (Adam and Kneeshaw 2008; Lyver et al. 2017). Fortunately, there is a slow, but growing, realization of the importance of considering Indigenous perspectives on forest health and in collaborative management of threats (CCFM 2006; Hill et al. 2021).

Tree mortality and forest decline are ecological processes that belong to the dynamical nature of forest ecosystems (Franklin et al. 1987). Interactions between abiotic (e.g., climate changes, pollutants, fire, storms) and biotic stressors (e.g., insect herbivores and diseases) are common (Skelly 1989) and have shaped forests for a long time (Waller 2013). However, the health of many forests is now increasingly being affected by the invasion not just of non-native insects and fungi (Santini et al. 2013; Roy et al. 2014) but also of non-native plants and trees, for example, those introduced as ornamentals by humans in novel environments (Felton et al. 2013). While non-native trees might temporarily escape in their introduced range from the insects or pathogens with which they have coevolved in their native range, there is a long-term potential for novel interactions (e.g., due to host shifts) with insects and pathogens with which these trees have not coevolved (Schmid et al. 2014). Alternatively, once established, invasive trees may harbor later introductions of pests from their native range and threaten plant health in the invaded regions. This is the case in North America currently with the well-established invasive tree-of-heaven, *Ailanthus altissima* (Mill.) Swingle, and spotted lanternfly, *Lycorma delicatula* (White) (Urban and Leach 2023). Proactive risk assessments relying on a clear understanding of host–herbivore or host–pathogen associations are therefore particularly important to safeguard the health of forests.

17.2 FOREST HEALTH-RELATED INDICATORS: INSECT AND PATHOGEN DAMAGE

Insects and pathogens are key biotic disturbance agents in forest ecosystems. They play essential roles in forest ecosystem processes such as facilitating removal of weakened trees, decomposition of dead or decaying trees, nutrient cycling, biodiversity maintenance, forest regeneration, and succession (Schowalter 1994; Castello et al. 1995; Ayres and Lombardero 2000). However, natural or anthropogenic changes to forests, accidental introduction of pest species into new habitats, and/or shifts to favorable climatic conditions may help insects or pathogens to transition to outbreak levels (Dale et al. 2001; Desprez-Loustau et al. 2006; Taylor et al. 2006). Outbreaks may cause severe tree mortality or growth reduction across scales over time and may facilitate rapid pest range expansion, especially in case of agents that have long-distance dispersal capabilities. For example, in the temperate and boreal forests of North America, large outbreaks of bark beetles such as the mountain pine beetle (*Dendroctonus ponderosae* Hopkins) and defoliators such as the eastern spruce budworm (*Choristoneura fumiferana* (Clemens)) cause extensive mortality of their respective host trees. Similarly, the European spruce bark beetle (*Ips typographus* (L.)) occasionally causes widespread mortality of spruce trees (*Picea* spp.) in Eurasia, although these infestations may improve the biodiversity conservation value of otherwise uniform and increasingly dense forests (Lehnert et al. 2013).

Invasions and spread of non-native forest insects and pathogens, driven by increasing global trade and climate change, are having far-reaching impacts on urban and natural forests (Weed et al. 2013;

Lovett et al. 2016). For example, the emerald ash borer (*Agrilus planipennis* Fairmaire; EAB) is a forest insect native to east Asia that is believed to have been introduced into North America through EAB-harboring woody materials such as crates and pallets (Cappaert et al. 2005). This non-native insect has killed millions of trees in north-eastern North America significantly altering urban environments and forested landscapes (Herms and McCullough 2014). EAB is presently spreading in European Russia and Ukraine and poses a serious threat to European ash species (e.g., *Fraxinus excelsior* L.) (Valenta et al. 2017; Drogvalenko et al. 2019). In the case of tree pathogens, some of the greatest impacts in terms of tree mortality, well exceeding those caused by non-native forest insects, have been caused by non-native pathogens (Lovett et al. 2006; Freer-Smith and Webber 2017; Fei et al. 2019). For example, chestnut blight caused by the ascomycete fungus *Cryphonectria parasitica* (Murrill) Barr has essentially extirpated the American chestnut (*Castanea dentata* (Marsh.) Borkh.) in the United States (Anagnostakis 1987). Similarly, butternut canker (*Ophiognomonia clavigignenti-juglandacearum* (V.M.G. Nair, Kostichka, & J.E. Kuntz) Broders & G.J. Boland) currently occurs throughout the range of butternut (*Juglans cinerea* L.) in North America and threatens the extinction of its host tree (Ostry and Woeste 2004; Sambaraju et al. 2018; Fei et al. 2019). The oomycete *Phytophthora ramorum* Werres, De Cock & Man in't Veld causes Sudden Oak Death (SOD), ramorum leaf blight, and ramorum shoot dieback, in Western North America and Sudden Larch Death (SLD) in the United Kingdom (UK) and some other European countries. This pathogen is responsible for the mortality of oaks (*Quercus* spp.) and tanoak (*Notholithocarpus densiflorus* (Hook. & Arn.) Manos, Cannon, & S.H. Oh) in California and Oregon, United States, and of Japanese larch (*Larix kaempferi* (Lamb.) Carrière) in the UK (Rizzo et al. 2005; Brasier and Webber 2010; Grünwald et al. 2012) and Ireland (Frankel and Palmieri 2014). Thus, both native and introduced insects and pathogens can potentially alter the forest ecosystem balance (Kizlinski et al. 2002; Hicke et al. 2012; Freer-Smith and Webber 2017).

Natural disturbances or forest management practices may impact critical elements of a forest ecosystem, such as organismal diversity, soil nutrient composition, carbon budgets, and water resources (Castello et al. 1995; Ballard 2000; Lovett et al. 2002). A holistic approach to studying the impacts of such disturbances on forest ecosystem health would ideally include the consideration of their stand-scale as well as landscape-scale impacts on the various structural, functional, and compositional components of an ecosystem (see Dale and Beyeler 2001). Due to the impracticality of measuring all the different levels of an ecosystem to determine the health status or sustainability of a forest ecosystem, a representative set of ecological indicators has been proposed. These are qualitative or quantitative attributes that would show trends when measured over time and which are related to the ecosystem elements to be sustained (referred to as "criteria," e.g., biodiversity) (FAO 2010; Woodall et al. 2011). Ecological indicators should be simple, feasible, reliable, measurable, and cost-effective in addition to being sensitive to small changes in the system caused by disturbances (Dale and Beyeler 2001). Ecological indicators are useful only when long-term baseline data concerning their natural range of variability in undisturbed habitats exist so that standards may be developed for comparisons with similar but disturbed habitats (O'Laughlin et al. 1994). However, frameworks to assess forest ecosystems using indicators under different criteria exist mainly at country scales under sustainable forest management (SFM) principles (Hall 2001; FAO 2010). To our knowledge, appropriate frameworks and standards to collect and assess such information at the stand/management unit scale are currently limited (FAO 2010).

With regard to forest health and vitality, several inter-governmental working groups (e.g., Montreal Process, Pan-European Forest Process) have identified indicators for monitoring and assessment at one or more spatial scales in those regions (FAO 2010). The forest health and vitality-related indicators proposed therein mainly concern the magnitude of stressors (e.g., "area and percent of forest affected by processes or agents beyond the range of historical variation"), damage severity (e.g., mortality, growth loss), occurrence and extent of regeneration, alterations in forest structure and composition, physical changes (e.g., soil nutrient balance, pH), and occurrence of non-native species. The United States Department of Agriculture-Forest Service (USDA-FS) uses "crown condition", "tree damage", and "tree mortality and standing dead trees" as indicators of forest ecosystem health impacted by stresses (e.g., insects, diseases) (Schomaker et al. 2007; Woodall et al. 2011; Randolph et al. 2020). As identification, development, and further refinement of normative ecological indicators (i.e., those with norms or baselines) with accompanying frameworks and standards, especially at stand/management-unit level, are emerging areas of investigation, other, more descriptive, indicators may be helpful to assess forest health across scales. Moreover, unlike insects, pathogen damage to forests, especially in boreal and temperate regions, may take years to manifest with or without tree mortality, which makes the utility of normative ecological indicators uncertain. Thus, alternative approaches such as using molecular detection technologies (e.g., PCR-DNA and Bioinformatics) for monitoring and early detection of a disease and, as necessary, application of relevant disease management options for maintaining forest health are critically important (Lamarche et al. 2016; Feau et al. 2019; Shamoun et al. 2019).

Descriptive indicators of forest condition, specifically with regard to heightened activity of insects or pathogens, may be used to assess forest health status at various scales. Visual indicators of incidences of insect infestations or disease symptoms (Riitters et al. 1992) in a susceptible habitat provide cues to later impacts if favorable conditions continue to persist. Qualitative indicators may include those that can be easily observed as a result of an insect infestation, or symptoms caused by a disease-causing agent. For example, in the case of insects such as bark beetles, occurrence of entry holes laden with resin and frass in the tree trunk, egg/larval galleries in the phloem and boring dust accumulation at the tree base are qualitative indicators of an infestation at the individual tree level (Christiansen and Bakke 1988; Safranyik and Carroll 2006; Raffa et al. 2008). Disease symptoms, such as leaf spots, leaf necrosis, rusts, and stem cankers, can serve as visual indicators of pathogen activity. Increased frequency of trees showing such infestations or symptoms, spatially and temporally, may signify an impending, potentially serious, damage to forest ecosystem health especially with regard to non-native pests. Small clusters of dead trees or localized infestations caused by insect activity, extensive foliage loss, and/or foliage discoloration (stand-scale) or huge areas of tree mortality (landscape-level) may also serve as visual indicators of biotic disturbance agents. Quantitative indicators suggesting changes in forest health condition as a result of insect or pathogen activity include percentage defoliation, percentage reduction in growth/productivity, or changes in area or percentage of tree mortality across different spatial scales and over time.

Insect and disease infestation data are generally collected by provincial, regional, and/or federal government departments (e.g., United States Forest Service-Forest Health Protection program) through ground-based monitoring plots, aerial surveys, and remote sensing technologies such as aerial photography (Fierke et al. 2011). Aerial monitoring surveys and Geographic Information System (GIS)-based mapping systems are especially suited to record insect infestations that can cause landscape-level tree mortality in boreal and temperate forests and which are readily recognized by an expert surveyor. Aerial surveys and/or remote sensing-based imagery may be used to monitor disease incidences (Kanaskie et al. 2005; Hansen et al. 2008; Hill et al. 2009). Ground surveys for collection of diseased plant samples, and subsequent isolation and laboratory testing of the disease-causing agents through culture-based or genomic approaches are common methods for identifying pathogens.

Model-based indices (Riitters et al. 1992) of risk or vulnerability developed by associating climate, host distribution, host density, and topography to some attribute related to an insect or disease (e.g., occurrence or magnitude) can be effective indicators for identifying forested regions potentially at risk. Other indicators of forest health include the amount and diversity of deadwood and of organisms (e.g., bryophytes, small mammals, woodpeckers) dependent on this key structural feature of healthy forests, lichen communities, vegetation structure and composition, soil and water quality, invasive plants, forest regeneration capacity, and forest fragmentation and urbanization (CCFM 2006; Frego 2007; Kumar et al. 2011; Nappi et al. 2015; Randolph et al. 2020). Some of these indicators are also relevant from the point of view of assessing the sustainability of forest management (Mäkelä et al. 2012).

In the following sections, a non-exhaustive account of important pest insects and pathogens of trees in boreal and temperate forests and forests in one tropical region are discussed to provide a broad overview of biotic threats faced by these forests. Major pests impacting forest health in these regions are described via select case studies, and different types of indicators are emphasized where applicable. Ecological and economic impacts of these agents as well as the forest management approaches being used to limit their impacts are also described.

17.3 BOREAL AND TEMPERATE FORESTS

17.3.1 Distribution

Boreal forest or taiga is the northernmost forest region, populated mainly by conifers (e.g., *Picea* spp., *Pinus* spp., *Abies* spp.) and certain deciduous trees such as *Populus* spp. and *Betula* spp. (Brandt 2009). This forest forms a circumpolar belt encircling the globe and covers nearly 27% of the world's forest area or about 1.1 billion hectares (FAO 2020). Boreal forest extends across higher latitudes in North America, Europe, and Asia; its northern limit being the southern arctic treeline going up to latitude 68°N in the Brooks Range of Alaska (Larsen 1980). Its southern limit is more difficult to define however, as the boundaries with the southern ecotones—broadleaf deciduous forest, parkland, or grassland—are not as sharp as with the tundra ecotone (Larsen 1980).

Temperate forests are mainly situated in the humid middle latitudes characterized by moderate climatic conditions, almost exclusively situated in the Northern Hemisphere, in eastern North America, north-eastern Asia, and western and central Europe. Temperate forests constitute 16% of world's forested area amounting to ~666 million ha (FAO 2020). This is an important biome in North America, present mainly in the eastern United States and southeastern Canada. These forests can be subdivided into different forest types depending on the seasonal distribution of rainfall: temperate conifer and evergreen broad-leaved forests, temperate dry conifer forests, temperate Mediterranean forests, temperate coniferous forests, and temperate broad-leaved rainforests (Kuennecke 2008). The following information mostly concerns the temperate conifer and evergreen broad-leaved forests.

17.3.2 Climatic Conditions

Boreal forests are characterized by subarctic climate with very large range of temperatures between seasons, with long, dry, and cold winters, warm and humid summers, and relatively low precipitation (400–1,000 mm), half of it occurring in summer. However, because of the cold temperature, evaporation rates are also low, keeping enough moisture to support plant growth (Kuennecke 2008). A short annual growing season is characteristic of boreal forests.

In contrast, temperate forests have a longer growing season of at least six months and are characterized by four distinct seasons, including warm to hot summers and cool to cold winters. Precipitation (750–1,250 mm) is evenly distributed throughout the year (Kuennecke 2008).

17.3.3 Soil Characteristics

Taiga's soil is thin and nutritively poor. The fall of conifer needles makes the soil acidic, allowing only lichens and some mosses to grow on it (Henry and Viney 2002). Because of this thin soil, trees usually have shallow roots. In addition, most of the taiga has permafrost—permanently frozen ground—chilling the soil above it and slowing the decomposition process. Poor soil tends to favor coniferous trees as these species can accumulate nutrients (Henry and Viney 2002). In contrast, the annual fall of deciduous tree foliage makes the soil of temperate forests much more fertile, deeply developed, and non-compacted (Kuennecke 2008).

17.3.4 Plant Diversity

Although the boreal forest appears homogeneous, its composition varies from region to region: the southern part is composed of closed canopy forest, with mossy ground cover and shrubs and wildflowers in clearings, whereas the northernmost part is composed of lichen woodland or sparse taiga, with spaced trees and lichen ground cover (Sayre 1994). Boreal tree diversity is low, and the main tree genera are spruces (*Picea* spp.), firs (*Abies* spp.), pines (*Pinus* spp.), and larches (*Larix* spp.). Two deciduous genera are also common, birches (*Betula* spp.) and aspens or poplars (*Populus* spp.) (Kuennecke 2008). In contrast, temperate forests are composed of more deciduous trees and have the highest vegetation richness in the middle latitudes. The richest ones are located in East Asia, followed by North America and Europe (Kuennecke 2008). The most important trees of this biome are in the beech family (Fagaceae)—beeches (*Fagus* spp.) and oaks—but also the birch family (Betulaceae)—hazels (*Corylus* spp.), hornbeams (*Carpinus* spp.), hop-hornbeams (*Ostrya* spp.)—the walnut family (*Juglans* spp.), and the maple family (Sapindaceae; *Acer* spp.). As these trees are deciduous, the canopy opens seasonally, allowing the growth and flowering of forbs and shrubs. This enables the temperate forests to develop a multilayer structure, from top to bottom: canopy layer, tree layer, shrub layer, herb layer, and moss layer (Kuennecke 2008).

17.3.5 Major Natural Disturbances

The major short-term destructive agent in the boreal forest is wildfire that shapes its composition and development (Payette 1992; Attiwill 1994). The fire cycle can vary depending on the climate, being around 50–100 years in drier regions such as western Canada and more than 200 years in more moist climates of eastern Canada (Bergeron et al. 2001; Henry and Viney 2002). Fires prevent the boreal forest from becoming homogeneous. Some boreal species, such as jack pine (*Pinus banksiana* Lamb.), lodgepole pine (*Pinus contorta* Dougl. ex. Loud. var. *latifolia* Engelm.), and to a lesser extent black spruce (*P. mariana* (Mill.) B.S.P.), are adapted to reproducing shortly after a fire has occurred. Their cones can be opened only by the heat of a fire. Other tree species have developed their own adaptations: for instance, aspen's roots mostly survive fires, allowing an abundant regrowth (Henry and Viney 2002). Seeds of understory plants buried in the leaf litter will germinate only when the canopy opens and the leaf litter gets removed, for example, by a fire (Henry and Viney 2002). On the coast where wildfires are less common, insects, diseases, storms, and wind are the major disturbances.

Compared to boreal forests, small-scale canopy disturbances are more frequent in temperate forests. The higher humidity makes forest fires less frequent compared to taiga. Windthrow is more important, generating small gap dynamic allowing regeneration (Bérard and Côté 1996). Certain native insect species (e.g., bark beetles) are natural disturbances in temperate conifer forests. Non-native invasive diseases and insects have historically caused negligible to, at times, very extensive tree mortality in this biome.

17.3.6 Overview of Major Forest Insects in Boreal and Temperate Regions

Native bark beetles (Coleoptera) and defoliators (Lepidoptera and Hymenoptera) are the most damaging insects in the boreal and temperate forests. Important examples of bark beetles in these forests include mountain pine beetle, *D. ponderosae* and spruce beetle, *D. rufipennis* Kirby (North America), and the European spruce bark beetle, *I. typographus* (Europe). Bark beetles colonize trees through a pheromone-mediated mass attack strategy. Adult beetles bore through the bark, feed and reproduce in the phloem, and in the process inoculate fungal associates that further aid the beetles in colonizing the host tree, eventually causing tree death. Optimal availability of host resource and favorable climatic conditions may quickly amplify the beetle numbers such that high tree mortality over a short time period may be observed. For example, spruce beetle outbreaks in

Alaska, United States, occur when abundant mature spruce occurs on the landscape and warm summer temperatures prevail (Berg et al. 2006). In case of defoliators, adults lay eggs on the tree crown (needles, branches, twigs etc.) and the emerging larvae feed on the foliage. Extensive defoliation by larval feeding affects tree growth and may potentially result in tree death over time. Some important examples of defoliators include the eastern spruce budworm and forest tent caterpillar (*Malacosoma disstria* Hübner) in northern North America and autumnal moth (*Epirrita autumnata* (Borkhausen)), winter moth (*Operophtera brumata* (L.)), and European pine sawfly (*Neodiprion sertifer* (Geoffroy)) in Europe.

In Canada, although there are numerous species of native insects that may be considered as problematic pests in the boreal and temperate forests and that may occasionally outbreak, impacts of many of these species are ephemeral over time and space or are currently of minor importance (Armstrong and Ives 1995). Nevertheless, there are several important species of eruptive insects of concern that can have huge impacts on forests, especially those forests influenced by anthropogenic or climate-induced changes. These include (1) bark beetles such as mountain pine beetle and spruce beetle and (2) defoliators such as forest tent caterpillar, eastern spruce budworm, large aspen tortrix (*Choristoneura conflictana* (Walker)), hemlock looper (*Lambdina fiscellaria* (Guenée)), and balsam fir sawfly (*Neodiprion abietis* (Harris)) (Table 17.1). Some species, such as the eastern

TABLE 17.1
List of Important Native Insect Pests in Canada's Forests

Common Name	Scientific Name	Order and Family	Primary Host(s)	Forest Area Affected Annually (× 1000 ha)[a]
Mountain pine beetle	*Dendroctonus ponderosae* Hopkins	Coleoptera: Curculionidae	Pines (*Pinus* spp., e.g., lodgepole pine, *Pinus contorta* Dougl. ex Loud. var. *latifolia* Engelm.; jack pine, *P. banksiana* Lamb.)	3,868.7 (189.6, 10,051.9)[b]
Forest tent caterpillar	*Malacosoma disstria* Hübner	Lepidoptera: Lasiocampidae	Various hardwood species (e.g., trembling aspen, *Populus tremuloides* Michx.; sugar maple, *Acer saccharum* Marsh.)	3,799.3 (89.6, 14,334.8)
Spruce budworm	*Choristoneura fumiferana* (Clemens)	Lepidoptera: Tortricidae	Balsam fir, *Abies balsamea* (L.) Mill.; spruces, *Picea* spp.	3,049.1 (721.5, 6,983.5)
Large aspen tortrix	*C. conflictana* (Walker)	Lepidoptera: Tortricidae	Trembling aspen (*P. tremuloides*); balsam poplar (*P. balsamifera* L.)	1058.4 (8.1, 6,013.5)
Jack pine budworm	*C. pinus pinus* Freeman	Lepidoptera: Tortricidae	Jack pine, *P. banksiana*	582.2 (1.0, 3,666.5)[c]
Spruce beetle	*D. rufipennis* Kirby	Coleoptera: Curculionidae	Spruces (*Picea* spp., e.g., white spruce, *P. glauca* (Moench) Voss)	222.3 (26.6, 528.2)
Balsam fir sawfly	*Neodiprion abietis* (Harris)	Hymenoptera: Diprionidae	Balsam fir, *A. balsamea*	29.4 (0, 68.7)
Hemlock looper	*Lambdina fiscellaria* (Guenée)	Lepidoptera: Geometridae	Balsam fir, *A. balsamea*; eastern hemlock, *Tsuga canadensis* (L.) Carrière	72.3 (0.2, 926.2)[d]

Notes:

[a] Gross estimate values represent Canada-wide annual averages (min, max in parentheses) of areas infested during the time period 2000–2020 (Source: National Forestry Database, http://nfdp.ccfm.org/insects/national_e.php, accessed May 2, 2023). Surveyed areas may include live trees (e.g., damage by defoliators).

[b] Value represents total area affected in the Provinces of British Columbia and Alberta.

[c] Figures not available from 2000 to 2003.

[d] Figures not available for 2019.

spruce budworm and forest tent caterpillar, show periodical outbreak patterns, which is evident from observing the trends in the average annual forest area affected by these insects (Figure 17.1). This national-level metric, an adaptation of the Montréal Process indicator (# 3.a.) concerning forest ecosystem health and vitality (Anon. 1995), is used as one of the indicators ("Area of forest disturbed by fire, insects, disease, and timber harvest") in the Criteria and Indicator (C&I) framework of the Canadian Council of Forest Ministers (CCFM) for assessing forest "ecosystem condition and productivity" (CCFM 2006). Similarly, several Canadian provinces (e.g., Québec, Ontario) use the annual extent and severity of insect infestations as one of the indicators for monitoring the overall health of forests in their respective jurisdictions.

Non-native insects are a huge concern in the boreal and temperate forests. Large numbers of invasive forest insect species are intercepted each year in North America and Europe, although only a few species cause significant harm after establishment (Allen and Humble 2002; Haack 2006; Roques and Auger-Rozenberg 2006). The invaded habitats are typically composed of trees that are not adapted to non-native insect invasions and lack the assemblage of natural enemies that regulate the population dynamics of these insects in their native range. Such naïve trees may quickly succumb to attacks by non-native insects due to lack of coevolutionary history; hence, these biological invasions have detrimental impacts on the affected forest ecosystems (Gandhi and Herms 2010). Examples of important non-native insect species include the Asian longhorned beetle (*Anoplophora glabripennis* (Motschulsky)), pine shoot beetle (*Tomicus piniperda* (L.)), brown spruce longhorn beetle (*Tetropium fuscum* (Fabr.)), European elm bark beetle (*Scolytus multistriatus* (Marsham)), spongy moth (*Lymantria dispar* L.), balsam woolly adelgid (*Adelges piceae* (Ratz.)), hemlock woolly adelgid (*A. tsugae* Annand), and the European woodwasp (*Sirex noctilio* Fabr.) (Allen and Humble 2002; Haack 2006).

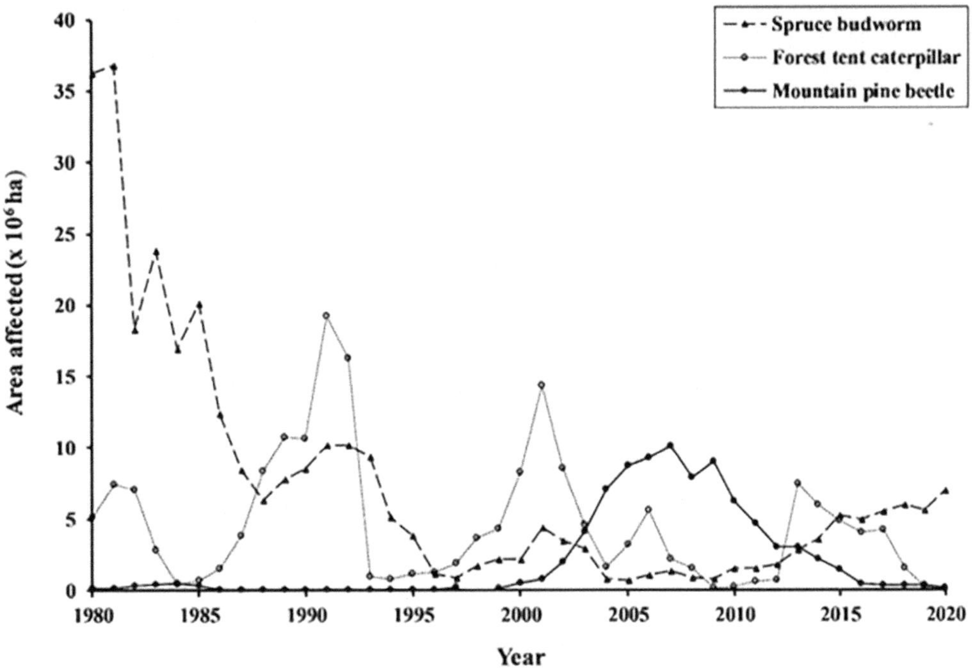

FIGURE 17.1 Forest area affected by three major native insect species in Canada from 1980 to 2020. Area infested by mountain pine beetle is not available for 1998.

(Source: National Forestry Database, http://nfdp.ccfm.org/index_e.php.)

17.3.6.1 Insect Case Study: Eastern Spruce Budworm

The spruce budworm (*C. fumiferana*; Lepidoptera: Tortricidae) is a native defoliator in northern North America. Its range coincides approximately with that of balsam fir (*Abies balsamea* (L.) Mill.), its preferred host. Other hosts in order of susceptibility include white (*P. glauca* (Moench) Voss), red (*P. rubens* Sarg.), and black spruce (Greenbank 1963; Hennigar et al. 2008). Several consecutive years of severe defoliation causes significant growth reduction of host trees and, eventually, mortality in defoliated stands. Although outbreaks mostly affect southeastern boreal forest in Canada, spruce budworm populations can also reach epidemic levels from Manitoba to British Columbia (east of the Rockies) and north of the 60th parallel in the Northwest Territories (Volney and Fleming 2007). Over the last 400 years, spruce budworm outbreak recurrence cycle has averaged 30–40 years (Boulanger and Arseneault 2004), with outbreaks lasting between five years and ten years on a regional scale. Variations in population levels may be synchronized over very large areas so that several millions of hectares may concurrently experience significant defoliation and mortality (Peltonen et al. 2002). During the last major outbreak in the 20th century (*ca.* 1965–1992), over 51 million ha were moderately or severely defoliated in just a single year (1975) (Canada's National Forestry Database, http://nfdp.ccfm.org/en/data/insects.php). The outbreak resulted in losses of 382 million m^3 of wood between 1977 and 1987 (Sterner and Davidson 1982b; Hall and Moody 1994). Widespread damage triggers high uncertainty about short- and long-term availability of forest resources (MacLean 1984).

First signs of budworm activity usually occur in late spring (May–June), with the defoliation of current-year shoots by third-instar larvae. When populations reach very high levels, late-instar larvae can consume previous-years needles (Piene 1989). Second-instar larvae may also feed on pollen produced by staminate flowers (Blais 1952), which increases larval development rate without affecting pupal weight and fecundity at low budworm density (Bauce and Carisey 1996). At high budworm density, the amount of current-year foliage produced by flowering branches is lower, and late-instar larvae are forced to feed on old foliage to complete development, which reduces pupal weight and fecundity (Bauce and Carisey 1996). Severely affected stands turn into a rust color, with dried needles held by strands of silk spun by larvae, and eventually turn gray in the fall when needles fall off. As within-stand defoliation levels can strongly vary among host species (Nealis and Régnière 2004b; Hennigar et al. 2008), the spruce budworm induces very high within-stand heterogeneity (Erdle and MacLean 1999).

Several multi-scale factors influence spruce budworm-related impacts during an infestation and can also be used as indicators of forest health and for improving forest management. These are essentially a function of host susceptibility and host vulnerability, which can be estimated through defoliation and mortality levels, respectively. Defoliation varies as a function of tree species, but environmental factors related to the host phenology and foliage quality (Mattson et al. 1983; Carisey and Bauce 1997; Nealis and Régnière 2004a) also influence it through their effects on larval growth rates and survival (Royama 1984; MacKinnon and MacLean 2003; Nealis and Régnière 2004b). Later bud break in spruce species, notably in black and red spruce, is known to impede early development of second-instar larvae in spring, increasing mortality rates (Lawrence et al. 1997). Larger trees may sustain larger populations because of greater foliage biomass resulting from increased crown exposure and thereby egg-deposition sites (Morris and Mott 1963), as well as more suitable microclimates for feeding larvae (Mott 1963). As a result, greater impacts are generally observed in mature and overmature stands with a high proportion of balsam fir (MacLean 1980). Moreover, mortality is consistently higher in mature than immature stands (MacLean 1980). Less vigorous trees, as those growing in poor, xeric or hydric sites, experience more budworm-related mortality (Osawa 1989; Dupont et al. 1991) and greater growth reductions (MacKinnon and MacLean 2004) than hosts growing on mesic sites. Recent studies have also shown that trees growing on mesic sites produce more foliage and have higher levels of defensive compounds (monoterpenes) than those growing on xeric sites (Fuentealba and Bauce 2012).

Host mortality is strongly related to cumulative defoliation over the preceding years (Blais 1958; MacLean and Ostaff 1989). Balsam fir mortality generally occurs after 4–5 years of severe defoliation, whereas mortality may be delayed by a few more years for spruce species (Blais 1981). In the most severely affected stands, near-complete mortality may be observed (MacLean 1980; Bergeron et al. 1995).

Vulnerability grids for the spruce budworm have been produced (Gagnon and Chabot 1988) and could serve as indicators of risk to forest health. They are based on variables available from Quebec provincial forest maps: forest composition, age, density, soil drainage, and surficial deposit. However, it is not clear how these generalities might be transposed over the entire range of spruce budworm outbreak. There are discrepancies in the literature with respect to the influence of soil drainage and surficial deposit (Bergeron et al. 1995; MacLean and MacKinnon 1997) that might be related to specific regional context. Such regional discrepancies advocate the development of better models linking spruce budworm defoliation rates and host mortality to environmental variables which would help defining specific vulnerability grids at this scale. Such grids might be particularly helpful in regions that were lightly affected in previous outbreaks but that might be much more at risk in the future as a result of climate change (Pureswaran et al. 2015).

Host mortality following spruce budworm infestations can have large impacts on stand renewal patterns. Depending on mortality patterns (gap size), established understory abundance and stand composition, post-outbreak stand characteristics may exhibit a range of responses (Kneeshaw and Bergeron 1998; Bouchard et al. 2005, 2006). Outbreaks may prompt self-replacement in balsam fir-dominated stands as they frequently contain an abundant seedling bank under their canopy (MacLean 1980; Morin 1994). However, large gap openings, particularly in mixed stands or where balsam fir is less ubiquitous, may favor the recruitment of non-host species, especially those that are shade-intolerant (Kneeshaw and Bergeron 1998; Bouchard et al. 2007). In mixed hardwood stands, outbreaks may trigger long-term changes in species composition where abundant and/or long-lived species (e.g., white cedar (*Thuja occidentalis* L.), yellow birch (*Betula alleghaniensis* Britton)) in the regeneration layer may subsequently increase their relative occurrence (Bouchard et al. 2005). In some situations, competitive exclusion of shrub species (e.g., mountain maple, *Acer spicatum* Lam.) or other hardwoods over balsam fir may occur depending on gap size, soil richness, and slope position (Kneeshaw and Bergeron 1998; Bouchard et al. 2006). As a consequence, successive spruce budworm outbreaks may have a profound and long-term influence on the forest composition and structure depending on the regional context (Bouchard et al. 2007).

Spruce budworm outbreaks are the most important natural disturbance regime in balsam fir forests of northeastern North America and thus are part of the regeneration cycle of this tree species. Black spruce, a secondary host of spruce budworm, is a major component of forests in northern, cooler latitudes. As climates warm, especially in more northern locations, spruce budworm could benefit from such favorable regimes as climate plays a key role in the life cycle and population dynamics of the spruce budworm. Spring and summer temperatures govern the ability of the spruce budworm to complete a one-year life cycle and as such, delimit the species' northern extent (Régnière 1987). Warmer temperatures during the overwintering period can contribute to exhaustion of energy reserves for the L2 larvae at the southern trailing edge of its range (Régnière 1990, Régnière and Duval 1998). These seasonal variables thus contribute to defining the climate envelope of the spruce budworm which, with climate change, is projected to be further displaced northward and to higher elevations (Régnière et al. 2012b, Boulanger et al. 2016). Such displacements appear to have already begun with outbreak foci occurring further north than in the past (Régnière et al. 2012b; Pureswaran et al. 2015). If severe defoliation and mortality of black spruce forests were to occur in future, it might result in a change in ecosystem state that may not be easily reversible (Pureswaran et al. 2015).

17.3.6.1.1 Management

Several management options are available to lower stand vulnerability to subsequent spruce budworm outbreaks. In most areas, management strategies that promote the occurrence of non-host (e.g., jack pine) or less vulnerable hosts (e.g., Norway spruce, *P. abies* (L.) H. Karst., black spruce) over preferential host species (balsam fir, white spruce) are to be favored. Such objectives may be achieved, for example, by planting, thinning, or smart harvest scheduling (MacLean 1996). As large-scale spraying operations during outbreaks may be very expensive, management strategies that alter tree composition represent an additional and effective option to reduce volume losses and overall costs in forest protection on a long-term perspective (MacLean et al. 2001).

Spatial decision-making regarding forest management in the context of SBW infestations may be assisted through the Spruce Budworm Decision Support System (SBWDSS). Using forest inventory, stand yield, stand growth models, and harvest schedule information, the SBWDSS can estimate losses in timber supply according to specific spruce budworm outbreak and forest management scenarios (MacLean et al. 2001). Consequently, such a tool is well suited to optimize forest management by evaluating the costs, benefits, and consequences of alternative scenarios (MacLean et al. 2001). Furthermore, the SBWDSS can be used to prioritize stands or spray blocks for protection to optimize the overall costs associated with spraying operations during an outbreak (MacLean et al. 2001). Currently, the SBWDSS has been implemented in New Brunswick, Canada, and Maine, United States (Hennigar et al. 2008; Hennigar et al. 2011).

Although preventive silvicultural approaches can be used to enhance the resilience of natural spruce and fir forests to spruce budworm outbreaks (Boulet et al. 2009), insecticidal intervention is often required to protect stands from extensive damage. Suppression of spruce budworm outbreaks typically involves insecticidal applications targeting high-value mature stands, once population densities have reached a level that could negatively affect tree growth and survival. This approach, referred to as the "foliage protection strategy," involves aerial surveys of defoliation at the end of the active budworm feeding stage. These are typically conducted using fixed-wing aircraft when large areas are affected. A permanent network of pheromone traps is also used to follow annual fluctuations of spruce budworm populations. In addition, sampling of branches in the fall is conducted to estimate the density of overwintering larvae (L2) at specific sites. These measurements produce various forest health indicators that are used to plan control operations against the budworm, to modify the harvesting schedule to pre-salvage severely affected stands and, when necessary, to prepare salvage logging plans. An alternative budworm management strategy aimed at suppressing or slowing the development of outbreaks is currently being evaluated in the Canadian province of New Brunswick (Johns et al. 2019). This "early intervention strategy" (EIS) involves maintaining population densities below a threshold at which natural enemies can keep populations in check. Monitoring approaches similar to those described here are used to identify loci of rising densities, which are then targeted for insecticidal applications.

The EIS approach requires understanding the dispersal behavior of spruce budworm adults, which can significantly impact the success of local mating and reproduction (MacLean et al. 2019). Indeed, massive dispersal events involving billions of moths can occur almost daily during outbreaks, with these swarms covering vast areas and traveling hundreds of kilometers in a single night (Boulanger et al. 2017; Garcia et al. 2022). Large-scale dispersal of moths can lead to increased mating success and fecundity in low-density populations, potentially surpassing a threshold where natural enemies can no longer control the rising populations (Miller et al. 1978; Régnière et al. 2012a). This can initiate new epicenters of outbreaks and favor outbreak synchrony (Peltonen et al. 2002). Key variables driving insect dispersal processes such as liftoff, landing, and atmospheric layering include individual characteristics (e.g., sex, mating status), as well as environmental factors (e.g., temperature, wind) and the physics of dispersal (Régnière et al. 2019b; Garcia et al. 2022; Rhainds et al. 2022). Weather surveillance radar can be combined with phenological models to assess the frequency, origin, destination, and density of long-dispersal events which can help assess the impacts of these key variables (Boulanger et al. 2017). Linking aerobiological principles to these variables, an agent-based model, the SBW-Atmospheric Transport Model, was developed to better simulate budworm flight behavior by assigning behavioral parameters to agents based on liftoff timing, flight strength, and descent (Sturtevant et al. 2013; Régnière et al. 2019b). Although these are yet to be implemented at large scale, these experimental tools are promising to better manage spruce budworm outbreaks.

Three pest-control products are currently registered in Canada for spruce budworm management. These include the microbial insecticide *Bacillus thuringiensis* var. *kurstaki* (Btk), the insect growth regulator tebufenozide (commercially known as Mimic®), and a pheromone-based mating disruptant (commercially known as Disrupt Micro-Flake SBW®). The choice of a suitable pest-control product for budworm management varies from one province to another; however, public pressure against the application of synthetic pesticides over forest land has led most provinces to adopt Btk as their main or sole weapon to fight budworm invasions.

Both Btk (Cadogan and Scharbach 1993) and tebufenozide (Cadogan et al. 1997) have been shown to provide adequate stand protection against the spruce budworm. While both products are specific to the Lepidoptera and must be ingested, their modes of action and half-life differ. Btk is a biopesticide inducing mortality by the formation of pores through the midgut as a result of the action of a bacterial toxin (Gill 1995), whereas tebufenozide, a synthetic non-steroidal analog of the molting hormone, induces a precocious molt that the insect cannot complete (Dhadialla et al. 1998). Btk has a half-life of 12 hours to a few days while tebufenozide has a half-life of 20–45 days (Sundaram A. et al. 1996; Sundaram K.M.S. et al. 1996). Tebufenozide has also been shown to display a carryover effect toward insects of the following generation (see Cadogan et al. 2002; Dallaire et al. 2004; Doucet et al. 2007). Finally, Btk and tebufenozide could also have detrimental effects on spruce budworm parasitoids, either directly or indirectly through their alternate hosts, although parasitized budworms feed less and are thus less likely to be affected (Nealis and van Frankenhuyzen 1990). For example, tebufenozide was shown to be more toxic to spruce budworm larvae than those of the oblique-banded leafroller, *Choristoneura rosaceana* (Harris), a known alternate host of several spruce budworm parasitoids (Dallaire et al. 2004). Negative impacts of both products on non-target caterpillars (potential alternate hosts for parasitoids) have been previously shown in some studies (Miller 1990; Butler et al. 1997; Rastall et al. 2003), although an increase in caterpillar richness, probably through an increase in foliage availability, was also shown after Btk treatments against outbreaking species (Scriber 2004). Nevertheless, neither product was shown to negatively affect Lepidoptera community or parasitism rates when used under EIS in New Brunswick (Glaus et al. 2023). Thus, while Btk and tebufenozide are considered environmentally benign (i.e., very limited non-target effects), each product's advantages and disadvantages must be taken into account in planning spray operations. Beyond Btk and tebufenozide, novel insect growth regulators targeting juvenile hormone production are currently being developed and may soon provide an additional control option to spruce budworm management specialists (Picard et al. 2021). Finally, biological control using the egg parasitoid *Trichogramma minutum* Riley (Hymenoptera: Trichogrammatidae) has also been shown to increase egg mortality and could be a complementary tool for sites where insecticidal applications are prohibited or difficult to conduct (Martel et al. 2021).

In Lepidoptera, virgin females release a species-specific sex pheromone that serves to attract males over long distances for copulation (Cardé and Baker 1984). Permeation of the atmosphere with synthetic pheromones has been used to disrupt mating by preventing conspecific males of the target species to locate females in the treated area (Cardé 1990). Over the past 40 years, the effectiveness of mating disruption for controlling spruce budworm populations has been assessed in more than 20 experimental trials using either ground or aerial applications of a broad variety of pheromone formulations, including the Disrupt Micro-Flake SBW®, registered in Canada since 2007. Although these assays have confirmed that mating disruption significantly reduces the number of males attracted to traps baited with either synthetic lures or virgin females and reduced the number of mated females, they did not reduce the mating success of feral females, the number of eggs laid by locally emerged females (apparent fecundity), or the number of overwintering larvae of the next generation (Rhainds et al. 2012; Régnière et al. 2019a). Massive immigration of mated females was hypothesized to be the cause for this lack of impact on the next generation in a recent replicated series of assays in Quebec (Régnière et al. 2019a). Similar results on population densities were recently reported in a study designed to test the effectiveness of aerial spray applications of the new microencapsulated formulation CONFOUND$_{SBW}$ in New Brunswick, even with little or no evidence of immigration events throughout the season (Roscoe et al. 2022). Although most studies clearly showed that pheromone treatment significantly reduced male captures in pheromone traps and/or mating success of virgin females held in individual cages, its inability to bring down population densities argues strongly against the use of mating disruption to effectively manage spruce budworm populations.

17.3.6.2 Insect case study: Mountain pine beetle

The mountain pine beetle (*D. ponderosae*; Coleoptera: Curculionidae) is a bark beetle infesting mature pines (*Pinus* spp.) in western North America (Amman and Cole 1983; Safranyik and Carroll 2006; Raffa et al. 2008; Sambaraju and Goodsman 2021). The beetle is normally present at endemic population densities and attacks trees that are weak, stressed, or diseased (Boone et al. 2011). Discrete beetle populations over a susceptible, pine-dominated landscape coalesce when favorable climatic conditions persist over several years resulting in large outbreaks that cause extensive tree mortality (Safranyik and Carroll 2006; Sambaraju et al. 2019). The outbreak that began in the late 1990s killed millions of hectares of pine forests in British Columbia, Canada (Figure 17.1). Lodgepole pine and ponderosa pine (*Pinus ponderosa* Douglas ex C. Lawson) are the main hosts in western Canada and the United States, although by the end of the outbreak, mountain pine beetle populations traversed the Rocky Mountains, expanding their range into the boreal forests of Alberta, now occurring in jack pine (*P. banksiana*) forests (Giroday et al. 2012). Climate change could further provide conditions favorable for beetle expansion (Sambaraju et al. 2012). Jack pine is widely distributed in the boreal forest of Canada and forms a bridge to pine species in eastern Canada (e.g., eastern white pine, *P. strobus* L.). Eastward progression of the mountain pine beetle through jack pine thus poses a risk to eastern pines as well as to the overall health of the boreal forest in Canada (Safranyik et al. 2010; Sambaraju and Goodsman 2021).

In early summer, beetles leave their brood trees and disperse in search of new hosts (Wood 1982). Thousands of beetles simultaneously mass attack a tree and must be able to kill it to successfully reproduce (Berryman et al. 1985). Signs of infestation occur on the bole of attacked trees: (1) Beetle entry holes are surrounded with pitch tubes that are composed of resin and boring dust. (2) Frass, which is a combination of boring dust and excrement, is seen as a reddish-brown powder in bark crevices and around the base of the tree. (3) Woodpecker marks are commonly seen on the bark. (4) Emergence holes where beetles have left brood trees are visible. Trees that are successfully attacked usually die, and the foliage fades to greenish-yellow and then turns red the following year (Amman 1982).

Beetles leaving a brood tree often attack pine trees in the vicinity. As population size grows, mountain pine beetle transitions from colonizing weakened trees to healthy and vigorous trees, eventually killing them. At this stage, affected areas are seen on the landscape as small clusters of reddened, dead trees known as "spot" infestations. Red tree crowns are used for aerial assessment and mapping of infestations over large areas of forest. These localized infestations grow into progressively larger patches as more trees come under beetle attack. Epidemics occur at the landscape level when there are large areas of susceptible host and triggering weather conditions (e.g., warmer winters and summers) that favor population growth and dispersal. Unprecedented outbreaks of the mountain pine beetle in the 2000s in western Canada were the result of such conditions (Aukema et al. 2006; Safranyik and Carroll 2006; Sambaraju et al. 2012). Mountain pine beetle outbreaks can have important ecological impacts such as on forest regeneration, carbon dynamics, soil biogeochemistry, interactions with other disturbances, and mid-term timber supply (Kurtz et al. 2008; Klutsch et al. 2011; Mikkelson et al. 2013; Dhar et al. 2016). Lessening the impacts of spread and damage caused by mountain pine beetle requires removal of dead trees, followed by afforestation or natural regeneration, both of which could result in a different vegetational progression that could affect forest resilience to future disturbances (Kashian et al. 2011; Rhoades et al. 2020).

17.3.6.1.2 Management

Different management options are available to limit pine mortality caused by the mountain pine beetle depending on tree attributes, site characteristics, and beetle population levels (Fettig et al. 2007; Shore and Safranyik 1992). For instance, although vast outbreaking populations cannot be suppressed by human interventions, measures can be taken to improve stand resilience to withstand a potential future outbreak. Therefore, long-term mitigation of epidemics involves management strategies that reduce the susceptibility of lodgepole pine over the landscape. Indirect control or preventive management can be applied by increasing stand vigor or decreasing the number of susceptible stands (Safranyik and Wilson 2006; Safranyik and Carroll 2006). Landscape susceptibility can be reduced if stands are managed to minimize conditions that favor population growth of the beetle. Beetle populations build up in high density, over-mature stands containing large diameter trees. Hence, creating a landscape with younger, more diverse age classes of pines planted in species mixes with low stand densities will help mitigate the development of outbreaks (Amman and Safranyik 1985). Silvicultural treatments have been used to increase tree vigor during stand development by reducing stand basal area via thinning and spacing (Safranyik et al. 2004).

Direct control includes mechanical treatments and manipulation of populations using pheromones. Prescribed fires have been used to ignite infested stands (Safranyik et al. 2001). Removal of either single infested trees or blocks of trees (sanitation logging) helps reduce the beetle population in an area. Infested logs are often submerged in water to kill developing brood before they emerge (Safranyik and Linton 1982).

Semiochemical-based management consists of manipulating insect behavior by using pheromones, which are produced by beetles during the attack process, to initiate and terminate attack on a tree, as well as volatiles emitted by trees that are attractive or repellent to beetles. These compounds are formulated in commercial devices and deployed in traps or on trees to manipulate beetle behavior. Integrated management using semiochemicals to lure beetles and concentrate attack on a target tree, then treating with insecticides, followed by sanitation logging is often employed (Borden 1995). Antiaggregation pheromones such as verbenone have been used in "push-pull" strategies where aggregation pheromones such as *trans*-verbenol "pull" beetles to a stand about to be treated and antiaggregation pheromones simultaneously "push" them away from a high value stand (Borden 1995; Gillette et al. 2012b). Area-wide aerial sprays of verbenone have been shown to be effective at lowering beetle pressure and reducing tree mortality (Gillette et al. 2012a).

The cryptic habit of mountain pine beetle necessitates topical insecticides applied to the bole that can penetrate the bark (e.g., ethylene bromide, lindane, carbaryl) (Smith et al. 1977) or systemic insecticides (e.g., copper sulphate, monosodium methanearsonate) (Bedard 1938; Maclauchlan et al. 1988) that are translocated into the bark of trees and then kill larvae that feed within the tree. Chemical treatments are used in conjunction with ecologically based treatments (e.g., pheromones) under the framework of integrated management to minimize mountain pine beetle impacts.

17.3.7 Overview of Major Forest Pathogens in Boreal and Temperate Regions

Tree fungal pathogens are an important component of boreal and temperate forest ecosystems. Pathogenic fungi colonize trees in a variety of ways, through various wounds, such as broken branches or injured roots, or through natural openings such as lenticels in the stem or stomata of the leaves. Some endophytic fungi can change their lifestyle and cause disease when trees become stressed due to drought, climate shifts, or other factors. Tree pathogens can also contribute to weakening trees and making them more vulnerable to drought and other stressors (Oliva et al. 2014). Although fire and insects are seen by the general public as more important than diseases in causing tree mortality (Gonthier 2013), the losses of timber caused by tree pathogens are often more important than those due to other disturbance agents (Tainter and Baker 1996). For example, losses caused by diseases in the Canadian boreal and temperate forests between 1982 and 1987 totaled 51 million m^3 per year (Hall and Moody 1994), very similar to the damage caused by insects.

The available data (CFS 2014; DesRochers and Allen 2014) show that four diseases cause major damage to natural forests and plantations in the boreal zone: Annosus root and butt rot (*Heterobasidion* spp. complex), Armillaria root rot (*Armillaria ostoyae* (Romagn.) Herink), white pine blister rust (*Cronartium ribicola* J.C. Fisch.), and western gall rust (*Cronartium harknessii* E. Meinecke). In Canada, Armillaria root rot affects 203 million hectares of forests (NRCAN 2014). Other important diseases that cause damage to trees in the northern boreal forests include violet root rot (*Helicobasidium mompa* Nobuj. Tanaka), blue stain fungus (*Endoconidiophora polonica* (Siemaszko) Z.W. de Beer, T.A. Duong & M.J. Wingf., comb. nov.), Hypoxylon canker (*Entoleuca mammata* (Wahlenb.) J.D. Rogers & Y.M. Ju), Scleroderris canker (*Gremmeniella abietina* (Lagerb.) M. Morelet), spruce needle cushion rust (*Chrysomyxa abietis* (Wallr.) Unger), and red banded needle blight (*Mycosphaerella pini* Rostr. ex Munk). While some diseases such as Annosus root rot, Armillaria root rot, white pine blister rust, or Hypoxylon canker are widespread throughout the circumboreal forest, others such as the western gall rust, the violet root rot, or the blue stain fungi are limited to some regions of the northern hemisphere.

Native diseases that cause some important damage to temperate forests include Annosus root rot and Armillaria root rot (*A. mellea* s.s.). Cankers, such as the Scleroderris canker on pines; the Lachnellula canker (*Lachnellula willkommii* (R. Hartig) Dennis) on larches; the Hypoxylon canker on poplars and the Ceratocystis canker (*C. platani* (J.M. Walter) Engelbr. & T.C. Harr.) on London plane (*Platanus* × *hispanica* Mill. ex Münchh.); and blue stain fungi (e.g., *E. polonica* and *E. laricicola* (Redfern & Minter) Z.W.de Beer, T.A. Duong & M.J. Wingf.) associated with insect vectors. Dwarf mistletoes (*Arceuthobium* spp.) including *A. americanum* Nutt. ex Engelm., *A. douglasii* Engelm., *A. laricis* (Piper) H. St. John, *A. pusillum* M. Peck, and *A. tsugense* (Rosend.) G.N. Jones are important in temperate forests of Canada and the United States.

Trees are normally able to cope with native fungal pathogens, given their long history of coevolution and the insurance and dilution effects provided by tree species and genetic diversity in semi-natural forests (Hantsch et al. 2014). A certain degree of tree mortality is now recognized as beneficial for the overall functioning of forest ecosystems, as well as the dynamics of forest biodiversity (Hansen 1999). However, when fungal pathogens are moved over long distances and infect naïve hosts with which they have not coevolved, it can take some time before some kind of equilibrium in the pathosystem is reached (Brunet et al. 2014). There are indeed several non-native tree pathogens of concern that have widespread impacts on temperate forests (Table 17.2).

Despite the growing threat posed by non-native diseases, there is an increased understanding among forest pathologists and ecologists of the ecological role played by native tree diseases (Ostry and Laflamme 2009). Also, in temperate forests, herbivores and diseases limit the abundance of dominant tree species, thereby allowing rare species to coexist (Yamazaki et al. 2009). Non-native tree diseases can be seen as effective tools to diversify homogeneous plantations of non-native tree species (Pautasso 2013b). Tree fungal pathogens can help restore deadwood supplies of managed forest ecosystems, where human actions have often reduced this important biodiversity habitat (Lonsdale et al. 2008). However, the introduction and spread of non-native invasive tree pathogens of common tree species can also have deleterious effects on forest ecosystems, affecting carbon and nitrogen cycles (Stenlid et al. 2011), as well as community dynamics and ecosystem services (Dietze and Matthes 2014). In conjunction with climate changes, non-native tree diseases, such as European ash dieback (*Hymenoscyphus fraxineus* (T. Kowalski) Baral, Queloz, & Hosoya), can lead to the decline of many host-associated species, although to differing degrees depending on the site conditions and host-replacing tree species (Ellis et al. 2014).

The co-occurrence in the coming decades of climate warming and an increase in trade of plants for planting and wood products is likely to facilitate the emergence of new tree health threats, not just in temperate forests (Pautasso et al. 2015). At the same time, climate change will induce changes in tree species distributions (Zolkos et al. 2015), which will shift areas suitable for many native and non-native tree pathogens. This process is expected to be facilitated by assisted migration of tree species, for instance when tree seedlings or saplings are moved together with their associated organisms (Garbelotto and Pautasso 2012). The complexity of interactions among the host tree, the

TABLE 17.2
Examples of Important Non-Native Pathogens in Temperate Forests

Pathogen	Main Host Species	Non-Native Distribution	Reference
Bretziella fagacearum (Bretz) Z.W. de Beer, Marinc., T.A. Duong & M.J. Wingf.	Red and white oaks (*Quercus* spp.)	North America	Juzwik et al. 2008
Cronartium ribicola J.C. Fisch.	Five needle pines such as *Pinus strobus* L. (eastern white pine)	North America and Europe	Mulvey and Hansen 2011; Kaitera et al. 2015
Cryphonectria parasitica (Murrill) Barr	*Castanea* spp.	North America and Europe	Prospero and Rigling 2012
Discula destructiva Redlin	*Cornus florida* L.	Eastern US and Europe	Holzmueller et al. 2010
Fusarium circinatum Nirenberg & O'Donnell	*Pinus* spp.	Worldwide	Wingfield et al. 2008
Hymenoscyphus fraxineus (T. Kowalski) Baral, Queloz, & Hosoya	*Fraxinus excelsior* L.	Europe	Gross et al. 2014
Neonectria faginata (M.L. Lohman, A.M.J. Watson & Ayers) Castl. & Rossman and *N. ditissima* (Tul. & C.Tul.) Samuels & Rossman (= *N. galligena*), associated with the scale insect, *Cryptococcus fagisuga* Lindinger.	American beech, *Fagus grandifolia* Ehrh.	Northern North America	Wieferich 2013; Cale et al. 2015
Ophiostoma spp.	*Ulmus* spp.	North America and Europe	Brasier 2001
Phytophthora ramorum Werres, De Cock & Man in't Veld	Variety of tree and shrub species	North America and Europe	Brasier and Webber 2010

disease, the prevailing climates, and competing biotic factors will ultimately determine the eventual range of forest diseases (Queloz et al. 2023). In the following section, we discuss two major diseases in greater detail to illustrate their patterns, trends, and impacts in the boreal and/or temperate forests and to provide insights into their management.

17.3.7.1 Disease Case Study: Annosus Root and Butt Rot in Boreal Forest

Annosus root and butt rot is an important tree disease that has been known since 1882 (Laflamme 2010). Damage caused by this disease has been recorded in several countries in the circumboreal forest zone since the first half of the 20th century, e.g., in Russia (Younitzky 1927), Finland (Tikka 1934), Sweden (Lagerberg 1936), Norway (Roll-Hansen 1940), Japan (Kamei and Hoshi 1948), and eastern Canada (Hord and Quirke 1955).

The causal agent of this disease has had numerous names previously, until it was renamed as *Heterobasidion annosum* (Fr.) Bref. However, it became obvious that some differences existed within this species. Recent taxonomic research investigations have re-classified *Heterobasidion annosum sensu lato (H. annosum s.l.)* into at least five different species: *H. annosum* (Fr.) Bref. *sensu stricto* (s.s.), *H. parviporum* Niemelä & Korhonen and *H. abietinum* Niemelä & Korhonen in Europe (Asiegbu et al. 2005), and *H. irregulare* Garbel. & Otrosina and *H. occidentale* Otrosina & Garbel. in North America (Otrosina and Garbelotto 2010). *H. parviporum* is also known to occur in Asia (Tokuda et al. 2009), and *H. irregulare* was introduced in Italy during the Second World War (Otrosina and Garbelotto 2010). Additional species have been described for Asia: *H. linzhiense* Y.C. Dai & Korhonen, *H. orientale* Tokuda, T. Hatt. & Y.C. Dai and *H. ecrustosum* Tokuda,

T. Hatt. & Y.C. Dai (Tokuda et al. 2009). Only *H. annosum* s.s., *H. parviporum*, *H. orientale*, *H. irregulare*, and *H. occidentale* are found in boreal and temperate forests (DesRochers et al. 2006; DesRochers and Allen 2014; DesRochers and Callan 2014; Shamoun et al. 2019). *Heterobasidion annosum* s.l. has been observed abundantly in the boreal forest before the recent description of new species within the *H. annosum* complex (DesRochers and Allen 2014). The distribution of recently identified *Heterobasidion* species among tree species and countries of boreal forest is listed in Table 17.3. Moreover, *H. annosum* s.l. was recorded on various firs (*Abies* spp., *A. alba,* A. balsamea, A. nobilis), spruces (*Picea* spp., *P. fennica* (Regel) Kom., *P. glauca, P. jezoensis* (Siebold & Zucc.) Carrière, *P. pungens* Engelm., *P. sitchensis* (Bong.) Carrière, *P. obovata* Ledeb.) and pines (*Pinus* spp., *P. cembra* L., *P. mugo* Turra, *P. sibirica* Du Tour). Signs and symptoms may vary according to *Heterobasidion* species. Fruiting bodies often grow on the tree butt and roots in the soil organic layer or in the butt itself (Otrosina and Garbelotto 2010). Their color and pore size may vary according to age and species. They may be sessile, effuse-reflex, or resupinate (Tokuda et al. 2009). As they are often variable and inconspicuous, fruiting bodies might not be an adequate indicator of *Heterobasidion* species. Circular patches of dead or dying trees following thinning activities however are indicative of root rot caused by *Heterobasidion spp.* (Myren et al. 1994).

Heterobasidion spores infect fresh stumps left in plantations after thinning (Laflamme 2010). The fungus can survive in these stumps, in the soil humus, and in soil from reclaimed agricultural land for long periods of time (Rennerfelt 1952, 1953; Laflamme 2010). Depending on host and pathogen species, Annosus root rot can infect roots only or cause butt rot as well. Following root rot, trees tend to dieback and die (Myren et al. 1994). They are prone to windfall. Butt rot reduces wood quality and volume (Huse et al. 1994; Myren et al. 1994). As the decay proceeds along the roots of these stumps, it infects neighboring trees by root contacts and root grafts (Laflamme 2010). Wound prevention, stump protection with chemical (borate, urea) or biological (*Phlebiopsis gigantea* (Fr.) Jülich) control products, stump removal, alternative species plantation, and winter harvesting are efficient control methods against Annosus root rot (Bendz-Hellgren et al. 1998; Laflamme 2010).

In the boreal forest of northern Europe, *H. parviporum* mainly occurs on Norway spruce while *H. annosum* s.s. frequently attacks Scots pine (*P. sylvestris* L.). *H. parviporum* is more aggressive than *H. annosum* s.s. (DesRochers and Allen 2014). Area impacted by Annosus root rot in the European boreal forest is either unknown or unpublished. Volume losses vary from 5.8% to 60%, 3.8 m^3 to 10.1 m^3 ha^{-1} and can total up to 1 Mm^3 $year^{-1}$ depending on tree species and country. Financial losses vary from 2.9% to 13% and can amount up to €54 million $year^{-1}$ (Younitzky 1927; Kallio 1972; Tuimala 1979; Bendz-Hellgren et al. 1998; Fiodorov 1998; Mattila and Nuutinen 2007).

In Asia (Russia and Japan), *H. annosum* s.s. is uncommon while *H. annosum* s.l., *H. parviporum*, and *H. orientale* are found most frequently. The first two were reported more frequently from Siberia, Russia, where they cause heavy damage (Younitzky 1927; DesRochers and Allen 2014), mostly on Siberian fir (*A. sibirica* Ledeb.) and Scots pine. *H. parviporum* and *H. orientale* are reported only from Hokkaido, Japan, where they act as decay fungi or saprobe decomposers on deadwood (Tokuda et al. 2007; Tokuda et al. 2009; Tokuda et al. 2011). Sakhalin fir (*A. sachalinensis* (F. Schmidt) Mast.) and Sakhalin spruce (*P. glehnii* (F. Schmidt) Mast.) are the most common hosts of Annosus root rot in Northern Japan. There is no information available on the losses caused by these fungi in Asia.

In North America, *Heterobasidion* spp. are thought to have migrated from Mexico northward (Dalman et al. 2010). In Alaska, *H. annosum* is restricted to the temperate forest (Shaw III 1981, 1989). In western Canada, the most northern collection of *H. annosum* s.l. was from latitude 54.6° N in the temperate forest (Callan 2014). *H. annosum* s.l. is subdivided in two species, *H. occidentale* and *H. irregulare.* In eastern Canada, following the first observation of *H. irregulare* in 1955 (Hord and Quirke 1955), the disease has constantly moved northward and eastward (Jorgensen 1961; Sterner and Davidson 1981, 1982a; Moody 1988; Innes and Gagnon 1998; Scarr et al. 2009; Scarr et al. 2011) from the province of Ontario to the province of Quebec (Figure 17.2). It was first collected in the boreal forest in Quebec in 1989 (DesRochers and Callan 2014). It is a threat to white pine (*P. strobus*) and red pine (*P. resinosa*) plantations. Moreover, *H. irregulare* is now a threat to jack pine in the boreal forest of Canada as jack pine has been found to be susceptible to this disease

TABLE 17.3
Distribution of *Heterobasidion* Species among Boreal and Temperate Forest Tree Species and within Countries of the Boreal Forest

Heterobasidion spp.	Host Tree Species	Countries
H. annosum s.s.	*Abies sibirica* Ledeb.	Finland
	Betula pendula Roth	Finland
	Picea abies (L.) H. Karst.	Finland, Sweden
	Pinus sylvestris L.	Finland, Russia, Sweden
	Larix kaempferi (Lamb.) Carrière	Sweden
H. parviporum	*Abies sibirica* Ledeb.	Finland, Russia
	Abies sachalinensis (F. Schmidt) Mast.	Japan
	Abies sachalinensis ssp. *mayriana* Miyabe & Kudô) Silba	Japan
	Abies veitchii Lindl.	Japan
	Betula pendula Roth	Finland
	Juniperus communis L.	Finland
	Picea abies (L.) H. Karst.	Finland, Russia, Sweden Norway
	Picea glehnii (F. Schmidt) Mast.	Japan
	Picea jezoensis (Siebold & Zucc.) Carrière	Japan
	Pinus contorta Dougl. ex Loud. var. *latifolia* Engelm.	Finland
	Pinus sylvestris L.	Finland, Russia, Sweden
H. orientale	*Abies sachalinensis* (F. Schmidt) Mast.	Japan
	Cryptomeria japonica (Thunb. ex L. f.) D. Don	Japan
	Fraxinus mandshurica Rupr.	Japan
	Picea glehnii (F. Schmidt) Mast.	Japan
	Picea jezoensis (Siebold & Zucc.) Carrière	Japan
	Picea maximowiczii Regel ex Mast.	Japan
	Pinus nigra J.F. Arnold	Japan
H. irregulare	*Picea rubens* Sarg.	Canada
	Pinus resinosa Ait.	Canada
	Pinus strobus L.	Eastern Canada
	Pinus sylvestris L.	Canada
	Pinus ponderosa Douglas ex C. Lawson	Western Canada
	Prunus spp.	Canada
H. occidentale	*Tsuga heterophylla* (Raf.) Sarg.	Western Canada & US
	Pseudotsuga menziesii (Mirb.) Franco	
	Picea sitchensis (Bong.) Carrière	
	Thuja plicata Donn ex D. Don	
	Sequoia sempervirens (D. Don) Endl.	
	Larix occidentalis Nutt.	
	Alnus rubra Bong.	

(Jorgensen 1961). Most recently, *H. irregulare* was discovered on ponderosa pine (*P. ponderosa*) and planted eastern white pine (*P. strobus*) in British Columbia (BC) forests (Shamoun et al. 2019). The presence of *H. irregulare* in eastern Canada forests, and the new discovery of its occurrence in BC forests, is of high phytosanitary concern. The European and Mediterranean Plant Protection Organization (EPPO) has listed *H. irregulare* on its A2 list of pests recommended for regulation as quarantine pests in 2015. The development of conventional and real-time PCR DNA assays resulted in multiple advantages: (1) current distribution of *Heterobasidion* spp. in North America

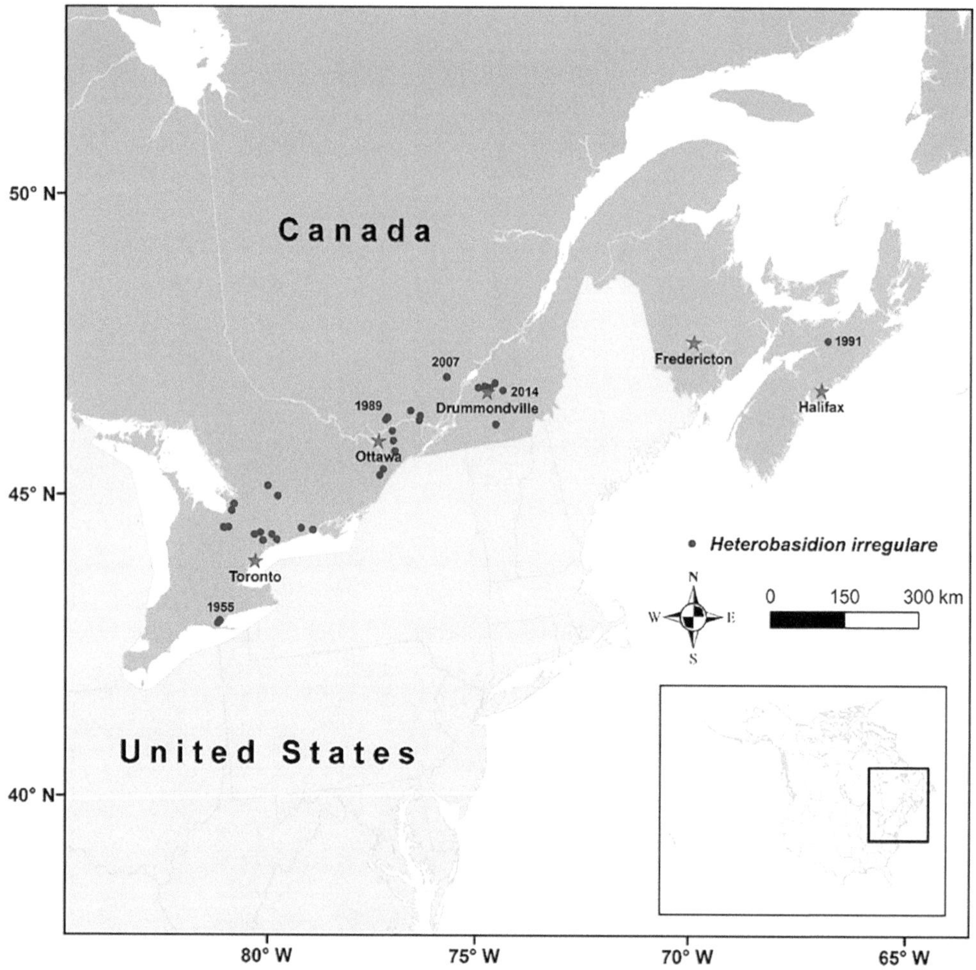

FIGURE 17.2 Northward and eastward propagation of *Heterobasidion irregulare* in eastern Canada.[1,2]

[1]Former names: *Fomes annosus* (Fr.) Karst. (Jorgensen 1956); *Heterobasidion annosum* (Fr.) Bref. (Laflamme and Blais 1995).

[2]Distribution data were collated from multiple sources including scientific publications (Jorgensen 1956; Punter and Cafley 1968; Laflamme and Blais 1995; Otrosina and Garbelotto 2010; Laflamme 2012), government reports (Sippell et al. 1968; MRNFQ 2016); Canadian Forest Service's Forest Insect and Disease Survey data, Global Biodiversity Information Facility (www.gbif.org; accessed on August 24, 2023), and MyCoPortal (www.mycoportal.org/portal/index.php, accessed on August 25, 2023).

could be mapped to species level; (2) wood for export can be tested specifically for presence of *Heterobasidion*; and (3) species-level diagnosis of *Heterobasidion* may help formulate better disease management strategies (Lamarche et al. 2016; Shamoun et al. 2019). Hence, volume losses may be expected in the future, although the extent of ecosystem-wide impacts is uncertain at this time, as they have not been measured yet. It is important to note that, although root rot may reduce the amount of timber available for extraction, it can contribute to diversify the structure of forests and to increase the amount of deadwood, a key feature of healthy forest ecosystems.

17.3.7.2 Disease Case Study: Ramorum Canker (Sudden Oak Death and Sudden Larch Death), Ramorum Blight, Ramorum (Shoot) Dieback

The causal agent of these non-native diseases is the oomycete *P. ramorum*. In Europe, the symptoms of this pathogen were first observed as twig dieback on rhododendrons (*Rhododendron* spp.)

in Germany and the Netherlands in 1993 (Werres et al. 2001). In North America, cankers and mortality were first reported on tanoak and coast live oak (*Q. agrifolia* Née) in coastal forests of California in 1995 (Garbelotto et al. 2001; Rizzo et al. 2005). The pathogen has a wide host range, more than 130 plant genera or species (APHIS 2013; Everhart et al. 2014). Infections have been found in the wild in California and Oregon in North America, and in Europe, damage was found mostly in limited areas in the UK until recently, but since 2009, an extensive outbreak has appeared in Japanese larch (*L. kaempferi*) plantations in the UK and in Ireland, where mortality frequently occurs (Frankel and Palmieri 2014). *Phytophthora ramorum* is likely to originate from Asia, from the laurisilva forests between Indochina and southwest Japan, and has been widely spread by the trade in ornamental plants (e.g., *Camellia* spp., *Magnolia* spp., *Pieris* spp., and *Rhododendron* spp.) (Xu et al. 2009; Goss et al. 2011; Jung et al. 2021).

17.3.7.2.1 Detection and Monitoring

Methods of detection of the pathogen include inspection, ground and aerial surveys, and stream monitoring. Suspicious material detected by these methods is sent to the laboratory to confirm the presence of the pathogen. Risk models based on several climatic and host parameters are also produced to map potential areas that are likely to favor disease development. The pathogen particularly thrives in nature on susceptible plants when mild temperatures (around 20°C) and saturated humid air (e.g., rain or fog) occur (Rizzo et al. 2005).

Visual inspection is difficult as early symptoms caused by *P. ramorum* can easily go undetected or be confused with symptoms caused by other agents. The pathogen may also be present within asymptomatic plants (Bienapfl et al. 2005; Denman et al. 2009; Redlin et al. 2014). Ground surveys are conducted in forests mainly to detect early signs of infection or simply to validate the data obtained by aerial surveys (Mai et al. 2006). Aerial surveys are used for mapping the disease or for detecting new focus of infection. For instance, in Oregon, the disease was first discovered on tanoak in the wild during an aerial survey conducted in July 2001 (Goheen et al. 2002). In California, the presence and abundance of bay laurel (*Umbellularia californica* (Hook. & Arn.) Nutt.), on which *P. ramorum* sporulates more abundantly, has also been identified as an indicator of the risk of infection on oak species as it drives the epidemic of sudden oak death disease (Kelly and Meentemeyer 2002; Davidson et al. 2005). Elsewhere in California and Oregon, the same role is attributed to tanoak, whereas in the UK the main hosts on which the pathogen sporulates are *Rhododendron ponticum* L. (common rhododendron) and Japanese larch, both non-native species in Britain (Brasier and Webber 2010).

Stream baiting using wounded rhododendron leaves enclosed in bait bags appears to be more efficient than ground surveys for early detection of *P. ramorum* and monitoring its distribution in forests (Oak et al. 2010). Using such baiting, the pathogen was first detected in several states outside of the known natural distribution of the disease on the west coast of the United States (Oak et al. 2010), but evidence from Oregon shows that new infestations in forests occur via dispersal of the pathogen through blowing fog and rain rather than from contaminated streams (Peterson et al. 2014).

The ultimate indicator of the presence of the pathogen is a laboratory confirmation by culturing the pathogen on selective media and/or by using molecular assays such as ELISA to screen for the presence of a *Phytophthora* species followed by tests with DNA markers specific for *P. ramorum* (Elliott et al. 2009; Kliejunas 2010; Feau et al. 2019).

Several risk models have been developed over the years to predict areas favorable for the establishment and spread of *P. ramorum*. For instance, using climatic parameters such as relative humidity and temperature and the susceptibility and sporulation potential of different plant species, Meentemeyer et al. (2004) have produced maps representing areas in California that are at risk to harbor *P. ramorum* and favor its dissemination. In Europe, a climate-based risk map using the system proposed by Meentemeyer et al. (2004) has been presented in the risk analysis carried out in 2009 (Sansford et al. 2009). Locations identified in the latter report that are at risk are mostly found in western coastal areas of countries such as the UK, Ireland, France, Portugal and Greece. It is important to include the trade in ornamental plants susceptible to *P. ramorum* in such risk models, as this is known to facilitate *P. ramorum* dispersal over long distances (Harwood et al. 2009).

17.3.7.2.2 Impacts

To date, *P. ramorum* has killed 30–45 million oak and tanoak trees in California and Oregon (Parke and Peterson 2019). In the UK and some other European countries, thousands of hectares of larch plantations, mainly Japanese larch (*L. kaempferi*) and European larch (*L. decidua*), have been lost to *P. ramorum* causing another phenomenon called sudden larch death (Brasier and Webber 2010). Increased fuel loads from dead trees killed by *P. ramorum* could have significantly boosted deadwood (although of a non-native tree species), had Japanese larch trees not been felled and removed in the UK. In California, however, tree mortality has a significant effect on fire severity, particularly in areas recently affected by the pathogen (Metz et al. 2011). In California's urban forest areas, costs associated with treatment, removal, and replacement of trees infected by *P. ramorum* and subsequent property value losses have been estimated at about $14 million/year over the decade 2010–2020 (Kovacs et al. 2011). Over a relatively small area in southwest Oregon, eradications efforts have cost $10 million from 2001 to 2012 (Kanaskie et al. 2013).

The disease is expected to influence insect, bird, or small mammal populations. It has been shown, for instance, that bird losses are predicted to be relatively high in certain areas in California, where *P. ramorum* is well established (Monahan and Koening 2006). The outbreaks of *P. ramorum* have many other ecological consequences, from decreasing forest carbon sequestration to shifts in species composition of forest assemblages (Lamsal et al. 2011; Metz et al. 2012; Wright and Dodd 2013). Yet, despite the high levels of tree and shrub mortality caused by this aggressive and generalist pathogen, can we conclude that *P. ramorum* has decreased the health of affected forest ecosystems in California and in the UK? In the latter case, *P. ramorum* has reduced the surface of plantations of non-native trees. Non-native trees such as Japanese larch are likely to host a limited number of native species in their canopy and roots. Were it not that *P. ramorum* also affects native species such as *Fagus sylvatica* L. (European beech), *Calluna vulgaris* (L.) Hull (Heather), and *Vaccinium myrtillus* L. (European blueberry), this pathogen could have been considered as a tool to control invasive non-native shrubs such as *R. ponticum*. On the West Coast of the United States, the main hosts on which the pathogen sporulates, California bay laurel and tanoak, are native species important for a range of other organisms, although they are considered as weeds by the forestry sector. However, also in this case, the impacts of *P. ramorum* go potentially beyond the levels of tree mortality experienced so far, because species such as Douglas-fir (*Pseudotsuga menziesii* (Mirb.) Franco) are also susceptible. However, at a time of rapid climate change, pathogens such as *P. ramorum* may ultimately at least in part help forest ecosystems increase their adaptability to further environmental shifts, by increasing mortality and thus also rejuvenation rates, thereby favoring selection for new conditions.

The nursery industry has also been seriously impacted in North America and in Europe by the presence of *P. ramorum*. For instance, in Canada, the pathogen has been found in a few nurseries in British Columbia, including those on mainland southern Vancouver and Vancouver Island. The impact and risk assessment of *P. ramorum* in Canada is based on policies in Canada and the United States for regulating and mitigating *P. ramorum*, which are quite similar between the two countries. *Phytophthora ramorum* has been under regulation in Canada and the United States since 2003. The Canadian Food Inspection Agency (CFIA) and the US Animal and Plant Health Inspection Service (USDA-APHIS) have implemented the Plant Protection Act and subsequently several Plant Protection Policy Directives (APHIS 2015; CFIA 2015). Nurseries in quarantine must be inspected annually for *P. ramorum*. In addition to annual inspections, pre-shipment inspections of plants are required before national movement. Eradication efforts are initiated if the pathogen is detected during the inspection process, which includes removal and destruction of the infected hosts, nursery stock, soil, wood packaging materials, and firewood. In British Columbia where *P. ramorum* is present only in nurseries, indirect economic losses due to domestic phytosanitary actions taken between 2003 and 2006 were estimated at $3 million for plant destruction and $500,000 for sanitation and disposal. In 2010, compensation payments to nurseries had reached $21.2 million (Canada Gazette 2010). Future consideration of indirect economic impact to trade should also be considered if *P. ramorum* ever became established in forested environments. For instance, although kiln drying may eliminate the pathogen in lumber, there could be potential impacts to log exports, valued in Canada at $580 million in 2012 (Shamoun et al. 2018).

17.3.7.2.3 Management

The best way of managing these diseases is to avoid their introduction. As diagnostic problems frequently occur and considering the growing globalization of trade with an increasing diversity of plant material, this is quite difficult to achieve. Moreover, introduction of *P. ramorum* to North America and Europe occurred when the pathogen was still unknown to science and society. Once established in a habitat, as it is the case for most non-native pathogens, it is really difficult to get rid of *P. ramorum*. Both in the UK, where *P. ramorum* is killing Japanese larch, and in California, eradication of the pathogen is no longer possible, and efforts are concentrated on containing or reducing the outbreak. Tree removal, early detection of the pathogen, and educational outreach seem to be the main management options to limit the outbreak. High density and/or homogenous populations of susceptible hosts, such as Japanese larch, tanoak, and California bay laurel, on which *P. ramorum* intensively sporulates, help propagate the disease in an uncontrollable way. For instance, in Oregon, despite attempts to eradicate the pathogen by cutting and burning all infected plants and potential hosts, and by using herbicides to prevent tanoak sprouting, the quarantine area has increased from 22 km^2 in 2001 to 505 km^2 in 2012 and then to 1,334 km^2 by 2019 (Kanaskie et al. 2013; COMTF 2020). Tree removal as a management option for *P. ramorum* appears not only ineffective but also counterproductive given that it removes deadwood from the affected forests (Pautasso 2013a). Deadwood is an important habitat for many organisms that has been strongly reduced by forest management (Cobb et al. 2012).

No chemical treatments are available to kill the pathogen in infected plants. Fungicides such as phosphite systemics can help prevent infection. These can be injected into the tree or applied directly to the bark (Garbelotto et al. 2007), but they are useful only at the tree level or over small areas such as urban forests.

17.4 TROPICAL FORESTS

Tropical forests in the world lie mostly in the area between the Tropic of Cancer (23.5°N latitude) and the Tropic of Capricorn (23.5°S latitude). These forests occupy 10% of the total land mass of the world above sea level and differ from temperate and boreal forests mainly by the distinct seasonality, 12 h day length, average temperature range of 20–25°C, continuous multilayered canopy allowing little light penetration, highly diverse flora with nearly 100 different species per square kilometer, rich faunal diversity, and rare occurrences of landscape or regional-scale forest pest outbreaks unlike boreal or temperate conifer forests. As it is out of the scope of this paper to describe all the major forest health issues related to insect infestations or disease incidences in the tropical forest worldwide, given the great diversity of tree species (relative to boreal and temperate forests) and the associated pest species, we restrict our observations to insect or pathogen issues impacting forest or plantation health in India. Geographically, India is under the Indo-Malayan Tropical region in which the forests are ancient in nature (Bhagwat et al. 2014). India is the tenth most forested nation in the world with ~79 million ha of forest and tree cover and approximately 24% of the country's geographical area is composed of forests (ISFR 2013). Forests in India are highly diverse and encompass several forest types such as evergreen tropical rainforests, dry alpine scrub types, semi-evergreen rainforests, deciduous monsoon forests, sub-tropical pine forests, and temperate montane forests (Lal 1989). In total, 16 major forest types occur (based on temperature and moisture conditions), which in turn are subdivided into 221 minor types (based on location-specific climatic and edaphic conditions) (Champion and Seth 1968). A list of important tree species in the natural forests and outside forests in India is presented in Tables 17.4 and 17.5. Together, these tree species form a total growing stock volume of

TABLE 17.4
Growing Stock in Forest for Top Ten Species in India[1]

Name of the Species	Total Volume[2] (%)	Total Stems[3] (%)
Shorea robusta Gaertn.	13.06	10.02
Tectona grandis L. f.	5.37	7.79
Pinus roxburghii Sarg.	3.40	2.02
Terminalia elliptica Willd.[4]	3.35	3.65
Picea smithiana (Wall.) Boiss.	3.24	0.30
Terminalia anogeissiana Gere & Boatwr.[5]	3.12	3.97
Pinus wallichiana (Wall. ex. D.Don) A.B.Jacks[6]	1.86	0.73
Lannea coromandelica (Houtt.) Merr.	1.84	2.46
Quercus semecarpifolia Sm.	1.81	0.64
Boswellia serrata Roxb.	1.71	1.28

Notes:

1 Source: India State of Forest Report (ISFR 2013).

2 Species-wise volume proportion as parameter estimated using local volume equations developed by Forest Survey of India (FSI).

3 Species-wise proportion as parameter of estimated growing stock calculated from available data on number of stems standing in field by FSI.

4 Synonym: *Terminalia crenulata* Roth.

5 Synonym: *Anogeissus latifolia* (Roxb. ex DC.) Wall. ex Guill. & Perr.

6 Synonym: *Pinus excelsa* Wall. ex D.Don.

TABLE 17.5
Growing Stock in Trees Outside Forest for Top Ten Species in India[1]

Name of the Species	Total Volume[2] (%)	Total Stems[3] (%)
Mangifera indica L.	11.41	8.25
Azadirachta indica A. Juss.	7.08	5.44
Cocos nucifera L.	6.08	6.69
Borassus flebellifer L.	4.95	2.16
Madhuca longifolia var. *latifolia* (Roxb.) A.Chev.	4.60	0.56
Pinus roxburghii Sarg.	2.65	2.71
Vachellia nilotica (L.) P.J.H.Hurter & Mabb.[4]	2.27	3.59
Bombax ceiba L.	2.23	0.90
Artocarpus heterophyllus Lam.	2.11	1.32
Eucalyptus spp.	2.09	2.63

Notes:

1 Source: India State of Forest Report (ISFR 2013).

2 Species-wise volume proportion as parameter estimated using local volume equations developed by Forest Survey of India (FSI).

3 Species-wise proportion as parameter of estimated growing stock calculated from available data on number of stems standing in field by FSI.

4 Synonym: *Acacia nilotica* (L.) Willd. ex Delile.

7,143 million m^3. Several species of trees occurring in India are also found in neighboring countries such as Pakistan, Bangladesh, Nepal, Burma, Malaysia, and Sri Lanka (Beeson 1941; Browne 1968).

We discuss important insects and diseases of trees in India in the following sections. We do not emphasize their likely ecological roles, for instance, in influencing tree diversity or in facilitating wildlife survival, because very limited published information exists on this topic. As noted by Gilbert (2005) regarding forest diseases, more work is needed to elucidate the potential ecological influences of damaging insects and pathogens in tropical forests in general.

17.4.1 An Overview of Forest Insects in India

Owing to an exponential rise in population pressure and resulting demands of forest products in a region with high dependency on forests for livelihood, it has long been a challenge to enhance productivity in terms of Minor Forest Produces (MFPs) and timber in the available areas. Insect pests are one of the most important biotic agents affecting forests in India. They have detrimental environmental impacts through destruction caused to the forest stands and economic impacts through losses to the seedlings in forest nurseries and end products such as timber and MFPs (Pathak et al. 1995; Thakur 2000; Kulkarni 2014). Insects attack trees at various life stages, such as during flowering or at the seed-bearing stage, as well as during storage (e.g., timber) (Beeson 1941; Pathak et al. 1995; FAO 2006, 2007). Unlike instances when epidemics occur, which may result in extensive product losses or cause tree mortality, small-scale or intermittent incidences of insect pests in large plantations or natural forests do not elicit any interventions. Nevertheless, young growing stock in forest nurseries is more prone to attack by insect pests that warrant management. A list of major insect species attacking forest or plantation trees in India is presented in Appendix 17A in Sambaraju et al. (2016). Case studies of important forest insects in India are presented here.

17.4.1.1 Insect Case Study: Sal Borer

Sal (*Shorea robusta* Gaertn.) is one of the most important natural forest tree species after teak (*Tectona grandis* L. f.), which is a plantation crop in India (Anon. 1972; Tewari 1995). It is a large deciduous tree, attaining 50 m in height and 4 m in girth with spreading crown. *Sal* occupies about 105,790 km^2 or approximately 10% of the total forest cover of India (ISFR 2013). It is mainly distributed in northern and central India with discrete pockets of trees in southern India. *Sal* forests have been classified by a few workers on the basis of annual rainfall and humidity (Troup 1921; Champion 1933; Champion and Seth 1968). Walter (1971) in his latest reclassification placed the *sal* bearing zones under semi-evergreen tropical forests. This study also indicates that *sal* forests are restricted to areas where annual rainfall ranges from 1,000–2,000 mm or above and where the dry period lasts less than five months.

The tree seeds yield a commercially important oleoresin, locally known as *dammer.* Thus, besides timber, fruits/seeds are one of the most important and valuable products having multiple socioeconomic importance, not only because they are eaten after roasting (in times of food scarcity) but also for their use in soap industry (Anon. 1972). They are mainly valued for yielding fatty oil and oil cake, which find use in various ways, viz., cooking and lighting by tribal populations inhabiting deep in forest, soap making after blending with other soft oils, in chocolate and confectionery after refining (Anon. 1972). Its principal use is in soap making. Owing to the economic importance, trade in *sal* seed has been nationalized (i.e., permission granted for its trade by the forest department) in the states of Madhya Pradesh, Chhattisgarh, and Bihar as an economic activity, which generates large-scale employment in rural areas.

Sal heartwood borer (*Hoplocerambyx spinicornis* Newman) (Coleoptera: Cerambycidae) is one of the most dreaded and tough-to-manage insect pests in this region. It has an annual life cycle. Newly hatched grubs of the pest enter the tree trunk during the rainy season after mating by the emerging beetles. They feed on the softwood and then the heartwood, making large tunnels inside the tree trunk, and emerge as adults during the next rainy season (June to August). The attacked tree succumbs to the injuries, dies, and has value merely as fuelwood. The dreaded borer has been causing havoc through reoccurring epidemics in *sal* forest areas throughout India (Joshi et al. 2006) in states such as Madhya Pradesh and Chhattisgarh influenced by various abiotic and biotic factors (Kulkarni et al. 2007) (see Appendix 17B in Sambaraju et al. 2016). An epidemic that occurred between 1996 and 2001 resulted in selective removal of damaged/attacked or dead *sal* trees to the tune of a million trees from the forest floor to maintain forest hygiene. A recent epidemic (2012–2015) in Madhya Pradesh and Chhattisgarh remained restricted to some pockets (Kulkarni 2017; Kulkarni and Chander 2022), possibly due to regular monitoring by the entomological team from Tropical Forest Research Institute in close collaboration with the concerned State Forest Departments. Attacked trees are marked and categorized into eight categories, viz., Category-I (completely dead) to Category-VIII (healthy tree) (in increasing order from dead to healthy), to inform decision-making regarding tree removal (Beeson 1941; Roonwal 1978). This categorization is mainly based on symptoms like top-dying effect and presence of frass extruded by the borers present inside the tree trunk. There is only one elaterid predator, *Cryptalaus sordidus* (Westwood) (Coleoptera: Elateridae) that is known and frequently collected from heavily infested logs (Joshi et al. 2006). However, its natural population is not sufficient to manage the pest insect and efforts to multiply this predator species have been unsuccessful to date.

After removing the badly affected trees, normally Category-I (completely dead) to III (half-crown dead), before the onset of rains, tree-trapping operations coinciding with the onset of rains are performed to trap adults emerging from rest of the categories of trees to prevent further oviposition on healthy trees. Tree-trapping is conducted using measured lengths of broken *sal* logs, which are beaten to loosen the bark and allow oozing out of cell sap during the evenings of initial months of rainy season (normally July–August) to attract beetles which are collected every morning before sunrise and killed during epidemic periods (Kulkarni 2014). Efforts have also been made to isolate and identify kairomonal compounds from the bark of the host trees for use as trap bait (Kulkarni et al. 2004), but that has not been very promising in the field and needs further research. However, investigations are underway for a deeper understanding of the field biology of the pest and to improvise the methods, so as to manage the pest population below threshold levels (Kulkarni and Chander 2022).

17.4.1.2 Insect Case Study: Teak Defoliator Complex

Teak (*T. grandis*) is native to south/southeast Asia, predominantly Burma, India, Indonesia, and Malaysia. In India, it is the second among the top ten growing stocks (ISFR 2013). It is a large broadleaf deciduous tree that can reach up to 40 m in height. It is found in a variety of habitats and climatic conditions, such as in areas where the annual precipitation varies from 500 mm to 5000 mm i.e., arid to moist areas. This tree species provides high-quality, durable timber, which is termite and pest resistant (Tewari 1992).

The complex of defoliators on one of the most valuable natural and plantation crop of tropical forests, teak, is mainly dominated by two oligophagous insect species: teak defoliator (*Hyblaea puera* (Cramer)) (Lepidoptera: Hyblaeidae) that consumes whole leaves leaving only the mid rib and teak skeletonizer (*Eutectona machaeralis* (Walker)) (Lepidoptera: Crambidae) that consumes only the mesophyll tissue leaving veins and midribs intact such that the leaves appear skeletonized. The incidence may reoccur every year in one or the other teak plantation areas and natural stands in India and the neighboring countries, where teak is grown at large. These two insect species, in succession or in overlapping generations, cause complete defoliation in teak that can affect the tree growth by up to 65% in plantations (Nair 2007) and up to 55% in forest nurseries (Joshi and Shukla 2001); however, defoliation-induced mortality is rare.

Management options for controlling defoliator populations include identifying resistant teak clones (Roychoudhury 2001) and using chemical insecticides and biocontrol agents (Joshi and Shukla 2001). Among the biological control options, Nuclear Polyhedrosis Viruses (NPV) have been isolated from the larvae of teak defoliator, *H. puera* (HpNPV) (Mohamed-Ali et al. 1992; Sudheendrakumar et al. 1988) and tested in field at Nilambur, Kerala, India. HpNPV is quick acting and causes significant mortality of larvae within three days (Nair et al. 1998). Foliar application of an HpNPV crude preparation at 10^5 Polyhedral Occlusion Bodies (POBs)/ml spray fluid in preliminary trials in Nilambur, Kerala, protected foliage by 70–76% and increased the basal area increment by 41% in protected trees (Nair et al. 1998). Out of seven types of formulations prepared, Sudheendrakumar et al. (2006) reported freeze-dried wettable powder formulation with activated charcoal as UV protectant to be the most efficient from the point of view of virulence, sunlight stability and field efficacy, and storage life of about 18 months with retention of a high percentage of original activity. This freeze-dried HpNPV formulation has been patented and is now a commercial product named Hybcheck®. This biopesticide is a new addition to the repertoire of measures undertaken in tropical forestry to reduce environmental hazards (Sajeev et al. 2007; Sudheendrakumar and Biji 2009). Ironically, HpNPV has not gained popularity across the country, possibly because of the difference in the environmental parameters affecting efficacy of the virus and impracticality of application in large plantations and natural forest areas in terms of process of spraying and economics.

Recently, Roychoudhury and Joshi (2009, 2010, 2011) and Roychoudhury (2012) have tested ivermectin against the larvae of *Ailanthus* webworm, *Atteva fabriciella* (Swederus) (Lepidoptera: Yponomeutidae), greater bamboo leaf roller, *Crypsiptya coclesalis* (Walker) (Lepidoptera: Crambidae), teak defoliator, *H. puera*, teak leaf skeletonizer, *E. machaeralis*, and *Albizia* foliage feeder, *Spirama retorta* (Clerck) (Lepidoptera: Erebidae). Ivermectin proved to be highly effective in inducing 100% larval mortality in the laboratory.

For handling larger, inapproachable areas and operational difficulties, classical biological control offers great promises. Collection and identification of native populations of potential insect parasitoids and evaluations are the first step toward classical biological control (Kulkarni et al. 2004; Yousuf et al. 2004). Sudheendrakumar et al. (1995) reported successful parasitization of *H. puera* eggs by *Trichogramma embryophagum* Hartig (Hymenoptera: Trichogrammatidae) and *T. dendrolimi* Matsumura under laboratory conditions. Further, Patil and Naik (1998) recorded higher parasitization of eggs of *H. puera* by *T. chilonis* Ishii under field conditions. At the Tropical Forest Research Institute, Jabalpur, India, field experiments were conducted using four non-native egg parasitoids, viz., *T. brasiliensis* (Ashmead), *T. chilonis*, *T. japonicum* Ashmead, and *T. pretiosum* Riley, to investigate the lowest highly effective quantity of wasps of the egg parasitoid species needed to minimize the attack by the teak skeletonizer. In this study, 25,000–300,000 parasitoid wasps/ha were released during the years 2001 and 2002 in teak forests (Joshi et al. 2007), but these non-native species did not prove promising. Experiments with native populations of *T. raoi* Nagaraja, showed that supplementing the egg parasitoid *T. raoi* alone in teak plantations or forests at 125,000 wasps/ha was sufficient to minimize at least half of the skeletonization incidences in released sites resulting in a reduction of half of the annual loss. Besides the egg parasitoid, *T. raoi*, investigations are ongoing to use native entomopathogenic nematode populations against the teak skeletonizers (Kulkarni, personal observation; Kulkarni et al. 2011) targeting young plantations and forest nurseries.

17.4.2 An Overview of Forest Diseases in India

Trees in the tropical forests are affected by diseases caused by biotic agents as well as environmental factors. Disease-causing pathogens affect every part of the tree, including the seeds, seedlings, saplings, leaves, branches, stems, roots, and even the whole tree. Major diseases affecting trees in natural forests and plantations in India, including the names of causative pathogens and the tree species affected, are presented in Appendix 17C in Sambaraju et al. (2016). Although comprehensive surveys of disease incidences and severities are not conducted in India and no centralized database of areas affected by forest diseases exists, foliar diseases, root rots, and heart rots generally comprise the major groups of diseases that impact forests and plantations in India.

Pathogenic foliage fungi destroy leaves in standing trees. For example, leaf spot caused by *Pseudocercospora anogeissi* U. Braun & Kamal in *Terminalia* spp. (syn. *Anogeissus* spp.; *Terminalia anogeissiana* Gere & Boatwr., known locally as *dhawda*, and *T. pendula* (Edgew.) Gere & Boatwr., known locally as *kardhai*) and *Rhytisma lagerstroemiae* Henn. tar spot in *Lagerstroemia* spp. (crape myrtle; known locally as *sendha*) are the most common foliage diseases in central India. Data on financial losses caused by these diseases, however, are not available. *Phomopsis tectonae* D.P. Tiwari, R.C. Rajak & Nikhra is a well-known pathogen causing dieback in teak. This pathogen was first reported as causing leaf spot in teak (Tiwari et al. 1981); later, this fungus was also found on twigs and branches. The pathogen currently attacks the main stem and causes top dying. Another fungal species belonging to the same genus, *P. variosporum* J.K. Sharma, C.N. Mohanan & Florence, also causes top dying of teak in southern India and in the state of Chhattisgarh in central India (Sharma et al. 1985; Verma et al. 2008). The occurrence of foliar disease outbreaks in various forest nurseries was studied by Pathak et al. (2015). Among the different types of diseases observed, rust (e.g., *Pedrocrousiella pongamiae* (Syd. & P. Syd.) Rajeshkumar, U. Braun & J.Z. Groenew., in Indian beech, *Pongamia pinnata* (L.) Pierre) and powdery mildew diseases (e.g., *Ovulariopsis sissoo* Shamsi, R. Sultana & Azad in Bombay blackwood, *Dalbergia latifolia* Roxb.) were found to be common in all nurseries in Indore, India. Another ethnomedicinally important plant called *Oxalis debilis* Kunth (large-flowered pink-sorrel) has been affected by foliar rust caused by *Puccinia oxalidis* Dietel & Ellis (Bansal et al. 2021). Recently, Negi et al. (2023) have reported a leaf spot disease of eucalyptus, *Eucalyptus* spp., caused by *Lasiodiplodia iraniensis* Abdollahz., Zare & A.J.L. Phillips in northern India.

Serious dieback-causing tree diseases have been reported in India. An economically important tree considered "Critically Endangered" under the International Union for Conservation of Nature (IUCN; www.iucnredlist.org/species/32056/2810130), the agarwood tree (*Aquilaria malaccensis* Lam.) is prized in the cosmetic industry and also used for medicinal purposes (Naziz et al. 2019). However, recently, the tree became highly susceptible to wilt and dieback disease caused by *Fusarium equiseti* (Corda) Sacc., in northeast India (Pandey et al. 2019). The affected tree loses its crown by drooping, wilting, and shedding of leaves, and severe dieback results in the complete death of the tree. Dieback of neem, *Azadirachta indica* A. Juss., caused by *Phomopsis azadirachtae* Sateesh, Shank. Bhat & Devaki, is a serious fungal disease in the southern states of India and affects all stages (leaves, twigs and the inflorescences) and ages of neem trees (Sateesh et al. 1997; Nagendra Prasad et al. 2010, 2011).

Bamboos (subfamily Bambusoideae; family Poaceae) are culturally and commercially important group of woody grasses in India comprising numerous species. Various pathogens infecting this group of plants have been reported. For example, Kainthola et al. (2021) studied fungi present in diseased culms (stems) and foliage of 61 Calcutta bamboo, *Dendrocalamus strictus* (Roxb.) Nees, provenances from 17 different states of India. Among the different phyla of fungi found, Ascomycota was the largest fungal group with 84.9% of total mycoflora infecting *D. strictus* followed by Basidiomycetes (8.2%) and Zygomycetes (5.5%) (Kainthola et al. 2021). Blight disease of another bamboo species, *Bambusa nutans* Wall. ex Munro, caused by *Sarocladium oryzae* (Sawada) W. Gams & D. Hawksw. is noteworthy. Heavy mortality in bamboo plantations has been reported in coastal areas in the state of Odisha in eastern India (Jamaluddin et al. 1992). This pathogen has been reported to cause a serious dieback of *B. balcooa* Roxb., *B. tulda* Roxb., and *B. vulgaris* Schrad. ex J.C. Wendl. in Bangladesh (Boa and Brady 1987; Rahman 1988).

Root rots are one of the most damaging diseases as they cause whole tree death. In nurseries, seedlings of several tree species such as neem, Indian rosewood (*Dalbergia sissoo* Roxb. ex DC.), and teak are affected by root rot or wilt pathogens generally caused by species belonging to the genera *Fusarium*, *Rhizoctonia*, and *Verticillium*. Under dry conditions in plantations, charcoal root rot caused by *Macrophomina phaseolina* (Tassi) Goid., affects several tree species such as *Acacia auriculiformis* A.Cunn. ex Benth., *Boswellia serrata* Roxb. Ex Colebr., *Eucalyptus* spp., and *Pinus caribaea* Morelet. *Ganoderma lucidum* (Curtis) P. Karst. causes root rot in young plantations, especially in leguminous trees such as cutch (*Senegalia catechu* (L.f.) P.J.H.Hurter & Mabb., syn. *Acacia catechu* (L.f) Willd.; locally known as *khair*) (Bakshi et al. 1976; Sankaran et al. 2005; Verma et al. 2013) and Indian rosewood (Harsh et al. 2011). This disease is a serious threat in 9–20-year-old plantations. In water-stressed nurseries and young plantations of teak, root rot is caused by *Helicobasidium longisporum* Wakef. (syn. *Helicobasidium compactum* (Boedijn) Boedijn) (Verma et al. 2013). This pathogen poses a potential threat to teak plantations as it has been reported recently from young plantations (Soni and Verma 2010). *Inonotus shoreae* (Wakef.) Ryvarden =*Polyporus shoreae* Wakef. causes root rot in dry and wet *sal* (*S. robusta*) forests in the states of Chhattisgarh and Madhya Pradesh in central India (Jamaluddin 1991). The affected trees can be recognized through epicormic branches with dark green leaves and signs of dying from the top as opposed to light green leaves and normal branching in healthy trees.

Heart rot fungi cause considerable wood damage in standing trees; however, heart rot-damaged trees may be beneficial to wildlife by providing nesting sites for birds that prefer to nest in tree cavities (Mudappa and Kannan 1997). Most heart rots are caused by fungi belonging to the genera *Hymenochaete* and *Phellinus* in the family Hymenochaetaceae and *Polyporus* and *Trametes* in the family Polyporaceae (Bakshi 1957; Bagchee 1959; Soni et al. 2010). *Phellinus badius* (Cooke) G. Cunn. and *P. rimosus* (Berk.) Pilát are wound parasites and infect cutch trees through injuries due to mechanical causes or damage due to animal feeding. The fungi attack heartwood and are responsible for hollowness in trees, whereas sapwood remains healthy. The infection can be confirmed from the development of fruiting bodies on the main stem at different heights (Verma et al. 2013). The damage caused by heart rot fungi cause the *khair* wood to become useless for the extraction of medicinally important (*kattha*) or industrially important (cutch) commercial products.

17.4.2.1 Disease Case Study: Root Rot of Cutch Tree in Central India

Cutch tree (*S. catechu*) is found throughout India, Sri Lanka, Pakistan, Southeast Asia, and southern China (Chakrabarty and Gangopadhyay 1996), up to an elevation of 1,500 m in the Himalayas and northeastern India, and up to 1,000 m in southern India. It is a small-to-medium sized deciduous tree 12–15 m in height, 60–90 cm in girth, and a clear bole of 2–3 m. The wood is hard to very hard, heavy with average weight of 1,010 kg/m^3 at 12% moisture content. Products derived from heartwood extracts of *S. catechu* (catechu, *kattha*, and cutch) are used as traditional medicine, as dyeing agents, and/or in the tanning industry. About 63,000 tons of wood is annually consumed for manufacture of cutch and catechu in several Indian states, such as Andhra Pradesh, Maharashtra, Gujarat, Madhya Pradesh, Chhattisgarh, Uttar Pradesh, and West Bengal. Root rot and heart rot are the main diseases that are responsible for serious losses in *S. catechu* cultivation. Root rot disease is described hereafter.

17.4.2.1.1 Pathogen Description

Ganoderma lucidum causes large-scale damage in young plantations through root rot (Bakshi et al. 1976). The affected plants exhibit pale foliage followed by drying. Young plants are killed soon after infection while mature trees die when most of the roots become affected. The fungus produces thin white mycelial mat between the bark and wood and causes white spongy rot in the sapwood. Fruit bodies or sporophores of this fungus commonly develop at the base of the affected plants or apparently on the ground, but in the latter case, fruit bodies are invariably attached to decaying root (Figure 17.3). The presence of sporophores is an indicator of infected plant. Sporophores are annual, usually stipitate (rarely sessile), corky, and later on woody. Stalk and upper surface are shiny laccate, smooth, first yellow but later turn red to almost black. Hymenial surface is white to cream when fresh, turning light brown with age, pore minute. Basidiospores truncate, light brown, 7–12 × 6–8μm. The spores are produced in abundance and are deposited on the adjoining weeds or grass as brownish red powdery mass.

17.4.2.1.2 Infection

Residual stumps and roots of infected trees are sources of infection (Figure 17.3). The fungus spreads from diseased to healthy plants by direct root contact. Therefore, planting cutch trees on sites with

FIGURE 17.3 *Ganoderma lucidum*, the causal agent of root rot of cutch (*khair*) tree in central India. Fruiting bodies develop on old stumps and roots and help propagate the disease over several years.

residual old stumps and roots containing mycelium increases the risk of infection. *Ganoderma lucidum* is normally endemic in natural forests where it does not cause any serious damage as some tree species may be resistant to the fungus while others may have varying degrees of susceptibility. In plantations, mortality starts from the third year and increases with age with up to 40% mortality in 7-year-old plantations (Bakshi 1976).

17.4.2.1.3 Status of Damage in Central India

Surveys were conducted by Verma and colleagues in central India (three forest *circles*, which are the forest areas demarcated by the forest department for easy operations and identification of sites, in Madhya Pradesh and two circles in Maharashtra) for recording mortality of cutch trees caused by *G. lucidum*. Out of the 21 sites visited by the study team, mortality was recorded in 17 sites, although extent of mortality greatly varied at different sites (Appendix 17D in Sambaraju et al. 2016). In Madhya Pradesh, the worst affected site was Palpur, which showed up to 93.3% mortality in some areas (Appendices 17D and 17E in Sambaraju et al. 2016) followed by Indore forest circle. In Maharashtra, one site in Yavatmal circle showed up to 60% mortality, however, in Thane circle, very low or no mortality was recorded. In 21 sites surveyed, on average, 30% mortality due to the disease was recorded (Verma et al. 2013).

17.4.2.1.4 Management

The disease can effectively be controlled by extraction of old stumps and cleaning of debris from the site, digging of isolation trenches in young plantations, planting of resistant species like *Bombax ceiba* L., *Bridelia retusa* (L.) A. Juss., and *Ailanthus excelsa* Roxb. and adopting mixed cropping (50:50) with resistant species to prevent rapid spread of the disease (Bakshi 1976).

17.5 SUMMARY

Forests are inherently complex and resilient ecosystems with multiple ecological hierarchies and feedback mechanisms. After humans, insects and pathogens are the major biotic disturbance agents in boreal, temperate, and tropical forests (Wohlgemuth et al. 2022). Native disturbance agents are essential components of a forest ecosystem. They play critical roles through their impacts on various ecological processes within a forest and thus help maintain system equilibrium. However, direct and indirect anthropogenic or abiotic influences could create conditions conducive for outbreaks of insects and pathogens that in turn could detrimentally impact forest ecosystem health. The concept, definition, and eventual classification of forest ecosystem health as "healthy" or "unhealthy" has been much debated and, in general, is subjective unlike human health. However, carefully selected normative ecological indicators may be helpful in assessing whether an ecosystem impacted by biotic disturbance agents is deviating from a normal or an expected state. Attempting to address this topic would require reliable data collection standards and determinations of system behavior (i.e., baselines and natural variabilities), which require significant time and resources. As broad-scale applicability of normative ecological indicators is hampered by the diversity of flora, fauna, topography, and climate in different regions within a forest type or among different forest types, more regional assessments and/or alternative assessment approaches are needed. Cross scale descriptive indicators of damage caused by insects and pathogens may be used as proxies for monitoring and assessing forest health, especially in cases where scientific, technical, and/or financial resource constraints exist. Several inter-governmental working groups have identified national-scale changes in the extent (or percentage) of area damaged by insects or diseases in forest ecosystems as a broad indicator of forest health. Visual indicators of insect or pathogen activity or occurrence of harmful, non-native pest species could serve as potential early warning indicators of significant later impacts on forest health.

Different approaches are needed to monitor and assess the overall health of forests impacted by problematic insects and diseases. A science-based understanding of the biology and ecology of insects and pathogens, the ecological roles that they play in forest ecosystems, and their interactions with other ecosystem components is imperative for the adoption of integrated management strategies that are ecologically sound and that pose minimal hazard to the environment. Such practices will improve the prospects of maintaining healthy and productive forest ecosystems worldwide in the decades and centuries to come.

17.6 ACKNOWLEDGMENTS

We would like to thank Sylvie Gauthier and Jean Bérubé (Canadian Forest Service-Laurentian Forestry Centre; CFS-LFC) and Ottmar Holdenrieder (Swiss Federal Institue of Technology) for their participation in helpful discussions leading up to the previous edition of this chapter. Our sincere appreciation to Chantal Côté (CFS-LFC) for collecting and summarizing data in Table 17.1 and for collating distribution data of *Heterobasidion irregulare*. We are deeply thankful to Guy R. Larocque (CFS-LFC) for inviting us to write this chapter.

REFERENCES

Adam, M.-C., and D. Kneeshaw. 2008. Local level criteria and indicator frameworks: A tool used to assess aboriginal forest ecosystem values. *For. Ecol. Manage.* 255:2024–2037.

Allen, E.A., and L.M. Humble. 2002. Nonindigenous species introductions: A threat to Canada's forests and forest economy. *Can. J. Plant Pathol.* 24:103–110.

Amman, G.D. 1982. Characteristics of mountain pine beetle reared in four pine hosts. *Environ. Entomol.* 11:590–593.

Amman, G.D., and W.E. Cole. 1983. Mountain pine beetle dynamics in lodgepole pine forests. Part II: population dynamics. *USDA For. Serv. Gen. Tech. Rep.* INT-145.

Amman, G.D., and L. Safranyik. 1985. Insects of lodgepole pine: Impacts and control. In *Lodgepole Pine, the Species and its Management*, eds. J.M. Baumgartner, R.G. Krebill, J.T. Arnott and G.F. Weetman, pp. 107–124. Pullman, WA: Washington State University, Cooperative Extension.

Anagnostakis, S.L. 1987. Chestnut blight: the classical problem of an introduced pathogen. *Mycologia.* 79:23–37.

Anon. 1972. *The wealth of India, raw materials (Vol. IX: Rh-So)*, 313–321. New Delhi, India: Publication and Information Directorate, Council of Scientific and Industrial Research.

Anon. 1995. *Criteria and indicators for the conservation and sustainable management of temperate and boreal forests.* The Montreal Process. Hull, Québec: Canadian Forest Service.

APHIS [Animal and Plant Health Inspection Service]. 2013. List of regulated hosts and plant proven or associated with *Phytophthora ramorum.* www.aphis.usda.gov/plant_health/plant_pest_info/pram/downloads/pdf_files/usdaprlist.pdf. Accessed October 3, 2023.

APHIS. 2015. *Phytophthora ramorum.* www.aphis.jungbinkellyusda.gov/plant-health/sod. Accessed July 26, 2023.

Armstrong, J.A., and W.G.H. Ives. 1995. *Forest insect pests in Canada.* Ottawa, Canada: Natural Resources Canada, Canadian Forest Service.

Asiegbu, F.O., A. Adomas, and J. Stenlid. 2005. Conifer root and butt rot caused by *Heterobasidion annosum* (Fr.) Bref. sl. *Mol. Plant Pathol.* 6:395–409.

Attiwill, P.M. 1994. The disturbance of forest ecosystems: the ecological basis for conservative management. *Forest Ecol. Manage.* 63:247–300.

Aukema, B.H., A.L. Carroll, J. Zhu, et al. 2006. Landscape level analysis of mountain pine beetle in British Columbia, Canada: Spatiotemporal development and spatial synchrony within the present outbreak. *Ecography* 29:427–441.

Ayres, M.P., and M.J. Lombardero. 2000. Assessing the consequences of global change for forest disturbance from herbivores and pathogens. *Sci. Total Environ.* 262:263–286.

Bagchee, K.D. 1959. The fungal diseases of sal (*Shorea robusta*) V. The heart rot of sal caused by *Trametes incerta. Indian Forest Record (NS). Mycology* 2:61–69.

Bakshi, B.K. 1957. Heart rots in relation to management of sal. *Indian For.* 83:651–661.

Bakshi, B.K. 1976. *Forest pathology: Principles and practices in forestry.* New Delhi, India: Controller of Publications.

Bakshi, B.K., M.A.R. Reddy, and S. Singh. 1976. Ganoderma root rot mortality in khair (*Acacia catechu* Willd.) in reforested stands. *Eur. J. For. Pathol.* 6:30–38.

Ballard, T.M. 2000. Impacts of forest management on northern forest soils. *For. Ecol. Manage.* 133:37–42.

Bansal, N., R. Bose, A. Pandey, et al. 2021. First report of rust caused by *Puccinia oxalidis* on *Oxalis debilis* in India. *J. Plant Pathol.* 104:453–453.

Bauce, E., and N. Carisey. 1996. Larval feeding behaviour affects the impact of flower production on the suitablity of balsam fir trees for spruce budworm. *Oecologia* 105:126–131.

Bedard, W.D. 1938. Control of mountain pine beetle by means of chemicals. *J. For.* 36:35–40.

Beeson, C.F.C. 1941. *The ecology and control of the forest insects of India and the neighbouring countries.* Dehradun, India: Vasant Press.

Bendz-Hellgren, M., K. Lipponen, H. Solheim, et al. 1998. Impact, control and management of *Heterobasidion annosum* root and butt rot in Europe and North America. The Nordic countries. In *Heterobasidion annosum: Biology, ecology, impact and control*, eds. S. Woodward, J. Stenlid, R. Karjalainen, and A. Hüttermann, pp. 333–345. Wallingford, UK: CAB International.

Bérard, J.A., and M. Côté. 1996. *Manuel de foresterie*. Québec, Canada: Université Laval Press.

Berg, E.E., J.D. Henry, C.L. Fastie, et al. 2006. Spruce beetle outbreaks on the Kenai Peninsula, Alaska, and Kluane National Park and Reserve, Yukon Territory: relationship to summer temperatures and regional differences in disturbance regimes. *For. Ecol. Manage.* 227:219–232.

Bergeron, Y., S. Gauthier, V. Kafka, et al. 2001. Natural fire frequency for the eastern Canadian boreal forest: consequences for sustainable forestry. *Can. J. For. Res.* 31:384–391.

Bergeron, Y., A. Luduc, H. Morin, et al. 1995. Balsam fir mortality following the last spruce budworm outbreak in northwestern Quebec. *Can. J. For. Res.* 25:1375–1384.

Berryman, A.A., B. Dennis, K.F. Raffa, et al. 1985. Evolution of optimal group attack with particular reference to bark beetles (Coleoptera: Scolytidae). *Ecology* 66:898–903.

Bhagwat, S.A., S. Nogué, and K.J. Willis. 2014. Cultural drivers of reforestation in tropical forest groves of the Western Ghats of India. *For. Ecol. Manage.* 329:393–400.

Bienapfl, J., J. Zanzot, S. Murphy, M. Garbelotto, et al. 2005. Isolation of a new lineage of *Phytophthora ramorum* from asymptomatic stems and roots of a commercial lot of rhododendron in California. *Phytopathology* 95:S9.

Blais, J.R. 1952. The relationship of the spruce budworm (*Choristoneura fumiferana*, Clem.) to the flowering condition of balsam fir (*Abies balsamea* (L.) Mill.). *Can. J. Zool.* 30:1–29.

Blais, J.R. 1958. Effects of 1956 spring and summer temperatures on spruce budworm populations (*Choristoneura fumiferana* Clem.) in the Gaspé Peninsula. *Can. Entomol.* 90:354–361.

Blais, J.R. 1981. Mortality of balsam fir and white spruce following a spruce budworm outbreak in the Ottawa River watershed of Quebec. *Can. J. For. Res.* 11:620–629.

Boa, E.R., and B.L. Brady. 1987. *Sarocladium oryzae* associated with blight of *Bambusa* species in Bangladesh. *Trans. Br. Mycol. Soc.* 89:161–166.

Boone, C.K., B.H. Aukema, J. Bohlman, et al. 2011. Efficacy of tree defense physiology varies with bark beetle population density: A basis for positive feedback in eruptive species. *Can. J. For. Res.* 41:1174–1188.

Borden, J.H. 1995. Development and use of semiochemicals against bark and timber beetles. In *Forest insects in Canada*, eds. J.A. Armstrong and W.G.H. Ives, pp. 431–449. Ottawa, Canada: Natural Resources Canada.

Bouchard, M., D. Kneeshaw, and Y. Bergeron. 2005. Mortality and stand renewal patterns following the last spruce budworm outbreak in mixed forests of western Quebec. *For. Ecol. Manage.* 204:297–313.

Bouchard, M., D. Kneeshaw, and Y. Bergeron. 2006. Forest dynamics after successive spruce budworm outbreaks in mixedwood forests. *Ecology* 87:2319–2329.

Bouchard, M., D. Kneeshaw, and C. Messier. 2007. Forest dynamics following spruce budworm outbreaks in the northern and southern mixedwoods of central Quebec. *Can. J. For. Res.* 37:763–772.

Boulanger, Y., and D. Arseneault. 2004. Spruce budworm outbreaks in eastern Quebec over the last 450 years. *Can. J. For. Res.* 34:1035–1043.

Boulanger Y., F. Fabry, A. Kilmabi, et al. 2017. The use of weather surveillance radar and high-resolution three dimensional weather data to monitor a spruce budworm mass exodus flight. *Agric. For. Meteorol.* 234–235:127–135.

Boulanger, Y., D.R. Gray, B.J. Cooke, et al. 2016. Model-specification uncertainty in future forest pest outbreak. *Global Change Biol.* 22:1595–1607.

Boulet, B., M. Chabot, L. Dorais, et al. 2009. Entomologie forestière. In *Ordre des ingénieurs forestiers du Québec, 'Manuel de foresterie'*, 2nd edn. Québec: Éditions MultiMondes.

Boyd, I.L., P.H. Freer-Smith, C.A. Gilligan, et al. 2013. The consequence of tree pests and diseases for ecosystem services. *Science* 342:1235773.

Brandt, J.P. 2009. The extent of the North American boreal zone. *Environ. Rev.* 17:101–161.

Brasier, C.M. 2001. Rapid evolution of introduced plant pathogens via interspecific hybridization hybridization is leading to rapid evolution of Dutch elm disease and other fungal plant pathogens. *Bioscience* 51:123–133.

Brasier, C., and J. Webber. 2010. Plant pathology: Sudden larch death. *Nature* 466:824–825.

Browne, F.G. 1968. *Pests and diseases of forest plantation trees: an annotated list of the principal species occurring in the British Commonwealth.* Oxford, UK: Clarendon Press.

Brunet, J., Y. Bukina, P.-O. Hedwall, et al. 2014. Pathogen induced disturbance and succession in temperate forests: Evidence from a 100-year data set in southern Sweden. *Basic Appl. Ecol.* 15:114–121.

Butler, L., V. Kondo, and D. Blue. 1997. Effects of tebufenozide (RH-5992) for gypsy moth (Lepidoptera: Lymantriidae) suppression on nontarget canopy arthropods. *Environ. Entomol.* 26:1009–1015.

Cadogan, B.L., A. Retnakaran, and J.H. Meating. 1997. Efficacy of RH5992, a new insect growth regulator against spruce budworm (Lepidoptera: Tortricidae) in a boreal forest. *J. Econ. Entomol.* 90:551–559.

Cadogan, B.L., and R.D. Scharbach. 1993. Efficacy of Foray 48B (*Bacillus thuringiensis* Berliner) applications against the spruce budworm, *Choristoneura fumiferana* (Clemens) (Lepidoptera: Tortricidae), timed for phenological development of balsam fir and black spruce. *Can. Entomol.* 125:479–488.

Cadogan, B.L., R.D. Scharbach, R.E. Krause, et al. 2002. Evaluation of tebufenozide carry-over and residual effects on spruce budworm (Lepidoptera: Tortricidae). *J. Econ. Entomol.* 95:578–586.

Cale, J.A., S.A. Teale, M.T. Johnston, et al. 2015. New ecological and physiological dimensions of beech bark disease development in aftermath forests. *For. Ecol. Manage.* 336:99–108.

Callan, B. 2014. *Pacific forestry centre's forest pathology herbarium—DAVFP collections database.* Natural Resources Canada, Canadian Forest Service, Pacific Forestry Centre, Victoria, British Columbia, Canada. http://cfs.nrcan.gc.ca/herbarium?lang=en_CA. Accessed October 4, 2023.

Canada Gazette. 2010. Part II, Statutory Instruments, SOR/2010–110 to 117 and SI/2010–42, Vol. 144, No. 12: 942–1015. www.registrelep-sararegistry.gc.ca/virtual_sara/files/orders/g2-14412_e.pdf. Accessed October 4, 2023.

Cappaert, D., D.G. McCullough, T.M. Poland, et al. 2005. Emerald ash borer in North America: a research and regulatory challenge. *Am. Entomol.* 51:152–165.

Cardé, R.T. 1990. Principles of mating disruption. In *Behavior-modifying chemicals for insect management: Applications of pheromones and other attractants*, eds. R.L. Ridgway, R.M. Silverstein, and M.N. Inscoe, pp. 47–71. New York, NY: Marcel Dekker, Inc.

Cardé, R.T. and T.C. Baker. 1984. Sexual communication with pheromones. In *Chemical ecology of insects*, eds. W.J. Bell and R.T. Cardé, 355–383. New York, NY: Springer.

Carisey, N. and E. Bauce. 1997. Intra-tree variation of balsam fir foliar chemistry and spruce budworm growth, development, food and nitrogen utilization. *J. Chem. Ecol.* 23:1963–1977.

Castello, J.D, D.J. Leopold, and P.J. Smallidge. 1995. Pathogens, patterns, and processes in forest ecosystems. *Bioscience* 45:16–24.

CCFM [Canadian Council of Forest Ministers]. 2006. *Criteria and indicators of sustainable forest management in Canada: National status 2005*, p. 154. Ottawa, Canada: Canadian Council of Forest Ministers.

CFIA [Canadian Food Inspection Agency]. 2015. Sudden oak death—*Phytophthora ramorum.* www.inspection.gc.ca/plants/plant-pests-invasive species/diseases/sudden-oak-death/eng/1327587864375/1327587972647. Accessed July 19, 2023.

CFS [Canadian Forest Service]. 2014. *Forest Insect and Disease Survey database.* New Brunswick, Canada: Natural Resources Canada, Atlantic Forestry Centre, Fredericton.

Chakrabarty, T. and M. Gangopadhyay. 1996. The genus *Acacia* P. Miller (Leguminosae: Mimosoideae) in India. *J. Econ. Tax. Bot.* 20:599–633.

Champion, H.G. 1933. *Regeneration and management of Sal* (*Shorea robusta*). Indian Forest Records (Old Series). Delhi, India: Manager of Publications.

Champion, H.G., and S.K. Seth. 1968. *A revised survey of the forest types of India.* Delhi, India: Manager of Publications.

Christiansen, E., and A. Bakke. 1988. The spruce bark beetle of Eurasia. In *Dynamics of forest insect populations*, ed. A.A. Berryman. New York, NY: Springer.

Clark, J.A., and K.R. Covey. 2012. Tree species richness and the logging of natural forests: A meta-analysis. *For. Ecol. Manage.* 276:146–153.

Cobb, R.C., M.N. Chan, R.K. Meentemeyer, et al. 2012. Common factors drive disease and coarse woody debris dynamics in forests impacted by sudden oak death. *Ecosystems* 15:242–255.

COMTF [California Oak Mortality Task Force]. 2020. *Monthly (February) reports to the California Oak Mortality Task Force.* www.suddenoakdeath.org/wp-content/uploads/2020/02/COMTF-Report-February.pdf. Accessed December 13, 2023.

Costanza, R. 1992. Toward an operational definition of ecosystem health. In *Ecosystem health: New goals for environmental management*, eds. R. Constanza, B.G. Norton, and B.D. Haskell, pp. 239–256. Washington, DC, USA: Island Press.

Costanza, R., and M. Mageau. 1999. What is a healthy ecosystem? *Aquat. Ecol.* 33:105–115.

Dale, V.H., and S.C. Beyeler. 2001. Challenges in the development and use of ecological indicators. *Ecol. Indic.* 1:3–10.

Dale, V.H., L.A. Joyce, S. McNulty, et al. 2001. Climate change and forest disturbances. *Bioscience* 51:723–734.

Dallaire, R., A. Labrecque, M. Marcotte, et al. 2004. The sublethal effects of tebufenozide on the precopulatory and copulatory activities of *Choristoneura fumiferana* and *C. rosaceana. Entomol. Exp. Appl.* 112:169–181.

Dalman, K., Å. Olson, and J. Stenlid. 2010. Evolutionary history of the conifer root rot fungus *Heterobasidion annosum* sensu lato. *Mol. Ecol.* 19:4979–4993.

Davidson, J.M., A.C. Wickland, H.A. Patterson, et al. 2005. Transmission of *Phytophthora ramorum* in mixed-evergreen forest in California. *Phytopathology* 95:587–596.

Denman, S., S.A. Kirk, E. Moralejo, et al. 2009. *Phytophthora ramorum* and *Phytophthora kernoviae* on naturally infected asymptomatic foliage. *EPPO Bulletin* 39:105–111.

Desprez-Loustau, M., B. Marçais, L. Nageleisen, et al. 2006. Interactive effects of drought and pathogens in forest trees. *Ann. For. Sci.* 63:597–612.

DesRochers, P., and E. Allen. 2014. *Pests on Canadian trees planted abroad database.* Natural Resources Canada, unpublished.

DesRochers, P., and B. Callan. 2014. *Fungal culture database of the Canadian Forest Service.* Natural Resources Canada, unpublished.

DesRochers, P., D. Lachance, M. Lebon, et al. 2006. *Herbier René-Pomerleau du Service canadien des forêts.* Ressources Naturelles Canada. https://apps-scf-cfs.rncan.gc.ca/collections/fr/herbier. Accessed October 6, 2023.

Dhadialla, T.S., G.R. Carlson, and D.P. Le. 1998. New insecticides with ecdysteroidal and juvenile hormone activity. *Annu. Rev. Entomol.*43:545–569.

Dhar, A., L. Parrott, and C.D.B. Hawkins. 2016. Aftermath of mountain pine beetle outbreak in British Columbia: Stand dynamics, management response and ecosystem resilience. *Forests* 7:1–19.

Dietze, M.C., and J.H. Matthes. 2014. A general ecophysiological framework for modelling the impact of pests and pathogens on forest ecosystems. *Ecol. Lett.* 17:1418–1426.

Döring, T.F., M. Pautasso, M.R. Finckh, et al. 2012. Concepts of plant health–reviewing and challenging the foundations of plant protection. *Plant Pathol.* 61:1–15.

Döring, T.F., A. Vieweger, M. Pautasso, et al. 2015. Resilience as a universal criterion of health. *J. Sci. Food Agric.* 95:455–465.

Doucet, D., C. Frisco, M. Cusson, et al. 2007. Diapause disruption with tebufenozide for early-instar control of the spruce budworm, *Choristoneura fumiferana. Pest Manag. Sci.* 63:730–736.

Drogvalenko, A.N., M.J. Orlova-Bienkowskaja, and A.O. Bieńkowski. 2019. Record of the emerald ash borer (*Agrilus planipennis*) in Ukraine is confirmed. *Insects*. 10:338.

Dupont, A., L. Belanger, and J. Bousquet. 1991. Relationships between balsam fir vulnerability to spruce budworm and ecological site conditions of fir stands in central Québec. *Can. J. For. Res.* 21:1752–1759.

Elliott, M., G. Sumampong, A. Varga, et al. 2009. PCR-RFLP markers identify three lineages of the North American and European populations of *Phytophthora ramorum. For. Pathol.* 39:266–278.

Ellis, C.J., S. Eaton, M. Theodoropoulos, et al. 2014. Response of epiphytic lichens to 21st Century climate change and tree disease scenarios. *Biol. Conserv.* 180:153–164.

Erdle, T.A., and D.A. MacLean. 1999. Stand growth model calibration for use in forest pest impact assessment. *For. Chron.* 75:141–152.

Everhart, S.E., J.F. Tabima, and N.J. Grünwald. 2014. *Phytophthora ramorum.* In *Genomics of plant-associated fungi and oomycetes: Dicot pathogens*, eds. R.A. Dean, A. Lichens-Park and C. Kole, pp. 159–174: Berlin, Germany: Springer.

FAO [Food and Agriculture Organization]. 2006. *Global Forest Resources Assessment 2005—progress towards sustainable forest management.* FAO Forestry Paper 147. FAO, Rome, Italy.

FAO. 2007. *Overview of the forest pests: India.* Forest Health & Biosecurity Working Papers, Working Paper FBS/18E, Forest Management Division. FAO, Rome, Italy. www.fao.org/docrep/012/al007e/al007e00.pdf. Accessed October 6, 2023.

FAO. 2010. *Criteria and indicators for sustainable forest management: A compendium.* Paper compiled by Froylan Castaneda, Christel Palmberg-Lerche and Petteri Vuorinen, May 2001. Forest Management Working Papers, Working Paper 5. Forest Resources Development Service, Forest Resources Division. FAO, Rome, Italy (unpublished). www.fao.org/docrep/004/ac135e/ac135e00.htm. Accessed August 24, 2023.

FAO. 2020. *Global forest resources assessment 2020: Main report.* FAO Forestry Paper 163. FAO, Rome, Italy. https://doi.org/10.4060/ca9825en. Accessed September 12, 2023.

Feau, N., D. Ojeda, S. Beauseigle, et al. 2019. Improved detection and identification of the sudden oak death pathogen *Phytophthora ramorum* and the Port Orford cedar root pathogen *Phytophthora lateralis. Plant Pathol.* 68:878–888.

Fei, S., R.S. Morin, C.M. Oswalt, et al. 2019. Biomass losses resulting from insect and disease invasions in US forests. *Proc. Natl. Acad. Sci. U.S.A.* 116:17371–17376.

Felton, A., J. Boberg, C. Björkman, et al. 2013. Identifying and managing the ecological risks of using introduced tree species in Sweden's production forestry. *For. Ecol. Manage.* 307:165–177.

Fettig, C.J., K.D. Klepzig, R.F. Billings, et al. 2007. The effectiveness of vegetation management practices for prevention and control of bark beetle infestations in coniferous forests of the western and southern United States. *For. Ecol. Manage.* 238:24–53.

Fierke, M., D. Nowak, and R. Hofstetter. 2011. Seeing the forest for the trees: forest health monitoring. In *Forest health: An integrated perspective*, eds. J.D. Castello and S.A. Teale, pp. 321–343. Cambridge, UK: Cambridge University Press.

Fiodorov, N. 1998. Eastern Europe and Baltic countries. In *Heterobasidion annosum: biology, ecology, impact and control*, eds. S. Woodward, J. Stenlid, R. Karjalainen and A. Hüttermann, pp. 387–403. Oxford, UK: CAB International.

Frankel, S.J. and K.M. Palmieri. 2014. Sudden Oak Death, *Phytophthora ramorum*: A persistent threat to oaks and other tree species. *International Oaks* 25:43–56.

Franklin, J.F., H.H. Shugart, and M.E. Harmon. 1987. Tree death as an ecological process. *Bioscience* 37:550–556.

Freer-Smith, P.H., J.F. Webber. 2017. Tree pests and diseases: the threat to biodiversity and the delivery of ecosystem services. *Biodivers. Conserv.* 26:3167–3181.

Frego, K.A. 2007. Bryophytes as potential indicators of forest integrity. *For. Ecol. Manage.* 242:65–75.

Fuentealba, A., and E. Bauce. 2012. Site factors and management influence short-term host resistance to spruce budworm, *Choristoneura fumiferana* (Clem.), in a species-specific manner. *Pest Manage. Sci.* 68:245–253.

Gagnon, G., and M. Chabot. 1988. Un système d'évaluation de la vulnérabilité des peuplements à la tordeuse des bourgeons de l'épinette: ses fondements, son implantation et son utilisation en aménagement forestier. *L'Aubelle.* 67:7–14.

Gandhi, K.J.K., and D.A. Herms. 2010. Direct and indirect effects of alien insect herbivores on ecological processes and interactions in forests of eastern North America. *Biol. Invasions.* 12:389–405.

Garbelotto, M., and M. Pautasso. 2012. Impacts of exotic forest pathogens on Mediterranean ecosystems: Four case studies. *Eur. J. Plant Pathol.* 133:101–116.

Garbelotto, M., D.J. Schmidt, and T.Y. Harnik. 2007. Phosphite injections and bark application of Phosphite+ Pentrabark™ control sudden oak death in coast live oak. *Arboric. Urban For.* 33:309–317.

Garbelotto, M., P. Svihra, and D.M. Rizzo. 2001. Sudden Oak Death syndrome fells three oak species. *Calif. Agric.* 55:9–19.

Garcia M., B.R. Sturtevant, R. Saint-Amant, et al. 2022. Modeling weather-driven long-distance dispersal of spruce budworm moths (*Choristoneura fumiferana*). Part 1: Model description. *Agric. For. Meteorol.* 315:108815.

Gauthier, S., P. Bernier, T. Kuuluvainen, et al. 2015. Boreal forest health and global change. *Science* 349:819–822.

Gilbert, G.S. 2005. Dimensions of plant disease in tropical forests. In *Biotic interactions in the tropics: their role in the maintenance of species diversity*, eds. D. Burslem, M. Pinard, and S. Hartley, pp. 141–164. Cambridge, UK: Cambridge University Press.

Gill, S.S. 1995. Mechanism of action of *Bacillus thuringiensis* toxins. *Memórias do Instituto Oswaldo Cruz* 90:69–74.

Gillette, N.E., E.M. Hansen, C.J. Mehmel, et al. 2012a. Area-wide application of verbenone-releasing flakes reduces mortality of whitebark pine *Pinus albicaulis* caused by the mountain pine beetle *Dendroctonus ponderosae*. *Agric. For. Entomol.* 14:367–375.

Gillette, N.E., C.J. Mehmel, S.R. Mori, et al. 2012b. The push–pull tactic for mitigation of mountain pine beetle (Coleoptera: Curculionidae) damage in lodgepole and whitebark pines. *Environ. Entomol.* 41:1575–1586.

Giroday, H-M.C. de la, A.L. Carroll, and B.H. Aukema. 2012. Breach of the northern Rocky Mountain geoclimatic barrier: Initiation of range expansion by the mountain pine beetle. *J. Biogeogr.* 39:1112–1123.

Glaus, V., A. Nisole, S. Edwards, et al. 2023. Non-target impacts of insecticide-based population control of eastern spruce budworm (Lepidoptera: Tortricidae) on non-target caterpillar communities and parasitism. *Can. Entomol.* 155:e8.

Goheen, E.M., E.M. Hansen, A. Kanaskie, et al. 2002. Sudden oak death caused by *Phytophthora ramorum* in Oregon. *Plant Dis.* 86:441–441.

Gonthier, P. 2013. Introduction to infectious forest diseases. In *Infectious forest diseases*, ed. P. Gonthier. Wallingford, UK: CAB International.

Goss, E.M., M. Larsen, A. Vercauteren, et al. 2011. *Phytophthora ramorum* in Canada: evidence for migration within North America and from Europe. *Phytopathology* 101:166–171.

Greenbank, D.O. 1963. The development of the outbreak.The dynámics of epidemic spruce budworm populations. *Entomol. Soc. Can. Mem.* 31:19–23.

Gross, A., O. Holdenrieder, M. Pautasso, et al. 2014. *Hymenoscyphus pseudoalbidus*, the causal agent of European ash dieback. *Mol. Plant Pathol.* 15:5–21.

Grünwald, N.J., M. Garbelotto, E.M. Goss, et al. 2012. Emergence of the sudden oak death pathogen *Phytophthora ramorum*. *Trends Microbiol.* 20:131–138.

Haack, R.A. 2006. Exotic bark-and wood-boring Coleoptera in the United States: recent establishments and interceptions. *Can. J. For. Res.* 36:269–288.

Hall, J.P. 2001. Criteria and indicators of sustainable forest management. *Environ. Monit. Assess.* 67:109–119.

Hall, J.P., and B.H. Moody. 1994. *Forest depletions caused by insects and diseases in Canada 1982–1987.* Information Report ST-X-8, 14 p. Ottawa, Canada: Natural Resources Canada, Canadian Forest Service, Headquarters, Science and Sustainable Development Directorate.

Hansen, E.M. 1999. Disease and diversity in forest ecosystems. *Australas. Plant Pathol.* 28:313–319.

Hansen, E.M., A. Kanaskie, S. Prospero, et al. 2008. Epidemiology of *Phytophthora ramorum* in Oregon tanoak forests. *Can. J. For. Res.* 38:1133–1143.

Hantsch, L., S. Bien, S. Radatz, et al. 2014. Tree diversity and the role of non-host neighbour tree species in reducing fungal pathogen infestation. *J. Ecol.* 102:1673–1687.

Harsh, N.S.K., S. Chandra, and K. Uniyal. 2011. Screening resistance of *Dalbergia sissoo* clones against *Ganoderma lucidum* root rot disease in field conditions. *For. Pathol.* 41:221–226.

Harwood, T.D., X. Xu, M.Pautasso, et al. 2009. Epidemiological risk assessment using linked network and grid based modelling: *Phytophthora ramorum* and *Phytophthora kernoviae* in the UK. *Ecol. Modell.* 220:3353–3361.

Hennigar, C.R., D.A. MacLean, D.T. Quiring, et al. 2008. Differences in spruce budworm defoliation among balsam fir and white, red, and black spruce. *For. Sci.* 54:158–166.

Hennigar, C.R., J.S. Wilson, D.A. MacLean, et al. 2011. Applying a spruce budworm decision support system to Maine: Projecting spruce-fir volume impacts under alternative management and outbreak scenarios. *J. For.* 109:332–342.

Henry, J.D., and M. Viney. 2002. *Canada's boreal forest.* Washington, DC, USA: Smithsonian Institution Press.

Herms, D.A., and D.G. McCullough. 2014. Emerald ash borer invasion of North America: history, biology, ecology, impacts, and management. *Annu. Rev. Entomol.* 59:13–30.

Hicke, J.A., C.D. Allen, A.R. Desai, et al. 2012. Effects of biotic disturbances on forest carbon cycling in the United States and Canada. *Global Change Biol.* 18:7–34.

Hill, L., E. Ashby, N. Waipara, R. Taua-Gordon, et al. 2021. Cross-cultural leadership enables collaborative approaches to management of Kauri dieback in Aotearoa New Zealand. *Forests* 12:1671.

Hill, R.J., B.A. Wilson, J.E. Rookes, et al. 2009. Use of high resolution digital multi-spectral imagery to assess the distribution of disease caused by *Phytophthora cinnamomi* on heathland at Anglesea, Victoria. *Australas. Plant Pathol.* 38:110–119.

Holzmueller, E.J., S. Jose, and M.A. Jenkins. 2010. Ecological consequences of an exotic fungal disease in eastern US hardwood forests. *For. Ecol. Manage.* 259:1347–1353.

Hord, H.H.V., and D.A. Quirke. 1955. Province of Ontario: Forest disease survey. In *Annual report of the forest insect and disease survey 1954*, pp. 56–69. Ottawa, Ontario: Department of Agriculture.

Huse, K.J., H. Solheim, and K. Venn. 1994. Stump inventory of root and butt rots in Norway spruce cut in 1992. *Rapport fra Skogforsk.* 23:26 pp.

Innes, L., and G. Gagnon. 1998. La maladie du rond. In *Direction de la conservation des forêts. 1998. Insectes et maladies des arbres Québec 1997*, pp. A1–A4. Québec: Ministère des Ressources naturelles.

ISFR [India State of Forest Report]. 2013. *India state of forest report 2013*, p. 252. Dehradun, India: Forest Survey of India, Ministry of Forest, Environment and Climate Change, Government of India.

Jamaluddin. 1991. Status of *Polyporus shoreae* causing root-rot in dry and wet sal forest. *J. Trop. For.* 7:342–344.

Jamaluddin, B.N. Gupta, S.C. Bohidar, et al. 1992. Mortality of bamboo (*Bambusa nutans* Wall.) in the coastal areas of Orissa. *J. Trop. For.* 8:252–261.

Johns, R.C., J.J. Bowden, R.D. Carleton, et al. 2019. A conceptual framework for the spruce budworm Early Intervention Strategy: can outbreaks be stopped? *Forests* 10:910.

Jorgensen, E. 1956. *Fomes annosus* (Fr.) Cke. on red pine in Ontario. *For. Chron.* 32:87–88.

Jorgensen, E. 1961. On the spread of *Fomes annosus* (Fr.) Cke. *Can. J. Bot.* 39:1437–1445.

Joshi, K.C., N. Roychoudhury, N. Kulkarni, et al. 2006. Sal heartwood borer in Madhya Pradesh. *Indian For.* 132:799–808.

Joshi, K.C., S. Sambath, M. Yousuf, et al. 2007. Evaluation of *Trichogramma* spp. to minimise the attack of teak leaf skeletonizer. *Indian For.* 133:527–533.

Joshi, K.C., and P.K. Shukla. 2001. *Recent trends in insect pest control to enhance forest productivity*. Jabalpur, India: Tropical Forest Research Institute.

Jung, T., M.H. Jung, J.F. Webber, et al. 2021. The destructive tree pathogen *Phytophthora ramorum* originates from the Laurosilva forests of East Asia. *J. Fungi* 7:226.

Juzwik, J., T.C. Harrington, W.L. MacDonald, et al. 2008. The origin of *Ceratocystis fagacearum*, the oak wilt fungus. *Annu. Rev. Phytopathol.* 46:13–26.

Kainthola, C., N. Harsh, A. Pandey, et al. 2021. New fungal records on *Dendrocalamus strictus* from India. *J. Bamboo Ratt.* 20:15–29.

Kaitera, J., R. Hiltunen, and J. Hantula. 2015. *Cronartium* rust sporulation on hemiparasitic plants. *Plant Pathol.* 64:738–747.

Kallio, T. 1972. An example of the economic loss caused by decay in growing spruce timber in S. Finland. *Silva Fenn.* 6:116–124.

Kamei, S., and S. Hoshi. 1948. On the brown root and butt rot of conifers in the national forest of Akan, Hokkaido. *Res. Bull. Coll. Exp. For. Hokkaido Univ.* 14:144–176.

Kanaskie, A., E.M. Hansen, E.M. Goheen, et al. 2013. Detection and eradication of *Phytophthora ramorum* from Oregon forests, 2001–2011. In *Proceedings of the sudden oak death fifth science symposium*, Tech. Coords. S.J. Frankel, J.T. Kliejunas, K.M. Palmieri, and J.M. Alexander, Gen. Tech. Rep. PSW-GTR-243, pp. 3–13. Albany, CA: US Department of Agriculture, Forest Service, Pacific Southwest Research Station.

Kanaskie, A., M. McWilliams, K. Sprengel, et al. 2005. *Swiss needle cast aerial surveys, 1996 to 2005*. Swiss Needle Cast Cooperative 2005 annual report, pp. 9–11. Corvallis, OR: College of Forestry, Oregon State University.

Kashian, D.M., R.M. Jackson, and H.D. Lyons. 2011. Forest structure altered by mountain pine beetle outbreaks affects subsequent attack in a Wyoming lodgepole pine forest, USA. *Can. J. For. Res.* 41:2403–2412.

Kelly, M., and R.K. Meentemeyer. 2002. Landscape dynamics of the spread of sudden oak death. *Photogramm. Eng. Remote Sensing* 68:1001–1010.

Kizlinski, M.L., D.A. Orwig, R.C. Cobb, and D.R. Foster. 2002. Direct and indirect ecosystem consequences of an invasive pest on forests dominated by eastern hemlock. *J. Biogeogr.* 29:1489–1503.

Kliejunas, J.T. 2010. Sudden oak death and *Phytophthora ramorum*: a summary of the literature, Gen. Tech. Rep. PSW-GTR-234, p. 181. USDA Forest Service, Pacific Southwest Research Station, Albany, CA. www.suddenoakdeath.org/wp-content/uploads/2010/03/psw_gtr234.pdf. Accessed October 6, 2023.

Klutsch, J.G., M.A. Battaglia, D.R. West, S.L. Costello, and J.F. Negrón. 2011. Evaluating potential fire behavior in lodgepole pine-dominated forests after a mountain pine beetle epidemic in north-central Colorado. *West. J. Appl. For.* 26:101–109.

Kneeshaw, D.D., and Y. Bergeron. 1998. Canopy gap characteristics and tree replacement in the southeastern boreal forest. *Ecology* 79:783–794.

Kolb, T.E., M.R. Wagner, and W.W. Covington. 1994. Utilitarian and ecosystem perspectives: Concepts of forest health. *J. For.* 92:10–15.

Kovacs, K., T. Václavík, R.G. Haight, et al. 2011. Predicting the economic costs and property value losses attributed to sudden oak death damage in California (2010–2020). *J. Environ. Manage.* 92:1292–1302.

Kuennecke, B.H. 2008. *Temperate forest biomes. Greenwood guides to biomes of the world.* Westport, USA: Greenwood Press.

Kulkarni, N. 2014. Status of potential biocontrol components for integrated management of forest insect pests in India. In *Biopesticides in sustainable agriculture: Progress and potential*, eds. O. Koul, G.S. Dhaliwal, S. Khokhar and R. Singh, pp. 389–419. New Delhi, India: Science Publishers.

Kulkarni, N. 2017. Integrated insect pest management in tropical forestry. In *Integrated pest management in tropical region*, eds. C. Rapisarda and G.E.M. Cocuzza, pp. 313–342. Oxfordshire, UK: CABI Publishing.

Kulkarni, N. and S. Chander. 2022. *Hoplocerambyx spinicornis* Newman: major heartwood borer of sal, *Shorea robusta* and its management in India. *In Science of wood degradation and its protection*, ed. R. Sundararaj, pp. 289–328. Singapore: Springer Nature.

Kulkarni, N., K.C. Joshi, and P.K. Shukla. 2004. Integrated insect pest management of forest insect pests. In *Contemporary trends in insect science*, ed. G.T. Gujar, pp. 370–410. New Delhi, India: Campus Books.

Kulkarni, N., S. Paunikar, and S.S. Hussaini. 2011. Susceptibility of teak skeletonizer, *Eutectona machaeralis* (Walker) to the EPN, *Heterorhabditis indica* Poinar. *World J. Zool.* 6:33–39.

Kulkarni, N., K.K. Soni, and K.C. Joshi. 2007. Biotic and abiotic factors responsible for outbreak of the sal heartwood borer, *Hoplocerambyx spinicornis* Newman in Achanakmar-Amarkantak biosphere reserve. In *Research needs for Achanakmar-Amarkantak biosphere reserve*, eds. K.C. Joshi and A.K. Mandal, pp. 115–126. Jabalpur, India: Tropical Forest Research Institute.

Kumar, R., G. Shahabuddin, and A. Kumar. 2011. How good are managed forests at conserving native woodpecker communities? A study in sub-Himalayan dipterocarp forests of northwest India. *Biol. Conserv.* 144:1876–1884.

Kurz, W.A., C.C. Dymond, G. Stinson, et al. 2008. Mountain pine beetle and forest carbon feedback to climate change. *Nature* 452:987–990.

Laflamme, G. 2010. Root diseases in forest ecosystems. Contribution to the symposium "Root Diseases: Challenges and Perspectives" held during the Canadian Phytopathological Society Annual Meeting, 22–25 June 2009, Winnipeg, Manitoba, Canada. *Can. J. Plant Pathol.* 32:68–76.

Laflamme, G. 2012. Spread of *Heterobasidion irregulare* in eastern Canada towards northern natural forests of *Pinus banksiana*. In *Proceedings of the XIII conference root and butt rot of forest trees.* IUFRO Working Party 7.02.01, Firenze (FI)—S. Martino di Castrozza (TN), Italy, 4–10 September 2012. eds. P. Capretti, C. Comparini, M. Garbelotto, N. La Porta, A. Santini, p. 162. Firenze, Italy: University Press.

Laflamme, G., and Blais, R. 1995. Détection du *Heterobasidion annosum* au Québec. *Phytoprotection* 76:39–43.

Lagerberg, T. 1936. Some aspect of care of standing timber and of wood/Nagra synpunkter pa bestandsvard och virkesvard. *Svenska skogdv Foren Tidskr.* 34:396–406 & 472–473.

Lal, J.B. 1989. *India's forests: Myth and reality*. Dehra Dun, India: Natraj Publishers.

Lamarche, J., A. Potvin, D. Stewart, et al. 2016. Real-time PCR assays for detection of *Heterobasidion irregulare, H. annosum sensu stricto* and *Heterobasidion annosum* complex. *Forest Pathol.* 47: e12321.

Lamsal, S., R.C. Cobb, J.H. Cushman, et al. 2011. Spatial estimation of the density and carbon content of host populations for *Phytophthora ramorum* in California and Oregon. *For. Ecol. Manage.* 262:989–998.

Larsen, J.A. 1980. *The boreal ecosystem*. New York, NY: Academic Press, Inc.

Lawrence, R.K., W.J. Mattson, and R.A. Haack. 1997. White spruce and the spruce budworm: Defining the phenological window of susceptibility. *Can. Entomol.* 129:291–318.

Lehnert, L.W., C. Bässler, R. Brandl, et al. 2013. Conservation value of forests attacked by bark beetles: Highest number of indicator species is found in early successional stages. *J. Nat. Conserv.* 21:97–104.

Lonsdale, D., M. Pautasso, and O. Holdenrieder. 2008. Wood-decaying fungi in the forest: conservation needs and management options. *Eur. J. For. Res.*127:1–22.

Lovett, G.M., C.D. Canham, M.A. Arthur, et al. 2006. Forest ecosystem responses to exotic pests and pathogens in eastern North America. *Bioscience* 56:395–405.

Lovett, G.M., L.M. Christenson, P.M. Groffman, et al. 2002. Insect defoliation and nitrogen cycling in forests. *Bioscience* 52:335–341.

Lovett, G.M., M. Weiss, A.M. Liebhold, et al. 2016. Nonnative forest insects and pathogens in the United States: Impacts and policy options. *Ecol. Appl.* 26:1437–1455.

Lyver, P.O'.B., P. Timoti, C.J. Jones, et al. 2017. An indigenous community-based monitoring system for assessing forest health in New Zealand. *Biodivers. Conserv.* 26:3183–3212.

MacKinnon, W.E., and D.A. MacLean. 2003. The influence of forest and stand conditions on spruce budworm defoliation in New Brunswick, Canada. *For. Sci.* 49:657–667.

MacKinnon, W.E., and D.A. MacLean. 2004. Effects of surrounding forest and site conditions on growth reduction of balsam fir and spruce caused by spruce budworm defoliation. *Can. J. For. Res.* 34:2351–2362.

Maclauchlan, L.E., J.H. Borden, J.M. D'Auria, et al. 1988. Distribution of arsenic in MSMA-treated lodgepole pines infested by the mountain pine beetle, *Dendroctonus ponderosae* (Coleoptera: Scolytidae), and its relationship to beetle mortality. *J. Econ. Entomol.* 81:274–280.

MacLean, D.A. 1980. Vulnerability of fir-spruce stands during uncontrolled spruce budworm outbreaks: a review and discussion. *For. Chron.* 56:213–221.

MacLean, D.A. 1984. Effects of spruce budworm outbreaks on the productivity and stability of balsam fir forests. *For. Chron.* 60:273–279.

MacLean, D.A. 1996. Forest management strategies to reduce spruce budworm damage in the Fundy Model Forest. *For. Chron.* 72:399–405.

MacLean D.A., P. Amirault, L. Amos-Binks, et al. 2019. Positive results of an Early Intervention Strategy to suppress a spruce budworm outbreak after five years of trials. *Forests* 10:448.

MacLean, D.A., T.A. Erdle, W.E. MacKinnon, et al. 2001. The spruce budworm decision support system: Forest protection planning to sustain long-term wood supply. *Can. J. For. Res.* 31:1742–1757.

MacLean, D.A., and W.E. MacKinnon. 1997. Effects of stand and site characteristics on susceptibility and vulnerability of balsam fir and spruce to spruce budworm in New Brunswick. *Can. J. For. Res.* 27:1859–1871.

MacLean, D.A., and D.P. Ostaff. 1989. Patterns of balsam fir mortality caused by an uncontrolled spruce budworm outbreak. *Can. J. For. Res.* 19:1087–1095.

Mai, J.A., W. Mark, L. Fischer, et al. 2006. Aerial and ground surveys for mapping the distribution of *Phytopthora ramorum* in California. In *The proceedings of the sudden oak death second science symposium: the state of our knowledge*. Tech. Coords. S.J. Frankel, P.J. Shea, and M.I. Haverty, Gen. Tech. Rep. PSW-GTR-196, pp. 345–360. Albany, USA: US Department of Agriculture, Forest Service, Pacific Southwest Research Station.

Mäkelä, A., M. del Rio, J. Hynynen, M.J. Hawkins, et al. 2012. Using stand-scale forest models for estimating indicators of sustainable forest management. *For. Ecol. Manage.* 285:164–178.

Martel, V., R.C. Johns, L. Jochems-Tanguay, et al. 2021. The use of UAS to release the egg parasitoid *Trichogramma* spp. (Hymenoptera: Trichogrammatidae) against an agricultural and a forest pest in Canada. *J. Econ. Entomol.* 114:1867–1881.

Mattila, U., and T. Nuutinen. 2007. Assessing the incidence of butt rot in Norway spruce in southern Finland. *Silva Fenn.* 41:29–43.

Mattson, W.J., S.S. Slocum, and C.N. Koller. 1983. Spruce budworm (*Choristoneura fumiferana*) performance in relation to foliar chemistry of its host plants. In *Forest defoliator-host interactions: A comparison between gypsy moth and spruce budworms*. eds. R.L. Talerico and M. Montgomery, Gen. Tech. Rep. NE-85, pp. 55–66. Broomall, PA: US Department of Agriculture, Forest Service, Northeastern Forest Experiment Station.

Meentemeyer, R.K., D.M. Rizzo, W. Mark, et al. 2004. Mapping the risk of establishment and spread of sudden oak death in California. *For. Ecol. Manage.* 200:195–214.

Metz, M.R., K.M. Frangioso, R.K. Meentemeyer, et al. 2011. Interacting disturbances: wildfire severity affected by stage of forest disease invasion. *Ecol. Appl.* 21:313–320.

Metz, M.R., K.M. Frangioso, A.C. Wickland, et al. 2012. An emergent disease causes directional changes in forest species composition in coastal California. *Ecosphere* 3:art86.

Mikkelson, K.M., L.A. Bearup, R.M. Maxwell, et al. 2013. Bark beetle infestation impacts on nutrient cycling, water quality and interdependent hydrological effects. *Biogeochemistry* 115:1–21.

Millar, C.I., and N.L. Stephenson. 2015. Temperate forest health in an era of emerging megadisturbance. *Science* 349:823–826.

Miller, C.A., D.O. Greenbank, and E.G. Kettela. 1978. Estimated egg deposition by invading spruce budworm moths (Lepidoptera: Tortricidae). *Can. Entomol.* 110:609–615.

Miller, J.C. 1990. Field assessment of the effects of a microbial pest control agent on nontarget Lepidoptera. *Am. Entomol.* 36:135–140.

Mohamed-Ali, M.I., R.V. Varma, and V.V. Sudheendrakumar. 1992. Microbial pathogens associated with forest insect pests in Kerala. In *Emerging Trends in Biological Control of Phytophagous Insects*, ed. T.N. Ananthakrishnan, pp. 131–141. New Delhi, India: Oxford and IBH Publishing.

Monahan, W.B., and W.D. Koening. 2006. Potential effects of sudden oak death on the oak woodland bird community of coastal California. In *The proceedings of the sudden oak death second science symposium: The state of our knowledge*, Tech. Coords. S.J. Frankel, P.J. Shea, and M.I. Haverty, Gen. Tech. Rep. PSW-GTR-196, pp. 195–209. Albany, CA: US Department of Agriculture, Forest Service, Pacific Southwest Research Station.

Moody, B.H. 1988. *Insectes et maladies des arbres au Canada 1987*, p. 92. Ottawa, Ontario: Forêts Canada.

Morin, H. 1994. Dynamics of balsam fir forests in relation to spruce budworm outbreaks in the boreal zone of Quebec. *Can. J. For. Res.* 24:730–741.

Morris, R.F., and D.G. Mott. 1963. Dispersal and the spruce budworm. *Mem. Entomol. Soc. Can.* 95:180–189.

Mott, D.G. 1963. The forest and the spruce budworm. *Mem. Entomol. Soc. Can.* 95:189–202.

MRNFQ [Ministère des Ressources naturelles et des Forêts du Québec]. 2016. *La maladie du rond*. https://mffp.gouv.qc.ca/les-forets/protection-milieu-forestier/protection-forets-insectes-maladies/maladies-fiches/maladie-rond/. Accessed October 5, 2023.

Mudappa, D.C., and R. Kannan. 1997. Nest-site characteristics and nesting success of the Malabar Gray Hornbill in the southern Western Ghats, India. *Wilson Bull.* 109:102–111.

Mulvey, R.L., and E.M. Hansen. 2011. *Castilleja* and *Pedicularis* confirmed as telial hosts for *Cronartium ribicola* in whitebark pine ecosystems of Oregon and Washington. *For. Pathol.* 41:453–463.

Myren, D.T., G. Laflamme, P. Singh, et al. 1994. *Tree diseases of eastern Canada*, pp. 119–121. Ottawa: Natural Resources Canada, Canadian Forest Service.

Nagendra Prasad, M.N., S.S. Bhat, and K. Girish. 2011. A survey of die-back disease of neem in Tamil Nadu, India and PCR-based confirmation of the isolates. *Arch. Phytopathol. Plant Prot.* 44:331–339.

Nagendra Prasad, M.N., S.S. Bhat, N. Haraprasad, et al. 2010. Study of die back disease incidence of neem in Karnataka, India and PCR based identification of the isolates. *Arch. Phytopathol. Plant Prot.* 43:446–453.

Nair, K.S.S. 2007. *Tropical forest insect pests: Ecology, impact, and management*. New York, USA: Cambridge University Press.

Nair, K.S.S., R.V. Varma, V.V. Sudheendrakumar, et al. 1998. Management of the teak defoliator (*Hyblaea puera*) using nuclear polyhedrosis virus (NPV). Kerala Forest Research Institute, Peechi, India, pp. 27.

Nappi, A., P. Drapeau, and A. Leduc. 2015. How important is dead wood for woodpeckers foraging in eastern North American boreal forests? *For. Ecol. Manage.* 346:10–21.

Naziz, P.S., R. Das, and S. Sen. 2019. The scent of stress: Evidence from the unique fragrance of agarwood. *Front. Plant Sci.* 10:840.

Nealis, V.G., and J. Régnière. 2004a. Insect-host relationships influencing disturbance by the spruce budworm in a boreal mixedwood forest. *Can. J. For. Res.* 34:1870–1882.

Nealis, V., and J. Régnière. 2004b. Fecundity and recruitment of eggs during outbreaks of the spruce budworm. *Can. Entomol.* 136:591–604.

Nealis, V., and K. van Frankenhuyzen. 1990. Interactions between *Bacillus thuringiensis* Berliner and *Apanteles fumiferanae* Vier. (Hymenoptera: Braconidae), a parasitoid of the spruce budworm, *Choristoneura fumiferana* (Clem.) (Lepidoptera: Tortricidae). *Can. Entomol.* 122:585–594.

Negi, N., R. Krishna, R.K. Meena, et al. 2023. First report of *Lasiodiplodia iraniensis* causing leaf spot disease of Eucalyptus in India. *Physiol. Mol. Plant Pathol.* 127:102113.

NRCAN [Natural Resources Canada]. 2014. *The state of Canada's forests—Annual report 2014,* 63 pp. Ottawa, Ontario, Canada: Canadian Forest Service

NRCAN. 2022. *The state of Canada's forests: Annual report 2022*, p. 96. Ottawa: Canadian Forest Service. https://natural-resources.canada.ca/sites/nrcan/files/forest/sof2022/SoF_Annual2022_EN_access.pdf. Accessed September 19, 2023.

Oak, S.W., J. Hwang, S.N. Jeffers, et al. 2010. *Phytophthora ramorum* in USA streams from the national early detection survey of forests. In *The proceedings of the sudden oak death fourth science symposium*, Tech. Coords. S.J. Frankel, J.T. Kliejunas, and K.M. Palmieri, Gen. Tech. Rep. PSW-GTR-229, pp. 353–354. Albany, CA: US Department of Agriculture, Forest Service, Pacific Southwest Research Station.

O'Laughlin, J., R.L. Livingston, R. Thier, et al. 1994. Defining and measuring forest health. *J. Sustainable For.* 2:65–85.

Oliva, J., J. Stenlid, and J. Martínez-Vilalta. 2014. The effect of fungal pathogens on the water and carbon economy of trees: Implications for drought-induced mortality. *New Phytol.* 203:1028–1035.

Oliver, C.D. 1980. Forest development in North America following major disturbances. *For. Ecol. Manage. 3*, 153–168.

Osawa, A. 1989. Causality in mortality patterns of spruce trees during a spruce budworm outbreak. *Can. J. For. Res.* 19:632–638.

Ostry, M.E., and G. Laflamme. 2009. Fungi and diseases—natural components of healthy forests. *Botany* 87:22–25.

Ostry, M.E., and K. Woeste. 2004. Spread of butternut canker in North America, host range, evidence of resistance within butternut populations and conservation genetics. In *Proceedings of the 6th Walnut Council Research Symposium.* eds. C.H. Michler, P.M. Pijut, J.W. Van Sambeek, et al., Gen. Tech. Rep. NC-243, pp. 114–120. St. Paul, MN: US Department of Agriculture, Forest Service, North Central Research Station.

Otrosina, W.J. and M. Garbelotto. 2010. *Heterobasidion occidentale* sp. nov. and *Heterobasidion irregulare* nom. nov.: A disposition of North American *Heterobasidion* biological species. *Fungal Biol.* 114:16–25.

Pandey, S., R.R.Rishi, R.S.C. Jayaraj, et al. 2019. *Fusarium equiseti* is associated with the wilt and dieback of *Aquilaria malaccensis* in Northeast India. *Forest Pathol.* 49:e12489.

Parke, J.L. and E.K. Peterson. 2019. *Sudden oak death, sudden larch death, and ramorum blight.* www.apsnet.org/edcenter/disandpath/oomycete/pdlessons/Pages/SuddenOakDeath.aspx. Accessed October 9, 2023.

Pathak, H., S. Maru, H.N. Satya, and S.C. Silawat. 2015. Fungal diseases of trees in forest nurseries of Indore, India. *J. Plant Pathol. Microbiol.* 6:297.

Pathak, S.C., N. Kulkarni, and P.B. Meshram. 1995. Forest insect control. In *Encyclopaedia of environmental biology*. San Diego, USA: Academic Press.

Patil, S.U., and M.I. Naik. 1998. Natural enemies of teak defoliator, *Hyblaea puera* Cramer and their seasonal incidence. *Indian J. For.* 21:253–255.

Pautasso, M. 2013a. *Phytophthora ramorum*-a pathogen linking network epidemiology, landscape pathology and conservation biogeography. *CAB Reviews* 8:1–14.

Pautasso, M. 2013b. Responding to diseases caused by exotic tree pathogens. In *Infectious Forest Diseases*, eds. P. Gonthier and G. Nicolotti, pp. 592–612. Wallingford, UK: CABI.

Pautasso, M., M. Schlegel, and O. Holdenrieder. 2015. Forest health in a changing world. *Microb. Ecol.*:1–17.

Payette, S. 1992. Fire as a controlling process in the North American boreal forest. In *A systems analysis of the global boreal forest*, eds. H.H. Shugard, R. Leemans, and G.B. Bonan, pp. 144–169. Cambridge, UK: Cambridge University Press.

Peltonen, M., A.M. Liebhold, O.N. Bjørnstad, et al. 2002. Spatial synchrony in forest insect outbreaks: roles of regional stochasticity and dispersal. *Ecology* 83:3120–3129.

Peterson, E., E.M. Hansen, and J. Hulbert. 2014. Source or sink? The role of soil and water borne inoculum in the dispersal of *Phytophthora ramorum* in Oregon tanoak forests. *For. Ecol. Manage.* 322:48–57.

Picard M-È., M. Cusson, S.E. Sen, et al. 2021. Rational design of Lepidoptera-specific insecticidal inhibitors targeting farnesyl diphosphate synthase, a key enzyme of the juvenile hormone biosynthetic pathway. *J. Pesticide Sci.* 46:7–15.

Piene, H. 1989. Spruce budworm defoliation and growth loss in young balsam fir: defoliation in spaced and unspaced stands and individual tree survival. *Can. J. For. Res.* 19:1211–1217.

Prospero, S. and D. Rigling. 2012. Invasion genetics of the chestnut blight fungus *Cryphonectria parasitica* in Switzerland. *Phytopathology* 102:73–82.

Punter, D. and J. Cafley. 1968. Two new hardwood hosts of *Fomes annosus. Plant. Dis. Rep.* 52:692–695.

Pureswaran, D.S., L. de Grandpre, D. Paré, et al. 2015. Climate-induced changes in host tree-insect phenology may drive ecological state-shift in boreal forest. *Ecology* 96:1480–1491.

Queloz, V., H. Jactel, B. Marçais, et al. 2023. Rôle des agents biotiques dans les crises sanitaires forestières. *Revue Forestiere Française* 74:133–143.

Raffa, K.F., B.H. Aukema, B.J. Bentz, et al. 2008. Cross-scale drivers of natural disturbances prone to anthropogenic amplification: The dynamics of bark beetle eruptions. *Bioscience* 58:501–517.

Rahman, M.A. 1988. Perspectives of bamboo blight in Bangladesh. *Indian For.* 114:726–736.

Randolph, K.C., R.S. Morin, K. Dooley, et al. 2020. *Forest ecosystem health indicators*, FS-1151, 28 pp. US Department of Agriculture, Forest Service, Northern Research Station.

Rapport, D.J. 1989. What constitutes ecosystem health? *Perspect. Biol. Med.* 33:120–132.

Rapport, D., R. Costanza, P.R. Epstein, et al. 1998a. Ecosystem health. Blackwell Science Inc., Victoria, Australia.

Rapport, D.J., R. Costanza, and A.J. McMichael. 1998b. Assessing ecosystem health. *Trends Ecol. Evol.*13:397–402.

Rastall, K., V. Kondo, J.S. Strazanac, et al. 2003. Lethal effects of biological insecticide applications on nontarget lepidopterans in two Appalachian forests. *Environ. Entomol.* 32:1364–1369.

Redlin, S.C., S. Werres, and T. Schröder. 2014. Invasive pathogens in plant biosecurity. Case study: *Phytophthora ramorum* Werres et al.: cause of Sudden Oak Death, ramorum leaf blight and ramorum dieback. In *The handbook of plant biosecurity*, eds. G. Gordh and S. McKirdy, pp. 593–611. Berlin, Germany: Springer.

Régnière, J. 1987. Temperature-dependent development of eggs and larvae of *Choristoneura fumiferana* (Clem.) (Lepidoptera: Tortricidae) and simulation of its seasonal history. *Can. Entomol.* 119:717–728.

Régnière, J. 1990. Diapause termination and changes in thermal responses during postdiapause development in larvae of the spruce budworm, *Choristoneura fumiferana. J. Ins. Physiol.* 36:727–735.

Régnière, J., J. Delisle, A. Dupont, et al. 2019a. The impact of moth migration on apparent fecundity overwhelms mating disruption as a method to manage spruce budworm populations. *Forests* 10:775.

Régnière, J., J. Delisle, D. Pureswaran, et al. 2012a. Mate-finding Allee effect in spruce budworm population dynamics. *Entomol. Exp. Appl.* 146:112–122.

Régnière J., J. Delisle, B.R. Sturtevant, et al. 2019b. Modeling migratory flight in the spruce budworm: temperature constraints. *Forests* 10:802.

Régnière, J., and P. Duval. 1998. Overwintering mortality of spruce budworm, *Choristoneura fumiferana* (Clem.) (Lepidoptera: Tortricidae), populations under field conditions. *Can. Entomol.* 130:13–26.

Régnière, J., R. St-Amant, and P. Duval. 2012b. Predicting insect distributions under climate change from physiological responses: spruce budworm as an example. *Biol. Invasions* 14:1571–1586.

Rennerfelt, E. 1952. Root-rot attack on Scots Pine/Om angrepp av rotrota pa tall. *Meddelanden fran Statens Skogsforskningsinstitut* 41:40.

Rennerfelt, E. 1953. On the infection of Scots Pine by root rot/Om angrepp av rotröta pu Tall. *Meddelanden fran Statens Skogsforskningsinstitut* 41:1–39.

Rhainds, M., E.G. Kettela, and P.J. Silk. 2012. Thirty-five years of pheromone-based mating disruption studies with *Choristoneura fumiferana* (Clemens) (Lepidoptera:Tortricidae). *Can. Entomol.* 144:379–395.

Rhainds, M., D. Lavigne, Y. Boulanger, et al. 2022. I know it when I see it: incidence, timing and intensity of immigration in spruce budworm. *Agric. For. Entomol.* 24:152–166.

Rhoades, C.C., R.M. Hubbard, K. Elder, et al. 2020. Tree regeneration and soil responses to management alternatives in beetle-infested lodgepole pine forests. *For. Ecol. Manage.* 468:118182.

Riitters, K.H., B.E. Law, R.C. Kucera, et al. 1992. A selection of forest condition indicators for monitoring. *Environ. Monit. Assess.* 20:21–33.

Rizzo, D.M., M. Garbelotto, and E.M. Hansen. 2005. *Phytophthora ramorum*: integrative research and management of an emerging pathogen in California and Oregon forests. *Annu. Rev. Phytopathol.* 43:309–335.

Roll-Hansen, F. 1940. Studies in *Polyporus annosus* Fr., especially in respect of its occurrence in Norway south of the Dovre Fell/Undersøkelser over *Polyporus annosus* Fr., særlig med henblikk på dens forekomst i det sennefjelske Norge. *Meddelelser fra det Norske Skogsforsoksvesen* 7:100 pp.

Roonwal, M.L. 1978. Biology, ecology and control of the sal heartwood borer, *Hoplocerambyx spinicornis*: a review of recent work. *Indian J. For.* 1:21–34.

Roques, A., and M.A. Auger-Rozenberg. 2006. Tentative analysis of the interceptions of non-indigenous organisms in Europe during 1995–2004. *EPPO Bulletin* 36:490–496.

Roscoe, L.E., W. MacKinnon, J. Régnière, et al. 2022. Use of sprayable sex pheromone formulation in landscape-level control of *Choristoneura fumiferana* populations. *Insects* 13:1175.

Roy, B.A., H.M. Alexander, J.M. Davidson, et al. 2014. Increasing forest loss worldwide from invasive pests requires new trade regulations. *Front. Ecol. Environ.* 12:457–465.

Royama, T. 1984. Population dynamics of the spruce budworm *Choristoneura fumiferana. Ecol. Monogr.* 54:429–462.

Roychoudhury, N. 2001. Tree resistance to insects: A novel approach of forest insect management. In *Recent trends in insect pest control to enhance forest productivity*, eds. P.K. Shukla and K.C. Joshi, pp. 28–60. Jabalpur, India: Tropical Forest Research Institute.

Roychoudhury, N. 2012. Toxicity of ivermectin in inducing larval mortality in bamboo leaf roller, *Crypsiptya coclesalis* Walker. *Bangladesh J. For. Sci.* 32:1–9.

Roychoudhury, N., and K.C. Joshi. 2009. Toxicity of ivermectin in inducing larval mortality in *Ailanthus* webworm, *Atteva fabriciella* Swederus (Lepidoptera: Yponomeutidae). *World Journal of Zoology*. 4:277–280.

Roychoudhury, N., and K.C. Joshi. 2010. Toxicity of ivermectin against the larvae of teak defoliator, *Hyblaea puera* Cramer (Lepidoptera: Hyblaeidae). *Indian For.* 136:38–44.

Roychoudhury, N., and K.C. Joshi. 2011. Effect of ivermectin on larval mortality and reproductive potential of *Albizia* foliage feeder, *Spirama retorta* Cramer (Lepidoptera: Noctuidae). *J. Pest Sci.* 83:117–122.

Safranyik, L., and A.L. Carroll. 2006. The biology and epidemiology of the mountain pine beetle in lodgepole pine forests. In *The Mountain pine beetle: A synthesis of biology, management, and impacts on lodgepole pine*, eds. L. Safranyik and B. Wilson, pp. 3–66. Victoria, Canada: Natural Resources Canada, Canadian Forest Service, Pacific Forestry Centre.

Safranyik, L., A.L. Carroll, J. Régnière, et al. 2010. Potential for range expansion of mountain pine beetle into the boreal forest of North America. *Can. Entomol.* 142:415–442.

Safranyik, L., and D.A. Linton. 1982. Mortality of spruce beetle bolts submerged in water. *J. Entomol. Soc. B.C.* 79:8–11.

Safranyik, L., D.A. Linton, T.L. Shore, et al. 2001. *The effects of prescribed burning on mountain pine beetle in lodgepole pine, Information Report BC-X-391*, p. 9. Victoria, Canada: Natural Resources Canada, Canadian Forest Service, Pacific Forestry Centre.

Safranyik, L., T.L. Shore, A.L. Carroll, et al. 2004. Bark beetle (Coleoptera: Scolytidae) diversity in spaced and unmanaged mature lodgepole pine (Pinaceae) in southeastern British Columbia. *For. Ecol. Manage.* 200:23–28.

Safranyik, L., and B. Wilson. 2006. *The mountain pine beetle: A synthesis of biology, management and impacts on lodgepole pine*. Victoria, Canada: Natural Resources Canada, Canadian Forest Service, Pacific Forestry Centre.

Sajeev, T.V., V.V. Sudheendrakumar, S.M. Helen, et al. 2007. Hybcheck-the biopesticide for managing teak defoliator (*Hyblaea puera* Cramer): an announcement. In *Processing and marketing of teak wood products of planted forests*, eds. K.M. Bhat, K.V. Balasundaran, E.M. Muralidharan, and P.K. Thulasidhas, pp. 173–177. Peechi, India: Kerala Forest Research Institute.

Sambaraju, K.R., A.L. Carroll, and B.H. Aukema. 2019. Multiyear weather anomalies associated with range shifts by the mountain pine beetle preceding large epidemics. *For. Ecol. Manage.* 438:86–95.

Sambaraju, K.R., A.L. Carroll, J. Zhu, et al. 2012. Climate change could alter the distribution of mountain pine beetle outbreaks in western Canada. *Ecography* 35:211–223.

Sambaraju, K.R., P. DesRochers, and D. Rioux. 2018. Factors influencing the regional dynamics of butternut canker. *Plant Dis.* 102:743–752.

Sambaraju, K., P. DesRochers, D. Rioux, et al. 2016. Forest ecosystem health and biotic disturbances: perspectives on indicators and management approaches. In *Ecological forest management handbook*, 1st edn, ed. G.R. Larocque, pp. 459–515. Boca Raton, FL: CRC Press.

Sambaraju, K.R., and D. Goodsman. 2021. Mountain pine beetle: an example of a climate-driven eruptive insect impacting conifer forest ecosystems. *CAB Reviews.* 16:1–18.

Sankaran, K., P.D. Bridge, and C. Gokulapalan. 2005. *Ganoderma* diseases of perennial crops in India–an overview. *Mycopathologia*. 159:143–152.

Sansford, C.E., A.J. Inman, R. Baker, et al. 2009. Report on the risk of entry, establishment, spread and socio-economic loss and environmental impact and the appropriate level of management for *Phytophthora ramorum* for the EU. Deliverable Report 28. EU Sixth Framework Project RAPRA. https://planthealthportal.defra.gov.uk/pests-and-diseases/uk-plant-health-risk-register/downloadExternalPra.cfm?id=3802. Accessed October 9, 2023.

Santini, A., L. Ghelardini, C. de Pace, et al. 2013. Biogeographical patterns and determinants of invasion by forest pathogens in Europe. *New Phytol.* 197:238–250.

Sateesh, M., S.S. Bhat, and N. Devaki. 1997. *Phomopsis azadirachtae* sp. nov. from India. *Mycotaxon*, 65:517–520.

Sayre, A.P. 1994. *Taiga*. Brookfield, CT: Twenty-First Century Books.

Scarr, T.A., K.L. Ryall, A. Piscopo, et al. 2011. *Forest health conditions in Ontario, 2009*, p. 116. Sault Ste. Marie, ON: Queen's Printer for Ontario, Ontario Ministry of Natural Resources.

Scarr, T., K. Ryall, and L. Thompson. 2009. *Forest health conditions in Ontario, 2008*, p. 71. Sault Ste. Marie, ON: Queen's Printer for Ontario, Ontario Ministry of Natural Resources.

Schmid, M., M. Pautasso, and O. Holdenrieder. 2014. Ecological consequences of Douglas fir (*Pseudotsuga menziesii*) cultivation in Europe. *Eur. J. For. Res.*133:13–29.

Schomaker, M.E., S.J. Zarnoch, W.A. Bechtold, et al. 2007. *Crown-condition classification: a guide to data collection and analysis, Gen. Tech. Rep. SRS-102*. Asheville, NC: US Department of Agriculture, Forest Service, Southern Research Station.

Schowalter, T.D. 1994. An ecosystem-centered view of insect and disease effects on forest health. In *Sustainable ecological systems: implementing an ecological approach to land management*, Tech. Coords. W.W. Covington and L.F. DeBano, Gen. Tech. Rep. RM-247, pp. 189–195. Fort Collins, CO: US Department of Agriculture, Forest Service, Rocky Mountain Forest and Range Experimental Station.

Scriber, J.M. 2004. Non-target impacts of forest defoliator management options: decision for no spraying may have worse impacts on non-target Lepidoptera than *Bacillus thuringiensis* insecticides. *J. Insect Conserv.* 8:241–261.

Shamoun, S.F., C. Hammett, G. Sumampong, et al. 2019. New taxon-specific *Heterobasidion* PCR primers detect and differentiate North American *Heterobasidion* spp. in various substrates and led to discovery of *Heterobasidion irregulare* in British Columbia, Canada. *Pathogens.* 8:156.

Shamoun, S.F., D. Rioux, B. Callan, et al. 2018. An overview of Canadian research activities on diseases caused by *Phytophthora ramorum*: results, progress and challenges. *Plant Dis.* 102:1218–1233.

Sharma, J.K., C. Mohanan, and E.J.M. Florence. 1985. *Disease survey in nurseries and plantations of forest tree species grown in Kerala*. KFRI Research Report 36. Peechi, India: Kerala Forest Research Institute.

Shaw III, C.G. 1981. Infection of western hemlock and Sitka spruce thinning stumps by *Fomes annosus* and *Armillaria mellea* in southeast Alaska. *Plant Dis.* 65:967–971.

Shaw III, C.G. 1989. Root disease threat minimal in young stands of western hemlock and Sitka spruce in southeastern Alaska. *Plant Dis.* 73:573–577.

Shore, T.L., and L. Safranyik. 1992. *Susceptibility and risk rating systems for the mountain pine beetle in lodgepole pine stands.* Information Report BC-X-336. Victoria, BC, Canada: Forestry Canada.

Sippell, W., H. Gross, and A. Rose. 1968. Ontario region: Important forest diseases. *Annual Report of the Forest Insect and Disease Survey* 66–76.

Skelly, J.M. 1989. Forest decline versus tree decline-the pathological considerations. *Environ. Monit. Assess.* 12:23–27.

Smith, R.H., G.C. Trostle, and W.F. McCambridge. 1977. Protective spray tests on three species of bark beetles in the western United States. *J. Econ. Entomol.* 70:119–125.

Soni, K.K., C.K. Tiwari, and R.K. Verma. 2010. Heart rot in Indian hard wood tree species. *J. Trop. For.* 26:15–21.

Soni, K.K., and R.K. Verma. 2010. *Helicobasidium* root rot of teak in central India and its control measures. *J. Trop. For.* 26:3–7.

Stenlid, J., J. Oliva, J.B. Boberg, et al. 2011. Emerging diseases in European forest ecosystems and responses in society. *Forests.* 2:486–504.

Sterner, T.E., and A.G. Davidson. 1981. *Insectes et maladies des arbres au Canada 1980*, p. 33. Ottawa, Ontario: Environnement Canada.

Sterner, T.E., and A.G. Davidson. 1982a. *Insectes et maladies des arbres au Canada 1981*, p. 40. Ottawa, Ontario: Environnement Canada.

Sterner, T.E., and A.G. Davidson. 1982b. *Forest insect and disease conditions in Canada*, p. 46. Forest Insect and Disease Survey Report. Ottawa: Canadian Forest Service.

Sturtevant, B.R., G.L. Achtemeier, J.J. Charney, et al. 2013. Long-distance dispersal of spruce budworm (*Choristoneura fumiferana* Clemens) in Minnesota (USA) and Ontario (Canada) via the atmospheric pathway. *Agric. For. Meteorol.* 168:186–200.

Sudheendrakumar, V.V., M.I.A. Ali, and R.V. Varma. 1988. Nuclear polyhedrosis virus of the teak defoliator, *Hyblaea puera*. *J. Invertebr. Pathol.* 51:307–308.

Sudheendrakumar, V.V., and C.P. Biji. 2009. Biocontrol of the teak defoliator, *Hyblaea puera*: prospects and constraints. *Envis Forestry Bulletin* 9:8–12.

Sudheendrakumar, V.V., S.K. Jalali, and S.P. Singh. 1995. Acceptance of the teak defoliator *Hyblaea puera* (Cramer) (Lepidoptera: Hyblaeidae) by two exotic species of *Trichogramma* (Hymenoptera: Trichogrammatidae). *J. Biol. Control* 9:43–44.

Sudheendrakumar, V.V., R.V. Varma, and T.V. Sajeev. 2006. *Demonstration of mass production, formulation and application of a baculovirus for management of the teak defoliator, Hyblaea puera*. KFRI Research Report 290. Peechi, India: Kerala Forest Research Institute.

Sundaram, A., K.M.S. Sundaram, and L. Sloane. 1996. Spray deposition and persistence of a *Bacillus thuringiensis* formulation (Foray® 76B) on spruce foliage, following aerial application over a northern Ontario forest. *J. Environ. Sci. Health, Part B*, 31:763–813.

Sundaram, K.M.S., A. Sundaram, and L. Sloane. 1996. Foliar persistence and residual activity of tebufenozide against spruce budworm larvae. *Pestic. Sci.* 47:31–40.

Tainter, F.H., and F.A. Baker. 1996. *Principles of forest pathology*. New York, NY: John Wiley.

Taylor, S.W., A.L. Carroll, R.I. Alfaro, et al. 2006. Forest, climate and mountain pine beetle outbreak dynamics in western Canada. In *The mountain pine beetle: A synthesis of biology, management, and impacts on lodgepole pine*, eds. L. Safranyik and B. Wilson, pp. 67–94. Victoria, Canada: Natural Resources Canada, Canadian Forest Service, Pacific Forestry Centre.

Tewari, D.N. 1992. *A monograph on teak* (*Tectona grandis* Linn. f.). Dehradun, India: International Book Distributors.

Tewari, D.N. 1995. *A monograph of sal* (*Shorea robusta* Gaertn. f.). Dehradun, India International Book Distributors.

Thakur, M.L. 2000. *Forest entomology*. Dehradun, India: Sai Publishers.

Tikka, P.S. 1934. On the butt rot of the conifer woods of North Finland/Über die Stockfäule der Nadelwälder Nord-Suomis (-Finnlands). *Acta For. Fenn.* 40:293–308.

Tiwari, D.P., R.C. Rajak, and K.M. Nikhra. 1981. A new species of *Phomopsis* causing leaf spot disease on *Tectona grandis* L. *Current Science* 50:1002–1003.

Tkacz, B., K. Riitters, and K.E. Percy. 2013. Forest monitoring methods in the United States and Canada: An overview. *Developments in environmental science* 12:49–73.

Tokuda, S., T. Hattori, Y.-C. Dai, et al. 2009. Three species of *Heterobasidion* (Basidiomycota, Hericiales), *H. parviporum*, *H. orientale* sp. nov. and *H. ecrustosum* sp. nov. from East Asia. *Mycoscience* 50:190–202.

Tokuda, S., Y. Ota, and T. Hattori. 2007. Root and butt rot of Todo fir (*Abies sachalinensis*) caused by *Heterobasidion annosum* sl in Hokkaido, Japan. *For. Pathol.* 37:155–166.

Tokuda, S., Y. Ota, T. Hattori, et al. 2011. The distribution of closely related large genets of *Heterobasidion parviporum* in a Todo fir (*Abies sachalinensis*) stand in Hokkaido, Japan. *For. Pathol.* 41:482–492.

Troup, R.S. 1921. *The silviculture of Indian trees (Vol. 1)*. Oxford, UK: Clarendon Press.

Tuimala, A. 1979. Changes in timber assortments and loss in stumpage prices caused by decay in standing spruce timber. *Silva Fenn.* 13:333–342.

Urban, J.M., and H. Leach. 2023. Biology and management of the spotted lanternfly, *Lycorma delicatula* (Hemiptera: Fulgoridae), in the United States. *Annu. Rev. Entomol.* 68:151–167.

Valenta, V., D. Moser, S. Kapeller, and F. Essl. 2017. A new forest pest in Europe: a review of emerald ash borer (*Agrilus planipennis*) invasion. *J. Appl. Entomol.* 141:507–526.

Verma, R.K., R. Hanwat, P.S. Rajput, et al. 2013. Status of khair mortality in central India. *Indian For.* 139:790–796.

Verma, R.K., N. Sharma, K.K. Soni, et al. 2008. *Forest fungi of Central India*. Lucknow, India: International Book Distributing Company.

Volney, W.J.A., and R.A. Fleming. 2007. Spruce budworm (Choristoneura spp.) biotype reactions to forest and climate characteristics. *Global Change Biol.* 13:1630–1643.

Waller, M. 2013. Drought, disease, defoliation and death: forest pathogens as agents of past vegetation change. *J. Quat. Sci.* 28:336–342.

Walter, H. 1971. *Ecology of tropical and sub-tropical vegetation*. New York, NY: Van Nostrand Reinhold Company.

Weed, A.S., M.P. Ayres, and J.A. Hicke. 2013. Consequences of climate change for biotic disturbances in North American forests. *Ecol. Monogr.* 83:441–470.

Werres, S., R. Marwitz, W.A. Man In'T Veld, et al. 2001. *Phytophthora ramorum* sp. nov., a new pathogen on *Rhododendron* and *Viburnum*. *Mycol. Res.* 105:1155–1165.

Wieferich, J.B. 2013. *Beech bark disease in Michigan: Distribution, impacts and dynamics*. Ph.D. dissertation. East Lansing, MI: Michigan State University.

Wingfield, M.J., A. Hammerbacher, R.J. Ganley, et al. 2008. Pitch canker caused by *Fusarium circinatum*-a growing threat to pine plantations and forests worldwide. *Australas. Plant Pathol.* 37:319–334.

Wohlgemuth, T., A. Jentsch, and R. Seidl. 2022. *Disturbance ecology*. Cham, Switzerland: Springer Nature.

Wood, D.L. 1982. The role of pheromones, kairomones and allomones in the host selection and colonization of bark beetles. *Annu. Rev. Entomol.* 27:411–446.

Woodall, C.W., M.C. Amacher, W.A. Bechtold, et al. 2011. Status and future of the forest health indicators program of the USA. *Environ. Monit. Assess.* 177:419–436.

Wright, J.W., and R.S. Dodd. 2013. Could tanoak mortality affect insect biodiversity? Evidence for insect pollination in tanoaks. *Madroño* 60:87–94.

Xu, X., T.D. Harwood, M. Pautasso, et al. 2009. Spatio-temporal analysis of an invasive plant pathogen (*Phytophthora ramorum*) in England and Wales. *Ecography* 32:504–516.

Yamazaki, M., S. Iwamoto, and K. Seiwa. 2009. Distance-and density-dependent seedling mortality caused by several diseases in eight tree species co-occurring in a temperate forest. *Plant Ecol.* 201:181–196.

Younitzky, A.A. 1927. Fungi attacking healthy and scorched trees in the forests of the Mari Region, and the damage they cause to young stands arising on areas devastated by tire or bark insects according to the observations of the 1926 expedition. *Kazan Inst. of Agric. and Silvicult. News* 3:1–21.

Yousuf, M., K.C. Joshi, and N. Kulkarni. 2004. *Trichogramma*: the noble egg parasitoids and their potential in biological control of insect pests. In *Contemporary trends in insect science*, ed. G.T. Gujar, pp. 307–369. New Delhi, India: Campus Books International.

Zolkos, S.G., P. Jantz, T. Cormier, et al. 2015. Projected tree species redistribution under climate change: Implications for ecosystem vulnerability across protected areas in the eastern United States. *Ecosystems* 18:202–220.

Section IV

Addressing Ecological and Socioeconomic Issues

18 Application of Surface Modeling for Large Regions

A Case Study for Forest Carbon Stocks in China

Tian-Xiang Yue, Yi-Fu Wang, and Guy R. Larocque

18.1 INTRODUCTION

Surface modeling refers to the process of simulating a surface by efficiently combining satellite remotely sensed data with ground-based observation data, such as a scattered point-form dataset, a line-form dataset and/or an area-form dataset. Surface modeling formulates an object in a grid system (Yue 2011). Each grid cell includes an estimate of the object that represents the characteristics for that particular location. There are four advantages to represent data in grid form (Martin and Bracken 1991; Deichmann 1996; Yue et al. 2008; 2009, 2010a, 2010b): (1) a regular grid can be easily regrouped into any new aerial arrangement; (2) ecological data organized in a grid form can facilitate compatibility among heterogeneous datasets; (3) multiresolution and multisource information can be amalgamated more easily when data are in a grid form, and (4) some of the problems caused by different types of boundaries may be avoided when data are converted into a grid form.

Surface modeling began to be used effectively in the 1960s when computers became more accessible (Lo and Yeung 2002). However, its development was limited until the 1990s because of the limited computational capacity of computers. The application of surface modeling requires powerful software and large storage capacities to handle spatially explicit data over large regions. Significant advances in surface modeling were intimately associated with developments in several areas, including trend surface analysis (TSA) (Ahlberg et al. 1967; Schroeder and Sjoquist 1976; Legendre and Legendre 1983), digital terrain modeling (Stott 1977), surface approximation (Long 1980), spatial simulation of wetland habitats (Sklar et al. 1985), spatial pattern matching (Costanza 1989), spatial prediction (Turner et al. 1989), and modeling of coastal landscape dynamics (Costanza et al. 1990). Technological advances in remote sensing (RS) and geographic information systems (GIS) and the accumulation of spatially explicit data have contributed to the extensive development and application of surface modeling to analyze and understand the spatial phenomena of ecological processes for the last three decades (Yue 2011). Landmark case examples can be found in Sklar et al. (1985) for the spatial simulation model of habitat changes in relation to marsh type, hydrology, subsidence, and sediment transport for a generalized coastal wetland area, Turner et al. (1989) for the comparison of spatial patterns and Costanza and Maxwell (1991) for application on workstations.

Then, methodological developments continued on several fronts, such as the simulation of ecosystems with spatial heterogeneity (Gao 1996), a pairwise potential modeling approach (Reich et al.

DOI: 10.1201/9781003431084-22

1997), a patch-based spatial modeling approach to analyze the effects of spatial heterogeneity on ecological processes (Wu and Levin 1997), a daily high-resolution weather generator (Friend 1998), and a GIS-based spatial modeling method for the simulation of the geographic distribution of wildlife populations (Ji and Jeske 2000). The global carbon cycle was examined using different surface models that were sensitive to spatial distributions of productivity, biomass, and soil organic matter (Haxeltine and Prentice 1996; Svirezhev 2002). Perry and Enright (2002) developed a spatially explicit grid-based model to examine the extent to which past and future modifications in disturbance regimes may affect landscape structure.

The surface modeling method in forestry applications has embodied the forest properties and parameters of surface simulation. The classical method for obtaining forest attributes and parameters includes forest resources surveys in which population estimates are derived using samples. The measured data from forest resources survey form a point-form dataset. However, forest attributes and parameters (e.g., forest type, stock volume, biomass, and carbon storage) should be expressed as grid or raster data to explain the distribution. Surface modeling is an efficient approach to convert point-form data into grid data and consists of an appropriate method to simulate the distribution of forest attributes and parameters based on different data amalgamations that may originate from remote sensing observations and ground-based observation data.

In this chapter, some methods of earth surface modeling are briefly introduced, with a focus on the method used for high accuracy surface modeling (YUE-HASM). For more details, readers are encouraged to consult Yue (2011) for a complete introduction on the subject. A case study of the application of surface modeling to quantify carbon stocks in China is presented.

18.2 METHODS OF SURFACE MODELING

18.2.1 Trend Surface Analysis

Trend surface analysis uses least-squares regression to fit a trend surface from sparse data that consist of spatial coordinates. Good approximations can be obtained when the trend surface is formulated as a "n" order polynomial surface:

$$z(x,y)=\sum_{i=0}^{n}\sum_{j=0}^{n-i}a_{i,j}\cdot x^{i}\cdot y^{j} \tag{18.1}$$

Where $z(x,y)$ is the value of the trend surface at the grid (x, y); x and y are the independent variables; $a_{i,j}(i = 1, \ldots, n; j = 1, \ldots, n)$ are parameters to be simulated; and n is the order of the polynomial surface. There is a major weakness with TSA. Details can be lost because the poor smoothing in some surface regions may affect the fit of other surface regions. Also, TSA assumes that the residuals are independent, but this assumption is violated most of the time, which has consequences on the estimated variance (Oliver and Webster 1990).

18.2.2 Inverse Distance Weighted Method

The inverse distance weighted method (IDW) can be formulated as (Shepard 1968)

$$z_{i,j}=\left(\sum_{k=1}^{m_{i,j}}\frac{1}{\left(d_{i,j,k}\right)^{a}}\right)^{-1}\sum_{k=1}^{m_{i,j}}\frac{z_k}{\left(d_{i,j,k}\right)^{a}} \tag{18.2}$$

where, $z_{i,j}$ indicates the estimate made at the location (i, j) for which an estimate is required because of the absence of measurements. The estimates are computed from the $m_{i,j}$ neighbor observations;

z_k is the k_{th} neighbor observation; $d_{i,j,k}$ is the distance from z_k to (i, j); and a is the parameter to be determined. The main shortcoming of IDW is the inability to integrate the spatial structure and information beyond the neighborhood (Zhao et al. 2005, Magnussen et al. 2007).

18.2.3 Triangulated Irregular Network

The triangulated irregular network (TIN) method is well recognized to represent surface models in GIS (Bengtsson and Nordbeck 1964), as it supports a simple data structure and can easily be rendered using common graphics hardware (Yang et al. 2000). TIN is composed of nodes, edges, triangles, topological information, and a hull. This method is based on linear interpolation in a triangulated irregular network and is used to fit irregularly distributed points into a surface of each triangle. It is one of the basic models used to represent digital terrain pieces of land on the Earth's surface (Tse and Gold 2004).

Even though TIN models of the terrain surface are well known in GIS, their main shortcoming is that they ignore nonlinear information and are unable to represent cliffs, caves, or holes (Tse and Gold 2004).

18.2.4 Kriging

Kriging, named after the South African mining engineer D.G. Krige (Kleijnen and van Beers 2005), is the term used for a family of generalized linear least-squares regression algorithms that interpolate values in a random field for unobserved locations using observations from neighborhood locations (Krige 1951). Considered as a fundamental method in geostatistics, Kriging includes ordinary kriging (OK), co-kriging, and disjunctive kriging (Kleijnen 2009).

18.2.5 Spline

There are different types of spline: uniform rational basis-spline, uniform non-rational basis-spline, non-uniform-non-rational basis-spline, and non-uniform rational basis-spline. These different methods are based on the basis-spline, which includes cubic polynomial regressions with any number of curve segments (Watt 2000). Each curve segment has four control points and each control point influences four curve segments, which is the local control property of the basis-spline curve. The entire set of curve segments, $Q(u)$, is defined as one basis-spline curve

$$Q(u) = \sum_{i=0}^{m} P_i \cdot B_i(u) \tag{18.3}$$

where i is a nonlocal control point number; u is a global parameter; P_i is the ith control point; m is total number of the control points; and $B_i(u)$ is the ith basis-spline.

18.2.6 YUE-HASM

The method used for high accuracy surface modeling (YUE-HASM) is a more complex approach than the previous ones. When the surface can be represented as $z = (x, y, f(x, y))$, then the first fundamental coefficients can be formulated as:

$$\begin{cases} E = 1 + f_x^2 \\ F = f_x f_y \\ G = 1 + f_y^2 \end{cases} \tag{18.4}$$

The second fundamental coefficients can be formulated as:

$$\begin{cases} L = \dfrac{f_{xx}}{\sqrt{1+f_x^2+f_y^2}} \\ M = \dfrac{f_{xy}}{\sqrt{1+f_x^2+f_y^2}} \\ N = \dfrac{f_{yy}}{\sqrt{1+f_x^2+f_y^2}} \end{cases} \tag{18.5}$$

These two coefficient sets must satisfy the following Gauss equation set,

$$\begin{cases} f_{xx} = \Gamma_{11}^1 f_x + \Gamma_{11}^2 f_y + \dfrac{L}{\sqrt{E+G-1}} \\ f_{xy} = \Gamma_{12}^1 f_x + \Gamma_{12}^2 f_y + \dfrac{M}{\sqrt{E+G-1}} \\ f_{yy} = \Gamma_{22}^1 f_x + \Gamma_{22}^2 f_y + \dfrac{N}{\sqrt{E+G-1}} \end{cases} \tag{18.6}$$

where $\Gamma_{11}^1 = \dfrac{GE_x - 2FF_x + FE_y}{2(EG-F^2)}$, $\Gamma_{12}^1 = \dfrac{GE_y - FG_x}{2(EG-F^2)}$, $\Gamma_{22}^1 = \dfrac{2GF_y - GG_x - FG_y}{2(EG-F^2)}$, $\Gamma_{11}^2 = \dfrac{2EF_x - EE_y - FE_x}{2(EG-F^2)}$, $\Gamma_{12}^2 = \dfrac{EG_x - FE_y}{2(EG-F^2)}$ and $\Gamma_{22}^2 = \dfrac{EG_y - 2FF_y + FG_x}{2(EG-F^2)}$ are the second kind of Christoffel symbols.

If $\{(x_i, y_j)\}$ is an orthogonal division of a computational domain and h the simulation step length, the central point of lattice $(x_{\rm i}, y_{\rm j})$ could be expressed as $(0.5h+(i-1)h, 0.5h+(j-1)h)$, where $i = 0,1,2,\ldots,I,I+1$ and $j = 0,1,2,\ldots,J,J+1$. If $f_{\rm i,j}^{(n)}$ $(n \geq 0)$ represents the iterants of $f(x,y)$ at $(x_{\rm i}, y_{\rm j})$ in the nth iterative step, the interpolations are represented by $\left\{f_{\rm i,j}^{(0)}\right\}$ based on sampling values $\left\{\bar{f}_{\rm i,j}\right\}$. In terms of numerical mathematics (Quarteroni et al. 2000), the iterative formulation of the YUE-HASM master equation set can be expressed as (Yue et al.,2013b; Zhao and Yue 2014),

$$\frac{-f_{i+2,j}^{(n+1)} + 16f_{i+1,j}^{(n+1)} - 30f_{i,j}^{(n+1)} + 16f_{i-1,j}^{(n+1)} - f_{i-2,j}^{(n+1)}}{12h^2} = \left(\Gamma_{11}^1\right)_{i,j}^{(n)} \frac{f_{i+1,j}^{(n)} - f_{i-1,j}^{(n)}}{2h} + \left(\Gamma_{11}^2\right)_{i,j}^{(n)} \frac{f_{i,j+1}^{(n)} - f_{i,j-1}^{(n)}}{2h} + \frac{L_{ij}^{(n)}}{\sqrt{E_{i,j}^{(n)} + G_{i,j}^{(n)} - 1}} \tag{18.7}$$

$$\frac{-f_{i,j+2}^{(n+1)} + 16f_{i,j+1}^{(n+1)} - 30f_{i,j}^{(n+1)} + 16f_{i,j-1}^{(n+1)} - f_{i,j-2}^{(n+1)}}{12h^2} = \left(\Gamma_{22}^1\right)_{i,j}^{(n)} \frac{f_{i+1,j}^{(n)} - f_{i-1,j}^{(n)}}{2h} + \left(\Gamma_{22}^2\right)_{i,j}^{(n)} \frac{f_{i,j+1}^{(n)} - f_{i,j-1}^{(n)}}{2h} + \frac{N_{ij}^{(n)}}{\sqrt{E_{i,j}^{(n)} + G_{i,j}^{(n)} - 1}} \tag{18.8}$$

$$\frac{f_{i+1,j+1}^{(n+1)} - f_{i+1,j}^{(n+1)} - f_{i,j+1}^{(n+1)} + 2f_{i,j}^{(n+1)} + f_{i-1,j}^{(n+1)} - f_{i,j-1}^{(n+1)} + f_{i-1,j-1}^{(n+1)}}{2h^2}$$

$$= \left(\Gamma_{12}^1\right)_{i,j}^{(n)} \frac{f_{i+1,j}^{(n)} - f_{i-1,j}^{(n)}}{2h} + \left(\Gamma_{12}^2\right)_{i,j}^{(n)} \frac{f_{i,j+1}^{(n)} - f_{i,j-1}^{(n)}}{2h} + \frac{M_{ij}^{(n)}}{\sqrt{E_{i,j}^{(n)} + G_{i,j}^{(n)} - 1}} \tag{18.9}$$

where $E_{\text{i,j}}^{(n)}$, $F_{\text{i,j}}^{(n)}$ and $G_{\text{i,j}}^{(n)}$ are the iterants of the first fundamental coefficients at the nth iterative step; $L_{\text{i,j}}^{(n)}$, $M_{\text{i,j}}^{(\text{n})}$ and $N_{\text{i,j}}^{(n)}$ represent the iterants of the second fundamental coefficients at the nth iterative step; $(\Gamma_{11}^1)_{i,j}^{(n)}$, $(\Gamma_{11}^2)_{i,j}^{(n)}$, $(\Gamma_{22}^1)_{i,j}^{(n)}$ and $(\Gamma_{22}^2)_{i,j}^{(n)}$ are the iterants of the Christoffel symbols of the second kind at the nth iterative step, which depend only upon the first fundamental coefficients and their derivatives.

The matrix formulation of YUE-HASM master equations can be respectively expressed as,

$$\mathbf{A} \cdot \mathbf{z}^{(n+1)} = \mathbf{d}^{(n)} \tag{18.10}$$

$$\mathbf{B} \cdot \mathbf{z}^{(n+1)} = \mathbf{q}^{(n)} \tag{18.11}$$

$$\mathbf{C} \cdot \mathbf{z}^{(n+1)} = \mathbf{p}^{(n)} \tag{18.12}$$

where $\mathbf{A}$, $\mathbf{B}$, and $\mathbf{C}$ represent coefficient matrixes of the first equation, the second equation, and the third equation; $\mathbf{d}^{(n)}$, $\mathbf{q}^{(n)}$ and $\mathbf{p}^{(n)}$ are right-hand side vectors of the three equations respectively; $\mathbf{z}^{(n+1)} = \left(f_{1,1}^{(n+1)},\ldots,f_{1,J}^{(n+1)},\ldots\ldots,f_{I,1}^{(n+1)},\ldots,f_{I,J}^{(n+1)}\right)^T = \left(z_1^{(n+1)},\ldots,z_J^{(n+1)},\ldots,z_{(I-1)\cdot J+1}^{(n+1)},\ldots,z_{I\cdot J}^{(n+1)}\right)^T$, $f_{i,j}^{(n)}$ is the value of the nth iteration of $f(x,y)$ at grid cell (x_i, y_i) ; $z_{(i-1)\cdot J+j}^{(n+1)} = f_{i,j}^{(n+1)}$ for $1 \le i \le I$, $1 \le j \le J$.

If $\overline{f}_{i,j}$ is value of $z = f(x,y)$ at the pth sampled point (x_i, y_j), $s_{p,(i-1)\times J+j} = 1$, $k_p = \overline{f}_{i,j}$. There is only one non-zero element, 1, in every row of the coefficient matrix, $\mathbf{S}$, making it a sparse matrix. The solution procedure of YUE-HASM, taking the sampled points as its constraints, can be transformed into solving the following linear equation set in terms of least squares principle.

$$\begin{bmatrix} \mathbf{A}^T & \mathbf{B}^T & \mathbf{C}^T & \lambda \cdot \mathbf{S}^T \end{bmatrix} \begin{bmatrix} \mathbf{A} \\ \mathbf{B} \\ \mathbf{C} \\ \lambda \cdot \mathbf{S} \end{bmatrix} \mathbf{z}^{(n+1)} = \begin{bmatrix} \mathbf{A}^T & \mathbf{B}^T & \mathbf{C}^T & \lambda \cdot \mathbf{S}^T \end{bmatrix} \begin{bmatrix} \mathbf{d}^{(n)} \\ \mathbf{q}^{(n)} \\ \mathbf{p}^{(n)} \\ \lambda \cdot \mathbf{k} \end{bmatrix} \tag{18.13}$$

The parameter λ is the weight of the sampling points and determines the contribution of the sampling points to the simulated surface. λ can be a real number, which means all sampling points have the same weight, or a sector, which means every sampling point has its own weight. An area affected by a sampling point in a complex region is smaller than in a flat region. Therefore, a smaller value of λ is selected in a complex region, and a bigger value of λ is selected in a flat region.

Let

$$\mathbf{W} = \begin{bmatrix} \mathbf{A}^T & \mathbf{B}^T & \mathbf{C}^T & \lambda \cdot \mathbf{S}^T \end{bmatrix} \begin{bmatrix} \mathbf{A} \\ \mathbf{B} \\ \mathbf{C} \\ \lambda \cdot \mathbf{S} \end{bmatrix} \text{ and,}$$

$\mathbf{v}^{(n)} = \begin{bmatrix} \mathbf{A}^T & \mathbf{B}^T & \mathbf{C}^T & \lambda \cdot \mathbf{S}^T \end{bmatrix} \begin{bmatrix} \mathbf{d}^{(n)} \\ \mathbf{q}^{(n)} \\ \mathbf{p}^{(n)} \\ \lambda \cdot \mathbf{k} \end{bmatrix}$, then YUE-HASM has the following formulation:

$$\mathbf{W} \cdot \mathbf{z}^{(n+1)} = \mathbf{v}^{(n)} \tag{18.14}$$

In terms of Gauss-Codazii equation set, iteration stopping criterion of YUE-HASM is formulated as,

$$\begin{aligned} &\left(\varphi_{1y} - \phi_{2x} - \varphi_2 P - \phi_1 Q\right)^2 + \\ &\left(\varphi_{2x} - \phi_{1y} - \varphi_1 Q - \phi_2 P\right)^2 \\ &\left(Q_x + P_y + \varphi_1 \varphi_2 - \phi_1 \phi_2\right)^2 < \text{EI} \end{aligned} \tag{18.15}$$

where $\varphi_1 = \frac{L}{\sqrt{E}}$; $\varphi_2 = \frac{N}{\sqrt{G}}$; $P = \frac{\sqrt{E}_y}{\sqrt{G}}$; $Q = \frac{\sqrt{G}_x}{\sqrt{E}}$; $\phi_1 = \frac{M}{\sqrt{G}}$; $\phi_2 = \frac{M}{\sqrt{E}}$; EI is the iteration stopping criterion of YUE-HASM determined by an application requirement for accuracy.

Many studies described the essential significance of both satellite remotely sensed data and ground-based observation data, but the challenge is how to combine these two kinds of information. YUE-HASM incorporates reinforced machine learning and provides a solution to this challenge (Haber 2021). It can be transformed into a large sparse linear system and combined with the HarrowHassidim-Lloyd (HHL) quantum algorithm to support quantum machine learning (Yue et al. 2022, 2023).

There are several applications of YUE-HASM, including the simulation of climate change (Yue et al. 2013a, 2013b, Zhao and Yue 2014), the development of digital elevation maps (Yue et al. 2007a, 2010a, 2010b, Yue and Wang 2010), and the interpolation of soil properties (Yue 2011, Shi et al. 2009, 2011).

18.3 CASE STUDY: SURFACE MODELING FOR FOREST CARBON STOCKS IN CHINA

By integrating photosynthesis, autotrophic respiration, and litterfall fluxes, biomass dynamics reflects the potential of vegetation to act as a long-term carbon sink (Thurner et al. 2014). Forest ecosystems cover more than 41 million km^2 of the Earth's land area and are thought to contain about half of the carbon in terrestrial biomes (Prentice et al. 2001). Many studies on forest carbon stocks have been conducted in China in recent years.

According to the Integrated Terrestrial Ecosystem C-budget model, China's forests were a source of 21.0±7.8 TgC yr^{1} from 1901 to 1949 due to human activities. This source increased by 122.3±25.3 TgCyr^{1} between 1950 and 1987 due to intensified human activities. However, forests became large sinks of 176.7±44.8 TgCyr^{1} from 1988 to 2001 because of large-scale plantation programs and forest regrowth in previously disturbed areas as well as climatic warming, atmospheric CO_2 fertilization, and N deposition (Wang et al. 2007). The latter factors contributed to increasing forest growth, resulting in more CO_2 uptake. Vegetation carbon sink was defined as carbon sequestration from the atmosphere (1.63 times NPP), vegetation carbon stock as the carbon content that aboveground vegetation holds, and soil carbon stock as the carbon content that soil organic matter holds. In terms of these definitions, the vegetation carbon stock was 1.58 Pg C and the soil carbon stock was 1.41 Pg C in forest ecosystems of China from 1981 to 2000 (Deng et al. 2011). Forest stands have great potential to sequester more biomass carbon in the future due to large fractions of young and middle-aged forests. According to available national-scale information, mean vegetation carbon in China was 36.98 Pg, and mean soil carbon was 100.75 Pg. Forests had mean vegetation carbon of 5.49 Pg, whereas grasslands had 1.41 Pg (Ni 2013).

According to the continuous biomass expansion factor (BEF) method, based on field measurements of forests plots in different age classes and forest inventory data, carbon density (CD) of the forests in the Pearl River Delta increased by 14.3% from 1.908 to 2.181 kg C m^{-2} during the period from 1989 to 2003 (Yang and Guan 2008). The factors mean biomass density (MBD), mean ratio (MR), and continuous biomass expansion factor (BEF) were applied to forest inventory data to estimate China's forest biomass carbon stocks and their changes from 1984 to 2003. In terms of the BEF, carbon stocks of forest biomass increased from 4.0 to 5.9 Pg C. According to the MR, the carbon stocks (CS) rose from 4.2 to 6.2 Pg, and according to the MBD, the carbon stocks grew from 5.7 to 7.7 Pg (Guo et al. 2010). A study based on seven forest inventories from 1973 to 2008 indicated that total biomass carbon stock (CS) for all forest types increased by 65% from 1973 to 2008 and recently reached 8.12 Pg in China (Zhang et al. 2013).

A satellite-based approach suggested that the averaged forest biomass CS and CD in northeastern China were respectively 2.10 Pg (1 Pg = 10^{15} g) and 4.465 kg m^{-2} between 1982 and 1999. The forest biomass CS increased by 7% with an annual rate of 0.0082 Pg (Tan et al. 2007). It was highlighted that the vegetation restoration in the subalpine coniferous forest was a large carbon sink in the western Sichuan province of China (Xian et al. 2009). The total forest biomass of China averaged 5.79 Pg C between 1981 and 1999, with an average biomass density of 4.531 kg C m-2. The total forest biomass CS increased from 5.62 Pg in the early 1980s to 5.99 Pg by the end of the 1990s, giving a total increase of 0.37 Pg and an annual sequestration rate of 19 kg yr^{1} (Piao et al. 2005). China's terrestrial ecosystems were a net carbon sink in the range of 0.19–0.26 Pg carbon per year and absorbed 28–37% of its cumulated fossil carbon emissions during the 1980s and 1990s. Northeast China is a net source of CO_2 due to overharvesting and forest degradation while southern China accounts for more than 65% of the carbon sink, which can be attributed to regional climate change, large-scale plantation programs initiated in the 1980s, and shrub recovery (Piao et al. 2009).

18.3.1 Data

18.3.1.1 Data Produced Using the National Forest Inventory Database

The national forest inventory database consists of seven datasets obtained from inventory data measured from 1984 to 1988 (termed as P1), 1989 to 1993 (P2), 1994 to 1998 (P3), 1999 to 2003 (P4), 2004 to 2008 (P5), 2009 to 2013 (P6), and 2014 to 2018 (P7). The data included 160,000 permanent sample plots and 90,000 temporary sample plots scattered over the whole land of China. The biomass density of each forest type in each province was calculated using linear relationships between a biomass expansion factor (BEF) and timber volume (Table 18.1):

$$BEF_{i,j} = a_i + \frac{b_i}{V_{i,j}} \tag{18.16}$$

The carbon density of each forest type in each province was calculated by multiplying biomass density by a carbon factor (CF) (Table 18.2). The CS by forest type in individual provinces was calculated by multiplying forest type area by CD. The total CS in China was the sum of the CS of all forest types in the 31 provinces of China, excluding Taiwan, Hong Kong, and Macao.

TABLE 18.1
Parameters Used to Calculate the Biomass Expansion Factor (BEF) for Different Forest Types in China (From Fang et al. 2007)

Forest Type	*a* (kg/m³)	*b* (kg/m²)	Number of Samples	R^2
Abies, Picea	551.9	4.8861	24	0.78
Tsuga, Cryptomeria, Keteleeria	349.1	3.9816	30	0.79
Larix	609.6	3.3806	34	0.82
P. koraiensis	572.3	1.6489	22	0.93
P. sylyestris var. mongolica	1,112	0.2695	15	0.85
P. tabulaefomis	869	0.9121	112	0.91
P. armandii	585.6	1.8744	9	0.91
P. massoniana, P. yunnanensis	503.4	2.0547	52	0.87
Cunninghamialanceolata	465.2	1.9141	90	0.94
Cypress	889.3	0.7397	19	0.87
Other pines and conifer forests	529.2	2.5087	19	0.86
Deciduous oaks	1,145.3	0.8547	12	0.98
Betula	1,068.7	1.0237	9	0.7
Mixed deciduous and Sassafras	978.8	0.5376	35	0.93
Eucalyptus	887.3	0.4554	20	0.8
Casuarina	744.1	3.2377	10	0.95
Populus	496.9	2.6973	13	0.92
Lucidophyllous forests	929.2	0.6494	24	0.83
Nonmerchantable woods	1,178.3	0.2559	17	0.95
Mixed conifer and deciduous	813.6	1.8466	10	0.99
Tropical forests	797.5	0.042	18	0.87

TABLE 18.2
Carbon Factor (CF) for Every Forest Type in China (From Li and Lei 2010)

Forest Type	CF	Forest Type	CF	Forest Type	CF
Pinuskoraiensis	0.5113	*Pinusyunnanensis*	0.5113	*Tilia*	0.4392
Abiesfabri	0.4999	*Pinuskesiya* var. langbianensis	0.5224	*Sassafras tzumu*	0.4848
Piceaasperata	0.5208	*Pinusdensata*	0.5009	*Eucalyptus robusta* Smith	0.5253
Tsugachinensis	0.5022	*Cunninghamialanceolata*	0.5201	*Casuarinaequisetifolia*	0.4980
Cupressusfunebris	0.5034	*Cryptomeriafortunei*	0.5235	*Populus*	0.4956
Larixgmelinii	0.5211	Metasequoiaglyptostroboides	0.5013	*Firmiana*	0.4695
*Pinussylvestris*var. mongolica	0.5223	Coniferous mixed forest	0.5101	Nonmerchantable woods	0.4834
Pinusdensiflora	0.5141	Broad-leaved and coniferous mixed forest	0.4978	Broad-leaved mixed forest	0.4900
Pinusthunbergii	0.5146	*Fraxinusmandschurica, Juglansmandshurica, Phellodendronamurense*	0.4827	Coppice	0.5000
Pinustabuliformis	0.5207	*Cinnamomumcamphora*	0.4916		
Pinusarmandii	0.5225	*Phoebe zhennan*	0.5030		
Keteleeriafortunei	0.4997	Oaks	0.5004		
Pinusmassoniana	0.4596	*Betula*	0.4914		

The following formulations were used to calculate the forest CS in China:

$$TCS = \sum_{i=1}^{M}\sum_{j=1}^{N} A_{i,j} \cdot BCD_{i,j} \cdot 10^{-12} \tag{18.17}$$

$$BCD_{i,j} = W_{i,j} \cdot CF_i \tag{18.18}$$

$$W_{i,j} = BEF_i \cdot V_{i,j} \tag{18.19}$$

where, TCS is the total forest carbon stocks of China (Pg); $BCD_{i,j}$ is the biomass carbon density of the i_{th} forest type in the j_{th} province, area weighted by mean forest carbon stocks (kg/m^2); $A_{i,j}$ is the area of the i_{th} forest type in the j_{th} province (m^2); M and N are respectively the number of forest types and number of provinces in China; $W_{i,j}$ is the area weighted mean forest biomass of the i_{th} forest type in the j_{th} province (kg m^{-2}); CF_i is the carbon factor of the i_{th} forest type; $V_{i,j}$ is the area weighted mean timber volume of the i_{th} forest type in the j_{th} province (m^3 m^{-2}); BEF_i is the biomass expansion factor of the i_{th} forest type (kg m^{-3}); a_i (kg m^{-3}) and b_i (kg m^{-2}) are constants for the i_{th} forest type to be simulated.

The CF_i value for coniferous mixed forests is an average of CF_i values for all coniferous forest types. For broad-leaved mixed forests and broad-leaved and coniferous mixed forests, the CF_i values were the averages of CF_i values for broad-leaved mixed forests and broad-leaved and coniferous forests, respectively.

18.3.1.2 Data Produced Using a Satellite-Based Approach (SBA)

The satellite data consisted of the Normalized Difference Vegetation Index (NDVI) from the Earth Observation System's moderate-resolution imaging spectroradiometer (EOS MODIS) at a spatial resolution of 1×1 km and 1-month interval. The spatial distribution map of China forests was obtained from the Vegetation Map of China published by the Institute of Botany, Chinese Academy of Sciences, in 2000. The carbon densities from forest inventory data were matched with NDVI data through the spatial distribution map of China forests.

The biomass carbon density was closely related to latitude, longitude, and the maximum value of the monthly averaged NDVI values during periods of national forest inventories, such as from 2004 to 2008 (Piao et al. 2005), i.e.,

$$\begin{aligned} BCD_j &= 93.351\ln(NDVI_j) - 2.96Lat_j - 21.388Lon_j \\ &\quad + 0.047Lat_j^{\,2} + 0.091Lon_j^{\,2} + 1339.03 \end{aligned} \tag{18.20}$$

where, $NDVI_j$ is the mean value of the maximum values of the monthly averaged NDVI values in a period of the national forest inventory in the j_{th} province; Lat_j and Lon_j are latitude and longitude of the center of the j_{th} province, respectively. The coefficient of correlation was 0.91 ($P<0.001$).

18.3.2 Validation

Cross-validation included four steps: (i) 5% of the control points of each forest type in each province were removed for validation prior to model creation; (ii) the spatial distribution of average forest CS in China from 2004 to 2008 was simulated at a spatial resolution of 5×5 km using the remaining 95% of the control points; (iii) mean absolute error (MAE) and mean relative error (MRE) were calculated using the 5% validation set; and (iv) the 5% validation set was returned to the pool of available stations for the next iteration, and another 5% validation set was removed. This process was repeated until all of the control points were used for validation at least one time, and the simulation error statistics for each control point could be calculated.

The MAE and MRE were respectively formulated as,

$$MAE = \frac{1}{n}\sum_{i}^{n}|o_i - s_i| \tag{18.21}$$

$$MRE = \frac{MAE}{\frac{1}{n}\sum_{i}^{n}|o_i|} \tag{18.22}$$

where, o_i represents forest carbon stocks at the i_{th} control point for validation, s_i is the simulated value at the i_{th} control point for validation, and n_i is the total number of control points for validation.

The cross-validation results indicated (Figure 18.1) that MAEs from YUE-HASM and SBA were 1.303 kg·m^{-2} and 1.919 kg·m^{-2}, respectively, while MREs were 33% and 49%. MRE for YUE-HASM was 16% lower than MRE for SBA. A regression analysis showed that the simulated CS using YUE-HASM was related to the observed CS more closely compared with the results obtained using SBA. The correlation coefficient between the YUE-HASM results and the observed data was 0.99 while the coefficient between the SBA results and the observed data was 0.89 (Figure 18.2). In other words, YUE-HASM had a much higher accuracy than SBA.

The annual mean carbon stock (AMCS) predicted using YUE-HASM for all forest types in China was 7.08 Pg between 2004 and 2008, of which AMCS for coniferous, broadleaf, and mixed forests were respectively 2.74, 3.95, and 0.39 Pg, respectively (Table 18.3). The annual mean carbon density (AMCD) was 4.55 kg/m^2 between 2004 and 2008, of which AMCD for coniferous forests, broadleaf forests, and mixed forests were 4.351, 4.742, and 4.198 kg/m^2, respectively. SBA underestimated AMCD for coniferous and broadleaf forests but overestimated AMCD for mixed forests.

18.3.3 Results

A zoning system dividing the land mass of China into nine regions with similar temperature, precipitation, and soil regimes was adopted to facilitate the analysis of the changes in forest carbon storage among different locations (Zhou et al. 1981). The nine regions were respectively termed as R_i (i=1 to 9) (Figure 18.3). The simulation results using YUE-HASM indicated that about 90% of AMCS was distributed in the R3, R5, R6, R7, and R9 regions, which accounted for 28.6%, 27.9%, 15.3%, 11.7%, and 6.1%, respectively, in the 2004–2008 period. The three largest AMCDs were in the regions R5 (Tibet plateau), R2 (arid area), and R3 (northeastern China), and the two smallest in R8 and R9 (Table 18.4 and Figure 18.1b). Results by forest types indicated that the AMCD of evergreen forests was greater than the AMCD for deciduous forests (Table 18.5). AMCS for evergreen coniferous forests had the highest proportion, accounting for 33.05%; which was followed by deciduous broad-leaved forests, accounting for 29.8%; and evergreen broad-leaved forests, accounting for 25.99% (Table 18.5). The broad-leaved and coniferous mixed forests and the deciduous coniferous forests accounted for the first two smallest proportions of the total CS, respectively 5.51% and 5.65%.

YUE-HASM simulation results indicated that AMCS rose 4.22 Pg from P1 to P7, mainly because of the increase in AMCD and the expansion of forest area (Table 18.6). AMCD increased by 1.035 kg m^{-2} from P1 to P7. The forest area increased by 0.3458 million km^2 in the same period. The regression analysis of the patterns of increase in AMCS, AMCD, and forest area resulted in significant relationships.

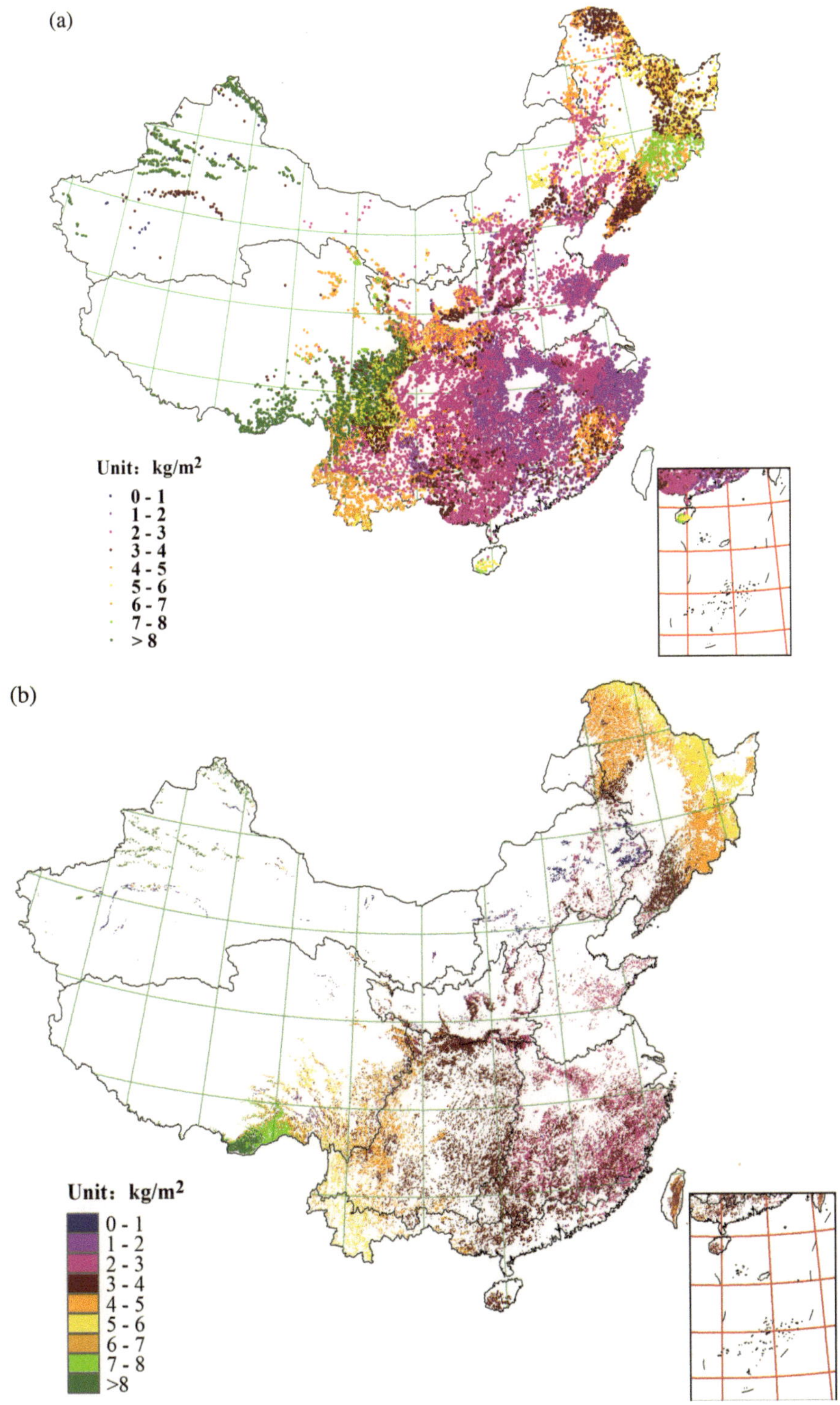

FIGURE 18.1 Spatial distribution of annual mean carbon density (AMCD) in forests during the 2004–2008 period in China: (a) observed, (b) using a satellite-based approach, and (c) YUE-HASM earth surface modeling methods.

(Continued)

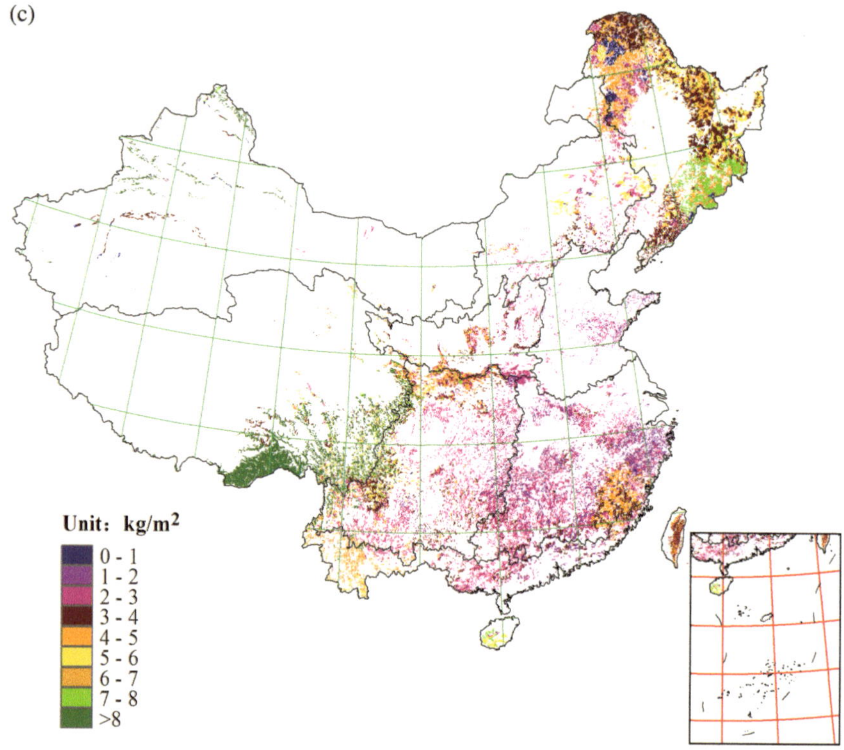

FIGURE 18.1 (*Continued*)

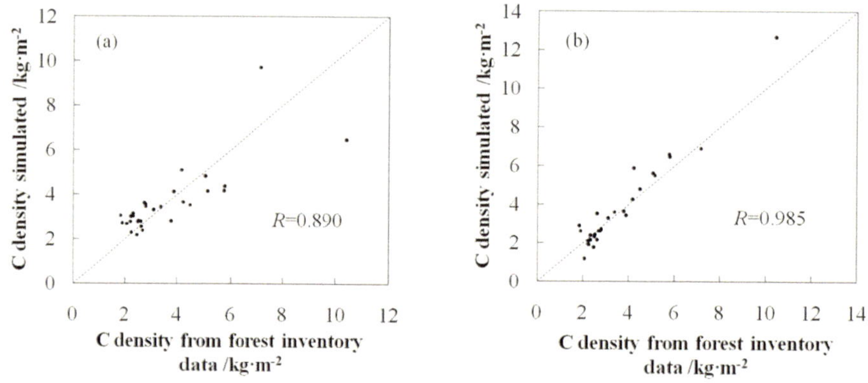

FIGURE 18.2 Comparison between observed and simulated results for annual mean carbon density (AMCD) in the 31 provinces obtained by (a) satellite-based approach and (b) YUE-HASM earth surface modeling methods.

TABLE 18.3
Comparison of Annual Mean Carbon Stocks (AMCS) and Annual Mean Carbon Density (AMCD) Using a Satellite-Based Approach (SBA) and YUE-HASM for Different Forest Types in China

	SBA		YUE-HASM	
Forest Type	**AMCS (Pg)**	**AMCD (kg/m²)**	**AMCS (Pg)**	**AMCD (kg/m²)**
Coniferous forests	2.48	3.937	2.74	4.351
Mixed forests	0.46	4.926	0.39	4.198
Broadleaf forests	3.61	4.338	3.95	4.742
Total	6.55		7.08	

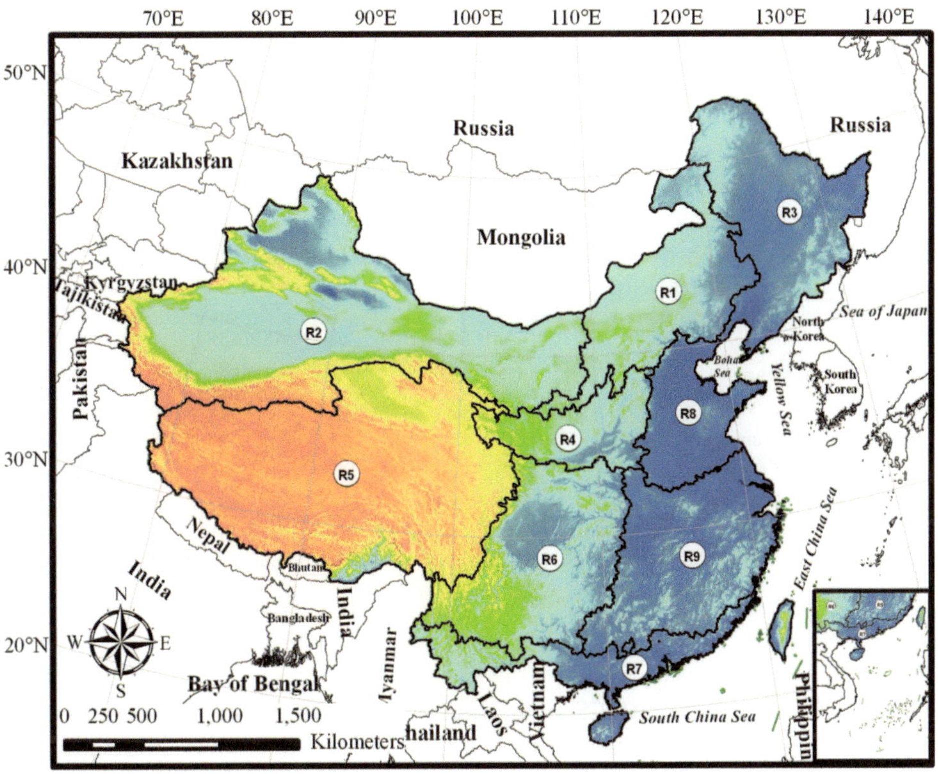

FIGURE 18.3 The nine regions of China characterized by similar temperature and precipitation conditions and soil regimes. R1 represents Inner Mongolia and the area along the Great Wall, R2 represents Arid area of China, R3 represents Northeastern China, R4 represents Loess plateau, R5 represents Tibet plateau, R6 represents Southwest China, R7 represents Southern China, R8 represents Huang-Huai-Hai River Plain, and R9 represents Middle and lower reaches of Yangtze River Basin.

TABLE 18.4
Annual Mean Carbon Stocks (AMCS) and Annual Mean Carbon Density (AMCD) for Forests in the Nine Regions of China from 2014 to 2018 and from 1984 to 1988

	2014–2018			1984–1988		CS Accumulation Rate
Regions	AMCD ($kg{\cdot}m^{-2}$)	AMCS (Pg)	Percentage (%)	AMCD ($kg{\cdot}m^{-2}$)	AMCS (Pg)	($Pg{\cdot}year^{-1}$)
R1	4.946	0.43	4.75	2.666	0.16	0.0090
R2	7.250	0.26	2.87	6.358	0.15	0.0037
R3	5.122	2.67	29.47	4.493	1.59	0.0360
R4	3.751	0.25	2.76	3.035	0.13	0.0040
R5	11.079	2.45	27.04	6.718	0.99	0.0487
R6	4.204	1.35	14.9	3.734	0.82	0.0177
R7	3.992	0.52	5.74	3.643	0.36	0.0053
R8	2.370	0.08	0.88	1.515	0.03	0.0017
R9	2.736	1.05	11.59	2.358	0.62	0.0143
Total		9.06	100		4.84	0.1407

TABLE 18.5
Annual Mean Carbon Density (AMCD) and Annual Mean Carbon Stocks (AMCS) for Different Forest Types in China from 2014 to 2018

Forest Types	AMCD ($kg\ m^{-2}$)	AMCS (Pg)	Percentage (%)
Deciduous coniferous forests	4.214	0.59	6.51
Evergreen coniferous forests	4.527	2.49	27.48
Broad-leaved and coniferous mixed forests	4.467	0.67	7.40
Deciduous broad-leaved forests	4.847	2.86	31.57
Evergreen broad-leaved forests	6.622	2.45	27.04
Total		9.06	100

TABLE 18.6
Annual Mean Carbon Stocks (AMCS) and Annual Mean Carbon Density (AMCD) of China in the Seven Study Periods (Excluding the Taiwan Province, Hong Kong, and Macao)

Periods	Area (million km^2)	AMCS (Pg)	AMCD ($kg\ m^{-2}$)
P1	1.2101	4.84	4.001
P2	1.2864	5.55	4.315
P3	1.2920	5.60	4.334
P4	1.4279	6.38	4.469
P5	1.5559	7.08	4.550
P6	1.6460	7.97	4.842
P7	1.7989	9.06	5.036

$$AMCS(t) = 0.6779t + 3.9286, R^2=0.96 \quad (18.23)$$

$$AMCD(t) = 0.1563t + 3.8817, R^2=0.95 \quad (18.24)$$

$$FA(t) = 0.0982t + 1.10668, R^2=0.96 \quad (18.25)$$

where, t corresponds to Pt, t = 1 to 7. $AMCS(t)$, $AMCD(t)$, and $FA(t)$ are respectively AMCS, AMCD, and forest area in the period of Pt.

Although AMCS increased in the nine regions from P1 to P7, it spatially varied over China, in concordance with forest distribution (Figure 18.4, Tables 18.4 and 18.7). In the regions R1 and R8, both AMCS and AMCD were characterized by a pattern of increase from P1. The smallest AMCS were obtained in R8, which accounted for only 0.88% of the total AMCS of China, and the smallest AMCD of 2.37 kg/m^2 in the period P7 as well as the lowest CS accumulation rate of 0.0017Pg·yr^{-1}. In R1, AMCS accounted for 4.75% of the total AMCS of China in the period P7; the CS accumulation rate was 0.0090 Pg·yr^{-1} on average.

The region R3 had the largest AMCS accounting for 29.47% of total AMCS of China in P7. The AMCS had a monotonically increasing trend since P1. The second largest AMCS appeared in the region R5. Its AMCS accounted for 27.04% of the total AMCS. It has the largest AMCD and the fastest CS accumulation rate of 0.0487 Pg year^{-1} (Table 18.7). In the regions R4 (loess plateau), R6, and R9, both AMCS and AMCD increased since P3. AMCS in R4, R6, and R9 accounted for 2.76%, 14.9%, and 11.59%, respectively, in the period P7. The CS accumulation rate was 0.0040, 0.0177, and 0.0143 Pg year-1. In R2, AMCS accounted for 2.87% in the period P7; AMCS and AMCD increased from the periods P4 to P7. The accumulation rate in CS was 0.0037 Pg year-1. In R7, AMCS accounted for 5.74%; AMCS and AMCD increased from the period P4 to P7. The CS accumulation rate was 0.0053 Pg year^{-1}.

In terms of forest types, evergreen broad-leaved forests had the fastest CS accumulation rate and the largest AMCD while deciduous broad-leaved forests had the biggest AMCS (Table 18.8). AMCS of broad-leaved forests increased during the seven periods while AMCS for evergreen broad-leaved forests increased from 0.63 to 2.45 Pg from P1 to P7, and it increased from 1.38 to 2.86 Pg in deciduous broad-leaved forests during the same period. Their patterns of increase had the following regression equations, respectively:

$$AMCS_1(t) = 0.305t + 0.3014, R^2=0.99 \quad (18.26)$$

$$AMCS_2(t) = 0.2539t + 0.9800, R^2=0.97 \quad (18.27)$$

where, t corresponds to Pt, t = 1 to 7; $AMCS_1(t)$ and $AMCS_2(t)$ are, respectively, AMCS for evergreen broad-leaved and deciduous broad-leaved forests in the period Pt.

AMCS for deciduous coniferous forests fluctuated among periods (Table 18.8). Evergreen coniferous and broad-leaved and coniferous mixed forests showed general patterns of increase in AMCS, which were formulated by the following regression equations,

$$AMCS_3(t) = 0.1521t + 1.5257, R^2=0.85 \quad (18.28)$$

$$AMCS_4(t) = 0.1046t - 0.1400, R^2=0.90 \quad (18.29)$$

where, t corresponds to Pt, t = 1 to 7; $AMCS_3(t)$ represents AMCS for evergreen coniferous forests in the period of Pt, and $AMCS_4(t)$ is AMCS for broad-leaved and coniferous mixed forest.

(a)

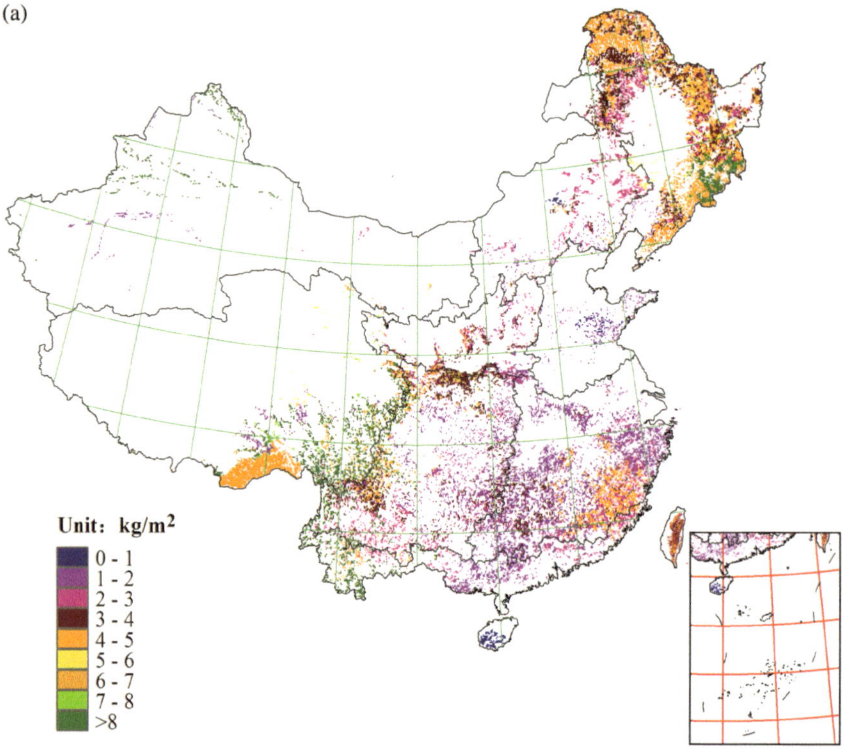

(b)

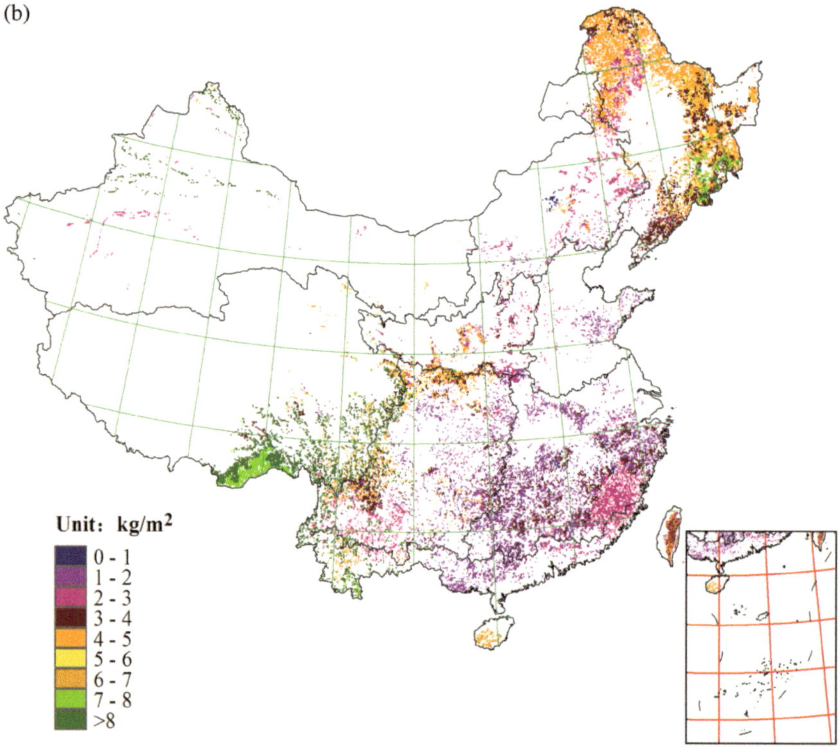

FIGURE 18.4 Spatial distribution of forest annual mean carbon density (AMCD) in China in the seven study periods: (a) 1984–1988 (P1), (b) 1989–1993 (P2), (c) 1994–1998 (P3), (d) 1999–2003 (P4), (e) 2004–2008 (P5), (f) 2009–2013 (P6), and (g) 2014–2018 (P7).

(Continued)

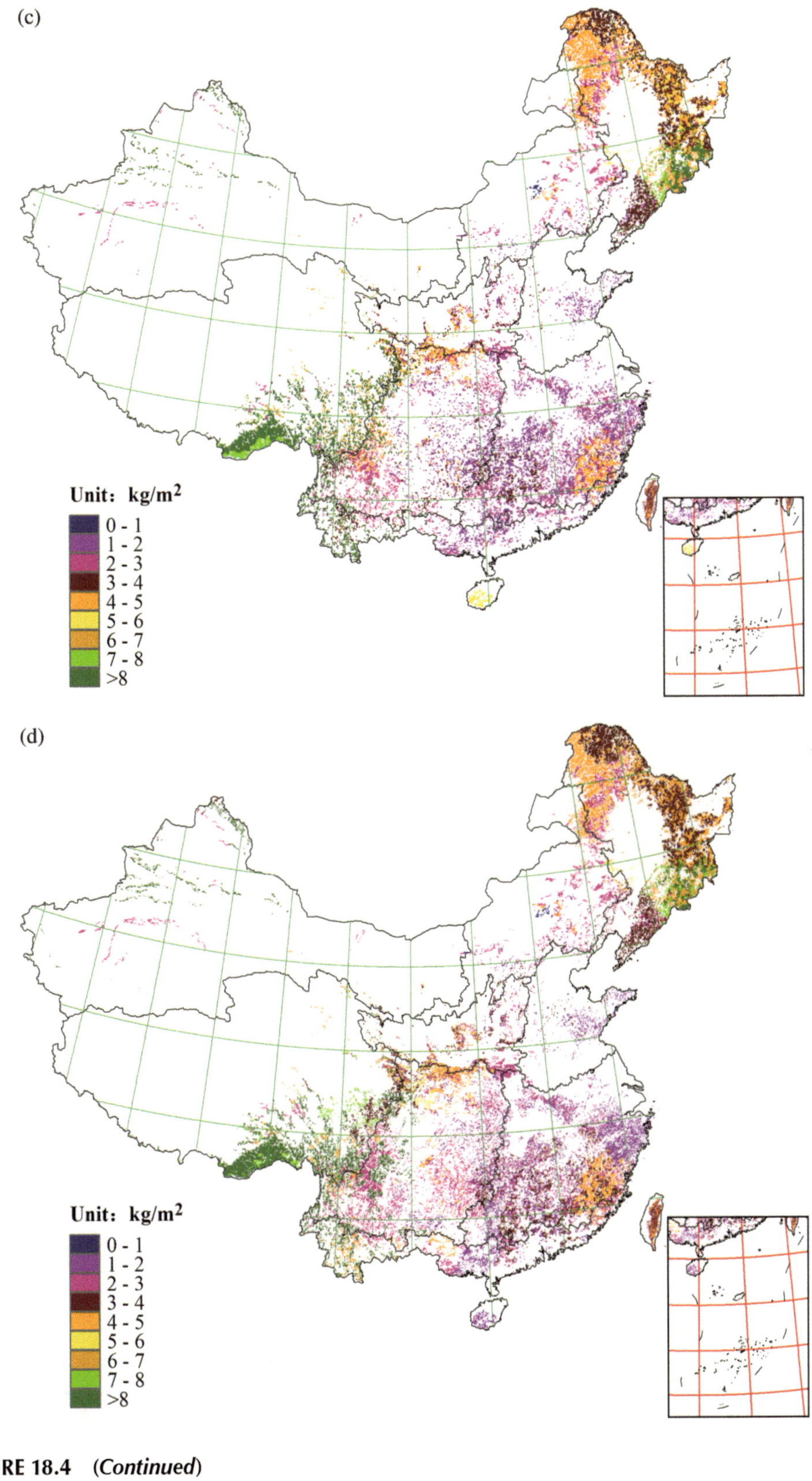

FIGURE 18.4 *(Continued)*

(e)

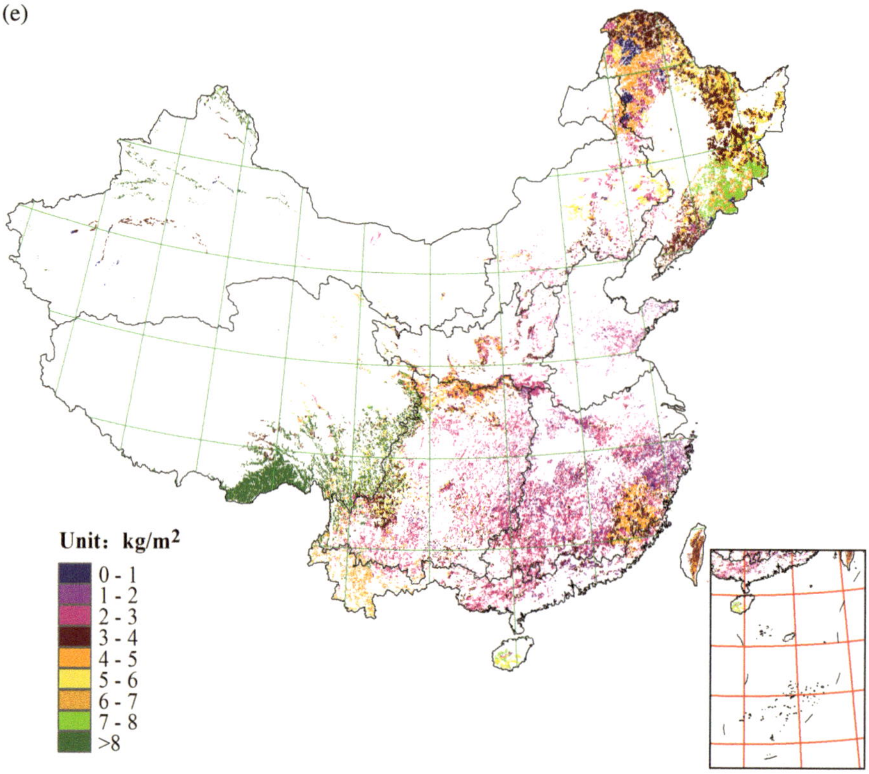

(f)

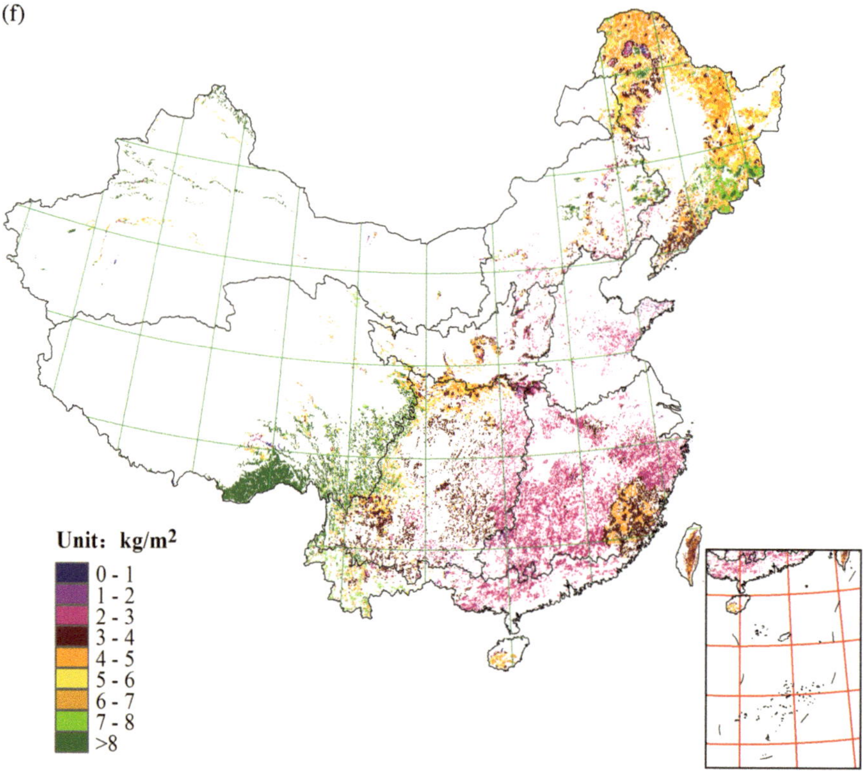

FIGURE 18.4 ***(Continued)***

(g)

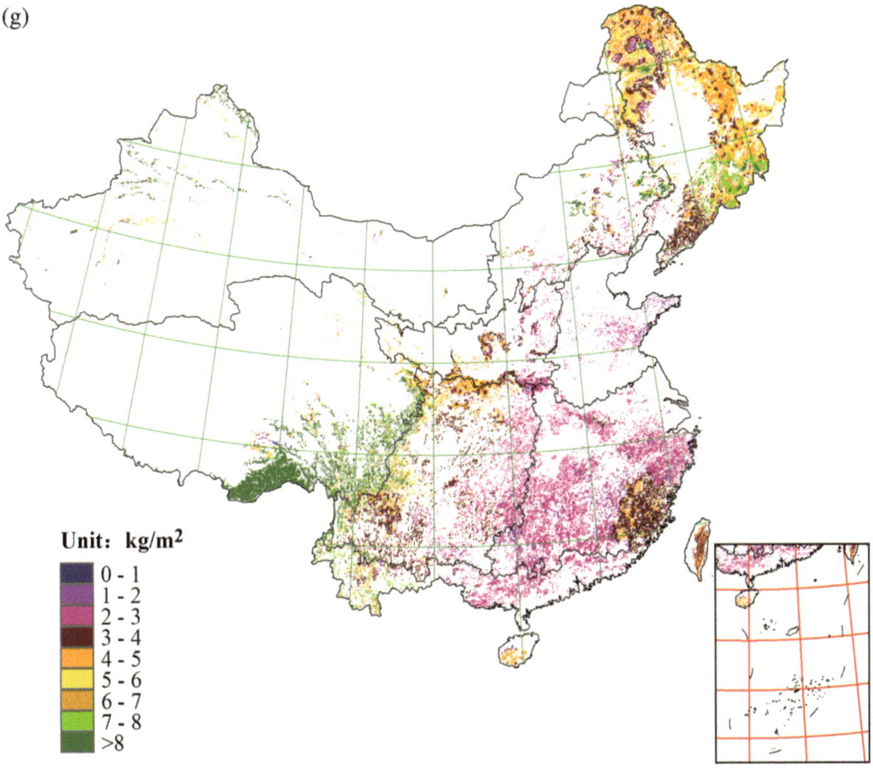

FIGURE 18.4 ***(Continued)***

18.3.4 Discussion

Ground national forest inventory is able to accurately estimate forest carbon stocks with high temporal resolution using sample plots, but these sample plots are too sparse to satisfy the spatial simulation of carbon stocks with required accuracy. SBA can supply spatially continuous information about the surface of forest carbon stocks, which is impossible to obtain from ground-based investigations, but remote sensing description has considerable uncertainties. YUE-HASM overcame the shortcomings of the ground national forest inventory and SBA by fusing information about the details of the carbon stocks observed on the Earth surface and the change in the carbon surface observed from outside the Earth surface. The cross-validation demonstrates that YUE-HASM is 16% more accurate than the SBA.

In terms of the results obtained from YUE-HASM, forest carbon stocks of China increased by 4.42 Pg from 1984 to 2018. The Grain for Green program made a great contribution to this carbon stock increase. This program aims at restoring the forests and grasslands of China to prevent soil erosion. Program designers made the steepness of slopes, one of the main criteria on which plots were selected, for inclusion into the Grain for Green program (Xu et al. 2006). The steepness criterion means that the program targets land with a slope greater than 25°.

The Grain for Green program was divided into three phases: the experimental phase from 1999 to 2001, the construction phase from 2002 to 2010, and the consolidation phase from 2011 to 2020. By the end of 2001, 1.206 million hectares of farmland had been converted into forestland or grassland, and 1.097 million hectares of barren land had been afforested. Before 2011, 14.667 million hectares of farmland would have been converted to forestland or grassland, for which almost all

TABLE 18.7
Annual Mean Carbon Stocks (AMCS) and Annual Mean Carbon Density (AMCD) for Nine Chinese Regions in the Seven Study Periods

Forest types	P1		P2		P3		P4		P5		P6		P7	
	AMCS (Pg)	AMCD (kg m^{-2})	AMCS (Pg)	AMCD (kg m^{-2})	AMCS (Pg)	AMCD (kg m^{-2})	AMCS (Pg)	AMCD (kg m^{-2})	AMCS (Pg)	AMCD (kg m^{-2})	AMCS (Pg)	AMCD (kg m^{-2})	AMCS (Pg)	AMCD (kg m^{-2})
R1	0.16	2.67	0.17	2.71	0.17	2.88	0.2	2.98	0.27	3.57	0.38	4.77	0.43	4.95
R2	0.15	6.36	0.16	6.27	0.15	6.25	0.18	6.23	0.2	6.35	0.24	7.12	0.26	7.25
R3	1.59	4.49	1.64	4.42	1.64	4.5	1.77	4.43	1.98	4.37	2.34	4.91	2.67	5.12
R4	0.13	3.04	0.15	3.23	0.14	3.13	0.16	3.15	0.19	3.25	0.22	3.55	0.25	3.75
R5	0.99	6.72	1.57	10.15	1.64	10.83	1.94	11.49	2.03	10.53	2.21	10.92	2.45	11.08
R6	0.82	3.73	0.87	3.78	0.82	3.66	0.96	3.88	1.09	3.9	1.18	4.01	1.35	4.2
R7	0.36	3.64	0.39	3.79	0.37	3.66	0.4	3.54	0.44	3.87	0.45	3.81	0.52	3.99
R8	0.03	1.52	0.04	1.64	0.04	1.89	0.05	1.89	0.06	2.16	0.06	2.18	0.08	2.37
R9	0.62	2.36	0.56	2.05	0.62	2.3	0.73	2.51	0.83	2.49	0.89	2.53	1.05	2.74

TABLE 18.8
Annual Mean Carbon Stocks (AMCS) and Annual Mean Carbon Density (AMCD) for All Forest Types during the Seven Study Periods

Forest Types	P1		P2		P3		P4		P5		P6		P7	
	AMCS (Pg)	AMCD (kg m^{-2})	AMCS (Pg)	AMCD (kg m^{-2})	AMCS (Pg)	AMCD (kg m^{-2})	AMCS (Pg)	AMCD (kg m^{-2})	AMCS (Pg)	AMCD (kg m^{-2})	AMCS (Pg)	AMCD (kg m^{-2})	AMCS (Pg)	AMCD (kg m^{-2})
Deciduous coniferous forests	0.41	4.35	0.39	4.28	0.44	4.2	0.47	4.37	0.4	3.77	0.46	3.83	0.59	4.21
Evergreen coniferous forests	1.5	3.81	1.8	4.13	2.23	4.09	2.19	4.4	2.34	4.47	2.39	4.51	2.49	4.53
Broad-leaved and coniferous mixed forests	0.06	3.08	0.09	3.75	0.07	3.03	0.19	5.18	0.39	4.2	0.48	4.36	0.67	4.47
Deciduous broad-leaved forests	1.38	3.75	1.44	3.77	1.66	3.87	1.97	3.89	2.11	3.93	2.55	4.55	2.86	4.85
Evergreen broad-leaved forests	0.63	4.35	0.87	5.65	1.2	6.35	1.57	7.49	1.84	6.22	2.09	6.33	2.45	6.62

TABLE 18.9
Areas of Converted Farmlands and Afforested Barren Lands in the Grain for Green Program in China (Million Hectares)

Year	Converted Farmlands	Afforestation on Bare Lands	Total
1999	0.381	0.066	0.448
2000	0.405	0.468	0.872
2001	0.420	0.563	0.983
2002	2.647	3.082	5.729
2003	3.367	3.767	7.133
2004	0.667	3.333	4.000
2005	1.114	1.321	2.435
1999–2010	14.667	17.333	32

farmland with a slope more than 25° would have been converted, and 2.667 million hectares of cultivated desertification land have been converted to grassland; 17.333 million hectares of barren land would have been afforested. During the consolidation phase, scientific management of the converted land and afforestation land will be strengthened to keep the achievement of the Grain for Green program, for which 72.933 billion kg of grains and 11.515 billion of RMB have been planned (Li 2006).

In 1999, the pilot program was first experimented in three provinces, Gansu, Shaanxi and Sichuan; 0.381 million hectares of farmland were converted into forestland, and 0.066 million hectares of barren land were afforested. In 2000, the experimental area was expanded to 17 provinces; the converted farmland and the afforested barren land were respectively 0.41 million hectares and 0.449 million hectares. In 2001, 20 provinces were involved in the experiment; 0.42 million hectares of farmland were converted into forestland, and 0.563 million hectares of barren land were afforested (Table 18.9). In 2002, the Grain for Green program was overall launched in China. By 2005, 9.001 million hectares had been withdrawn from farmland and planted with trees or converted to permanent grassland.

REFERENCES

Ahlberg, J.H., E.N. Nilson and J.L. Walsh. 1967. *The theory of splines and their application*. New York: Academic Press.

Baffetta F, P. Corona and L. Fattorini. 2012. A matching procedure to improve k-NN estimation of forest attribute maps. *For. Ecol. Manage.* 272:35–50.

Bengtsson, B.E. and S. Nordbeck. 1964. Construction of isarithms and isarithmic maps by computers. *BIT* 4:87–105.

Costanza, R. 1989. Model goodness of fit: A multiple resolution procedure. *Ecol. Modell.* 47:199–215.

Costanza, R. and T. Maxwell. 1991. Spatial ecosystem modeling using parallel processors. *Ecol. Modell.* 58:159–183.

Costanza, R., F.H. Sklar, and M.L. White. 1990. Modeling costal landscape dynamics. *BioScience* 40(2):91–107.

Deichmann, U. 1996. *A review of spatial population database design and modeling*. Technical Report 96–3, National Center for Geographic Information and Analysis, USA.

Deng H.B., P. Zheng, T.X. Liu, et al. 2011. Forest ecosystem services and eco-compensation mechanisms in China. *Environ. Manage.* 48:1079–1085.

Fang, J.Y., Z.D. Guo, S.L. Piao, et al. 2007. Terrestrial vegetation carbon sinks in China, 1981–2000. *Sci. China Ser. D. Earth Sci.* 50:1341–1350.

Friend, A.D. 1998. Parameterisation of a global daily weather generator for terrestrial ecosystem modeling. *Ecol. Modell.* 109:121–140.

Gao, Q. 1996. Dynamic modeling of ecosystems with spatial heterogeneity: A structured approach implemented in windows environment. *Ecol. Modell.* 85:241–252.

Guo, Z., J. Pang, Y. Pan, et al. 2010. Inventory-based estimates of forest biomass carbon stocks in China: A comparison of three methods. *For. Ecol. Manage.* 269:1225–1231.

Haber, W. 2021. Eco-environmental surface modelling requires integration of both extrinsic and intrinsic informations. *Sci China Earth Sci.* 64:185-186.

Haxeltine, A. and I.C. Prentice. 1996. BIOME3: An equilibrium terrestrial biosphere model based on ecophysiological constraints, resource availability, and competition among plant functional types. *Global Biogeochem. Cycles*10:693–710.

Ji, W. and C. Jeske. 2000. Spatial modeling of the geographic distribution of wildlife populations: A case study in the lower Mississippi River region. *Ecol. Modell.* 132:95–104.

Kleijnen, J.P.C. 2009. Kriging metamodeling in simulation: A review. *Eur. J. Oper. Res.* 192:707–716.

Kleijnen, J.P.C. and W.C.M. van Beers. 2005. Robustness of kriging when interpolating in random simulation with heterogeneous variances: Some experiments. *Eur. J. Oper. Res.* 165(3):826–834.

Krige, D.G. 1951. A statistical approach to some basic mine valuation problems on the Witwatersrand. *J. Chem. Metall. Min. Soc. S. Afr.* 52:119–139.

Legendre, L. and P. Legendre. 1983. *Numerical ecology*. Amsterdam: Elsevier Scientific Pul. Co.

Li, S.D. 2006. *Research on the optimized models of the conversion of farmland to forests in China*. Beijing: China Environmental Science Press (in Chinese).

Li, H. and Y.C. Lei. 2010. *Estimation and evaluation of forest biomass carbon storage in China*. Beijing: China Forestry Press (in Chinese).

Lo, C.P. and A.K.W. Yeung. 2002. *Concepts and techniques of geographic information systems*. Upper Saddle River, NJ: Prentice-Hall.

Long, G.E. 1980. Surface approximation: A deterministic approach to modeling spatially variable systems. *Ecol. Modell.* 8:333–343.

Magnussen, S., E. Naesset, and M.A. Wulder. 2007. Efficient multi-resolution spatial predictions for large data arrays. *Remote Sens. Environ.* 109:451–463.

Martin, D. and I. Bracken. 1991. Techniques for modeling population-related raster databases. *Environ. Plan.* 23:1069–1075.

Ni, J. 2013. Carbon storage in Chinese terrestrial ecosystems: approaching a more accurate estimate. *Clim. Change* 119:905–917.

Oliver, M.A. and R. Webster. 1990. Kringing: A method of interpolation for geographical information systems. *J. Geogr. Inf. Sci.* 4:313–332.

Perry, G.L.W. and N.J. Enright. 2002. Spatial modeling of landscape composition and pattern in a maquis-forest complex, Mont Do, New Caledonia. *Ecol. Modell.* 152:279–302.

Piao S.L., J.Y. Fang, B. Zhu, et al. 2005. Forest biomass carbon stocks in China over the past 2 decades: Estimation based on integrated inventory and satellite data. *J. Geophys. Res.* 110(G1). https://doi.org/10.1029/2005JG000014.

Piao S.L., J.Y. Fang, P. Ciais, et al. 2009. The carbon balance of terrestrial ecosystems in China. *Nature* 458:1009–1014.

Prentice I.C., L.O. Bengtsson, and C.U. Hammer. 2001. Interactions of climate change and the terrestrial biosphere. In *Geosphere-biosphere interactions and climate*, eds. L.O. Bengtsson and C.U. Hammer, pp. 176–195. Cambridge: Cambridge Books Online.

Quarteroni, A., R. Sacco, and F. Saleri. 2000. *Numerical mathematics*. New York: Springer.

Reich, R.M., C.D. Bonham, and K.L. Metzger. 1997. Modeling small-scale spatial interaction of shortgrass prairie species. *Ecol. Modell.* 101:163–174.

Schroeder, L.D. and D.L. Sjoquist. 1976. Investigation of population density gradients using trend surface analysis. *Land Econ.* 52:382–392.

Shepard, D. 1968. A two-dimensional interpolation function for irregularly-spaced data. In *Proceedings of the 1968 23rd ACM national conference*, pp. 517–524. New York: Association for Computing Machinery.

Shi, W.J., J.Y. Liu, Z.P. Du, et al. 2011. Surface modelling of soil properties based on land use information. *Geoderma* 162:347–357.

Shi, W.J., J.Y. Liu, Z.P. Du, et al. 2009. Surface modelling of soil ph. *Geoderma* 150(1–2):113–119.

Sklar, F.H., R. Costanza, and J.W. Day, Jr. 1985. Dynamic spatial simulation modeling of coastal wetland habitat succession. *Ecol. Modell.* 29:261–281.

Stott, J.P. 1977. Review of surface modeling. In *Proceedings of surface modeling by computer*, pp. 1–8. London: A conference jointly sponsored by the Royal Institution of Chartered Surveyors and the Institution of Civil Engineers, 1976.

Svirezhev, Y.M. 2002. Simple spatially distributed model of the global carbon cycle and its dynamic properties. *Ecol. Modell.* 155:53–69.

Tan, K., S.L. Piao, C.H. Peng, et al. 2007. Satellite-based estimation of biomass carbon stocks for northeast China's forests between 1982 and 1999. *For. Ecol. Manage.* 240:114–121.

Thurner, M., C. Beer, M. Santoro, et al. 2014. Carbon stock and density of northern boreal and temperate forests. *Glob. Ecol. Biogeogr.* 23:297–310.

Tse, R.O.C. and C. Gold. 2004. TIN meets CAD—Extending the TIN concept in GIS. *Future Gener. Comput. Syst.* 20:1171–1184.

Turner, M.G., R. Costanza, and F.H. Sklar. 1989. Methods to evaluate the performance of spatial simulation models. *Ecol. Modell.*48:1–18.

Wang, S., J.M. Chen, W.M. Ju, et al. 2007. Carbon sinks and sources in china's forests during 1901–2001. *J. Environ. Manage.* 85:524–537.

Watt, A. 2000. *3D Computer graphics*. New York: Addison-Wesley.

Wu, J.G. and S.A. Levin. 1997. A patch-based spatial modeling approach: conceptual framework and simulation scheme. *Ecol. Modell.*101:325–346.

Xian J.R., Y.B. Zhang, K.Y. Wang, et al. 2009. Carbon stock and its allocation in five forest ecosystems in the subalpine coniferous forest zone of western Sichuan province, southwest China. *Chin. J. Plant Ecol.* 33:283–290.

Xu ZG, J.T. Xu, X.Z. Deng, et al. 2006. Grain for Green versus grain: conflict between food security and conservation set-aside in China. *World Development* 34:130–148.

Yang, K. and D.S. Guan. 2008. Changes in forest biomass carbon stock in the Pearl River Delta between 1989 and 2003. *J. Environ. Sci.* 20:1439–1444.

Yang, Q.H., J.P. Snyder, and W.R. Tobler. 2000. *Map projection transformation*. New York, NY: Taylor &Francis.

Yue, T.X., C.C. Wu, Y. Liu, et al. 2023. HASM quantum machine learning. Science China-Earth *Sciences* 66(9):1937–1945.

Yue, T.X., Y. Liu, Z.P. Du, et al. 2022. Quantum machine learning of eco-environmental surfaces. *Sci. Bull.* 67:1031–1033.

Yue, T.X. 2011. *Surface Modelling: High accuracy and high speed methods*. New York: CRC Press.

Yue, T.X., S. Jorgensen, and G. Larocque. 2011. Progress in global ecological modelling. *Ecol. Modell.* 222:2172–2177.

Yue, T.X., C.F. Chen, and B.L. Li. 2010a. An adaptive method of high accuracy surface modeling and its application to simulating elevation surfaces. *Trans. GIS* 14:615–630.

Yue, T.X., D.J. Song, Z.P. Du, et al. 2010b. High-accuracy surface modelling and its application to dem generation. *Int. J. Remote Sens.* 31 2205–2226.

Yue, T.X., N. Zhao, H. Yang, et al. 2013b. A multi-grid method of high accuracy surface modeling and its validation. *Trans. GIS* 17:943–952.

Yue, T.X., N. Zhao, R.D. Ramsey, et al. 2013a. Climate change trend in China, with improved accuracy. *Clim. Change* 120:137–151.

Yue, T.X. and S.H. Wang. 2010. Adjustment computation of HASM: A high-accuracy and high-speed method. *Int. J. Geogr. Inf. Sci.* 24:1725–1743.

Yue, T.X., Y.A. Wang, and Z.M. Fan. 2009. Surface modeling of population distribution. In *Handbook of ecological modeling and informatics*, eds. S.E. jOrgensen, T.S. Chon, and F. Recknagel, pp. 71–98. Ashurst Lodge, Ashurst, Southampton S040 7AA, UK: WIT Press.

Yue, T.X., Z.M. Fan, and J.Y. Liu. 2007b. Scenarios of land cover in China. *Glob. Planet. Change* 55:317–342.

Yue, T.X., Z.P. Du, D.J. Song, et al. 2007a. Anew method of surface modeling and its application to DEM construction. *Geomorphology* 91:161–172.

Yue, T.X., Z.P. Du, and Y.J. Song. 2008. Ecological models: Spatial models and geographic information systems. In *Encyclopedia of ecology*, eds. S.E. Jorgensen and B. Fath, pp. 3315–3325. Oxford: Elsevier Limited.

Zhang C.H., W.M. Ju, J.M. Chen, et al. 2013. China's forest biomass carbon sink based on seven inventories from 1973 to 2008. *Clim. Change* 118:933–948.

Zhao, N. and T.X. Yue. 2014. A modification of HASM for interpolating precipitation in China. *Theor. Appl. Clim.* 116:273–285.

Zhao, C.Y., Z.G. Nan, and G.D. Cheng. 2005. Methods for modeling of temporal and spatial distribution of air temperature at landscape scale in the southern Qilian mountains, China. *Ecol. Modell.* 189:209–220.

Zhou L.S., H. Sun, Y.Q. Shen, et al. 1981. *Comprehensive agricultural planning of China*. Beijing: China Agricultural Press. (in Chinese)

19 Comprehensive Analysis of Land Tenure Effects on the Structure and Services of Regional Forest Ecosystems

Patrick Morin, Luc Sirois, and Luc Bouthillier

19.1 INTRODUCTION

When considering the optimal conditions for ecological forest management, the first variables that may come to mind as being determinants are intimately linked to silvicultural practices such as harvest method, machinery used, rotation period, and retention. But a fundamental characteristic that is often overlooked is the type of governance that a given patch of exploited forest is under (Dietz et al. 2003; Ostrom 2007; Agrawal et al. 2008). In a forest governance system, tenures are key policy instruments to adjust when facing challenges (Vertinsky and Luckert 2009).

Forest tenure creates a context in which forest management decisions emerge, and different tenures are conducive to different sets of motivations for forest management. Land tenure is defined "as the societal institutions (organizations, rules, rights, and restrictions) that control the allocation and use of land and its associated resources by people" (Cumming and Barnes 2007). The predominant forest tenure categories encountered are private and public (White and Martin 2002). Generally, public tenure refers to forests owned by central, regional, or local governments, while private forests are owned by first nations or local community groups, private individuals, or firms.

There has been a certain level of controversy over the tenure model that best optimizes the management of the forest resources. In his seminal "The tragedy of the commons," Hardin (1968) suggests that the solution to the overexploitation of common-pool resources is either private or public property. From that private–public dichotomy, more than 45 years of scientific research have not yet provided a comprehensive assessment of the tenure that is most likely to provide a sustainable outcome.

Privatization is argued by some as a key to improve productivity and avoid overexploitation of forests (Desrochers 2002; McElligott 2003). In a free-market economy, private property is associated with a high degree of freedom and closer control, which increases competitiveness. A belief that private belongings are better tended than collective goods also exists (Kindscher and Scott 1977). On the other hand, the mere thought of privatizing public resources raises a massive outcry from public ownership advocates (Luckert and Vertinsky 2007). For them, keeping natural resources as public wealth is paramount, and government control provides some level of safeguard in that respect. However, when a sign of unsustainability is detected under either forms of ownership, advocates of either sides claim that changing the tenure is the solution (e.g., Desrochers 2002).

Some studies help us understand the link—or the absence of—between levels of sustainability and tenure types (e.g., Berkes et al. 1989; Lovett-Doust and Kuntz 2001; Dietz et al. 2003; Folke 2007; Ostrom 2007). Although there is a certain level of consensus in the scientific literature that sustainability does not solely hinge on property rights, the fierce debate in the decision-making arena is often polarized around which tenure guarantees a responsible intendance (Beckley 1998; Worrell and Appleby 2000; Haley and Nelson 2007; Luckert and Vertinsky 2007). Furthermore,

 DOI: 10.1201/9781003431084-23

the literature on tenures appears to suffer from the paucity of sound comparative studies that could enlighten the debate; this is the goal of this chapter.

19.2 UNIQUELY TENURED FOREST ECOSYSTEM

The Bas-Saint-Laurent (BSL) region of the Québec Province, Canada, provides a unique study model. It comprises a peculiar juxtaposition of private and public tenures within the same ecological region (Figure 19.1) in nearly equal proportions. This is in sharp contrast to the rest of the province and other Canadian regions where, respectively, 89% and 93% of forests are on public land (MRN 2002; NRCAN 2006). Since this forest region has been under this specific tenure regime for many decades, it offers a unique opportunity to gauge the respective imprint these tenures had on the forests and communities via environmental and socioeconomic metrics.

19.2.1 Study Area

The BSL is an administrative region of the Province of Québec, Canada. It is located on the south shore of the St. Lawrence River estuary and covers an area of 22,637 km^2, of which 19,320 km^2 (85%) are covered by forests. The BSL region is distinctive because

- 51% (11,574 km^2) of the total land area is under public ownership (or 59% of forested land);
- 49% (11,105 km^2) is privately owned land (or 41% of forests), including roughly 5% of large industrially owned forests.

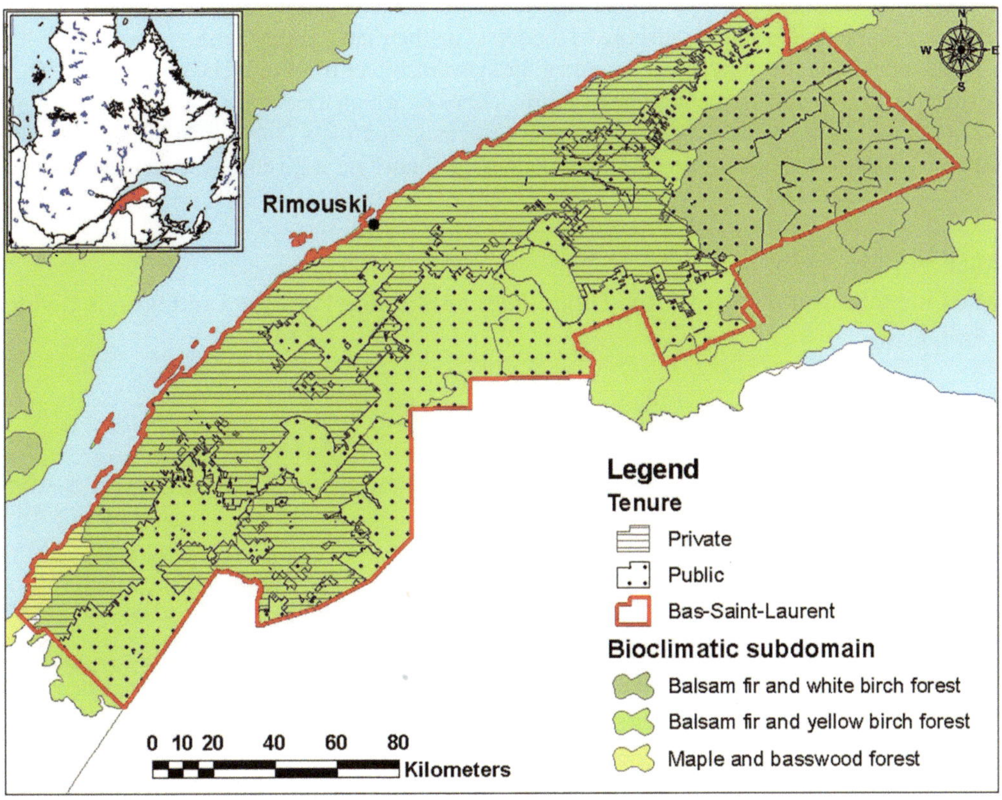

FIGURE 19.1 Map of private and public forests of the administrative region of Bas-Saint-Laurent in relation to the bioclimatic subdomain.

The study area is mostly 90% within the eastern balsam fir (*Abies balsamea* (L.) Mill.) and yellow birch (*Betula alleghaniensis* Britt.) forest ecoregion at the northern limit of the northern temperate zone (Robitaille and Saucier 1998). These mixed forests, at the transition between temperate and boreal forests, include other species, such as red maple (*Acer rubrum* L.), sugar maple (*Acer saccharum* Marsh), white birch (*Betula papyrifera* Marsh), white spruce (*Picea glauca* (Moench) Voss), and eastern white cedar (*Thuya occidentalis* L.) (Archambault et al. 2006).

The natural disturbance regime is driven by recurrent spruce budworm (*Choristoneura fumiferana* (Clemens)) outbreaks at a 40-year interval on average (Boulanger and Arseneault 2004), coupled with windthrow and few rare fires. Together, these disturbances produce a naturally occurring small gap dynamics (Archambault et al. 2006), conducive to uneven-aged/irregular forests. However, since the onset of logging activities in the late 1800s, anthropogenic forms of disturbance now account for most of the changes in the region's forest dynamics (Archambault et al. 2006; Boucher et al. 2006, 2009).

19.2.2 Historical and Current Tenures

Colonization into private woodlots, which began around 1850 (Fortin and Lechasseur 1993), occurred from the shores of the St. Lawrence toward the uplands and also along two major valleys to the southeast (Figure 19.1). Some logging took place before in the early 1800s, but essentially occurred in the region's scattered forests dominated by red and white pines for the purpose of shipbuilding in Great Britain (Fortin and Lechasseur 1993).

There are currently about 18,735 individually owned private woodlots—or nonindustrial private forests—in the BSL of 25.7 ha in average (Table 19.1). Roughly, 70% of the woodlots are under a management plan developed by a regional agency of private forests (OFBSL 2004a), but ultimately the landowners have the right to dispose of the forest on their land as they wish.

Public forests, on the other hand, are under the stewardship of the provincial government, which also determines the management regime and guidelines by which users must abide. Private companies hold timber supply and forest management agreement contracts (Table 19.1), which give them exclusive rights to harvest predetermined volumes of timber on a given forest management unit

TABLE 19.1
Basic Characteristics of Private and Public Tenures Showing How the Land Base Is Divided in Each Case

	Private	Public
Total forest area (ha)	689,598	1,095,103
Management units	Woodlots[a]	Common areas[b]
Number of management units in the study area	~18,735	15–18
Mean area (ha)	25.7	65,787
(min–max)	(0.01–178)	(6,524–259,972)
Users	Woodlot owners	TSFMA holder[c]
Number of users	~8,400	33–36

Note: Since the study covers a 10-year time period (1990–1999), the number of management units as well as the number of users may have changed during that time, explaining that some metrics are presented as intervals or approximations.

[a] *Definition*: Territorial units, established during the English regime, that are organized along rows within townships. A typical lot measures 1 mile (1.6 km) long by 13 chains (260 m) wide.

[b] *Definition*: TSFMA reference territorial unit on Québec public land in the management of forest resources that was effective until March 31, 2008. Since this date, the *forest management unit* replaces the common areas.

[c] *Definition*: Timber supply and forest management agreement. In Québec, a contract that entitles its holder to harvest each year a volume of round timber, provided that the holder undertakes silvicultural treatments to attain the annual yield indicated in the agreement.

(Bouthillier 2001b). Since these companies most often use this wood to supply wood or pulp and paper industrial mills, this tenure will also be referred to as industrial public forest.

19.2.3 Sampling Strategy

Watersheds were used to subdivide the BSL study area into sampling units to facilitate direct comparison between tenures and replicability of the whole study. Watersheds averaging 50 km^2 were selected. Those smaller than 50 km^2 were grouped together and larger ones divided to obtain units of approximately 50 km^2. A total of 436 such watersheds were obtained. The watershed map was overlaid on the tenures map to select watersheds that contain both public and private properties in proportions averaging between 35% and 65% of their total area. Only the watersheds that met these criteria were retained for our analyses, yielding a sample size of 22 watersheds (Figure 19.2).

To ensure that differences in the structure and composition of the private and public shares of the forests within these watersheds were a consequence of tenure, the similarity of permanent variables (surface deposits, slope, drainage) that typically drive forest ecosystem composition (Grondin et al. 1999) was verified across sides. Since no significant differences were found, we were confident that the differences in forest metrics under the private and the public portion of each watershed in our sample were due to the management under each tenure.

19.2.4 Silvicultural and Socioeconomic Data

The respective impacts of the forest tenures were compared according to a set of biological indicators:

- Habitat **loss** and **fragmentation**
- **Heterogeneity** or size of forest patches
- **Age** and **composition** of forest stands

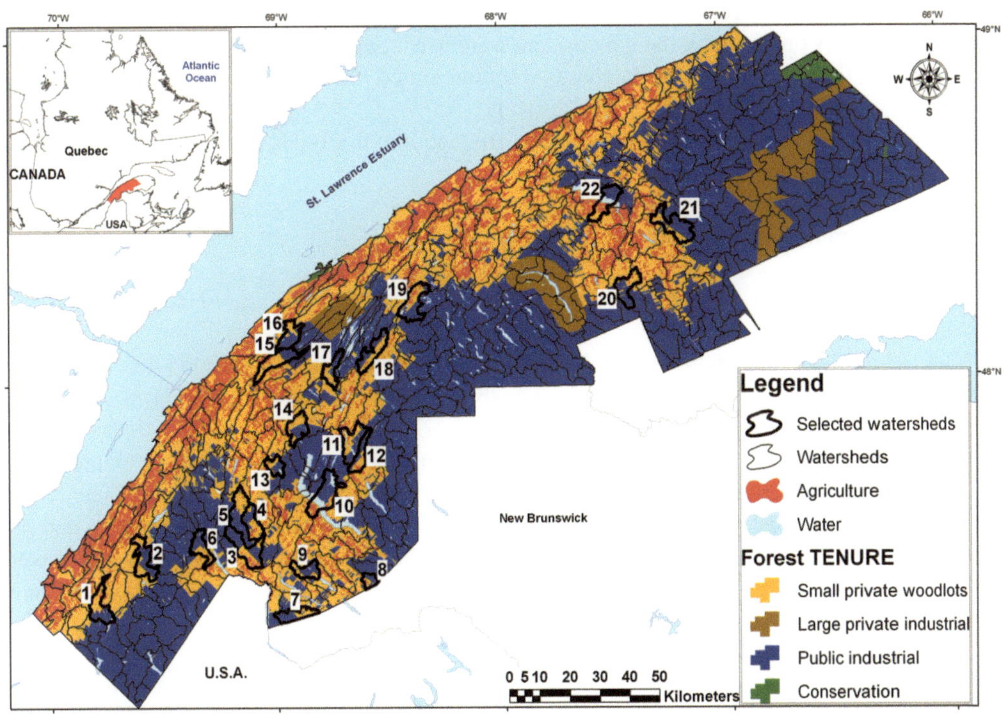

FIGURE 19.2 Map of selected watersheds in the Bas-Saint-Laurent region. The 22 selected watersheds are within 35% and 65% under each tenure.

19.2.4.1 Habitat Fragmentation, Loss, and Heterogeneity

Fragmentation is used here to define a "process through which formerly continuous forest expanses turn into forest patches of varying size, isolated from each other by tracts of nonforest land" (Haila 1999). Habitat fragmentation also involves habitat loss when a formerly forested land is converted into non-forested land (e.g., roads, agriculture, human settlements) (see also Haila 2002; Fahrig 2003; Jaeger et al. 2007). In this study, we compare how many forest habitats were lost to other land covers under each tenure and assess differences in the levels of fragmentation by comparing the size of remaining cont iguous forest patches. In addition, we compared forest heterogeneity that is the extent to which the forest is fractioned into patches of different ages, covers, heights, and stand densities.

19.2.4.2 Age and Composition

These data were obtained through the Québec's forest department 2001 maps (MRN 2000), which provide ecological and forest data. These maps were further scaled down to a 1:40,000 scale (mapping patches as small as 4 ha) to survey small privately owned woodlots and better represent the small-scale operations that occur under such forms of tenure (OFBSL 2004b).

19.2.4.3 Silvicultural Treatments, Jobs, and Economic Benefits

To provide a complete assessment of the differential impact of tenure, a set of socioeconomic variables were used, namely

- Type and size of silvicultural treatments applied
- Number of **jobs** that were created to realize these treatments on the field
- **Economic** benefits induced by government financial incentives to silviculture

It was possible to obtain the amount and size of total and partial cuts directly from the map databases. Yet, for all other treatments, the databases that compile realized hectares admissible to financial incentives provided by the government were used. Separate incentive systems exist under private and public forests. Once all databases were integrated, the sole period that allowed a robust comparison spanned from 1990 to 1999 inclusively.

All 196 possible silvicultural treatments were grouped into five types of treatment based on their characteristics (Bérard and Côté 1996; Côté 2003):

(1) Site preparation before plantation
(2) Tree plantation
(3) Stand education, which includes pruning, precommercial thinning, sanitation, or improvement cuttings
(4) Partial cutting of stands, such as selection cutting or commercial thinning
(5) Total cutting of a stand, where all merchantable stems are harvested (clearcut)

To better discriminate the number of job creation within each tenure, only the jobs created in the forest were considered, excluding sawmills. Indeed, there were many interactions between tenures when the number of jobs created in transformation plants was considered. For example, wood processing facilities use wood coming from both tenures without keeping track of the origin of wood during the transformation and retailing steps. Moreover, professionals and technicians in forest management businesses work routinely on either tenure. Hence, jobs created in the forest were a more discriminating indicator of respective job creation on each tenure.

The amount of work hours induced by every hectare of silvicultural treatment realized was derived from productivity rates. These well-documented rates establish the relation between the silvicultural treatment, the type of machinery used, and the time required to complete the work on each hectare. On this basis, the area (ha) of realized treatments was converted into jobs. One day-person (d-p) corresponds to 8 h of work and 5 d-p in one week-person (w-p).

As an indicator of economic incentives, the financial incentives given to support silviculture were used. Funds came from both the Canadian government in the case of the *Programme de développement forestier de l'Est du Québec* (Service canadien des forêts 1994) and the Québec government in the case of the regional agency for private forests.

All data were compiled and assembled in a Microsoft ACCESS® database to perform analyses. Since the silvicultural treatments, data were a complete inventory rather than a sample, and results were compared directly without the use of statistics.

19.2.5 Spatial Analysis

To run our analyses, the private and public maps were converted from vector to raster format at a 20 m × 20 m resolution using IDRISI Andes (Clark Labs, www.clarklabs.org). For each indicator chosen (land cover, age, composition etc.), two metrics were calculated: patch size and the proportion in each watershed. Statistical analyses were performed on all watersheds separately as well as an overall structure obtained from the average for each age class over all 22 watersheds. The same procedure was used for cover, height, and density.

Since the data were not normally distributed, a Wilcoxon test was used to perform nonparametric, paired comparisons. The private and public share of every watershed constituted the pairs used.

19.3 RESULTS AND DISCUSSION

19.3.1 Habitat Loss and Fragmentation

The habitat loss caused by roads and villages as well as forest conversion into agricultural lands was greater on private lands. This explains the significantly smaller amount of productive forest on private land (Figure 19.3). Conversely, water bodies and roads covered a significantly greater proportion of private land than public land. Agriculture was present only on private land. Habitat loss induced a fragmentation of contiguous patches of productive forests. On average, contiguous patches of private productive forests were 109.80 ± 9.35 ha in size, which was significantly smaller than public forests, where patches averaged at 328.14 ± 41.85 ha ($p < 0.001$). Therefore, the private forests were more fragmented than the public forests.

19.3.2 Heterogeneity

Contiguous patches of forests belonging to the same age class were systematically smaller in size within private forests than within public forests (Figure 19.4). That difference was statistically significant in three of the eight age classes (Figure 19.4). There was clearly a higher heterogeneity in private forests. Furthermore, for all treatment types, the size of silvicultural treatment patches was significantly smaller under private tenure (Figure 19.5). The size of total cut patches was not the same as the observed forest patch size, total cuttings (Figure 19.5) being roughly half the size of forest patch size under public tenure (Figure 19.4), while private total cuttings (Figure 19.5) were roughly eight times smaller than forest patches (Figure 19.4). This suggests that multiple total cuts aggregated over time with neighbors into larger patches.

In short, smaller cuts in private forests yielded smaller forest patches and consequently a higher level of heterogeneity.

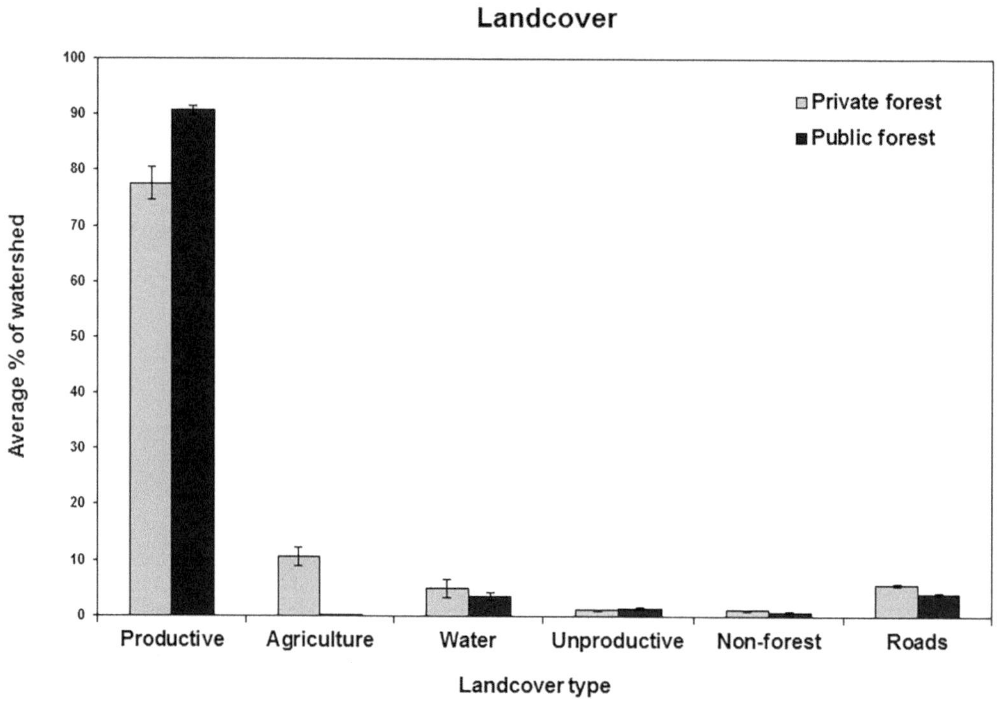

FIGURE 19.3 Mean proportion, for the 22 sample watersheds, of land cover types, surface deposits, and slope and drainage classes under each tenure. Significance levels of the difference between tenures ($p < 0.05$) are shown above each bar.

19.3.3 Age and Composition

The age structure that was observed in the BSL forests is best understood when considering some of its history. The last spruce budworm outbreak (1975–1992) was quite severe in the region (Boulanger and Arseneault 2004) and induced extensive salvage cutting. With the combined epidemic and ongoing logging effort, large expenses of forests were pushed into early seral stages, so drastically away from historical characteristics (Boucher et al. 2009). This resulted in the observed overabundance of young forests, even-aged (class 10) and young uneven aged (Jin), and concomitant lack of old-growth forests, either even-aged (classes 90 and 120) or old uneven aged (Vin), under both forms of tenures (Figure 19.4).

Where tenures differed the most in their age class was in the commercially mature stands, which for the BSL conifer-based industry corresponded to stands in 50 and 70 years age classes. Private forests had 34% of their productive stands in the 50 years age class (or 40–60 years of age, on our 2001 map) and very little in older stands (Figure 19.4). This matched with the last settlement episode in the region, around 1930 and 1940. Logging and slash burning were performed to convert land to agriculture, and many uncontrolled wildfires resulted (Blanchet 2003). A major portion of the deciduous stands observed in private forests originated from uncontrolled wildfires. This characteristic, induced in the private forests of the BSL, is similar to what has been observed in a similar tenure setting in the boreal forest of Saskatchewan (Weir and Johnson 1998). On the other hand, commercially mature forests under public tenure were equally distributed between the 50 and 70 age classes (Figure 19.4).

Each tenure displayed different forest covers, where public forests had more conifers than private forests, while private forests had more deciduous stands (Figure 19.4). Considering that, historically,

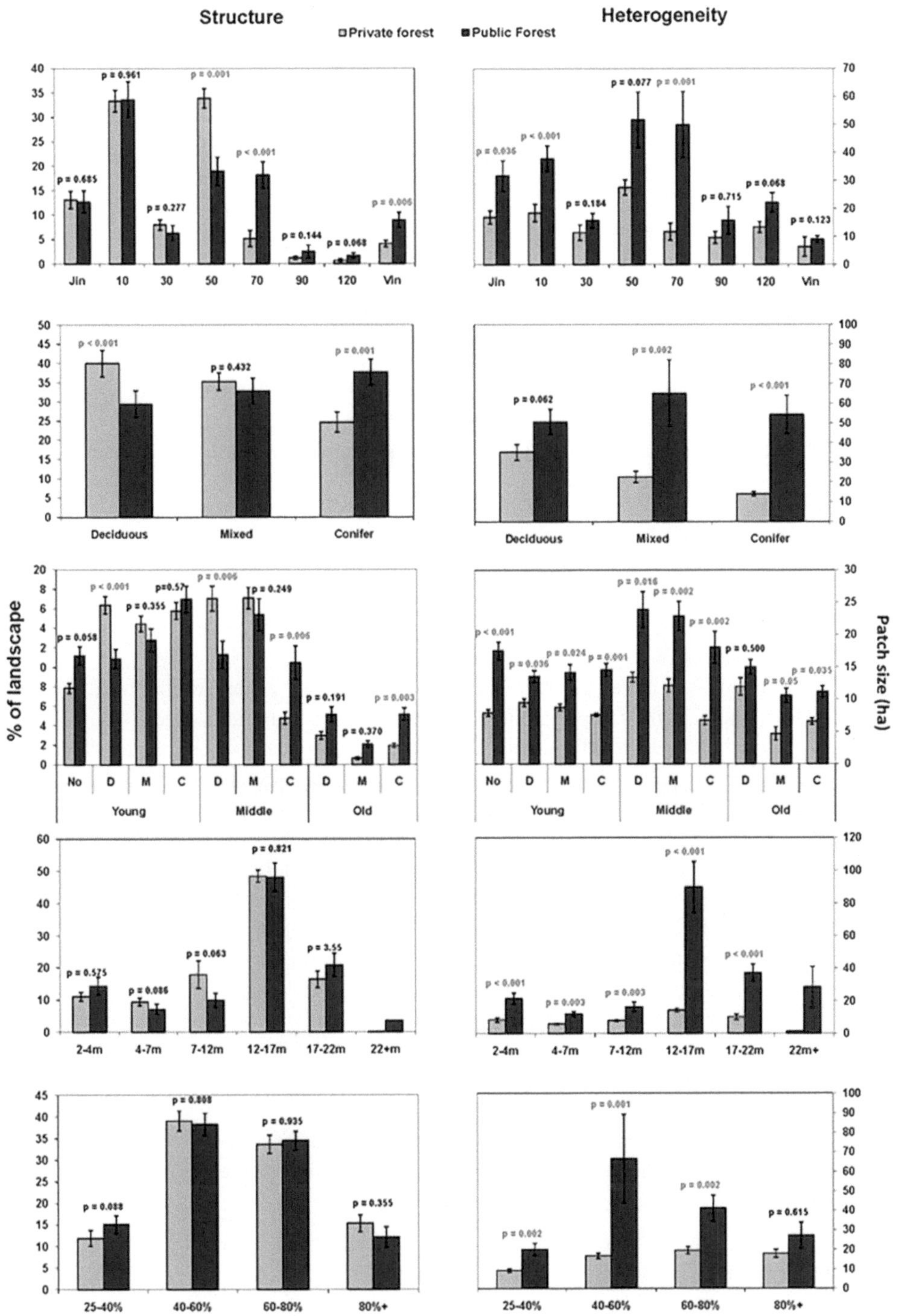

FIGURE 19.4 Composite graph showing age, cover, age cover, height, and density classes for the 22 sampled watersheds. Density classes represent the percent of the ground in a stand that is covered by the projection of tree crowns. Structure is the mean proportion of productive forest in each variable class. Heterogeneity is the size of contiguous patches of the same class for each variable (age, cover, age cover, height and density). Significance levels of the difference between tenures, as p values, are shown above each bar. *Note*: Jin, young uneven-aged; Vin, old uneven-aged.

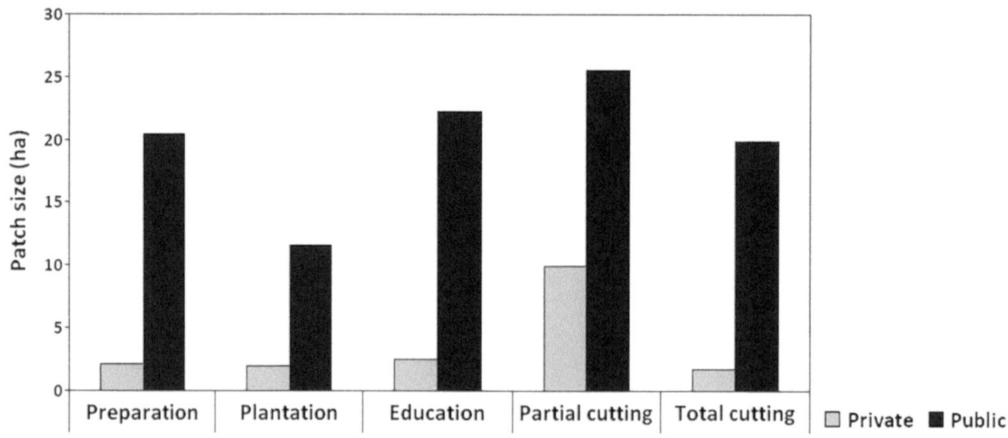

FIGURE 19.5 Mean patch size of realized silvicultural treatments for private and public tenures.

the regional landscape was dominated by a coniferous forest cover (Boucher et al. 2009), our observations showed that private forests have shifted to be dominated by deciduous and mixed stands. In contrast, public forests, which were more similar to historical forests with their conifer dominance, still showed a high degree of change in favor of deciduous and mixed stands when compared to 1930 records (Boucher et al. 2009).

With the combined age cover variable (Figure 19.4), it can further be observed that the predominance of deciduous stands in private forests was in young and middle-age classes. So the historical human disturbances factor discussed earlier induced the establishment of deciduous forests. Trembling aspen (*Populus tremuloides*) was the predominant deciduous species in these deciduous stands, which is characteristic of forests frequently disturbed by logging and fire (Weir and Johnson 1998).

In essence, forests under both tenures have undergone considerable changes in terms of their structure and composition, away from historical patterns. Since there was no undisturbed forest in the BSL that could have served as a control to describe natural dynamics, historical data were reliable alternatives. As suggested by Kneeshaw et al. (2000), the contrasts between historical and current age structure and stand composition (e.g., Hessburg et al. 1999; Boucher et al. 2009) can serve in a coarse-filter strategy to assess threats to biodiversity. The watersheds examined had a drastically different age and cover structure than natural forests (Boucher et al. 2006; Dupuis et al. 2011) and even more so under private tenure. In their study of the effects of landownership on biodiversity, Lovett-Doust and Kuntz (2001) also found that private land held lower levels of biodiversity than public land. Considering the similarities between their case study and the BSL situation, the concerns about highly fragmented and disturbed forests in the private woodlots of BSL can be further emphasized.

19.3.4 Silvicultural Treatments

The greatest differences between the private and public management recipes were in the harvest operations. Partial cutting was performed annually on a proportion (0.86%) slightly greater than total cutting (0.7%) on nonindustrial private forests, but strikingly higher than partial cutting in public forests (1.5%) (Figure 19.6).

This trend most probably originated from the shared concept that it is in public interest to invest in woodlot management (Cubbage 1997). This opinion was clearly brought forward in the reports of advisory committees (Paillé et al. 1978; Lortie et al. 1984; Audet 1988; Service canadien des

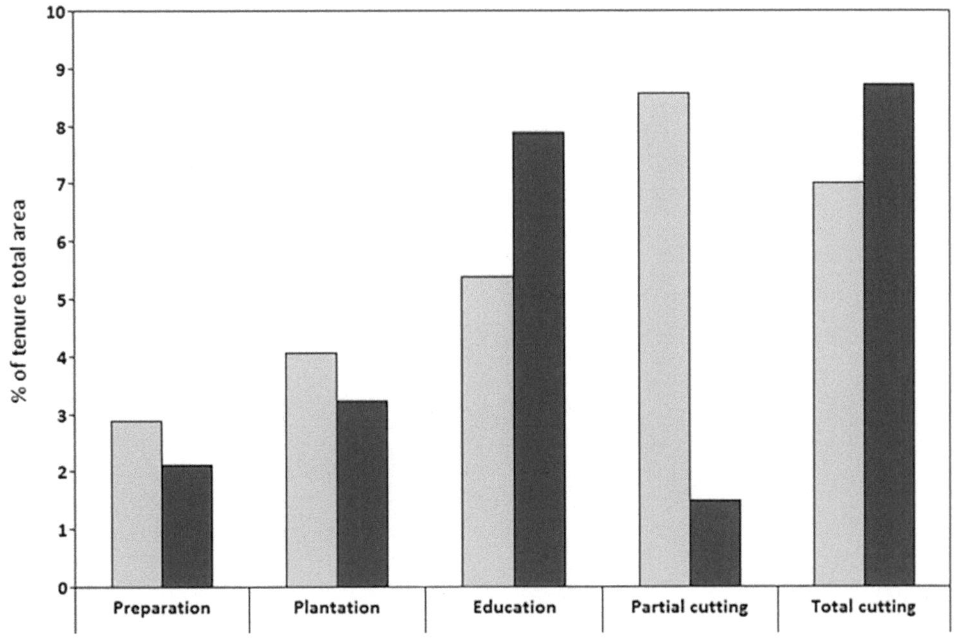

FIGURE 19.6 Silvicultural intervention "baskets," shown as the percentage of total forest area under private and public tenures that was submitted to each of the five types of silvicultural treatments. In total, respectively, 28% and 23% of total private and public forest area have been submitted to a silvicultural treatment between 1990 and 1999.

forêts 1994), who supported an increased nurturing of private forests. This idea implied to maintain a forest cover, which can provide a constant source of revenues to the landowner, an objective that partial cutting is particularly suited for (Nyland 2005). Furthermore, these policies aimed to enroot populations on this territory to ensure its occupation (Bouthillier and Dionne 1995) and thereby the resilience of communities (Bouthillier 2001a). Our case study suggested that private forest practices were coherent with the structure results observed as well as with this orientation.

19.3.5 Jobs and Economic Benefits

In the forest sector, most silvicultural treatments (planting, site preparation, precommercial thinning) were performed seasonally, meaning that forest workers must have relied more frequently on employment insurance (Stanek 1997; Comité Sectoriel de Main-D'œuvre en Aménagement Forestier 2010). Thus, the number of jobs created by forest operations was crucial.

When looking at the job creation per hectare of silvicultural treatment realized (Figure 19.7), the most noticeable trend was that private tenure yields more jobs per hectare for all types of treatments examined. Thus, every hectare of private forest yielded on average 68% more week-persons of work than if the same hectare was managed using a typical public recipe. In other words, to create one week-person job per year required 101.8 ha of private forest compared to 171.3 ha of public forest. This strategic job creation advantage could be brought forward, especially because private forests created more jobs per dollar invested. In short, our case study supports the idea that public investments in private forests are profitable (Bouthillier 2001a). This job creation difference was due to the level of mechanization, which was lower in private forests according to our case study. The main objective of mechanization being to increase efficiency, it inevitably caused a decrease in the number of day-persons required to achieve one hectare of treatment. Partial cutting, education, and

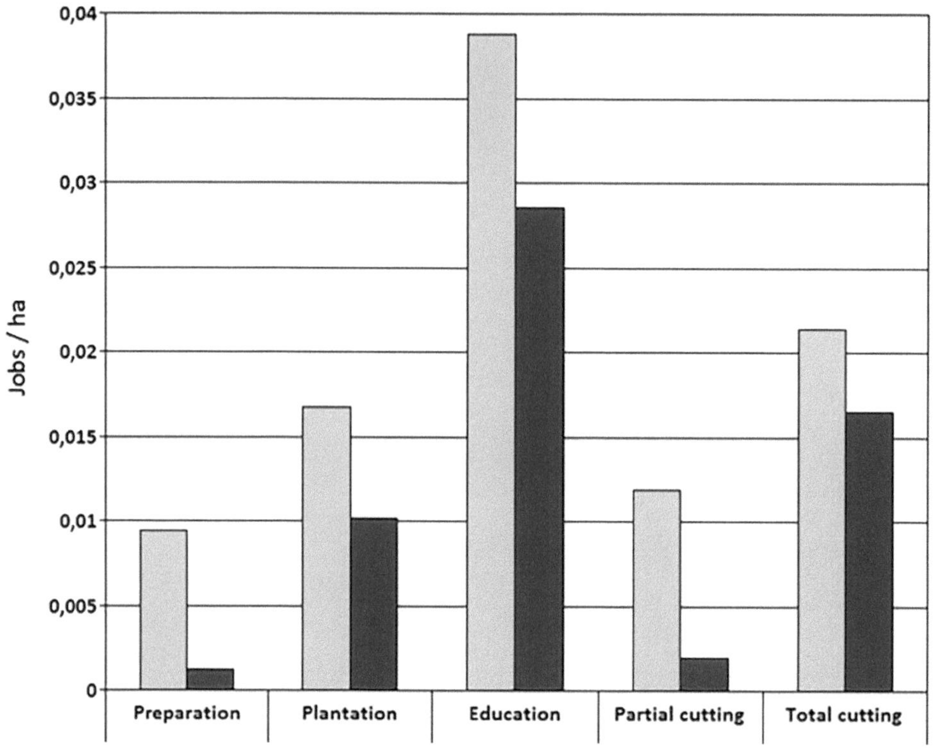

FIGURE 19.7 Number of jobs created (week-person) per hectare (ha) for each of the five types of silvicultural treatments in private and public tenures.

site preparation were the treatments for which this difference was most noticeable (Figure 19.7). The small size of woodlots, and consequently of interventions (Figure 19.5), encouraged the trend toward less mechanized operations in private forests.

Education treatments were the intervention type that generated the greatest number of jobs (Figure 19.7). At first conceived to increase allowable cuts, it has been shown that this effect did not materialize in other regions of Québec, especially in the case of pre commercial thinning (CEGFPQ 2004). Furthermore, the data from the present study showed that public incentives for education treatments were higher under public tenure but generated more jobs under private tenure (Figure 19.7).

Maintaining manual harvest methods, such as chainsaw felling, was thus coherent with the policies that accompanied the private institutional arrangement that aimed to favor the enrooting of local population and land occupancy (Paillé et al. 1978; Lortie et al. 1984; Audet 1988; Bouthillier and Dionne 1995; CEGFPQ 2004). However, since the recruitment of chainsaw operators declined markedly in recent years, a certain level of mechanization appears inevitable (Stanek 1997; Fréchette 2007).

19.4 MANAGEMENT IMPLICATIONS

The main results of the case study indicated that

- Public forests held a greater biodiversity conservation potential.
- Nonindustrial private forests tenure yielded higher socioeconomic benefits.

		Private	Public
1	Age	+ young	+ old
	Cover	+ deciduous	+ conifer
	Heterogeneity	+	-
	Fragmentation	+	-
	Size of cuttings	Small	Large
2	Sylvicultural treatments	+Preparation/plantation + Partial cuttings	+ Education + Total cuttings
	Jobs	+	-
	Financial incentives	Create more jobs	Education
3	*Status quo*	↑ gap with natural forest	↑ gap
	Ecosystem-based management	↓ gap ↓ jobs and incentives	↓ gap ↓ jobs and incentives
	Privatization	–	↑ gap ↑ jobs

FIGURE 19.8 Synthesis of the main differences between tenures for each of the variables examined in our case study and simulation model. (From Morin, P. et al., Modeling the effect of private and public forest management recipes on the social-ecological system of a southern boreal forest. In Analyse comparative des modes de tenure du territoire forestier en fonction d'indicateurs environnementaux et socio-économiques: Une étude de cas à l'interface tempérée-boréale de l'est canadien, PhD thesis, Université du Québec à Rimouski, Rimouski, Québec, Canada, 2012.)

The simulation modeling experiment performed using these data (Morin et al. 2012) demonstrated that

- Adjusting the actual practices by implementing ecosystem-based management improved societal benefits better than maintaining the status quo or privatization of public forests.

We suggest that managers must realize that private and public forest tenures are complementary (Figure 19.8). Acknowledging this fact allows us to reconcile that old debate on the subject and instead search for more efficient ways to coordinate management decisions between tenures to invest under the tenure that will optimally yield the desired societal benefits. For example, in the BSL region, after this case study was performed, many stakeholders worked together for the creation of protected areas in public forest: the forest industry was motivated by its FSC certification while the regional and provincial governments were responding to policy objectives. During the public hearings, it was demonstrated that the reduction in allowable cut under public tenure could be compensated by wood available under private tenure.

So we think that our study demonstrated the potential to achieve complementarity from the regional to national (or international) levels.

19.5 OUTREACH

This study contributed to a corpus of studies that also contradict the proposed superiority of privatization with different approaches, through the study of local forest management (Agrawal 1996), the study of cost-efficiency (Grebner and Amacher 2000), or the meta-analysis of natural resource governance studies (Ostrom 2003; Dietz et al. 2003; Gueneau and Tozzi 2008). Furthermore, decision-making could be better oriented by the inclusion of ecosystem services valuation (Daily et al. 2000; Heal 2000; Kremen and Ostfeld 2005) in the process. This way, in addition to current economic indicators, such as the widespread yet problematic GDP (Turner and Fisher 2008; Holt and Hattam 2009), a more thorough estimate of our assets, including built capital as well as natural capital (Costanza et al. 1997; Vallée 2002; Mead 2011) would be accessible to maximize human well-being. Private and public forests can participate in this welfare buildup, given that the appropriate tools to fine-tune the governance of our forests are used, to make our social–ecological system more resilient (Holling 2001).

REFERENCES

Agrawal, A. 1996. The community vs. the market and the state: Forest use in Uttarakhand in the Indian Himalayas. *J. Agric. Environ. Ethic.* 9(1):1–15.

Agrawal, A., A. Chhatre, and R. Hardin. 2008. Changing governance of the world's forests. *Science* 320(5882):1460–1462.

Archambault, L., C. Delisle, G.R. Larocque, L. Sirois, and P. Belleau. 2006. Fifty years of forest dynamics following diameter-limit cuttings in balsam fir—Yellow birch stands of the lower St. Lawrence region, Québec. *Can. J. For. Res.* 36(11):2745–2755.

Audet, J. 1988. *La forêt privée: un potentiel à développer.* Rapport du Comité Audet sur l'orientation de la politique relative à la forêt privée, 84pp. Québec City, Québec: Ministère de l'énergie et des ressources.

Beckley, T.M. 1998. Moving toward consensus-based forest management: A comparison of industrial, co-managed, community and small private forests in Canada. *For. Chron.* 74(5):736–744.

Bérard, J., and M. Côté. 1996. *Manuel de foresterie.* Ordre des ingénieurs forestiers du Québec. Québec, Canada: Presses de l'Université Laval, Sainte-Foy.

Berkes, F., D. Feeny, B.J. McCay, and J.M. Acheson. 1989. The benefits of the commons. *Nature* 340:91–93.

Blanchet, P. 2003. *Feux de forêt: L'histoire d'une guerre.* Montréal, QC: Éditions Trait d'Union.

Boucher, Y., D. Arseneault, and L. Sirois. 2006. Logging-induced change (1930–2002) of a pre-industrial landscape at the northern range limit of northern hardwoods, eastern Canada. *Can. J. For. Res.* 36:505–517.

Boucher, Y., D. Arseneault, L. Sirois, and L. Blais. 2009. Logging pattern and landscape changes over the last century at the boreal and deciduous forest transition in eastern Canada. *Landsc. Ecol.* 24(2):171–184.

Boulanger, Y. and D. Arseneault. 2004. Spruce budworm outbreaks in eastern Québec over the last 450 years. *Can. J. For. Res.* 34(5):1035–1043.

Bouthillier, L. 2001a. *L'impact des investissements publics en forêt privée*, 24pp. Rimouski, Québec, Canada: La Forêt Modèle du BasSaint-Laurent.

Bouthillier, L. 2001b. Québec: Consolidation and the movement towards sustainability. In *Canadian forest policy*, ed. M. Howlett. Toronto, Ontario, Canada: University of Toronto Press.

Bouthillier, L. and H. Dionne. 1995. *La forêt à habiter: la notion de "forêt habitée" et ses critères de mise en oeuvre: rapport final au Service canadien des forêts (Région du Québec).* Rimouski, Québec, Canada: Université Laval, Sainte-Foy, Québec, Université du Québec à Rimouski.

CEGFPQ. 2004. *Rapport: Commission d'étude sur la gestion de la forêt publique québécoise*, 261pp. Québec, Canada: Commission d'étude sur la gestion de la forêt publique Québécoise.

Comité sectoriel de main-d'œuvre en aménagement forestier. 2010. *Fiche sectorielle—Aménagement forestier*, 5pp. Québec, Qc: Chantier sur la saisonalité.

Costanza, R., R. D'Arge, R.D. Groot, et al. 1997. The value of the world's ecosystem services and natural capital. *Nature* 387(6630):253–260.

Côté, M. 2003. *Dictionnaire de la foresterie.* Éd. spéciale 12e Congrès forestier mondial ed. Sainte-Foy, Québec, Canada: Ordre des ingénieurs forestiers du Québec.

Cubbage, F.W. 1997. The public interest in private forests: Developing regulations and incentives. In *Creating a forestry for the 21st century: The science of ecosystem management*, eds. K.A. Kohm and J.F. Franklin. Washington, DC: Island Press.

Cumming, G.S. and G. Barnes. 2007. Characterizing land tenure dynamics by comparing spatial and temporal variation at multiple scales. *Landsc. Urban Plan.* 83(4):219–227.

Daily, G.C., T. Soderqvist, S. Aniyar, K. Arrow, P. Dasgupta, P.R. Ehrlich, C. Folke et al. 2000. Ecology—The value of nature and the nature of value. *Science* 289(5478):395–396.

Desrochers, P. 2002. *Comment assurer le développement durable de nos forêts?* Note économique n 24, 4pp. Montréal, Québec, Canada: Institut économique de Montréal.

Dietz, T., E. Ostrom, and P.C. Stern. 2003. The struggle to govern the commons. *Science* 302(5652):1907–1912.

Dupuis, S., D. Arseneault, and L. Sirois. 2011. Change from pre-settlement to present-day forest composition reconstructed from early land survey records in eastern Québec, Canada. *J. Veg. Sci.* 22(3):564–575.

Fahrig, L. 2003. Effects of habitat fragmentation on biodiversity. *Annu. Rev. Ecol. Evol. S.* 34:487–515.

Folke, C. 2007. Social–ecological systems and adaptive governance of the commons. *Ecol. Res.* V22(1):14–15.

Fortin, J.-C. and A. Lechasseur. 1993. *Histoire du Bas-Saint-Laurent*. Québec, Canada: Institut québécois de recherche sur la culture.

Fréchette, A. 2007. The sustainable governance of common-pool resources: Property rights institutions and the social, political, and economic conditions affecting resource use outcomes. In *Rapport de synthèse environnementale présenté comme exigence partielle du doctorat en sciences de l'environnement*, 57pp. Montréal, Qc: Université du Québec à Montréal.

Grebner, D.L. and G.S. Amacher. 2000. The impacts of deregulation and privatization on cost efficiency in New Zealand's forest industry. *For. Sci.* 46(1):40–51.

Grondin, P., J. Blouin, and P. Racine. 1999. *Rapport de classification écologique de sous-domaine bioclimatique de la sapinière à bouleau jaune de l'est*, 198pp. Québec, Canada: Ministère des Ressources naturelles du Québec, Direction des inventaires forestiers.

Gueneau, S. and P. Tozzi. 2008. Towards the privatization of global forest governance? *Int. For. Rev.* 10(3):550–562.

Haila, Y. 1999. Islands and fragments. In *Maintaining biodiversity in forest ecosystems*, ed. M.L. Hunter. Cambridge, UK: Cambridge University Press.

Haila, Y. 2002. A conceptual genealogy of fragmentation research: From island biogeography to landscape ecology. *Ecol. Appl.* 12(2):321–334.

Haley, D. and H. Nelson. 2007. Has the time come to rethink Canada's crown forest tenure systems? *For. Chron.* 83(5):630–641.

Hardin, G. 1968. The tragedy of the commons. *Science* 162:1243–1248.

Heal, G. 2000. Valuing ecosystem services. *Ecosystems* 3(1):24–30.

Hessburg, P.F., B.G. Smith, and R.B. Salter. 1999. Detecting change in forest spatial patterns from reference conditions. *Ecol. Appl.* 9(4):1232–1252.

Holling, C.S. 2001. Understanding the complexity of economic, ecological, and social systems. *Ecosystems* 4(5):390–405.

Holt, A.R. and C. Hattam. 2009. Capitalizing on nature: How to implement an ecosystem approach. *Biol. Lett.* 5(5):580–582.

Jaeger, J.A.G., H.G.S. Raumer, H. Esswein, et al. 2007. Time series of landscape fragmentation caused by transportation infrastructure and urban development: A case study from Baden-Wurttemberg, Germany. *Ecol. Soc.* 12(1):22. www.ecologyandsociety.org/vol12/iss1/art22/.

Kindscher, K. and N. Scott. 1977. Land ownership and tenure of the largest land parcels in the flint Hills of Kansas, USA. *Nat. Area J.* 17:131–135.

Kneeshaw, D.D., A. Leduc, P. Drapeau, et al. 2000. Development of integrated ecological standards of sustainable forest management at an operational scale. *For. Chron.* 76(3):481–493.

Kremen, C. and R.S. Ostfeld. 2005. A call to ecol+ogists: Measuring, analyzing, and managing ecosystem services. *Front. Ecol. Environ.* 3(10):540–548.

Lortie, M., F. Côté, H. Filion, L. Laneville, and M. Pleau. 1984. L'avenir de la forêt privée. Rapport du comité de consultation sur l'avenir de la forêt privée au Québec. In *Rapport du comité de consultation sur l'avenir de la forêt privée au Québec*, 35pp. Québec, Canada: Ministère de l'énergie et des ressources.

Lovett-Doust, J. and K. Kuntz. 2001. Land ownership and other landscape-level effects on biodiversity in southern Ontario's Niagara Escarpment Biosphere Reserve, Canada. *Landsc. Ecol.* 16(8):743–755.

Luckert, M. and I. Vertinsky. 2007. Forest privatization should not be censored from debates about forest tenure policies. *For. Chron.* 83(6):790–791.

McElligott, P. 2003. Creating win-wins in BC's forests . . . Taking on the sacred cows. *For. Chron.* 79(3):659–663.

Mead, H.L. 2011. *L'indice de progrès véritable du Québec: quand l'économie dépasse l'écologie*. Québec, Canada: Éditions MultiMondes.

Morin, P., L. Sirois, and L. Bouthillier. 2012. *Modeling the effect of private and public forest management recipes on the social-ecological system of a southern boreal forest. In Analyse comparative des modes de tenure du territoire forestier en fonction d'indicateurs environnementaux et socio-économiques: Une étude de cas à l'interface tempérée-boréale de l'est canadien.* PhD thesis, Université du Québec à Rimouski, Rimouski, Québec, Canada.

MRN. 2000. *Système d'information écoforestière (SIEF)*, 16pp. Québec, Canada: Ministère des Ressources Naturelles, Forêt Québec, Direction des inventaires forestiers.

MRN. 2002. *Rapport sur l'état des forêts québécoises 1995–1999*, 276pp. Charlesbourg, Québec, Canada: Ministère des Ressources Naturelles, Gouvernement du Québec.

NRCAN. 2006. *The State of Canada's Forests 2005–2006: Forest industry competitiveness*, 79pp. Ottawa, Ontario, Canada: Natural Resources Canada, Canadian Forest Service.

Nyland, R.D. 2005. Diameter-limit cutting and silviculture in Northern Hardwoods. In *Proceedings of the conference on diameter-limit cutting in northeastern forests*, eds. L.S. Kenefic and R.D. Nyland. Gen. Tech. Rep. NE-341. Newtown Square, PA, U.S. Forest Service, Northeastern Research Station, Amherst, MA.

OFBSL. 2004a. *Déclaration sur l'État de la forêt privée au Bas-Saint-Laurent*, 106pp. Québec, Canada: Observatoire de la Foresterie du Bas-Saint-Laurent, Rimouski. http://ofbsl.net/Rapport%20final%20-%20etat%20foret%20privee.pdf.

OFBSL. 2004b. *Mise à jour de la carte écoforestière en forêt privée (POLYGIF)*. Rimouski, Québec, Canada: Observatoire de la Foresterie du Bas-Saint-Laurent.

Ostrom, E. 2003. How types of goods and property rights jointly affect collective action. *J. Theor. Polit.* 15(3):239–270.

Ostrom, E. 2007. A diagnostic approach for going beyond panaceas. *Proc. Natl. Acad. Sci. U S A* 104(39):15181–15187.

Paillé, G., A.-C. Leamy, Y. Barrettte, et al. 1978. *Rapport du groupe de travail sur la politique forestière relative aux forêts privées du Québec*, 246pp. Québec, Canada: Gouvernement du Québec, Ministère des Terres et Forêts Québec.

Robitaille, A. and J.-P. Saucier. 1998. *Paysages régionaux du Québec méridional*. Sainte-Foy, Québec, Canada: Direction de la gestion des stocks forestiers, et direction des Relations publiques, Ministère des Ressources naturelles du Québec. Publications du Québec.

Service Canadien des Forêts. 1994. *Le programme de développement forestier de l'est du Québec, 1983–1993: 10 ans d'appui aux communautés rurales*. Sainte-Foy, Québec, Canada: Ressources naturelles Canada, Service canadien des forêts, Région du Québec.

Stanek, O. 1997. *Les travailleurs forestiers du Bas-Saint-Laurent*, 255pp. Rimouski, Québec, Canada: UQAR-Grideq.

Turner, R.K. and B. Fisher. 2008. Environmental economics: To the rich man the spoils. *Nature* 451(7182).1067–1068.

Vallée, A. 2002. *Économie de l'environnement, Points*. Paris, France: Éditions du Seuil.

Vertinsky, I. and M. Luckert. 2009. *The future of forest tenure systems in Canada. Tomorrow's Forest.* Sustainable Forest Management Network. Edmonton, Alberta, Canada: University of Alberta.

Weir, J.M.H. and E.A. Johnson. 1998. Effects of escaped settlement fires and logging on forest composition in the mixedwood boreal forest. *Can. J. For. Res.* 28(3):459–467.

White, A. and A. Martin. 2002. Who owns the world's forests? In *Forest Trends*, 32 pp. Washington, DC: Center for International Law.

Worrell, R, and M.C. Appleby. 2000. Stewardship of natural resources: Definition, ethical and practical aspects. *J. Agric. Environ. Ethic.* 12(3):263–277

20 Forest Management and Climate Change

Adaptive Measures for the Temperate–Boreal Interface of Eastern North America

Laurent Gagné, Luc Sirois, and Luc Lavoie

20.1 INTRODUCTION

Forests cover about 30% (four billion ha) of the Earth's continental surfaces (FAO 2010; Lindquist et al. 2012), 17% of which are in North America (Cohen and Miller 2001). There are four main forest biomes on the planet. Among these, the deciduous temperate forest biome covers 1.04 billion ha (Fischlin et al. 2007). The boreal biome covers 1.27 billion ha (33% of the planet's forested surfaces) and constitutes the second largest forest biome in the world after the tropical forests, which cover 1.75 billion ha (Fischlin et al. 2007, 2009). This biome extends primarily to Canada, Alaska, Fennoscandia, and Russia (FAO 2001, Fischlin et al. 2007, 2009) and beyond the polar circle in some countries. This is the forest biome with the lowest average annual temperatures and the lowest number of tree species (Fischlin et al. 2009).

These two biomes are present alongside one another, both in Eurasia and in America, where they support a sustained silvicultural activity (Amiro et al. 2006; Fischlin et al. 2009). For the boreal forest only, about 600 million m^3 of wood is harvested annually, which corresponds to 20% of the planet's entire harvest (Fischlin et al. 2009). For the boreal and temperate forests, this proportion should increase by 25% by 2030 (Turner et al. 2006). Between 2005 and 2010, the total C stored in the world's forests has decreased by 0.5 gigaton (Gt) annually (FAO 2010), mostly due to human activities, that is, forest exploitation and forest clearance for agriculture (Heimann 2008). The C sequestration function of forest ecosystems is a major component of the terrestrial C balance that is highly influenced by natural disturbances and forest management.

20.1.1 Climate–Forest Relations

Climate influences forests and forests influence the climate on a global and on a regional scale. The Holocene/Anthropocene global warming is the major cause of the northern migration of tree species that had taken refuge south of the continental ice cap (MacDonald 1993; King and Herstrom 1997; McLachlan et al. 2005). Species would have moved northward at a pace of 10–50 km century^{-1} (McLachlan et al. 2005; Feurdean et al. 2013) to form the forests of the temperate and boreal biomes, with early successional species migrating more rapidly than late-successional species (Feurdean et al. 2013). The global warming of 0.6°C in 30 years (Hansen et al. 2006) is much more rapid than the one of 0.6°C between 11,300 BP and 9,500 BP during the Holocene (Marcott et al. 2013); accordingly, the current and future northern migration rate of tree species should occur faster

DOI: 10.1201/9781003431084-24

than the one observed during the Holocene so that species can follow the displacement of their thermal zone (Feurdean et al. 2013).

The current geographical distribution of tree species, along with regional climate patterns, allows for the delimitation of zones that are defined by macroclimatic conditions and assemblages of relatively homogeneous species, variably designated under terms such as ecoregions (Bailey 1980), ecozones (MRNF 2009), and forest provinces (Bailey 1980). From south to north of the temperate–boreal interface in eastern North America, there is a succession pattern of these types of zones. In the northeastern United States and southeastern Canadian forests, species such as white pine, hickories, oaks, and maples are found (Bailey 1980; Burns et Honkala 1990a, 1990b). Annual average temperatures are between 16°C and 6°C, and precipitations are at least 500 mm (Walter 1979). Progressing toward the north, the balsam fir zone and then the spruce zone with more boreal species, such as balsam fir, spruces, poplars, and birches, are observed. Annual average temperatures vary from 5°C to –5°C, and there is a precipitation gradient from 1,300 mm in the balsam fir zone to 300 mm in the more northerly spruce zone (Fischlin et al. 2009).

The role of forests in climate regulation has just recently been recognized (Cohen et Miller 2001; Bonan 2002; Seppälä et al. 2009), notably in context with the interest in the global C cycle (Rollinson 2008; Lindquist et al. 2012). Globally, the planet's forests constitute a 1,200 GtC reservoir, including the C from the soil compared to the 762 GtC for the whole atmosphere (Dixon et al. 1994; Freer-Smith et al. 2008; Rollinson 2008). Boreal forests account for more than 30% of the planet's C (Post et al. 1982; Manies et al. 2005; Symon et al. 2005) and 13% of all the C contained in forests (Jobbagy and Jackson 2000), with 20% as living biomass and 80% stored in woody debris, litter, and organic C in the soil (Melillo et al. 1990; Dixon et al. 1994; Goulden et al. 1998; Apps et al. 1999). Meanwhile, temperate deciduous forests store 14% (168 GtC) of the C of all forest biomes from around the world (Dixon et al. 1994). During the growth process, the C sequestration rate can vary from 1 ton to 26 tons of CO_2 ha^{-1} $year^{-1}$, depending on the site's productivity (Fensham and Guymer 2009); about a third of this amount is transported to the root system to be used in respiration and root growth or to be lost to the soil (Norby and Jackson 2000).

The influence of forests on climate is closely related to their successional status and to the disturbance regime. Following a severe disturbance at the local scale, forests become a C source because the decomposition of organic matter emits more C than the amount sequestrated in tree growth (Kirschbaum 2000; Seppälä 2008). Of the 40 GtC from human origin emitted annually in the atmosphere (Lemprière et al. 2008), 5.9 GtC comes from surface areas exploited for timber production and agriculture (Heimann 2008). This trend is progressing because 60,000 km^2 $year^{-1}$ of forest areas are modified by industrial activities, primarily by forest exploitation (FAO 2010). At the same time, the net surface area of global forests decreased by 1.7% between 1990 and 2005 (Lindquist et al. 2012). Between 1920 and 1980, the boreal forest of Canada was a C sink (Kurz and Apps 1999), but several natural disturbances by insects and fires have reduced the C stock accumulated in the Canadian forest for more than half between 1990 and 2007, the largest decline among forest biomes of the world (Pan et al. 2011). Forest practices associated with even-aged management (Box 20.1), in single or few species stands, also have a significant impact on C balance on a large scale because of the types of logging practices, the area of harvested surfaces, and the intervals between them (Hunter 1999; Drever et al. 2006; Taylor et al. 2008; Innes et al. 2009; Lucier et al. 2009; Coursolle et al. 2012). Because of the associated greenhouse gas emissions, classic silvicultural practices may prove to be a catalyst for ongoing climate change (Taylor et al. 2008; Nunery and Keeton 2010). By comparison, the changeover from an even-aged to an uneven-aged management regime in the boreal forest biome could store up to 6% more CO_2 over a period of 140 years (Taylor et al. 2008) and up to 100% over a longer period in the boreal forest and deciduous temperate biome (Harmon et al. 2009; Nunery and Keeton 2010). Silvicultural practices that aim to maintain a permanent cover with frequent partial cuts promote long-term C sequestration capacity higher than unmanaged forest or forestry approach clearcutting practices (Harmon et al. 2009). If adaptation measures are applied, the sequestration potential from the planet's forest ecosystems is evaluated to 4 Pg C $year^{-1}$

BOX 20.1 EVEN-AGED AND UNEVEN-AGED SILVICULTURAL REGIMES

The forest stands can be managed to get or maintain an even-aged or an uneven-aged structure (Schütz 1997). At the temperate-boreal forest interface, the even-aged forest management regime is mainly applied in stands dominated by short-living species such as balsam fir and intolerant hardwood. The even-aged management regime can also be applied in plantations of long-living softwood when the objective is essentially timber production. The silvicultural sequence usually involves a thinning in the case of naturally occurring stands and two or even three commercial thinnings in the case of plantations. The final step of this silvicultural regime is generally clearcutting.

The uneven-aged management regime is primarily applied in natural stands or plantations with a component of long-living species such as spruces, northern white cedar, sugar maples, oaks, ash, bitternut hickory, black walnut, butternut, and black cherry. This silvicultural regime can be implemented from an even-aged stand where two or three commercial thinnings are performed prior to irregular shelterwood cuts that keep permanent cover. This is the type of silvicultural system for creating more complex structures compared to an even-aged forest management plan (O'Hara and Ramage 2013).

while currently it is 1.1 Pg C year^{-1} (Settele et al. 2014). This chapter proposes proactive silvicultural adaptation measures that could substantially contribute to reaching goals in the overall reduction of greenhouse gases emissions and that are applicable at the landscape and stand levels. These measures are mainly proposed to forest managers and foresters of the temperate–boreal interface of northeastern North America.

20.2 IMPACTS OF CLIMATE CHANGE ON FOREST DYNAMICS

Global warming should increase the impact of climate-related disruptive agents. In certain regions, we expect longer droughts, fires that are more severe and more frequent (Flannigan et al. 2005a, 2005b; Bond-Lamberty et al. 2009; Allen et al. 2010), more violent wind episodes, and more devastating ice storms (Johnston et al. 2010). Insect infestations and diseases should be more severe, especially if the hosts are abundant, vulnerable to climate change, and located at the limit of their range of distribution (Volney and Fleming 2000; Johnston et al. 2010) and if they form forests with low tree species diversity (Moreau et al. 2006; Jactel and Brockerkoff 2007).

Temperature, CO_2 concentration, light, water, and access to nutrients constitute limiting factors of prime importance to the productivity of forests (Price et al. 1999; Boisvenue and Running 2006; Karnosky et al. 2008). An increase in global average temperature of 4°C by 2100 could lead to a northern movement of isotherms 500 km in the northern hemisphere (Thuiller 2007). Most models predict that a warmer climate will lead to increased productivity in temperate (Alcamo et al. 2007; Field et al. 2007; Alo and Wang 2008) and boreal (Fischlin et al. 2007; Kurz et al. 2008; Fischlin et al. 2009) forests, and certain empirical studies confirm this trend (Pretzsch et al. 2014). Productivity has a direct link with C accumulation at the stand level (Man et al. 2013; Schütz and Pommerening 2013). C accumulation appears to be linked to species and to drought conditions prevailing during or before the growth period (Barr et al. 2004; Boisvenue and Running 2006; Fischlin et al. 2009; Huang et al. 2010; Lloyd et al. 2013). However, the productivity increase associated with the lengthening of the growth season could be canceled due to longer drought periods (Kirschbaum 2000; Barr et al. 2004; Kirschbaum 2004; Boisvenue and Running 2006; Huang et al. 2010), particularly in the central part of the North American boreal forest (Meehl et al. 2007; Fischlin et al. 2009; Girardin et al. 2009).

The geographical position, the tree species diversity, and the current extent of each ecological unit might be altered in the course of the present century (Malcolm et al. 2002; McKenney et al. 2007; Iverson et al. 2008; Allen et al. 2010; Chambers et al. 2013; Duveneck et al. 2014). The planet's global warming should provoke a northern migration and a southern decline for some species (McLachlan et al. 2005; Fischlin et al. 2007). The northern migration of species at the same rate than the northern displacement of isotherms induced by an increase in temperature of 4°C by 2100 would result in a displacement rate of 5 km year^{-1}, which is ten times faster than what was observed during the Holocene. A rapid and sustained warming of annual average temperatures may cause phenological changes in species, which will make them more vulnerable to disruptive agents (Colombo 2008) and less able to regenerate naturally (Chambers et al. 2013; Fisichelli et al. 2013a, 2013b). It is expected that temperate and boreal forests will be particularly affected by the ecosystem disturbances caused by global warming because temperature increase is greater in these regions and more particularly in the boreal forests (Plummer et al. 2006; Christensen et al. 2007; Colombo 2008; Lemmen et al. 2008; Fischlin et al. 2009; Lucier et al. 2009). For example, by the end of this century, in the northeastern United States and neighboring regions of Canada, a decline in conifer species, such as fir and spruces, and their gradual replacement by hardwood species, such as oaks, maples, and hickories, is expected (Iverson et al. 2008). Therefore, most of the balsam fir zone will become within a few decades a sensitive forest region and the lack of proactive adaptation measures may reduce the productivity of these ecosystems (Duveneck et al. 2014).

20.3 ADAPTED FOREST MANAGEMENT

Although the effects of global warming on forest ecosystems have been measured in the last few decades (Boisvenue and Running 2006; Johnston et al. 2006; Innes et al. 2009), the development of a silviculture adapted to this phenomenon is only beginning (Spittlehouse 2005; Innes et al. 2009; Magruder et al. 2013). Adapted forest management enhances forest productivity and resilience (Folke et al. 2004; Spittlehouse 2005), which in turn improve their capacity to sequester C originating from the use of fossil fuel. The design of adapted silvicultural approaches will be inspired by new emerging practices focused on the following.

(1) Ecosystem-based forest management (Hunter 1999; Bergeron et al. 1999; Gauthier et al. 2008; Vaillancourt et al. 2008)
(2) The complex structure principle (Steenberg et al. 2011; Puettmann 2011; Buongiorno et al. 2012; Schütz and Pommerening 2013)
(3) Species diversity (Cavard et al. 2010; Schütz et al. 2012; O'Hara and Ramage 2013; Puettmann and Tappeiner 2014)
(4) Adaptability by genetic diversity (Stokes and Kerr 2009; Kuparinen et al. 2010; Küchli 2013)

The use of an uneven-aged silviculture in mixed-species stands is presently imposing itself as the most favorable silvicultural regime since it allows for increased adaptation capacities to changes in water and thermal regime induced by global warming (Evans and Perschel 2009) and would have positive effects on C accumulation compared to even-aged management (Seidl et al. 2007; Taylor et al. 2008) or unmanaged stands (Harmon et al. 2009).

The general principle/objective of silvicultural adaptation measures to global warming should be the maintenance or increase of C sequestration by the forest ecosystem (Drever et al. 2006; Vallet 2008; Bradford and Kastendick 2010; Liu et al. 2011; Wang et al. 2013). These measures are notably based on the following.

(1) The reforestation of abandoned farmlands and of poorly regenerated forested areas (Harmon and Marks 2002; Spittlehouse and Stewart 2003; Anderson et al. 2006; Foote and Grogan 2010; Tremblay and Ouimet 2013);

(2) The implementation of partial cuts (or selective felling) to maintain optimal microclimatic conditions (Brassard and Chen 2010; Horner et al. 2010; Magruder et al. 2013) favorable to growth (Sohn et al. 2013) and the natural regeneration of adapted tree species (Fisichelli et al. 2013a, 2013b);
(3) The creation of seedbeds suitable for establishing regeneration (Fisichelli et al. 2013a, 2013b);
(4) The control of the size and frequency of clearcuts (Brassard and Chen 2010) to move toward a stable age structure at the landscape scale (Pregitzer and Euskirchen 2004; Magnani et al. 2007; Bradford and Kastendick 2010), given the longevity of the tree species and vulnerability of the stands (Wang et al. 2013);
(5) The restoration of a small gap regime to establish a regeneration of adapted and rarefied tree species (O'Hara 2001; Fahey and Puettmann 2007; Franklin et al. 2007);
(6) The structural conversion of regular stands toward uneven-aged/irregular ones to restore a structural, floristic (Puettmann 2011; Schütz and Pommerening 2013; Puettmann and Tappeiner 2014), and genetic complexity (Stokes and Kerr 2009; Küchli 2013).

20.4 ADAPTATION MEASURES AT THE REGIONAL AND LANDSCAPE SCALES

Current policies that govern forest management worldwide remain primarily focused on the production of woody material (Puettmann and Tappeiner 2014), which leaves the practitioner few silvicultural options specifically designed to prepare forests for ongoing climate change (Macdonald et al. 2010; Patry et al. 2013). In western Canada, a few approaches have been initiated to integrate new practices adapted to climate change. For example, adaptation measures have been included in management plans, such as assisted migration and forest logging management, to decrease the risks of fires (Yamasaki et al. 2008) or epidemics (Johnston et al. 2006).

In addition to assisted migration, control of the age structure, reforestation of abandoned farmlands, and poorly regenerated forested sectors are other examples of adaptation measures to climate change applicable at a large scale. Forest landscapes composed predominantly of stands aged between 30 and 120 years should display a positive C balance (Pregitzer and Euskirchen 2004; Magnani et al. 2007), whereas surface areas in regeneration should be minimized (Pregitzer and Euskirchen 2004; Coursolle et al. 2012; Tremblay and Ouimet 2013) along with areas of poorly productive stands that represent a C source (Pregitzer and Euskirchen 2004). Even if old forests represent a C source, their role is to store very large amounts of C over a long period (Kurz and Apps 1999; Harmon and Marks 2002; Pregitzer and Euskirchen 2004), hence the importance in conserving them. The choice of a silvicultural regime, even-aged or uneven-aged or both at the same time, will depend on the natural disturbance regime for a given region. For example, in regions where large-scale, stand-replacing natural disturbances dominate, or where the fire cycle would be less than 100 years, the even-aged silvicultural regime might be suitable because after large-scale disturbance natural regeneration will result in even-aged stand in any event (Etheridge and Kayahara 2013). However, in regions dominated by secondary natural disturbances of medium and small sizes (<10 ha) with long intervals (150 years and more), such as insect attacks, storms, or windthrow, the uneven-aged regime should be favored (Harvey et al. 2002; Etheridge and Kayahara 2013). In addition to these two alternatives, a mixed situation could be well suited in a region where the entire range of natural disturbances is represented. In such situations, both silvicultural regimes can be applied in concomitance and could constitute the best orientation to take in most regions of northeastern North America.

The reforestation of abandoned farmlands represents another adaptation measure at the regional and landscape levels. Several regions of eastern North America were affected by the rapid expansion of agricultural lands in the early 19th century that were followed by their massive abandonment a few decades later (Parson 1999; McLauchlan 2006). The abandonment of marginal farmlands is a continuing trend (Smith et al. 2005; McLauchlan 2006), and the surface area of abandoned agricultural lands is probably underestimated (Vouligny and Gariépy 2008). The rehabilitation of fallow lands by silvicultural interventions that favor a natural succession or by plantation establishment can contribute to sequester large quantities of C (Harmon and Marks 2002; Anderson et al. 2006; Foote and Grogan 2010; Tremblay and Ouimet 2013). Natural regeneration has the advantage of being inexpensive and may contribute to preserving the genetic heritage and maintaining high diversity (Tremblay and Ouimet 2013). Naturally regenerated forests generally represent a greater stand density than a plantation (Peterken 1996). Natural regeneration reduces forest management costs (Tremblay and Ouimet 2013) but lowers the forester's control over the composition of regeneration. As a consequence, vulnerable tree species, such as balsam fir (Duveneck et al. 2014), can germinate and colonize perturbed sites and form stands that are not well adapted to climate change (Tremblay and Ouimet 2013). It is estimated that natural regeneration established on abandoned agricultural lands in Ontario can sequester 5% of the total annual CO_2 emissions of this province, which are estimated at 55 Tg of C (Foote and Grogan 2010).

The establishment of plantations represents a complement or an alternative to natural regeneration but remains an expensive option. The production of plants in a nursery, field preparation, reforestation, and work involved in plantation clearing represent additional costs compared to natural regeneration. However, plantations are generally more productive and sequester more C than naturally regenerated stands (Anderson et al. 2006; van Minnen et al. 2008; Coursolle et al. 2012; Tremblay and Ouimet 2013). If plantations of adapted species had been established at a rate of 73,000 ha year^{-1} in Ontario between 1990 and 2012, this would have allowed an additional sequestration of 1.6 million metric tons of total C when compared to natural regeneration (Colombo et al. 2005).

The assisted migration of certain forest species or genotypes is already applied in certain regions (Rehfeldt et al. 1999; Beaulieu and Rainville 2005; O'Neill et al. 2008a) and should increase considerably in magnitude in the future (Pedlar et al. 2012). This measure is considered as a necessity to ensure that forests retain optimal productivity and CO_2 sequestration capacity (Savva et al. 2007; O'Neill et al. 2008a; Pedlar et al. 2012) under a climate regime that would have modified itself at a pace that neither gene flow nor natural migration could have compensated for.

There are three types of assisted migration: (1) the assisted expansion of populations, (2) the assisted expansion of the species' range, and (3) the introduction of exotic tree species.

(1) The assisted expansion of populations is already applied when seedlings produced from the seeds of external sources are planted in regions where the genotype is believed to be better adapted to the climate that will prevail during the tree's lifespan. For several years now, statistical models have been used to determine the migration distance that should be respected to optimize plantation productivity (McKenney et al. 2009). For white spruce, one of the species most frequently planted in eastern Canada, it is estimated that the seed source transfer protocol used in the Province of Québec will still be valid for the next 50 years (Beaulieu and Rainville 2005). However, this information remains unknown for many species.

(2) The assisted expansion of a species outside of its current range consists in introducing a species originating from a neighboring region into one where it has not yet migrated. Generally, the further away from the original seed source, the greater is the reduction in volume growth or survival rate (Beaulieu and Rainville 2005). Methods have been developed to assess, from climate projections, the ideal migration distance to obtain optimal growth for plantations in the future (McKenney et al. 2009). In British Columbia, for example, the displacement distance of climate zones was estimated for the next 25 years to plan the establishment of plantations of species outside their current distribution areas (O'Neill et al. 2008a; Rehfeldt and Jaquish 2010). Although climate conditions are suboptimal at this time, they still allow seedlings to settle (McKenney et al. 2009),

and a progressive enhancement of growth conditions during the tree's lifespan is expected (O'Neill et al. 2008a).

(3) Finally, the introduction of non-native species is the third type of assisted migration. Norway spruce (*Picea abies* (L.) Karst.), some species of larches (e.g., *Larix decidua* Mill., *Larix kaempferi* (Lamb.) Carrière), and hybrid poplars are examples of species that were introduced in several regions of eastern North America. Although this option is available to the practitioner, it remains a very controversial one (Aubin et al. 2011; Pedlar et al. 2012) and should be carefully monitored (Hunter 2007).

20.5 FOREST MANAGEMENT ADAPTED TO THE STAND LEVEL

In the eastern region of North America, at the temperate–boreal interface, there are no silvicultural measures specifically designed for the adaptation of stands to climate change (Patry et al. 2013). In the following section, we propose a sequence of silvicultural treatments that can contribute to mitigating the economic and ecological threats resulting from climate change. These silvicultural treatments are inspired by those that already exist here or elsewhere in the world but have sometimes been adapted according to the implications of climate change. These silvicultural treatments may be applied to stands of different ages and composition and according to whether an even- or uneven-aged strategy is followed. For each silvicultural treatment, precisions are given on its application, advantages, and particularities in relation to climate change.

20.5.1 Educational Silvicultural Treatments

20.5.1.1 Stands in Regeneration

During the regeneration stage, the maintenance of site productivity and the restoration of forest biodiversity are the main silvicultural objectives for adaptation to climate change (Folke et al. 2004; Steffen et al. 2004; Fischlin et al. 2007; Schütz and Pommerening 2013; Puettmann and Tappeiner 2014). At this stage, in both even-aged and uneven-aged management regimes, reforestation and precommercial thinning will be used. The reforestation of poorly regenerated forest sectors and fallow lands (van Minnen et al. 2008; Tremblay and Ouimet 2013) with adapted species (Table 20.1) represents the main adaptation measures that can be used by the forester (Figure 20.1). A fallow land reforested with softwood species (2,500 stems ha^{-1}) cumulates a total of ~ 30 tC ha^{-1} after 50 years compared to 21tC ha^{-1} for naturally regenerated fallow lands (Tremblay and Ouimet 2013). The forester must have a good knowledge of the ecological traits of the species used for reforestation or replacement planting. A species can be well adapted for a given region as long as its shade tolerance and optimal thermal zone are respected (Table 20.1). The range of temperatures is well known for most of the species used for reforestation, except for the white pine's minimum temperatures (Table 20.1). In this case, the range of temperatures is considered to be similar to the one observed for red pine.

Precommercial thinning is the silvicultural treatment that allows the stand to maintain optimum productivity by managing the density and composition. This silvicultural treatment results in changing the structure of the stand (Schneider et al. 2013) and making more water available for residual trees, thus reducing mortality risks caused by prolonged droughts (Sohn et al. 2013). Stands located on rich soils should be primarily selected because of their capacity of reaction to thinning (Johnston 2009). Modulations in the application of precommercial thinning are possible, and its resulting form depends on the composition of the stand. In the temperate–boreal transition zone, precommercial thinning prescriptions must aim at conserving all adapted tree species such as eastern white cedar, pines, spruces, and tolerant or semi-tolerant hardwoods along with other biodiversity components (fruit trees, noncommercial species, snags, etc.) (Figure 20.1). For stands composed of species sensitive to insects (Cappucino et al. 1998; Jactel and Brockerhoff 2007), diseases, or other factors such as overgrazing, the objective is to reduce their vulnerability. In certain regions,

TABLE 20.1
Temperature Range within the Distribution Area and Silvicultural Tolerances for Some Native Species of North America at the Seedling Stage

Scientific Name of Species	Average Temperature[18a,b] (Minimum and Maximum)	Shade Tolerance
Betula alleghaniensis	−18°C to 16°C[a]	Intermediate[3]
Acer saccharum	−18°C to 16°C[a]	Very tolerant[4]
Pinus strobus	−18°C to 21°C[a]	Intermediate[5]
Pinus resinosa	−18°C to 21°C[a]	Intolerant-intermediate[6]
Quercus rubra	4°C to 16°C[b]	Intermediate[7]
Quercus alba	−18°C to 21°C[a]	Intermediate to tolerant[8]
Carya ovata	4°C to 21°C[b]	Intermediate to tolerant[9]
Juglans nigra	7°C to 19°C[b]	Intolerant[10]
Juglans cinerea	4°C to 16°C[b]	Intolerant[11]
Prunus serotina	−8°C to 28°C[a]	Intolerant[12]
Fraxinus americana[1]	−14°C to 18°C[a]	Tolerant to intolerant[13]
Thuja occidentalis[2]	−12°C to 16°C[a]	Intermediate to tolerant[14]
Picea glauca	−29°C to 18°C[a,15a]	Intolerant-intermediate[15a–c]
Picea mariana	−20°C to 7°C[a]	Tolerant[16]
Picea rubens	−7°C to 14°C[a]	Tolerant[17]

1 *White ash seedlings are very tolerant to shade and can survive with less than 5% of light.*
2 *At the seedling stage, the eastern white cedar (Thuja occidentalis) has a rather intermediate tolerance to shade.*
3 *Erdmann (1990); Kneeshaw and Prévost (2007); Shields et al. (2007).*
4 *Godman et al. (1990).*
5 *Wendel and Smith (1990); Burgess and Wetzel (2002) except for the minimum temperature.*
6 *Rudolf (1990).*
7 *Sander (1990).*
8 *Rogers (1990).*
9 *Graney (1990).*
10 *Williams (1990).*
11 *Rink (1990).*
12 *Marquis (1990).*
13 *Schlesinger (1990).*
14 *Johnston (1990).*
15 *(a) Nienstaedt and Zasada (1990); (b) Comeau et al. (2008); (c) Hébert et al. (2013).*
16 *Viereck and Johnston (1990).*
17 *Blum (1990).*
18 *(a) Represents the average temperature for the coldest month (minimum = January) at the northern limit and for the warmest month (maximum = July) at the southern limit of its current distribution. (b) Mean annual temperature minimum and maximum.*

damages caused by specific faunal species can induce decreased productivity, especially by excessive grazing (Anderson et al. 2002; Danell et al. 2003; Frelich and Reich 2010; Fisichelli et al. 2012).

20.5.2 Commercial Silvicultural Treatments

Commercial silvicultural treatments recommended in the temperate–boreal transition zone of eastern Canada can be applied in several other regions that contain species that are particularly vulnerable to climate change, as is the case for the balsam fir (Ogden 2006; Williamson et al. 2009; Duveneck et al. 2014). Adaptation measures should consider its progressive replacement as soon

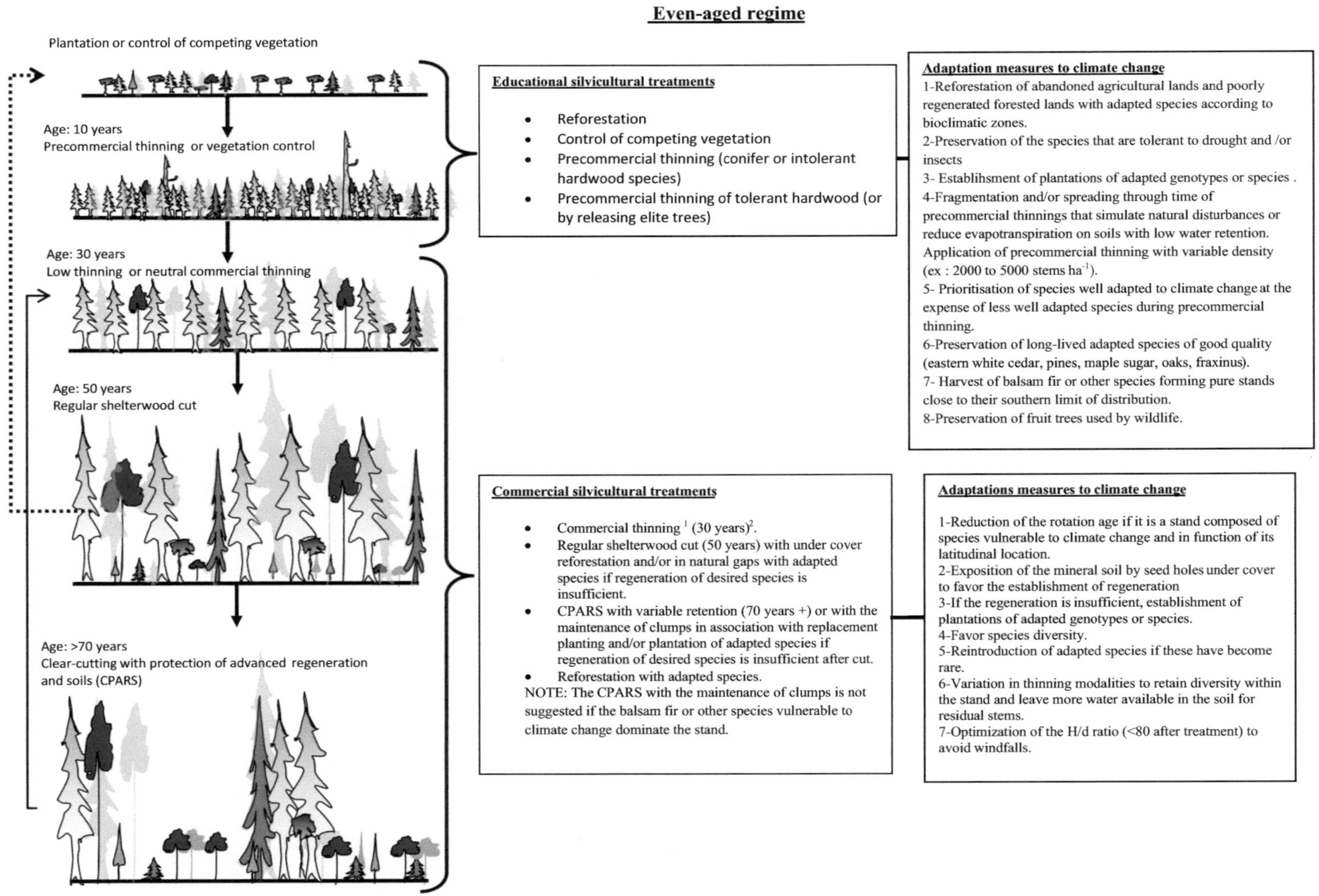

FIGURE 20.1 Sequence of silvicultural treatments adapted to climate change for the even-aged regime.

[1] Can include thinning modalities by the bottom, by the top or neutral with the releasing of 150, 200, or 300 elite trees ha^{-1} with or without gaps or a combination of these modalities.

[2] The age between parentheses is indicative and is function of the stand, its species composition, and site productivity.

as possible within the stand's lifespan, and this is also the case for all other species, such as white spruce (Fisichelli et al. 2013a; Duveneck et al. 2014), aspen, yellow birch, white birch, black spruce, and jack pine, judged as currently vulnerable or that will become so in the future (Duveneck et al. 2014). Subsequent silvicultural treatments will aim at maintaining or increasing stand productivity by the diversification of composition, structure, or both at the same time.

20.5.2.1 ADAPTATION MEASURES FOR TRANSITIONAL AND MATURE STANDS

20.5.2.1.1 Even-Aged Forest Regime (Single or Few Species)

Recommended silvicultural measures depend on the developmental stage of the stand from transitional to mature stages. When the competitive exclusion process is engaged after about 30 years or so of development, commercial thinning becomes one of the main silvicultural treatments that can be applied (Sohn et al. 2013; Magruder et al. 2013) (Figure 20.1). Commercial thinning offers several advantages, notably the delivery of forest products and the increase in C reserves (Harmon and Marks 2002). This intervention provides the possibility of harvesting species that are judged vulnerable while favoring the establishment of seedlings from long-lived species. Most long-lived species having small seeds, such as spruces (Ruel and Pineau 2002; Simard et al. 2003), pines (Cornett et al. 1998), eastern white cedar (Cornett et al. 2000, 2001; Larouche 2006), and yellow birch (Erdmann 1990; Shields et al. 2007), generally require specific germination conditions frequently associated with the presence of mineral soil (Cornett et al. 1998; Greene and Johnson 1998; Cornett et al. 2000) or rotten wood (Minore 1972; Greene and Johnson 1998; Cornett et al. 2000; Simard et al. 2003; Robert et al. 2012).

In subsequent developmental stages, contrasted silvicultural approaches between the south and north of the distribution area of species will have to be considered to account for the effects of climate change. For example, in a monospecific stand of balsam fir of 50 years or so located in the south of the species distribution area, stand recovery should be considered, particularly in the case of a spruce budworm outbreak or other insect pests (Williamson et al. 2009) (Figure 20.1, dotted arrow). When the balsam fir regenerates in abundance, a clearcut should be applied followed by field preparation and reforestation using adapted species. A balsam fir seedling that germinates now will develop under growth conditions that will become less favorable with time, particularly at the south of its distribution area (Johnston et al. 2009; Chambers et al. 2013). However, in the northern sector of its distribution area, the treatment generally recommended will be a shelterwood cut followed by a regeneration protection cut 10– 15 years later (Figure 20.1). When spruces, eastern white cedar, or other adapted species constitute an important component of the arborescent stratum, efforts should be devoted to the establishment of natural regeneration by exposing the mineral soil and maintaining a sufficient quantity of large woody debris on the soil (Simard et al. 2003). The progressive reduction of balsam fir proportion or of any other species judged vulnerable should be done for the benefit of adapted species.

20.5.2.1.2 Uneven-Aged Forest Regime

The structural conversion appears as an innovative adaptation approach, particularly in even-aged monospecific stands or those consisting of species vulnerable to climate change (Steenberg et al. 2011) (Box 20.2). Since this is a silvicultural approach that spans several decades (Schütz 1997, 2001; O'Hara 2001), the structural conversion provides the opportunity for managing at the same pace than the evolution of anticipated climate conditions. The type of intervention recommended to initiate structural conversion depends on the age of the stand.

In 30-year-old stands, it is suggested to apply a commercial thinning by releasing elite trees from competitors and create small gaps (Figure 20.2). In the context of silvicultural adaptation to climate change, an elite tree is a long-lived tree species, from the dominant or codominant class

BOX 20.2 STRUCTURAL CONVERSION

Structural conversion is a silvicultural treatment that requires a long period (Reininger 1987; Schütz 1997, 2001; Hanewinkel and Pretzsch 2000; O'Hara 2001). This silvicultural approach consists of transforming an even-aged stand into an uneven-aged, irregular stand without using clearcutting, except for creating small gaps (Figure 20.2). There are four steps to be followed (O'Hara 2001; Schütz 2001). No step should be omitted at the risk of jeopardizing the success of subsequent steps. The first step is to ensure that the stand is differentiated. The main differentiation factors can be represented by competition indices between stems such as the tree surface area, the dominant height/diameter ratio, and the Hart-Becking coefficient. A stand basal area higher than 25 m^2 ha^{-1} indicates that the competition has begun. However, this threshold may vary depending on the species and site productivity. The height/diameter ratio of dominant trees for the main conifer species should be maintained below 80 (Becquey 1986). The targeted value of the Hart-Becking coefficient at the time of the first thinning varies in function of the stand's age and species composition (Pardé 1961). For coniferous species such as eastern North American spruces, this coefficient must be between 15 and 20 (Pardé 1961; Riou-Nivert 1984; Prégent 1998). Beyond a value of 20, the stand does not have a density problem, but when below 15, the stand is potentially vulnerable to wind. For pines, the value of the Hart-Becking coefficient at the first thinning is around 22 (Riou-Nivert 1984). The second step consists in first establishing, as early as possible in the life of the stand, a regeneration of long-lived species in quantity and quality and then to create a heterogeneous horizontal structure. It's the abundance of regeneration (Schütz 1997; Hanewinkel and Pretzsch 2000; Schütz 2001) of long-lived species (Schütz 2001) that enables the initiation of the conversion process. For the regeneration to establish, interventions must be frequent to allow seedlings to receive sufficient light (Schütz 1997; 2001; Davies et al. 2008; Susse et al. 2011). The third step is to make sure that the regeneration is well established and that it will gradually form at first a bi-storied stand, and then an uneven, irregular stand. If this sequence is respected, the conversion process is underway. The fourth step is to practice uneven-aged management interventions to preserve the irregular structure of the stand. The absence of long-lived species in the original stand can be compensated by replacement planting in gaps or undercover if understory light conditions are favorable. Species such as pines, white spruce, eastern white cedar, red spruce, yellow birch, sugar maple, and oaks can be used for replacement planting based on the regions and their ecological requirements.

GAPS

Gaps have been used for a long time in silviculture of tolerant hardwood (O'Hara 2001) and coniferous (Anderson 1951; Franklin et al. 2007) stands. They have several silvicultural and ecological roles notably to allow natural or artificial seedlings to receive more light and to emulate natural disturbances (O'Hara 2001; Franklin et al. 2007). This is the reason why we associate gaps to structural conversion since it helps in orienting stands toward an irregular structure from the first thinning and also in subsequent silvicultural interventions. In natural conditions, their surface area can vary from a few tens to 500–1,000 m^2. These gaps are aimed at emulating small disturbances such as windfall or mortality by small groups of trees. The size of the gap influences local environmental conditions (Canham and Marks 1985; Bradshaw 1992; Fahey and Puettmann 2007), hence the importance of knowing the requirements of each species.

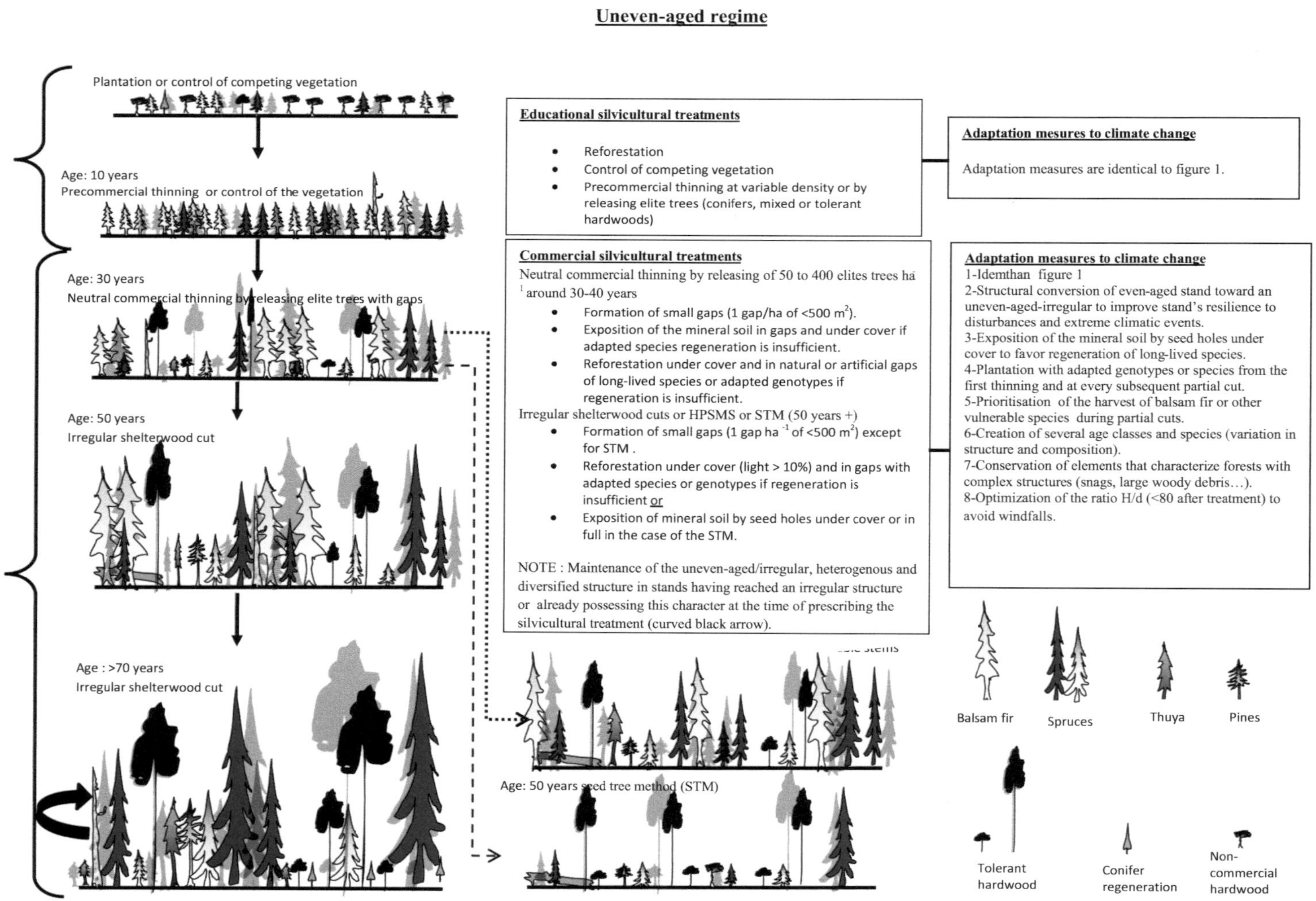

FIGURE 20.2 Sequence of silvicultural treatments adapted to climate change for the uneven-aged regime.[3] The modalities of commercial thinning can be applied under the condition that the stand basal area is ≥28 $m^2\ ha^{-1}$. Commercial thinning can be applied twice before moving to an irregular shelterwood cut regime.

with a straight bole. The number of elite trees that should be released can be modulated in function of the stand's characteristics (initial stem density, quality of dominant and codominant stems, average diameter) and can vary from 50 stems to 400 stems ha^{-1} (Schütz 1997, 2001; Hanewinkel and Pretzsch 2000; Mason and Kerr 2004; Balleux and Ponette 2006; Davies et al. 2008; Susse et al. 2011; Sanchez 2012; Sohn et al. 2013). The choice of elite trees must be made at the first thinning (Davies et al. 2008) without however completely releasing them. A certain number of elite trees can be released during the second thinning. Thinning should be conducted along with the exposition of the mineral soil in small patches and filling plantations with adapted species in natural or artificial gaps. An undercover filling plantation is possible in all types of stands provided that at least 25% of incident light reaches the forest floor at the time of plantation of shade-tolerant species (Greene et al. 2002). To promote plant growth undercover in young conifer stands, thinnings of light to moderate intensity must be frequent. Mixed stands dominated by hardwood species are more suitable for undercover plantation since they generally provide more light for seedlings during the growth period (Greene et al. 2002; Comeau et al. 2008).

The formation of small gaps emulates the effect of small windfalls to create heterogeneity in stands (Anderson 1951; Frelich and Lorimer 1991), whether these are conifer, mixed (Fahey and Puettmann 2007; Franklin et al. 2007; Hébert et al. 2013; Laarmann et al. 2013), or hardwood stands (O'Hara 2001), while allowing the natural or artificial regeneration to settle. In young stands, gaps should not exceed 0.05 ha, especially on rich sites, to halt the invasion of competing species and at the same time to allow the optimal growth of seedlings (Hébert et al. 2013). Gaps must be dispersed in such a way as to constitute 9–30% of the stand's surface area (Kneeshaw and Prévost 2007). For each silvicultural intervention, the existing gaps must be enlarged by the harvest of several peripheral trees to allow seedlings to receive at least 10% of incident light for tolerant hardwood (Anderson 1951; Kerr et al. 2010) and 25% for tolerant conifer species (Greene et al. 2002).

Ten years later, a second commercial thinning by the release of elite trees can be performed. However, the number of elite trees to be released during the second intervention will be inversely proportional to the number of elite trees released in the first thinning

(Hanewinkel and Pretzsch 2000; Mason and Kerr 2004; Balleux and Ponette 2006). The number of elite trees can also be constant through time, whereas the number of competitive trees harvested can change (Susse et al. 2011; Sohn et al. 2013).

In 50-year-old stands, two situations are possible: (1) if the stand contains at least 50% of long-lived species (spruces, eastern white cedar, oaks, yellow birch, pines, sugar maple, hickories, ashes), the irregular shelterwood cut system should be applied (Figure 20.2). The irregular shelterwood cut allows a progressive establishment of regeneration and creates an irregular structure with time (Grassi et al. 2003). The harvest of 30%–40% of the stand basal area at intervals of 15–25 years and moderate interventions provide for a steady flow of light under the forest cover necessary for the establishment of regeneration (Grassi et al. 2003). The creation of gaps of 200–500 m^2 and infilling plantations are also suitable silvicultural tools to promote the establishment of regeneration; (2) when the stand consists mostly of balsam fir (or of another vulnerable species), while long-lived species occupy more than 30% of stand basal area, the harvesting with protection of small merchantable stems (Figure 20.2, dotted arrow) is recommended. For stands dominated by long-lived hardwood species, the seed tree method (STM) is proposed (Figure 20.2, broken arrow). The STM should be applied only when long-lived species dominate the stand (Figure 20.2).

In stands that are 70 years old or older with a long-lived species component greater than 50% (Figure 20.2), the regime of irregular shelterwood cuts at intervals of 15–25 years is the most appropriate silvicultural approach to maintain structure and control composition (Figure 20.2, curved arrow). In this type of stand, the structure is already uneven aged or irregular, and gaps are already present. If more gaps are required at the time of the application of the treatment, they must not exceed a size of 0.2 ha (Kneeshaw and Prévost 2007; Kneeshaw et al. 2008) to avoid the invasion of competing species.

The challenge for today's forest specialists is to implement adaptive measures in an emergency context that will have positive impacts much later on (Brooks et al. 1998; Lindner et al. 2000; Chapin et al. 2004; Goldblum and Rigg 2005; van Minnen et al. 2008; Williamson et al. 2009). Current silvicultural practices must be improved (Johnston et al. 2009; Magruder et al. 2013; Patry et al. 2013; Wang et al. 2013) to allow the maintenance of site productivity at an optimal level in a context of climate change. Silvicultural treatments applied in the last decades were primarily established for economic reasons. Although being still relevant today, this trend is being reconsidered for environmental reasons. Climate change is largely responsible for this shift in favor of a forest management approach aimed at improving the adaptability of forest ecosystems to changing environmental conditions.

The costs to improve the adaptability of forests toward climate changes will be very high if we do not start as of now to modify our silvicultural practices (Maclean et al. 2010). Large-scale implementation of the mitigative measures proposed in this chapter can generate important social, economical, and ecological benefits, and this is at a lesser cost. Recent studies have shown that a managing strategy focused on diversity could lead to greater resilience in forest ecosystems and generate a greater and more regular income compared to a management strategy that is solely based on woody matter production (Davis 2010; Anonyme 2013; Dymond et al. 2014). Managing strategies can differ from one region to another and will depend on the state of forests or on their vulnerability to disturbances (Anonyme 2013). Adapted practices that will be integrated in management plans will significantly reduce the vulnerability of forests to epidemics or to other types of disturbances after only a few decades (Spittlehouse 2005; Bunnell and Kremsater 2012).

Forests provide many products and services such as timber, wildlife habitat, and environmental services, which are values that are essential to a collective well-being. Forest economy and governmental policies should, as of now, consider all of these values when calculating income, particularly in the context of global changes such as those that will be imposed by climate changes (Dymond et al. 2014). The complete life cycle of wood, from its production to its processing, use, and eventual decomposition, should be modeled to estimate more precisely its role in sequestering C from the atmosphere and in mitigating greenhouse effect. Studies will be needed to analyze in greater detail the long-term costs and benefits of the mitigation measures that are proposed in this chapter.

20.6 CONCLUSION

Vast regions bordering the temperate northern forest and the southern part of the boreal forest might strongly react to ongoing climate change because this part of the globe is experiencing a rapid increase in temperature (Plummer et al. 2006; Christensen et al. 2007; Lemmen et al. 2008; Fischlin et al. 2009; Lucier et al. 2009). In this chapter, we presented proactive measures to improve forest adaptability to climate change that included several of the practices associated with both ecosystem and forest management aimed at structural diversity. Concurrent to these, reactive management measures that respond to ecological disturbances could also increase the adaptability of forests to climate change (Kuparinen et al. 2010). Several of these practices are already applied locally in North America. However, elements that specifically aim at adapting to climate change as it is proposed in this chapter must still be integrated in management guides (Patry et al. 2013). Most silvicultural guides in North America do not consider climate change in their planning strategies. Yet climate change offers an exceptional scientific challenge to develop an adapted silviculture. Adapted management contributes to maintaining very high productivity of forests increasing the capacity of forests to sequester C and maintaining sustainable socioeconomic values (Maclean et al. 2010). Considering the unprecedented pace at which climate change is taking place, there is an urgent need for immediate modifications of our forest management practices (Smit et al. 1996; Smit et al. 2000; Lazar 2005; McKinnon and Webber 2005; Lemprière et al. 2008).

REFERENCES

Alcamo, J., J.M. Moreno, B. Nováky, et al. 2007. Europe. In *Climate change 2007: Impacts, adaptation and vulnerability*, eds. M.L. Parry, O.F. Canziani, J.P. Palutikof, P.J. van der Linden, and C.E. Hanson, pp. 541–580. Contribution of Working Group II to the Fourth Assessment Report of the Intergovernmental Panel of Climate Change (IPCC). Cambridge, UK: Cambridge University Press.

Allen, C.D., A.K. Macalady, H. Chenchouni, et al. 2010. A global overview of drought and heatinduced tree mortality reveals emerging climate change risks for forests. *For. Ecol. Manage.* 259:660–684. https://doi.org/10.1016/j.foreco.2009.09.001.

Alo, C.A. and G.L. Wang. 2008. Potential future changes of the terrestrial ecosystem based on climate projections by eight general circulation models. *J. Geophy. Res. Biogeosci.* 113:1–16 (G01004). https://doi.org/10.1029/2007JG000528.

Amiro, B.D., A.G. Barr, T.A. Black, et al. 2006. Carbon, energy and water fluxes at mature and disturbed forest sites, Saskatchewan, Canada. *Agric. For. Meteorol.* 136:237–251.

Anderson, C.E., K. Chapman, M. White, et al. 2002. Effects of browsing control on establishment and recruitment of Eastern white pine (*Pinus strobus* L.) at Cathedral Grove, Lake Superior Highlands, Minnesota, MN. *Nat. Areas J.* 22:202–210.

Anderson, K.J., A.P. Allen, J.F. Gillooly, et al. 2006. Temperature-dependence of biomass accumulation rates during secondary succession. *Ecol. Lett.* 9:673–682. https://doi.org/10.1111/j.1461-0248.2006.00914.x

Anderson, M.L. 1951. Spaced group-planting and irregularity of stand structure. *Empire For. Rev.* 30:328–341.

Anonyme. 2013. *Climate mitigation potential of British Columbian forests: Growing carbon sinks*, 29p. Victoria, British Columbia, Canada: Ministry of forests, Lands and Natural Resource Operations. www.for.gov.bc.ca/het/index.htm.

Apps, M.J., W.A. Kurz, S.J. Beukema, et al. 1999. Carbon budget of the Canadian forest product sector. *Environ. Sci. Policy* 2:25–41.

Aubin, I., C.M. Garbe, S. Colombo, et al. 2011. Why we disagree about assisted migration: Ethical implications of a key debate regarding the future of Canada's forests. *For. Chron.* 87(6):755–765.

Bailey, R.G. 1980. *Description of the ecoregions of the United States*, 77pp. U.S. Department of Agriculture, Forest Service, in cooperation with U.S. Fish and Wildlife Service. Miscellaneous Publication No. 1391.

Balleux, P. and Ponette, Q. 2006. Dispositif d'éclaircie en épicea, principaux enseignements de trente années d'expérience. *Forêt Walonne* no. 83—Juillet/Août, 21p. www.foretwallonne.be/images/stories/pdffolder/fw83_3–21[libin].pdf.

Barr, A.G., T.A. Black, E.H. Hogg, et al. 2004. Inter-annual variability in the leaf area index of boreal aspen-hazelnut forest in relation to net ecosystem production. *Agric. For. Meteorol.* 126:237–255.

Beaulieu, J. and A. Rainville. 2005. Adaptation to climate change: Genetic variation is both a shortand a long term solution. *For. Chron.* 81(5):704–709.

Becquey, J. 1986. Hauteur et facteur d'élancement: un équilibre à respecter. *Forêt-Entreprise* 34:14–21.

Bergeron, Y., B. Harvey, A. Leduc, and S. Gauthier. 1999. Forest management guidelines on natural disturbance dynamics: Stand-and forest-level considerations. *For. Chron.* 75:49–54.

Blum, B.M. 1990. Red spruce. In *Sylvics of North America*. Vol. 1. Softwoods, eds. R.M. Burns and B.H. Honkola. U.S. Department of Agriculture, Forest Service. *Agriculture Handbook 654*. Washington, DC. http://na.fs.fed.us/spfo/pubs/silvics_manual/Volume_1/vol1_Table_of_ contents.htm.

Boisvenue, C. and S.W. Running. 2006. Impacts of climate change on natural forest productivity—Evidence since the middle of the 20th century. *Glob. Chang. Biol.* 12:862–882. https://doi.org/10.1111/j.1365-2486.2006.01134.x

Bonan, G. 2002. *Ecological climatology: Concepts and applications*, 678pp. New York: Cambridge University Press.

Bond-Lamberty, B., S.D. Peckham, et al. 2009. Effects of fire on regional evapotranspiration in the central Canadian boreal forest. *Glob. Chang. Biol.* 15(5):1–13. https://doi.org/10.1111/j.1365-2486.2008.01776.x.

Bradford, J.B. and D.N. Kastendick. 2010. Age-related patterns of forests complexity and carbon storage in pine and aspen-birch ecosystems of northern Minnesota, USA. *Can. J. For. Res.* 40:401–409. https://doi.org/10.1139/X10-002

Bradshaw, F.J. 1992. Quantifying edge effect and patch size for multiple-use silviculture—A discussion paper. *For. Ecol. Manage.* 48:249–264.

Brassard, B.W. and H.Y.H. Chen. 2010. *Dynamique de structure et de composition des peuplements en forêt boréale mixte: Implications pour l'aménagement forestier*, 21p. Edmonton, Alberta, Canada: Réseau de gestion durable des forêts.

Brooks, J.R., L.B. Flanagan, and J.R. Ehleringer. 1998. Responses of boreal conifers to climate fluctuations: Indications from tree-ring widths and carbon isotope analyses. *Can. J. For. Res.* 28:524–533.

Bunnell, F.L. and L.L. Kremsater. 2012. Migrating like a herd cats: Climate change and emerging forest in British Columbia. *J. Ecosyst. Manage.* 13(2):1–24. http://jem.forrex.org/index.php/jem/article/viewfile/131/122.

Buongiorno, J., E.A. Halvorsen, O.M. Bollandsas, et al. 2012. Optimizing management regimes for carbon storage and other benefits in uneven-aged stands dominated by Norway spruce, with a derivation of the economic supply of carbon storage. *Scand. J. For. Res.* 27(5):460–473. http://dx.doi.org/10.1080/02827581.2012.657671.

Burgess, D. and S. Wetzel. 2002. Recruitment and early growth of eastern white pine (*Pinus strobus*) regeneration after partial cutting and site preparation. *Forestry* 75(4):419–423.

Burns, R.M. and B.H. Honkola. 1990a. *Sylvics of North America.* Vol. 2, ed. U.S. Department of Agriculture, Forest Service, *Agriculture Handbook 654.* Washington, DC. http://na.fs.fed.us/spfo/pubs/silvics_manual/volume_2/vol2_Table_of_contents.htm.

Burns, R.M. and B.H. Honkola. 1990b. *Sylvics of North America.* Vol. 1. Softwoods, ed. U.S. Department of Agriculture, Forest Service, *Agriculture Handbook 654.* Washington, DC. http://na.fs.fed.us/spfo/pubs/silvics_manual/Volume_1/vol1_Table_of_contents.htm.

Canham, C.D. and P.L. Marks. 1985. The responses of woody plants to disturbance: Patterns of establishment and growth. In *The ecology of natural disturbance and patch dynamics*, eds. S.T.A. Pickett and P.S. White, pp. 197–217. New York: Academic Press.

Cappucino, N., D. Lavertu, Y. Bergeron, and J. Regniere. 1998. Spruce budworm impact, abundance and parasitism rate in a patchy landscape. *Oecologia* 114:236–242.

Cavard, X., Y. Bergeron, H.Y.H. Chen, and D. Paré. 2010. Mixed-species effect on tree aboveground carbon pools in the east-central boreal forests. *Can. J. For. Res.* 40:37–47. https://doi.org/10.1139/X09-171.

Chambers, D., C. Périé, N. Casajus, et al. 2013. Challenges in modelling the abundance of 105 tree species in eastern North America using climate, edaphic, and topographic variables. *For. Ecol. Manage.* 291:20–29. http://dx.doi.org/10.1016/j.foreco.2012.10.046.

Chapin, F.S. III, T.V. Callaghan, Y. Bergeron, et al. 2004. Global change and the boreal forest: Thresholds shifting states or gradual change? *Ambio* 33(6):361–365.

Christensen, J.H., B. Hewitson, A. Busuioc, et al. 2007. Regional climate projections. In *Climate change 2007: The physical science basis*, eds. S. Salomon, D. Qin, M. Manning, et al. Contribution of Working Group I to the Fourth Assessment Report of the Intergovernmental Panel of Climate Change (IPCC), pp. 847–940. Cambridge, UK: Cambridge University Press.

Cohen, S. and K. Miller. 2001. North America. In *Climate change 2001: Impacts, adaptations, and vulnerability*, eds. J.J. McCarthy, O.F. Canziani, N.A. Leary, et al. Contribution of Working Group II to the Third Assessment Report of the Intergovernmental Panel of Climate Change (IPCC) (Chapter 15). Cambridge, UK: Cambridge University Press.

Colombo, S.J. 2008. *Ontario's forests and forestry in a changing climate,* 22 p. Climate change research report CCRR-12. Ontario Forest Research Institute. Ontario Ministry of Natural Resources, Sault Ste. Marie, Ontario, Canada. ISBN: 978-1-4249-7495-5 (PDF).

Colombo, S.J., W.J. Parker, N. Luckai, et al. 2005. *The effects of forest management on carbon storage in Ontario's forests.* Climate change research report; CCRR-03. Toronto, Ontario, Canada: Ministry of Natural Resources.

Comeau, P.G., C.N. Filipescu, R. Kabzems, et al. 2008. Growth of white spruce underplanted beneath spaced and unspaced aspen stands in northeastern B.C.—10 years results. *For. Ecol. Manage.* 257:1087–1094.

Cornett, M.W., L.E. Frelich, K.J. Puettmann, et al. 1998. Canopy type, forest floor, predation, and competition influence conifer seedling emergence and early survival in two Minnesota conifer deciduous forests. *Can. J. For. Res.* 28:196–205.

Cornett, M.W., K.J. Puettmann, L.E. Frelich, et al. 2001. Comparing the importance of seedbed and canopy type in the restoration of upland *Thuja occidentalis* forests of northeastern Minnesota. *Rest. Ecol.* 9(4):386–396.

Cornett, M.W., P.B. Reich, K.J. Puettmann, et al. 2000. Seedbed and moisture availability determine safe sites for early *Thuja occidentalis* (Cupressaceae) regeneration. *Am. J. Bot.* 87(12):1807–1814.

Coursolle, C., M-A. Giasson, H.A. Margolis, et al. 2012. Moving towards carbon neutrality: CO_2 exchange of a black spruce forest ecosystem during the first 10 years of recovery after harvest. *Can. J. For. Res.* 42:1908–1918. https://doi.org/10.1139/x2012-133.

Danell, K., R. Bergström, L. Edenius, et al. 2003. Ungulates as drivers of tree population dynamics at module and genet levels. *For. Ecol. Manage.* 181:67–76.

Davies, O., J. Haufe, and A. Pommerening. 2008. *Silvicultural principles of continuous cover forestry. A guide to best practice. Forestry Commission Wales*, 111p. Bangor, UK: Bangor University.

Davis, R. 2010. *Timber supply analysis report.* Merritt Timber supply review #4. Forest analysis and inventory branch. Victoria, British Columbia, Canada: Ministry of Forests and Range, 111p. www.for.gov.bc.ca/hts/.

Dixon, R.K., S. Brown, R.A. Houghton, et al. 1994. Carbon pools and flux of global forest ecosystems. *Science* 263:185–190.

Drever, C.R., G. Peterson, C. Messier, et al. 2006. Can forest management based on natural disturbances maintain ecological resilience? *Can. J. For. Res.* 36:2285–2299.

Duveneck, M.J., R.M. Scheller, M.A. White, et al. 2014. Climate change effects on northern Great Lake (USA) forests: A case for preserving diversity. *Ecosphere* 5(2):23 1–26. http://dx.doi.org/10.1890/ES13-00370.1.

Dymond, C.C., T. Sinclair, D.L. Spittlehouse, et al. 2014. Diversifying managed forests to increase resilience. *Can. J. For. Res.* 44:1196–1205. dx.doi.org/10.1139/cjfr-2014-0146.

Erdmann, G.G. 1990. *Betula alleghaniensis.* In *Sylvics of North America.* Vol. 2. Hardwoods, eds. R.M. Burns and B.H. Honkola, U.S. Department of Agriculture, Forest Service, *Agriculture Handbook 654.* Washington, DC. http://na.fs.fed.us/spfo/pubs/silvics_manual/volume_2/vol2_Table_ of_contents. htm.

Etheridge, D.A. and G.J. Kayahara. 2013. Challenges and implications of incorporating multi-cohort management in northeastern Ontario, Canada: A case study. *For. Chron.* 89(3):315–326.

Evans, A.M. and R. Perschel. 2009. A review of forestry mitigation and adaptation strategies in the Northeast U.S. *Clim. Change* 96:167–183.

FAO. 2001. *Global forest resource assessment 2000—Main report*, 479p. FAO Forestry Paper 140, Rome, Italy: Food and Agriculture Organization of the United Nations.

FAO. 2010. *Global forest resources Assessment 2010—Main report.* Rome, Italy: FAO Forestry Paper No. 163. www.fao.org/docrep/013/i1757e/i1757e00.htm.

Fahey, R.T. and K.J. Puettmann. 2007. Ground-layer disturbance and initial conditions influence gap partitioning of understorey vegetation. *J. Ecol.* 95:1098–1109.

Fensham, R.J. and G.P. Guymer. 2009. Carbon accumulation through ecosystem recovery. *Environ. Sci. Policy* 12:367–372. https://doi.org/10.1016/j.envsci.2008.12.002

Feurdean, A., S.A. Bhagwat, K.J. Willis, et al. 2013. Tree migration-rates: Narrowing the gap between inferred post-glacial rates and projected rates. *PLoS ONE* 8(8):e71797. https://doi.org/10.1371/journal.pone.0071797.

Field, C.B., L.D. Mortsch, M. Brklacich, et al. 2007. North America. In *Climate change 2007: impacts, adaptation and vulnerability*, eds. M.L. Parry, O.F. Canziani, J.P. Palutikof, et al., pp. 617–652, Contribution of Working Group II to the Fourth Assessment Report of the Intergovernmental Panel of Climate Change (IPCC). Cambridge, UK: Cambridge University Press.

Fischlin, A., M. Ayres, D. Karnosky, et al. 2009. Future environmental impacts and vulnerabilities In *Adaptation of forests and people to climate change—a global assessment report*, eds. Seppala, R., Buck, A., and Katila, P., World Series Volume 22, 224p. Helsinki, Finland: International Union of Forest Research Organizations (IUFRO).

Fischlin, A., G.F. Midgley, J.T. Price et al. 2007. In *Ecosystems, their properties, goods, and services. Climate change 2007: Impacts, adaptation and vulnerability*, eds. M.L. Parry, O.F. Canziani, J.P. Palutikof et al., pp. 211–272. Contribution of working Group II to the Fourth Assessment Report of the Intergovernmental Panel on Climate Change. Cambridge, UK: Cambridge University Press.

Fisichelli, N., L.E. Frelich, and P.B. Reich. 2012. Sapling growth responses to warmer temperatures 'cooled' by browse pressure. *Glob. Change Biol.* 18:3455–3463.

Fisichelli, N., L.E. Frelich, and P.B. Reich. 2013a. Temperate tree expansion into adjacent boreal forest patches facilitated by warmer temperatures. *Ecography* 37:152–161. https://doi.org/10.1111/j.1600-0587.2013.00197.x.

Fisichelli, N., L.E. Frelich, and P.B. Reich. 2013b. Climate and interrelated tree regeneration drivers in mixed temperate–boreal forests. *Landsc. Ecol.* 28:149–159.

Flannigan, M.D., B.D. Amiro, K.A. Logan, et al. 2005a. Forest fires and climate change in the 21st century. *Mitig. Adapt. Strat. Glob. Change* 11(4):847–859. https://doi.org/10.1007/s11027-005-9020-7.

Flannigan, M.D., K.A. Logan, B.D. Amiro, et al. 2005b. Future area burned in Canada. *Clim. Change* 72:1–16.

Folke, C., S. Carpenter, B. Walker, et al. 2004. Regime shifts, resilience, and biodiversity in ecosystem management. *Ann. Rev. Ecol. Evol. Syst.* 35:557–581.

Foote, R.L. and P. Grogan. 2010. Soil carbon accumulation during temperate forest succession on abandoned low productivity agricultural lands. *Ecosystems* 13:795–812. https://doi.org/10.1007/s10021-010-9355-0.

Franklin, J.F., R.J. Mitchell, and B.J. Palik 2007. *Natural disturbance and stand development principles for ecological forestry*, 44p. General Technical Report NRS-19. Newtown Square, PA: U.S. Department of Agriculture, Forest Service, Northern Research Station.

Freer-Smith, P.H., M.S.J. Broadmeadow, and J.M. Lynch 2008. Forests and climate change: The knowledge-base for action. In *Forestry and climate change* (Chapter 2), eds. P.H. Freer-Smith, M.S.J. Broadmeadow, and J.M. Lynch, 253p. Oxfordshire, UK: CABI.

Frelich, L.E. and C.G. Lorimer 1991. Natural disturbance regimes in hemlock-hardwood forests of the upper Great Lakes Region. *Ecol. Monogr.* 61(2):145–164.

Frelich, L.E. and P.B. Reich. 2010. Will environmental changes reinforce the impact of global warming on the prairie-forest border of central North America? *Front. Ecol. Environ.* 7:371–378.

Gauthier, S., M-A. Vaillancourt, D. Kneeshaw, et al. 2008. Aménagement forestier écosystémique. Origines et fondements. In *Aménagement écosystémique en forêt boréale*, eds. S. Gauthier, M.-A. Vaillancourt, A. Leduc et al., Chapitre 1, pp. 13–40, 567 p. Montréal, Québec, Canada: Presses de l'Université du Québec.

Girardin, M., A.A. Ali, C. Carcaillet, et al. 2009. Heterogeneous response of circumboreal wildfire risk to climate change since the early 1900s. *Glob. Change Biol.* 15:2751–2769. https://doi.org/10.1111/j.1365-2486.2009.01869.x.

Godman, R.M., H.W. Yawney, and C.H. Tubbs. 1990. Sugar maple. In *Sylvics of North America.* Vol. 2. Hardwoods, eds. R.M. Burns and B.H. Honkola. U.S. Department of Agriculture, Forest Service, *Agriculture Handbook 654.* Washington, DC. http://na.fs.fed.us/spfo/pubs/silvics_ manual/volume_2/vol2_Table_of_contents.htm.

Goldblum, D. and L.S. Rigg 2005. Tree growth response to climate change at the deciduous-boreal forest ecotone, Ontario, Canada. *Can. J. For. Res.* 35(11):2709–2718.

Goulden, M.L., S.C. Wofsy, J.W. Harden, et al. 1998. Sensitivity of boreal forest carbon balance to soil thaw. *Science* 279(5348):214–217.

Graney, D.L. 1990. *Carya ovata.* In *Sylvics of North America.* Vol. 2. Hardwoods. eds. R.M. Burns and B.H. Honkola. U.S. Department of Agriculture, *Agriculture Handbook 654.* Forest Service, Washington, DC. http://na.fs.fed.us/spfo/pubs/silvics_manual/volume_2/vol2_Table_of_ contents.htm.

Grassi, G., G. Minotta, R. Giannini, et al. 2003. The structural dynamics of managed uneven-aged conifer stands in the Italian eastern Alps. *For. Ecol. Manage.* 185:225–237.

Greene, D.F and E.A. Johnson 1998. Seed mass and early survivorship of tree species in upland clearings and shelterwoods. *Can. J. For. Res.* 28:1307–1316.

Greene, D.F., D.D. Kneeshaw, C. Messier, et al. 2002. Modelling silvicultural alternatives for conifer regeneration in boreal mixedwood stands (aspen/white spruce/balsam fir). *For. Chron.* 78(2):281–295.

Hanewinkel, M. and H. Pretzsch. 2000. Modelling the conversion from even-aged to uneven-aged stands of Norway spruce (*Picea abies* L. Karst.) with a distance-dependent growth simulator. *For. Ecol. Manage.* 134:55–70.

Hansen, J., M. Sato, R. Ruedy, et al. 2006. Global temperature change. *PNAS* 103(39):14288–14293.

Harmon, M.E. and B. Marks. 2002. Effects of silvicultural practices on carbon stores in Douglas fir-western hemlock forests in the Pacific Northwest, USA: Results from a simulation model. *Can. J. For. Res.* 32:863–877.

Harmon, M.E., A. Moreno, and J.B. Domingo. 2009. Effects of partial harvest on the carbon stores in Douglas fir/western hemlock forests: A simulation study. *Ecosystems* 12:777–791.

Harvey, B.D., A. Leduc, S. Gauthier, et al. 2002. Stand-landscape integration in natural disturbance based management of the southern boreal forest. *For. Ecol. Manage.* 155:369–385.

Hébert, F., V. Roy, I. Auger, and M.-M. Gauthier. 2013. White spruce *(Picea glauca)* restoration in temperate mixedwood stands using patch cuts and enrichment planting. *For. Chron.* 89(3):392–400.

Heimann, M. 2008. Present and future carbon sources and sinks. In *Forestry and climate change*, eds. P.H. Freer-Smith, M.S.J. Broadmeadow, and J.M. Lynch (Chapter 3), 253p. Oxfordshire, UK: CABI.

Horner, G.J., P.J. Baker, R. Mac Nally, et al. 2010. Forest structure, habitat and carbon benefits from thinning floodplain forests: Managing early stand density makes a difference. *For. Ecol. Manage.* 259:286–293.

Huang, J., J.C. Tardif, Y. Bergeron, et al. 2010. Radial growth response of four dominant boreal tree species to climate along a latitudinal gradient in the eastern Canadian boreal forest. *Glob. Change Biol.* 16:711–731. https://doi.org/10.1111/j.1365-2486.2009.01990.x.

Hunter, M.L. Jr. 1999. *Maintaining biodiversity in forest ecosystems*, 698pp. Cambridge, UK: Cambridge University Press.

Hunter, M.L. Jr. 2007. Climate change and moving species: Furthering the debate on assisted colonization. *Conserv. Biol.* 21(5):1356–1358.

Innes, J., L.A. Joyce, S. Kellomäki et al. 2009. Management for adaptation. In *Adaptation of forest and people to climate change—a global assessment report*, eds. R. Seppälä, A. Buck, and P. Katila, 224p. (Chapter 6). IUFRO World Series Vol. 22. Helsinki, Finland.

Iverson, L., A. Prasad, and S. Matthews 2008. Modeling potential climate change impacts on the trees of the northeastern United States. *Mitig. Adapt. Strat. Glob. Change* 13:487–516.

Jactel, H. and E.G. Brockerhoff. 2007. Tree diversity reduces herbivory by forest insects. *Ecol. Lett.* 10:835–848. https://doi.org/10.1111/j.1461-0248.2007.01073.x.

Jobbagy, E.G. and R.B. Jackson. 2000. The vertical distribution of soil organic carbon and its relation to climate and vegetation. *Ecol. Appl.* 10(2):423–436.

Johnston, M.H., M. Campagna, P. Gray, et al. 2009. *Vulnérabilité des arbres du Canada aux changements climatiques et propositions de mesures visant leur adaptation.* Un aperçu destiné aux décideurs et aux intervenants du monde forestier. Conseil canadien des ministres des forêts. ISBN: 978-1-100-92687-2 www.ccfm.org/pdf/TreeSpecies_web_f.pdf 47p.

Johnston, M., T. Williamson, A. Munson, et al. 2010. *Climate change and forest management in Canada: Impacts, adaptive capacity and adaptation options*, 60p. A State of Knowledge report. Sustainable forest management network, Edmonton, Alberta. http://sfmnetwork.ca. ISBN: 978-1-55261-235-4.

Johnston, M.H., T. Williamson, D. Price et al. 2006. *Adapting forest management to the impacts of climate change in Canada.* A BIOCAP Research integration program. Synthesis paper. 96p. www.for.gov.bc.ca/hre/pubs/pubs/1390.htm.

Johnston, W.F. 1990. Northern White-cedar. In *Sylvics of North America.* Vol. 1. Softwoods, eds. R.M. Burns and B.H. Honkola. U.S. Department of Agriculture, Forest Service, *Agriculture Handbook 654.* Washington, DC. http://na.fs.fed.us/spfo/pubs/silvics_manual/Volume_1/vol1_Table_of_contents.htm.

Karnosky, D.F., M. Tallis, J. Darbah, and Taylor, G. 2008. Direct effects of elevated carbon dioxide on forest tree productivity. In *Forestry and climate change*, eds. Freer-Smith, P.H., M.S.J. Broadmeadow, and J.M. Lynch (Chapter 17), 253p. Wallingford: CABI.

Kerr, G., G. Morgan, J. Blyth, and V. Stokes. 2010. Transformation from even-aged plantations to an irregular forest: The world's longest running trial area at Glentress, Scotland. *Forestry* 83(3):329–344. https://doi.org/10.1093/forestry/cpq015.

King, G.A. and A.A. Herstrom. 1997. Holocene tree migration rates objectively determined from fossil pollen data. In: *Past and future environmental changes: The spatial and evolutionary responses of terrestrial biota*, eds. B. Huntley, W. Cramer, A.V. Marsa et al. pp. 91–102. New York: Springer-Verlag.

Kirschbaum, M.U.F. 2000. Forest growth and species distribution in a changing climate. *Tree Physiol.* 20:309–322.

Kirschbaum, M.U.F. 2004. Assessing long-term trend in forest productivity under changing climatic conditions. In *IUFRO 4.01.09 process based models for forest management*—International Conference on Modeling Forest Production. Vienna, Austria: University of Natural Resources and Applied Life Sciences, Department of Ecology.

Kneeshaw, D., È. Lauzon, A. de Römer, et al. 2008. Appliquer les connaissances sur les régimes de perturbations naturelles pour développer une foresterie qui s'inspire de la nature dans le sud de la péninsule gaspésienne In *Aménagement écosystémique en forêt boréale.* (Chapter 9), pp. 215–240. Montréal, Québec, Canada: Presses de l'Université du Québec.

Kneeshaw, D. and M. Prévost. 2007. Natural canopy gap disturbance and their role in maintaining mixed-species forests on central Québec, Canada. *Can. J. For. Res.* 37:1534–1544.

Küchli, C. 2013. The Swiss experience in forest sustainability and adaptation. *Unasylva* 64:12–18.

Kuparinen, A., O. Savolainen, and F.M. Schurr 2010. Increased mortality can promote evolutionary adaptation of forest trees to climate change. *For. Ecol. Manage.* 259:1003–1008.

Kurz, W.A. and M.J. Apps. 1999. A 70-year retrospective analysis of carbon fluxes in the Canadian Forest Sector. *Ecol. Appl.* 9(2):526–547.

Kurz, W.A., C.C. Dymond, G. Stinson, et al. 2008. Mountain pine beetle and forest carbon feedback to climate change. *Nature* 452(7190):987–990.

Laarmann, D., H. Korjus, A. Sims et al. 2013. Initial effects of restoring natural forest structures in Estonia. *For. Ecol. Manage.* 304:303–311.

Larouche, C. 2006. Rarefaction du thuya. *Chapitre 5 (addenda) du document: Les enjeux de biodiversité relatifs à la composition forestière*, eds. P. Grondin and A. Cimon, 32p. Ministère des Ressources naturelles, de la Faune et des Parcs, Direction de la recherche forestière et Direction de l'environnement forestier. www.mrnf.gouv.qc.ca/forets/connaissances/connaissances-activites-diversite.jsp.

Lazar, A. 2005. A proud record of leadership in addressing and adapting to climate change. *For. Chron.* 81(5):631–632.

Lemmen, D.S., F.J. Warren, J. Lacroix, et al., eds. 2008. *Vivre avec les changements climatiques au Canada: édition 2007*, 448p. Ottawa, Ontario, Canada: Gouvernement du Canada. [en ligne] www.adaptation.nrcan.gc.ca/assess/2007/index_f.php.

Lempriere, T.C., P.Y. Bernier, A.L. Carroll, et al. 2008. *L'importance d'adapter le secteur forestier aux changements climatiques*. Ressources naturelles du Canada, Rapport d'information NOR-X416F. Centre de foresterie du Nord, Edmonton (Alberta). Service canadien des forets, 64p. ISBN: 978-0-662-04836-7. http://librairie.scf.rncan.gc.ca/.

Lindner, M., P. Lasch, and M. Erhard. 2000. Alternative forest management strategies under climatic change—Prospects for gap model applications in risk analyses. *Silva Fennica* 34(2):101–111.

Lindquist, E.J., R. D'Annunzio, A. Gerrand, et al. 2012. *Global forestland-use change*. FAO Forestry Paper No. 169. Food and Agriculture Organization of the United Nations and European Commission Joint Research Centre. Rome, Italy: FAO.

Liu, S., B. Bond-Lamberty, J.A. Hicke, et al. 2011. Simulating the impacts of disturbances on forest carbon cycling in North America: Processes, data, models, and challenges. *J. Geophysical Res. Biogeosci* 116: G00K08. https://doi.org/10.1029/2010jg001585.

Lloyd, A.H., P.A. Duffy, and D.H. Mann. 2013. Nonlinear responses of white spruce growth to climate variability in interior Alaska. *Can. J. For. Res.* 43:331J343. https://dx.doi.org/10.1139/cjfr-2012-0372.

Lucier, A., M. Ayres, D. Karnosky, et al. 2009. Chap. 2. Forest responses and vulnerabilities to recent climate change. In *Adaptation of forest and people to climate change—a global assessment report*, eds. R. Seppala, A. Buck, and P. Katila, 224p. IUFRO World Series Vol. 22. Helsinki, Finland. ISBN: 978-3-901347-80-1. www.iufro.org/

Macdonald, E., N. Lecomte, Y. Bergeron, et al. 2010. *Répercussions écologiques découlant de changements dans la composition des forêts boréales mixtes*, 52p. Edmonton, Alberta, Canada: Rapport sur l'état des connaissances. Réseau de gestion durable des forêts.

MacDonald, G.M. 1993. Fossil pollen analysis and the reconstruction of plant invasions. *Adv. Ecol. Res.* 24:67–110.

MacLean, D.A., L.J. Amos-Binks, C.R. Hennigar, et al. 2010. Stratégie d'entreposage du carbone dans les forêts et les produits ligneux. Réseau GDF. Série de note de recherche No. 64. Réseau de gestion durable des Forêts. http://sfmnetwork.ca.

Magnani, F., M. Mencuccini, M. Borghetti, et al. 2007. The human footprint in the carbon cycle of temperate and boreal forests. *Nature* 447(7146):848–850. https://doi.org/10.1038/nature05847.

Magruder, M., S. Chhin, B. Palik, et al. 2013. Thinning increases climatic resilience of red pine. *Can. J. For. Res.* 43:878–889. dx.doi.org/10.1139/cjfr-2013–0088.

Malcolm, J.R., A. Markham, R.P. Neilson, et al. 2002. Estimated migration rates under scenarios of global climate change. *J. Biogeogr.* 29:835–849.

Man, C.D., K.C. Lyons, J.D. Nelson, et al. 2013. Potential of alternate forest management practices to sequester and store Carbon in two forest estates in British Columbia, Canada. *For. Ecol. Manage.* 305:239–247.

Manies, K.L., J.W. Harden, B.P. Bond-Lamberty, and K.P. O'Neill. 2005. Woody debris along an upland chronosequence in boreal Manitoba and its impact on long-term carbon storage. *Can. J. For. Res.* 35(2):472–482. https://doi.org/10.1139/X04-179.

Marcott, S.A., J.D. Shakun, P.U. Clark, et al. 2013. A reconstruction of regional and global temperature for the last 13,000 years. *Science* 339:1198–1201.

Marquis, D.A. 1990. Prunus serotina. In *Sylvics of North America*. Vol. 2. Hardwoods, eds. R.M. Burns and B.H. Honkola. U.S. Department of Agriculture, Forest Service, *Agriculture Handbook 654*. Washington, DC. http://na.fs.fed.us/spfo/pubs/silvics_manual/volume_2/vol2_Table_ of_contents.htm.

Mason, B. and G. Kerr. 2004. *Transforming even-aged conifer stands to continuous cover management*, 8 p. Edinburgh: Information note. Forestry Commission. www.forestry.gov.uk. ISBN: 0-85538-620-7.

McKenney, D.W., J.H. Pedlar, K. Lawrence, et al. 2007. Potential impacts of climate change on the distribution of North American trees. *BioScience* 57(11):939–948.

McKenney, D.W., J.H. Pedlar, and G. O'Neill. 2009. Climate change and forest seed zones: Past trends, future prospects and challenges to ponder. *For. Chron.* 85(2):258–266.

McKinnon, G.A. and S.L. Webber 2005. Climate change impacts and adaptation in Canada: Is the forest sector prepared? *For. Chron.* 81(5):653–654.

McLachlan, J.S., J.S. Clark, and P.L. Manos. 2005. Molecular indicators of tree migration capacity under rapid climate change. *Ecology* 86(8):2088–2098.

McLauchlan, K. 2006. The nature and longevity of agricultural impacts on soil carbon and nutrients: A review. *Ecosystems* 9:1364–1382. https://doi.org/10.1007/s10021-005-0135-1.

Meehl, G.A., C. Covey, T. Delworth, et al. 2007. The WCRP CMIP3 Multimodel Dataset, a new era in climate change research. *Bull. Am. Meteorol. Soc.* 88(9):1383–1394.

Melillo, J.M., T.V. Callaghan, F.I. Woodward, et al. 1990. Effects on ecosystems. In *Climate change: The IPCC scientific assessment*, eds. Houghton, J.T., G.J. Jenkins, and J.J. Ephraums, pp. 283–310 Cambridge, UK: Cambridge University Press.

Minore, D. 1972. Germination and early growth of coastal tree species on organic seed beds. USDA Forest Service Research Paper PNW-135. Pacific northwest forest and range experiment station, Portland, Oregon. 18p.

Moreau, G., E.S. Eveleigh, C.J. Lucarotti, et al. 2006. Stage-specific responses to ecosystem alteration in an eruptive herbivorous insect. *J. Appl. Ecol.* 43:28–34.

MRNF, Ministère des Ressources naturelles et de la faune du Québec. 2009. *Ressources et industries forestières*. Portrait statistique édition 2009. Québec, Canada: Direction du développement de l'industrie des produits forestiers. ISBN: 978-2-550-55569-8 (PDF). www.mrnf.gouv. qc.ca/forets/connaissances/connaissances-statistiques.jsp.

Nienstaedt, H. and J.C. Zasada 1990. White spruce. In *Sylvics of North America*. Vol. 1. *Softwoods*, eds. R.M. Burns and B.H. Honkola. U.S. Department of Agriculture, *Agriculture Handbook*. http://na.fs.fed.us/spfo/pubs/silvics_manual/Volume_1/vol1_Table_of_contents.htm.

Norby, J.R. and R.B. Jackson. 2000. Root dynamics and global change: Seeking an ecosystem perspective. *New Phytol.* 147:3–12.

Nunery, J.S. and W.S. Keeton. 2010. Forest Carbon storage in the northeastern United States: Net effects of harvesting frequency, postharvest retention, and wood products. *For. Ecol. Manage.* 259:1363–1375.

Ogden, A.E. 2006. *Forest management in a changing climate: Building the environmental information base for southwest Yukon: Overview report*, 35p. University of British Columbia. Whitehorse, Yukon, Canada: Northern Climate ExChange. http://yukon.taiga.net/swyukon/.

O'Hara, K.L. 2001. The silviculture of transformation—A commentary. *For. Ecol. Manage.* 151:81–86.

O'Hara, K.L. and S.R. Ramage 2013. Silviculture in an uncertain world: Utilizing multi-aged management systems to integrate disturbance. *Forestry* 86:401–410. https://doi.org/10.1093/forestry/cpt012.

O'Neill, G.A., N. Ukrainetz, M. Carlson et al. 2008a. *Assisted migration to address climate change in British Columbia*. Recommendations for interim seed transfer standards. Victoria, British Columbia, Canada: B.C. Min. For. Range, Res. Br., Tech. Rep. 048. www/for.gov.bc.ca/hfd/pubs/Docs/Tr/Tro48.htm.

Pan, Y., R.A. Birdsey, J. Fang, et al. 2011. A large and persistent carbon sink in the world's forest. *Science* 333:988–993.

Pardé, J. 1961. Comment préciser l'intensité d'une éclaircie? *Revue Forestière Française* 8–9:551–557.

Parson, H.E. 1999. Regional trends of agricultural restructuring in Canada. *Can. J. Reg. Sci.* 22:343–356.

Patry, C., D. Kneeshaw, S. Wyatt, et al. 2013. Forest ecosystem management in North America: From theory to practice. *For. Chron.* 89(4):525–537.

Pedlar, J.H., D.W. McKenney, I. Aubin et al. 2012. Placing forestry in the assisted migration debate. *BioScience* 62(9):835–842.

Peterken, G.F. 1996. *Natural woodland; Ecology and conservation in northern temperate regions*, 20p. Cambridge, UK: Cambridge University Press.

Plummer, D.A., D. Caya, A. Frigon, et al. 2006. Climate and climate change over North America as simulated by the Canadian RCM. *J. Clim.* 19(13):3112–3132.

Post, W.M., W.R. Emanuel, P.J. Zinke, et al. 1982. Soil carbon pools and world life zones. *Nature*. 298:156. https://doi.org/10.1038/298156a0.

Prégent, G. 1998. *L'éclaircie des plantations*. Mémoire de recherche no. 133. Direction de la recherche forestière. Service de l'amélioration des arbres, Forêt Québec, Ministère des Ressources Naturelles. RN98–3093, 39p. ISBN: 2-550-33778-6. www.mffp.gouv.qc.ca/publications/forets/connaissances/recherche/Pregent-Guy/Memoire133.pdf.

Pregitzer, K.S. and E.S. Euskirchen. 2004. Carbon cycling and storage in world forests: Biome patterns related to forest age. *Glob. Change Biol.* 10(12):2052–2077. https://doi.org/10.1111/j.1365-2486.2004.00866.x.

Pretzsch, H., P. Biber, G. Schütze, et al. 2014. Forest stand growth dynamics in Central Europe have accelerated since 1870. *Nat. Commun.* 5:4967. https://doi.org/10.1038/ncomms5967.

Price, D.T., D.H. Halliwell, M.J. Apps, et al. 1999. Adapting a patch model to simulate the sensitivity of Central-Canadian boreal ecosystems to climate variability. *J. Biogeogr.* 26:1101–1113.

Puettmann, K.J. 2011. Silvicultural challenges and options in the context of global change: Simple fixes and opportunities for new management approaches. *J. Forest.* 109:321–331.

Puettmann, K.J. and J.C. Tappeiner. 2014. Multi-scale assessments highlight silvicultural opportunities to increase species diversity and spatial variability in forests. *Forestry* 87:1–10. https://doi.org/10.1093/forestry/cpt050.

Rehfeldt, G.E. and B.C. Jaquish. 2010. Ecological impacts and management strategies for western larch in the face of climate-change. *Mitig. Adapt. Strat. Glob. Change* 15:283–306. https://doi.org/10.1007/s11027-010-9217-2.

Rehfeldt, G.E., C.C. Ying, D.L. Spittlehouse, et al. 1999. Genetic responses to climate in Pinus Contorta: Niche breadth, climate change and reforestation. *Ecol. Monogr.* 69(3):375–407.

Reininger, H. 1987. *ZielstaÈrken-Nutzung oder die Plenterung des Altersklassenwaldes*, 163pp. Wien, Austria: OÈ sterreichischer Agrarverlag.

Rink, G. 1990. Juglans cinerea. In *Sylvics of North America.* Vol. 2. *Hardwoods*, eds. R.M. Burns and B.H. Honkola. U.S. Department of Agriculture, *Agriculture Handbook.* http://na.fs.fed.us/spfo/pubs/silvics_manual/volume_2/vol2_Table_of_contents.htm.

Riou-Nivert, P. 1984. Le facteur d'espacement: un guide pour les premières éclaircies dans les peuplements résineux. *Forêt-Entreprise* 20:18–25.

Robert, E., S. Brais, B.D. Harvey, et al. 2012. Seedling establishment and survival on decaying logs in boreal mixedwood stands following a mast year. *Can. J. For. Res.* 42:1446–1455.

Rogers, R. 1990. White Oak. In *Sylvics of North America.* Vol. 2. Hardwoods, eds. R.M. Burns and B.H. Honkola. U.S. Department of Agriculture, Forest Service, *Agriculture Handbook* 654. Washington, DC. http://na.fs.fed.us/spfo/pubs/silvics_manual/volume_2/vol2_Table_of_ contents.htm.

Rollinson, T.J.D. 2008. Forests and climate change: Conclusions and the way forward. In: *Forestry and Climate Change* (Chapter 28), eds. P.H. Freer-Smith, M.S.J. Broadmeadow, and J.M. Lynch, 253p. Oxfordshire, UK: CABI.

Rudolf, P.O. 1990. Red pine. In *Sylvics of North America.* Vol. 1. *Softwoods*, eds. R.M. Burns and B.H. Honkola. U.S. Department of Agriculture, *Agriculture Handbook.* http://na.fs.fed.us/spfo/pubs/silvics_manual/Volume_1/vol1_Table_of_contents.htm.

Ruel, J.-C. and M. Pineau 2002. Windthrow as an important process for white spruce regeneration. *For. Chron.* 78(5):732–738.

Sanchez, C. 2012. Synthèse des méthodes d'irrégularisation des pessières pour la Wallonnie. Vers une sylviculture durable, des écosystemes diversifiés et des revenus soutenus. *Forêt Wallonne asbl*, 44p.

Sander, I.L. 1990. Northern red Oak. In *Sylvics of North America.* Vol. 2. *Hardwoods*, eds. R.M. Burns and B.H. Honkola. U.S. Department of Agriculture, *Agriculture Handbook.* http://na.fs.fed.us/spfo/pubs/silvics_manual/volume_2/vol2_Table_of_contents.htm.

Savva, Y, B. Denneler, A. Koubaa, et al. 2007. Seed transfer and climate change effects on radial growth of jack pine populations in a common garden in Petawawa, Ontario, Canada. *For. Ecol. Manage.* 242:636–647.

Schütz, J.-P. 1997. *Sylviculture 2. La gestion des forêts irrégulières et mélangées.* Eds. Les Presses polytechniques et universitaires romandes. Lausanne. Collection Gérer l'environnement. 176p. Première édition. ISBN: 2-88074-349-4.

Schütz, J.-P. 2001. Opportunities and strategies of transforming regular forests to irregular forests. *For. Ecol. Manage.* 151:87–94.

Schütz, J.-P. and A. Pommerening. 2013. Can Douglas fir (*Pseudotsuga menziesii* (MIRB.) FRANCO) sustainably grow in complex forest structures? *For. Ecol. Manage.* 303:175–183.

Schütz, J.P., T. Pukkala, P.J. Donoso, and K. von Gadow. 2012. Historical emergence and current application of CCF. In *Continuous cover forestry*, 2nd edn, eds. T. Pukkala and K. von Gadow, pp. 1–28. Dordrecht, The Netherlands: Springer.

Schlesinger, R.C. 1990. White ash. In *Sylvics of North America.* Vol. 2. *Hardwoods*, eds. R.M. Burns and B.H. Honkola. U.S. Department of Agriculture, Agriculture Handbook. http://na.fs.fed. us/spfo/pubs/silvics_manual/volume_2/vol2_Table_of_contents.htm.

Seidl, R., W. Rammer, D. Jäger, et al. 2007. Assessing trade-offs between carbon sequestration and timber production within a framework of multi-purpose forestry in Austria. *For. Ecol. Manage.* 248:64–79.

Seppälä, R. 2008. Global forest sector: Trends, threats and opportunities. In *Forestry and climate change*, eds. P.H. Freer-Smith, M.S.J. Broadmeadow, and J.M. Lynch (Chapter 4), 253 p. Wallingford: CABI.

Seppälä, R., A. Buck, and P. Katila, eds. 2009. *Adaptation of forest and people to climate change—A global Assessment Report (Executive summary)*, 224p. Helsinki, Finland: IUFRO World Series Vol. 22.

Settele, J., R. Scholes, R. Betts, et al. 2014: Terrestrial and inland water systems. In *Climate change 2014: Impacts, adaptation, and vulnerability.* part A: Global and Sectoral Aspects. Contribution of Working Group II to the Fifth Assessment Report of the Intergovernmental Panel on Climate Change (IPCC), eds. C.B. Field, V.R. Barros, D.J. Dokken et al., pp. 271–359. Cambridge, UK: Cambridge University Press.

Shields, J.M., C.R. Webster, and L.M. Nagel. 2007. Factors influencing tree species diversity and *Betula alleghaniensis* establishment in silvicultural openings. *Forestry* 80(3):293–307.

Simard, M.-J., Y. Bergeron, and L. Sirois. 2003. Substrate and litterfall effects on conifer seedling survivorship in southern boreal stands of Canada. *Can. J. For. Res.* 33:672–681.

Smit, B., I. Burton, R.J.T. Klein, and J. Wandel. 2000. An anatomy of adaptation to climate change and variability. *Clim. Change* 45:223–251.

Smit, J. and S.S. Lenhart. 1996. Climate change adaptation policy options. *Clim. Res.* 6:193–201.

Smith, P., O. Andren, T. Karlsson, et al. 2005. Carbon sequestration potential in European croplands has been overestimated. *Glob. Change Biol.* 11:2153–2163.

Sohn, J.A., T. Gebhardt, C. Ammer, et al. 2013. Mitigation of drought by thinning: Short-term and long-term effects on growth and physiological performance of Norway spruce (*Picea abies*). *For. Ecol. Manage.* 308:188–197.

Solano-Rivera, C. 2010. *La détermination de la valeur des forêts: une mesure économique*. Réseau de gestion durable des forêts. Notes de recherches no. 61. Université de l'Alberta, Edmonton (AB). http://sfmnetwork.ca.

Spittlehouse, D.L. 2005. Integrating climate change adaptation into forest management. *For. Chron.* 81(5):691–695.

Spittlehouse, D.L. and R.B. Stewart. 2003. Adaptation to climate change in forest management. B.C. *J. Ecosys. Manage.* 4(1):1–11. www.forrex.org/jem/2003/vol4/no1/art1.pdf.

Steenberg, J.W.N., P.N. Duinker, and P.G. Bush 2011. Exploring adaptation to climate change in the forests of central Nova Scotia, Canada. *For. Ecol. Manage.* 262:2316–2327. https://doi.org/10.1016/j.foreco.2011.08.027.

Steffen, W., A. Sanderson, P.D. Tyson, et al. 2004. *Global Change and the Earth System: A Planet under Pressure*. Heidelberg, Germany: Springer-Verlag.

Stokes, V. and G. Kerr. 2009. *The evidence supporting the use of CCF in adapting Scotland's forests to the risks of climate change*. Surrey, UK: Report to the Forestry Commission of Scotland by Forest Research.

Symon, C., L. Arris, and B. Heal, eds. 2005. *Arctic climate impact assessment (ACIA)—Scientific report*, 1042p. Cambridge, UK: Cambridge University Press.

Taylor, A.R., J.R. Wang, and W.A. Kurz. 2008. Effects of harvesting intensity on Carbon stocks in eastern Canadian red spruce (*Picea rubens*) forests: An exploratory analysis using the CBMCFS3 simulation model. *For. Ecol. Manage.* 255:3632–3641.

Thuiller, W. 2007. Climate change and the ecologist. *Nature* 448:550–552.

Tremblay, S. and R. Ouimet. 2013. White spruce plantations on abandoned agricultural land: Are they more effective as C sinks than natural succession? *Forests* 4:1141–1157. https://doi.org/10.3390/f4041141.

Turner, J.A., J. Buongiorno, and S.S. Zhu. 2006. An economic model of international wood supply, forest stock and forest area change. *Scand. J. For. Res.* 21(1):73–86.

Vaillancourt, M-A., L. De Granpré, S. Gauthier et al. 2008. Comment les perturbations naturelles peuvent-elles constituer un guide pour l'aménagement forestier écosystémique? In *Aménagement écosystémique en forêt boréale*, Chapter 2, pp. 41–60. Montréal, Québec, Canada: Presses de l'Université du Québec.

Vallet, P. 2008. Impact de différentes stratégies sylvicoles sur la fonction puits de carbone des peuplements forestiers. *Forêt Walonne* 95:57.

van Minnen, J.G., B.J. Strengers, B. Eickhout, et al. 2008. Quantifying the effectiveness of climate change mitigation through forest plantations and carbon sequestration with an integrated land-use model. *Carbon Balance Manage.* 3(3):1–20. https://doi.org/10.1186/1750-0680-3-3.

Viereck, L.A. and W.F. Johnston. 1990. Black spruce. In *Sylvics of North America*. Vol. 1. *Softwoods*, eds. Burns, R.M. and Honkola, B.H. U.S. Department of Agriculture, Forest Service, *Agriculture Handbook* 654. Washington, DC. http://na.fs.fed.us/spfo/pubs/silvics_manual/Volume_1/vol1_Table_of_contents.htm.

Volney, W.J.A. and R. Fleming. 2000. Climate change and impacts of boreal forest insects. *Agri. Ecosyst. Environ.* 82:283–294.

Vouligny, C. and S. Gariépy. 2008. *Abandoned farmland in Québec—Status and development options*. Ottawa, Ontario, Canada: Agriculture and Agri-Food Canada.

Walter, H. 1979. *Vegetation of the Earth and Ecological Systems of the Geobiosphere*, 274p. New York: SpringerVerlag.

Wang, W., C. Peng, D.D. Kneeshaw, et al. 2013. Modeling the effects of varied forest management regimes on carbon dynamics in jack pine stands under climate change. *Can. J. For. Res.* 43:469–479. dx.doi.org/10.1139/cjfr-2012–0320.

Wendel, G.W. and H.C. Smith. 1990. Eastern White pine. In *Sylvics of North America.* Vol. 1. Softwoods, eds. R.M. Burns and Honkola, B.H. U.S. Department of Agriculture, Forest Service, *Agriculture Handbook 654.* Washington, DC. http://na.fs.fed.us/spfo/pubs/silvics_manual/Volume_1/vol1_Table_of_contents.htm.

Williams, R.D. 1990. Juglans nigra. In *Sylvics of North America.* Vol. 2. *Hardwoods*, eds. Burns, R.M. and Honkola, B.H. U.S. Department of Agriculture, Forest Service, *Agriculture Handbook 654.* Washington, DC. http://na.fs.fed.us/spfo/pubs/silvics_manual/volume_2/vol2_Table_of_ contents.htm.

Yamasaki, S.H., R. Duchesneau, F. Doyon, et al. 2008. Making the case for cumulative impacts assessment: Modelling the potential impacts of climate change, harvesting, oil and gas, and fire. *For. Chron.* 84(3):349–368.

Index

Note: Page numbers in *italic* indicate a figure and page numbers in **bold** indicate a table on the corresponding page

C

D

E

F

G

H

I

O

P

U

V

W

X

Y

Z